Atlas of the
Vacuum Ultraviolet Emission Spectrum
of Molecular Hydrogen

Journal of
Physical and Chemical Reference Data

Jean W. Gallagher, Editor

The **Journal of Physical and Chemical Reference Data** (ISSN 0047-2689) is published bimonthly by the American Chemical Society (1155 16th St., N.W., Washington, DC 20036-9976) and the American Institute of Physics (500 Sunnyside Blvd., Woodbury, NY 11797-2999) for the National Institute of Standards and Technology. Second-class postage paid at Washington, DC and additional mailing offices. POSTMASTER: Send address changes to *Journal of Physical and Chemical Reference Data*, Membership and Subscription Services, P.O. Box 3337, Columbus, OH 43210-0012.

The objective of the Journal is to provide critically evaluated physical and chemical property data, fully documented as to the original sources and the criteria used for evaluation. Critical reviews of measurement techniques, whose aim is to assess the accuracy of available data in a given technical area, are also included. Papers which report new estimation, or prediction techniques and test these techniques against existing databases are also accepted. The journal is not intended as a publication outlet for original experimental measurements such as those that are normally reported in the primary research literature, nor for review articles of a descriptive or primarily theoretical nature.

Monographs are published at irregular intervals and are not included in subscriptions to the Journal. They contain compilations which are too lengthy for a journal format.

The Editor welcomes appropriate manuscripts for consideration as Journal articles or monographs. Potential contributors who are interested in preparing a compilation are invited to submit an outline of the nature and scope of the proposed compilation, with criteria for evaluation of the data and other pertinent factors, to:

Jean W. Gallagher, Editor
J. Phys. Chem. Ref. Data
Bldg. 221/A323
National Institute of Standards and Technology
Gaithersburg, MD 20899-0001

One source of contributions to the Journal is The National Standard Reference Data System (NSRDS), which was established in 1963 as a means of coordinating on a national scale the production and dissemination of critically evaluated reference data in the physical sciences. Under the Standard Reference Data Act (Public Law 90-396) the National Institute of Standards and Technology of the U.S. Department of Commerce has the primary responsibility in the Federal Government for providing reliable scientific and technical reference data. The Standard Reference Data Program of NIST coordinates a complex of data evaluation centers, located in university, industrial, and other Government laboratories as well as within the National Institute of Standards and Technology, which are engaged in the compilation and critical evaluation of numerical data on physical and chemical properties retrieved from the world scientific literature. The participants in this NIST-sponsored program, together with similar groups under private or other Government support which are pursuing the same ends, comprise the National Standard Reference Data System.

The primary focus of the NSRDS is on well-defined physical and chemical properties of well-characterized materials or systems. An effort is made to assess the accuracy of data reported in the primary research literature and to prepare compilations of critically evaluated data which will serve as reliable and convenient reference sources for the scientific and technical community.

Information for Contributors

Manuscripts submitted for publication must be prepared in accordance with *Instructions for Preparation of Manuscripts for the Journal of Physical and Chemical Reference Data*, available on request from the Editor.

New and renewal subscriptions should be sent with payment to the Office of the Controller at the American Chemical Society, 1155 Sixteenth Street, N.W., Washington, DC 20036-9976. **Address changes,** with at least six weeks advance notice, should be sent to *Journal of Physical and Chemical Reference Data,* Membership and Subscription Services, American Chemical Society, P.O. Box 3337, Columbus, OH 43210-0012. Changes of address must include both old and new addresses and ZIP codes and, if possible, the address label from the mailing wrapper of a recent issue. Claims for missing numbers will not be allowed: if loss was due to failure of the change-of-address notice to be received in the time specified; if claim is dated (a) North America: more than 90 days beyond issue date, (b) all other foreign: more than one year beyond issue date.

Members of AIP member and affiliate societies requesting member subscription rates should direct subscriptions, renewals, and address changes to American Institute of Physics, Circulation and Fulfillment Division, 500 Sunnyside Blvd., Woodbury, NY 11797-2999.

Subscription Prices (1994)

	U.S.A.	Canada and Mexico	Other Foreign	Air Freight — Europe, Mideast, Asia Africa and Oceania
Members (of ACS, AIP, or affiliated society)	$90.00	$100.00	$100.00	$110.00
Regular rate	$460.00	$470.00	—	$480.00†

†Regular rate subscriptions to Europe, Mideast, Asia, Africa, Oceania, and Central and South America include air freight services.

Rates above do not apply to nonmember subscribers in Japan, who must enter subscriptions orders with Maruzen Company Ltd., 3-10 Nihonbashi 2-chrome, Chuo-ku, Tokyo 103, Japan. Tel. (03) 272-7211.

Current 1994 single-issue price: $85.00.

Back numbers are available at a cost of $85 per single copy of 1990 and later issues, $130 for single copies of 1989 and previous years, and $460 per volume.

Orders for reprints, supplements, monographs and back numbers should be addressed to the American Chemical Society, 1155 Sixteenth Street, N.W., Washington, DC 20036-9976. Prices for reprints and supplements are listed at the end of this issue

Copying Fees: The code that appears on the first page of articles in this journal gives the fee for each copy of the article made beyond the free copying permitted by AIP. (See statement under "Copyright" elsewhere in this journal.) If no code appears, no fee applies. The fee for pre-1978 articles is $0.25 per copy. With the exception of copying for advertising and promotional purposes, the express permission of AIP is not required provided the fee is paid through the *Copyright Clearance Center, Inc. (CCC), 222 Rosewood Dr., Danvers, MA 01923.* Contact the CCC for information on how to report copying and remit payment.

Microfilm subscriptions of the *Journal of Physical and Chemical Reference Data* are available on 16mm and 35 mm. This journal also appears in Sec. I of *Current Physics Microfilm* (CPM) along with 32 other journals published by the American Institute of Physics and its member societies. A *Microfilm Catalog* is available on request.

Journal of
**Physical and
Chemical
Reference Data**

Monograph No. 4

Atlas of the Vacuum Ultraviolet Emission Spectrum of Molecular Hydrogen

Jean-Yves Roncin

*Laboratoire Traitement du Signal et Instrumentation
CNRS URA 842
23 rue Dr. P. Michelon
42023 Saint-Etienne Cedex 2, France*

Françoise Launay

*Observatoire de Paris, Section de Meudon
DAMAP and CNRS URA 812
5 Place Jules Janssen
92195 Meudon Cedex, France*

  AIP

Published by the **American Chemical Society**
and the **American Institute of Physics** for
the **National Institute of Standards and Technology**

Library of Congress Catalog Card Number 94-70936

International Standard Book Number
1-56396-339-6

American Institute of Physics

500 Sunnyside Boulevard

Woodbury, New York 11797-2999

Printed in the United States of America

Foreword

As a spectroscopist with a long-standing interest in molecular hydrogen, I welcome, with enthusiasm, the publication of the vacuum ultraviolet emission lines of H_2. This atlas will be of immense help to all of those who want to study further the H_2 spectrum in the VUV or identify it in observed spectra both in the laboratory and in space.

The authors have greatly increased the usefulness of the atlas by their simultaneous publications with colleagues of the detailed analysis of the Lyman and Werner bands. Further assignments of so far unassigned lines will be made much easier and will likely lead to many new results.

For a long time, this atlas will stand as the last word on the vacuum ultraviolet spectra of H_2.

Gerhard Herzberg
Ottawa, Canada

Contents

1. Introduction 1

2. Experimental 1

   2.1. Photographic Recording 1

   2.2. Photoelectric Recording 2

3. Description of the Spectrum 2

4. Acknowledgments 3

5. References 3

165 Sets of Tables and Plates 5

1. Introduction

Following the work of Schumann (1903), the first list of vacuum ultraviolet (VUV) spectral lines emitted by an electric discharge in molecular hydrogen (H_2) was published at the beginning of the century by Lyman (1904 and 1906). The spectrum was photographed at low resolution and no identification was possible at that time. Later on, Werner (1926) gave the first tentative classification of some lines. Other pioneering investigations were also performed, always at low resolution (Dieke and Hopfield, 1927; Hori, 1927; Hyman, 1930; Jeppesen, 1933). No rotational analysis was given yet in the list of characteristic band wavelengths of diatomic molecules published in 1952 by the "International Tables of Selected Constants" under the direction of Rosen (1952).

Although the absorption spectrum of H_2 has been extensively studied, relatively less study was done in emission at high as well as low resolution simply because the source of light used was an ordinary glow discharge, the pressure of which (about 100 Pa) was high enough to completely prevent emission at short wavelength by radiation trapping. Regarding absorption many references before 1977 can be found in Huber and Herzberg's table (1979). The most useful are Wilkinson (1968), Namioka (1964a and b), Dabrowski and Herzberg (1974), and Takezawa (1970).

The first VUV emission study of H_2 at high resolution was published by Herzberg and Howe (1959), concerning the Lyman band system $B\,^1\Sigma_u^+ \rightarrow X\,^1\Sigma_g^+$, down to 118 nm. These data were used by Junkes, Salpeter, and Milazzo (1965) to list about 1500 lines of the Lyman band system that they published, together with 32 lines of the Werner band system $C\,^1\Pi_u \rightarrow X\,^1\Sigma_g^+$ at wavelengths longer than 113.5 nm, in an atlas of VUV atomic emission lines. Pictures of high resolution spectra were also given.

In 1980 Larzillière, Launay, and Roncin (1980) published preliminary results about the $D\,^1\Pi_u \rightarrow X\,^1\Sigma_g^+$ band system at short wavelength around 80 nm. The spectrum was obtained at Meudon Observatory with a low pressure electric discharge where self-absorption is much reduced at short wavelength. It was the first observation of emission from levels above the ionization limit. Two spectrally more extended studies were reported in 1984. The first one, by Dabrowski (1984), was devoted to the analysis of the Lyman and Werner band systems down to 100 nm. The second one, by Roncin, Launay, and Larzillière (1984) and Larzillière, Launay, and Roncin (1985), extended the analysis of the four unperturbed band systems C, D, D', and $D''\,^1\Pi_u \rightarrow X\,^1\Sigma_g^+$ down to 78 nm.

At first the analysis of the very crowded emission spectrum rested on absorption data and on calculations which did not take into account the rotational coupling. However, the temperature at which the emission spectrum occurs is much higher than that for the absorption spectrum, so that the emission spectrum exhibits lines of high rotational quantum number J impossible to ascribe unambiguously except for the lowest v' (unperturbed) bands of the Lyman system and the Q branches of the $^1\Pi_u \rightarrow X\,^1\Sigma_g^+$ band systems.

From the theoretical point of view, Allison and Dalgarno (1970) first reported emission probabilities from each vibrational level of the $B\,^1\Sigma_u^+$ and $C\,^1\Pi_u$ states towards all vibrational levels of the electronic ground state, neglecting the rotational coupling between B and C. In the mid-1970s, Julienne (1973) and Ford (1974 and 1975) performed the first calculations of line emission probabilities of the Lyman and Werner band systems of H_2 taking into account the rovibronic coupling between the excited states. But at that time no complete experimental high resolution emission spectrum was available to check the calculation so that the problem was passed over. Recently several refined calculations have been performed. Abgrall *et al.* (1987) calculated emission probabilities for the Lyman and Werner band systems and found very good agreement with intensity measurements. The method of calculation is what we call semi-*ab initio* in the sense that the potential energy curves were fitted to the experimental results for each (v', $J=0$ or 1) before solving the system of two Schrödinger equations. An extended list of level energies and oscillator strengths has been subsequently published by Abgrall and Roueff (1989), while Senn, Quadrelli, and Dressler (1988) obtained level energies, up to $J=6$, of the B, B', C, and D states, calculated completely *ab initio* by including the four excited states in the interaction. Finally a semi-*ab initio* calculation, including the four excited states, has allowed Abgrall *et al.* (1993a, 1993b, 1993c, and 1994) to publish a complete analysis of the Lyman, Werner, $B' \rightarrow X$, and $D \rightarrow X$ band systems.

2. Experimental

2.1. Photographic Recording

The spectra have been recorded by means of the Meudon Observatory Eagle mounting VUV 10-m spectrograph. The instrument is fitted with a concave (R=10.685 m), 3600 lines/mm Jobin-Yvon holographic grating which gives a plate factor of 0.026 nm/mm in the first order. The slitwidth was 30 µm, giving a resolving limit of 8×10^{-4} nm throughout the whole range of observation (0.8 cm^{-1} at 100 nm).

The presence of a magnetic field of about 0.1 tesla in the water cooled electric discharge lamp (ANVAR), made of a brass anode and two aluminum cathodes, makes it possible to run the source at a pressure as low as a few pascals, so that self-absorption is much reduced at short wavelength. With hydrogen flowing in the source, the lamp was operated at 350 V, 200 mA.

Exposure time on Kodak SWR plates varied from 5 mm to 13 h, depending on the spectral range photographed.

Reference lines (Cu II, Ge II, Si II and O I, N I, Cl I, Cl II as impurities) were obtained by flowing helium in an ancillary water cooled copper hollow cathode discharge lamp operated at 500 V, 200 mA. Small pieces of germanium and silicon were introduced inside the cathode.

Due to the difference of the grating illumination by the two lamps, one finds a random shift between the two spectra (Kaufman and Edlén 1974, Freeman and King 1977). For each plate the shift is determined by means of the impurity lines common to the spectra emitted by both lamps.

In the shortest wavelength range (78-89 nm), the hollow cathode spectrum was recorded in the same zone of the photographic plate as the H_2 spectrum, and extra Ar I reference lines were also obtained by flowing argon in the low pressure discharge lamp itself.

The plates were measured on the Meudon Observatory photoelectric comparator giving an accuracy of 1 μm on line position (Launay 1975). The measuring machine is now connected to a PC compatible micro-computer. The standard deviation of our polynomial fit for the hollow cathode reference lines is typically 7×10^{-5} nm and the wavelength accuracy of the hydrogen unblended lines is estimated to be $\pm 15 \times 10^{-5}$ nm.

2.2. Photoelectric Recording

Relative intensity measurements have been performed by replacing the plate-holder of the spectrograph by a scanning photoelectric device moving along the Rowland circle and including a slit and either a photomultiplier with a MgF_2 window (Hamamatsu R1459) or a channeltron (RTC X919BL), depending whether the measured range is above or below 120 nm. The width of entrance and exit slits was 30 μm.

A set of 29 scans was obtained, each section of the spectrum being scanned at least twice. The relative intensities derived from raw data were scaled in order to get a consistent set of values over the whole spectral range (100 nm) but the instrumental response was not taken into account.

3. Description of the Spectrum

The spectrum contains 12 265 lines, including impurity and reference lines, extending from 78.60 to 171.35 nm. It has been divided into 165 sections covering about 0.58 nm each. Each section comprises the following:

(i) A picture of the plate enlarged ten times where emission lines appear in white on the dark background. H_2 lines look always slightly broad due to the Doppler effect, whereas atomic lines are generally sharp and readily recognized. It has to be noted that, due to different exposure times, blackening varies from plate to plate so that it may happen that weak high-J lines show up on a plate, whereas strong lower-J lines of the same band are not seen on the next plate.

(ii) The corresponding microdensitometer trace in which the strong lines are saturated while even very weak lines show up. Each line is marked by a vertical tick below the trace. Longer ticks indicate lines whose number is a multiple of five, thicker ones those whose number is a multiple of ten. The numbering of the lines is marked above the trace every five lines. A wavelength scale in nanometers and a wavenumber scale in reciprocal centimeters are indicated at the bottom and at the top of the trace, respectively.

The appearance of the reference atomic lines on the microdensitometer traces depends on which part of the spectrum has been scanned. Plot of photoelectric recording would have been more realistic but more difficult to display due to the dynamic range of the existing data acquisition unit.

(iii) A table listing all the lines appearing in the corresponding picture and densitometer trace.

- In these tables, lines are numbered in the first column. A few atomic lines, and very close H_2 lines, appear in the table not in the same order as in the picture. This is because reference lines are not emitted by the same source as the H_2 lines so that they may be shifted with respect to the H_2 spectrum as explained in the experimental section. Sometimes a standard line appears twice in the list, which means that the line is emitted by both the low pressure discharge and the hollow cathode.

- The second column contains the relative peak intensities of H_2 lines, measured from photoelectric recording but not corrected for instrumental response. Intensities are reported for H_2 lines only. However no value of intensity is given for the shoulder lines indicated "sh" in the comment and also for the reversed lines, indicated "r" because for the latter lines the intensity depends strongly on the pressure in the source at the time of the recording. An intensity of zero is given to lines barely seen in the picture and the trace, for avoiding values smaller than one. It should also be remarked that, as intensities have been scaled throughout the spectrum, the figures given in the tables are rounded and may not correspond strictly to what is seen on the traces. Also the running conditions of the source may have been different at the time of the photoelectric recording from what it was when the picture was taken.

- The third and fourth columns display, respectively, the wavelength (in nm) in ascending order and the corresponding wavenumber (in cm^{-1}).

- The fifth column indicates the abbreviated assignment of H_2 lines. In the VUV range, H_2 emission lines can be ascribed to seven band systems, namely, $2p\sigma B\,^1\Sigma_u^+ \to X\,^1\Sigma_g^+$ (Lyman), $2p\pi C\,^1\Pi_u^+ \to X\,^1\Sigma_g^+$ (Werner), as well as $3p\sigma B'$ and $4p\sigma B''\,^1\Sigma_u^+ \to X\,^1\Sigma_g^+$ and $3p\pi D$, $4p\pi D'$ and $5p\pi D''\,^1\Pi_u^+ \to X\,^1\Sigma_g^+$. We have also tentatively identified a few lines of the band system $6p\pi\,^1\Pi_u \to X\,^1\Sigma_g^+$. As all VUV transitions terminate on the ground state $X\,^1\Sigma_g^+$ we have omitted X everywhere. For example, $C2$-7P5 designates the line $J''=5$ in the P-branch of the transition $C\,^1\Pi_u(v'=2) \to X\,^1\Sigma_g^+(v''=7)$.

- The last column is devoted to the identification of reference and impurity lines and may contain comments about H_2 lines. For the self-reversed lines exhibiting clearly two components, the average wavenumber is repeated in the comments of the two components. Such a value is very accurate because the two components are always sharp. Finally "b" stands for blended lines, "d" for diffuse lines, and "s" for sharp unidentified lines.

Some molecular impurity lines due to N_2 and CO appear on a few pictures. Their wavelengths are indicated and they are identified in the comment.

At longer wavelengths, most of the unascribed lines have negligible intensities. They may be due to the presence of some impurities. At shorter wavelengths a number of unascribed lines have appreciable intensities. Most of these are certainly due to transitions from still unidentified high-J levels of the states B', D, B'', D', and D''. These lines cannot be ascribed because the calculation does not fit with the lines predicted from absorption data well enough for a few levels of B' and D, and no calculation is yet available for B'', D', and D''.

For atomic lines appearing in the reference spectrum we have indicated only the wavelength unless the lines are good standards with 5 decimals. In this case we give only the literature wavelength, in the comment. All the wavelengths of reference atomic lines are extracted either from a paper by Kaufman and Edlén (1974) or from Kelly's table (1987). In the latter case the value is followed by Ke.

In a first stage of the work the lines could be identified from the values of level energies published long ago (Wilkinson 1968, Namioka 1964a and b, Dabrowski and Herzberg 1974, and Takezawa 1970), and from more recent identifications of small v', high-J levels of the Lyman band system by Dabrowski (1984). The semi-*ab initio* calculations of Abgrall *et al.* (1993a, 1993b, 1993c, and 1994) give emission probabilities and wavenumbers of most lines to an accuracy of about 0.5 cm^{-1}. Such a calculation was necessary, in order to assign unambiguously many high-J lines. In that way about 96% of the lines have at least one assignment. The absence of most lines missing in the tables of Abgrall *et al.* (1993b, 1993c, and 1994) is well explained either by radiation trapping or a weak emission probability. Typical cases have been illustrated by Abgrall *et al.* (1993a). Most mistakes, but probably not all of them, have been tracked down by several error tests included in the file handling programs used for editing the tables of Abgrall *et al.* (1993b, 1993c, and 1994).

It has to be pointed out that, due to interactive work with theoreticians, the present atlas is more than a simple catalog of spectral lines, as 96% of molecular lines are identified, most of them for the first time. It is hoped that this book will be useful to scientists in many fields and particularly to:

- astrophysicists who study the interstellar medium as well as those who are receiving, from spacecrafts, VUV spectra emitted by planetary atmospheres where H_2 is the major constituent,
- physicists who study emission of H_2 excited by electron impact in connection with planetary atmosphere processes,
- laser physicists, who need a very good knowledge of molecular line wavelengths,
- theoreticians for testing any improvement of the model.

4. Acknowledgments

The present work is dedicated to Dr. Gerhard Herzberg, who did so much for the knowledge of H_2 and many other molecules. It was initially undertaken under the auspices of the International Tables of Selected Constants (Paris) and CNRS (Centre National de la Recherche Scientifique), which provided useful advice and encouragement. Over the years, many people helped the authors to complete this long term project.

First of all credit must be given to Dr. Henri Damany for designing the low pressure gas discharge lamp, twenty five years ago, when he was at the Laboratoire des Hautes Pressions (Meudon-Bellevue). Thanks are due to Dr. Christian Jungen (Orsay, France) and to Dr. Michel Larzillière (University Laval, Québec, Canada) who contributed at the early stage of the work. We appreciated the help of Maurice Benharrous in taking pictures and photoelectric recordings. Microdensitometer traces were carefully done by Monique Clemino. Professor Jean-Louis Subtil (University of Saint-Etienne) wrote the file handling software required by the great number of data. Rosario Battaglia assisted in entering and checking the data. We also acknowledge the help of Dr. Stéphane Mottin (University of Saint-Etienne) in the task of editing the tables in the proper format. The greatest contribution came certainly from Dr. Hervé Abgrall (Meudon Observatory), who performed the semi-*ab initio* calculations of emission probabilities and line positions that were so necessary to identify many features of the spectrum. Finally we would like to thank all of those who encouraged us to pursue this project in spite of the disregard of some people for what they term "grandpa's spectroscopy."

5. References

Abgrall, H., F. Launay, E. Roueff, and J.-Y. Roncin (1987), *J. Chem. Phys.* **87**, 2036.

Abgrall, H. and E. Roueff (1989), *Astron. and Astrophys. Suppl. Series* **79**, 313.

Abgrall, H., F. Launay, E. Roueff, J.-Y. Roncin, and J.-L. Subtil (1993a), *J. Molec. Spectrosc.* **157**, 512.

Abgrall, H., F. Launay, E. Roueff, J.-Y. Roncin, and J.-L. Subtil (1993b), *Astron. and Astrophys. Suppl. Series* **101**, 273.

Abgrall, H., F. Launay, E. Roueff, J.-Y. Roncin, and J.-L. Subtil (1993c), *Astron. and Astrophys. Suppl. Series* **101**, 323.

Abgrall, H., F. Launay, E. Roueff, and J.-Y. Roncin (1994), *Can. J. Phys.* (submitted).

Allison, A. C. and A. Dalgarno (1970), *At. Data* **1**, 289.

ANVAR (Agence Nationale pour la Valorisation de la Recherche) License Nb 7 296 800. Manufactured by Instruments SA, France.

Dabrowski, I. and G. Herzberg (1974), *Can. J. Phys.* **52**, 1110.

Dabrowski, I. (1984), *Can. J. Phys.* **62**, 1639.

Dieke, G. H. and J. J. Hopfield (1927), *Phys. Rev.* **30**, 400.

Ford, A. L. (1974), *J. Mol. Spectrosc.* **53**, 364.

Ford, A. L. (1975), *J. Mol. Spectrosc.* **56**, 251.

Freeman, G. H. C. and W. H. King (1977), *J. Phys. E* **10**, 895.

Herzberg, G. and L. L. Howe (1959), *Can. J. Phys.* **37**, 636.

Hori, T. (1927), *Z. Phys.* **44**, 834.

Huber, K.-P. and G. Herzberg, *Molecular Spectra and Molecular Structure* Vol. IV, *Constants of Diatomic Molecules*, Van Nostrand-Reinhold, New York, 1979.

Hyman, H. H. (1930), *Phys. Rev.* **36**, 187.

Jeppesen, C. R. (1933), *Phys. Rev.* **44**, 165.

Julienne, P. S. (1973), *J. Mol. Spectrosc.* **48**, 508.

Junkes, J., E. W. Salpeter, and G. Milazzo, *Atomic Spectra in the Vacuum Ultraviolet from 2250 to 1100 Å*, Specola Vaticana, Città del Vaticano, 1965.

Kaufman, V. and B. Edlén (1974), *J. Phys. Chem. Ref. Data* **3**, 825.

Kelly, R. L. (1987), *J. Phys. Chem. Ref. Data* **16**, Suppl. 1.

Larzillière, M., F. Launay, and J.-Y. Roncin (1980), *J. Phys. (Paris)* **41**, 1431.

Larzillière, M., F. Launay, and J.-Y. Roncin (1985), *Can. J. Phys.* **63**, 1416.

Launay, F., *Proceedings of the International Conference on Image Processing Techniques in Astronomy*, Utrecht, March 1975 (Reidel, Dordrecht, Holland).

Lyman, Th. (1904), *Astrophys. J.* **19**, 263.

Lyman, Th. (1906), *Astrophys. J.* **23**, 181.

Namioka, T. (1964a), *J. Chem. Phys.* **40**, 3154.

Namioka, T. (1964b), *J. Chem. Phys.* **41**, 2141.

Roncin, J.-Y., F. Launay, and M. Larzillière (1984), *Can. J. Phys.* **62**, 1686.

Tables de Constantes et Données Numériques. 5 Atlas des longueurs d'onde caractéristiques des bandes d'émission et d'absorption des molécules diatomiques. Edit. B. Rosen, Hermann, Paris 1952.

Schumann, V. (1903), *Smithsonian contributions*, Nb 1413.

Senn, P., P. Quadrelli, and K. Dressler (1988), *J. Chem. Phys.* **89**, 7401.

Takezawa, S. (1970), *J. Chem. Phys.* **52**, 2575.

Werner, S. (1926), *Proc. R. Soc. (London)* **A113**, 107.

Wilkinson, P. G. (1968), *J. Molec. Spectrosc.* **46**, 1225.

Tables and Plates

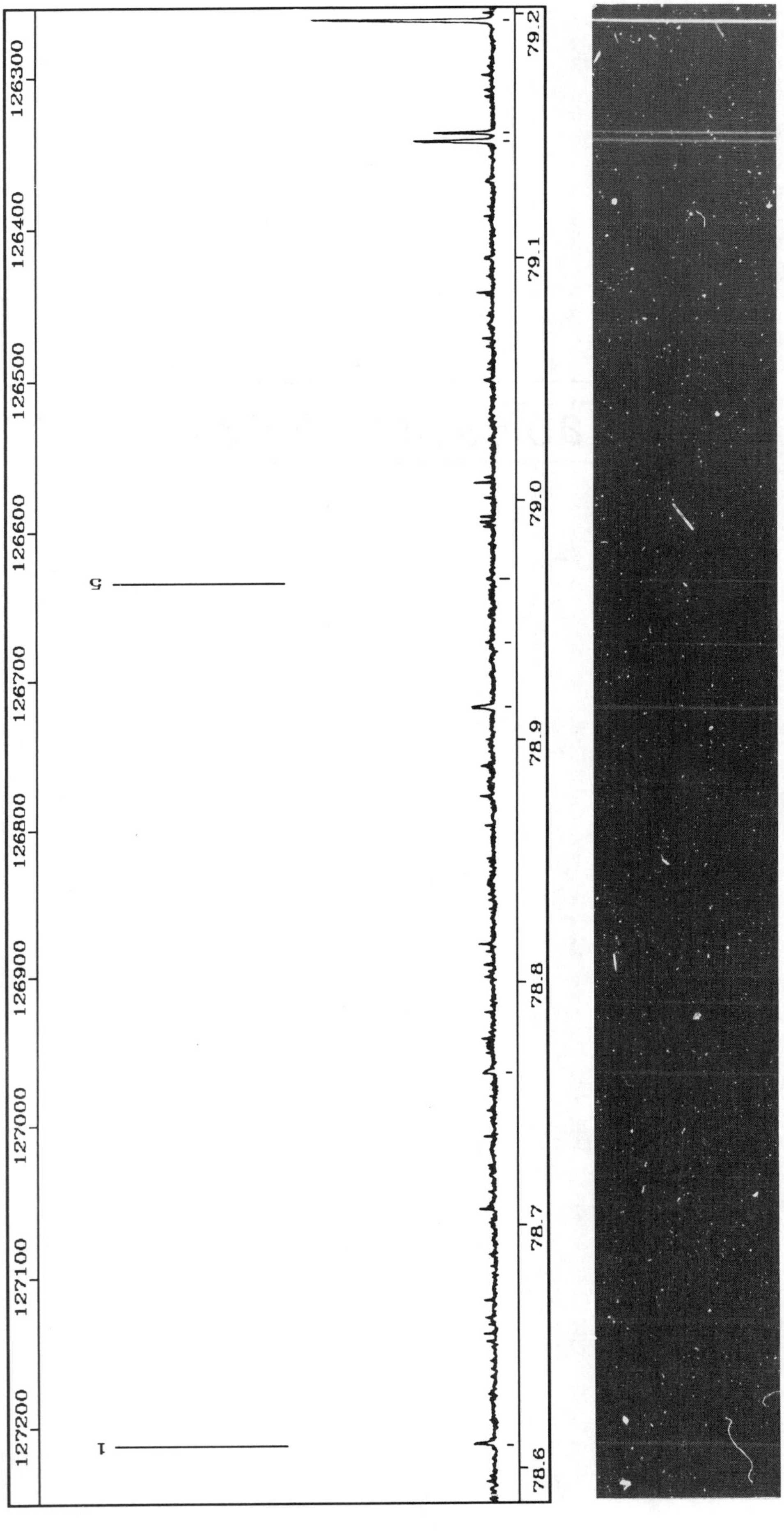
126300
126400
126500
126600
126700
126800
126900
127000
127100
127200
79.2
79.1
79.0
78.9
78.8
78.7
78.6
5
1

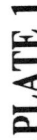

PLATE 1

TABLE 1

Nb	I	λ (nm)	σ (cm^{-1})	Assignment	Comment
1	0	78.6097	127210.72	D12-1Q1	
2	0	78.7619	126965.01	D8-0Q2	
3	0	78.9142	126719.97	D8-0Q3	
4		78.9397			
5		78.9659			
6	0	79.1489	126344.18	D11-1Q1	OI 79.15136
7					OI 79.19732
8					

Nb	I	λ (nm)	σ (cm^{-1})	Assignment	Comment

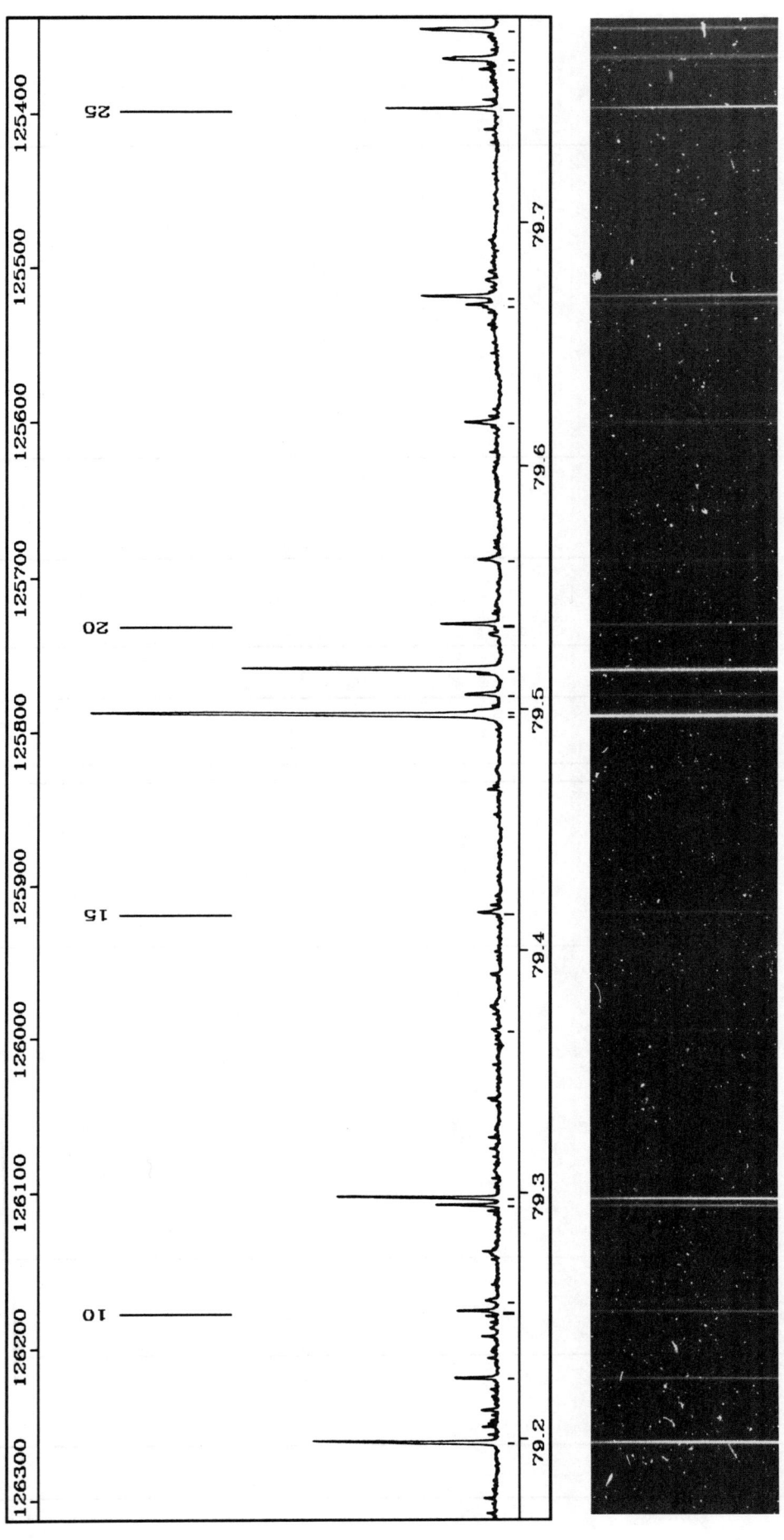

PLATE 2

TABLE 2

Nb	I	λ (nm)	σ (cm-1)	Assignment	Comment
9					OI 79.22330
10					OI 79.25063
11	0	79.2555	126174.18	D11-1Q2	
12					OI 79.29381
13					OI 79.29671
14	0	79.3661	125998.44	D8-0Q5	
15	1	79.4155	125920.04	D11-1Q3	
16	1	79.4963	125792.03		
17		79.4981	125789.20		sh
18	0	79.5048	125778.54		
19	1	79.5152	125762.15		
20	0	79.5338	125732.71		
21	1	79.5605	125690.59	D'4-0Q2	
22	1	79.6172	125601.08	D7-0Q2	
23		79.6647			s
24		79.6681			CuII 79.74552
25					
26		79.7622			
27	1	79.7670	125365.10	D7-0Q3	
28	1	79.7786	125346.88	D10-1Q1	

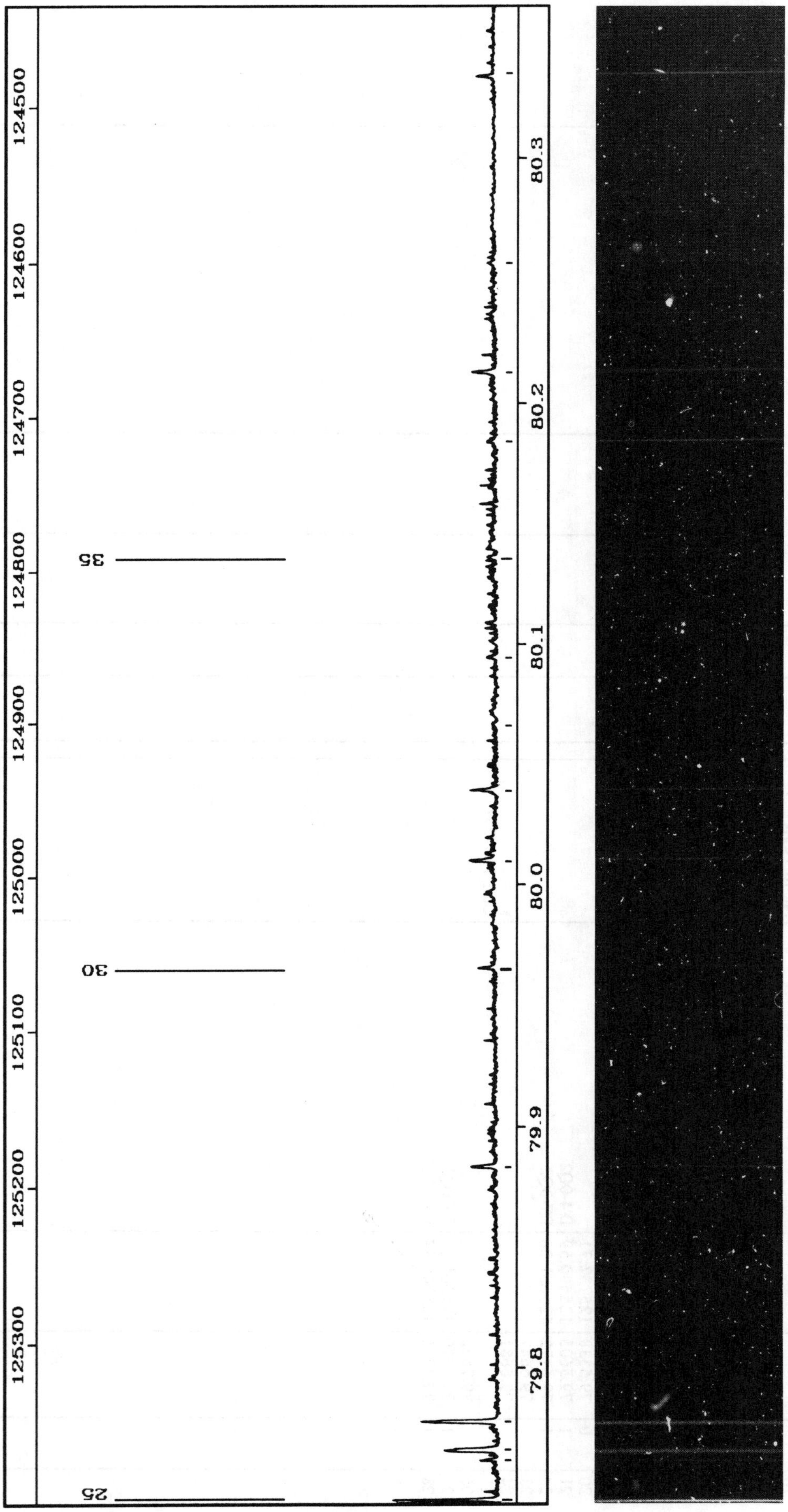
124500
124600
124700
124800
124900
125000
125100
125200
125300
35
30
25
80.3
80.2
80.1
80.0
79.9
79.8

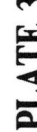

PLATE 3

TABLE 3

Nb	I	λ (nm)	σ (cm⁻¹)	Assignment	Comment
29	0	79.8831	125182.86	D10-1Q2	
30	0	79.9658	125053.41	D7-0Q4	
31	0	80.0103	124983.91		
32	0	80.0396	124938.14	D10-1Q3	
33	0	80.0671	124895.31		
34	0	80.0946	124852.42		
35	0	80.1361	124787.66		s
36		80.1826			
37	0	80.2125	124668.90	D7-0Q5	s
38	0	80.2580	124598.14	D14-2Q1	
39		80.3340			

Nb	I	λ (nm)	σ (cm⁻¹)	Assignment	Comment

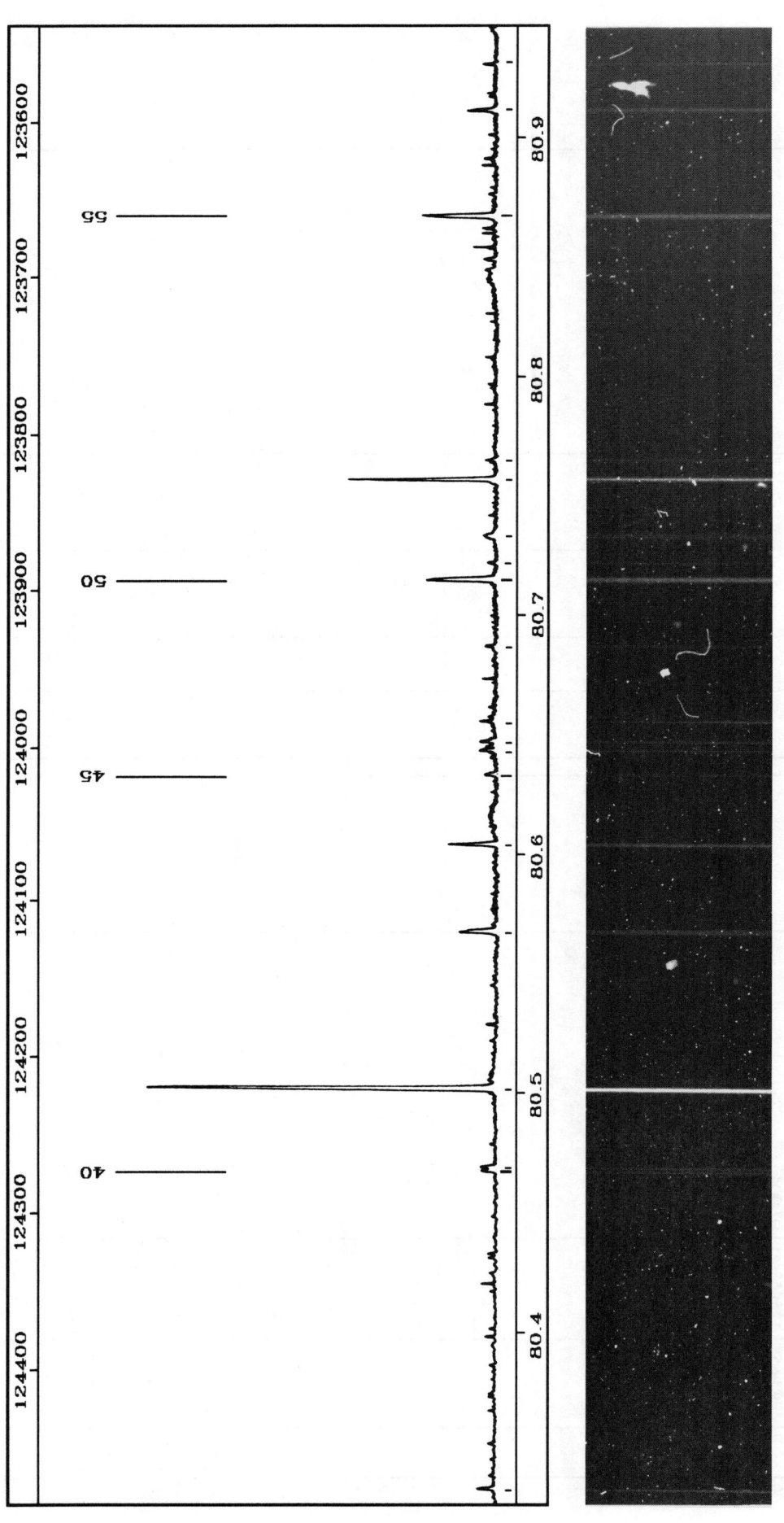

PLATE 4

TABLE 4

Nb	I	λ (nm)	σ (cm-1)	Assignment	Comment
40		80.4667	124274.96	D6-0Q1	r 124273.72
41		80.4683	124272.48	D6-0Q1	r 124273.72
42	3	80.5011	124221.87	D9-1Q1	r
43		80.5669	124120.44	D6-0Q2	
44	1	80.6040	124063.37	D9-1Q2	
45	0	80.6336	124017.84	D13-2Q1	
46	0	80.6437	124002.26	D3-0Q1	
47					ArI 80.64711 Ke
48					CuII 80.65472
49					ArI 80.68689 Ke
50	1	80.7150	123892.69	D6-0Q3	
51					ArI 80.72184 Ke
52		80.7334	123864.45	D3-0Q2	r
53	1	80.7573	123827.88	D9-1Q3	
54					ArI 80.76529 Ke
55		80.8670	123659.84	D3-0Q3	r
56	1	80.9114	123591.92	D6-0Q4	
57		80.9297			s

Nb	I	λ (nm)	σ (cm-1)	Assignment	Comment

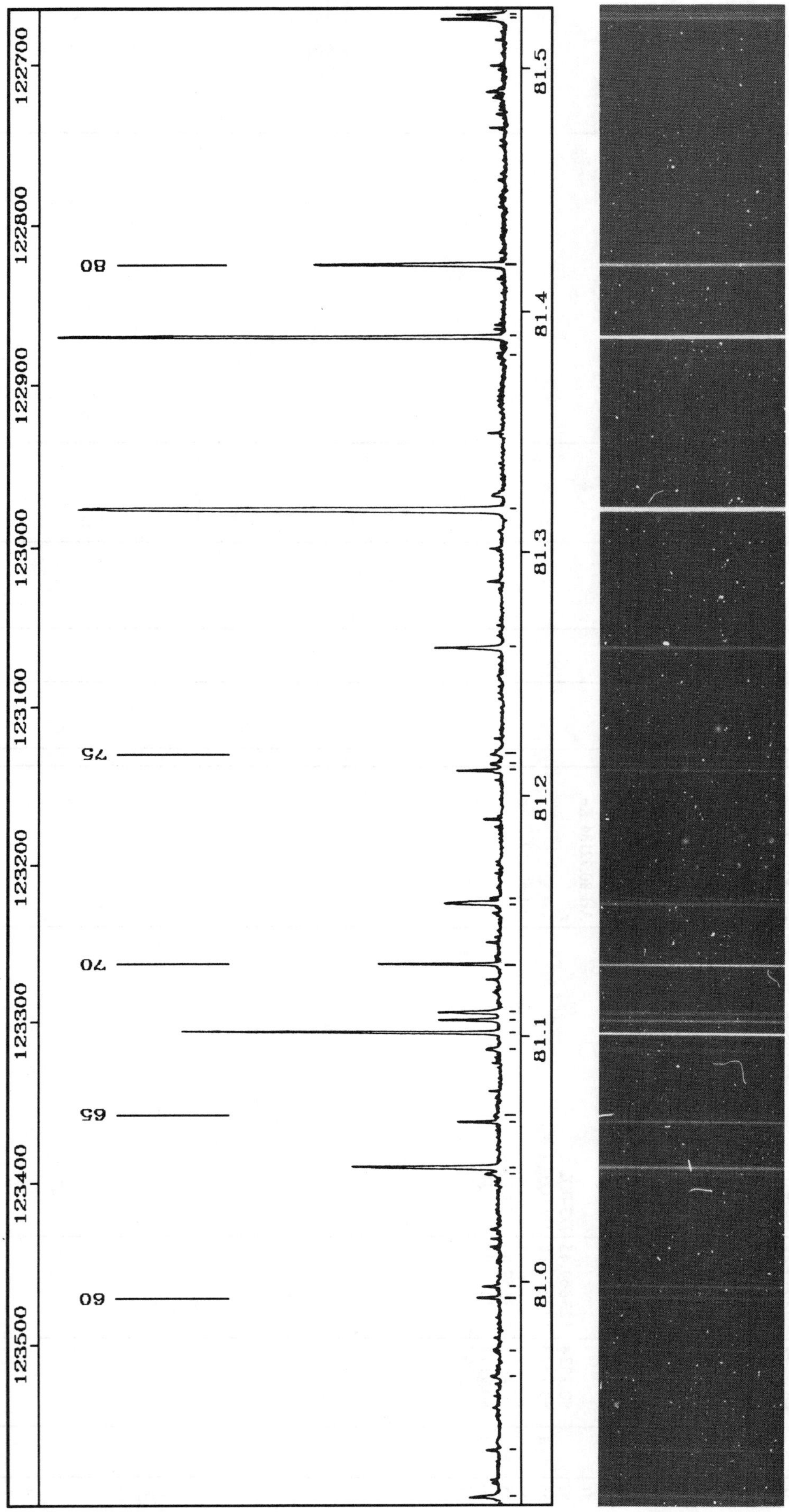
122700
122800
122900
123000
123100
123200
123300
123400
123500
80
75
70
65
60
81.5
81.4
81.3
81.2
81.1
81.0

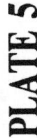

PLATE 5

TABLE 5

Nb	I	λ (nm)	σ (cm-1)	Assignment	Comment
58	0	80.9607	123516.67	D9-1Q4	
59		80.9710			s
60					ArI 80.99266 Ke
61		80.9966			s
62	0	81.0433	123390.88	D'3-0Q4	
63	2	81.0459	123386.95	D'5-1Q1	CuII 81.0635 Ke
64					OI 81.06650
65		81.0637			
66	0	81.0941	123313.53		CuII 81.09984
67					OI 81.10512
68					
69	1	81.1095	123290.08	D12-2Q1	
70		81.1286			CuII 81.129 Ke
71	0	81.1550	123221.06	D6-0Q5	
72		81.1561			
73					OI 81.20936
74	0	81.2132	123132.77	D9-1Q5	
75	0	81.2183	123124.98	D12-2Q2	
76	1	81.2612	123059.93	D'3-0Q5	d
77	9	81.3182	122973.75	D8-1Q1	
78	0	81.3820	122877.24	D12-2Q3	
79					CuII 81.38834
80	2	81.4191	122821.24	D8-1Q2	r 122667.45
81		81.5203	122668.77	D5-0Q1	r 122667.45
82		81.5221	122666.12	D5-0Q1	

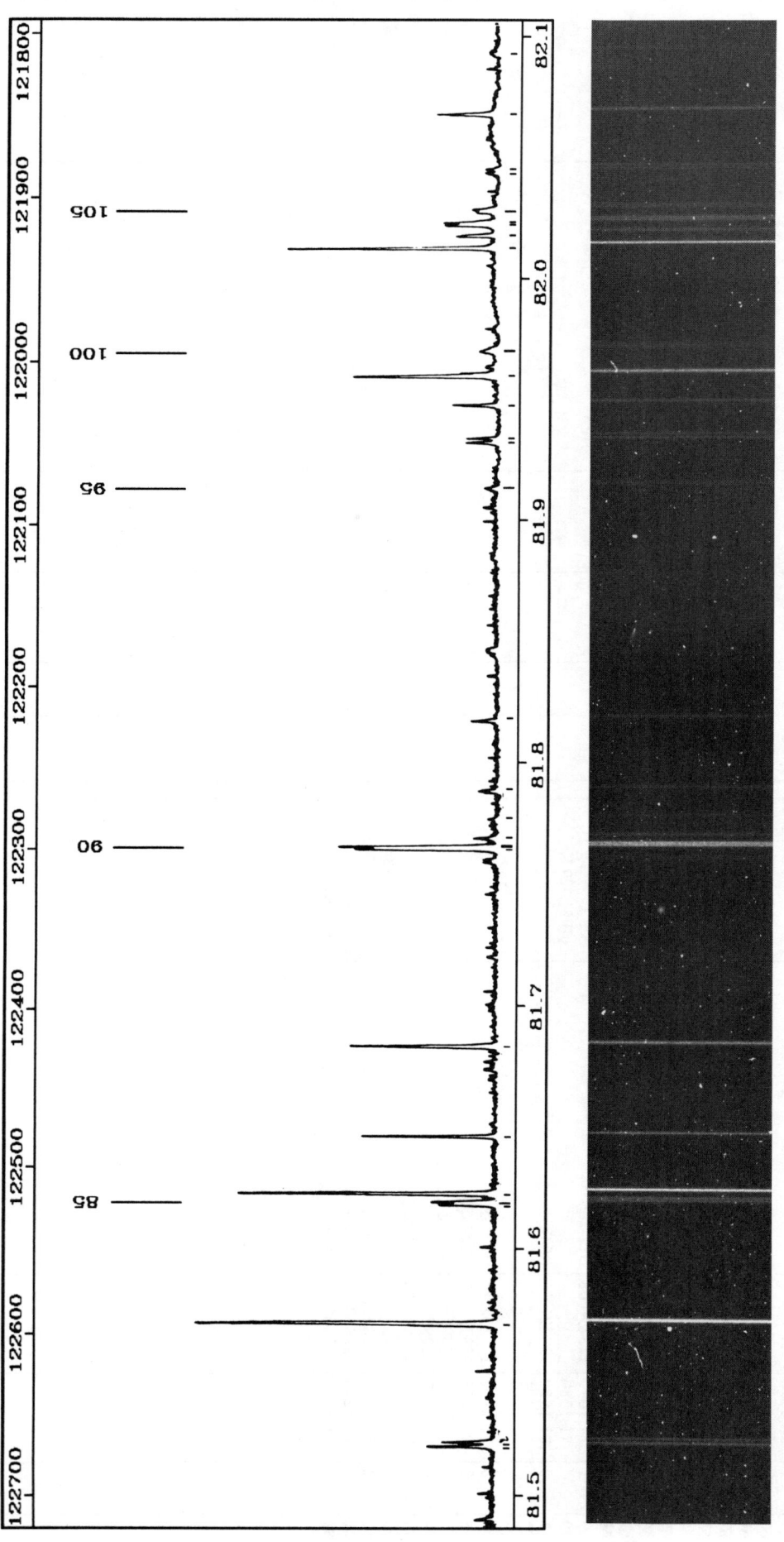

PLATE 6

TABLE 6

Nb	I	λ (nm)	σ (cm⁻¹)	Assignment	Comment

Nb	I	λ (nm)	σ (cm⁻¹)	Assignment	Comment
83	2	81.5700	122594.15	D8-1Q3	
84		81.6187	122520.88	D5-0Q2	r 122520.07
85		81.6198	122519.25	D5-0Q2	r 122520.07
86					ArI 81.62320
87					ArI 81.64640
88	1	81.6834	122423.96	D11-2Q1	
89		81.7653	122301.23	D5-0Q3	r 122300.47
90		81.7663	122299.71	D5-0Q3	r 122300.47
91	0	81.7699	122294.42	D8-1Q4	
92	0	81.7779	122282.50	D6-0Q7	
93	0	81.7893	122265.32	D11-2Q2	
94	0	81.8184	122221.96		
95		81.9141	122079.06		
96	0	81.9327	122051.41	D2-0Q1	r 122050.08
97		81.9345	122048.75	D2-0Q1	r 122050.08
98		81.9480	122028.59	D11-2Q3	
99	1	81.9599	122010.84	D5-0Q4	
100		81.9701	121995.69	D"1-0Q3	r
101					ArI 82.01235
102	0	82.0178	121924.83	D8-1Q5	r 121917.95
103		82.0219	121918.64	D2-0Q2	r 121917.95
104		82.0228	121917.26	D2-0Q2	r
105		82.0278	121909.83		
106		82.0434	121886.75		r 121885.71
107		82.0448	121884.67		r 121885.71
108	1	82.0675	121850.98		
109	0	82.0921	121814.39		

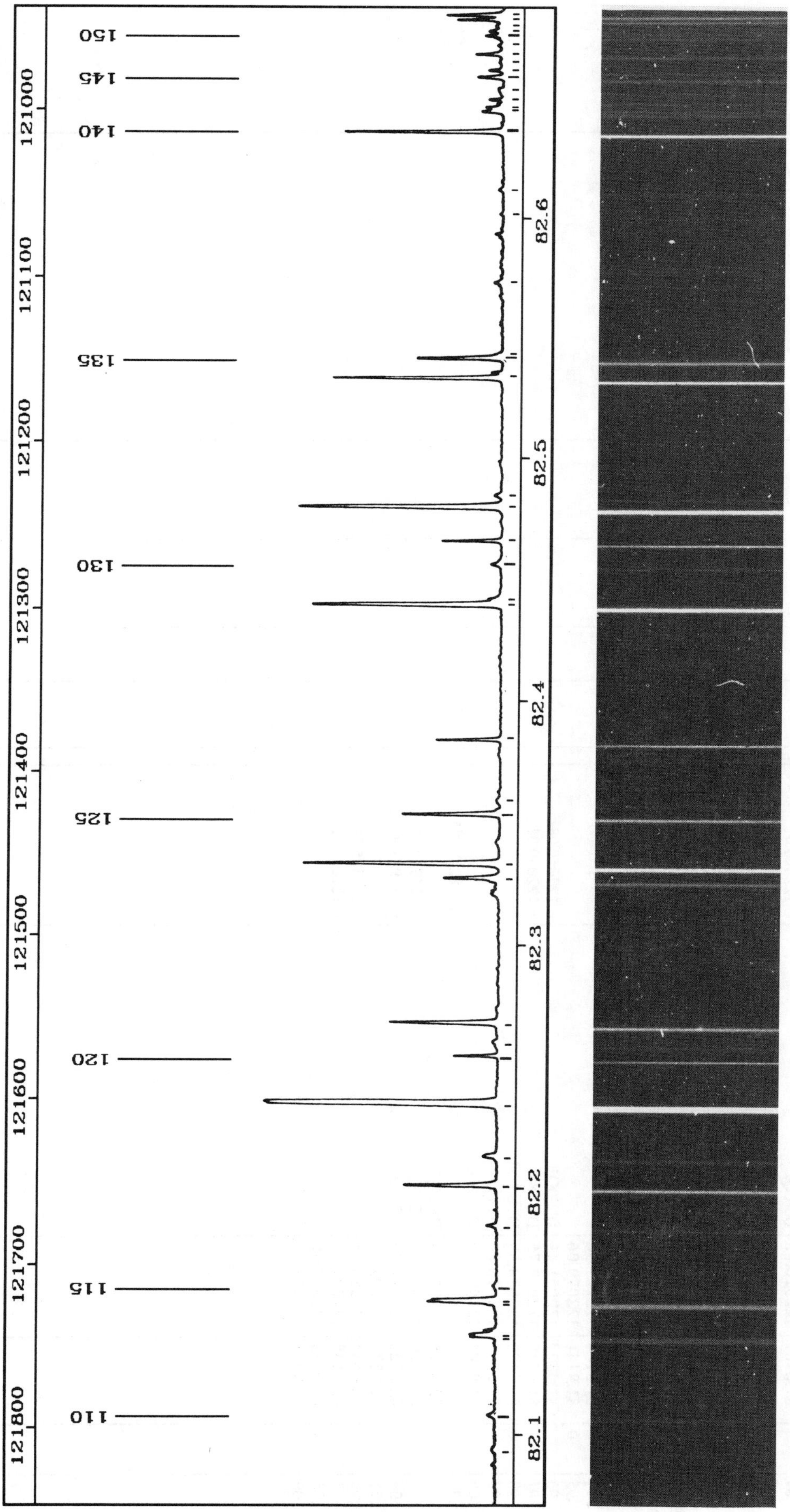
121000
121100
121200
121300
121400
121500
121600
121700
121800
110
115
120
125
130
135
140
145
150
82.1
82.2
82.3
82.4
82.5
82.6

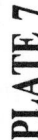

PLATE 7

TABLE 7

Nb	I	λ (nm)	σ (cm-1)	Assignment	Comment
110	0	82.1063	121793.29		
111		82.1386	121745.51		r 121744.73
112		82.1396	121743.94		r 121744.73
113		82.1529	121724.23	D'2-0Q3	r 121723.66
114		82.1537	121723.08	D'2-0Q3	r 121723.66
115	0	82.1595	121714.53		
116		82.1843	121677.76	D'4-1Q1	
117	1	82.2006	121653.63	D'5-0Q5	
118	1	82.2125	121636.05		d
119	20	82.2339	121604.42	D'7-1Q1	s
120		82.2525	121567.10		
121	0	82.2591	121555.44		
122	3	82.2670	121466.23	D'2-0Q4	
123	2	82.3274	121457.67	D'7-1Q2	
124	4	82.3332	121426.89	D'10-2Q1	
125	3	82.3541	121418.65		
126	0	82.3597	121298.80		
127		82.4411	121274.21		CuII 82.38378 Ke
128	5	82.4416			sh
129	0	82.4578	121239.22	D'10-2Q2	
130		82.4665			s
131	4	82.4816	121232.37	D'7-1Q3	
132	0	82.4862	121149.56	D'5-0Q6	ArI 82.53460
133					sh
134	2	82.5426	121147.17		
135		82.5442	121103.45		
136		82.5740	121061.84	D'2-0Q5	
137		82.6024	121047.06	D'6-0Q9	
138		82.6125		D'10-2Q3	
139					
140					ArI 82.63649
141		82.6447	120999.91		N2 (0,1)R5
142		82.6465	120997.30		N2 R4
143		82.6496	120992.75		N2 R2
144	1	82.6526	120988.29	D'8-1Q7	+N2 R0
145		82.6589	120979.12		N2 P1
146		82.6618	120974.85		N2 P2
147		82.6651	120969.99		N2 P3
148		82.6685	120965.04		N2 P4
149		82.6722	120959.59		N2 P5
150		82.6762	120953.76		N2 P6
151	0	82.6781	120951.03	D'7-1Q4	N2 P7
152		82.6804	120947.66		
153		82.6829	120943.96	D'4-0Q1	r 120942.58
154		82.6848	120941.20	D'4-0Q1	r 120942.58

Nb	I	λ (nm)	σ (cm-1)	Assignment	Comment

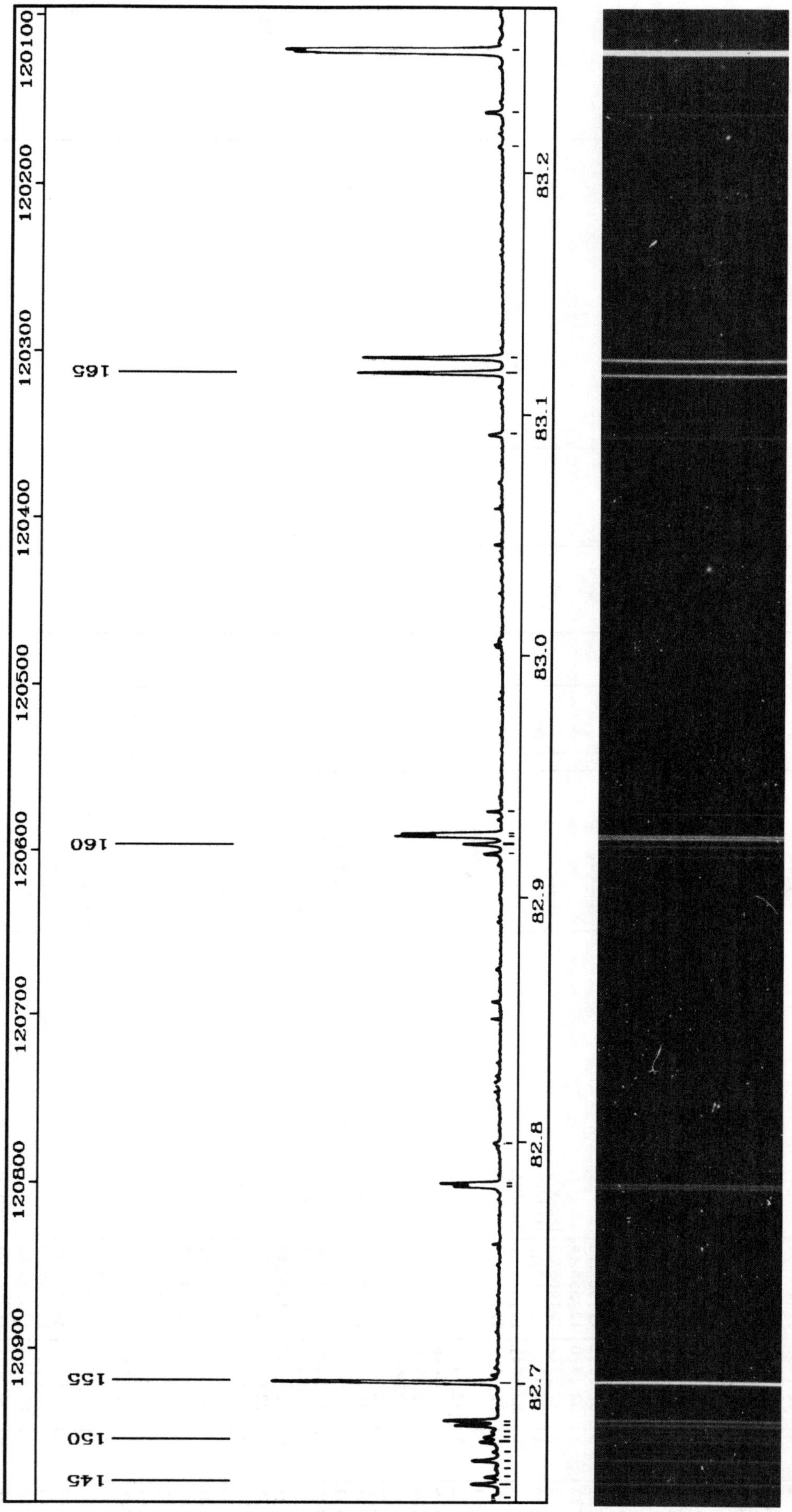
120100
120200
120300
120400
120500
120600
120700
120800
120900
83.2
83.1
83.0
82.9
82.8
82.7
165
160
155
150
145

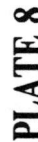

PLATE 8

TABLE 8

Nb	I	λ (nm)	σ (cm-1)	Assignment	Comment
155					CuII 82.69961
156		82.7804	120801.60	D4-0Q2	r 120800.67
157		82.7816	120799.75	D4-0Q2	r 120800.67
158	0	82.7980	120775.84		
159	0	82.9176	120601.67		
160	0	82.9218	120595.61	D7-1Q5	
161		82.9253	120590.46	D4-0Q3	r 120589.56
162		82.9265	120588.69	D4-0Q3	r 120589.56
163		82.9351			s
164	1	83.0921	120348.39	D4-0Q4	
165	2	83.1179	120311.03	D9-2Q1	
166	2	83.1242	120301.97	D7-1Q6	
167	0	83.2113	120176.04	D9-2Q2	
168	0	83.2256	120155.35	D6-1Q1	
169	17	83.2512	120118.45		

Nb	I	λ (nm)	σ (cm-1)	Assignment	Comment

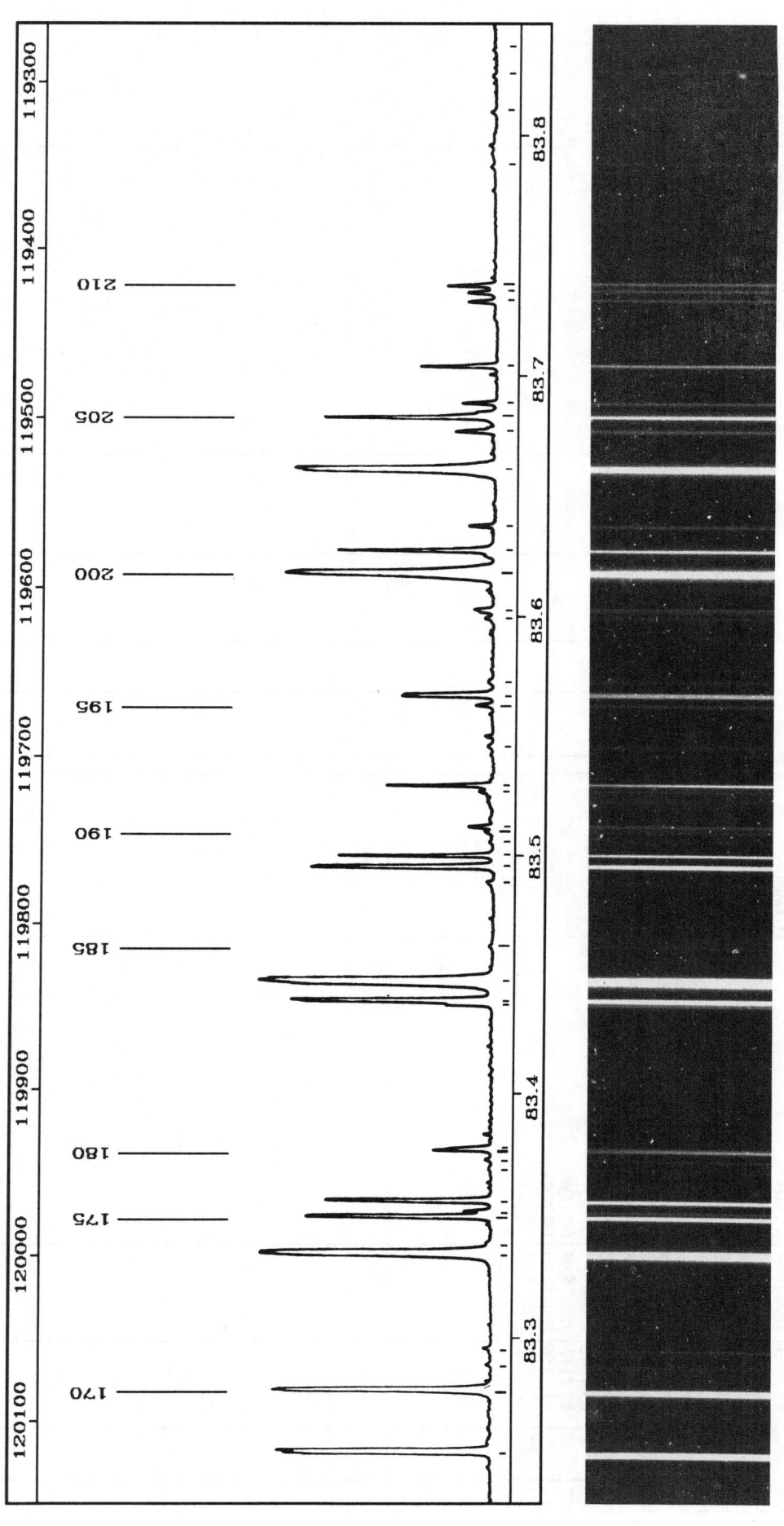

PLATE 9

TABLE 9

Nb	I	λ (nm)	σ (cm-1)	Assignment	Comment

Nb	I	λ (nm)	σ (cm-1)	Assignment	Comment
170		83.2759			OII 83.2762 Ke
171	0	83.2871	120066.66		
172		83.2930			OIII 83.2927 Ke
173		83.3329			OII 83.3332 Ke
174	0	83.3378	119993.60	D"0-0Q2	
175	4	83.3492	119977.23	D6-1Q2	
176	1	83.3513	119974.13	D'1-0Q1	
177	3	83.3557	119967.84	D4-0Q5	
178	0	83.3684	119949.57		
179		83.3716			
180		83.3754			sh
181		83.3772	119936.84	D9-2Q3	r 119849.07
182		83.4374	119850.37	D'1-0Q2	r 119849.07
183		83.4392	119847.77	D'1-0Q2 D'3-1Q1	+ArI 83.43918
184	0	83.4464			OII 83.4462 Ke
185	0	83.4616	119815.57	D"0-0Q3	
186	0	83.4883	119777.23		
187	4	83.4955	119766.92	D6-1Q3	
188		83.5062	119751.56		ArI 83.50021
189	0	83.5098			OIII 83.5096 Ke
190	0	83.5128	119742.10		
191	0	83.5275	119721.12	D'3-1Q2	
192		83.5292			OIII 83.5292 Ke
193					
194	0	83.5459	119694.65	D7-1Q7	
195	1	83.5631	119669.99		
196		83.5675	119663.73	D'1-0Q3	
197		83.5731	119655.73	D5-0Q9	r
198	0	83.5997	119617.67		
199	0	83.6030	119612.93		+CuII 83.60278 Ke
200		83.6185			NII 83.6187 Ke
201		83.6279			NII 83.6279 Ke
202	0	83.6381	119562.78	D4-0Q6	
203		83.6619			NII 83.6616 Ke
204	1	83.6776	119506.32		
205		83.6835			NII 83.6837 Ke
206	1	83.6893	119489.63	D6-1Q4	
207	1	83.7049	119467.29	D5-2Q1	
208	1	83.7324	119428.04	6pπ1-1Q1	
209	1	83.7363	119422.58	D'1-0Q4	
210	1	83.7392	119418.40		
211		83.7894	119346.80		
212	0	83.8122	119314.43	6pπ1-1Q2	
213	0	83.8276	119292.50	D9-2Q5	
214	0	83.8384	119277.11		

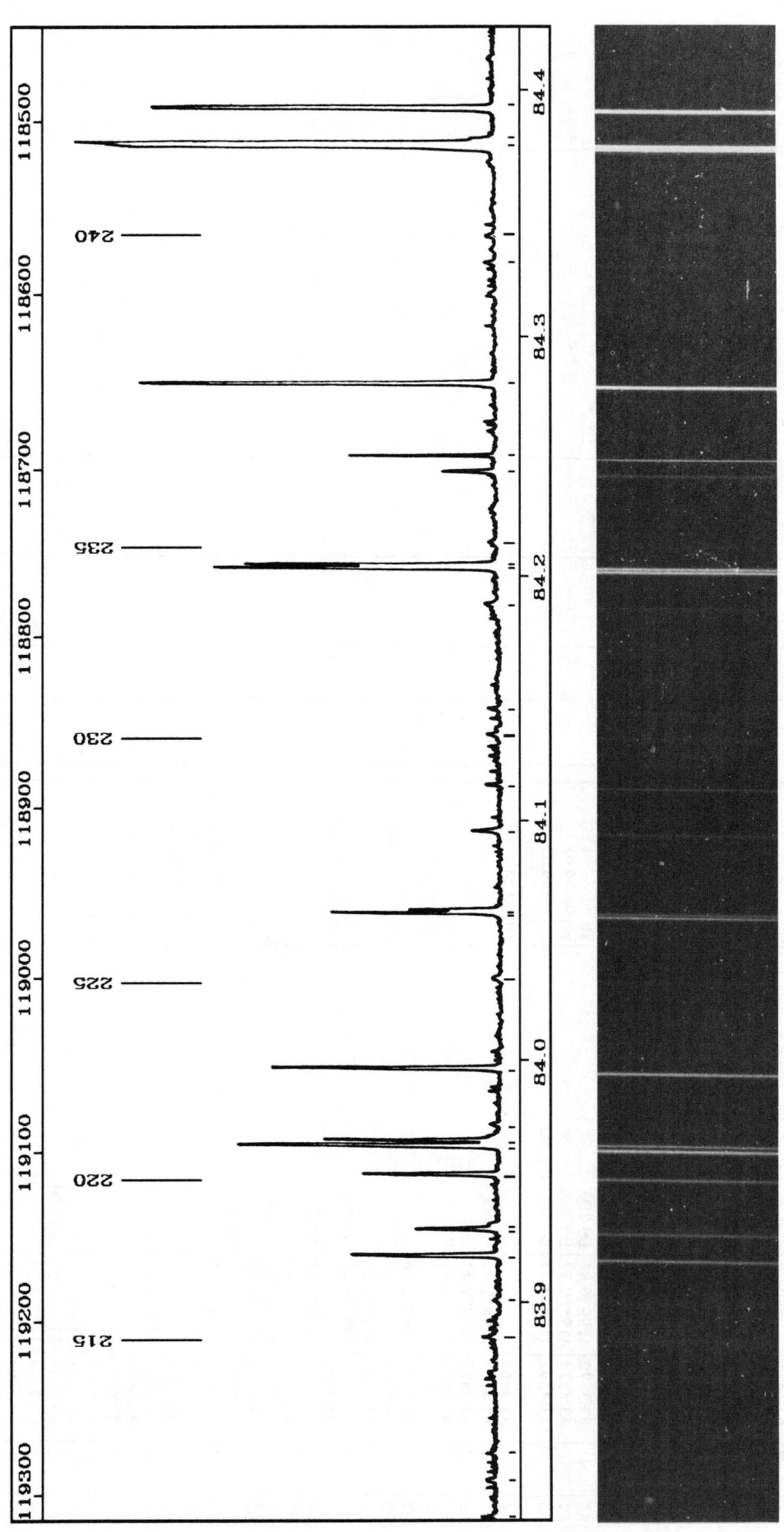

PLATE 10

TABLE 10

Nb	I	λ (nm)	σ (cm^{-1})	Assignment	Comment
215	1	83.8857	119209.86		
216	0	83.9013	119187.71		
217	1	83.9185	119163.28		
218	1	83.9294	119147.75	D6-1Q5	
219	0	83.9315	119144.81	6pπ1-1Q3	
220	1	83.9518	119115.93	D1-0Q5	
221		83.9638	119098.90	D3-0Q1 D4-0Q7	r 119097.39
222		83.9660	119095.87	D3-0Q1	r 119097.39
223		83.9724	119086.69		
224	0	83.9956	119053.84	D8-2Q1	
225	2	84.0333	119000.37		
226	0	84.0604	118962.14	D3-0Q2	r 118961.19
227		84.0617	118960.24	D3-0Q2	r 118961.19
228	0	84.0951	118912.95	D8-2Q2	CuII 84.11346 Ke
229	0				
230	0	84.1351	118856.49		
231	0	84.1458	118841.36		
232	0	84.1892	118780.15		
233		84.2039	118759.30	D3-0Q3	r 118758.34
234		84.2053	118757.37	D3-0Q3	r 118758.34
235	0	84.2143	118744.71	D6-1Q6	
236	0	84.2439	118702.94	D8-2Q3	CuII 84.24964 Ke ArI 84.28051
237					
238	0	84.3304	118581.17	D4-0Q8	
239	0	84.3417	118565.36	D7-1Q9	
240	22	84.3793	118512.45	D5-1Q1	sh
241		84.3820	118508.76		
242	4	84.3945	118491.18	D3-0Q4	
243					

Nb	I	λ (nm)	σ (cm^{-1})	Assignment	Comment

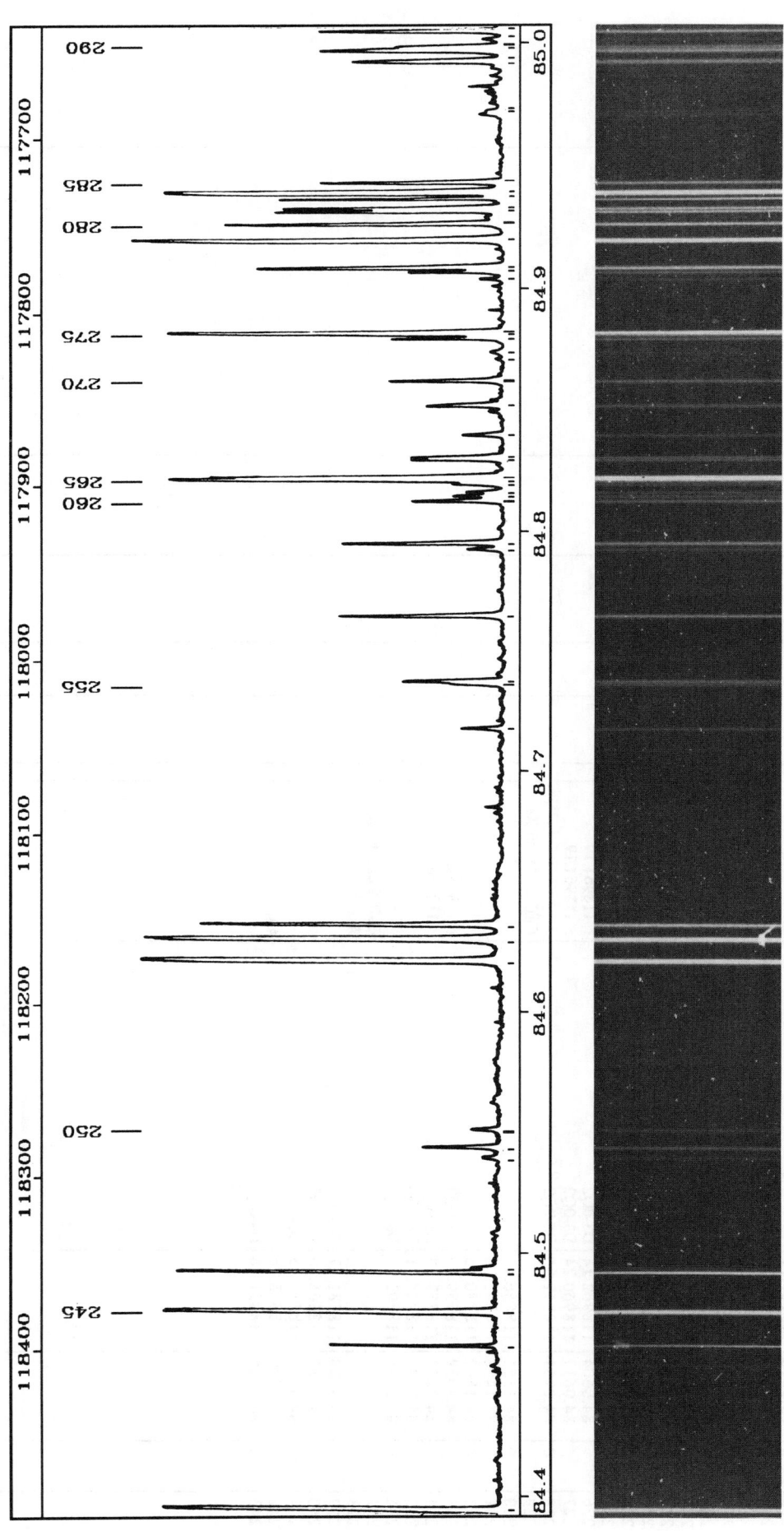

PLATE 11

TABLE 11

Nb	I	λ (nm)	σ (cm⁻¹)	Assignment	Comment

Nb	I	λ (nm)	σ (cm⁻¹)	Assignment	Comment
244					
245	5	84.4761	118376.72	D5-1Q2	CuII 84.46128
246					CuII 84.49122
247		84.4937	118352.06		sh
248	0	84.5390	118288.66	D5-0Q11	
249	1	84.5431	118282.86	D6-1Q7	
250	0	84.5504	118272.60		
251	8	84.6205	118174.74	D5-1Q3	
252	5	84.6296	118161.99	D3-0Q5	
253	4	84.6357	118153.49	D"1-1Q1	
254	1	84.7172	118039.81	D"1-1Q2	
255		84.7359	118013.78		sh
256	1	84.7370	118012.23	D4-0Q9	
257	2	84.7642	117974.36		
258	1	84.7917	117936.13	B34-0P1	
259	3	84.7942	117932.57		
260	0	84.8116	117908.44	D5-1Q4	
261		84.8137	117905.54	D0-0R1	r 117904.53
262		84.8151	117903.52	D0-0R1	r 117904.53
263		84.8185	117898.81	D0-0R0	r 117897.95
264		84.8197	117897.08	D0-0R0	r 117897.95
265	12	84.8211	117895.15	D2-1Q1	
266		84.8287	117884.60	B'5-0R0	r 117883.94
267		84.8297	117883.27	B'5-0R0	r 117883.94
268	1	84.8389	117870.50	D"1-1Q3	
269	1	84.8511	117853.46	D0-0R2	
270	1	84.8615	117839.10		
271	0	84.8706	117826.38	B6-0P2	
272	0	84.8733	117822.73	B'5-0R1	r 117813.70
273		84.8789	117814.90	B'5-0R1	r 117813.70
274		84.8806	117812.50	B'5-0R1	CuII 84.88075
275					
276		84.9027	117776.56	D0-0Q1 D2-1Q2	
277		84.9065	117773.91	D0-0Q1 D2-1Q2 D3-0Q6	r 117775.23
278		84.9085	117758.14	D4-2Q1	r 117775.23
279	9	84.9198	117748.81	D0-0R3	
280	3	84.9266	117741.70	B'5-0P1	r 117740.78
281		84.9317	117739.84	B'5-0P1	r 117740.78
282		84.9330			CuII 84.93594
283					
284	12	84.9395	117730.83		
285	2	84.9435	117725.27		
286	0	84.9715	117686.46		
287	2	84.9728	117684.70	D7-2Q1	
288	2	84.9934	117656.20	D0-0Q2	
289	2	84.9979	117649.94		
290	1	84.9995	117647.71	B'5-0R2 D"1-1Q4	N2 c₆'(0-1)R7
291					+N₂ R6
292	2	85.0059	117638.87	D4-2Q2	

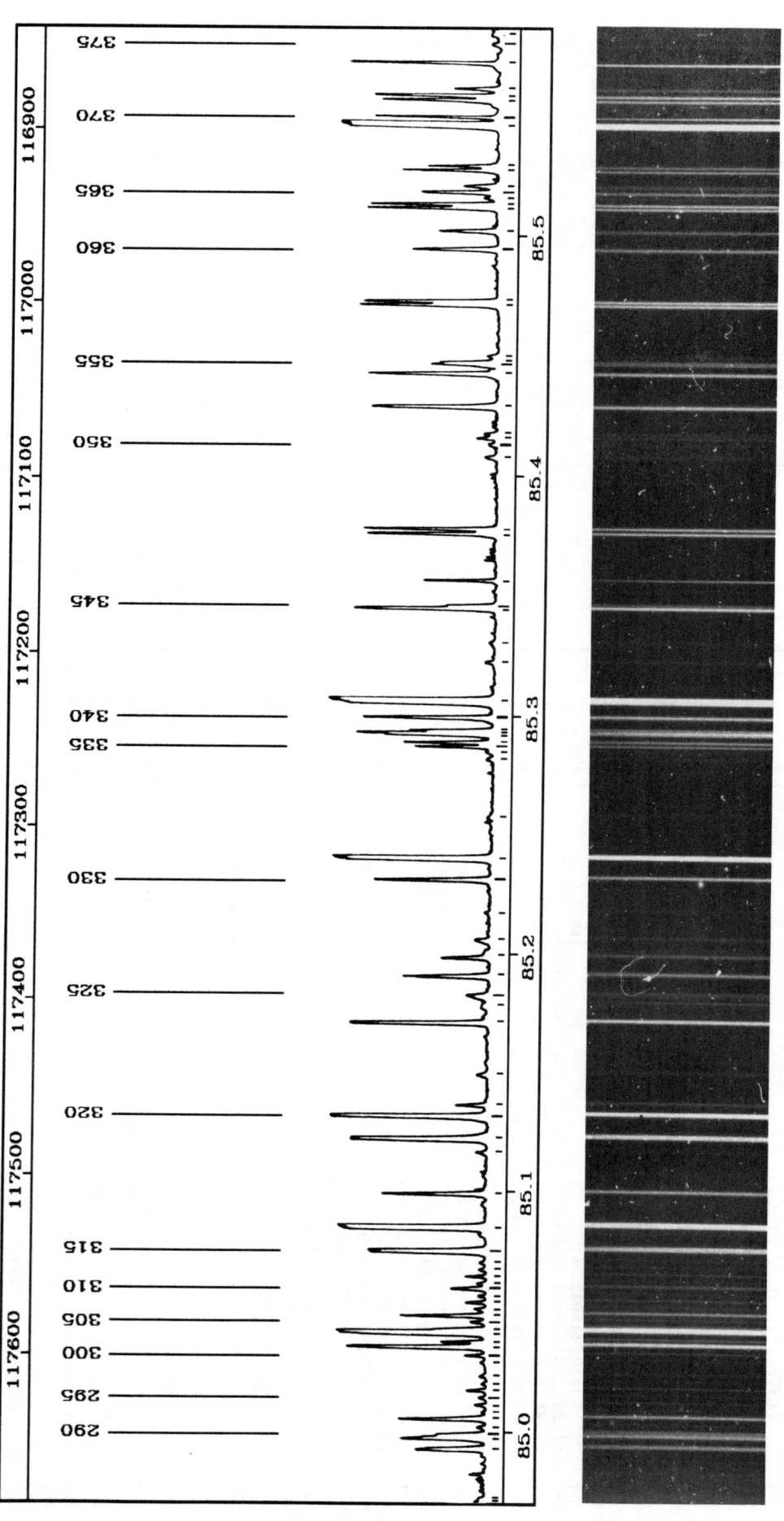

PLATE 12

TABLE 12

Nb	I	λ (nm)	σ (cm^{-1})	Assignment	Comment
293		85.0087	117635.06		N_2 c_6'(0-1)R5
294	4	85.0118	117630.80		N_2 R4
295	0	85.0147	117626.66		N_2 R3
296	9	85.0177	117622.54		N_2 R2
297		85.0205	117618.68		N_2 R1
298		85.0235	117614.50		N_2 R0
299		85.0292	117606.60		N_2 P1
300		85.0321	117602.63		N_2 P2
301	4	85.0355	117597.90	D'2-1Q3	+N_2 P3
302	0	85.0376	117595.07		+N_2 P4
303	9	85.0415	117589.65	D'0-0R4	
304		85.0428	117587.82		sh +N_2 P6
305		85.0457	117583.78		N_2 P7
306	3	85.0484	117580.11	D5-1Q5	+N_2 P8
307		85.0511	117576.39		N_2 P9
308		85.0539	117572.53		N_2 P10
309		85.0566	117568.79		N_2 P11
310		85.0596	117564.64		N_2 P12
311		85.0617	117561.71		N_2 P13
312		85.0646	117557.67		N_2 P14
313	0	85.0675	117553.71	C12-0Q2	+N_2 P15
314	0	85.0703	117549.77	D7-2Q2	+N_2 P16
315		85.0752	117543.14	D'0-0P2	r
316		85.0850	117529.53	B'5-0P2	r
317	1	85.0988	117510.42		r
318	0	85.1160	117486.74	B'6-0P3	
319		85.1217	117478.82	D'0-0Q3	
320					CuII 85.13027
321	1	85.1358	117459.37	D4-2Q3	r
322	0	85.1489	117441.37		
323	5	85.1700	117412.22		CuII 85.17714 Ke
324					
325	1	85.1813	117396.69	D4-0Q10	b
326	1	85.1897	117385.14	B'5-0R3	
327	1	85.1974	117374.46	D'1-1Q5	b
328	0	85.2047	117364.46	D2-1Q4 D'0-0R5	
329	2	85.2166	117348.07	D7-2Q3	
330	10	85.2302	117329.36	D3-0Q7	
331	0	85.2389	117317.34	D'0-0P3	
332	0	85.2569	117292.63	B32-0P1	
333	0	85.2806	117259.95	C12-0Q3	
334	0	85.2838	117255.62		
335		85.2859	117252.75	D2-0R0	r 117251.54
336		85.2876	117250.33	D2-0R0	r 117251.54
337		85.2909	117245.78	D'0-0Q4 D2-0R1	r 117244.66
338					CuII 85.29061
339		85.2926	117243.53	D'0-0Q4 D2-0R1	r 117244.66
340	2	85.2980	117236.11	6pπ0-1Q1	
341	24	85.3049	117226.56	B'5-0P3	
342	0	85.3213	117204.01	C12-0P3	
343	0	85.3294	117192.97	D5-1Q6	
344		85.3436	117173.42	D2-0R2	r 117172.60
345		85.3448	117171.78	D2-0R2	r 117172.60
346		85.3543			s
347		85.3747	117130.70	D2-0Q1 6pπ0-1Q2 ?	r 117129.34
348		85.3767	117127.97	D2-0Q1 6pπ0-1Q2 ?	r 117129.34
349	1	85.4068	117086.68		
350	0	85.4121	117079.49		
351		85.4148	117075.69	D'2-1Q5	r 117074.49
352		85.4166	117073.29	D'2-1Q5	r 117074.49
353	5	85.4278	117057.95	D'0-0P4	
354	2	85.4417	117038.94	D2-0R3	r 117032.98
355		85.4457	117033.45	D2-0R3 C11-0Q1	r 117032.98
356		85.4464	117032.50		
357	0	85.4491	117028.79	D2-0Q2	r 116998.97
358		85.4701	117000.01	D2-0Q2	r 116998.97
359		85.4716	116997.93	6pπ0-1Q3	
360	1	85.4935	116967.97	D'0-0Q5 ?	
361	1	85.5010	116957.68	B"0-0R0	r 116943.68
362		85.5105	116944.74	B"0-0R0	r 116943.68
363		85.5120	116942.62		
364	0	85.5148	116938.81		
365	2	85.5171	116935.71	B'4-0R0	
366	0	85.5196	116932.23	B"0-0R1	r 116921.95
367		85.5263	116923.15	B"0-0R1	r 116921.95
368		85.5280	116920.75	D2-0P2	
369		85.5452	116897.23		r
370					CuII 85.54762
371		85.5555	116883.17	B'4-0R1	r 116881.86
372		85.5574	116880.56	B'4-0R1	r 116881.86
373	1	85.5603	116876.61		CuII 85.57002
374					
375	0	85.5786	116851.59	C11-0Q2	
376		85.5832	116845.34		r

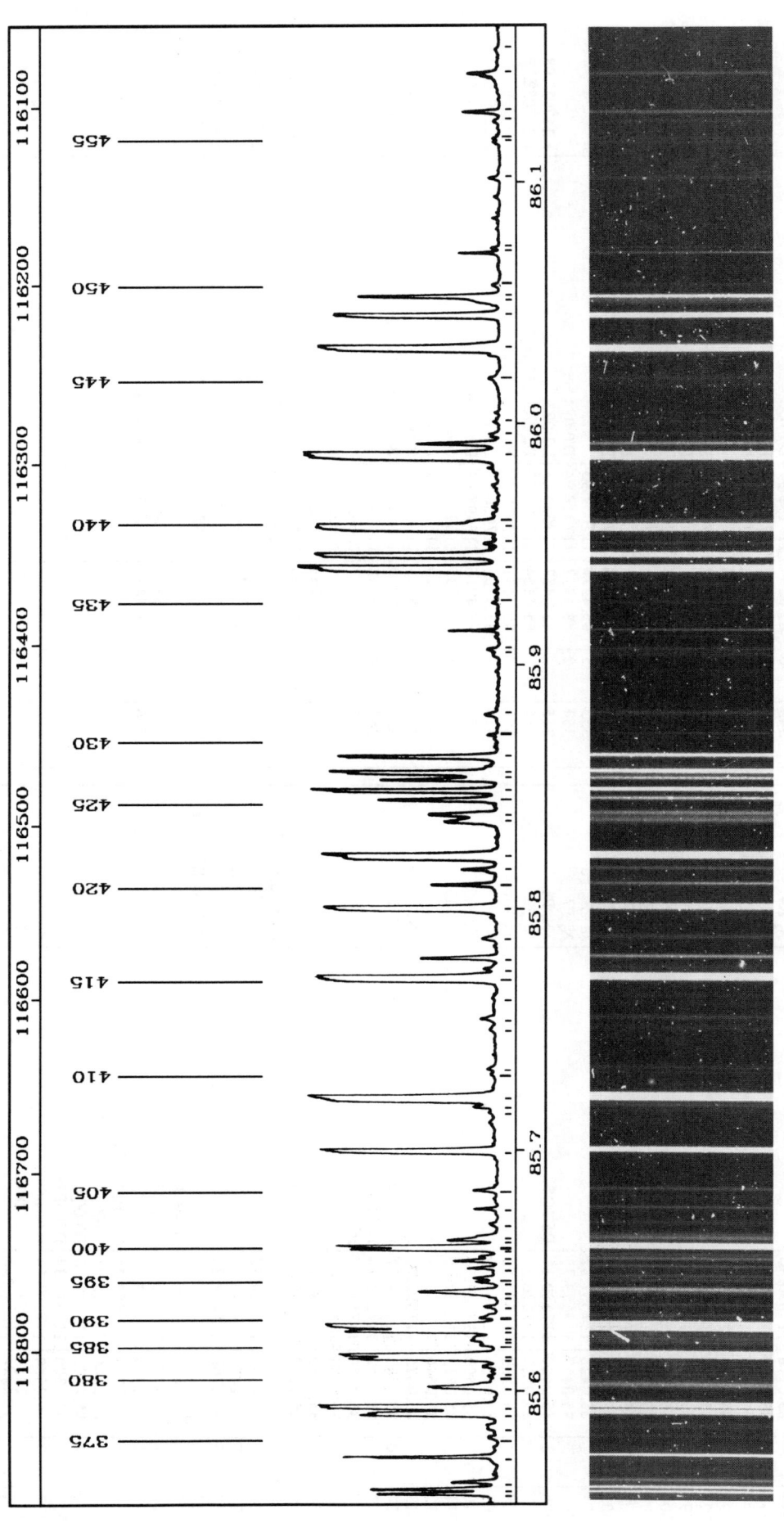

PLATE 13

TABLE 13

Nb	I	λ (nm)	σ (cm⁻¹)	Assignment	Comment
377	7	85.5890	116837.43	B"0-0R2	r
378	3	85.5919	116833.53	D3-0Q8 B'5-0P4	b
379		85.6003	116822.06	D2-0R4	+N₂ c₄(6-0)R-head
380		85.6049	116815.78		N₂ R2
381		85.6066	116813.38		N₂ R1
382		85.6089	116810.31		N₂ R0
383		85.6122	116805.80	D2-0Q3	r 116804.90
384		85.6135	116804.00	D2-0Q3	r 116804.90
385		85.6178	116798.12		N₂ P2
386	1	85.6197	116795.61	C11-0P2 D5-0Q13	
387		85.6212	116793.46		N₂ P3
388		85.6237	116790.08	B'4-0P1	r 116788.69
389		85.6258	116787.30	B'4-0P1 D4-1Q1	r 116788.69
390		85.6294	116782.29		N₂ P5
391		85.6339	116776.22		N₂ P6
392		85.6390	116769.24		N₂ P7 sh
393	1	85.6402	116767.63	B"0-0P1	
394		85.6440	116762.45		N₂ P8
395	0	85.6458	116759.92		
396	0	85.6498	116754.49	D7-2Q5	+N₂ P9
397	1	85.6529	116750.23	D5-1Q7	
398		85.6547	116747.83		N₂ P10
399		85.6577	116743.74	B'4-0R2	r 116742.84
400		85.6590	116741.94	B'4-0R2	r 116742.84
401		85.6618	116738.15		N₂ P11
402		85.6630	116736.46	D4-0Q11	sh
403	0	85.6682	116729.37	D2-1Q6	+N₂ P12
404		85.6745	116720.82		N₂ P13
405	0	85.6823	116710.29		
406	7	85.6986	116688.02	B"0-0R3	
407	0	85.7149	116665.86	C11-0R3	
408	0	85.7172	116662.69		
409	34	85.7210	116657.56	D4-1Q2 D2-0P3	
410	0	85.7309	116644.13		
411	0	85.7332	116641.00		
412	0	85.7499	116618.21		
413	0	85.7542	116612.38	D'0-0Q6 D9-3Q1	
414	0	85.7633	116599.98		
415	23	85.7712	116589.22	B"0-0P2	
416	0	85.7749	116584.21		
417	1	85.7796	116577.78	B'4-0P2	
418	0	85.7880	116566.47	C11-0Q3	
419		85.8005	116549.41	D2-0Q4	r
420					CII 85.80918
421	1	85.8164	116527.84	B'4-0R3	r
422		85.8221	116520.10	B'6-0P5	
423	0	85.8368	116500.15	C11-0P3	
424	1	85.8396	116496.30		
425	1	85.8458	116487.98	B'5-0R5	

Nb	I	λ (nm)	σ (cm⁻¹)	Assignment	Comment
426	1				CuI 85.84869
427		85.8542	116476.58	B"0-0R4 D9-3Q2	CII 85.85590
428					+CuII 85.85667 Ke
429	5	85.8638	116463.54	D4-1Q3	
430		85.8720			s
431	0	85.8814	116439.65	B30-0P1	
432	0	85.9060	116406.35	B31-0R3	
433	0	85.9081	116403.50		
434					CuII 85.91509 Ke
435	0	85.9278	116376.80		
436		85.9417	116357.94	D2-0P4	r
437	8	85.9473	116350.38	B'5-0P5	
438	0	85.9521	116343.87		
439	36	85.9588	116334.82	B"0-0P3 D2-1Q7	sh
440		85.9611	116331.64	B31-0P3	
441	51	85.9884	116294.76	B'4-0P3	
442	2	85.9934	116287.98	D3-0Q9	
443	0	85.9976	116282.38		
444	0	86.0025	116275.68	D9-3Q3	
445	0	86.0208	116250.95		
446	15	86.0332	116234.27	D2-0Q5	
447	6	86.0469	116215.65	B4-0R4 B30-0P2	
448		86.0526	116207.95	D4-1Q4	sh
449	3	86.0549	116204.92	B"0-0R5	
450	0	86.0595	116198.66	C11-0Q4	
451					CuII 86.07217 Ke
452	0	86.0751	116177.56	C10-0R1	
453	0	86.1041	116138.45	D12-4Q1	
454		86.1188	116118.69	C10-0Q1	r 116117.59
455		86.1204		C10-0Q1	r 116117.59
456	0	86.1275	116106.88	C11-0P4	
457	1	86.1313	116101.83		
458	0	86.1473	116080.28		
459	0	86.1580	116065.83		

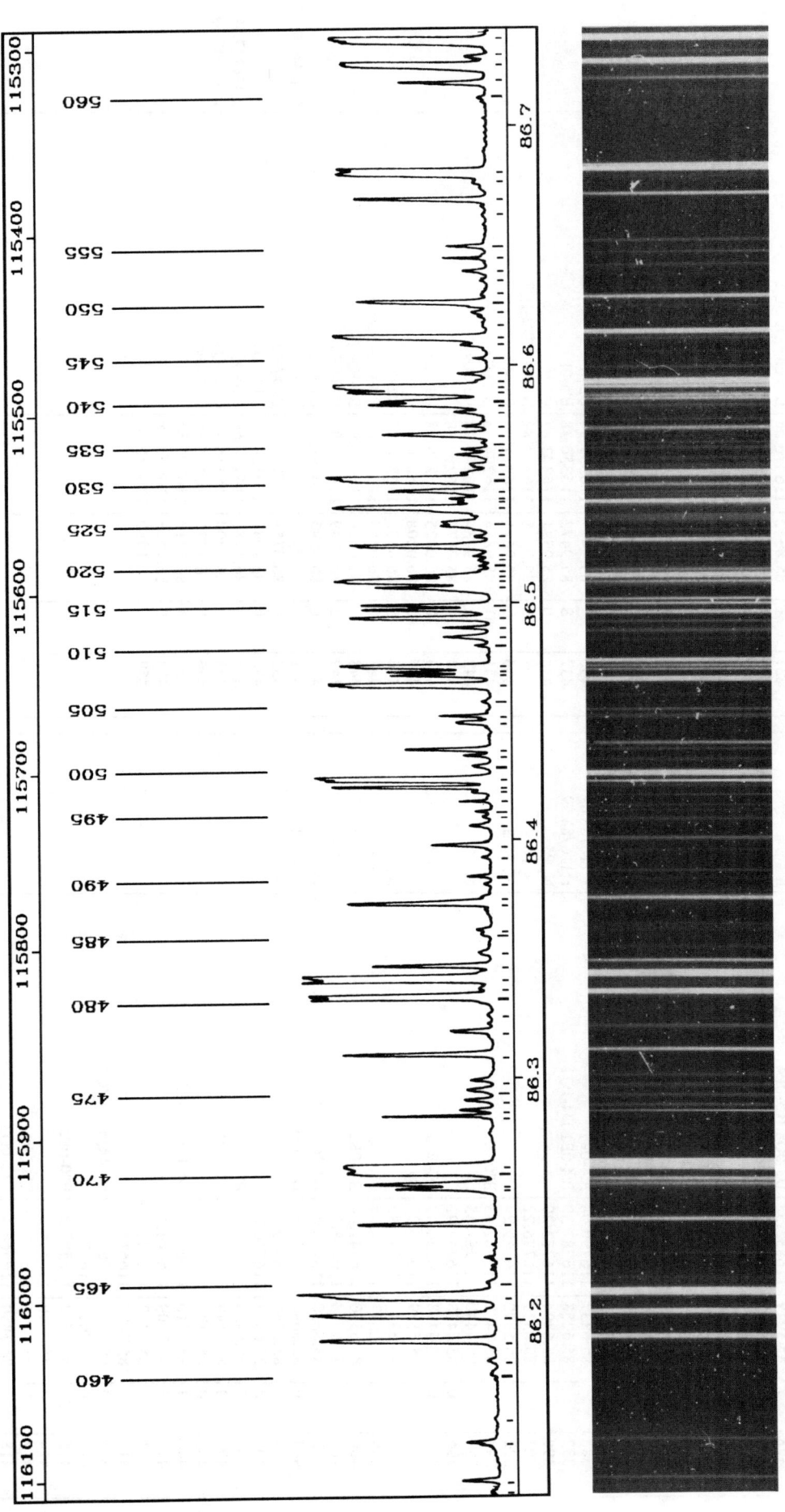

PLATE 14

TABLE 14

Nb	I	λ (nm)	σ (cm^{-1})	Assignment	Comment
460	0	86.1769	116040.32	D4-0Q12	
461	0	86.1842	116030.55	C10-0R2	
462	7	86.1898	116023.06	B"0-0P4	
463					CuII 86.19936
464	30	86.2084	115998.03	D2-0P5 D5-0Q14	
465	0	86.2140	115990.48		
466					GeII 86.22339
467	5	86.2383	115957.72	D"0-1Q1 B29-0P1	
468	2	86.2531	115937.88		
469	6	86.2551	115935.24		
470	23	86.2605	115927.97	B'4-0P4	
471	34	86.2620	115925.94	D3-2Q1	
472					CuII 86.28226 Ke
473	0	86.2858	115893.99	D4-1Q5	
474	0	86.2897	115888.70	B30-0P3 D'2-1Q8	
475	0	86.2939	115883.05	C10-0P2	
476	0	86.2980	115877.49	D2-0Q6 B"0-0R6	
477	3	86.3089	115862.92	D"0-1Q2	
478	2	86.3188	115849.68		
479	0	86.3256	115840.56		
480		86.3330	115830.54	B'4-0R5	r
481	0	86.3367	115825.59	D'1-1Q1	
482	3	86.3409	115820.03	D3-2Q2	r
483	0	86.3463	115812.74		
484	0	86.3519	115805.17		
485	0	86.3598	115794.67	B29-0R2	
486	0	86.3616	115792.15	C10-0R3	
487	6	86.3727	115777.39	D5-3Q1 ?	
488	0	86.3753	115773.89	D'0-0Q8	sh
489	0	86.3809	115766.31		
490	0	86.3843	115761.78		s
491	0	86.3929	115750.21		
492	2	86.3979	115743.57	C11-0Q5	
493	0	86.4058	115732.95	B29-0P2	
494	0	86.4098	115727.65		
495	0	86.4119	115724.75		
496					CuII 86.41546 Ke
497	0	86.4204	115713.39		
498					CuII 86.42138 Ke
499	17	86.4257	115706.31	D'1-1Q2	
500	0	86.4315	115698.63	D3-0Q10	
501	0	86.4358	115692.75		s
502	2	86.4384	115689.30	D"0-1Q3	
503	0	86.4497	115674.19	C10-0Q3	
504	0	86.4526	115670.28	D5-3Q2	s
505	0	86.4586	115662.32	B"0-0P5	
506	8	86.4653	115653.28	B'3-0R0	
507		86.4694	115647.86	B'3-0R0	r 115646.83
508		86.4709	115645.79	B'3-0R0	r 115646.83
509	7	86.4731	115642.86	D'3-2Q3	
510	0	86.4821	115630.89	C11-0P5	
511	0	86.4858	115625.92		
512	0	86.4897	115620.77		s
513	4	86.4936	115615.52		
514	0	86.4973	115610.56	B'3-0R1	r 115609.32
515	0	86.4992	115608.07	B'3-0R1	r 115609.32
516	3	86.5064	115598.42		
517	16	86.5090	115594.87		
518	2	86.5114	115591.77	C10-0P3	
519	0	86.5135	115588.96		
520	0	86.5157	115585.97		
521	0	86.5180	115582.93		
522	0	86.5203	115579.83		
523	3	86.5239	115574.95		
524	0	86.5282	115569.26	B28-0R0 D2-0P6	
525	0	86.5333	115562.46		
526		86.5424	115550.32		r CuII 86.53902
527		86.5438	115548.39		r 115549.36 ?
528		86.5471	115544.01		r 115549.36 ?
529	4	86.5518	115537.81		
530	25	86.5548	115533.80	D'1-1Q3	
531	0	86.5580	115529.42		
532	0	86.5623	115523.67		
533	0	86.5647	115520.56	D4-1Q6	sh
534	0	86.5673	115517.01		
535	0	86.5708	115512.36		
536	2	86.5744	115507.57	6pπ1-2Q1	
537	0	86.5804	115499.64		
538	0	86.5833	115495.75	B'3-0R2	r 115494.85
539		86.5846	115493.95	B'3-0R2 B29-0R3	r 115494.85
540		86.5884	115488.96	B'3-0P1 B31-0R5	r 115487.52
541		86.5905	115486.07	B'3-0P1 B4-0P5 B28-0R1	r 115487.52
542	0	86.5966	115477.94		
543	0	86.6030	115469.49		
544	0	86.6087	115461.86	C10-0R4 B30-0P4	
545	0	86.6116	115457.92		
546	0	86.6168	115451.11		
547	20	86.6223	115443.70		
548	0	86.6265	115438.13		
549	0	86.6300	115433.41	B28-0P1	
550	5	86.6356	115426.01	D2-0Q7	
551	0	86.6399	115420.26		s
552	0				s
553	1	86.6502	115406.50	D'3-2Q4	CuII 86.64427
554	0				
555	1	86.6502	115406.50	B29-0P3 6pπ1-2Q2	
556	0	86.6636	115388.71	B"0-0P5	
557	4	86.6699	115380.32	B'3-0R0	
558	0	86.6769	115370.94	B31-0P5	

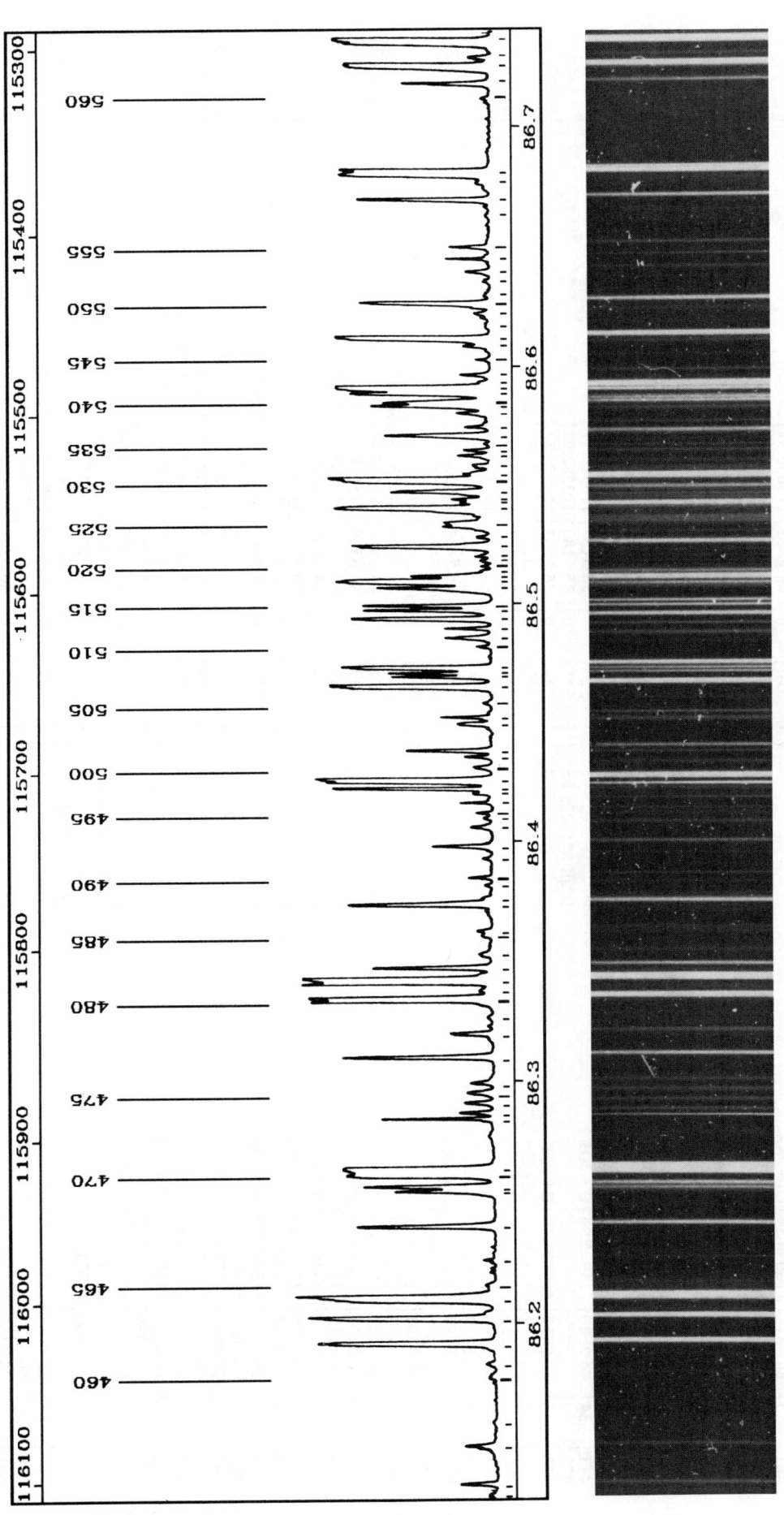

PLATE 14

TABLE 14 *continued*

Nb	I	λ (nm)	σ (cm⁻¹)	Assignment	Comment
559	9	86.6819	115364.40	D8-3Q1	+ArI 86.6800
560	0	86.7125	115323.64	C10-0Q4	
561	2	86.7186	115315.46	D1-1Q4	
562	7	86.7257	115306.13	B'3-0R3 D4-0Q13	
563	0	86.7296	115300.86		r
564	0	86.7359	115292.46	B'3-0P2	

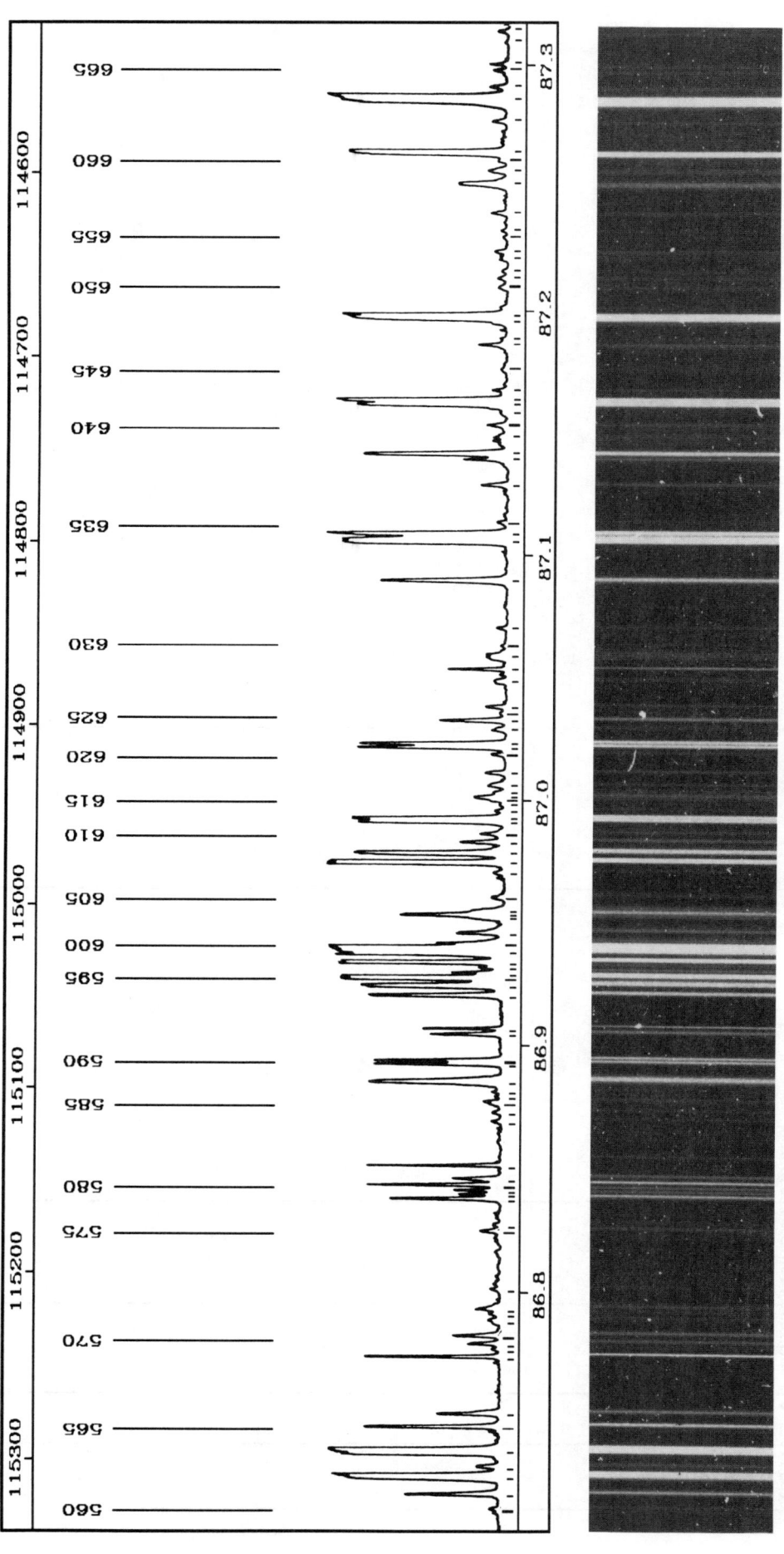

PLATE 15

TABLE 15

Nb	I	λ (nm)	σ (cm⁻¹)	Assignment	Comment
565	4	86.7460	115279.10	D0-0Q9	
566	3	86.7513	115272.04	D11-4Q1	
567					CuII 86.77336
568	0	86.7772	115237.64		
569	1	86.7797	115234.39	D8-3Q2	
570	0	86.7829	115230.10	B"0-0P6	
571	0	86.7886	115222.49	B28-0P2 D"0-1Q5	
572	0	86.7918	115218.27		b
573	0	86.7937	115215.74	C10-0P4	
574	0	86.8015	115205.38	C11-0Q6	
575	0	86.8249	115174.36		
576	0	86.8274	115171.07	D5-0Q15	
577		86.8381	115156.89	D1-0R0	r 115155.74
578		86.8398	115154.59	D1-0R0	r 115155.74
579		86.8414	115152.46	D1-0R1	r 115151.01
580		86.8436	115149.56	D1-0R1	r 115151.01
581	1	86.8461	115146.17	D3-2Q5	
582	0	86.8540	115135.79	D11-4Q2	
583	0	86.8694	115115.28		
584	0	86.8730	115110.56		
585	0	86.8771	115105.16	B'6-0P7	
586	0	86.8796	115101.87	C10-0R5	
587	0	86.8816	115099.17	D4-1Q7	
588	0	86.8858	115093.66	B29-0R4	+NI 86.8860 Ke
589	1	86.8929	115084.24	D1-0R2	r 115083.30
590	0	86.8943	115082.36	D1-0R2	r 115083.30
591	3	86.9050	115068.14	C11-0P6 D3-0Q11	
592					CuII 86.90641
593	5	86.9211	115046.84	B'3-0R4	
594	4	86.9251	115041.60	D1-1Q5 D8-3Q3	
595		86.9282	115037.49	D1-0Q1	r 115036.03
596		86.9304	115034.56	D1-0Q1 C9-0Q1	r 115036.03
597		86.9324	115031.90	C9-0Q1	+r 115033.22
598	60	86.9396	115022.42	B'3-0P3	CuII 86.93360
599		86.9424	115018.67		
600	0	86.9468	115012.86		
601	0	86.9528	115004.87		
602	0	86.9542	115003.06		
603	0	86.9558	115000.97		
604	0	86.9612	114993.76	B28-0R3	
605	0	86.9712	114980.58	B29-0P4	
606	6	86.9757	114974.67		+ArI 86.97541
607	3	86.9796	114969.51	B'4-0P6	
608	1	86.9840	114963.62	D2-0Q8	
609	0	86.9873	114959.35		
610	0				
611		86.9925	114952.48	D1-0R3	r 114951.68
612	0	86.9937	114950.88	D1-0R3	r 114951.68
613	0	86.9967	114946.89	B27-0R1	
614	0	86.9994	114943.31	D3-1Q1	
615	0	87.0011	114941.09	B30-0P5	
616	0	87.0024	114939.38		
617	0	87.0044	114936.72		
618	0	87.0071	114933.10	D11-4Q3	
619	0	87.0124	114926.22		
620	0	87.0199	114916.22		r 114911.27
621		87.0229	114912.29	D1-0Q2	r 114911.27
622	0	87.0245	114910.24	D1-0Q2	
623		87.0302	114902.64	B27-0P1	
624	1	87.0339	114897.77	B28-0P3	
625	0	87.0369	114893.77		
626	0	87.0395	114890.41	C10-0Q5	
627	0	87.0500	114876.52		CuII 87.05389
628					r
629	0	87.0600	114863.30	C9-0Q2	
630	0	87.0645	114857.38		
631	0	87.0716	114848.08		
632	3	87.0910	114822.43		
633	19	87.1073	114801.01	D1-0P2	ArIII 87.110 Ke
634	0	87.1099	114797.58		
635	0	87.1147	114791.18		
636	1	87.1301	114770.90	C10-0P5	
637	0	87.1405	114757.19	D1-0R4	
638	6	87.1429	114754.10	C9-0R3	
639	0	87.1496	114745.29	D0-0Q10	
640	0	87.1546	114738.62		r 114726.08
641	0	87.1599	114731.70	D1-0Q3	r 114726.08
642	0	87.1634	114727.01	D1-0Q3	
643	0	87.1649	114725.14		
644	0	87.1688	114719.92		
645	0	87.1772	114708.94		
646	0	87.1874	114695.53		
647	0	87.1897	114692.46		
648	36	87.1963	114683.77	B'3-0P4 B27-0P2	
649	0	87.1989	114680.40	D11-4Q4	
650	0	87.2108	114664.65		
651	0	87.2150	114659.24		
652	0	87.2182	114654.95		
653	0	87.2232	114648.37		
654	0	87.2256	114645.26		
655	0	87.2319	114636.92		
656	0	87.2345	114633.51	D3-1Q3	
657	0	87.2415	114624.41	D4-1Q8	
658	0	87.2532	114609.00	C9-0Q3	
659	0	87.2586	114601.92	B28-0R4	
660	0	87.2628	114596.35	B29-0R5	r
661	10	87.2659	114592.31	D5-2Q1	

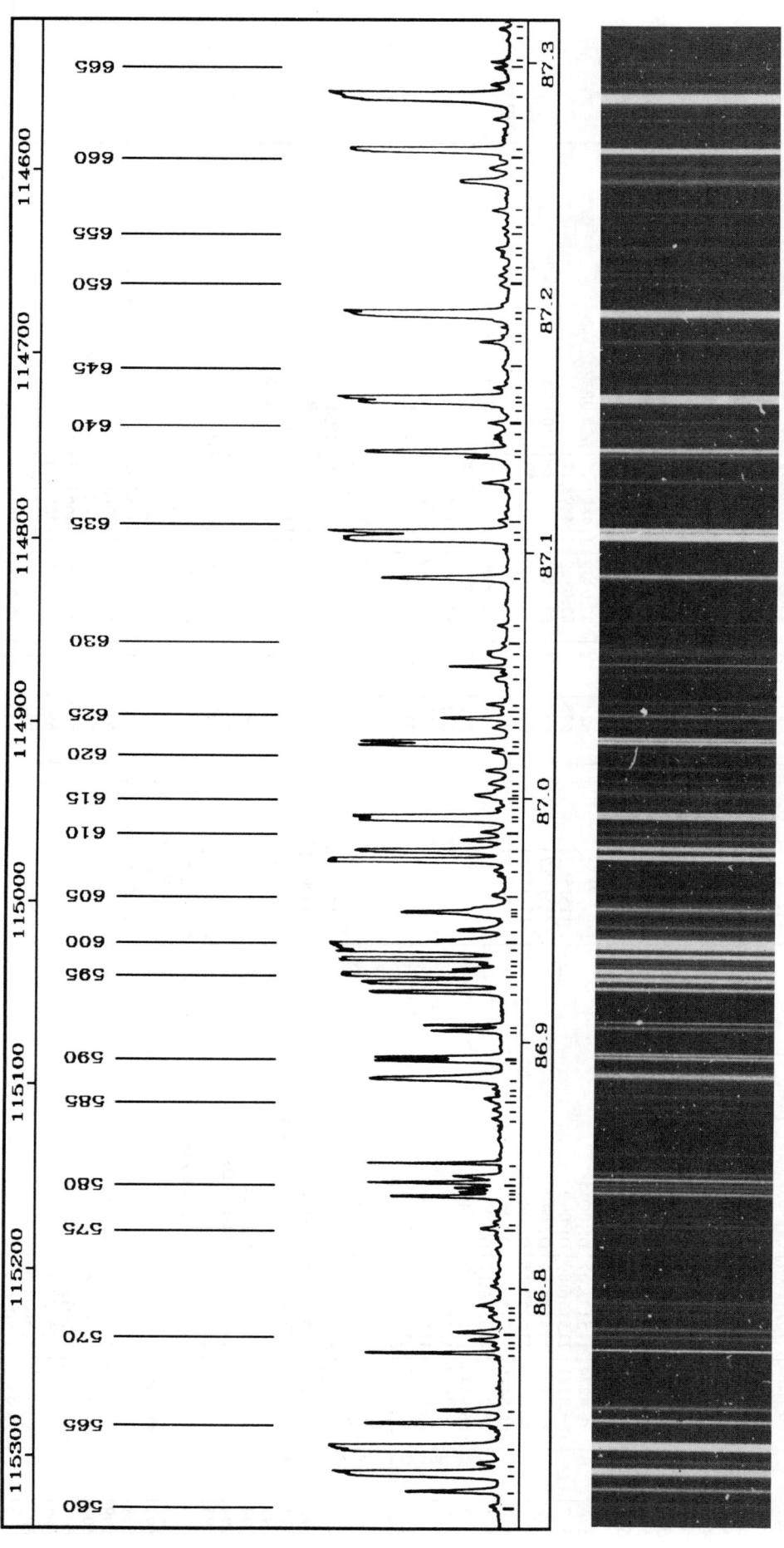

PLATE 15

TABLE 15 *continued*

Nb	I	λ (nm)	σ (cm⁻¹)	Assignment	Comment
662	0	87.2786	114575.61		
663	44	87.2875	114563.95	D1-0P3	
664	0	87.2927	114557.17		
665	0	87.2993	114548.52		
666	0	87.3021	114544.83	D4-0Q14	
667	0	87.3109	114533.29		
668	0	87.3153	114527.52		

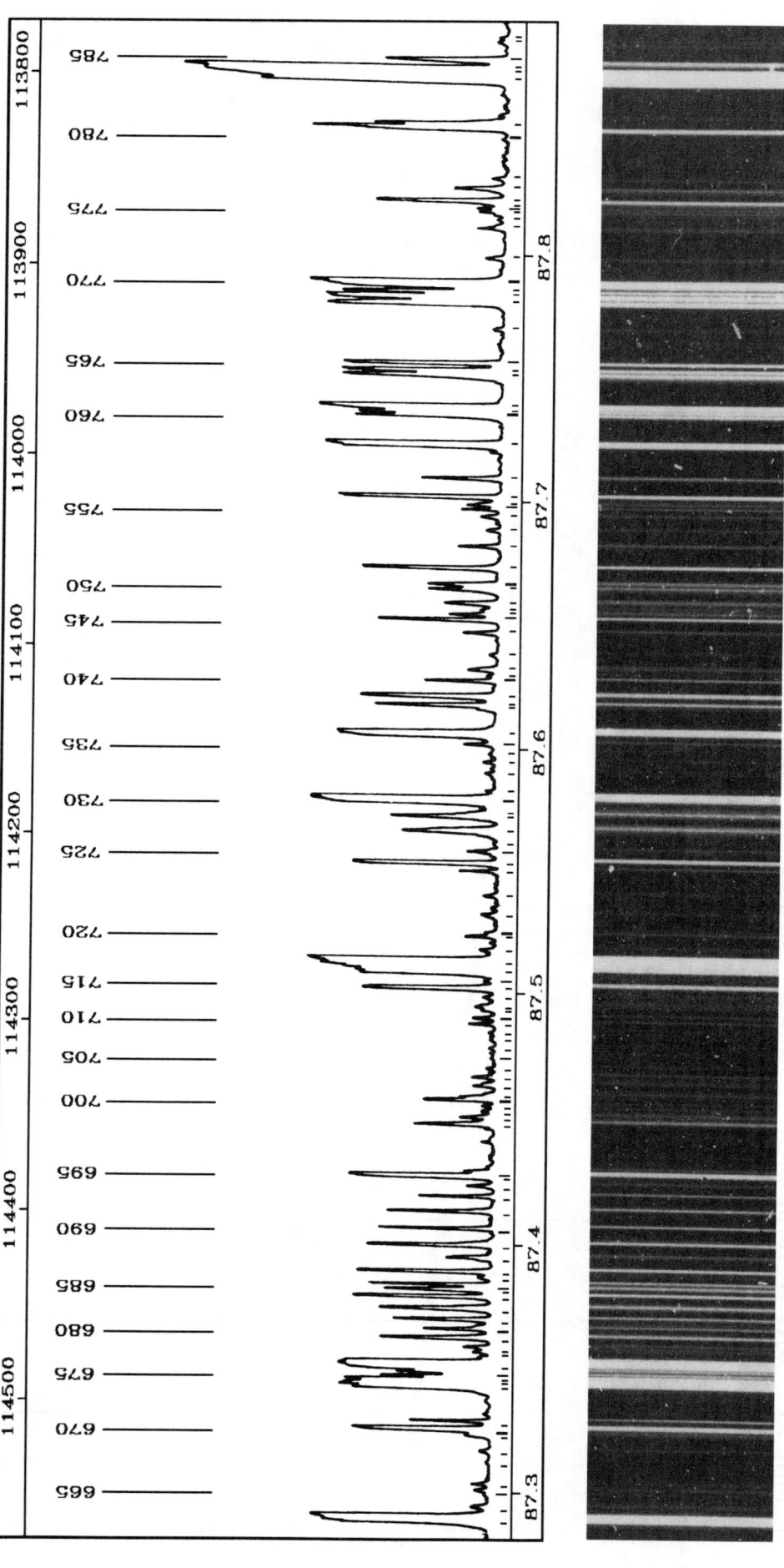

PLATE 16

TABLE 16

Nb	I	λ (nm)	σ (cm⁻¹)	Assignment	Comment
669	0	87.3216	114519.19		
670	2	87.3239	114516.24	C9-0P3	
671		87.3414	114493.25		CuII 87.32629
672	0	87.3437	114490.27		+N₂ c₄'(6-1)R-head b
673		87.3451	114488.40	D5-1Q11	N₂R4
674		87.3468	114486.14		N₂R3
675		87.3495	114482.59		N₂R2
676	19	87.3550	114475.44	D1-0Q4 D1-0R5	+N₂R1 b
677	0	87.3569	114472.98	B28-0P4	
678	0	87.3605	114468.26		
679	0	87.3639	114463.70	D5-2Q2	+N₂P2
680		87.3678	114458.66		N₂P3
681	0	87.3722	114452.95	B27-0R3	N₂P4
682		87.3767	114446.99		+N₂P5
683		87.3796	114443.20	D2-0Q9	N₂P6
684	2	87.3816	114440.59		
685		87.3846	114436.71	C9-0R4	N₂P7
686	0	87.3869	114433.67		
687		87.3926	114426.19		N₂P8
688	0	87.3976	114419.63		
689		87.4042	114410.94		N₂P10
690	1	87.4110	114402.05	D3-0Q12	N₂P11
691		87.4172	114393.98		+N₂P12
692	0	87.4216	114388.22	D3-1Q4	N₂P13
693	3	87.4255	114383.13	B4-0P7	
694	2	87.4274	114380.63	B27-0P3	
695					
696		87.4466	114355.53		
697	0	87.4495	114351.74		
698	0	87.4520	114348.44		
699	0	87.4542	114345.63		N₂ b'(19-1)R4
700		87.4580	114340.60		
701	0	87.4590	114339.26	B26-0P1 D11-4Q5	N₂R5
702		87.4626	114334.58		+N₂P3
703	0	87.4657	114330.60		N₂R6
704		87.4691	114326.07		N₂P4
705		87.4733	114320.60		N₂R7
706		87.4774	114315.19		N₂P5
707		87.4818	114309.43		N₂R8
708	0	87.4846	114305.87		
709		87.4875	114302.11		N₂P6
710	0	87.4898	114299.03		
711		87.4932	114294.66		N₂R9
712	0	87.4945	114292.95		
713		87.4984	114287.80		N₂P7
714	0	87.5018	114283.40	D5-2Q3	
715	2	87.5048	114279.39		
716	5	87.5087	114274.38	C9-0Q4 D10-4Q1	
717	59	87.5119	114270.23	D1-0P4 B'3-0P5	

Nb	I	λ (nm)	σ (cm⁻¹)	Assignment	Comment
718	0	87.5177	114262.67		
719	0	87.5233	114255.25		
720	0	87.5247	114253.49		
721	0	87.5316	114244.43		
722	0	87.5398	114233.78	D"1-2Q1	GeII 87.54927
723		87.5535	114215.85		+ArIII 87.553 Ke
724	0	87.5581	114209.83	D14-5Q1	+NI 87.5656 Ke
725	0	87.5611	114205.97		NI 87.5721 Ke
726	0	87.5667	114198.61		sh
727	0	87.5727			+NI 87.5791 Ke
728		87.5743	114188.72		
729	20	87.5793	114182.21	D1-0Q5	
730	0	87.5831	114177.23		
731	0	87.5899	114168.44		
732	0	87.5948	114162.02		
733	0	87.5982	114157.64		
734	0	87.6020	114152.59	C9-0P4 D0-0Q11 ?	+ArI 87.60577
735	7	87.6062	114147.17		+NI 87.6066 Ke
736		87.6161	114134.31	D"1-2Q2 B'3-0R6	sh
737	3	87.6176	114132.27		
738	1	87.6213	114127.55	B'2-0R0	r 114117.96
739		87.6276	114119.24	B'2-0R0	r 114117.96
740		87.6296	114116.67	B26-0P2	
741	0	87.6322	114113.29	D4-1Q9	
742	0	87.6384	114105.23		
743	1	87.6472	114093.83	B'2-0R1 D3-1Q5	r 114085.59
744		87.6524	114087.05	B'2-0R1	r 114085.59
745		87.6546	114084.12		
746	2	87.6563	114081.89	C13-1Q1	
747	1	87.6591	114078.24		NI 87.6645 Ke
748		87.6646			
749	1	87.6667	114068.46	D4-3Q1	CuII 87.67227
750		87.6822	114048.30	D5-2Q4	
751	1	87.6887	114039.74	C9-0R5	
752	0	87.6943	114032.48		
753	0	87.6963			
754		87.6991			s
755					NI 87.6987 Ke
756					CuII 87.70121
757					
758	2	87.7098	114012.37	D7-3Q1	
759	16	87.7234	113994.66		
760		87.7352	113979.35	B'2-0R2 D"1-2Q3	r 113978.37
761		87.7367	113977.38	B'2-0R2	r 113978.37
762	18	87.7386	113974.96	D2-2Q1	
763		87.7518	113957.83	B'2-0P1	r 113956.63
764		87.7536	113955.44	B'2-0P1	r 113956.63
765					CuII 87.75548

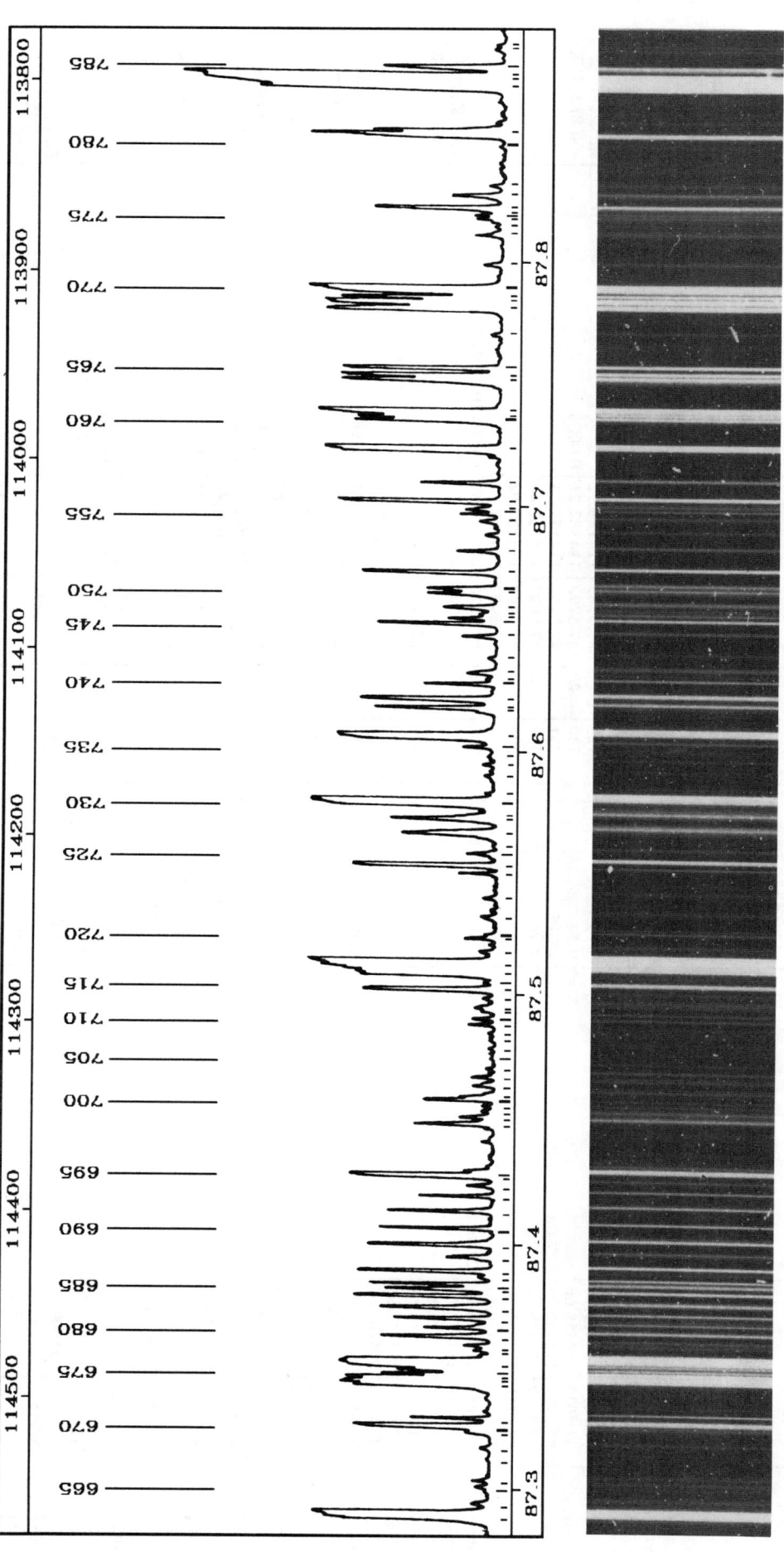

PLATE 16

TABLE 16 *continued*

Nb	I	λ (nm)	σ (cm⁻¹)	Assignment		Comment
766	0	87.7701	113934.03	B27-0P4		
767						OI 87.77983
768	11	87.7836	113916.46	D1-0P5		
769						CuII 87.78471
770						OI 87.78787
771	0	87.7991	113896.42	B26-0R3		
772	0	87.8114	113880.42	D2-0Q10		
773	0	87.8146	113876.23			
774	0	87.8170	113873.18			
775		87.8181	113871.78	B25-0R0 D7-3Q2		r 113870.94
776		87.8194	113870.09	B25-0R0		r 113870.94
777	2	87.8225	113866.09	D2-2Q2		
778	0	87.8276	113859.49	C9-0Q5		
779	0	87.8316	113854.27			
780	0	87.8474	113833.81			
781	3	87.8517	113828.16	D1-0Q6		
782						CuII 87.86986
783	9	87.8742	113799.03	D4-3Q3		+ ArII 87.873 Ke
784		87.8760	113796.77	C8-0Q1 B'2-0R3		r 113794.54
785		87.8794	113792.31	C8-0Q1 B26-0P3		r 113794.54
786		87.8867	113782.92	B25-0R1		r 113781.82
787		87.8884	113780.71	B25-0R1		r 113781.82

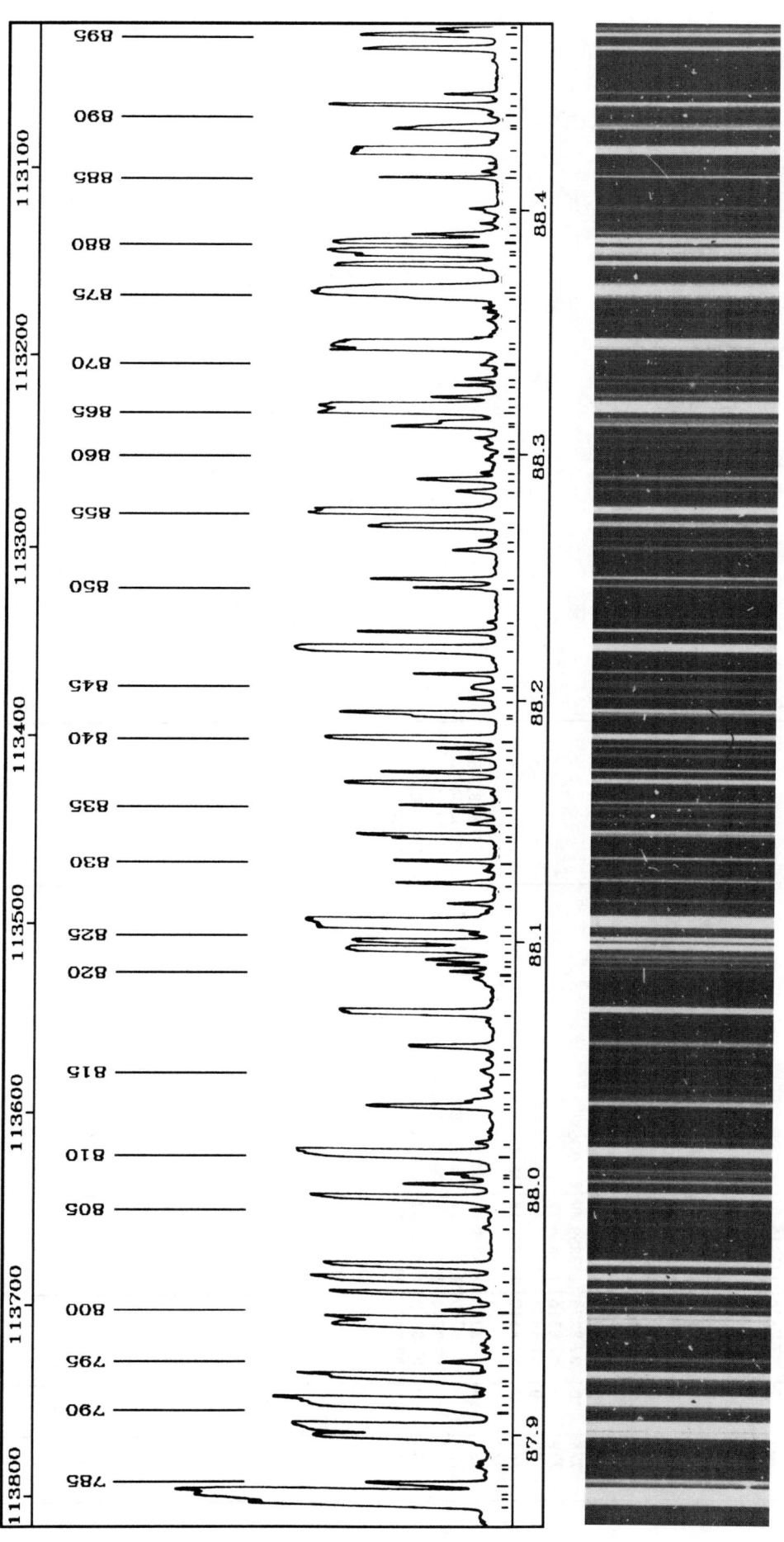

PLATE 17

TABLE 17

Nb	I	λ (nm)	σ (cm⁻¹)	Assignment	Comment
788					OI 87.89720
789	5	87.9017	113763.47	B'2-0P2	r +OI 87.90194
790	0	87.9129	113748.94	D0-1R1	OI 87.91001
791	0	87.9164	113744.49		
792	20	87.9204	113739.25	B25-0P1 D5-2Q5	
793	1	87.9222	113736.91	D0-1R0	
794	0	87.9283	113728.99		
795	0	87.9333	113722.51	B'5-1R0	
796	7	87.9356	113719.53		
797	5	87.9431	113709.86	D0-1R2 C9-0P5 C12-1R0	
798	1	87.9457	113706.55	D2-2Q3	
799		87.9492	113702.01	D3-0Q13	
800					OI 87.95507
801	4	87.9615	113686.09	D7-3Q3	
802	6	87.9672	113678.75		+ ArIII 87.962 Ke
803	0	87.9829	113658.50	B'5-1R1	
804					CuII 87.98912 Ke
805	7	87.9947	113643.22		+ArI 87.99466
806	1	88.0003	113635.96	C12-1R1	
807		88.0033	113632.05	C8-0Q2	
808		88.0046	113630.39	C8-0Q2	r 113631.21
809		88.0125	113620.22	D0-1Q1	r 113631.21
810		88.0176	113613.69	B33-1R0	r
811	0	88.0321	113594.90	C12-1Q1	
812	3	88.0346	113591.66	B25-0R2	
813	0	88.0392	113585.74	B'5-1P1 D4-3Q4	
814	0	88.0470	113575.70	B27-0R5	
815	0	88.0513	113570.20	C8-0P2	
816	0	88.0564	113563.59	B'2-0R4	
817	3	88.0704	113545.48		
818	8	88.0847	113527.11		
819	0	88.0874	113523.64		
820	0	88.0900	113520.31		s
821	1	88.0921	113517.62	B25-0P2 B33-1R1 D0-0Q12 ?	s
822	21	88.0963	113512.17	D0-1Q2	
823	3	88.0996	113507.91	D1-0P6	
824	0	88.1029	113503.66	B'5-1R2	
825	51	88.1069	113498.53	B'2-0P3	+N₂ c₃'(0-1)R8+P4
826		88.1155	113487.45		N₂ R9+P5
827		88.1237	113476.92		N₂ R10+P6
828	0	88.1289	113470.13	C12-1R2	
829		88.1330	113464.85		N₂ R11+P7
830		88.1430	113451.99		N₂ P8
831		88.1439	113450.82		N₂ R12
832		88.1487	113444.64		N₂ P9
833		88.1537	113438.26		N₂ R13
834					
835	0	88.1561	113435.17	D7-3Q4	
836	6	88.1649	113423.78	D1-0Q7	

Nb	I	λ (nm)	σ (cm⁻¹)	Assignment	Comment
837		88.1694	113418.03		N₂ R14
838	1	88.1755	113410.20	C12-1Q2	N₂ P11
839		88.1794	113405.21		
840	15	88.1833	113400.17		r 113387.48
841		88.1926	113388.25	D0-1P2	r 113387.48 +N₂ P12
842		88.1938	113386.72	C8-0Q3	N₂ R16
843		88.1998	113378.85	C8-0Q3	
844	0	88.2041	113373.42	B26-0P4	
845	0	88.2051	113372.09	C12-1P2	
846		88.2097	113366.24		N₂ P13
847	38	88.2198	113353.21	D0-1Q3	
848		88.2267	113344.39		N₂ P14
849		88.2310	113338.81		N₂ R18
850		88.2450	113320.88		N₂ P15
851	3	88.2484	113316.53		s
852	0	88.2606	113300.83	B25-0R3	
853		88.2644	113295.96		N₂ P16
854		88.2707	113287.93	C8-0P3	r
855	11	88.2764	113280.51	B'3-0P7 D2-0Q11	N₂ P17
856		88.2852	113269.24		OI 88.28895
857					
858	0	88.2926	113259.81	B32-1R0 B'5-1R3	
859	0	88.2981	113252.72	C8-0R4	
860	0	88.2997	113250.67		
861	0	88.3035	113245.77		
862		88.3069	113241.40		N₂ P18
863	0	88.3118	113235.18		
864		88.3136	113232.79	D2-2Q5	
865		88.3180	113224.65		ArII 88.318 Ke
866	12	88.3200	113219.69	B'2-0R5	CuII 88.32800 Ke
867	0	88.3239			
868					
869	0	88.3313	113210.15	C12-1R3	d
870	0	88.3376	113202.14		sh
871		88.3435	113194.47	B25-0P3	
872	46	88.3456	113191.85	D0-1P3 C9-0P6	
873	0	88.3552	113179.56		sh
874		88.3647	113167.43	B32-1R1	
875	37	88.3671	113164.35	B'2-0P4	sh
876		88.3693	113161.43		
877	5	88.3782	113150.13	D9-4Q1	
878	2	88.3823	113144.78	D7-3Q5	
879	4	88.3838	113142.86	D0-1Q4	
880	7	88.3877	113137.90	B32-1P1	
881	1	88.3907	113134.14	C12-1Q3	
882		88.3952	113128.35	D5-1Q13	
883	0	88.3997	113122.57		
884	0	88.4011	113120.75		
885				B24-0P1	CuII 88.41332

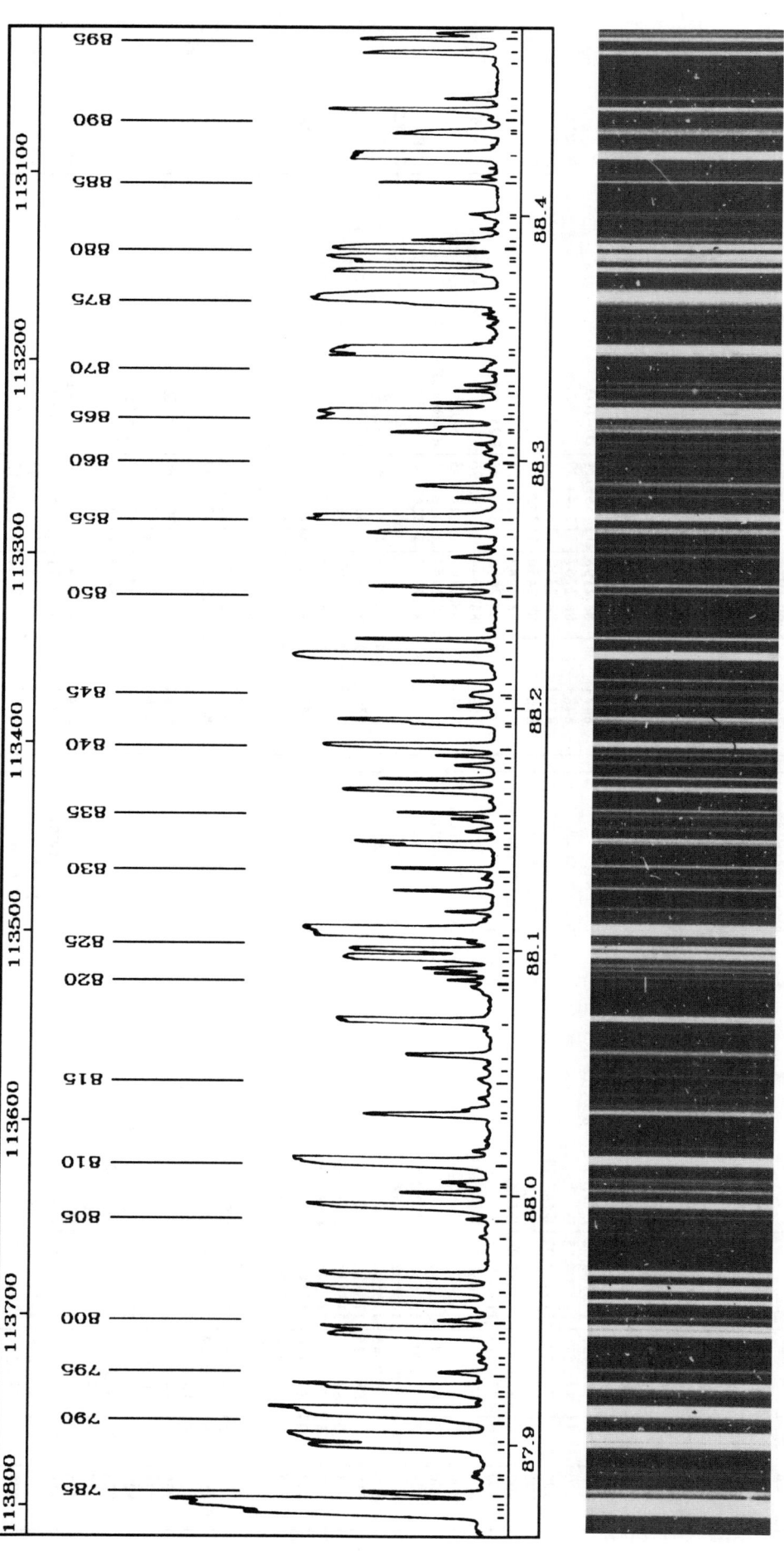

PLATE 17

TABLE 17 *continued*

Nb	I	λ (nm)	σ (cm⁻¹)	Assignment	Comment
886	0	88.4167	113100.81	B'5-1P3	
887	28	88.4252	113089.97	D2-1R0 D2-1R1	
888	2	88.4344	113078.19	C12-1P3	
889		88.4354	113076.88		sh
890	0	88.4399	113071.12		
891	1	88.4443	113065.58	C8-0Q4	+CuII 88.44346
892	1	88.4486	113059.98		
893	0	88.4621	113042.72	D1-0P7	
894	2	88.4672	113036.22	D2-1R2	
895	3	88.4727	113029.17	D9-4Q2	
896	1	88.4751	113026.10		

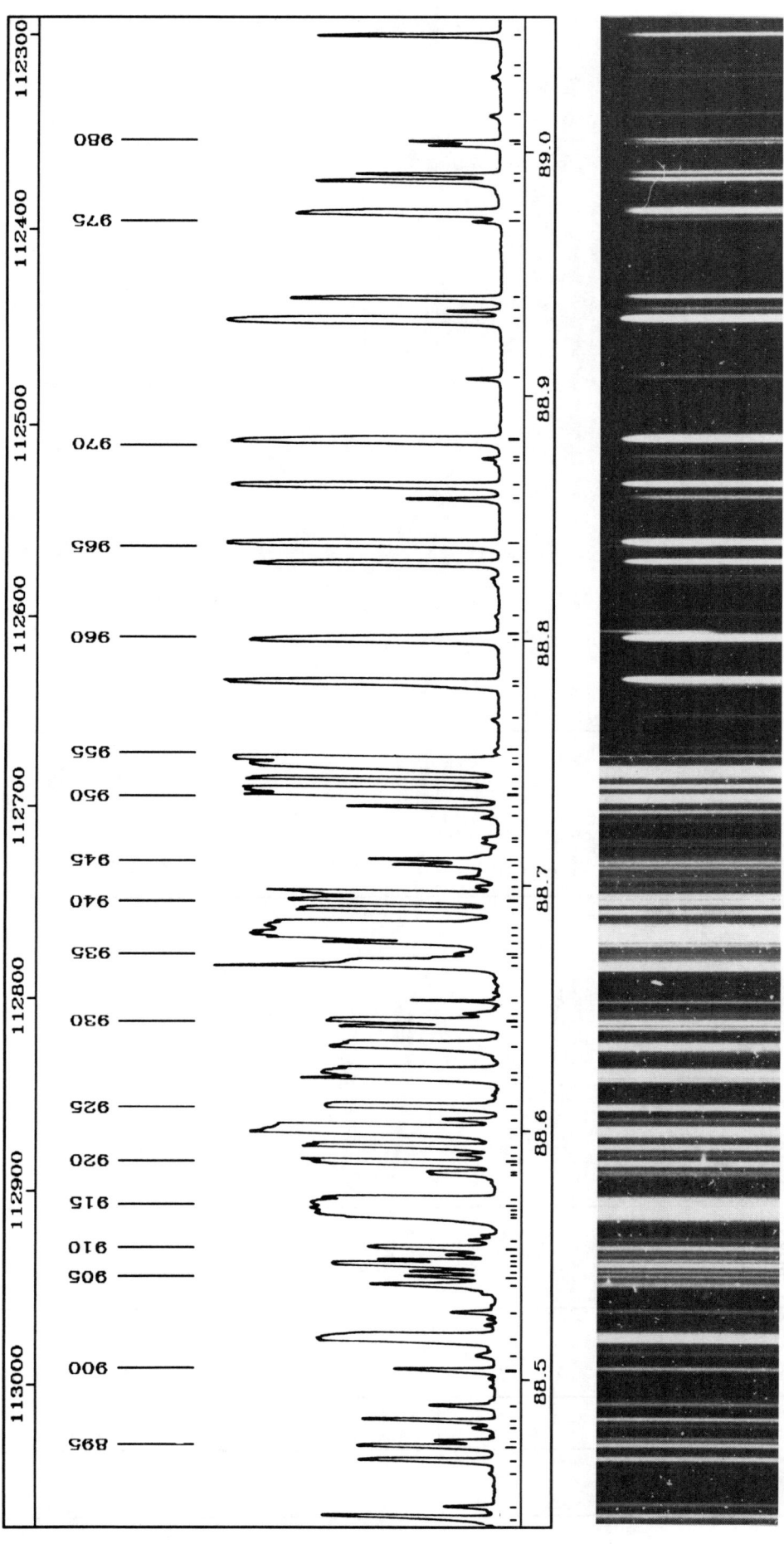

PLATE 18

TABLE 18

Nb	I	λ (nm)	σ (cm-1)	Assignment	Comment
897	0	88.4805	113019.25		
898	1	88.4832	113015.80		
899	1	88.4893	113008.03		
900	2	88.5036	112989.73	C11-1R0	
901	0	88.5096	112982.06	D5-2Q7	
902	52	88.5159	112974.00	D2-1Q1 D3-0Q14	
903	1	88.5267	112960.27		
904	1	88.5373	112946.77		
905		88.5405	112942.65	D0-0R1	r 112941.40
906		88.5425	112940.15	D0-0R1	r 112941.40
907		88.5451	112936.86	D0-0R0 D0-1P4	r 112935.70
908		88.5469	112934.53	D0-0R0 C8-0P4	r 112935.70
909	1	88.5494	112931.30		
910	4	88.5519	112928.07	C11-1R1	
911	0	88.5555	112923.60	B32-1P2 D'2-2Q6	
912	0	88.5651	112911.28		
913		88.5657			NI 88.5656 Ke P
914		88.5667			NI 88.5668 Ke P
915	2	88.5685	112907.01	D2-1R3	
916		88.5698			NI 88.5704 Ke P
917		88.5816	112890.26	D0-0R2	r 112889.73
918		88.5824	112889.20	D0-0R2	r 112889.73
919					CuII 88.58472
920	4	88.5865	112883.97	D0-1Q5	
921	0	88.5896	112880.03	C8-0R5	
922	8	88.5927	112876.08	C11-1Q1	
923	54	88.5996	112867.35	D4-2Q1	
924	0	88.6036	112862.23	B26-0P5	s
925	11	88.6089	112855.49	B31-1R0 D2-1Q2	
926	2	88.6201	112841.23	D9-4Q3	
927	3	88.6228	112837.73	B'2-0R6	+NI 88.6226 Ke
928		88.6334			NI 88.6332 Ke
929		88.6411	112814.43	D0-0Q1	r 112813.18 +NI 88.6428 Ke
930		88.6431	112811.93	D0-0Q1	r 112813.18 +NI 88.6465 Ke
931	0	88.6467	112807.38	C9-0Q7	CuII 88.65111
932					
933		88.6663	112782.43	D0-0R3 B"0-1R0	r 112780.69
934		88.6690	112778.95	D0-0R3 C11-1R2	r 112780.69
935	1	88.6707	112776.81	B25-0P4	
936	5	88.6753	112770.97	B4-1R0 D12-5Q2	
937	114	88.6786	112766.84	B"0-1R1	
938	51	88.6814	112763.23	B31-1R1 B'2-0P5	
939	8	88.6888	112753.85	D2-1P2	
940	14	88.6927	112748.84	D4-2Q2	
941					CuII 88.69434
942	0	88.6979	112742.25		
943	0	88.7019	112737.17		
944	0	88.7068	112730.95	B31-1P1	
945	0	88.7089	112728.26		
946	0	88.7165	112718.63		
947	0	88.7182	112716.39		
948	0	88.7262	112706.31		
949	0	88.7300	112701.52	C11-1Q2	
950		88.7343	112695.98	D0-0Q2 C12-1P4	r 112695.06
951		88.7358	112694.14	D0-0Q2 B"0-1R2	r 112695.06
952		88.7401			ArIII 88.740 Ke
953		88.7458			NI 88.7457 Ke
954	19	88.7475	112679.22	D2-1Q3 B28-0P7	
955	1	88.7553	112669.35	C8-0Q5	
956	1	88.7682	112652.90	C11-1P2	
957	2	88.7815	112636.13		
958	25	88.7832	112633.88	B4-1P1	
959	42	88.8002	112612.32	B"0-1P1	
960		88.8023			NI 88.8022 Ke
961	0	88.8103	112599.51	B4-1R2	
962		88.8240	112582.17	D0-0P2	r 112580.99
963		88.8259	112579.81	D0-0P2	r 112580.99
964	18	88.8317	112572.46	B31-1R2 D4-2Q3	
965	55	88.8397	112562.29	B"0-1R3	
966	5	88.8582	112538.83	C11-1R3	
967	28	88.8636	112532.02	D2-1P3	
968		88.8729	112520.21	D0-0Q3	r 112519.04
969		88.8748	112517.86	D0-0Q3 B31-1P2	r 112519.04
970	31	88.8820	112508.69	D6-3Q1	
971	4	88.9078	112476.12	D1-0Q9	
972	55	88.9317	112445.87	D2-1Q4 B"0-1P2	
973	4	88.9357	112440.85	C11-1Q3	
974	9	88.9408	112434.31	B4-1P2	
975	2	88.9724	112394.47	B4-1R3	
976	10	88.9759	112389.97	B'2-0R7 D6-3Q2	
977	8	88.9887	112373.82	B"0-1R4	
978	5	88.9915	112370.28	C11-1P3	
979		89.0034	112355.25	D0-0P3	r 112354.28
980		89.0049	112353.31	D0-0P3	r 112354.28
981	3	89.0154	112340.16	D4-2Q4	
982	1	89.0310	112320.38	B30-1R1	
983	1	89.0350	112315.33	C12-1Q5	
984	8	89.0477	112299.31	B'2-0P6	

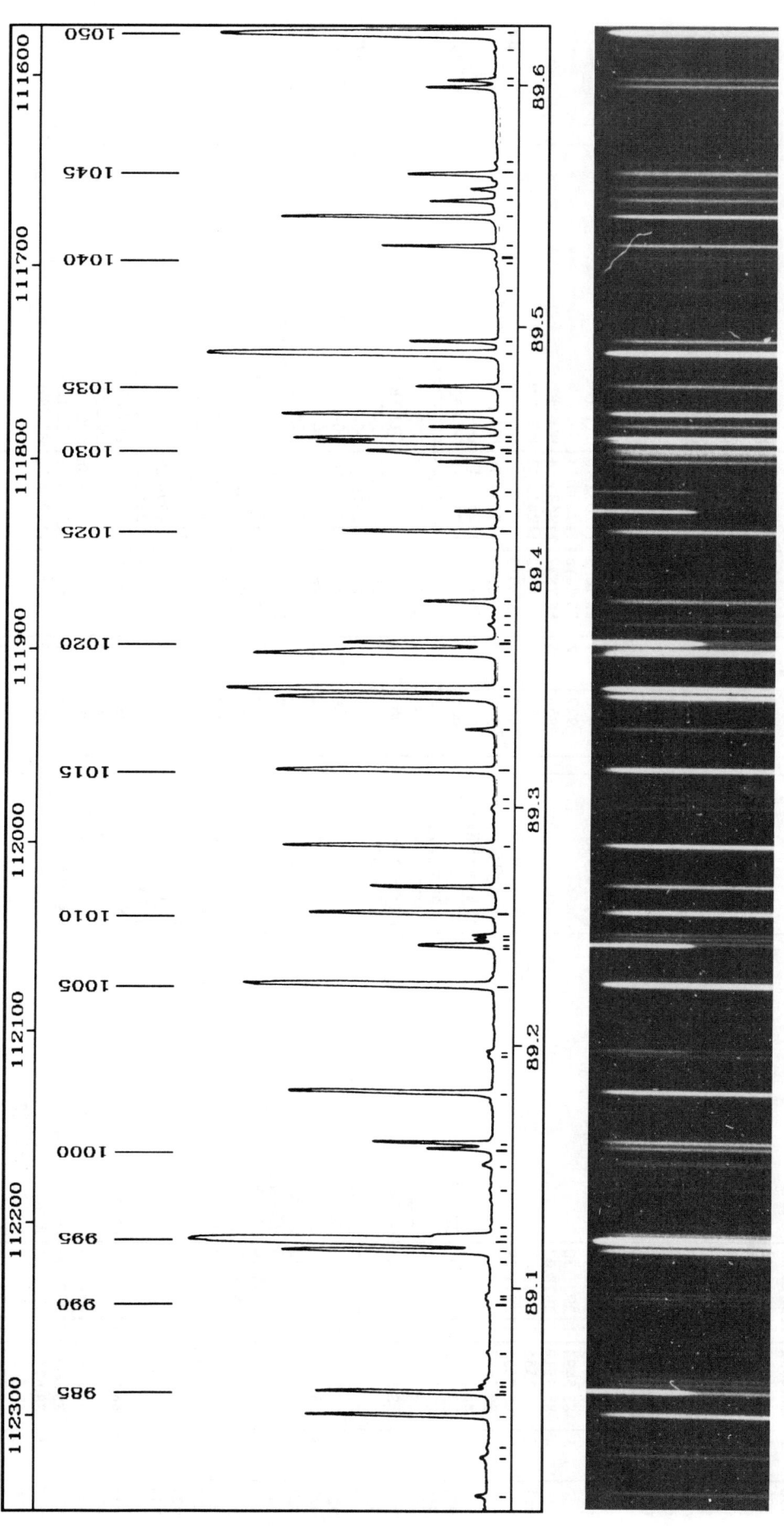

PLATE 19

TABLE 19

Nb	I	λ (nm)	σ (cm⁻¹)	Assignment	Comment
985	9	89.0572	112287.37	D0-0Q4	
986		89.0600	112283.84		CuII 89.05669
987	2	89.0622	112281.11	B30-1P1	
988	3	89.0735	112266.82	B31-1R3	
989	1	89.0937	112241.38	B25-0P5	
990		89.0960	112238.47	B1-0P1	r 112239.93
991	1	89.0977	112236.35	B1-0P1	r 112239.93
992	1	89.1124	112217.78	D3-3Q1	
993	1	89.1153	112214.14	D3-0Q15	
994	9	89.1194	112209.05	D6-3Q3	
995	130	89.1222	112205.55	B"0-1P3	
996	3	89.1258	112200.96	B31-1P3	
997	1	89.1410	112181.86	B32-1P4 C8-0Q6	
998	4	89.1517	112168.31	B4-1P3	
999	1	89.1579	112160.59	D2-1Q5	
1000	3	89.1604	112157.44	B"0-1R5	
1001	5	89.1812	112131.28	B1-0R3	r 112110.88
1002	12	89.1968	112111.71	B1-0R3	r 112110.88
1003		89.1981	112110.05	D0-0P4 C10-1R0	
1004		89.2257	112075.28		CuII 89.24144
1005	21	89.2429	112053.76	D4-2Q5	
1006	3	89.2450	112051.12	B1-0P2	r 112050.02
1007		89.2467	112048.92	B1-0P2	r 112050.02
1008	18	89.2559	112037.41	D'0-2Q1 D11-5Q1	
1009	10	89.2671	112023.37	C10-1R1	
1010	13	89.2841	112001.99	D0-0Q5	
1011	1	89.3000	111982.03	D6-3Q4 D'3-3Q3 C9-0P8	
1012	2	89.3038	111977.31	D2-0Q13	
1013	19	89.3159	111962.14	C10-1Q1 C8-0R7	
1014	5	89.3328	111940.99	D'0-2Q2 D1-0Q10	
1015	9	89.3460	111924.39	D2-1P5	
1016	15	89.3493	111920.34	B"0-1P4	
1017	11	89.3641	111901.76	D8-4Q1	
1018		89.3685	111896.21	C7-0P3	
1019	2	89.3760	111886.84	C10-1R2	
1020	1	89.3793	111882.69	B'2-0R8	CuII 89.36777
1021	10	89.3858	111874.61	B1-0R4	r
1022	8	89.4149	111838.23	B29-1R1 C11-1R5	
1023					CuII 89.42274
1024					ArI 89.43102
1025					sh
1026					
1027					
1028	3	89.4442	111801.57	B29-1P1	
1029		89.4472	111797.81	D1-2Q2 D"0-2Q3	
1030	6	89.4485	111796.14	C10-1Q2	
1031		89.4522	111791.55	B1-0P3	r 111790.56
1032		89.4538	111789.58	B1-0P3	r 111790.56
1033	4	89.4587	111783.49	D8-4Q2	
1034	6	89.4638	111777.03	B'2-0P7	
1035	4	89.4754	111762.60	B30-1P3	
1036	26	89.4889	111745.74	D0-0P5	
1037	4	89.4939	111739.41	C10-1P2	
1038	3	89.5154	111712.62	C7-0Q4	
1039	0	89.5260	111699.42	B32-1P5	
1040	1	89.5284	111696.39	D6-3Q5	
1041	5	89.5337	111689.81		
1042	18	89.5457	111674.87		
1043	5	89.5526	111666.23	D0-0Q6 C10-1R3	
1044	1	89.5575	111660.17	C8-0Q7	
1045	4	89.5638	111652.28	B29-1R2	
1046	1	89.5684	111646.51	D1-2Q3	
1047	2	89.5998	111607.38	D8-4Q3	
1048	2	89.6027	111603.84		
1049	2	89.6143	111589.28	B29-1P2	
1050	13	89.6220	111579.70	B"0-1P5	

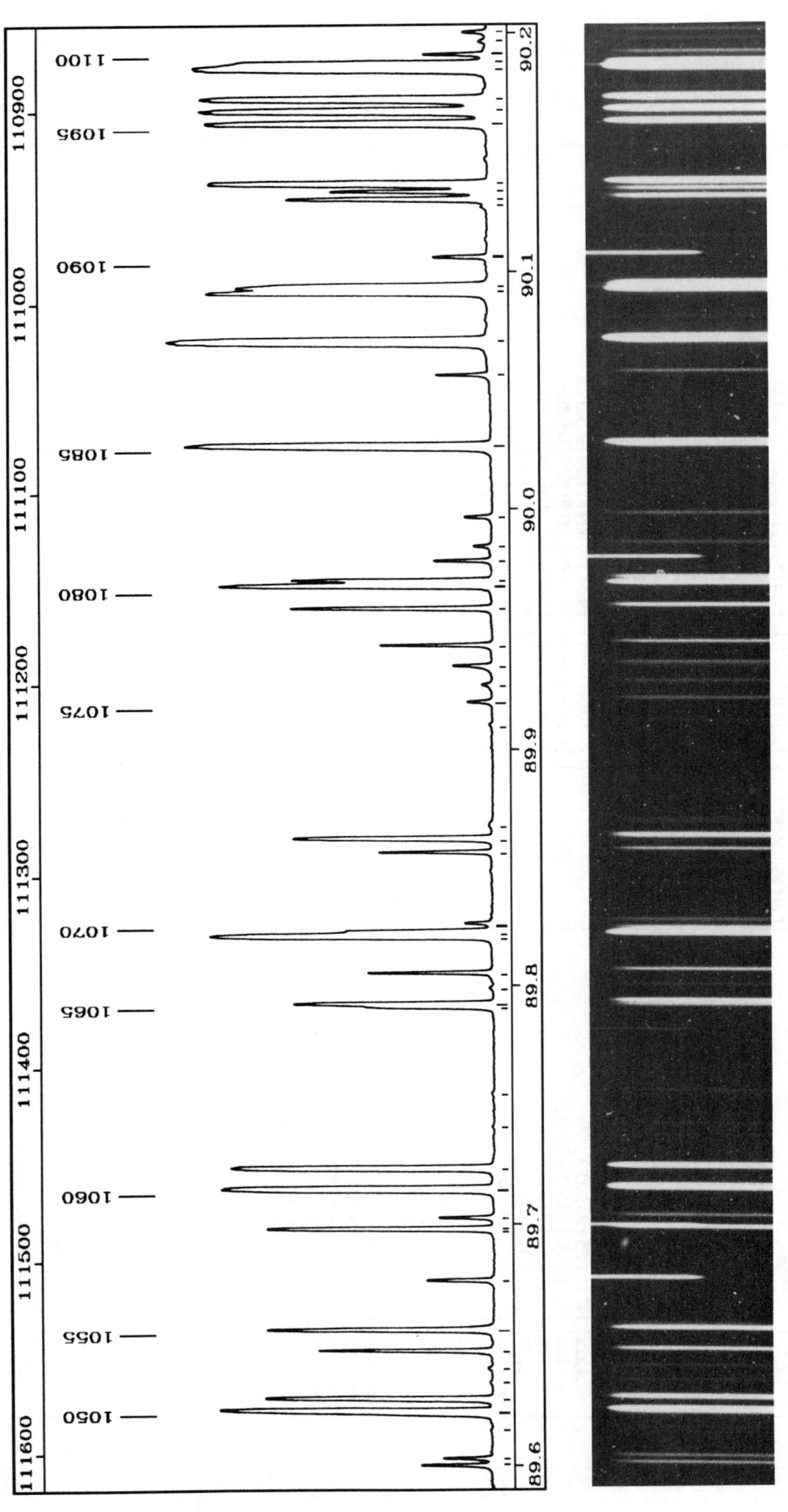

PLATE 20

TABLE 20

Nb	I	λ (nm)	σ (cm-1)	Assignment	Comment
1051	9	89.6271	111573.35	B'1-0R5	
1052	1	89.6340	111564.80		
1053	1	89.6401	111557.23	C11-1P5	
1054	4	89.6472	111548.44	C10-1Q3	
1055	3	89.6556	111537.89	D9-4Q7	
1056					CuII 89.67588
1057	7	89.6975	111485.87	B'3-1R0	
1058					CuII 89.69762
1059	5	89.7024	111479.73	C9-0Q9	
1060	39	89.7142	111465.04	B'1-0P4 C10-1P3	
1061	26	89.7229	111454.25	B'3-1R1	
1062	1	89.7406	111432.28	C8-0P7	
1063	1	89.7545	111415.05	B28-1R0 B31-1R5	
1064	5	89.7910	111369.75		
1065	20	89.7921	111368.37	B29-1R3 D0-0P6 D1-0Q11	
1066	2	89.7991	111359.68	B30-1P4 C10-1R4	
1067	5	89.8056	111351.63	B'3-1R2	
1068	38	89.8210	111332.56	C7-0Q5 B'3-1P1	
1069	6	89.8229	111330.25	B28-1R1 D4-2Q7	
1070	3	89.8260	111326.29	B'2-0R9	
1071	4	89.8564	111288.73	B28-1P1	
1072	6	89.8622	111281.50	D0-0Q7 B29-1P3	r
1073		89.8687	111273.50	B22-0P3	
1074	0	89.9111	111221.02	C10-1Q4	
1075	2	89.9202	111209.75	B'1-0R6	
1076	1	89.9275	111200.71	B'2-0P8	
1077	2	89.9355	111190.86	B''0-1P6	
1078	2	89.9432	111181.33	B'3-1R3 ?	
1079	6	89.9595	111161.15	C7-0P5	
1080	17	89.9693	111149.07	B'3-1P2	
1081	7	89.9714	111146.39	B28-1R2	
1082					CuII 89.97922 Ke
1083	2	89.9860	111128.43	C10-1P4	
1084	2	89.9982	111113.31	B'1-0P5	
1085	33	90.0282	111076.27	B'3-1R4 D0-0P7	
1086	2	90.0583	111039.15	D10-5Q1	
1087	105	90.0723	111021.93	D3-2Q1	
1088	121	90.0932	110996.19	D1-1R1	
1089		90.0953	110993.64	D1-1R0 C9-1R0	sh CuII 90.10731
1090				B30-1R5	
1091	1	90.1295	110951.46	C9-1R1	
1092	6	90.1323	110948.03	B'3-1R4 D0-0P7	
1093	5	90.1356	110943.98	D1-1R2	
1094	30	90.1387	110940.10	D3-2Q2	
1095	25	90.1640	110909.03	D5-3Q1	
1096	40	90.1692	110902.60	B'3-1P3	
1097	43	90.1741	110896.56	D1-1Q1	
1098	201	90.1879	110879.64	B29-1P4 C9-1Q1	
1099	40	90.1891	110878.13		
1100	3	90.1932	110873.10	B27-1R0	
1101	2	90.1988	110866.28	B28-1R3 B30-1P5	
1102	2	90.2025	110861.66	B22-0P4	

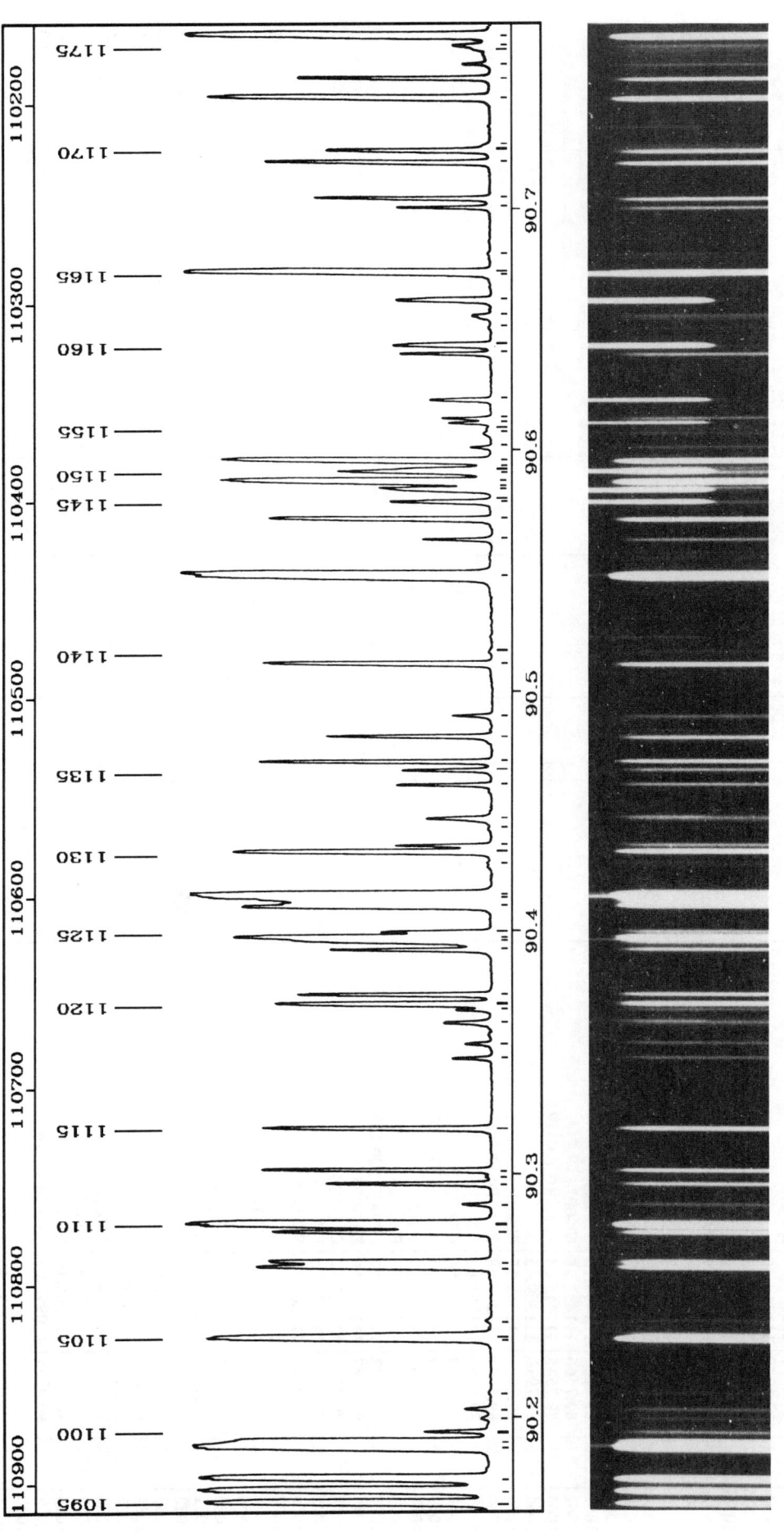

PLATE 21

TABLE 21

Nb	I	λ (nm)	σ (cm⁻¹)	Assignment	Comment
1103	1	90.2094	110853.21	D0-0Q8	
1104	39	90.2315	110826.01	D1-1R3	
1105	1	90.2320	110825.39	C9-1R2	sh
1106		90.2390	110816.84	C10-1Q5	
1107	13	90.2611	110789.72	D5-3Q2	
1108	16	90.2631	110787.30	B27-1R1 B'1-0R7	
1109	7	90.2757	110771.78	B28-1P3	
1110	71	90.2789	110767.89	D1-1Q2	
1111	3	90.2872	110757.71	C11-1R7	
1112	6	90.2957	110747.31	B27-1P1	
1113	3	90.2989	110743.37	D10-5Q3	
1114	27	90.3008	110740.94	D3-2Q3 B21-0R3	
1115	11	90.3184	110719.45	C9-1Q2 C9-0Q10	
1116	2	90.3475	110683.77	C10-1P5	
1117	2	90.3535	110676.35	C7-0P6	
1118					CII 90.36235
1119	3	90.3678	110658.95	C6-0P2 D1-1P2	
1120	24	90.3701	110656.13	C9-1P2	
1121	5	90.3740	110651.36	D1-1R4	
1122	5	90.3926	110628.50	B'1-0P6	
1123					CII 90.39616
1124	14	90.3978	110622.11	D5-3Q3	
1125	3	90.4002	110619.18	C9-1R3	
1126	15	90.4113	110605.61	B27-1R2 D4-4Q1	
1127					CII 90.41416
1128	103	90.4153	110600.74	D1-1Q3	
1129	23	90.4284	110584.72	B21-0P3	
1130	3	90.4340	110577.88	B'3-1P4	
1131	3	90.4367	110574.55	B2-0P9	
1132	2	90.4438	110565.87	D2-0Q15	r
1133					CII 90.44801
1134	6	90.4620	110543.69	D'1-3Q1	
1135	3	90.4680	110536.30	B27-1P2	
1136	11	90.4713	110532.27	D7-4Q1	
1137	6	90.4813	110520.04	D3-2Q4	
1138	3	90.4900	110509.46	D4-4Q2 D10-5Q4	
1139	12	90.5116	110483.14	C9-1Q3	
1140	1	90.5167	110476.86		
1141	148	90.5484	110438.15	D1-1P3	
1142	4	90.5635	110419.82	D7-4Q2	
1143	10	90.5719	110409.52	D1-1R5	
1144	8	90.5789	110400.99	D5-3Q4	
1145		90.5786			NI 90.5787 Ke
1146		90.5835			NI 90.5839 Ke
1147	18	90.5862	110392.09	C6-0P3	
1148	12	90.5876	110390.37	C9-1P3	
1149	6	90.5913	110385.71	B29-1P5	
1150		90.5913			NI 90.5916 Ke
1151	3	90.5933	110383.44	D0-0Q9	
1152	18	90.5962	110379.87	D1-1Q4	
1153	2	90.6018	110373.03	B28-1P4	
1154	4	90.6077	110365.96	C7-0Q7	
1155	1	90.6095	110363.69	D'4-4Q3	CuII 90.61134
1156					
1157	7	90.6139	110358.37	B22-0P5	
1158		90.6208			NI 90.6206 Ke
1159	5	90.6401	110326.41	B27-1R3	NI 90.6433 Ke
1160		90.6432			
1161	1	90.6514	110312.74	B21-0R4	
1162	2	90.6562	110306.86	B26-1R0	b
1163		90.6619			NI 90.6617 Ke
1164		90.6731			NI 90.6730 Ke
1165	51	90.6739	110285.28	D'2-3Q1	
1166	8	90.6811	110276.57	B'0-0R3	
1167	6	90.7010	110252.39	D7-4Q3	
1168	7	90.7047	110247.90	D3-2Q5	
1169	7	90.7196	110229.73	B27-1P3	
1170	7	90.7245	110223.77	B26-1R1 D10-5Q5	
1171	5	90.7264	110221.46	C6-0Q4	
1172	16	90.7464	110197.23	B3-1P5	
1173	8	90.7542	110187.79	D'2-3Q2	
1174	6	90.7601	110180.57	B21-0P4	
1175	11	90.7639	110175.94	B'0-0P2	
1176	3	90.7678	110171.29	C9-1Q4	
1177	32	90.7717	110166.52	D1-1P4	

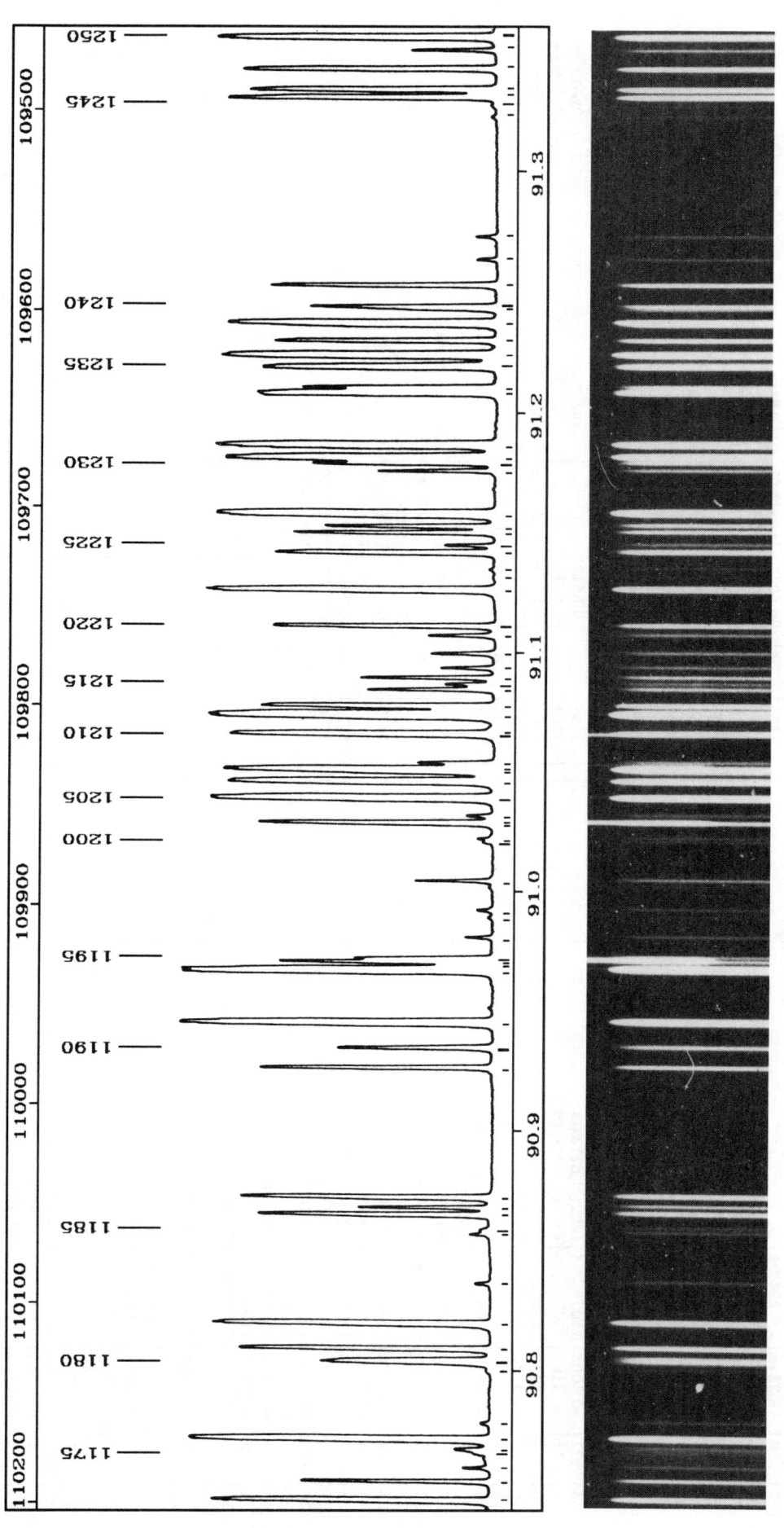

PLATE 22

TABLE 22

Nb	I	λ (nm)	σ (cm⁻¹)	Assignment	Comment
1178	1	90.7787	110158.06	C13-2Q1	
1179	2	90.8005	110131.50	D1-0Q13	
1180	8	90.8041	110127.14	D5-3Q5 B1-0P7 C7-0P7	b
1181	17	90.8092	110120.97		
1182	19	90.8196	110108.37	D1-1Q5	
1183	2	90.8363	110088.10		
1184	2	90.8570	110063.11		
1185	3	90.8589	110060.73	B0-0R4	
1186	10	90.8649	110053.53	C6-0P4	
1187	4	90.8678	110049.94	C9-1P4	
1188	13	90.8719	110044.98	D2-3Q3 B26-1R2	
1189	7	90.9258	109979.74		
1190	6	90.9341	109969.71	B26-1P2 C13-2Q2	
1191	34	90.9446	109957.10	B2-1R0	
1192	99	90.9666	109930.46	B2-1R1	
1193		90.9701	109926.28	B0-0P3 D3-2Q6	r 109925.28
1194					NI 90.96976 Ke / r 109925.28
1195		90.9717	109924.28	B0-0P3	
1196	1	90.9802	109913.98	D9-5Q1	
1197	1	90.9888	109903.67	C9-0Q11 B'2-0P10	
1198	2	90.9918	109899.96		
1199	1	91.0039	109885.41	B28-1P5 D10-5Q6	
1200		91.0204	109865.48	B20-0P3	r 109864.53
1201		91.0220	109863.57	B20-0P3	r 109864.53
1202	10	91.0283	109855.90	D2-3Q4 C6-0Q5	
1203					NI 91.02785 Ke
1204	2	91.0315	109852.07	B21-0R5	
1205	33	91.0387	109843.43	D1-1P5	
1206	29	91.0457	109834.95	B2-1R2	
1207	48	91.0506	109829.05	D0-2R1	
1208					CuII 91.05185
1209					NI 91.06456 Ke / +NI 91.06456 Ke
1210	4	91.0536	109825.41	D0-2R0	
1211	0	91.0656	109811.01	D0-2R2 B'2-1P1 D9-5Q2	
1212	96	91.0736	109801.35	B'5-2R0	
1213	9	91.0773	109796.83	D1-1Q6 C7-0Q8 C10-1Q7	
1214	4	91.0842	109788.49	C9-1Q5	
1215	3	91.0865	109785.74	B0-0R5 C12-2R0	
1216	7	91.0892	109782.51	B'1-0R9	
1217	2	91.0934	109777.46	B26-1R3 B22-0P6	
1218	2	91.0993	109770.36	B3-1P6 D7-4Q5	
1219	2	91.1068	109761.30	C8-1R0	
1220	9	91.1108	109756.53	B'5-2R1	
1221	21	91.1259	109738.30		
1222	1	91.1309	109732.23		
1223	0	91.1346	109727.81		
1224	14	91.1414	109719.60	C8-1R1	
1225	2	91.1446	109715.81	C12-2R1	
1226	8	91.1497	109709.64	B25-1R0	
1227	6	91.1525	109706.23	B21-0P5	
1228	106	91.1577	109700.00	D0-2Q1	
1229	3	91.1756	109678.49	C12-2Q1	
1230	5	91.1788	109674.63	B2-1R3	
1231	44	91.1812	109671.74	B5-2P1 B26-1P3	
1232	36	91.1864	109665.53	C8-1Q1	
1233	46	91.2082	109639.30	C9-1P5	
1234	5	91.2104	109636.60	B25-1R1 C6-0P5	sh
1235	28	91.2191	109626.22	B2-1P2 B20-0R4	
1236	59	91.2239	109620.42	B0-0P4	
1237	14	91.2299	109613.19	D0-2Q2	
1238	39	91.2375	109604.12	B33-2R1	
1239		91.2428	109597.75	B5-2R2	
1240	4	91.2442	109596.02	B25-1P1	
1241	9	91.2531	109585.31	B1-0P8 B33-2P1	
1242	2	91.2637	109572.61	C12-2R2 D3-2Q7	
1243	2	91.2732	109561.14	C12-2Q2	
1244	0	91.3231	109501.37	B27-1R5	
1245	0	91.3280	109495.48	D0-2P2	
1246	19	91.3311	109491.72	C8-1Q2	
1247	15	91.3345	109487.71	B5-2P2	
1248	13	91.3430	109477.49	D1-1P6	
1249	6	91.3508	109468.14	D0-2Q3	
1250	47	91.3562	109461.67		

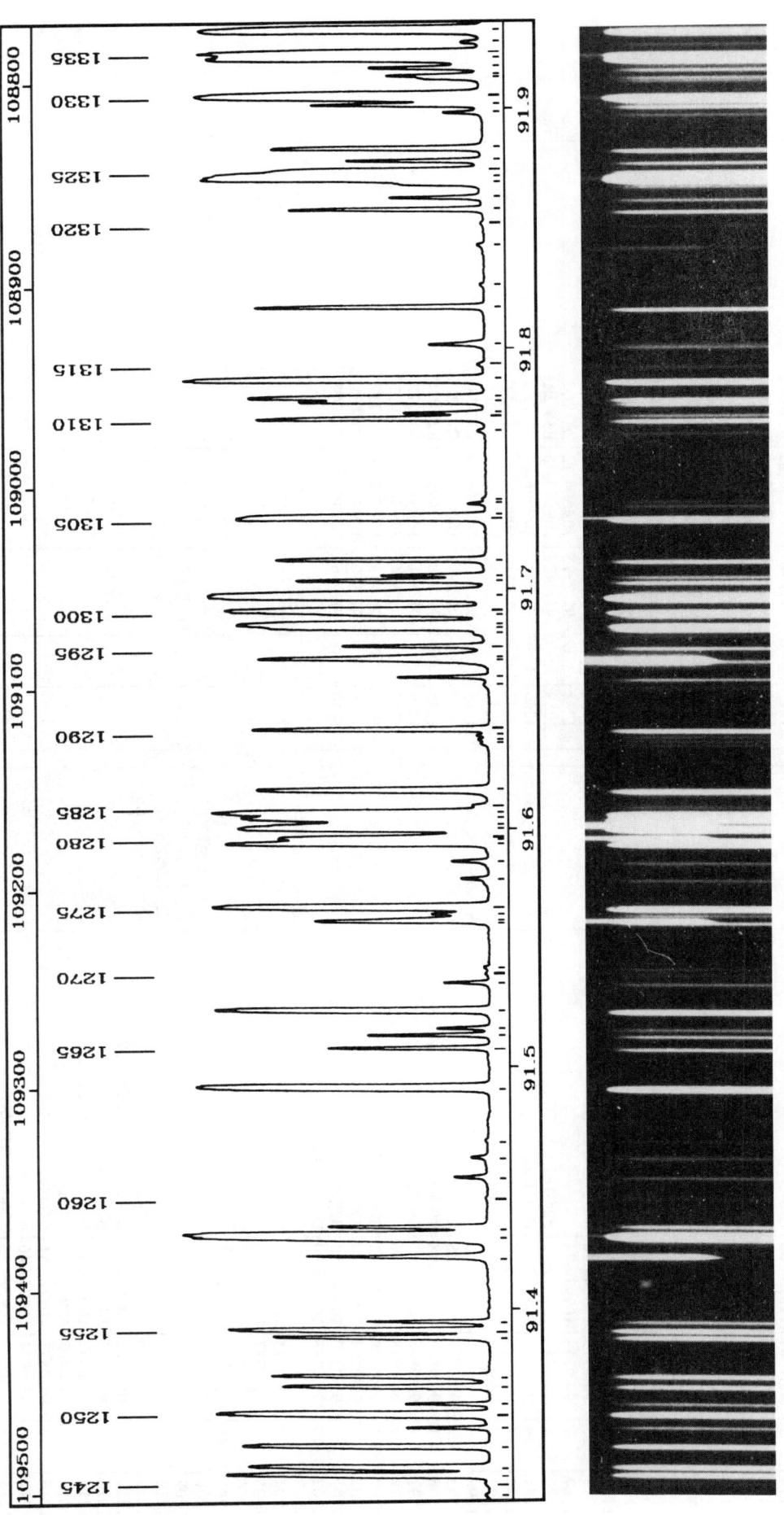

PLATE 23

TABLE 23

Nb	I	λ (nm)	σ (cm⁻¹)	Assignment	Comment
1251	3	91.3607	109456.31	B20-0P4	
1252	9	91.3677	109447.93	B25-1R2	
1253	7	91.3719	109442.83	B'2-1R4	
1254	8	91.3882	109423.35	D1-1Q7 C6-0Q6	
1255	12	91.3910	109419.92	C8-1P2	
1256	6	91.3947	109415.52	C8-1R3	CuII 91.42133
1257					
1258	134	91.4303	109372.92	B'2-1P3 B25-1P2	
1259	3	91.4345	109367.97	B33-2P2 B'5-2R3	
1260	0	91.4460	109354.21	B34-2P3	
1261	4	91.4546	109343.84	B27-1P5	
1262	4	91.4627	109334.20	B32-2R0 D0-0Q11	
1263	1	91.4696	109325.89		
1264	68	91.4913	109300.06	D0-2P3 B21-0R6	
1265	4	91.5077	109280.39	B'3-1P7	
1266	4	91.5130	109274.07	D0-2Q4?	
1267	2	91.5158	109270.70	B26-1P4	
1268	22	91.5233	109261.76	C8-1Q3	
1269	2	91.5351	109247.66	B32-2R1	
1270		91.5388	109243.26	C5-0Q1 B'0-0P5	r 109241.55
1271		91.5417	109239.86	C5-0Q1 B'0-0P5	r 109241.55
1272	7	91.5609	109216.89	B32-2P1	
1273		91.5614			NII 91.5612
1274	2	91.5643	109212.86	C6-0P6	
1275	26	91.5672	109209.44	B'5-2P3	
1276	2	91.5790	109195.39	B'2-0P11	
1277	2	91.5861	109186.85	C12-2P3	
1278	23	91.5939	109177.59	D4-3Q1	
1279	11	91.5960	109175.10	B25-1R3	
1280		91.5963			NII 91.5962
1281	44	91.6009	109169.19	D2-2R1	
1282		91.6019			NII 91.6015
1283	25	91.6049	109164.42	D2-2R0	
1284	32	91.6069	109162.05	C8-1P3	
1285	1	91.6099	109158.56	B20-0R5	
1286	11	91.6165	109150.62	C8-1R4 B'2-1R5 C9-1P6 C7-0Q9	
1287	0	91.6358	109127.68	B33-2R3	
1288		91.6376	109125.56	B19-0P3	r 109124.47
1289		91.6394	109123.38	B19-0P3	r 109124.47
1290	6	91.6418	109120.46	D2-2R2	
1291		91.6599	109098.95	C5-0Q2 B22-0P7	r 109096.88
1292		91.6634	109094.81	C5-0Q2 B24-1R0	r 109096.88 +NII 91.6704
1293	29	91.6709	109085.85		NII 91.6704
1294		91.6710			
1295	5	91.6765	109079.18	D'5-5Q1	
1296	1	91.6818	109072.94	D7-4Q7	
1297		91.6844	109069.85	D4-3Q2	
1298	17	91.6858	109068.12	B25-1P3 B33-2P3 D'0-2P4	sh
1299		91.6905	109062.55		sh
1300	20	91.6916	109061.23	B'2-1P4	CuII 91.73058 Ke
1301	106	91.6978	109053.93	D2-2Q1	r 109007.23
1302	4	91.7040	109046.53	D6-4Q1	r 109007.23
1303	6	91.7063	109043.82	D'0-2Q5?	
1304	6	91.7128	109036.00	D1-1P7	
1305	14	91.7304	109015.17	B24-1R1 B32-2P2 D1-1Q8 D2-2R3	
1306					
1307		91.7360	109008.41	C5-0P3 C11-2R1	
1308		91.7380	109006.05	C5-0P2	
1309	1	91.7664	108972.41	B'1-0P9	
1310	9	91.7714	108966.37	B24-1P1	
1311	3	91.7738	108963.57	C8-1Q4	
1312	5	91.7785	108958.00	B20-0P5 B26-1R5	
1313	10	91.7803	108955.90	C11-2Q1	
1314	25	91.7878	108946.89	D2-2Q2	
1315	1	91.7942	108939.40	D6-4Q2	
1316	3	91.8024	108929.59	B31-2R0 C6-0Q7	b
1317	5	91.8181	108910.96	D4-3Q3	
1318	0	91.8276	108899.74	C12-2Q4 C7-0P9	
1319		91.8434	108881.03	C5-0Q3	r
1320	0	91.8532	108869.39	C11-2R2	
1321	3	91.8591	108862.38	B'5-2P4	
1322	1	91.8641	108856.48	B''0-2R0 D'5-5Q3	
1323	1	91.8697	108849.86	D2-2R4	
1324	73	91.8728	108846.12	D2-2P2 B''0-2R1	sh
1325	1	91.8743	108844.34	B4-2R0	
1326	3	91.8794	108838.35	B24-1R2	
1327	6	91.8845	108832.28	C8-1P4	
1328	1	91.8994	108814.61	B'0-0P6	
1329	1	91.9027	108810.76	B25-1R4 B31-2P1	
1330	88	91.9061	108806.68	B4-2R1 C9-1Q7 C8-1R5	
1331	1	91.9135	108797.91	B'2-1R6	
1332		91.9147	108796.48		
1333	3	91.9179	108792.71	C11-2Q2	
1334	28	91.9224	108787.37	D2-2Q3 B26-1P5	
1335	84	91.9241	108785.38	D0-1R1 B''0-2R2	
1336	1	91.9286	108780.06	D6-4Q3	
1337	76	91.9338	108773.89	D0-1R0 D'3-4Q1	

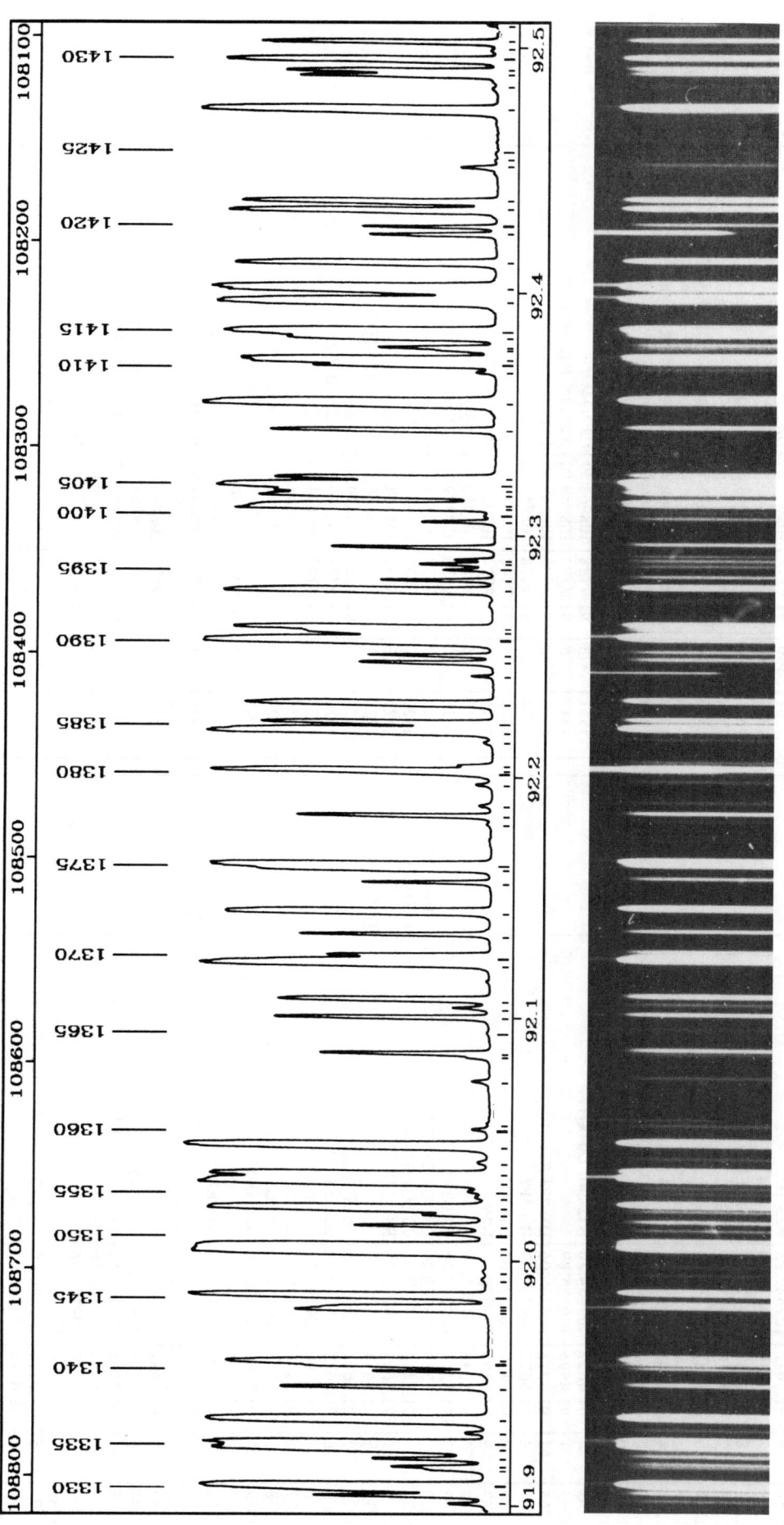

PLATE 24

TABLE 24

Nb	I	λ (nm)	σ (cm⁻¹)	Assignment	Comment
1338	5	91.9470	108758.27	B24-1P2	
1339		91.9540	108750.00	C5-0P3	r 108749.29
1340		91.9552	108748.57	C5-0P3	r 108749.29
1341	19	91.9573	108746.10	D0-1R2 C11-2P2	
1342					ArII 91.97810
1343		91.9789	108720.60	B19-0P4	r 108719.99
1344		91.9799	108719.38	B19-0P4	r 108719.99
1345	22	91.9847	108713.72	B4-2P1	
1346		91.9919	108705.21	D3-2Q9	
1347	0	91.9954	108701.10	D4-3Q4	
1348	45	92.0032	108691.82	B'4-2R2 B"0-2P1	
1349	26	92.0049	108689.84	B2-1P5	
1350	2	92.0104	108683.37	D3-4Q2	
1351	2	92.0139	108679.22	B28-1P7 C6-0P7	
1352	1	92.0183	108674.06	B25-1P4	
1353	31	92.0210	108670.88	B"0-2R3	
1354	2	92.0257	108665.27	B21-0R7	
1355	1	92.0279	108662.75	B31-2R2	
1356		92.0321	108657.76	D0-1Q1	r
1357	28	92.0347	108654.71	D0-1R3	
1358	1	92.0400	108648.47	C11-2R3	
1359	58	92.0467	108640.56	D2-2P3	
1360	1	92.0535	108632.45		
1361					GeII 92.05537
1362	1	92.0735	108608.93	B31-2P2	
1363		92.0831	108597.54	C5-0Q4	r 108596.63
1364		92.0847	108595.71	C5-0Q4 C8-1Q5	r 108596.63
1365	1	92.0932	108585.69	C9-1P7	
1366	4	92.0995	108578.22	D2-2Q4	
1367	4	92.1037	108573.29	B'1-0R11	
1368	6	92.1071	108569.30	B24-1R3 D1-1Q9	
1369	74	92.1225	108551.15	C11-2Q3 C5-0R5 D0-1Q2	
1370	6	92.1256	108547.50	D3-4Q3	
1371	6	92.1342	108537.28	B"0-2P2	
1372	27	92.1440	108525.75	B4-2P2	
1373	4	92.1562	108511.40	D0-1R4	
1374	7	92.1617	108504.91	B"0-2R4	
1375	34	92.1633	108503.06	B'4-2R3	
1376		92.1806	108482.72	B30-2R0	
1377	3	92.1838	108478.92	C11-2P3	
1378	2	92.1883	108473.60	C12-2Q5	
1379	3	92.1971	108463.29	C7-0Q10	
1380					CuII 92.20190
1381	13	92.2026	108456.80	B24-1P3	
1382	1	92.2050	108453.97	B23-1R0	
1383	1	92.2146	108442.73	D4-3Q5	
1384	56	92.2193	108437.23	D0-1P2 B'5-2P5	
1385	8	92.2227	108433.20	C8-1P5	
1386	17	92.2306	108423.84	C5-0P4	
1387					CuII 92.24161
1388	2	92.2477	108403.80	B19-0R5 B21-0P7	
1389	2	92.2508	108400.13	B30-2R1	
1390	101	92.2569	108393.00	D0-1Q3	
1391	5	92.2602	108389.11	B31-2R3 B2-1R7	
1392	12	92.2620	108387.04	D2-2P4 C12-2P5	
1393	23	92.2775	108368.78	D3-4Q4 C7-1R0	
1394	2	92.2818	108363.69	B30-2P1	
1395	1	92.2862	108358.56	B25-1R5	r 108354.77
1396		92.2885	108355.92	B18-0P3 B26-1R6 B22-0P8	r 108354.77
1397	7	92.2904	108353.62	B18-0P3	
1398	3	92.2953	108347.91	D"0-3Q1	
1399	0	92.3060	108335.27	B'0-0P7	
1400		92.3087	108332.15	B32-2P4	r 108327.94
1401	11	92.3118	108328.51	C7-1R1 B'1-0P10	r 108327.94
1402	7	92.3128	108327.36	C7-1R1	
1403	75	92.3170	108322.46	B23-1P1	
1404	5	92.3190	108320.12	D2-2Q5	
1405	8	92.3210	108317.67	D0-1R5 B"0-2P3	
1406	70	92.3240	108314.19	B31-2P3	
1407	2	92.3438	108290.97	B'0-2R5	
1408	5	92.3552	108277.62	B'4-2P3	
1409		92.3678	108262.85	D"0-3Q2	
1410	5	92.3707	108259.43	B2-1P6	r
1411		92.3727	108257.06	C7-1Q1	sh
1412	3	92.3768	108252.28	C7-1R2	
1413		92.3779	108250.98		
1414	10	92.3818	108246.38	C5-0Q5	
1415	44	92.3845	108243.33	B'1-1R0	
1416	128	92.3971	108228.53	D0-1P3 C11-2Q4 B19-0P5	
1417	153	92.4023	108222.41	B'1-1P3 B26-1P6	
1418	35	92.4127	108210.20	B24-1R4 D1-3Q1	
1419					CuII 92.42386
1420	1	92.4278	108192.51	B25-1P5	
1421	22	92.4346	108184.55	D0-1Q4	
1422	17	92.4383	108180.33	B23-1R2	
1423	2	92.4527	108163.45	B30-2P2 D0-0Q13	
1424	0	92.4554	108160.32	C8-1Q6	
1425	1	92.4582	108157.00		
1426	61	92.4762	108135.94	B'1-1R2 C11-2P4 D"0-3Q3	
1427	1	92.4853	108125.30	B18-0R4	
1428	7	92.4903	108119.48	D1-3Q2	
1429	6	92.4923	108117.13	B23-1P2	
1430	26	92.4970	108111.60	C7-1Q2	
1431	1	92.5010	108106.94		
1432	8	92.5041	108103.30	C10-2R1	
1433					CuII 92.50992 Ke

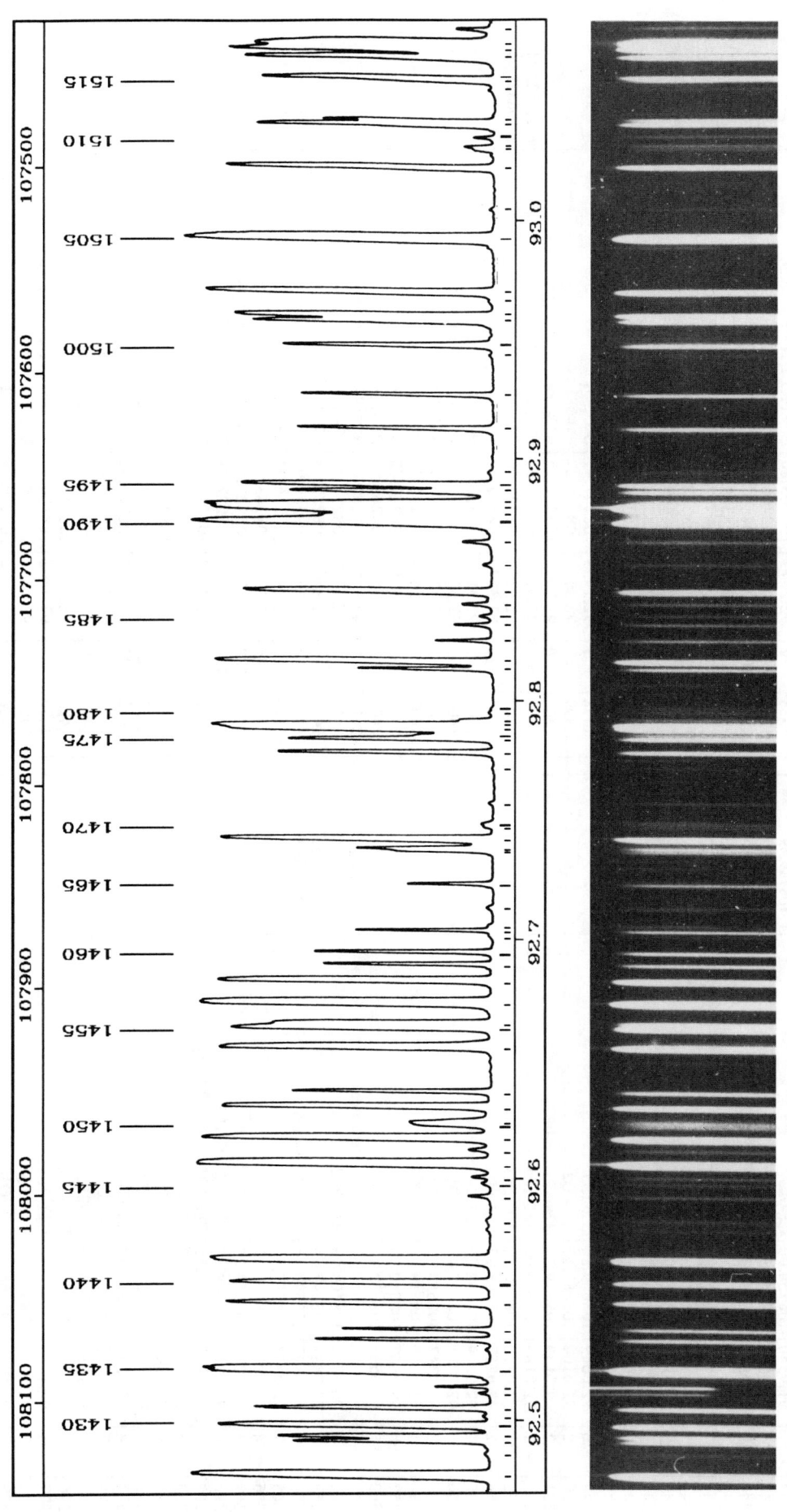

PLATE 25

TABLE 25

Nb	I	λ (nm)	σ (cm⁻¹)	Assignment	Comment
1434	154	92.5201	108084.62		CuII 92.51263 Ke
1435	1	92.5289	108074.37	B'1-1P1 D2-2P5	
1436	1	92.5328	108069.78		
1437	5	92.5372	108064.71	C7-1R3	
1438	3	92.5482	108051.80	B24-1P4	
1439	16	92.5566	108042.03	B"0-2P4	
1440	20	92.5657	108031.33	C10-2Q1	
1441	36	92.5784	108016.51	C7-1P2 C5-0P5	
1442	1	92.5823	108011.96	D2-2Q6 B"0-2R6	
1443	1	92.5923	108000.35		
1444	1	92.5975	107994.24	B29-2R0	
1445	1	92.6002	107991.17		
1446	3	92.6055	107984.95		
1447	101	92.6114	107978.07	B'1-1R3 D1-3Q3 B20-0R7	
1448	46	92.6161	107972.60	C10-2R2	
1449	2	92.6217	107966.11	D0-1P4 D"0-3Q4	HI 92.6226 Ke
1450		92.6227	107964.90	C8-1P6	
1451	18	92.6291	107957.40		
1452	8	92.6355	107949.96	B30-2R3 B'4-2P4	
1453	37	92.6542	107928.20	B18-0P4	
1454	22	92.6622	107918.92	B'5-2P6 D0-1Q5	
1455	8	92.6640	107916.77	B23-1R3 B29-2R1	
1456	111	92.6728	107906.55	B4-2R5	
1457	43	92.6823	107895.49	B'1-1P2	
1458	5	92.6892	107887.46	C7-1Q3	
1459	5	92.6942	107881.65	C10-2Q2	
1460	0	92.7014	107873.21	B29-2P1	
1461	7	92.7033	107871.07	B17-0P2	
1462	1	92.7050	107869.02	B30-2P3	
1463	0	92.7134	107859.30	B19-0R6	
1464	3	92.7230	107848.06	B32-2P5	
1465	3				
1466	3	92.7366	107832.34	C5-0Q6	
1467	22	92.7378	107830.92	C10-2P2 C11-2Q5	
1468	1	92.7421	107825.87	B23-1P3	
1469	2	92.7473	107819.80	C7-1R4	
1470	2	92.7486	107818.30		
1471	0	92.7575	107808.03	D1-3Q4	
1472	8	92.7719	107791.23	D4-3Q7	
1473	7	92.7784	107783.67	B22-1R0	
1474		92.7842	107777.02	B2-1P7	
1475	20	92.7847	107776.42	C10-2R3	sh
1476	29	92.7885	107772.00	B'1-1R4	
1477	2	92.7897	107770.56	C7-1P3 D"0-3Q5	
1478	1	92.7923	107767.56	C6-0Q9	
1479	0	92.7954	107763.96	B24-1R5	
1480	4	92.7975	107761.50	B26-1R7	
1481	23	92.8134	107743.03	B29-2R2	
1482		92.8166	107739.38	B"0-2P5	
1483	2	92.8254	107729.09	D2-2P6	
1484	2	92.8322	107721.30	C5-0R7 C7-0Q11	
1485	1	92.8359	107716.93	C11-2P5	
1486	1	92.8408	107711.32	B20-0P7	
1487	17	92.8459	107705.38	B22-1R1	
1488	1	92.8573	107692.15		
1489	2	92.8668	107681.11	B29-2P2	
1490	53	92.8744	107672.34	D0-1P5	
1491	6	92.8777	107668.48	B19-0P6	
1492	225	92.8809	107664.72	B21-0P8 B'1-1P3	
1493		92.8835	107661.71	C8-1Q7	sh
1494	9	92.8875	107657.10	C10-2Q3	
1495	13	92.8903	107653.87	B22-1P1	
1496		92.8957	107647.61	B'1-0P11	
1497	8	92.9139	107626.56	D0-1Q6 B25-1P6	
1498	6	92.9279	107610.26	C7-1Q4	
1499	2	92.9448	107590.75	D1-3Q5	
1500	6	92.9486	107586.32	B24-1P5	
1501	13	92.9589	107574.42	C10-2P3	
1502	22	92.9613	107571.69	B'4-2P5 C9-1Q9	
1503	3	92.9674	107564.54	B23-1R4	
1504	21	92.9713	107560.06	B22-0P9 B'3-2R0	
1505	70	92.9939	107533.89	B22-1R2 B'3-2R1	
1506	1	93.0063	107519.63	B'5-2R6	
1507	20	93.0235	107499.72	B'1-1R5 C7-1R5	
1508	1	93.0311	107490.89	B30-2P4 C10-2R4	
1509	1	93.0326	107489.25	B28-2R0	
1510	1	93.0364	107484.80		
1511	9	93.0412	107479.22	C7-1P4	
1512	4	93.0433	107476.86	B29-2R3	r 107460.05
1513		93.0564	107461.71	C4-0Q1	r 107460.05
1514		93.0593	107458.40	C4-0Q1	
1515	11	93.0606	107456.83	B18-0P5	
1516	13	93.0693	107446.78	B22-1P2	
1517	21	93.0726	107442.96	B'3-2R2	+HI 93.07483
1518	18	93.0749	107440.37	D5-4Q1	
1519	1	93.0817	107432.54	C8-1P7	

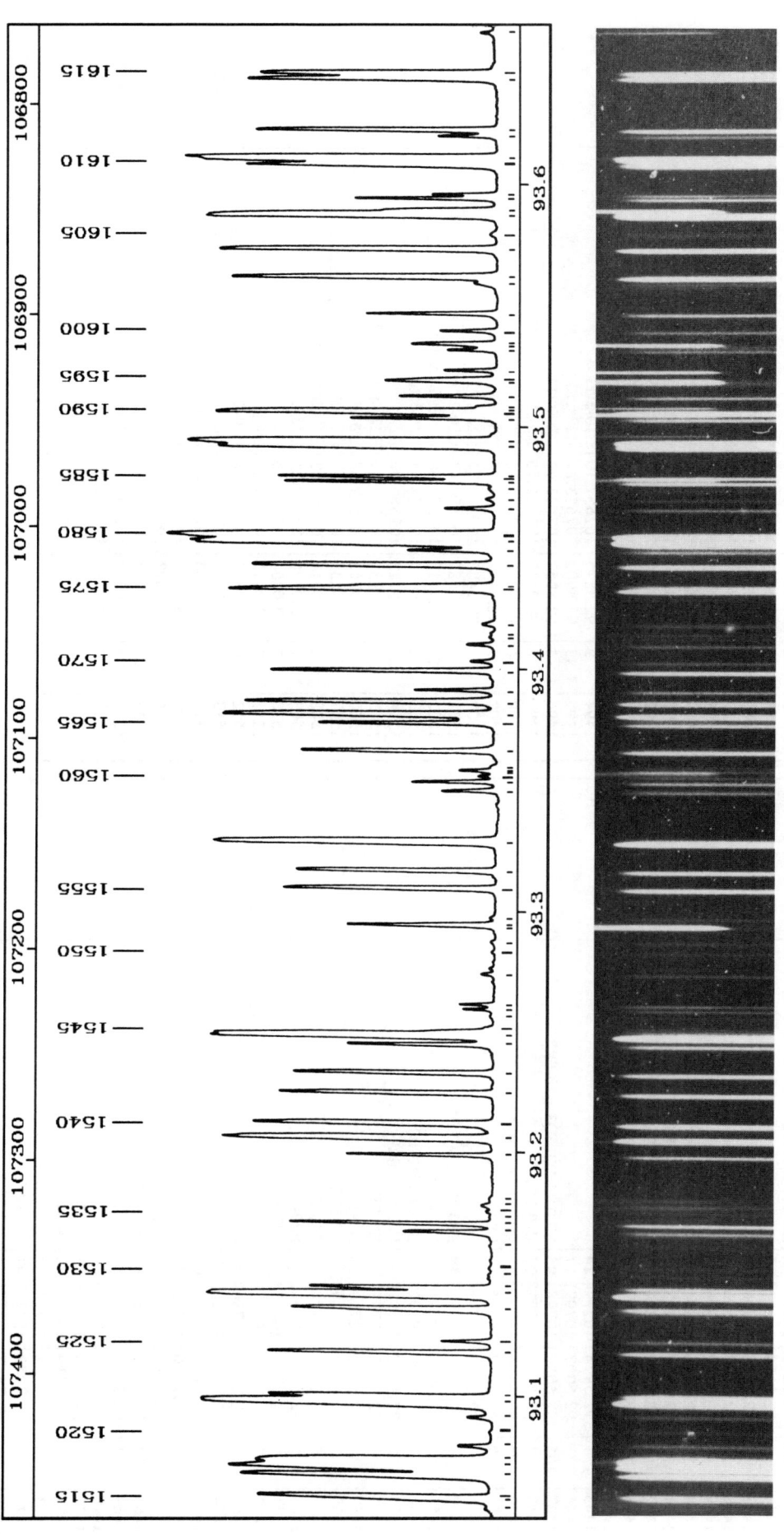

PLATE 26

TABLE 26

Nb	I	λ (nm)	σ (cm⁻¹)	Assignment	Comment
1520	1	93.0873	107426.07	C6-0P9	
1521	2	93.0934	107419.04	B'2-1R9	
1522	74	93.0989	107412.69	B'3-2P1	
1523	7	93.1013	107409.86	B28-2R1	
1524	10	93.1190	107389.46	B29-2P3	
1525	3	93.1237	107384.02	B"0-2P6 C12-2Q7 C7-0P11	
1526	7	93.1369	107368.83	B28-2P1	
1527	44	93.1426	107362.23	B'1-1P4	
1528	8	93.1454	107359.01	C5-0Q7 C11-2Q6	
1529	2	93.1507	107352.90	C10-2Q4	
1530	2	93.1537	107349.48	B17-0R4	
1531	1	93.1627	107339.06	D5-4Q2	
1532	0	93.1685	107332.38	D3-3Q1	
1533		93.1717	107328.75	B16-0R1 D0-1P6	r 107326.91
1534		93.1749	107325.06	B16-0R1	r 107326.91
1535		93.1768	107322.83	C4-0Q2	r 107321.22
1536		93.1796	107319.62	C4-0Q2 C4-0R3	r 107321.22
					+ r 107318.07
1537		93.1823	107316.51	C4-0R3	r 107318.07
1538	4	93.1997	107296.48	D7-5Q1	
1539	43	93.2064	107288.72	B'3-2R3	
1540	15	93.2123	107281.96	D0-1Q7	
1541	11	93.2249	107267.51	B22-1R3	
1542	9	93.2330	107258.18	C7-1Q5	
1543	6	93.2446	107244.86	B19-0R7 B24-1R6 C10-2P4 B'2-1P8	
1544	36	93.2482	107240.63	B'3-2P2	
1545		93.2502	107238.42	B28-2R2	sh
1546	0	93.2565	107231.17	D3-3Q2	
1547		93.2594	107227.75	C4-0P2	r 107226.56
1548		93.2615	107225.37	C4-0P2	r 107226.56
1549		93.2741	107210.86		
1550	1	93.2834	107200.23		
1551	1	93.2876	107195.34	D7-5Q2	
1552	0	93.2946	107187.39	C5-0R8 D5-4Q3	CuII 93.29387 Ke
1553		93.2980	107183.48		
1554	1	93.3096	107170.08	B'1-1R6 B28-2P2 B21-0R9	
1555	7	93.3170	107161.57	B17-0P4	
1556	9	93.3293	107147.51	B22-1P3	
1557	25	93.3504	107123.22	B29-2R4 B'4-2P6	
1558	1	93.3542	107118.94	B23-1R5 D12-7Q1	
1559	2	93.3569	107115.86	C4-0Q3	
1560		93.3578	107113.55	C4-0Q3	r 107114.70
1561		93.3589	107111.74	B30-2R5	r 107114.70
1562	1	93.3605	107104.43	C5-0P7 C7-1R6	
1563	1	93.3668	107091.49	B21-1R0 C4-0R4	
1564	5	93.3781	107087.34	C7-1P5	
1565	8	93.3817	107081.39	D"1-4Q1	
1566	14	93.3869			
1567	14				
1568	2	93.3917	107075.89	B'3-2R4	
1569	3	93.3997	107066.74	C9-2R0	
1570	2	93.4037	107062.12		
1571	2	93.4107	107054.11		
1572		93.4137	107050.73	B16-0P2	r 107049.68
1573		93.4155	107048.63	B16-0P2	r 107049.68
1574	2	93.4192	107044.34	D7-5Q3	
1575	10	93.4335	107027.98	C9-2R1	
1576		93.4348	107026.56	B24-1P6 B30-2P5	sh
1577	10	93.4435	107016.57	B21-1R1	
1578	1	93.4497	107009.50	B29-2P4 D12-7Q2	
1579	71	93.4534	107005.21	B'3-2P3	
1580	51	93.4556	107002.74	D"1-4Q2 B'1-1P5	
1581	2	93.4670	106989.67	B"0-2P7	r 106975.86
1582	0	93.4710	106985.03	B25-1P7	
1583	1	93.4751	106980.42	D4-3Q9	r 106975.86
1584		93.4779	106977.11	C4-0P3	
1585		93.4789			
1586		93.4801	106974.61	C4-0P3	
1587	24	93.4927	106960.19	B21-1P1	
1588	29	93.4945	106958.19	C9-2Q1	
1589	2	93.5041	106947.16	B27-2R0	CuII 93.50577 Ke
1590					
1591	14	93.5069	106944.00	D0-1P7	CuII 93.50855 Ke
1592					
1593	2	93.5132	106936.76	B23-1P5	OI 93.51930
1594		93.5193	106929.86		OI 93.51930
1595					CuII 93.52325
1596					
1597	0	93.5323	106914.98	C9-2R2 B17-0R5	
1598	1	93.5340	106913.04	B22-1R4	CuII 93.53434
1599					
1600	2	93.5401	106905.99	D9-6Q1	
1601	4	93.5475	106897.59	D0-1Q8	
1602	2	93.5601	106883.16	D"1-4Q3	
1603	11	93.5623	106880.62	B28-2P3 B18-0P6	
1604	11	93.5739	106867.38	B27-2R1	
1605	0	93.5802	106860.21	D2-2Q9	
1606	37	93.5881	106851.17	B21-1R2 B21-0P9 C6-1R0	
1607					
1608		93.5949	106843.44	C4-0Q4 C10-2P5	CuII 93.58977
1609		93.5966	106841.50	C4-0Q4	r 106842.47
1610	6	93.6090	106827.34	B27-2P1	r 106842.47
1611	57	93.6116	106824.37	C6-1R1	
1612	2	93.6208	106813.86	C11-2Q7 B'3-2R5	
1613	6	93.6233	106811.01	C9-2Q2	
1614	9	93.6446	106786.77	B'1-1R7	
1615	10	93.6467	106784.33	C4-0R5	
1616					OI 93.66295

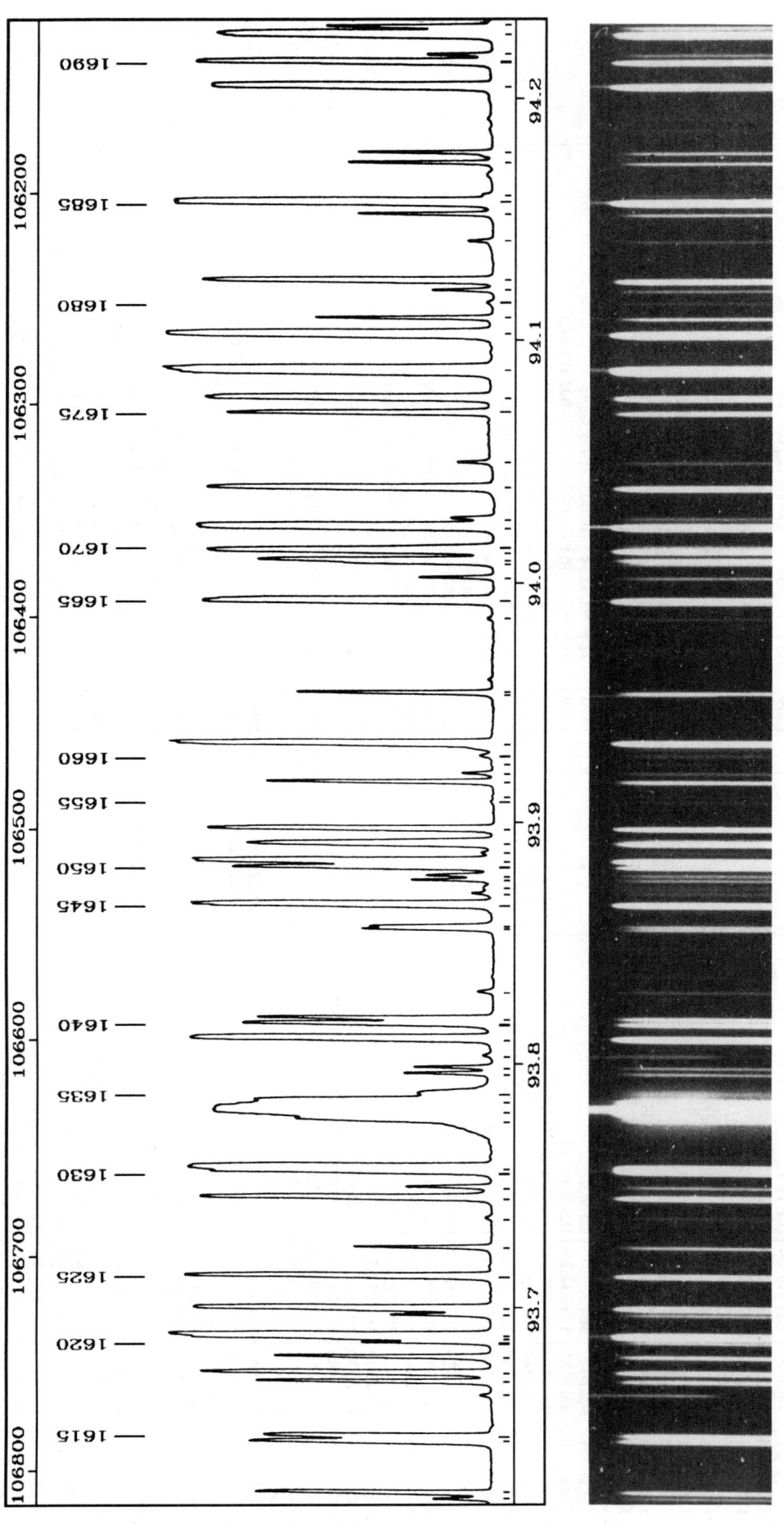

PLATE 27

TABLE 27

Nb	I	λ (nm)	σ (cm⁻¹)	Assignment	Comment
1617	4	93.6690	106758.93	B22-1P4	
1618	9	93.6726	106754.79	B21-1P2	
1619	3	93.6791	106747.37	C9-2P2	
1620		93.6850	106740.66	B16-0P3	r 106739.64
1621		93.6868	106738.61	B16-0P3	r 106739.64
1622	99	93.6876	106737.72	D2-4Q2 C6-1Q1	
1623	2	93.6964	106727.64	D"1-4Q4 C9-2R3	
1624	18	93.6991	106724.62	C6-1R2	
1625	10	93.7125	106709.41	B3-2P4	
1626	2	93.7242	106696.06	B27-2R2 C11-2P7	
1627	0	93.7364	106682.11	B29-2R5	
1628	10	93.7444	106673.07	B17-0P5	
1629	2	93.7488	106668.03	B2-1P9	
1630		93.7552	106660.81	C4-0P4	r 106660.08
1631		93.7564	106659.34	C4-0P4	r 106660.08
1632	7	93.7763	106636.71	C7-1P6 D12-7Q4	
1633					HI 93.78035 Ly ε
1634	13	93.7843	106627.61	B27-2P2	
1635	3	93.7871	106624.44		
1636	2	93.7953	106615.14	B4-2P7	
1637	4	93.7978	106612.25	D2-4Q3	
1638					OI 93.80200
1639	36	93.8097	106598.75	C6-1Q2	
1640	8	93.8158	106591.84	C9-2Q3 B23-1R6	
1641	5	93.8184	106588.96	B'1-1P6	
1642	1	93.8294	106576.45	B24-1R7	
1643	1	93.8556	106546.66	D1-2P3	
1644	1	93.8566	106545.47	B29-2P5	
1645	34	93.8658	106535.11	C6-1R3	
1646	1	93.8706	106529.67	C5-0P8	
1647	2	93.8730	106526.87	B16-0R4	
1648	2	93.8761	106523.44	D4-3Q10	
1649	2	93.8780	106521.24		
1650	10	93.8812	106517.60		
1651	24	93.8839	106514.58	C6-1P2	
1652	0	93.8872	106510.75	B18-0R7	
1653	21	93.8911	106506.38	C4-0Q5	
1654	15	93.8975	106499.12	C9-2P3	
1655	0	93.9088	106486.34	C8-1Q9	
1656	1	93.9110	106483.86	C4-0R6	
1657	6	93.9168	106477.21	D0-1Q9	
1658	1	93.9205	106473.05	B22-1R5	
1659	0	93.9238	106469.34	C13-3Q1	
1660	1	93.9278	106464.71	C9-2R4	
1661	29	93.9329	106458.97	B21-1P3	
1662					CuII 93.95232
1663	4	93.9539	106435.23	B27-2R3	
1664	0	93.9848	106400.24	C6-0Q11	
1665	42	93.9923	106391.71	C6-1Q3	
1666	1	94.0018	106380.96	B26-2R0 B23-1P6	
1667		94.0080	106373.98	D12-7Q5	sh
1668	8	94.0091	106372.67	B20-1R0	
1669	33	94.0131	106368.13	B'0-1R0	sh
1670		94.0142	106366.91	C7-1Q7	
1671	99	94.0229	106357.05	C6-1R4 B'0-1R1 B'3-2P5	
1672	8	94.0264	106353.08	B'1-1R8	
1673	15	94.0390	106338.81	B16-0P4 B27-2P3	
1674	1	94.0495	106326.99	B17-0R6 B21-0R10	
1675	7	94.0701	106303.68	C9-2Q4 B26-2R1	
1676	26	94.0764	106296.60	B20-1R1	
1677	71	94.0879	106283.65	B'0-1R2 C4-0P5 B22-1P5	
1678	78	94.1028	106266.72	C6-1P3	
1679	6	94.1094	106259.28	B26-2P1	
1680	1	94.1159	106251.94	D11-7Q1	
1681	5	94.1210	106246.20	C5-0Q9	
1682	28	94.1250	106241.68	B20-1P1 C8-1P9 C10-2R7	
1683	1	94.1416	106223.01	B18-0P7	
1684	5	94.1527	106210.43	B21-1R4	
1685		94.1578	106204.65	B15-0P2 B'0-1P1	r 106202.95
1686		94.1608	106201.26	B15-0P2	r 106202.95
1687	5	94.1741	106186.37	C7-1R8	
1688	4	94.1783	106181.56	C9-2P4	
1689	59	94.2056	106150.77	B'0-1R3 D11-7Q2	
1690	31	94.2158	106139.37	D0-3R1	
1691	2	94.2188	106135.96	B26-2R2	
1692	19	94.2272	106126.46	C7-1P7 B20-1R2	
1693	5	94.2301	106123.17	D'0-3R2	

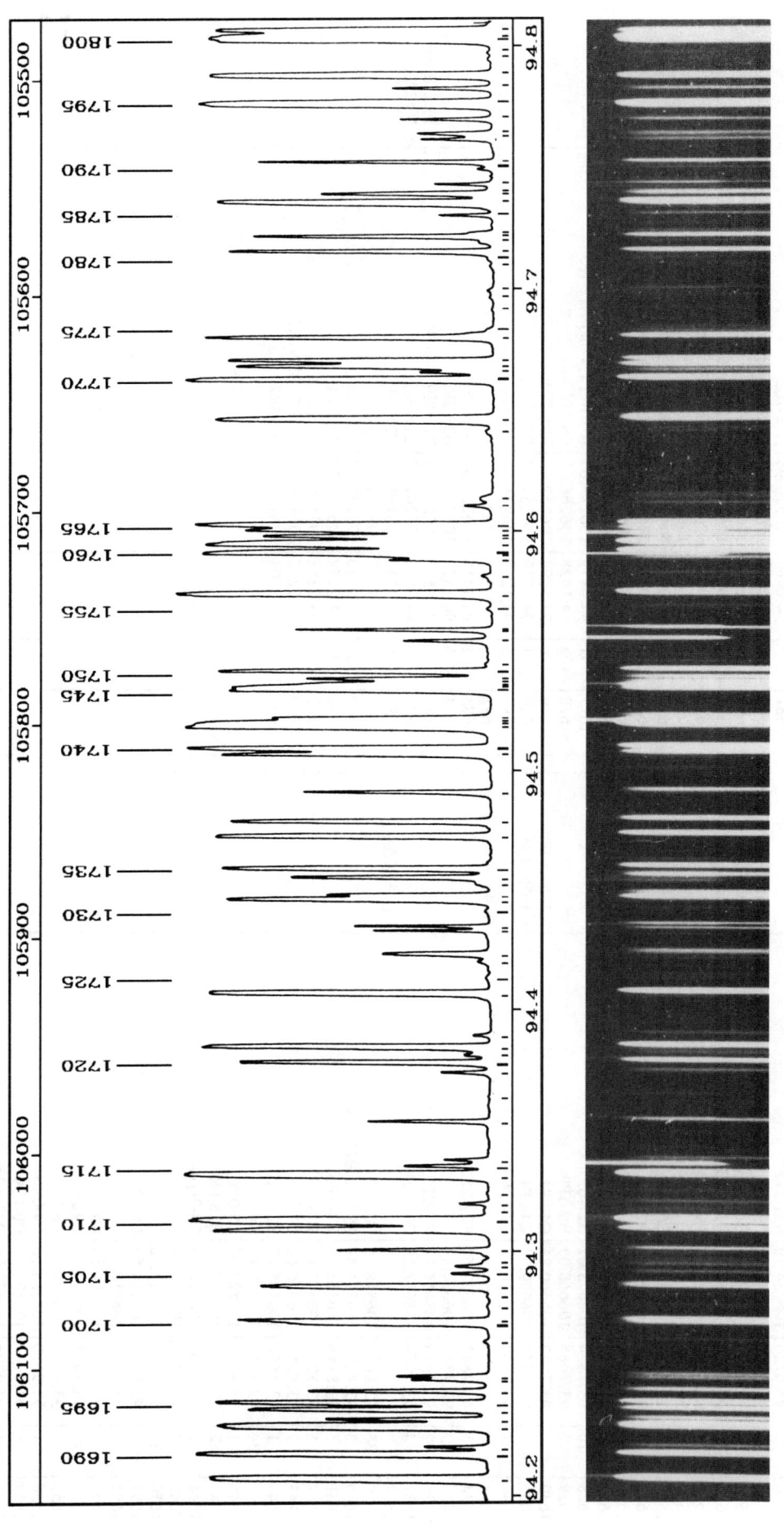

PLATE 28

TABLE 28

Nb	I	λ (nm)	σ (cm⁻¹)	Assignment	Comment
1694	9	94.2340	106118.79	C6-1Q4 C9-2R5	
1695	19	94.2367	106115.74	D0-3R0	
1696	5	94.2416	106110.21	C4-0Q6	
1697	6	94.2463	106104.95	C7-0Q13	
1698	6	94.2477	106103.41	B17-0P6	
1699	0	94.2619	106087.38	B27-2R4 C12-3R0	
1700	4	94.2692	106079.24	B16-0R5	
1701	6	94.2703	106077.94	B21-1P4	
1702	0	94.2761	106071.38		
1703	0	94.2804	106066.62	D12-7Q6	
1704	7	94.2843	106062.23	B26-2P2	
1705	0	94.2898	106056.04	C4-0R7	
1706	0	94.2928	106052.66	C6-1R5	
1707	0	94.2949	106050.24	C6-0P11 B'2-1P10	
1708	0	94.2993	106045.32	B28-2P5	
1709	16	94.3078	106035.73	B20-1P2 D4-3Q11	
1710	71	94.3116	106031.46	B0-1P2 B'2-2R0	
1711	3	94.3161	106026.45	C12-3R1	
1712	2	94.3189	106023.24	D0-1Q10	
1713		94.3226			
1714	91	94.3303	106010.47	D0-3Q1 B'2-2R1	CuII 94.33348
1715	1	94.3367	106003.34	D1-2P5	
1716	4	94.3524	105985.62	B23-1R7 C12-3Q1	
1717		94.3623			
1718	1	94.3729	105962.52	B27-2P4	
1719	12	94.3769	105958.15	B0-1R4	
1720	1	94.3803	105954.34	B22-1R6 B'3-2P6	
1721	15	94.3832	105951.08	C6-1P4	
1722	1	94.3881	105945.55	C9-2Q5	
1723					
1724	21	94.4057	105925.81	D0-3Q2 B'2-2R2	
1725	0	94.4123	105918.45	D2-2Q11	
1726	3	94.4193	105910.54	B33-3R1 C5-0P9	
1727		94.4222	105907.33	B15-0P3	r 105895.09
1728		94.4321	105896.19	B15-0P3	r 105895.09
1729	0	94.4341	105893.99	C10-2P7	
1730	9	94.4404	105886.93	B'2-2P1 B33-3P1	
1731	3	94.4453	105881.42	B26-2R3	
1732		94.4471	105879.41		
1733		94.4518			
1734	7	94.4544	105871.19	B'1-1R9	
1735	14	94.4582	105866.98	B20-1R3	
1736	11	94.4721	105851.26	B16-0P5	
1737	9	94.4784	105844.31	C4-0P6	
1738	4	94.4903	105830.92	C8-2R0 C7-1Q8	
1739	11	94.5060	105813.42	D0-3P2	
1740	14	94.5085		D6-5Q1	
1741		94.5177	105800.32	D0-3Q3	sh
1742	190	94.5191	105798.73	B0-1P3 C8-2R1	+CI 94.5191 Ke
1743	7	94.5183	105796.63	C9-2P5	CI 94.5191 Ke
1744	7	94.5210	105783.69	B25-2R0	
1745	11	94.5325	105782.25	C6-1Q5	+CI 94.5338 Ke
1746		94.5338			CI 94.5338 Ke
1747		94.5333			sh
1748	4	94.5350	105780.92	B'2-2R3	
1749	9	94.5371	105778.53	B21-1R5	
1750	0	94.5403	105774.95	B26-2P3	
1751	0	94.5435	105771.45	B19-0R9	
1752					CuII 94.55249
1753		94.5575	105755.72		CuII 94.5579 Ke
1754					CuII 94.5579 Ke
1755	0	94.5668	105745.33	B23-1P7	
1756	43	94.5726	105738.87	B20-1P3	
1757	0	94.5806	105729.98	B22-1P6	
1758	3	94.5869	105722.94	B17-0R7	CuII 94.58769
1759	28	94.5901	105719.29	C8-2Q1	
1760	55	94.5937	105715.25	D4-4Q1 D6-5Q2	
1761	4	94.5966	105712.02	B'2-2P2	CuII 94.59648
1762	16	94.5995	105708.85	B0-1R5	
1763	24	94.6016	105706.48	B25-2R1	
1764	0	94.6098	105697.34	C8-2R2	
1765	1	94.6128	105694.00	B15-0R4	
1766					
1767					
1768	3	94.6411	105662.33	D1-2P6	d
1769	14	94.6456	105657.32	C4-0Q7 D8-6Q1	d
1770	35	94.6622	105638.74	B32-3R0 D0-3P3	
1771	4	94.6654	105635.17	D0-3Q4	
1772	10	94.6677	105632.66	B21-1P5	
1773	7	94.6701	105629.93	B19-1R0	
1774	9	94.6796	105619.42	D4-4Q2	
1775	1	94.6830	105615.55	C5-0Q10	
1776	1	94.6912	105606.44	C7-1R9	
1777		94.6968	105600.20	B14-0R1	r 105598.75
1778		94.6994	105597.29	B14-0R1	r 105598.75
1779		94.7097	105585.81	C3-0R2	r 105584.19
1780		94.7124	105582.77	C3-0R2	r 105584.19
1781	4	94.7156	105579.24	C8-2Q2	
1782	1	94.7196	105574.78	C4-0R8	
1783	5	94.7219	105572.26	D6-5Q3	
1784	1	94.7234	105570.54		
1785	21	94.7309	105562.15	D8-6Q2	sh
1786	5	94.7360	105556.51	B19-1R1 B32-3R1	
1787		94.7397	105552.43	C6-1P5	
1788		94.7408	105551.16	C3-0Q1	
1789	4	94.7436	105548.06	C3-0Q1 B16-0R6	
1790		94.7504	105540.50	B14-0P1 B25-2R2 D0-1Q11	r 105549.61 sh
1791	11	94.7528	105537.82	B26-2R4 C8-1P10 D'3-5Q1	r 105549.61

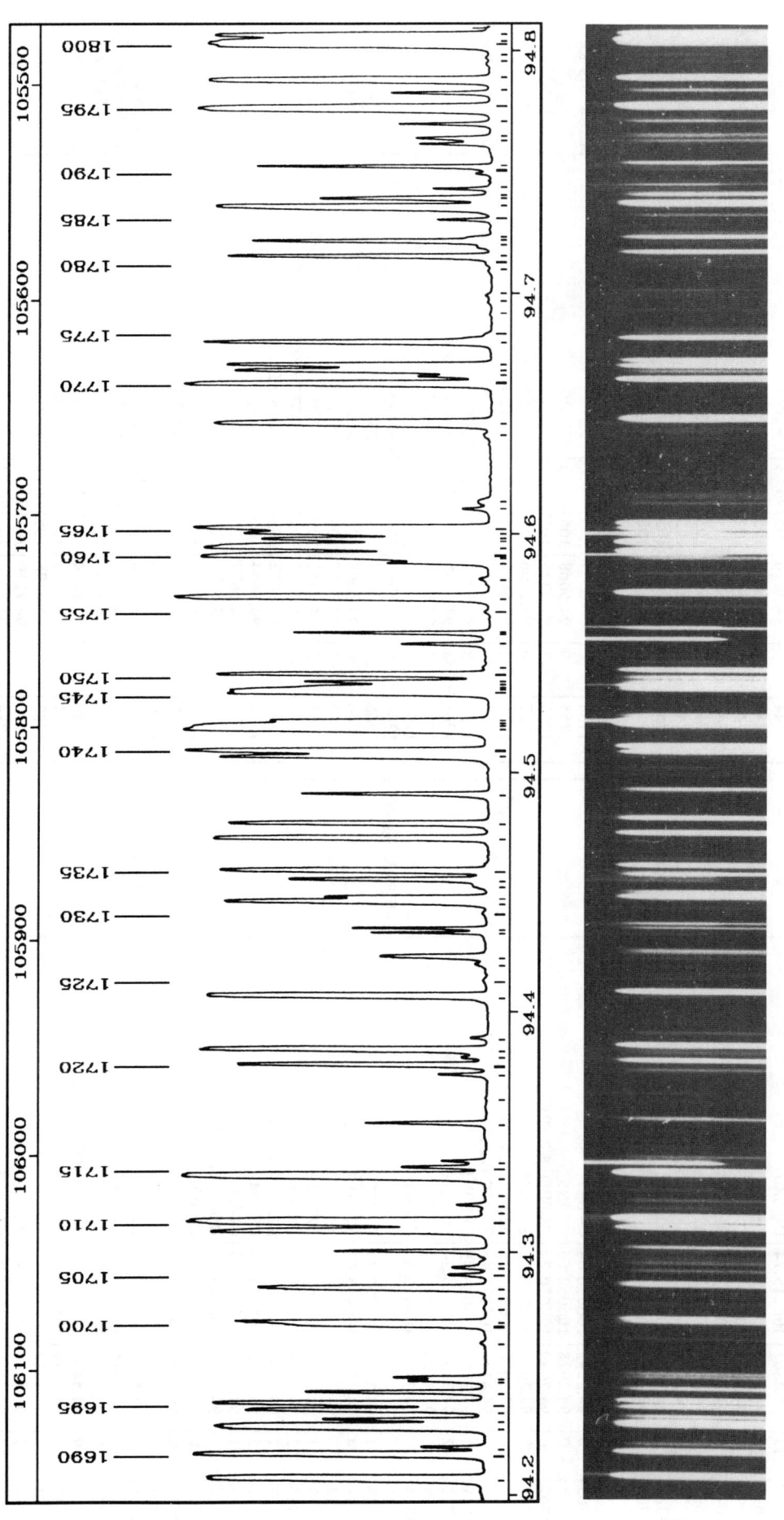

PLATE 28

TABLE 28 *continued*

Nb	I	λ (nm)	σ (cm⁻¹)	Assignment	Comment
1792	1	94.7621	105527.38	B32-3P1	
1793	2	94.7646	105524.68	C8-2R3 C9-2Q6	
1794	2	94.7704	105518.19	B20-1R4	
1795	36	94.7770	105510.79	C8-2P2 B'0-1P4	
1796	2	94.7834	105503.71	B27-2P5	
1797	31	94.7888	105497.70	B19-1P1 B15-0P4	
1798	0	94.7934	105492.55	B18-0P8	
1799		94.7948			sh
1800		94.8035	105481.35	B'2-2P3	
1801	63	94.8057	105478.85	D2-3R1	
1802	12	94.8075	105476.90	D4-4Q3	

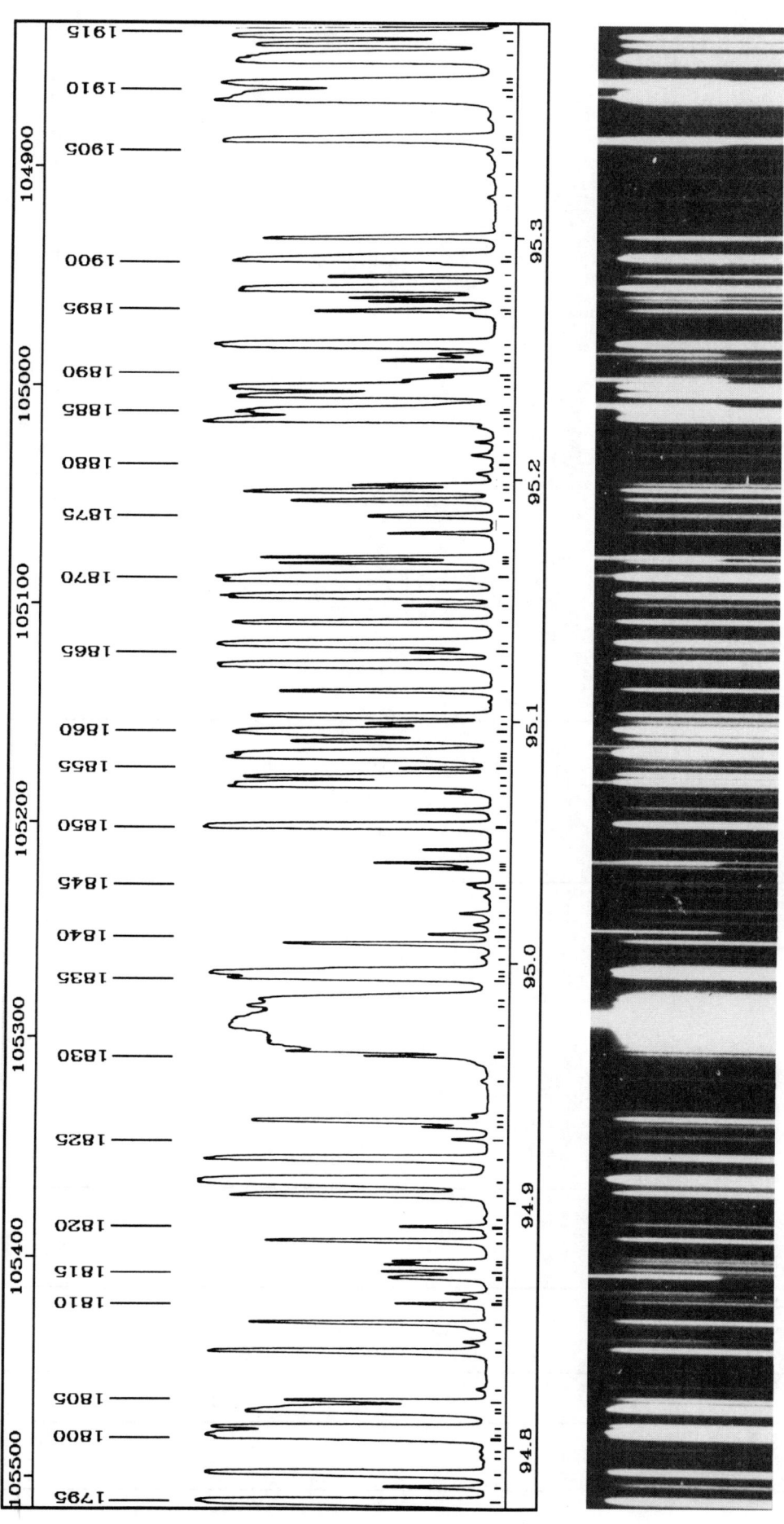

PLATE 29

TABLE 29

Nb	I	λ (nm)	σ (nm)	Assignment	Comment
1803	33	94.8148	105468.72	D2-3R0	
1804		94.8163	105467.06	B17-0P7 C8-1R11	sh
1805	3	94.8185	105464.70	B25-2P2 B33-3R3	
1806	3	94.8238	105458.79	D3-5Q2	
1807	8	94.8388	105442.10	D2-3R2	
1808		94.8419	105438.61	C3-0R3	r
1809	7	94.8505	105429.04	D0-3P4	
1810	1	94.8586	105420.07	D8-6Q3	
1811		94.8602	105418.26	C3-0Q2	r 105416.72
1812		94.8630	105415.19	C3-0Q2	r 105416.72
1813					OI 94.86855
1814	1	94.8699	105407.48	B33-3P3	
1815	3	94.8717	105405.53	B'0-1R6	
1816	2	94.8746	105402.29	B26-2P4	
1817	2	94.8760	105400.71	B'2-1P11	
1818	5	94.8844	105391.46	B19-1R2	
1819	0	94.8879	105387.52	B32-3R2	
1820	3	94.8903	105384.82	D6-5Q4 C6-1Q6	
1821	0	94.8938	105380.92	B29-2P7	
1822	9	94.9032	105370.52	C8-2Q3	
1823	128	94.9087	105364.42	D2-3Q1	
1824	13	94.9181	105353.98	B20-1P4 B21-0R11 D2-3R3	
1825	1	94.9260	105345.17	C9-2P6	
1826	2	94.9311	105339.56	D3-5Q3	
1827	1	94.9334	105337.02	C4-0P7	
1828	1	94.9354	105334.83	B14-0P7 B20-0P10 B32-3P2	
1829	2	94.9504	105318.18	C11-3R1	
1830		94.9603	105307.22	C3-0P2 B23-1R8	r 105306.25
1831		94.9620	105305.28	C3-0P2	r 105306.25
1832					HI 94.97430 Ly δ
1833	22	94.9810	105284.20	B25-2R3	
1834	11	94.9835	105281.48	B16-0P6 C8-2R4	
1835	17	94.9933	105270.56	C8-2P3 C10-2P8	
1836	38	94.9951	105268.63	D2-3Q2 D1-2P7	sh
1837		94.9968	105266.72	C11-3Q1	
1838	0	95.0021	105260.84	B21-1R6	
1839	6	95.0075	105254.87	B15-0R5 D10-7Q1	
1840					OI 95.01121
1841	1	95.0158	105245.64	C7-1Q9	
1842	1	95.0206	105240.33	C6-1R7	
1843	0	95.0273	105232.95	B31-3R0 D8-6Q4	
1844		95.0310	105228.84	C3-0R4	r 105228.21
1845		95.0321	105227.58	C3-0R4	r 105228.21
1846		95.0385	105220.53	C3-0Q3	r 105219.04
1847		95.0398			
1848		95.0412	105217.56	C3-0Q3	r 105219.04
1849	1	95.0469	105211.25	D2-3R4	
1850	38	95.0567	105200.40	C5-1R0	
1851	1	95.0634	105192.98	C11-3R2	
1852	1	95.0705	105185.12	D3-5Q4	
1853	57	95.0741	105181.13	C5-1R1	
1854	9	95.0773	105177.58	B25-2P3	
1855		95.0806	105173.90	B14-0R3 C6-1P6	r 105172.42
1856		95.0833	105170.93	B14-0R3	r 105172.42 sh
1857	52	95.0862	105167.74	D2-3P2 B0-1P5	
1858					OI 95.08846
1859	4	95.0921	105161.22	B"0-3R0	
1860	57	95.0959	105157.02	B"0-3R1	
1861	1	95.0992	105153.37	B31-3R1 D6-5Q5	
1862	8	95.1024	105149.86	B4-3R0 C4-0Q8	
1863	6	95.1126	105138.52	B19-1R3	
1864	47	95.1235	105126.45	D2-3Q3	
1865	1	95.1285	105120.98	B31-3P1	
1866	38	95.1321	105116.99	B4-3R1 B26-2R5 C4-0R9	
1867	15	95.1411	105107.10	B"0-3R2	
1868	1	95.1481	105099.35	B22-1P7 C8-1Q11	
1869	26	95.1523	105094.64	C5-1R2 C8-2Q4 B24-2R1	
1870	120	95.1601	105086.02	B20-1R5 C5-1Q1	
1871		95.1660	105079.52	C3-0P3	r 105078.20
1872		95.1671			r 105078.20
1873		95.1684	105076.87	C3-0P3	
1874	2	95.1783	105065.97	C11-3P2 B'1-1P9	
1875	2	95.1854	105058.18	B32-3P3 D4-4Q5	
1876	6	95.1920	105050.87	B'0-1R7	
1877	7	95.1961	105046.36	B24-2P1	
1878	3	95.1985	105043.66	C7-1R10	
1879	1	95.2032	105038.50	B23-1P8 B21-0P11 C9-2Q7	
1880	1	95.2067	105034.62	B17-0R8	
1881	1	95.2109	105030.04	D0-1Q12	
1882	0	95.2163	105024.04	B4-3P1	
1883	0	95.2226	105017.04	D10-7Q3	
1884	25	95.2266	105012.68	B14-0P3 B4-3R2	
1885	22	95.2293	105009.69	B"0-3R3	
1886					NI 95.23037 Ke
1887	26	95.2357	105002.61	B"0-3P1 D8-6Q5	
1888	43	95.2391	104998.85	B19-1P3	
1889					NI 95.24151 Ke
1890	1	95.2440	104993.50	C7-1P9 D3-5Q5	
1891	3	95.2502	104986.69	B18-0R9 C11-3R3	
1892					NI 95.25231 Ke
1893	59	95.2569	104979.26	D2-3P3	
1894		95.2692	104965.69	C8-2R5	
1895	2	95.2711	104963.67	C8-2P4 B'5-3R5	
1896		95.2749	104959.44	C3-0Q4	r 104958.47
1897		95.2767	104957.50	C3-0Q4	r 104958.47
1898	44	95.2804	104953.39	C5-1Q2 C3-0R5	
1899	2	95.2855	104947.76	B26-2P5	
1900		95.2907	104942.02	B25-2R4	sh

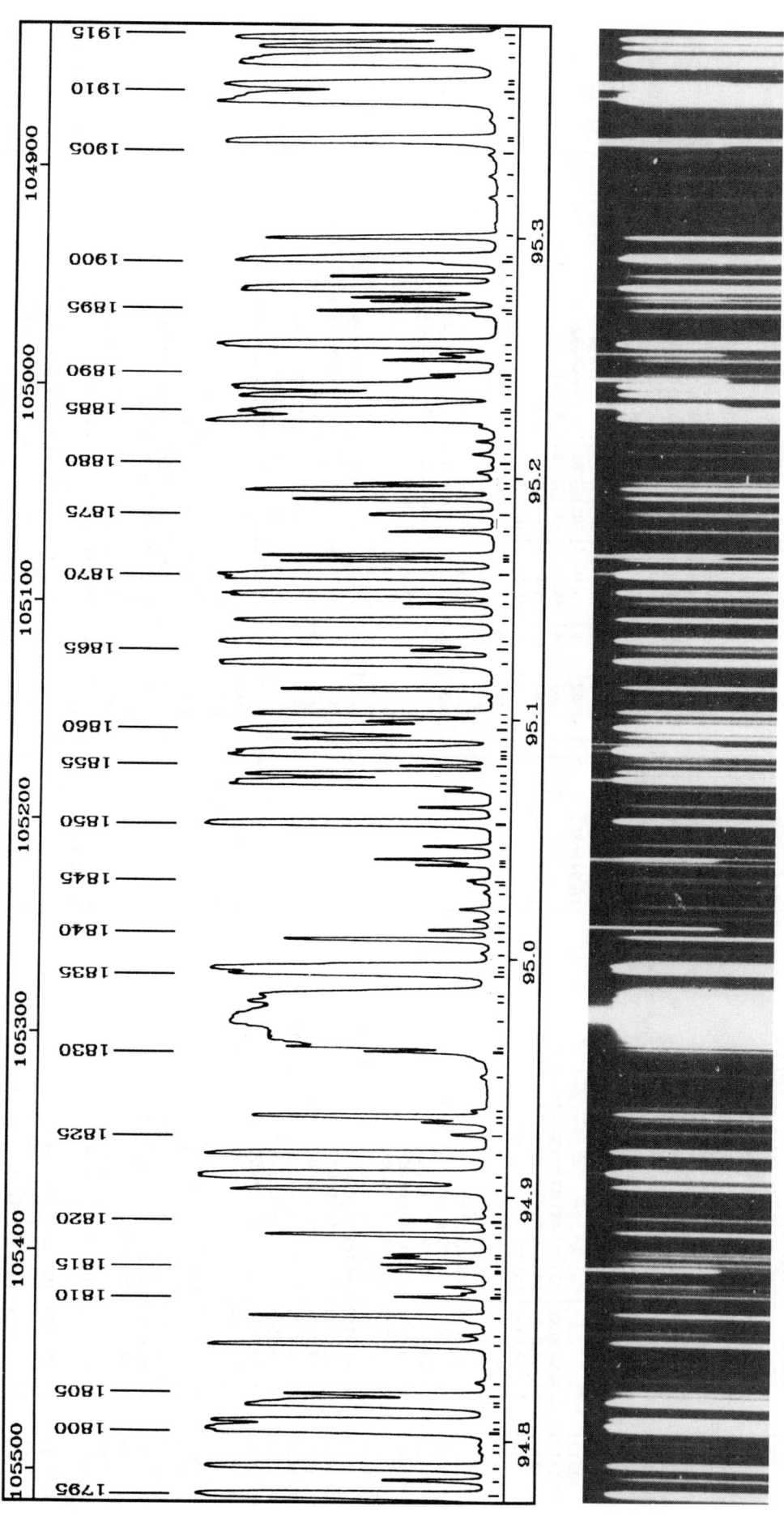

PLATE 29

TABLE 29 *continued*

Nb	I	λ (nm)	σ (nm)	Assignment	Comment
1901	28	95.2930	104939.49	B16-0R7 C5-0Q11 C5-1R3 D2-3Q4	
1902	6	95.3017	104929.89	B24-2R2 B31-3P2 C6-1Q7	
1903	0	95.3181	104911.84	B28-2P7	
1904	1	95.3273	104901.73		
1905	0	95.3361	104892.08		NI 95.34152
1906					
1907	15	95.3429	104884.60	D"0-4Q1 B20-1P5	
1908	0	95.3509	104875.74	B27-1P11	
1909	150	95.3594	104866.43	D0-2R1 B"0-3R4	
1910	33	95.3620	104863.59	C5-1P2	
1911	35	95.3656	104859.57	B"0-3P2 B18-1R0 C6-0Q13	NI 95.36549
1912					
1913	95	95.3762	104847.94	B4-3P2 D0-2R0	
1914	12	95.3819	104841.67	B4-3R3	
1915	28	95.3856	104837.60	D0-2R2	

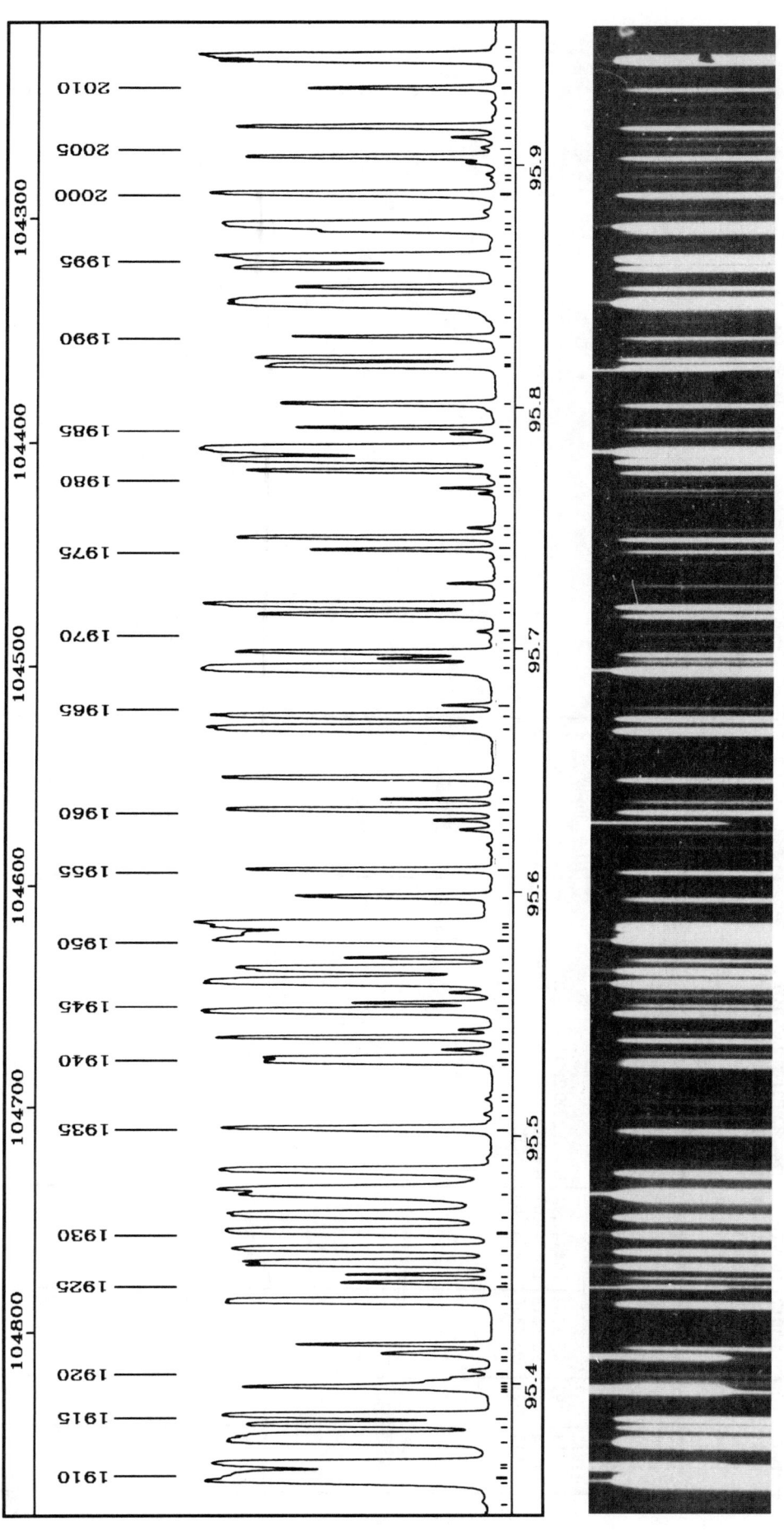

PLATE 30

TABLE 30

Nb	I	λ (nm)	σ (cm⁻¹)	Assignment	Comment
1916	27	95.3970	104825.07	B'5-3P5	
1917					
1918		95.3998	104821.98	B14-0R4	NI 95.39699
1919		95.4010	104820.70	B14-0R4	r 104821.34
1920	1	95.4039	104817.50	C11-3P3 C9-2P7	r 104821.34
1921	0	95.4092	104811.75	D"0-4Q2 B19-1R4	
1922					NI 95.41040 Ke
1923	6	95.4144	104806.02	B25-2P4	
1924	35	95.4322	104786.48	B18-1R1 B30-3R0 D4-4Q6	
1925					CuII 95.43830
1926		95.4396	104778.27	B13-0R0	r 104776.54
1927		95.4428	104774.81	B13-0R0	r 104776.54
1928		95.4474	104769.70	C3-0P4	r
1929	30	95.4533	104763.25	C8-1P11 D0-2R3	
1930	51	95.4605	104755.36	C5-1Q3 C8-2Q5	
1931	69	95.4673	104747.93	D1-4Q1 D2-3P4	
1932	360	95.4765	104737.84	D0-2Q1	
1933	41	95.4852	104728.28	B18-1P1 B31-3R3	
1934	0	95.4901	104722.85	B20-0R11	
1935	5	95.5023	104709.53	B30-3R1 D2-3Q5	
1936	2	95.5084	104702.87	D"0-4Q3	
1937	1	95.5146	104696.08	B18-0P9	
1938	0	95.5171	104693.27	B32-3P4	
1939	7	95.5297	104679.49	B"0-3R5 C6-1P7	
1940	4	95.5311	104677.95	B24-2R3	
1941	1	95.5346	104674.12	B30-3P1 B33-3R5	
1942	18	95.5397	104668.52	D1-4Q2	
1943	1	95.5429	104664.99	B21-1R7	
1944	45	95.5508	104656.34	B"0-3P3	
1945	1	95.5542	104652.70	B31-3P3	
1946	2	95.5584	104648.04	B"0-1R8	
1947	119	95.5632	104642.81	D0-2Q2 D0-2R4	
1948	45	95.5683	104637.25	B22-1R8 C3-0Q5	
1949	5	95.5728	104632.34	B16-0P7	
1950	102	95.5809	104623.45	C5-1P3	
1951	15	95.5832	104620.87	B18-1R2	
1952	20	95.5858	104618.02	B19-1P4 B14-0P4	
1953	15	95.5871	104616.58	B'4-3P3	
1954	6	95.5978	104604.93	C3-0R6 B'4-3R4	
1955	11	95.6088	104592.94	C4-0Q9 C8-2P5	
1956	0	95.6157	104585.38	B33-3P5	
1957	0	95.6195	104581.18	D12-8Q1	
1958	2	95.6256	104574.46	C5-0P11	CuII 95.62903
1959					
1960	15	95.6339	104565.41	B24-2P3 B23-1R9	
1961	2	95.6383	104560.68	D"0-4Q4	
1962	25	95.6472	104550.90	B26-2R6 D'1-4Q3	
1963	79	95.6675	104528.76	D0-2P2 B23-2R0	
1964	24	95.6724	104523.39	B18-1P2	
1965	1	95.6769	104518.39	B25-2R5	
1966	185	95.6922	104501.76	D0-2Q3	
1967	5	95.6963	104497.30	D0-1Q13	
1968	10	95.6989	104494.43	C5-1Q4	
1969	0	95.7033	104489.62	C6-0P13	
1970	0	95.7074	104485.09	B30-3P2 D12-8Q2	
1971	9	95.7145	104477.37	D4-4Q7 C5-1R5 D0-2R5	
1972	19	95.7184	104473.15	D2-3P5	
1973	2	95.7271	104463.65	B23-2R1	
1974	1	95.7365	104453.36		
1975	3	95.7411	104448.41	B15-0P6	
1976	7	95.7460	104443.04	C7-2R0	
1977	2	95.7499	104438.73	D2-3Q6	
1978		95.7640	104423.34	B13-0P2	r 104422.04
1979		95.7664	104420.74	B13-0P2 C6-1Q8	r 104422.04
1980	0	95.7707	104416.07	B26-2P6 B21-0R12	
1981	8	95.7736	104412.92	B'0-3P4	
1982	32	95.7782	104407.88	C7-2R1	
1983	108	95.7822	104403.52	C3-0P5 B23-2P1	
1984	2	95.7887	104396.41	D'1-4Q4	
1985	3	95.7914	104393.55		
1986	9	95.8013	104382.76	B14-0R5	CuII 95.81542
1987					
1988	12	95.8171	104365.47	B18-1R3	
1989	15	95.8202	104362.13	D4-6Q1	
1990	5	95.8290	104352.51	C8-2Q6 B25-2P5 C10-3Q1 C7-1R11	
1991		95.8348			
1992	175	95.8436	104336.67	B20-1P6 C7-2Q1 B'0-1P7 D0-2P3	
1993	7	95.8495	104330.19	B19-1R5	
1994	22	95.8578	104321.19	C5-1P4	
1995	44	95.8619	104316.73	B1-2R0 D0-2Q4	
1996	1	95.8674	104310.75	C7-1P10	
1997	3	95.8726	104305.06	B4-3R5 B29-3R0	
1998	105	95.8751	104302.41	B1-2R1	
1999	1	95.8805	104296.49	B30-3R3	
2000	23	95.8880	104288.39	D4-6Q2 D7-6Q1	
2001		95.8935	104282.41	B13-0R3	r 104281.16
2002		95.8957	104279.92	B13-0R3	r 104281.16
2003	1	95.9005	104274.79	B17-0R9	
2004	7	95.9031	104271.96	B23-2R2 C6-1R9	
2005	2	95.9067	104267.97		
2006	2	95.9109	104263.46	C3-0R7	
2007	11	95.9153	104258.61	C3-0Q6	
2008	2	95.9192	104254.38	B16-0R8	
2009	2	95.9249	104248.24	B32-3P5	
2010	6	95.9311	104241.48	C4-0P9	
2011	2	95.9375	104234.55		
2012	70	95.9433	104228.26	B18-1P3 B1-2R2 B29-3R1	
2013	1	95.9473	104223.86	C5-0Q12	

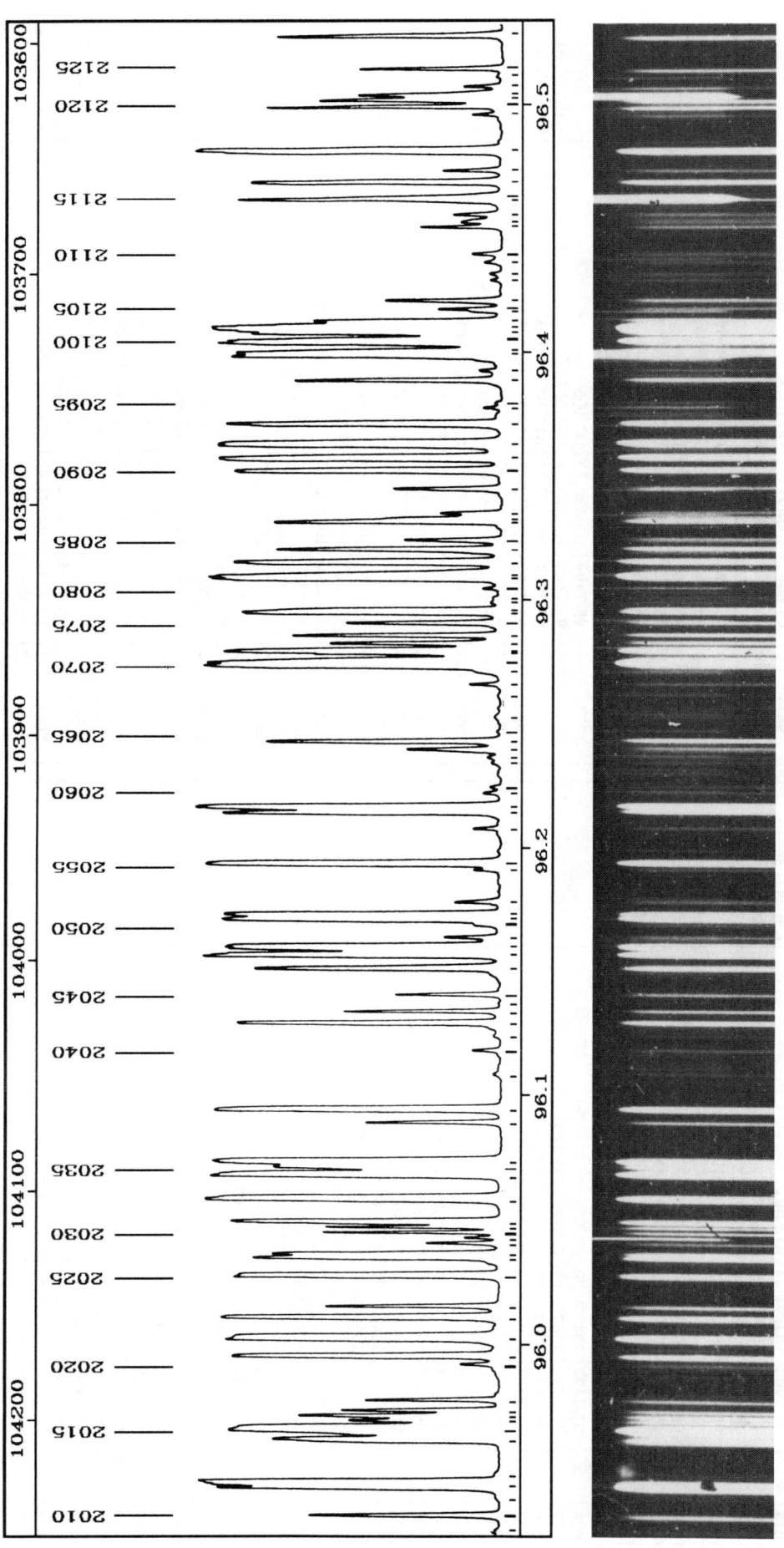

PLATE 31

TABLE 31

Nb	I	λ (nm)	σ (cm⁻¹)	Assignment	Comment
2014	6	95.9610	104208.97	B23-2P2 B30-3P3 C10-3Q2	
2015	49	95.9653	104204.38	C7-2Q2 D5-5Q1 D1-4Q5	
2016	5	95.9686	104200.71	B'0-1R9	
2017	4	95.9704	104198.75	D7-6Q2	
2018	2	95.9726	104196.45	B24-2P4	
2019	3	95.9767	104191.98	B29-3P1	
2020	2	95.9909	104176.55	D4-6Q3	
2021	15	95.9944	104172.70	C5-1Q5	
2022	54	96.0019	104164.65	B'1-2P1	
2023	16	96.0100	104155.77	B19-1P5 D12-8Q4	
2024	4	96.0143	104151.12	D2-3P6	
2025	15	96.0268	104137.62	B14-0P5	
2026	10	96.0340	104129.84	D9-7Q1 D2-3Q7	
2027	7	96.0354	104128.23	B"0-3P5 C5-1R6	
2028	2	96.0396	104123.72	C7-2P2	
2029					CuII 96.04135
2030		96.0439	104119.02	B13-0P3	r 104117.85
2031		96.0461	104116.67	B13-0P3	r 104117.85
2032	15	96.0484	104114.19	D5-5Q2	
2033	45	96.0577	104104.13	D0-2P4	
2034	47	96.0674	104093.62	B'1-2R3	
2035	9	96.0705	104090.27	B15-0R7	
2036	42	96.0725	104088.09	D0-2Q5	
2037	8	96.0878	104071.44		
2038	18	96.0932	104065.60	B17-1R0 D7-6Q3 B29-3R2	
2039	4	96.1072	104050.53	B19-0R11	
2040	2	96.1167	104040.17	D9-7Q2	
2041	0	96.1227	104033.68	C6-0Q14	
2042	9	96.1280	104027.94	B23-2R3	
2043	2	96.1326	104022.95	B18-1R4	
2044	1	96.1364	104018.93		
2045	3	96.1394	104015.62	B25-2R6	
2046	11	96.1500	104004.14	B29-3P2 C7-2Q3	
2047	26	96.1557	103997.96	B21-1R8 B'1-2P2	
2048	45	96.1588	103994.62	B17-1R1 B26-2R7	
2049	4	96.1630	103990.15	C4-0Q10	
2050	1	96.1681	103984.62	B20-1R7	
2051	18	96.1705	103981.99	C3-0P6	
2052	18	96.1718	103980.56	D5-5Q3	
2053	4	96.1772	103974.80	C5-0R13	
2054	1	96.1905	103960.36	B4-3P5	
2055	26	96.1933	103957.30	C5-1P5	
2056	0	96.2077	103941.82		
2057	11	96.2147	103934.27	B23-2P3 B13-0R4	
2058	40	96.2163	103932.48	B17-1P1	
2059	4	96.2219	103926.43	C7-1Q11	
2060	0	96.2244	103923.73	B24-2R5 D12-8Q5	
2061	0	96.2343	103913.01	C10-3P3	
2062	0	96.2362	103910.96	B16-0P8	

Nb	I	λ (nm)	σ (cm⁻¹)	Assignment	Comment
2063	1	96.2397	103907.19	B22-1R9 D9-7Q3	b
2064	3	96.2430	103903.69	B'1-2R4	
2065		96.2464			
2066	0	96.2527	103893.22	C8-2Q7	
2067	1	96.2554	103879.06	C7-2P3	
2068		96.2658			
2069		96.2702			
2070	120	96.2745	103869.69	D3-4Q1 C4-0R11	
2071	18	96.2793	103864.54	B'3-3R0	
2072		96.2798			
2073	4	96.2825	103861.07	C6-1Q9 B14-0R6	
2074	3	96.2856	103857.67	B22-2R0	sh
2075	5	96.2906	103852.28	B'0-1P8	
2076	13	96.2952	103847.32	B18-1P4	
2077		96.2964	103846.00	B12-0R0 D"1-5Q1	r 103844.59
2078		96.2991	103843.18	B12-0R0	r 103844.59
2079	0	96.3007	103841.38	B18-0P10	
2080		96.3043			
2081	70	96.3096	103831.82	B17-1R2 D0-2P5	
2082		96.3117	103829.60	B19-1R6	
2083	21	96.3155	103825.43	C3-0Q7	
2084	8	96.3207	103819.84	D0-2Q6 B25-2P6	
2085	2	96.3244	103815.81	B26-2P7 B29-3R3 C5-0P12	
2086	5	96.3321	103807.54	B15-0P7	
2087		96.3330	103806.59	C3-0R8	sh
2088	3	96.3355	103803.86	D5-5Q4	
2089	3	96.3454	103793.25	C5-1Q6 B28-3R0	
2090	14	96.3529	103785.20	B22-2R1	
2091	30	96.3581	103779.59	D3-4Q2	
2092	50	96.3638	103773.45	B'1-2P3	
2093	13	96.3719	103764.71	B'3-3R2	
2094		96.3773			
2095		96.3794			
2096	3	96.3893	103745.94	B24-2P5	NI 96.39903
2097	1	96.3932	103741.81	C7-2Q4	
2098		96.3994			
2099	20	96.3994	103735.12	B22-2P1	
2100	25	96.4007	103733.73	B17-1P2 B29-3P3	
2101	83	96.4053	103728.78	B13-0P4 B'3-3P1	sh
2102	4	96.4102	103723.48	B28-3R1 C5-1R7	
2103		96.4131	103720.40		
2104		96.4165			
2105	2	96.4182	103714.83	D11-8Q1	
2106	3	96.4217	103711.12	B20-1P7	r 103701.04
2107		96.4297	103702.46	B12-0P1	r 103701.04
2108		96.4324	103699.63	B12-0P1	
2109	2	96.4371	103694.59	B23-2R4	
2110	1	96.4405	103690.93	C8-2R8	
2111	3	96.4516	103678.92	B28-3P1	

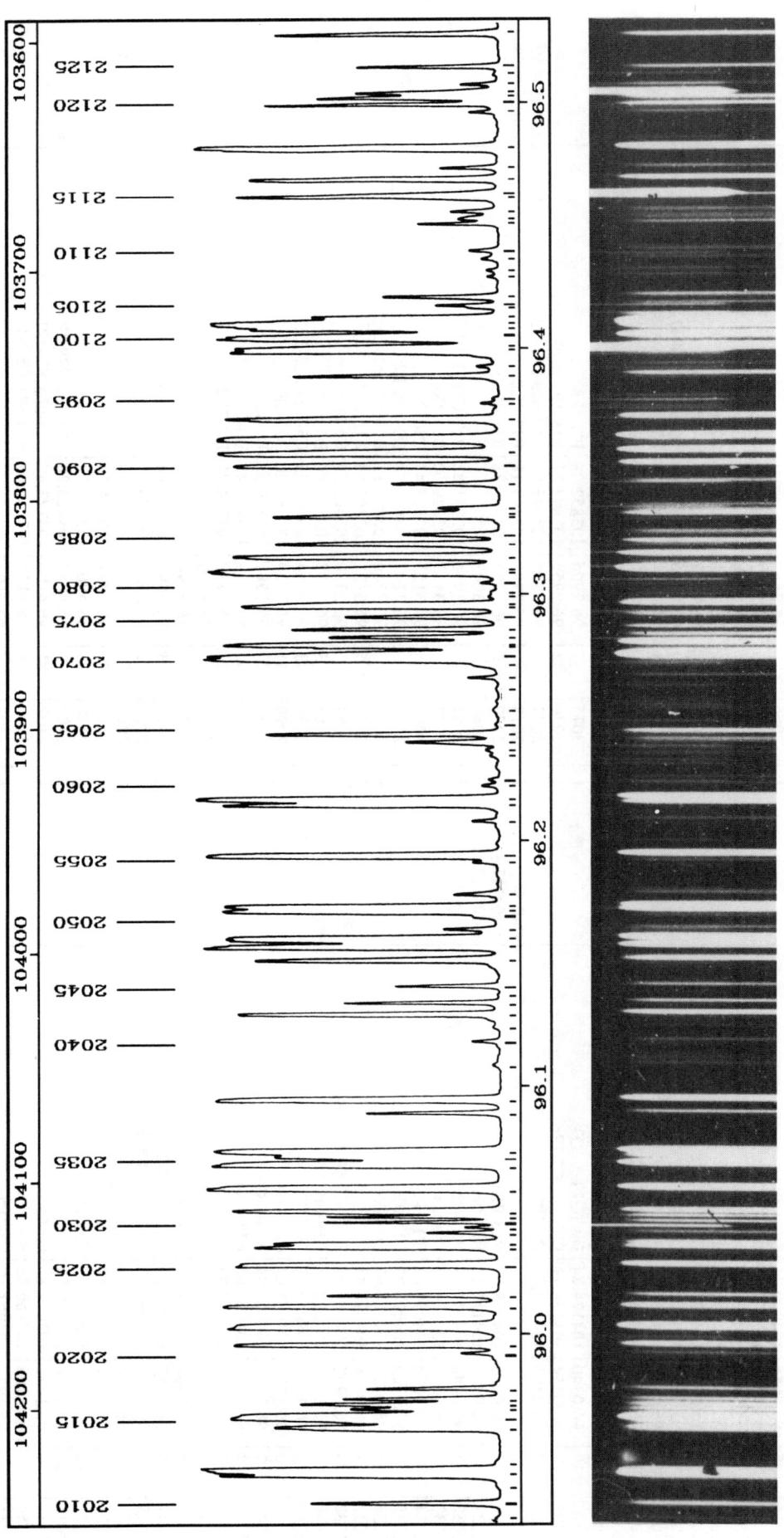

PLATE 31

TABLE 31 *continued*

Nb	I	λ (nm)	σ (cm⁻¹)	Assignment	Comment
2112	2	96.4541	103676.22	D"1-5Q3	b
2113	2	96.4569	103673.29	D7-6Q5	
2114					NI 96.46256
2115	20	96.4627	103666.98		
2116	9	96.4698	103659.34	B'1-2R5 C7-2R5	
2117	1	96.4745	103654.33	B22-1P9 C6-1R10	
2118	41	96.4826	103645.60	D3-4Q3	
2119	5	96.4969	103630.24	B19-1P6	
2120	11	96.5001	103626.89	B'3-3R3 D11-8Q2	
2121	7	96.5027	103624.09	B22-2R2	
2122					NI 96.50413
2123	5	96.5081	103618.25		
2124		96.5112		C7-2P4	
2125	2	96.5151	103610.70	B18-1R5 C4-0P10	
2126	5	96.5284	103596.48		

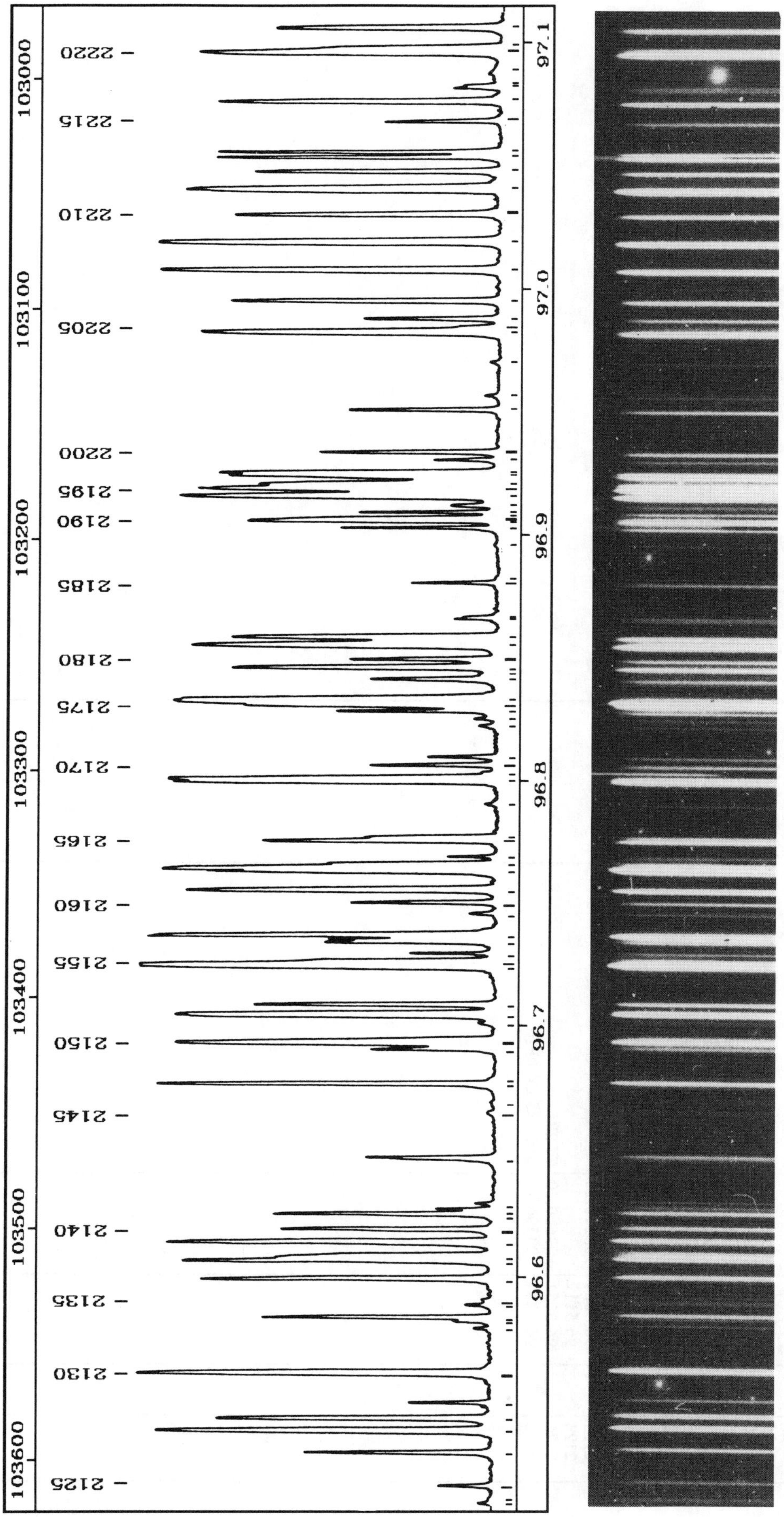
103000
103100
103200
103300
103400
103500
103600
97.1
97.0
96.9
96.8
96.7
96.6
2220
2215
2210
2205
2200
2195
2190
2185
2180
2175
2170
2165
2160
2155
2150
2145
2140
2135
2130
2125

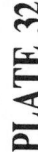

PLATE 32

TABLE 32

Nb	I	λ(nm)	σ(cm⁻¹)	Assignment	Comment
2127	59	96.5372	103587.01	C7-1P11 D5-5Q5 D2-5Q1	
2128	14	96.5418	103582.09	B17-1R3	
2129	1	96.5482	103575.26	B14-0P6	
2130	39	96.5601	103562.47	B33-3R7 B'3-3P2	
2131		96.5780	103543.21	C2-0R2	r 103541.83
2132		96.5806	103540.44	C2-0R2	r 103541.83
2133	8	96.5823	103538.68	B22-2P2 B23-2P4	
2134	3	96.5875	103533.03		
2135	3	96.5896	103530.81		
2136	11	96.5979	103521.88	C5-1P6 D0-2P6	
2137	14	96.6053	103514.01	D0-2Q7 D2-5Q2	
2138	14	96.6073	103511.80		
2139	21	96.6127	103506.06	C3-0P7	
2140	5	96.6179	103500.51	B13-0R5	
2141	8	96.6241	103493.86	B'1-2P4 D11-8Q3	
2142		96.6263	103491.48	B12-0P2	r 103490.39
2143		96.6283	103489.30	B12-0P2	r 103490.39
2144	4	96.6467	103469.66	C5-0Q13 D3-4Q4	b
2145	1	96.6651	103449.98	B17-0R10	
2146	1	96.6694	103445.37	B33-3P7	
2147		96.6766	103437.62	B17-1P3 B25-2R7 C2-0R3	r 103436.36
2148		96.6790	103435.10	C2-0R3	r 103436.36
2149	6	96.6907	103422.52	B15-0R8	
2150	34	96.6933	103419.78	C4-1R0 D7-6Q6	
2151	0	96.7011	103411.44	D2-3Q9	
2152	47	96.7045	103407.71	B21-0R13 C4-1R1	
2153	13	96.7086	103403.38	D2-5Q3	
2154	113	96.7250	103385.92	D1-3R1	
2155		96.7265	103384.26	C2-0Q2	r 103382.79
2156		96.7293	103381.32	C2-0Q2	r 103382.79
2157	5	96.7341	103376.17	B22-2R3	
2158	53	96.7367	103373.39	D1-3R0 D0-1Q15	
2159	1	96.7461	103363.39	B'1-2R6	
2160	3	96.7505	103358.68	C5-1Q7	
2161	18	96.7556	103353.23	D1-3R2	
2162	63	96.7641	103344.15	B'3-3P3 C4-0Q11	
2163		96.7662	103341.82	B12-0R3 C3-0Q8	r 103340.46
2164		96.7688	103339.09	B12-0R3	r 103340.46
2165	9	96.7757	103331.69	C4-1R2 D5-5Q6	
2166	5	96.7772	103330.09	B'0-1P9	
2167	1	96.7909	103315.51	B28-3R3	
2168	111	96.8008	103304.95	C4-1Q1 B18-0R11	CuII 96.80416
2170	4	96.8065	103298.82	C3-0R9	
2171	6	96.8099	103295.22	C7-2R6	
2172	2	96.8224	103281.89		
2173	2	96.8252	103278.89	B23-2R5	
2174		96.8287	103275.16	C2-0P2	r 103273.93
2175	2	96.8311	103272.61	C2-0P2 D1-3R3	r 103273.93
2176	220	96.8328	103270.83	D1-3Q1	
2177	4	96.8413	103261.74	B14-0R7 B21-1R9	
2178	0	96.8434	103259.52	D2-5Q4	
2179	19	96.8463	103256.36	B22-2P3	
2180	7	96.8497	103252.74	D3-4Q5 B27-3R0	
2181	43	96.8556	103246.45	B17-1R4 C7-2P5 B13-0P5 B19-1R7	
2182	23	96.8587	103243.14	B16-1R0	
2183		96.8660	103235.36	C2-0R4	r 103234.90
2184		96.8669	103234.45	C2-0R4	r 103234.90
2185	3	96.8810	103219.45	B28-3P3 B24-2P6	
2186	1	96.8831	103217.20	B25-2P7	
2187	0	96.8968	103202.54	B'3-3R5 C7-1Q12	
2188		96.9041	103194.80	C2-0Q3	r 103194.00
2189		96.9056	103193.21	C2-0Q3	r 103194.00
2190	22	96.9066	103192.14	C4-1R3	
2191		96.9078	103190.82	B12-0P3	r 103189.68
2192		96.9099	103188.53	B12-0P3	r 103189.68
2193	4	96.9128	103185.58	B'0-1R11	
2194	54	96.9169	103181.17	D1-3Q2	
2195	45	96.9200	103177.91	C4-1Q2 B27-3R1	
2196	21	96.9217	103176.10	D0-2P7 C6-0Q15	
2197	60	96.9246	103172.92	D0-2Q8	sh
2198	3	96.9265	103170.98	B16-1R1	
2199	3	96.9313	103165.81	B21-2R0	
2200	6	96.9346	103162.39	B'1-2P5	
2201	5	96.9518	103144.06	D1-3R4	
2202	1	96.9576	103137.88	B27-3P1	
2203	1	96.9711	103123.49	B16-0P9	
2204	41	96.9839	103109.87	B16-1P1	
2205	2	96.9859	103107.73	B15-0P8	
2206	6	96.9890	103104.52	C5-1P7 C8-2R9	
2207	12	96.9964	103096.65	B21-2R1 B23-2P5 C5-0P13	
2208	36	97.0091	103083.14	C4-1P2	
2209	41	97.0204	103071.15	B'3-3P4 D1-3P2	
2210	11	97.0316	103059.16	B17-1P4	
2211	63	97.0422	103047.96	D1-3Q3	
2212	12	97.0494	103040.34	B21-2P1	
2213		97.0552	103034.15	C2-0P3	r 103033.02
2214		97.0573	103031.89	C2-0P3	r 103033.02
2215	4	97.0698	103018.62	B27-3R2 B20-1P8 B'1-2R7	
2216	18	97.0779	103010.09	B16-1R2	
2217		97.0832	103004.42	B12-0R4	r 103003.88
2218		97.0842	103003.33	B12-0R4	r 103003.88
2219	0	97.0890	102998.30	D3-4Q6	
2220	53	97.0978	102988.94	C4-1Q3 C4-1R4	
2221	2	97.0996	102987.02	B13-0R6 C4-0P11	
2222	14	97.1076	102978.52	C2-0R5	sh

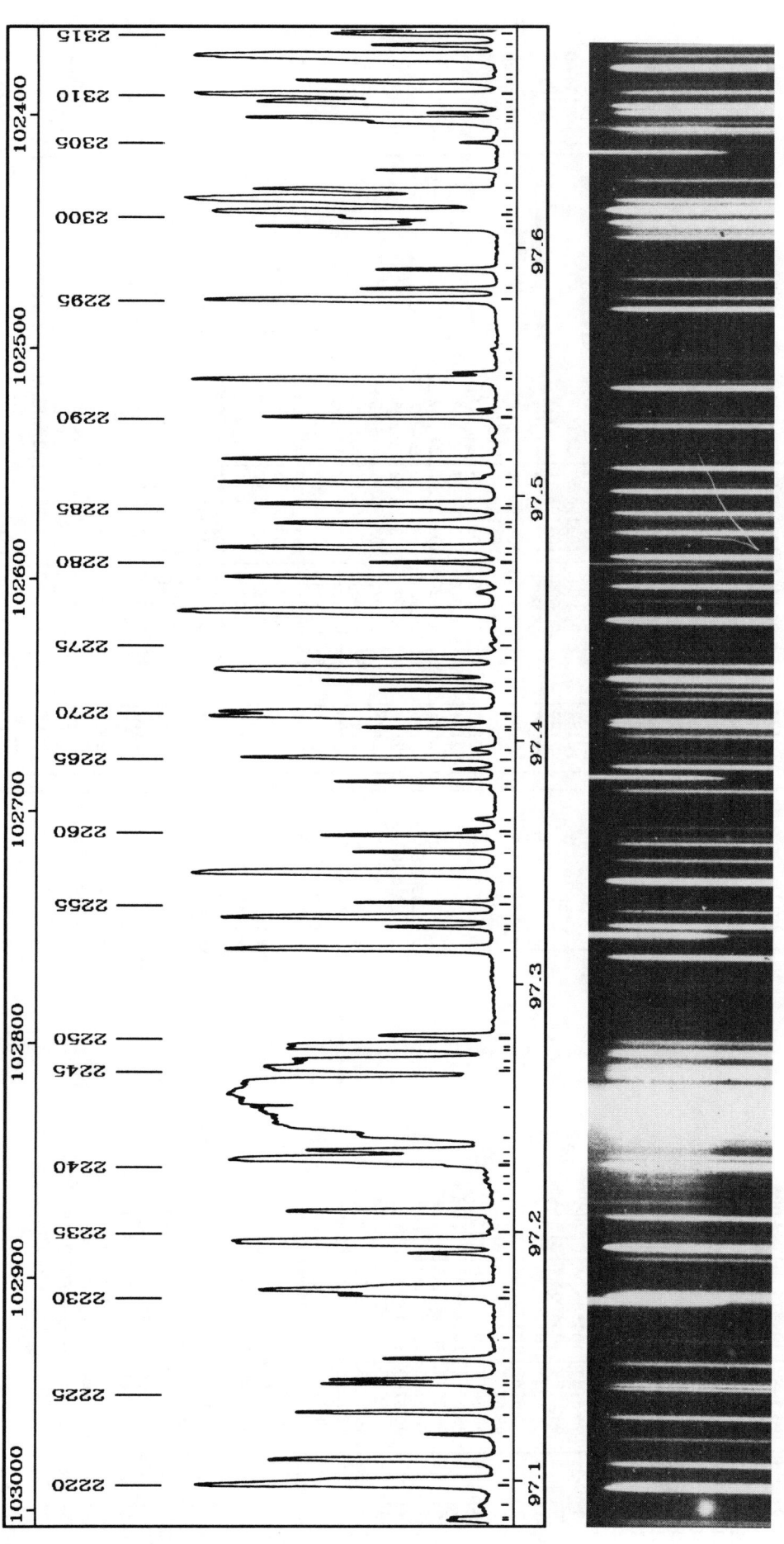

PLATE 33

TABLE 33

Nb	I	λ (nm)	σ (cm⁻¹)	Assignment	Comment
2223	3	97.1178	102967.75	C3-0P8	
2224	8	97.1267	102958.30	D1-3R5 B19-1P7	
2225	0	97.1350	102949.56	B27-3P2	
2226		97.1381	102946.25	C2-0Q4 C6-1P10	r 102945.25
2227		97.1398	102944.40	C2-0Q4 B21-1P9	r 102945.25
2228	3	97.1483	102935.44	B14-0P7	
2229	1	97.1576	102925.61	C6-2R0	
2230					OI 97.17381
2231	21	97.1757	102906.40	B16-1P2	
2232		97.1773	102904.71	C6-2R1	sh
2233	1	97.1906	102890.63	B22-2P4	
2234	61	97.1954	102885.58	D1-3P3	
2235	0	97.1998	102880.89	D8-7Q1	
2236	8	97.2075	102872.71	D1-3Q4 C5-1Q8	HeII 97.2088 Ke P
2237		97.2085			
2238	0	97.2194	102860.17	B18-0P11	
2239	0	97.2226	102856.76	B23-1R11	
2240	0	97.2262	102852.89	C4-1P3	
2241	38	97.2289	102850.06	B21-2P2	
2242	3	97.2324	102846.43	B18-1P6 B17-1R5	sh
2243		97.2385	102839.89		HI 97.25368 Ly γ
2244					
2245	23	97.2653	102811.60	C3-0Q9 C8-2Q9	
2246	28	97.2665	102810.32	B12-0P4 B24-2R7	
2247	30	97.2692	102807.47	D6-6Q1	
2248	23	97.2738	102802.61	D0-2Q9 C3-0R10	
2249	22	97.2752	102801.16	D8-7Q2	
2250	5	97.2789	102797.18	B16-1R3	
2251	30	97.3139	102760.28		
2252					OI 97.32342
2253					OI 97.32342
2254	33	97.3233	102750.32	B'3-3P5	
2255		97.3270	102746.37	B11-0P1 C4-1Q4	r 102738.78
2256		97.3328	102740.26	B11-0P1	r 102738.78
2257	93	97.3356	102737.29	C2-0P4	
2258	5	97.3452	102727.20	D6-6Q2 D10-8Q1	
2259	5	97.3538	102718.10	C4-1R5	
2260	3	97.3609	102710.65	D3-4Q7 B'3-3R6	
2261	1	97.3631	102708.31		
2262	1	97.3676	102703.60		
2263	4	97.3816	102688.75	B13-0P6	
2264		97.3828	102687.52		OI 97.38852
2265	16	97.3927	102677.09	B27-3P3 D'0-4R1 B15-0R9	
2266	3	97.3965	102673.15	D8-7Q3 D'0-4R2	
2267	5	97.4052	102663.91	C2-0R6	
2268		97.4065	102662.51	C4-0Q12	sh
2269	25	97.4098	102659.05	D1-3P4	
2270	16	97.4114	102657.40	D1-3Q5	
2271	6	97.4205	102647.79	D'0-4R0	
2272	8	97.4243	102643.80	B16-0R10 C6-2R3	
2273	75	97.4287	102639.13	C2-0Q5	
2274	9	97.4340	102633.54	B'5-4R0 D10-8Q2 B22-2R5	
2275	0	97.4389	102628.41	B'1-2R8	
2276	0	97.4447	102622.36	B24-2P7	
2277	105	97.4525	102614.12	B16-1P3	
2278	1	97.4604	102605.76	C6-1Q11	
2279	16	97.4665	102599.36	B17-1P5 C7-2Q7	
2280	8	97.4726	102592.95	D6-6Q3	
2281					CuII 97.47589
2282	27	97.4786	102586.58	B19-1R8 B'5-4R1	
2283	20	97.4886	102576.15	B12-0R5	
2284		97.4902	102574.41	C5-1P8	sh
2285	3	97.4948	102569.56	B26-3P1 B17-0R11	
2286	18	97.4966	102567.69	B21-2P3	
2287	26	97.5056	102558.20	C4-1P4	
2288	1	97.5082	102555.44	B20-1R9	
2289	33	97.5152	102548.15	D0-4Q1	
2290		97.5326	102529.84	B11-0P2 D'3-6Q1 C4-0R13	r 102528.25
2291		97.5356	102526.65	B11-0P2	r 102528.25
2292	29	97.5482	102513.45	B'5-4P1	
2293	3	97.5508	102510.67	B18-1R7 D10-8Q3	
2294	1	97.5604	102500.63	C6-2Q3	
2295	31	97.5805	102479.46	D4-5Q1	
2296	7	97.5850	102474.76	D0-4Q2	
2297	11	97.5927	102466.72	B'5-4R2	
2298	11	97.6097	102448.87	C3-0P9	
2299	4	97.6116	102446.84	B20-2R0	
2300	9	97.6138	102444.57	B22-2P5 C4-1R6	
2301	61	97.6157	102442.51	B'0-2R0	
2302	218	97.6209	102437.13	C7-1Q13 B'0-2R1	
2303	13	97.6248	102432.95	C4-1Q5	
2304	5	97.6330	102424.35	B16-1R4	
2305					OI 97.64481
2306	5	97.6526	102403.87	D1-3Q6	
2307	5	97.6540	102402.39	B11-0R3	r 102401.05
2308		97.6565	102399.71	B11-0R3 D0-2Q10	r 102401.05
2309	21	97.6609	102395.16	D4-5Q2 B15-1R0	
2310	36	97.6635	102392.37	D1-3P5	
2311	10	97.6692	102386.42	D3-4Q8	
2312	1	97.6721	102383.39	B26-3P2	
2313	76	97.6793	102375.88	B20-2R1 C6-2P3 B'0-2R2 B'3-3P6	
2314	5	97.6841	102370.76		
2315	11	97.6887	102366.01	D'0-4Q3	

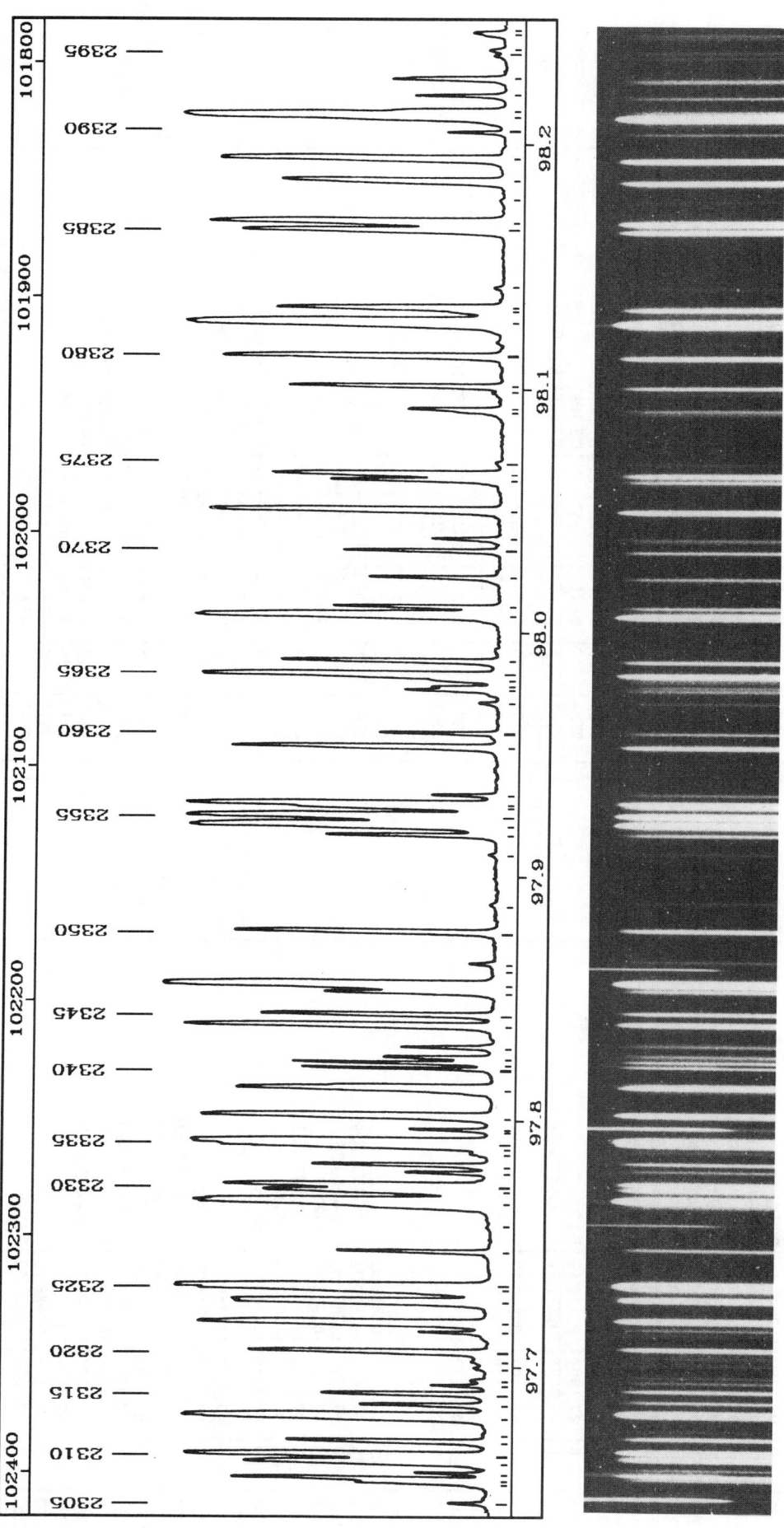

PLATE 34

TABLE 34

Nb	I	λ (nm)	σ (cm⁻¹)	Assignment	Comment
2316	4	97.6922	102362.34	D'0-4P2	
2317	3	97.6943	102360.08	D3-6Q3	
2318	0	97.6995	102354.72		
2319	0	97.7025	102351.48	B21-0R14	
2320	13	97.7058	102348.11	B'5-4P2	
2321	3	97.7116	102341.95	B21-2R4	
2322	48	97.7138	102339.67	C5-1Q9	
2323	54	97.7173	102336.08	B'2-3R0	
2324		97.7260	102326.93	B15-1R1	sh
2325	190	97.7316	102321.05	B'2-3R1 B20-2P1	
2326	5	97.7464	102305.55	B12-0P5	
2327					CuII 97.75674
2328	218	97.7663	102284.77	B'0-2P1	b
2329	17	97.7707	102280.13	C2-0R7	
2330	28	97.7728	102277.93	C2-0Q6	
2331	5	97.7781	102272.40	B'5-4R3	
2332	10	97.7812	102269.11	D4-5Q3	
2333	0	97.7862	102263.89	B20-1P9	
2334	75	97.7896	102260.38	B15-1P1	
2335	120	97.7904	102259.57	B'0-2R3	
2336	3	97.7956	102254.05	C6-1P11 C5-0P14	
2337					OI 97.79594
2338	59	97.8011	102248.34	B'2-3R2	
2339	21	97.8127	102236.17	B16-1P4 C3-0Q10	
2340		97.8208	102227.77	B11-0P3 D3-6Q4	r 102226.67
2341		97.8229	102225.56	B11-0P3 D6-6Q5 B14-0P8	r 102226.67
2342	6	97.8252	102223.17	D'0-4Q4 B18-1P7	
2343	5	97.8294	102218.73	B20-2R2	
2344	36	97.8384	102209.34	C4-1P5 B21-2P4	
2345	19	97.8429	102204.63	D'0-4P3	
2346	12	97.8523	102194.83	C3-0R11 B'1-2R9	
2347	170	97.8552	102191.80	B'2-3P1	
2348					OI 97.86170
2349	5	97.8645	102182.13	B'0-1P11	
2350	23	97.8775	102168.58	B15-1R2	
2351	5	97.8888	102156.75		
2352	3	97.9092	102135.47	C8-3R0	
2353	9	97.9169	102127.44	B20-2P2	
2354	138	97.9211	102123.04	B'0-2P2	
2355	80	97.9248	102119.22	B'2-3R3	
2356	14	97.9285	102115.27	D1-3Q7	
2357	46	97.9298	102113.92	B'5-4P3 B26-3P3	
2358	3	97.9335	102110.14	C8-3R1	
2359	22	97.9531	102089.73	C7-1P13 B'0-2R4	
2360	6	97.9587	102083.80	D1-3P6	
2361	3	97.9712	102070.83	C4-1Q6	
2362	4	97.9764	102065.41	B12-0R6	
2363	5	97.9777	102064.04	D'3-6Q5 B17-1P6	
2364	13	97.9803	102061.35	B11-0R4	
2365	47	97.9826	102058.92	B15-1P2	
2366	11	97.9881	102053.18	B13-0P7	
2367	100	98.0068	102033.69	D3-4Q9 B'2-3P2	
2368	8	98.0105	102029.93	C8-3Q1	
2369	9	98.0225	102017.37	D'0-4P4 D2-4R1	
2370	7	98.0335	102005.91	B16-1R5	
2371	5	98.0385	102000.76	D2-4R0 C5-1P9	
2372	42	98.0503	101988.46	C2-0P6	
2373	10	98.0625	101975.79	B20-2R3 B25-3P1	
2374	19	98.0650	101973.21	B'3-3P7 D0-2Q11	
2375	4	98.0695	101968.47	B24-2P8	
2376	3	98.0907	101946.44		
2377	9	98.0917	101945.48	C4-0Q13	
2378	0	98.0986	101938.29	B21-2R5	
2379	15	98.1011	101935.63	B'2-3R4	
2380	30	98.1135	101922.80	B15-1R3 B22-2P6	
2381	403	98.1276	101908.10	B'0-2P3	
2382	20	98.1317	101903.82	C2-0R8	
2383		98.1336	101901.92	D4-5Q5 D2-4Q1 C8-3Q2 C11-4R0	sh
2384	4	98.1423	101892.90	B27-3P5	
2385	43	98.1656	101868.68	B'0-2R5 C5-0Q15	
2386	54	98.1691	101865.09	C2-0Q7 B18-1R8 B19-1R9	
2387	2	98.1779	101855.94	C11-4R1	
2388	31	98.1855	101848.00	B20-2P3 B14-0R9	
2389	73	98.1948	101838.34	B11-0P4 D'0-4Q6 ?	
2390	8	98.2055	101827.27	C3-0P10	
2391	287	98.2122	101820.31	B'2-3P3	
2392		98.2142	101818.27	D2-4Q2	sh
2393	6	98.2204	101811.90	B'5-4P4	
2394	9	98.2273	101804.66	C4-1P6 D'0-4P5 C11-4Q1	
2395	2	98.2367	101794.92	D1-3Q8	
2396	3	98.2390	101792.60	B21-2P5	
2397		98.2450	101786.32	B25-3P2	sh
2398	5	98.2462	101785.15	B16-0R11	

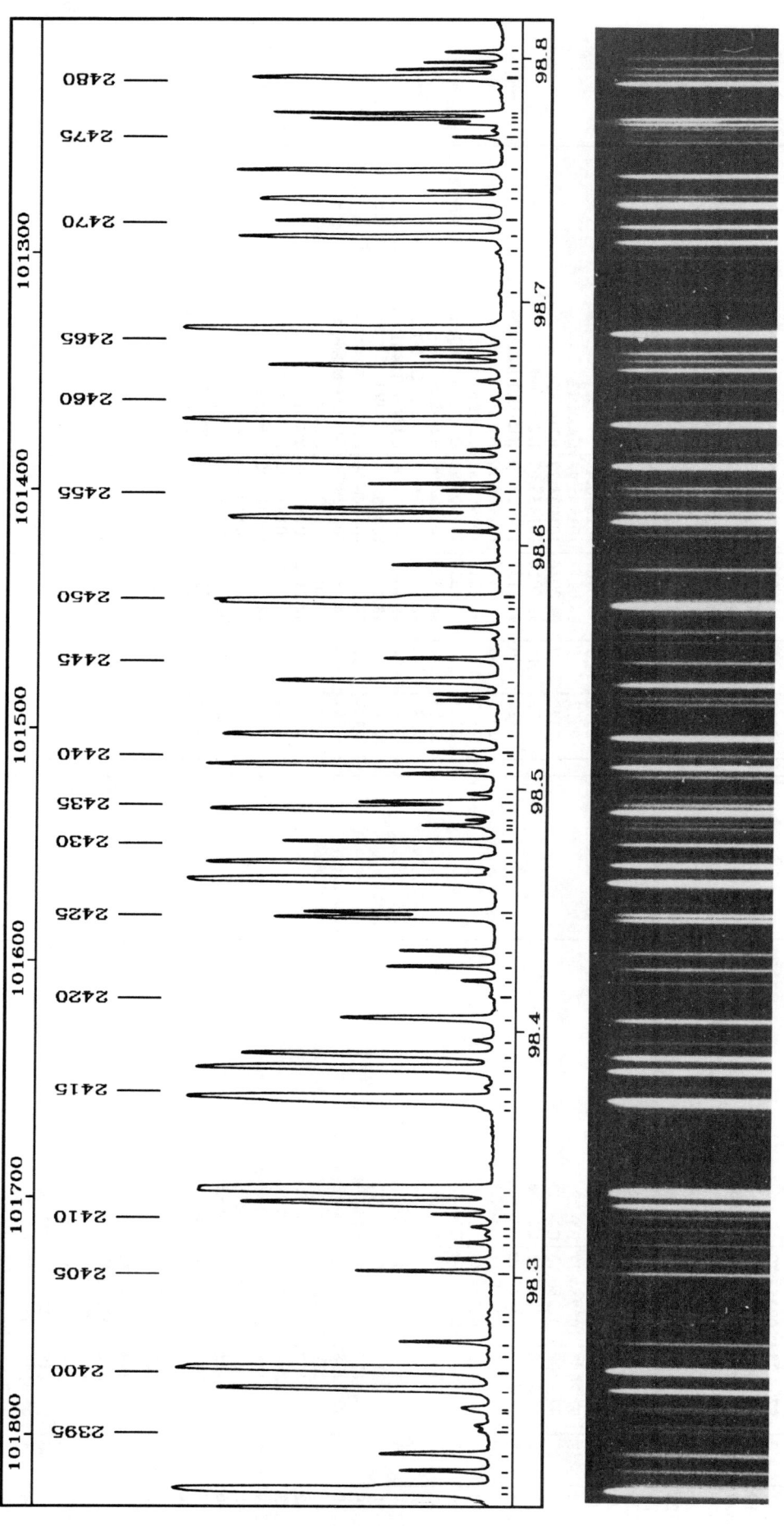

PLATE 35

TABLE 35

Nb	I	λ (nm)	σ (cm⁻¹)	Assignment	Comment
2399	31	98.2533	101777.73	B16-1P5	
2400	150	98.2615	101769.27	B15-1P3	
2401	1	98.2676	101762.98	B26-3P4 C5-1Q10	
2402	6	98.2725	101757.86	B12-0P6	
2403		98.2820	101748.05	B10-0P1	r 101746.58
2404		98.2848	101745.11	B10-0P1	r 101746.58
2405	8	98.3011	101728.27	D1-3P7	
2406	5	98.3062	101722.94	B17-1R7	
2407	4	98.3129	101716.01	D2-4P2	
2408	1	98.3164	101712.41	C6-2P5	
2409	2	98.3195	101709.26	C8-3Q3	
2410		98.3244	101704.16	B13-0R8 B19-2R0	r 101701.83
2411		98.3289	101699.49	B13-0R8 B2-3R5	r 101701.83 b
2412	76	98.3343	101693.87	B"0-4R0 B"0-4R1	b
2413	2	98.3679	101659.18	B19-0P13	
2414	95	98.3716	101655.38	B"0-4R2 B'4-4R1 D3-4Q10	
2415	5	98.3767	101650.01	B20-2R4	
2416	99	98.3844	101642.18	B'0-2P4	b
2417	35	98.3897	101636.65	B11-0R5 B19-2R1	
2418	3	98.3958	101630.40		
2419	18	98.4050	101620.84	C3-0Q11	
2420	2	98.4159	101609.62	C8-3P3	
2421	4	98.4206	101604.73		
2422	8	98.4263	101598.90	B'0-2R6	
2423	6	98.4329	101592.10	B15-1R4 D9-8Q1	
2424	16	98.4467	101577.83	B19-2P1	
2425	14	98.4489	101575.52	B"0-4R3	
2426	143	98.4621	101561.92	B'4-4P1 B'4-4R2	
2427	9	98.4657	101558.18	B21-1R11	
2428	44	98.4695	101554.28	B'2-3P4 C11-4R3	
2429	0	98.4719	101551.80		
2430	11	98.4785	101545.05	D2-4P3	
2431	5	98.4832	101540.16	C3-0R12	
2432		98.4851	101538.23	B10-0P2 B"0-4P1	r 101536.96
2433		98.4874	101535.79	B10-0P2	r 101536.96
2434	58	98.4918	101531.30	C2-0P7	
2435	9	98.4948	101528.24	B14-1R0	
2436	6	98.4981	101524.77	D0-2Q12 B19-1P9 D'4-7Q2 ?	
2437	8	98.5064	101516.28	B25-3P3	
2438	101	98.5103	101512.27	D1-5Q1	
2439	4	98.5137	101508.71	B16-1R6	
2440	8	98.5152	101507.20	C6-1P12	
2441	82	98.5226	101499.55	C3-1R0	
2442	9	98.5363	101485.47	B20-2P4	
2443	7	98.5387	101482.98	B19-2R2 B24-2R9	
2444	22	98.5444	101477.15	B12-0R7 C4-0P13	
2445	7	98.5535	101467.78	B17-1P7	
2446		98.5617	101459.28	C1-0R0 C1-0R1	r 101457.30
2447		98.5656	101455.31	C1-0R0 C1-0R1 C8-3Q4 B"0-4R4	r 101457.30
2448	3	98.5738	101446.88	C11-4Q3 B21-2R6	
2449	198	98.5769	101443.64	B14-0P9	
2450		98.5789	101441.58	B14-1R1 D'1-5Q2 D1-3Q9	sh
2451	9	98.5920	101428.12	C3-1R2 B5-4P5	
2452	5	98.6061	101413.63	C2-0R9 D11-9Q1 D'4-7Q3 ?	
2453	89	98.6116	101407.95	B'2-3R6	
2454	13	98.6153	101404.12	B'4-4R3 B"0-4P2	
2455		98.6236	101395.61	C2-0Q8	r 101394.85
2456		98.6251	101394.10	C1-0R2 B'4-4P2 B15-1P4	r 101394.85
2457	99	98.6347	101384.17	C1-0R2 C3-1Q1	
2458	3	98.6397	101379.08	B14-1P1 B19-2P2 C11-4P3	
2459	96	98.6520	101366.38	B16-0P11	
2460	3	98.6609	101357.28	B11-0P5	
2461	3	98.6683	101349.65	B24-3P1	
2462	31	98.6745	101343.29	B13-0P8	
2463		98.6781	101339.61	D1-5Q3	r 101337.86
2464		98.6815	101336.12	C1-0Q1	r 101337.86
2465	0	98.6862	101331.25	B26-3P5 C1-0Q1 D2-4P4	
2466	123	98.6894	101327.97	B22-2P7	
2467	2	98.7039	101313.14	B'0-2P5	
2468	2	98.7209	101295.72	C3-1R3	
2469	37	98.7271	101289.31	B"0-4R5	
2470	16	98.7335	101282.74	B14-1R2	
2471		98.7424	101273.62	B'0-2R7	r 101271.21
2472		98.7470	101268.80	C1-0R3 C3-1Q2	r 101271.21
2473		98.7545	101261.18	C1-0R3	
2474	16	98.7624	101253.13	C5-2R1	
2475	1	98.7682	101247.14	D3-4Q11	
2476	2	98.7710	101244.31	B19-2R3	
2477	2	98.7740	101241.23	B20-2R5 B21-1P11	
2478	3	98.7758	101239.40	B16-1P6	r 101238.30
2479		98.7779	101237.19	B10-0P3	r 101238.30
2480	30	98.7926	101222.13	B10-0P3	
2481		98.7961	101218.61	B"0-4P3	r 101217.20
2482		98.7988	101215.79	C1-0Q2	r 101217.20
2483	7	98.8030	101211.46	D1-5Q4	

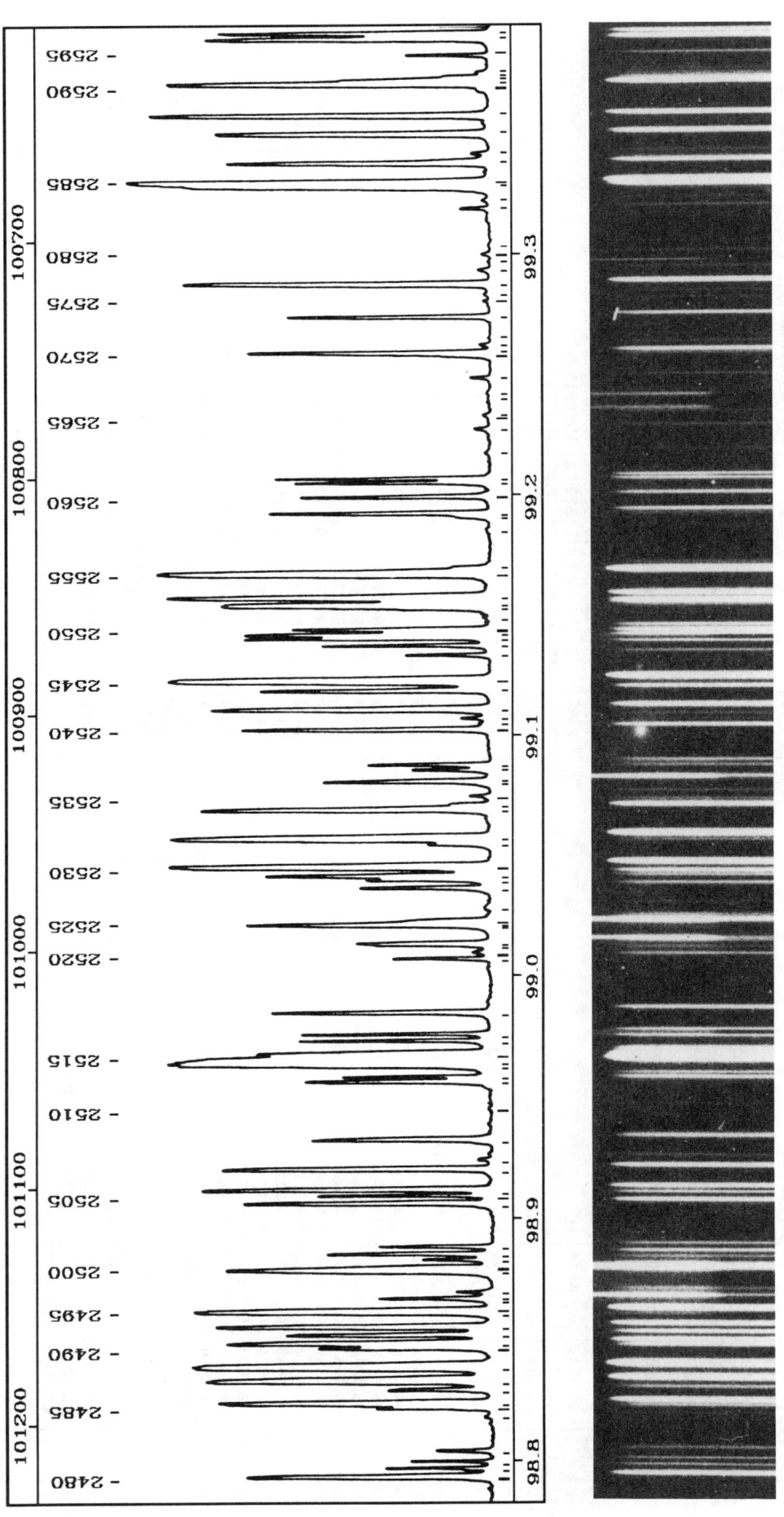

PLATE 36

TABLE 36

Nb	I	λ (nm)	σ (cm⁻¹)	Assignment	Comment
2484	2	98.8170	101197.17	C4-0Q14	
2485	6	98.8204	101193.72	B4-4R4 C4-1Q8	
2486	36	98.8225	101191.53	B14-1P2 C6-1Q13	
2487	6	98.8279	101186.04	C5-2R2	
2488	76	98.8318	101182.01	B15-1R5 B17-0P12 B'4-4P3	
2489	215	98.8374	101176.37	D0-3R1	
2490	12	98.8451	101168.44	C3-0P11	
2491	31	98.8470	101166.44	C5-2Q1 B25-3P4	
2492	14	98.8504	101162.93	C3-1P2	
2493	39	98.8539	101159.43	D0-3R2	
2494	116				OI 98.85778 Ke
2495		98.8602	101152.97	D0-3R0	
2496					OI 98.86549
2497		98.8655	101147.52		OI 98.86549
2498	5	98.8682	101144.73	C8-3Q5 C5-1Q11	
2499					OI 98.87734
2500		98.8773	101135.43		OI 98.87734
2501	4	98.8814	101131.24	B11-0R6	
2502	9	98.8838	101128.79	B12-0P7	
2503	5	98.8868	101125.71	C3-1R4 C1-0R4	
2504	19	98.9047	101107.47	B19-2P3	
2505		98.9078	101104.28	C1-0P2	r 101102.99
2506		98.9103	101101.71	C1-0P2 D0-3R3	r 101102.99
2507	19	98.9186	101093.23	C3-1Q3	
2508	3	98.9227	101089.02	D2-4P5 D11-9Q4	
2509	10	98.9307	101080.82	B18-1R9 B'2-3R7	
2510	0	98.9435	101067.78	D1-3Q10	
2511		98.9550	101056.05	B10-0R4 D0-2Q13	r 101055.27
2512		98.9565	101054.49	B10-0R4	r 101055.27
2513	410	98.9626	101048.25	D0-3Q1 D'1-5Q5 C5-2R3	
2514		98.9661	101045.90	B14-1R3	sh
2515	16	98.9661	101044.71	C5-2Q2 B20-2P5 C4-0R15	
2516		98.9715	101039.19	C1-0Q3	r 101037.75
2517		98.9743	101036.32	C1-0Q3	r 101037.75
2518	17	98.9835	101026.97	C2-0P8 C5-0Q16	
2519	5	99.0059	101004.06	D0-3R4	
2520	2	99.0088	101001.12	B"0-4P4	
2521	3	99.0122	100997.71	B15-0R11	+OI 99.01269 b
2522					OI 99.01269
2523					OI 99.02043
2524	3	99.0204	100989.33	C10-4R0	
2525	5	99.0219	100987.73	C8-3P5	
2526	8	99.0277	100981.85	B13-0R9	
2527	8	99.0356	100973.84	C3-0Q12	
2528	11	99.0390	100970.35	C3-0Q12	
2529	17	99.0407	100968.56	B'0-2P6	
2530	110	99.0446	100964.66	D0-3Q2	
2531	4	99.0539	100955.13	C5-2P2	
2532	96	99.0566	100952.41	C3-1P3 C10-4R1	
2533	31	99.0686	100940.15	B15-1P5	
2534	4	99.0708	100937.97	B16-1R7	
2535	2	99.0743	100934.38	B18-2R0	
2536					OI 99.08010
2537	7	99.0802	100928.37	B'0-2R8 C3-0R13	OI 99.08010
2538	8	99.0848	100923.65	B4-4R5	
2539	14	99.0870	100921.39	B4-4P4 D11-9Q5	
2540	0	99.1015	100906.66	B24-3P3	
2541	4	99.1037	100904.44	C2-0R10	
2542		99.1062	100901.91	B21-2R7 C2-0Q9	
2543	33	99.1099	100898.09	C10-4Q1 C4-1P8	
2544	15	99.1177	100890.20	B14-1P3	
2545	163	99.1219	100885.87	B'2-3P6	
2546	6	99.1323	100875.26	C1-0P3 C1-0R5	r 100869.52
2547		99.1364	100871.11	C1-0P3	r 100869.52
2548	20	99.1395	100867.93	B18-2R1 D0-3R5	
2549	10	99.1410	100866.49	C5-2Q3	
2550	1	99.1431	100864.32	B16-0R12	
2551		99.1481	100859.20	C3-1Q4	sh
2552	84	99.1516	100855.62	B10-0P4	
2553	65	99.1534	100853.88	D0-3P2	
2554	175	99.1565	100850.66	D0-3Q3	
2555	10	99.1665	100840.55	D"1-6Q1 C5-2R4	
2556	2	99.1691	100837.82	C6-1P13	
2557	5	99.1846	100822.14	B12-0R8	
2558	12	99.1900	100816.63	B11-0P6	
2559		99.1915	100815.10	B18-2P1	
2560	13	99.1984	100808.09	C1-0Q4	r 100801.34
2561		99.2042	100802.22	C1-0Q4 D2-4P6	r 100801.34
2562		99.2059	100800.46	B18-1P9	
2563	1	99.2165	100789.66	C5-1P11 D"1-6Q2	
2564	4	99.2264	100779.64	B23-3R1	
2565	1	99.2320	100773.96		
2566		99.2336			HeII 99.2338 Ke P
2567		99.2390			HeII 99.2391 Ke P
2568	0	99.2411	100764.68	B20-2R6	
2569	3	99.2474	100758.28	C10-4Q2	
2570		99.2565	100747.72	B19-2P4	sh
2571	15	99.2578	100744.28	C7-3R0 C4-0P14	
2572	3	99.2612	100741.35	B"0-4P5	
2573	1	99.2641	100732.35	B25-3P5	
2574	12	99.2730	100725.81	B9-0P1	r 100724.34
2575		99.2794		B9-0P1	r 100724.34
2576		99.2823	100722.88	B14-1R4 C7-3R1	
2577	37	99.2866	100712.56	B18-2R2 B23-3P1	
2578	1	99.2925			
2579					CuII 99.29532
2580	2	99.2991	100705.86	B15-1R6 B'2-3R8	
2581	0	99.3032	100701.68	C10-4P2	

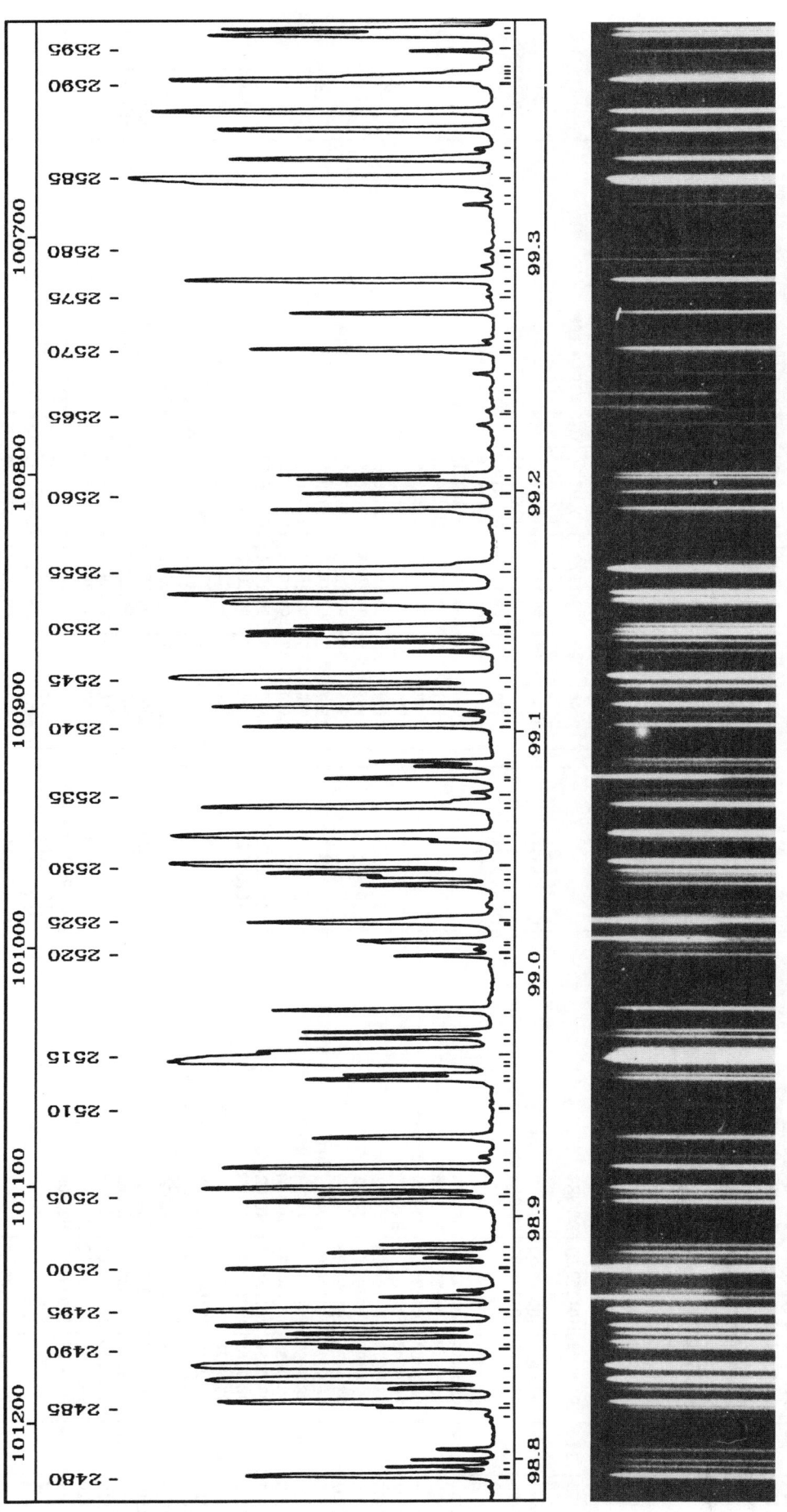

PLATE 36

TABLE 36 *continued*

Nb	I	λ (nm)	σ (cm-1)	Assignment	Comment
2582	0	99.3181	100686.54	C4-1Q9	
2583	0	99.3214	100683.21	B21-1R12	
2584		99.3273	100677.25	D0-3Q4	sh
2585	185	99.3288	100675.75	D0-3P3	
2586	21	99.3372	100667.18	C3-1P4 D1-3Q11 C7-3R2	
2587	4	99.3412	100663.17		
2588	37	99.3492	100655.03	B10-0R5	
2589	57	99.3569	100647.31	C7-3Q1	
2590	0	99.3671	100636.90	B21-2P7 C5-2R5	
2591	50	99.3700	100633.96	D3-5Q1	sh
2592		99.3715	100632.53	B16-1P7	
2593	5	99.3729	100631.09		
2594	5	99.3749	100629.03	B17-0R13	
2595	4	99.3818	100622.08	B1-3R0	
2596	31	99.3886	100615.14	B18-2P2 B13-1R0	
2597	16	99.3909	100612.79	B1-3R1	

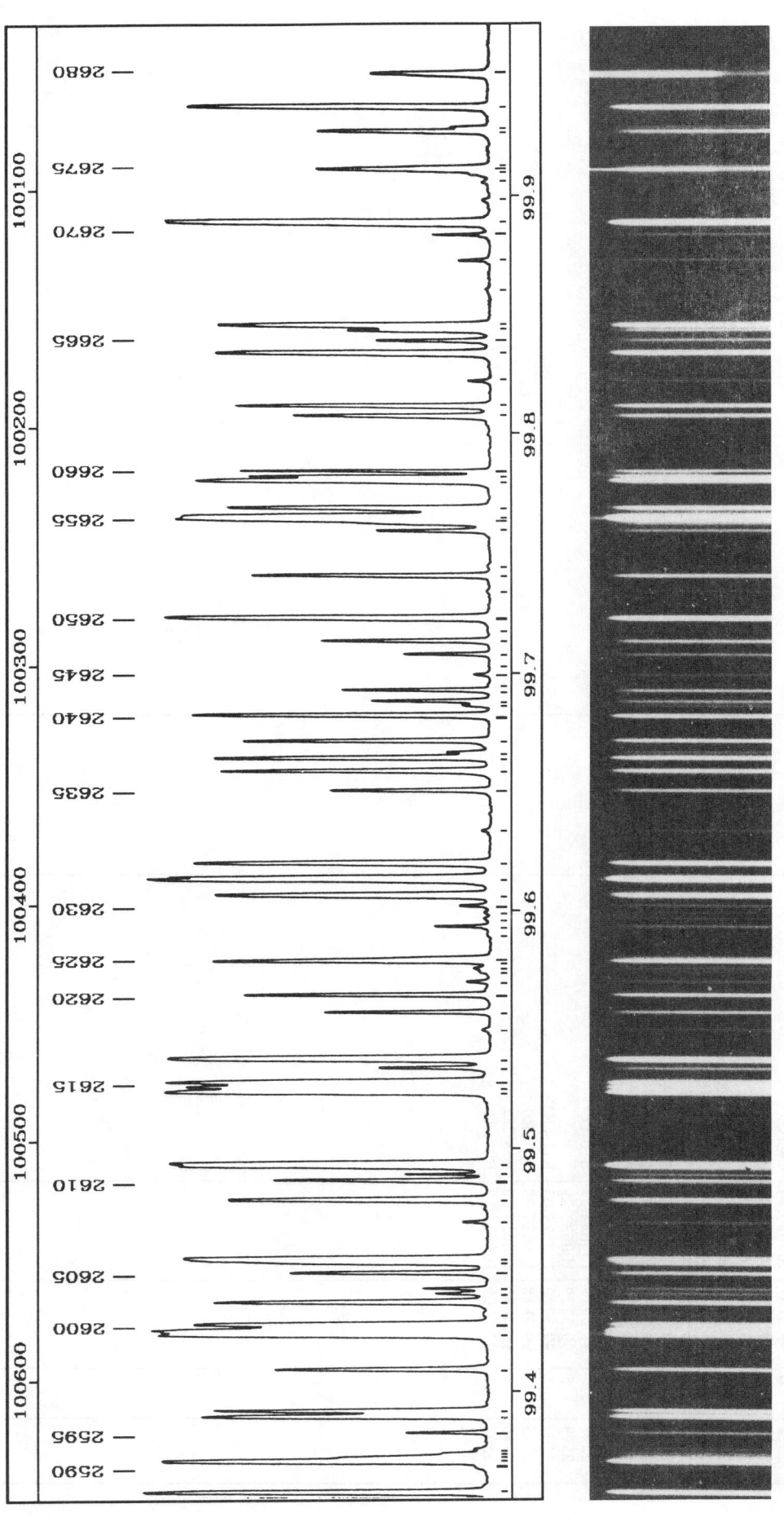

PLATE 37

TABLE 37

Nb	I	λ (nm)	σ (cm⁻¹)	Assignment	Comment
2598	13	99.4083	100595.26	B13-0P9	
2599	220	99.4238	100579.58	C1-0P4 D'2-6Q1	
2600	28	99.4266	100576.72	C1-0R6 B'4-4P5	
2601	4	99.4311	100572.13	C5-0P16 D0-2Q14	
2602	25	99.4359	100567.32	B'0-2P7	
2603	5	99.4394	100563.75	C3-1Q5	
2604	5	99.4415	100561.68	C10-4Q3	
2605	12	99.4480	100555.11	D3-5Q2	
2606		99.4524	100550.58	B11-0R7	sh
2607	86	99.4538	100549.20	B13-1R1 B'1-3R2	
2608	6	99.4687	100534.17	C3-0P12	
2609	29	99.4779	100524.83	C7-3Q2 B'0-2R9	
2610		99.4861	100516.61	B9-0P2 C7-3R3 D'2-6Q2 B14-1P4	r 100515.32
2611	150	99.4885	100514.11	B9-0P2	r 100515.32
2612	104	99.4925	100510.11	C1-0Q5	
2613		99.5233	100478.94	B13-1P1 C10-4P3	
2614	42	99.5251	100477.15	D0-3Q5	
2615	41	99.5267	100475.55	B18-2R3 B'1-3P1	
2616	7	99.5329	100469.25	B2-3P7	
2617	50	99.5366	100465.54	D0-3P4 C2-0P9	
2618	3	99.5491	100452.93	C5-2P4	
2619	10	99.5565	100445.38	C7-3P2	
2620	17	99.5639	100438.00	D3-5Q3	
2621	4	99.5695	100432.39	B'1-3R3	
2622	2	99.5734	100428.43	B12-0P8	
2623	3	99.5750	100426.86		
2624	44	99.5781	100423.68	D2-6Q3	
2625		99.5792	100422.61	C4-0Q15	
2626	0	99.5899	100411.84		sh
2627	4	99.5926	100409.04	B15-1P6	
2628		99.5959	100405.72	B9-0R3	r 100404.38
2629		99.5986	100403.05	B9-0R3	r 100404.38
2630		99.6013	100400.32		
2631	25	99.6056	100395.96	B13-1R2 B'3-4R0	
2632	131	99.6124	100389.07	B10-0P5	
2633	32	99.6190	100382.47	B3-4R1 C2-0R11	
2634	3	99.6331	100368.21	B17-1R9	
2635	11	99.6500	100351.28	C2-0Q10	
2636	20	99.6582	100342.96	C7-3Q3	
2637	18	99.6636	100337.50	B18-2P3 B18-1R10	
2638	4	99.6659	100335.24	C4-1P9	
2639	14	99.6709	100330.22	C3-1P5	
2640	28	99.6812	100319.79	B'1-3P2 C5-2R6	
2641	4	99.6853	100315.71	B19-2P5	
2642	10	99.6872	100313.74	C7-3R4 B'3-4R2	
2643	6	99.6919	100309.08	B14-1R5	
2644	5	99.6941	100306.85	C5-1R13	
2645	5	99.6985	100302.45	C10-4Q4 D'2-6Q4	
2646	0	99.7021	100298.77	B16-1R8	
2647	6	99.7070	100293.91	B'2-3R9	
2648	17	99.7126	100288.26	C3-0Q13	
2649	1	99.7167	100284.09	D3-5Q4	
2650	51	99.7222	100278.55	B13-1P2	
2651	1	99.7334	100267.30		
2652	14	99.7400	100260.70	B'3-4P1	
2653	1	99.7443	100256.36	D10-9Q3	
2654	7	99.7588	100241.82	D0-3Q6	
2655		99.7622	100238.41		sh
2656	285	99.7641	100236.51	C1-0P5	
2657	19	99.7682	100232.31	C1-0R7	
2658	47	99.7797	100220.78	D0-3P5	
2659		99.7814	100219.04	B9-0P3 C3-1Q6	r 100217.86
2660		99.7838	100216.69	B9-0P3 B20-2R7	r 100217.86
2661	12	99.8068	100193.48	C10-4P4 B'3-4R3 B'4-4P6	
2662	19	99.8111	100189.25	B11-0P7	
2663	5	99.8217	100178.61	B13-0R10 B17-0P13	
2664	38	99.8335	100166.83	C1-0Q6	
2665	12	99.8388	100161.51	B14-0R11 C5-0Q17	
2666	9	99.8429	100157.31	B10-0R6	
2667	30	99.8450	100155.24	B13-1R3 B18-2R4	
2668	2	99.8602	100140.04	B17-2R0	
2669	5	99.8723	100127.85	B'0-2P8	
2670	3	99.8830	100117.10	C5-2P5	
2671	86	99.8880	100112.10	B'1-3P3 B'3-4P2	
2672	2	99.8972	100102.87	C7-3Q4	
2673	4	99.9047	100095.36	D3-5Q5	
2674	5	99.9078	100092.32	B15-1R7	
2675					GeII 99.91011
2676	7	99.9102	100089.87	B12-0R9	
2677	20	99.9258	100074.28	B17-2R1	
2678	9	99.9275	100072.54	B9-0R4 D0-2Q15	
2679	41	99.9360	100064.07	B14-1P5	
2680		99.9498	100050.26		OI 99.94974
2681					OI 99.94974

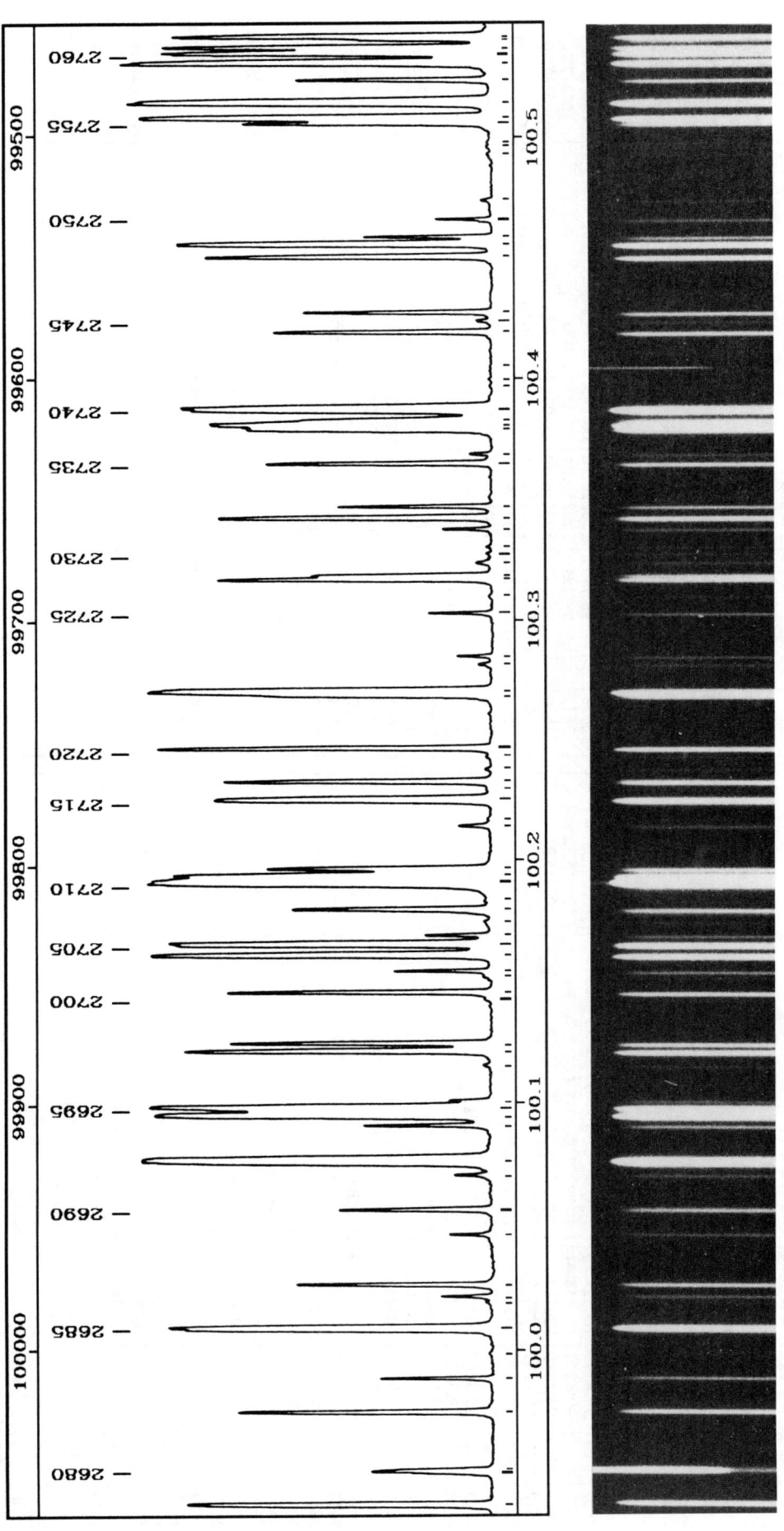

PLATE 38

TABLE 38

Nb	I	λ (nm)	σ (cm⁻¹)	Assignment	Comment
2682	19	99.9738	100026.18	D6-7Q1	
2683	5	99.9875	100012.46	B17-2P1	
2684	0	99.9980	100001.98	B17-1P9	
2685	109	100.0080	99991.99	B13-1P3	
2686	2	100.0188	99981.22	C10-4Q5	
2687	3	100.0211	99978.93	B18-2P4	
2688	10	100.0259	99974.10	D0-3Q7	
2689	2	100.0463	99953.68	C5-2R7 D6-7Q2	
2690	5	100.0564	99943.63	B20-2P7 C3-1P6 D0-3P6	
2691	3	100.0702	99929.86	C4-0P15	
2692	175	100.0762	99923.89	D1-4R1 B17-2R2 C2-0P10	
2693	5	100.0902	99909.84	B'3-4P3	
2694	135	100.0946	99905.54	D1-4R0	
2695	58	100.0974	99902.73	D1-4R2 C9-4R0	
2696	2	100.1003	99899.85	B23-3P4	
2697	3	100.1151	99885.06	B16-0R13	
2698	33	100.1209	99879.29	C1-0P6	
2699	16	100.1240	99876.12	C9-4R1	
2700	0	100.1426	99857.57		
2701	14	100.1453	99854.91	B'1-3P4	
2702	1	100.1518	99848.42	C10-4P5	
2703	5	100.1542	99846.06	D6-7Q3	
2704	51	100.1611	99839.18	D1-4R3	
2705	69	100.1655	99834.74	B9-0P4	
2706	3	100.1691	99831.21	B13-1R4	
2707	1	100.1749	99825.44	C3-1Q7	
2708	8	100.1800	99820.33	B17-2P2 C1-0R8	
2709	4	100.1836	99816.75	B'3-4R5	
2710	495	100.1918	99808.61	D1-4Q1	
2711	25	100.1925	99807.85	D1-3Q13 C9-4Q1 C7-3Q5 B15-1P7	r 99755.30
2712	18	100.1968	99803.55	C3-0P13 C5-1Q13	r 99755.30
2713	5	100.2149	99785.51	C9-4R2	sh
2714	3	100.2179	99782.58	B27-4R0 C2-0R12	
2715	58	100.2258	99774.75	C1-0Q7	
2716	0	100.2303	99770.19	C13-5Q1	
2717	26	100.2332	99767.32	C2-0Q11	
2718	1	100.2385	99762.12	B'4-4P7	
2719		100.2436	99757.00	B8-0R1 B18-2R5	
2720		100.2470	99753.59	B8-0R1 D'3-7Q1	
2721		100.2689	99731.79	D1-4R4	
2722	144	100.2705	99730.19	D1-4Q2	
2723	3	100.2820	99718.81	B14-0P11	
2724	4	100.2853	99715.53	C5-2P6 B27-4R1	
2725	12	100.3031	99697.79	D'3-7Q2	
2726	1	100.3103	99690.65	B17-2R3	
2727	18	100.3176	99683.39	B12-1R0	
2728	9	100.3190	99681.98	C9-4Q2	
2729	2	100.3245	99676.59	D0-3Q8	
2730		100.3289	99672.19	B8-0P1	r 99671.16
2731	2	100.3310	99670.12	B8-0P1	r 99671.16
2732		100.3381	99663.04	B12-0P9	
2733	24	100.3428	99658.37	B9-0R5	
2734	7	100.3474	99653.76	B'0-2P9	
2735	11	100.3651	99636.19	C7-3P5 D0-3P7	
2736	2	100.3690	99632.36	C9-4R3	
2737	95	100.3794	99621.99	B12-1R1 B13-1P4 D3-5Q7	
2738		100.3812	99620.26	D1-4P2	sh
2739		100.3829	99618.60	C9-4P2	sh
2740	226	100.3877	99613.76	D1-4Q3 D'3-7Q3	r 99603.20
2741		100.3968	99604.81	B8-0R2	r 99603.20
2742		100.4000	99601.59	B8-0R2	CuII 100.40554
2743					
2744	12	100.4195	99582.28	B10-0R7	
2745	5	100.4240	99577.79	C3-0Q14	
2746	10	100.4276	99574.22	D1-4R5	
2747	25	100.4508	99551.24	B'1-3P5 C4-1Q11	
2748	59	100.4562	99545.86	B12-1P1 B17-2P3	
2749	4	100.4591	99543.01	B18-2P5 B2-3P9	
2750	3	100.4663	99535.84	B14-1P6	
2751	3	100.4743	99527.94		
2752	0	100.4935	99508.96	B18-1R11	
2753	1	100.4964	99506.03	C3-1P7	
2754	1	100.4983	99504.15	D'3-7Q4	
2755	12	100.5062	99496.33	C9-4Q3	
2756	64	100.5085	99494.03	C4-2R0 B11-0P8	
2757	121	100.5150	99487.69	C4-2R1	
2758	11	100.5240	99478.71	B12-1R2	
2759	87	100.5313	99471.48	D4-6Q1	
2760	39	100.5353	99467.51	C2-1R0 B15-1R8	r 99463.73
2761		100.5376	99465.27	B8-0P2 C2-1R1	r 99463.73 sh
2762		100.5407	99462.19	B8-0P2	
2763	23	100.5420	99460.89	D1-4Q4	

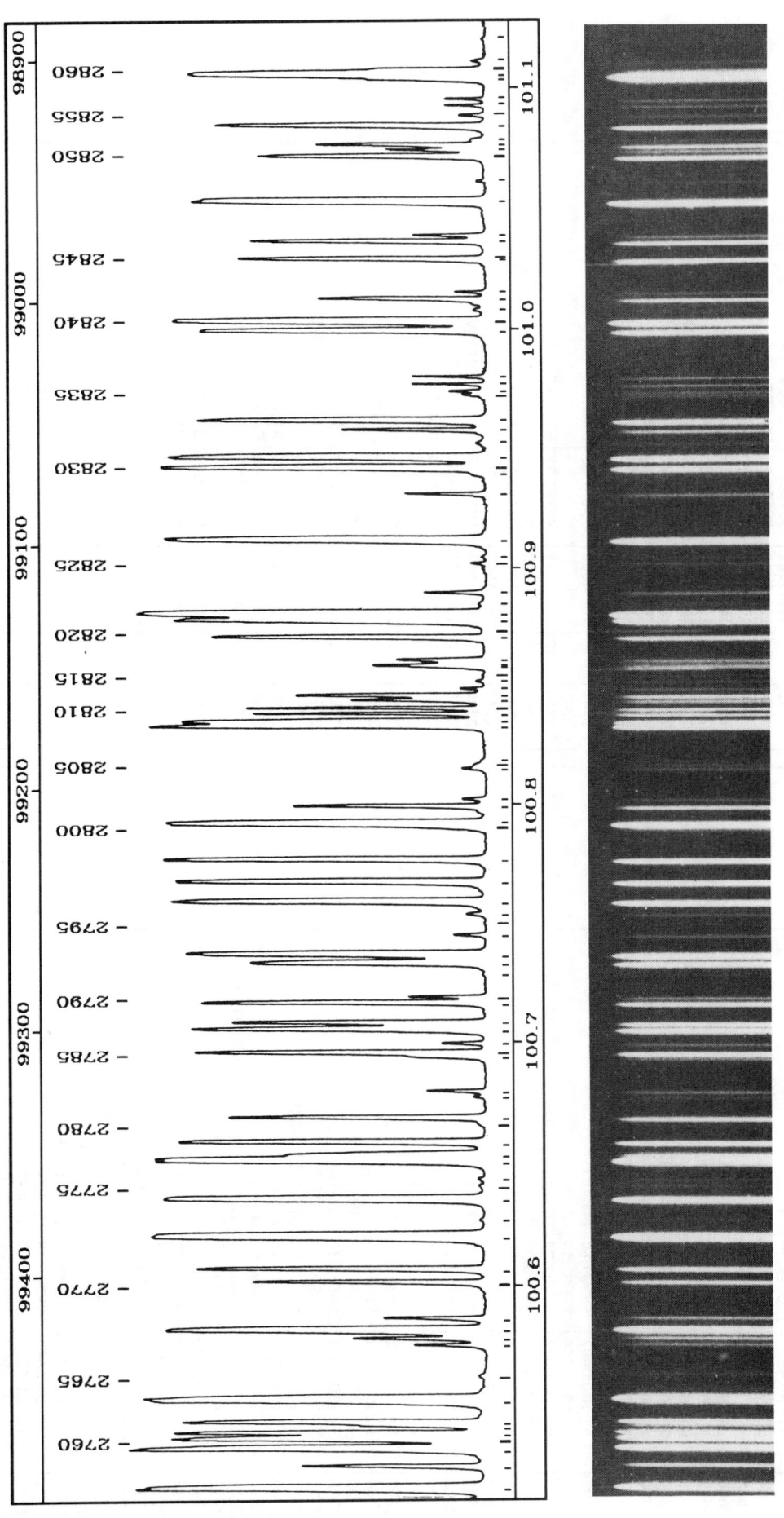

PLATE 39

TABLE 39

Nb	I	λ (nm)	σ (cm⁻¹)	Assignment	Comment
2764	295	100.5516	99451.46	D1-4P3	
2765	270	100.5619	99441.27	D0-5R1 C10-4P6 C4-0R17	
2766	8	100.5741	99429.20	C5-1P13 D0-5R3	
2767	8	100.5766	99426.68	B13-1R5	
2768	84	100.5802	99423.19	C4-2R2 C1-0P7	
2769	7	100.5851	99418.35	C1-0R9	
2770	17	100.5999	99403.72	C9-4P3	
2771	33	100.6054	99398.27	D4-6Q2 C2-1R2	
2772	270	100.6189	99384.95	C4-2Q1 C3-1Q8	
2773	0	100.6259	99377.99	B17-2R4 C12-5R0	
2774	116	100.6342	99369.81	D3-7Q5 B9-0P5	
2775		100.6400	99364.08	B8-0R3	r 99362.82
2776		100.6425	99361.56	B8-0R3 B3-4P5	r 99362.82
2777	198	100.6499	99354.26	C2-1Q1	
2778		100.6519	99352.32	D0-3Q9	sh
2779	33	100.6575	99346.83	B12-1P2	
2780		100.6648	99339.60	B27-4R3	sh
2781	16	100.6675	99336.96	C1-0Q8	
2782	2	100.6765	99328.03	C12-5R1	
2783	5	100.6786	99325.97	B13-0R11	
2784	7	100.6926	99312.16	D0-5Q1	
2785	26	100.6944	99310.39	C2-1R3	
2786	4	100.6984	99306.43	D9-9Q1	
2787	31	100.7043	99300.68	C4-2R3	
2788	23	100.7069	99298.06	C2-0P11	
2789	27	100.7154	99289.64	D4-6Q3	
2790	6	100.7180	99287.16	C12-5Q1	
2791	0	100.7276	99277.68	B5-5P1	
2792	38	100.7323	99273.01	C4-2Q5	
2793	68	100.7358	99269.55	B14-1R7	
2794	4	100.7442	99261.35	B14-0R12	
2795	1	100.7498	99255.83	C9-4Q4 C7-3P6	
2796	4	100.7534	99252.25	D1-4P4	
2797	71	100.7583	99247.42	C2-1Q2	
2798	46	100.7667	99239.15	C6-3R0	
2799	45	100.7759	99230.06	B16-1P9	sh
2800	113	100.7898	99216.39	C6-3R1 B12-1R3	
2801	10	100.7914	99214.81	B10-0P7 C12-5R2	
2802		100.7986	99207.75	B'1-3P6 B17-1P10	
2803	3	100.8018	99204.62	C4-1P11 C2-0R13	
2804	3	100.8146	99191.98	B16-2P1 B17-2P4	
2805	3	100.8161	99190.47	B19-2P7	
2806	3	100.8181	99188.57	C4-2P2	
2807	44	100.8324	99174.47	B13-1P5	r 99168.44
2808	33	100.8340	99172.88	B8-0P3	r 99168.44
2809		100.8373	99169.61	B8-0P3	
2810		100.8397	99167.26	B9-0R6	
2811	5	100.8430	99164.03	B11-0R9	
2812	9	100.8450	99162.13	B11-0R9	

Nb	I	λ (nm)	σ (cm⁻¹)	Assignment	Comment
2813		100.8481	99159.01	C0-0R1 B17-0P14 D0-5Q3	r 99157.46
2814		100.8513	99155.91	C0-0R1	r 99157.46
2815		100.8536	99153.63	C0-0R0	r 99151.62
2816					CuII 100.85688
2817		100.8577	99149.61	C0-0R0 C2-0Q12	r 99151.62
2818	8	100.8601	99147.24	C4-0P16 C12-5Q2 D4-6Q4 D9-9Q3	
2819	20	100.8696	99137.88	C6-3R2 D0-5P2	
2820					CuII 100.87284
2821	35	100.8769	99130.77	C9-4P4 C2-1P2	
2822	178	100.8794	99128.32	C6-3Q1	
2823	2	100.8839	99123.81	C9-4R5 B'5-5P2	
2824	5	100.8886	99119.23	C4-2R4	
2825		100.9010	99107.10	C0-0R2 B26-4P1	r 99105.74
2826		100.9037	99104.37	C0-0R2	r 99105.74
2827	94	100.9106	99097.63	C4-2Q3	
2828	8	100.9298	99078.74	B15-0R13	
2829	5	100.9382	99070.56		
2830	58	100.9406	99068.16	C2-1Q3	
2831	68	100.9450	99063.84	B12-1P3	
2832	3	100.9512	99057.73	C7-3Q7	
2833	7	100.9564	99052.69	D1-4Q6	
2834	61	100.9604	99048.78	D4-8Q1	
2835		100.9712	99038.13	B8-0R4 D3-5Q9	r 99037.45
2836		100.9726	99036.77	B8-0R4 D0-5Q4	r 99037.45
2837		100.9755	99033.91	C0-0Q1 C10-4P7	r 99032.36
2838		100.9787	99030.82	C0-0Q1	r 99032.36
2839	44	100.9978	99012.06	C6-3Q2	
2840	103	101.0018	99008.10	D1-4P5	
2841	2	101.0068	99003.21	D0-3Q10	
2842		101.0116	98998.51	C0-0R3 B16-2P2 D4-8Q2	r 98997.23
2843		101.0142	98995.96	C0-0R3	r 98997.23
2844					CuII 101.02690
2845	21	101.0281	98982.38	C6-3R3	
2846	18	101.0355	98975.13	D7-8Q1	
2847	3	101.0381	98972.54	D4-6Q5	
2848	110	101.0522	98958.80	C4-2P3	
2849	0	101.0609	98950.25	C9-4Q5	
2850	14	101.0713	98940.08	C12-5Q3 C1-0P8	
2851	6	101.0740	98937.45	C0-0Q2	
2852	6	101.0760	98935.46	B10-0R8 B14-1P7	
2853	8	101.0782	98933.30	C1-0R10	
2854	4	101.0839	98927.68	C6-3P2	
2855	24	101.0886	98923.15	D4-8Q3	
2856	6	101.0931	98918.73	C0-0Q2	r 98918.03
2857		101.0945	98917.32	C0-0Q2	r 98918.03
2858		101.1032	98908.87	D7-8Q2	sh
2859		101.1049	98907.16	C2-1P3	sh
2860	125	101.1067	98905.39	C2-1R5 B21-3P3 B'5-5P3	

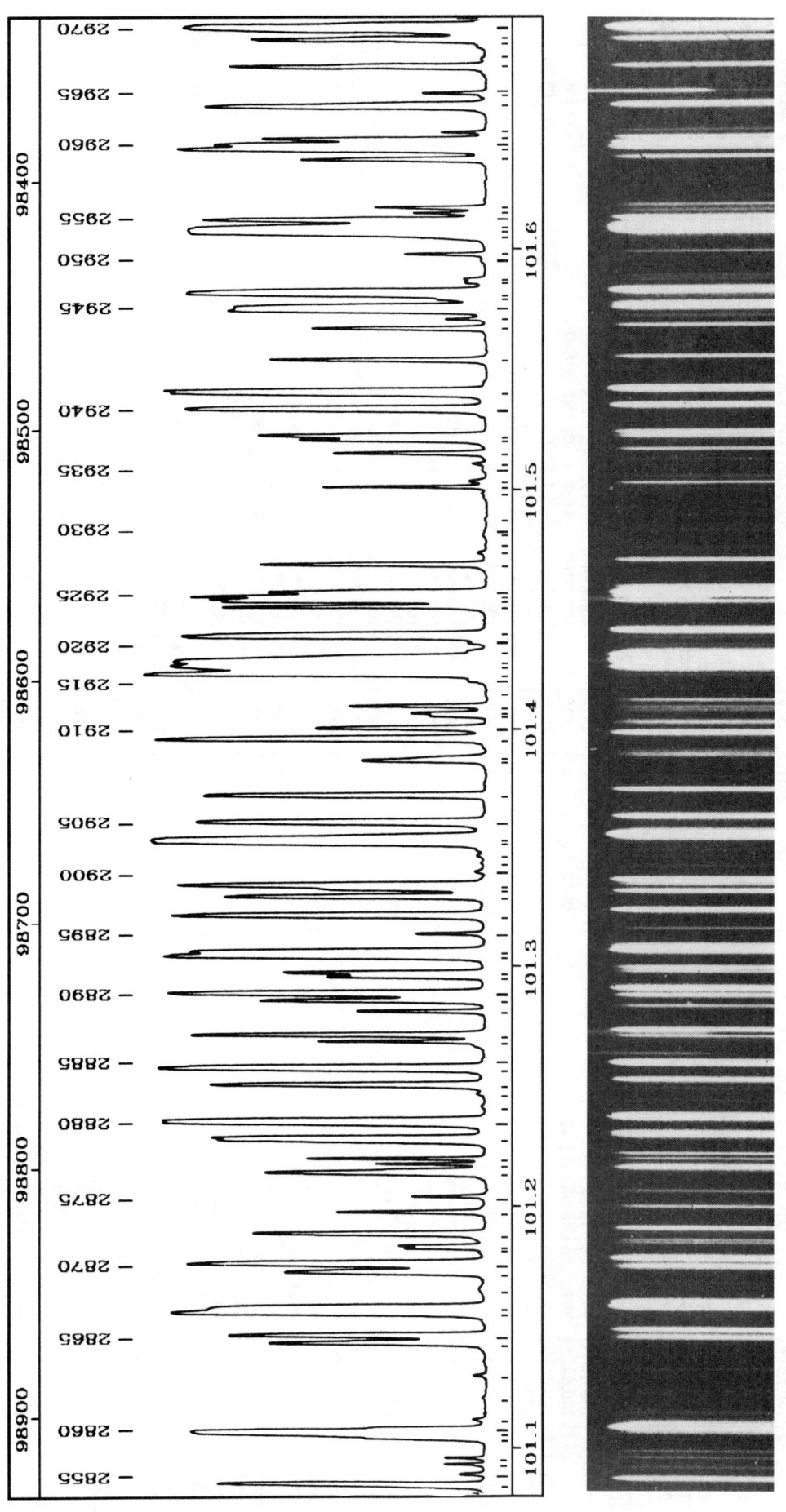

PLATE 40

TABLE 40

Nb	I	λ (nm)	σ (cm⁻¹)	Assignment	Comment
2861	3	101.1110	98901.24	B12-1R4	
2862	2	101.1196	98892.77	B26-2P13	
2863	2	101.1288	98883.79	C12-5P3	
2864	9	101.1418	98871.11	B16-0R14 C4-2Q4 C4-2R5	
2865	22	101.1449	98868.02	B2-4R0	
2866	98	101.1547	98858.46	B2-4R1	
2867	35	101.1563	98856.93	C1-0Q9	
2868	0	101.1642	98849.20	C6-3R4	
2869	18	101.1712	98842.39	C2-1Q4 C3-0Q15	
2870	42	101.1744	98839.20	C6-3Q3 B12-0P10	r 98832.41
2871		101.1808	98832.94	C0-0R4	r 98832.41
2872		101.1819	98831.89	C0-0R4	
2873	16	101.1870	98826.92	B9-0P6 D4-8Q4	
2874	4	101.1959	98818.22	C7-3P7	
2875	3	101.2023	98811.94	D7-8Q3	
2876	9	101.2122	98802.31	D1-4Q7 C9-4P5 C4-0Q17	b
2877		101.2157	98798.86	C0-0P2	r 98797.63
2878		101.2182	98796.41	C0-0P2	r 98797.63
2879		101.2262	98788.62	B8-0P4	
2880	63	101.2334	98781.59	D2-5R1	
2881	150	101.2399	98775.31	B17-1R11	
2882	0	101.2450	98770.28	B15-1R9	
2883	15	101.2488	98766.57	D2-5R2 D4-6Q6	
2884	84	101.2559	98759.71	B17-2P5 D2-5R0	
2885		101.2666	98749.29	C0-0Q3	CuII 101.25971
2886		101.2696	98746.27	C0-0Q3 B0-3R0 B0-3R1	r 98747.78
2887		101.2794	98736.80	B11-0P9	r 98747.78
2888	7	101.2839	98732.37	D1-4P6	
2889	11	101.2869	98729.48	B2-4P1	
2890	51	101.2940	98722.57	B16-2P3	
2891	4	101.2955	98721.06	B11-1R0	
2892	11	101.3023	98714.44	C6-3P3	
2893	55	101.3043	98712.49	D2-5R3	
2894	23	101.3119	98705.08	B12-1P4	
2895	3	101.3198	98697.38	B0-3R2	
2896	47	101.3278	98689.65	C4-2P4	
2897	15	101.3308	98686.68	B20-3R1	sh
2898		101.3324	98685.09	B26-4P3	
2899	27	101.3383	98679.37	B7-0R1	r 98674.17
2900	0	101.3420	98675.77	B7-0R1	r 98674.17
2901		101.3453	98672.57	C8-4R0 D2-5Q1	
2902		101.3518	98666.23	C12-5Q4	sh
2903	390	101.3534	98658.60	C11-5R0 B11-1R1	
2904	36	101.3596	98647.82	C8-4R1 B13-1P6 B8-0R5	
2905	22	101.3707	98633.58	C2-0P12	sh
2906		101.3854	98632.58	B14-1R8 B20-3P1 D0-3Q11	
2907	6	101.3864	98624.67	C2-1P4 C2-1R6	
2908		101.3945			sh
2909	32				
2910	4	101.3989	98620.37	C11-5R1 B25-4R0	
2911	3	101.4040	98615.47	B'0-2P11	
2912	4	101.4052	98614.29	D2-5R4	
2913	4	101.4080	98611.54	C6-3Q4 C7-3Q8	
2914	2	101.4129	98606.77	B15-0P13	
2915	10	101.4186	98601.27	C6-3R5	
2916	66	101.4218	98598.11	B'0-3R3	
2917	180	101.4250	98595.01	B0-3P1 C0-0R5 C9-4Q6 B9-0R7	
2918	115	101.4270	98593.11	D2-5Q2 C4-2Q5 C12-5P4	
2919		101.4311	98589.12	B7-0P1 D''0-6Q2	r 98587.65
2920		101.4341	98586.18	B7-0P1	r 98587.65
2921	51	101.4374	98583.01	B11-1P1 B'2-4P2	
2922		101.4493	98571.41	C0-0P3	r 98570.31
2923		101.4515	98569.22	C0-0P3 C11-5Q1	r 98570.31
2924	100	101.4532	98567.61	C8-4Q1 C8-4R2	
2925	12	101.4553	98565.60	C2-1Q5	
2926	17	101.4667	98554.52	B25-4R1	
2927	0	101.4716	98549.72	B17-0R15	
2928	0	101.4747	98546.75	D12-11Q1	
2929	0	101.4794	98542.20	C3-1P9	
2930	0	101.4810	98540.63	B20-3R2	
2931	0	101.4856	98536.20	D4-6Q7	
2932		101.4963	98525.76	C0-0Q4 B7-0R2	r 98524.37
2933		101.4992	98522.98	C0-0Q4 B7-0R2	r 98524.37
2934	1	101.5018	98520.44	B10-0P8	
2935	0	101.5065	98515.85	D''0-6Q3 C11-5R2	
2936	2	101.5090	98513.48	B25-4P1	
2937	7	101.5133	98509.31	B11-1R2	
2938	19	101.5188	98503.92	D1-6Q1 C10-4P8	
2939	22	101.5205	98502.27	B12-1R5 C2-0Q13	
2940	67	101.5317	98491.40	D2-5P2	
2941	122	101.5387	98484.63	D2-5Q3	
2942	25	101.5522	98471.55		
2943	11	101.5653	98458.81	B''0-5R1	
2944	2	101.5693	98455.00	B18-2P7 B31-5R1	
2945	39	101.5736	98450.79	C8-4Q2 B'0-3R4 B''0-5R0	
2946	0	101.5775	98447.03	C9-4R7 D'1-6Q2	
2947	91	101.5798	98444.73	B'0-3P2 C6-3P4	
2948	2	101.5845	98440.24	B'4-5R0	
2949	2	101.5858	98438.94	C11-5Q2	
2950	1	101.5935	98431.47	B''0-5R2	
2951	2	101.5962	98428.91	C8-4R3	
2952	80	101.6028	98422.48	B13-0R12 B16-1P10 B31-5P1 B'5-5R5	sh
2953	80	101.6052	98420.16	D5-7Q1	
2954	36	101.6065	98418.92	B'4-5R1 C1-0P9 D''0-6Q4 ?	b
2955	15	101.6135	98415.04	C9-4P6 D1-4P7 D12-11Q3	
2956	2	101.6158	98412.15	D10-10Q1	
2957	4	101.6185	98409.92	B25-4R2 B13-1R7	
2958	13	101.6360	98390.30	C11-5P2 C1-0R11	

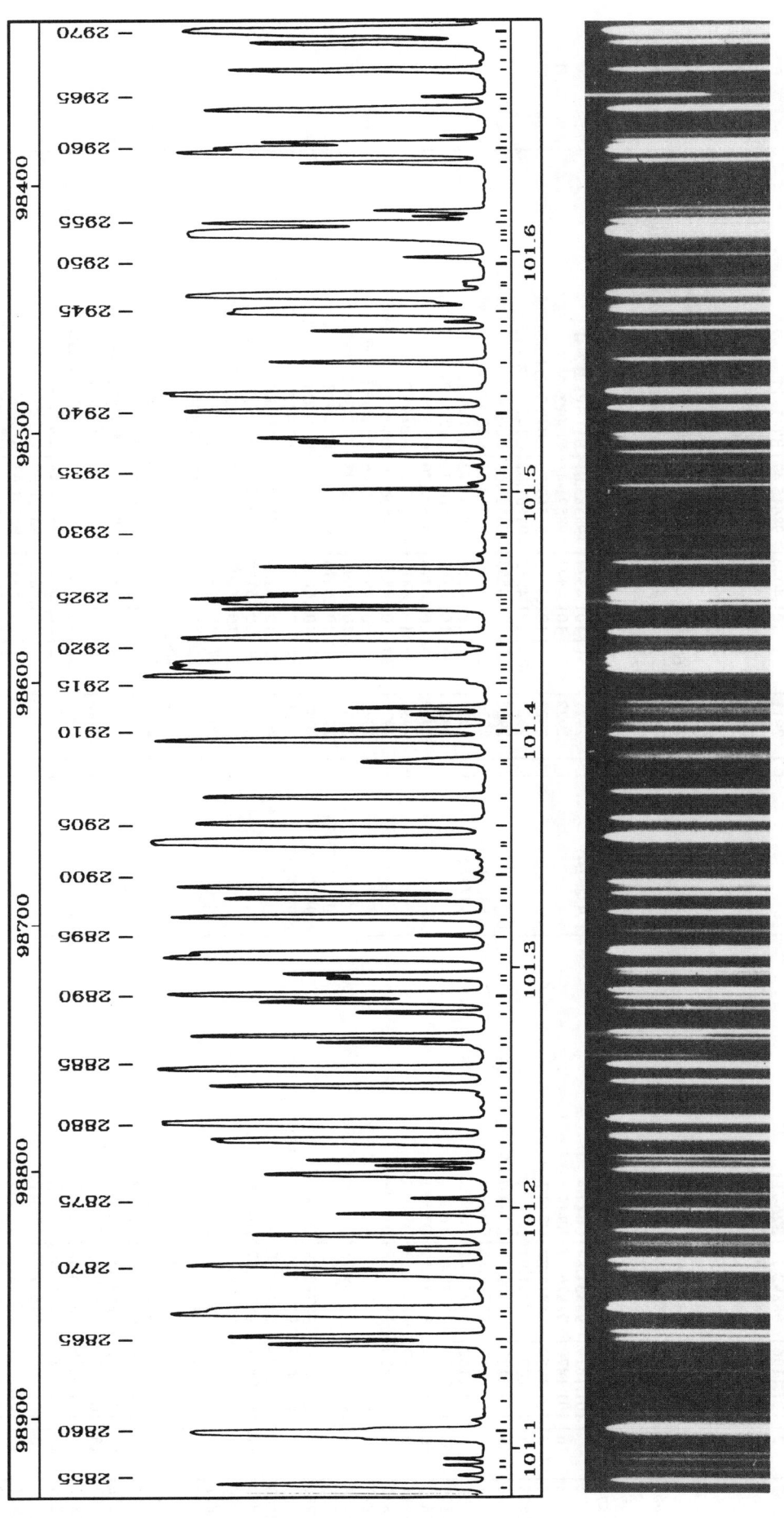

PLATE 40

TABLE 40 *continued*

Nb	I	λ (nm)	σ (cm⁻¹)	Assignment	Comment
2959	58	101.6403	98386.22	B'2-4P3	
2960	27	101.6421	98384.40	B11-1P2 B11-0R10	r 98380.84
2961	1	101.6444	98382.22	B7-0P2 C8-4P2	r 98380.84
2962		101.6472	98379.46	B7-0P2	
2963	22	101.6580	98369.04	C4-2P5 B16-2P4	
2964	2	101.6635	98363.77	D'1-6Q3	
2965					GeII 101.66377
2966	16	101.6745	98353.05	D5-7Q2 C0-0R6	
2967	0	101.6779	98349.84	B17-1P11	
2968	1	101.6827	98345.12	C11-5R3	
2969	17	101.6858	98342.11	C5-1Q15 D2-5Q4	
2970	178	101.6907	98337.46	B'4-5R2 D2-5P3 C1-0Q10	

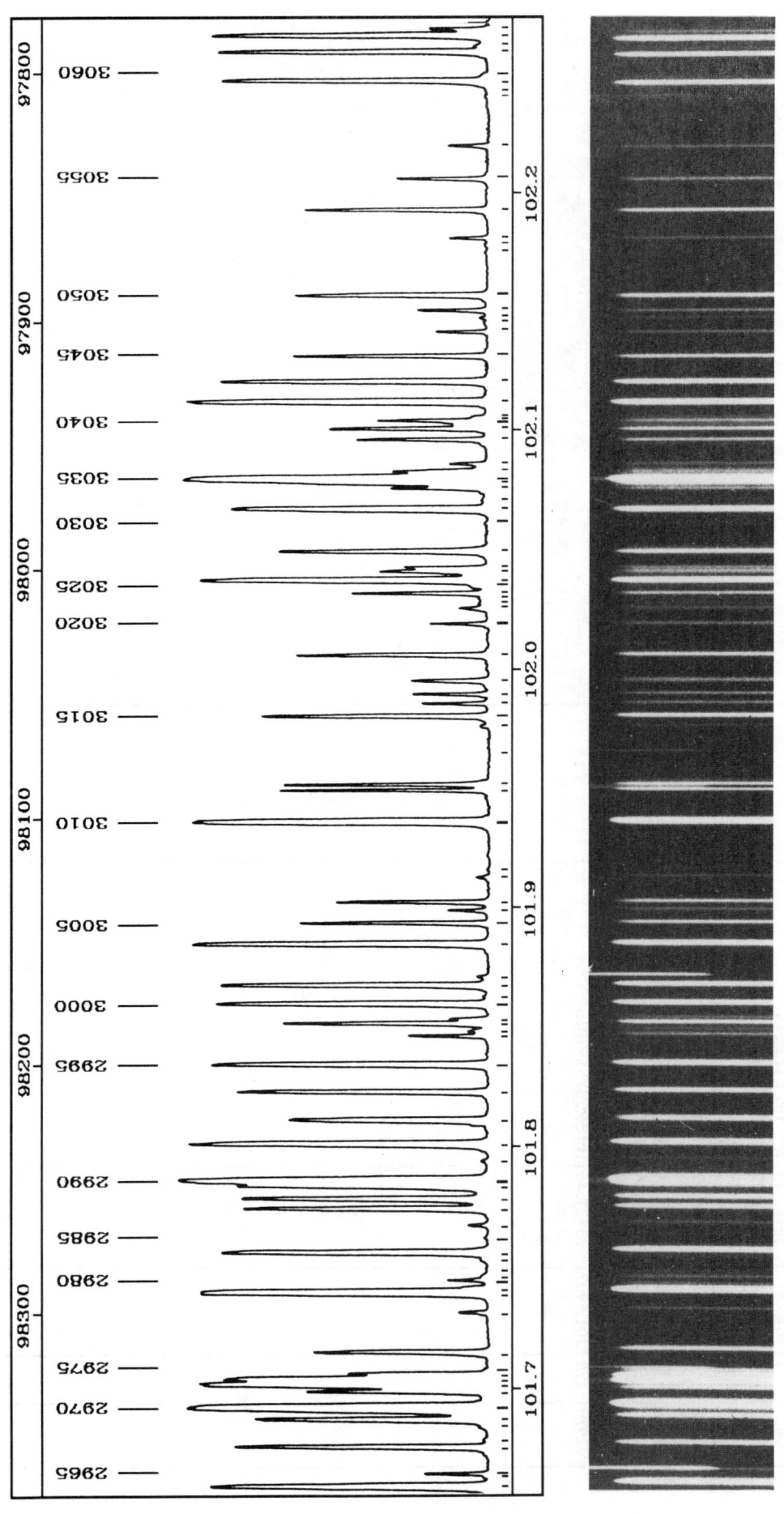

PLATE 41

TABLE 41

Nb	I	λ (nm)	σ (cm⁻¹)	Assignment	Comment
2971	9	101.6974	98330.87	C6-3Q5	
2972	116	101.7003	98328.09	B8-0P5 C7-3P8 B21-3R5	
2973	60	101.7024	98326.10	B4-5P1	
2974	10	101.7046	98324.01	C3-0P15	
2975					GeII 101.70600
2976	11	101.7138	98315.07	B20-3R3 B'2-4R5	
2977	3	101.7305	98298.94	B14-0R13	
2978	121	101.7385	98291.24	C0-0P4	
2979		101.7407	98289.08	B7-0R3	r 98287.52
2980		101.7439	98285.95	B7-0R3 C4-0P17	r 98287.52
2981	1	101.7461	98283.84	B'5-5P5	
2982	1	101.7491	98280.93	C2-1R7	
2983	0	101.7525	98277.70	D4-6Q8 C4-1Q13	
2984	26	101.7553	98275.03	C8-4Q3 B11-1R3	
2985	0	101.7620	98268.52	B'0-5R4 B14-1P8	
2986	3	101.7666	98264.10	C4-2Q6	
2987	20	101.7733	98257.62	B0-3R5	
2988	21	101.7773	98253.75	D5-7Q3 D1-6Q4	
2989		101.7838	98247.44	C0-0Q5	sh
2990	275	101.7844	98246.87	B'0-3P3 C11-5Q3	
2991	2	101.7931	98238.49	C2-1Q6	
2992	56	101.8000	98231.84	C8-4R4 B12-1P5	
2993	15	101.8099	98222.29	B10-0R9 D1-4Q9	
2994	20	101.8217	98210.93	B9-0P7	
2995	35	101.8332	98199.80	B4-5R3	
2996	2	101.8451	98188.30	B25-4R3	
2997	3	101.8473	98186.22	C9-4Q7 B20-3P3	
2998	15	101.8503	98183.33	B'0-5P2	
2999	3	101.8523	98181.43	B21-3P5	
3000	39	101.8587	98175.27	C4-8P3 C11-5P3	
3001	0	101.8621	98171.92	B4-5P2	
3002	22	101.8666	98167.57	D2-5Q5	
3003					CuII 101.87073
3004	32	101.8840	98150.84	D2-5P4	
3005	9	101.8929	98142.25	B'2-4P4	
3006	4	101.8985	98136.88	C10-4Q9 B"0-5R5	
3007	7	101.9019	98133.61	B8-0R6	
3008	2	101.9125	98123.44	D5-7Q4	
3009	0	101.9150	98121.01	C7-3Q9	
3010	76	101.9357	98101.02	B11-1P3 C6-3P5	
3011		101.9491	98088.20	B7-0P3	r 98087.07
3012		101.9514	98085.93	B7-0P3 B31-5R3 C3-0Q16	r 98087.07
3013					CuII 101.96545
3014	2	101.9761	98062.16	D'1-7Q1 B'2-4R6	
3015	14	101.9803	98058.19	D8-9Q1	
3016	2	101.9857	98052.97	B13-1P7	
3017	3	101.9896	98049.19		
3018	4	101.9953	98043.78	C8-4Q4	
3019	9	102.0061	98033.38	C0-0R7	

Nb	I	λ (nm)	σ (cm⁻¹)	Assignment	Comment
3020	3	102.0191	98020.83	B'0-3R6	
3021	1	102.0257	98014.54	B"0-5P3 D'1-7Q2	
3022	0	102.0277	98012.56	B30-5R1	
3023	0	102.0298	98010.58	B20-3R4 B31-5P3	
3024	4	102.0320	98008.51	B4-5R4 B19-3R0	sh
3025	43	102.0349	98005.64	B24-4R0	
3026		102.0375	98003.20	B0-3P4	
3027	6	102.0412	97999.65	C6-3Q6 D4-6Q9	
3028	5	102.0426	97998.34	C4-2P6 D8-9Q2	
3029	17	102.0495	97991.62	B16-1R11 C11-5Q4 C2-0P13	
3030	0	102.0611	97980.54		
3031	20	102.0671	97974.73	B4-5P3 B30-5P1	
3032	1	102.0715	97970.50	C8-4K5	
3033		102.0760	97966.26	B7-0R4	r 97965.27
3034		102.0780	97964.28	B7-0R4 C9-4P7 D5-7Q5	r 97965.27 sh
3035	191	102.0798	97962.55	C0-0P5 D2-5Q6	
3036	5	102.0829	97959.65	B11-1R4 B12-0P11	
3037	2	102.0862	97956.39	B9-0R8	
3038	7	102.0963	97946.70	B19-3R1 C2-1R8	
3039	8	102.1009	97942.33	B24-4R1 B14-1R9	
3040	2	102.1027	97940.60	C10-4P9	
3041	5	102.1044	97939.01	B11-0P10	
3042	1	102.1058	97937.65	B16-2P5	
3043	44	102.1123	97931.41	D2-5P5	
3044	22	102.1209	97923.16	C0-0Q6	
3045	9	102.1316	97912.91	C8-4P4 D8-9Q3	
3046	2	102.1419	97903.05		
3047	0	102.1452	97899.87	C11-5P4	
3048	1	102.1479	97897.24	D1-4Q10	
3049	3	102.1509	97894.41	B24-4P1	
3050	10	102.1572	97888.34	C4-2Q7 B19-3P1	
3051	0	102.1753	97871.01	C1-0R12	
3052	0	102.1783	97868.13	C7-3P9 D13-12Q1	
3053	2	102.1815	97865.10	C2-1Q7	
3054	6	102.1932	97853.83	B'2-4P5	
3055	4	102.2063	97841.31	D0-3Q13 C1-0P10	
3056	5	102.2198	97828.39	C2-0Q14	
3057	1	102.2401	97809.01	B30-5P2	Cl I 102.24143
3058	0				
3059	28	102.2467	97802.70	D2-7Q1	
3060	1	102.2487	97800.75	B24-4R2	
3061	15	102.2586	97791.33	B8-0P6	
3062	1	102.2610	97788.98	C6-3P6	
3063	21	102.2653	97784.85	C1-0Q11	
3064	5	102.2681	97782.22	B16-0R15	

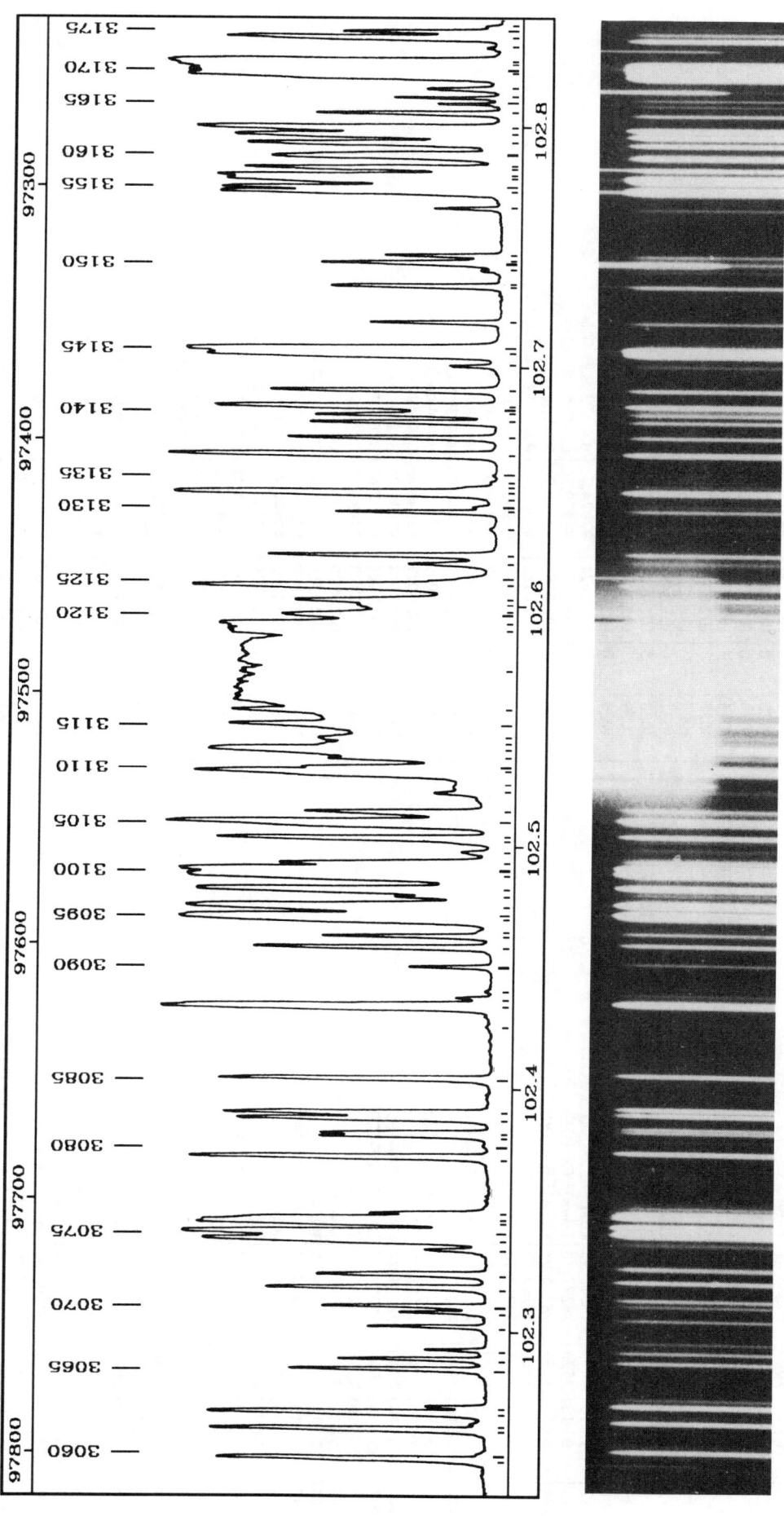

PLATE 42

TABLE 42

Nb	I	λ (nm)	σ (cm^{-1})	Assignment	Comment
3065	6	102.2832	97767.75	B10-0P9 B'2-4R7	
3066	4	102.2874	97763.74	B4-5R5 D13-12Q3	
3067	2	102.2913	97759.99	C8-4Q5	
3068	6	102.3008	97750.98	D2-7Q2	
3069	3	102.3067	97745.35	C10-5R0 B13-1R8	
3070	7	102.3092	97742.97	B'0-3R7	
3071	9	102.3167	97735.75	B11-1P4	
3072	8	102.3222	97730.47	B10-1R0 D2-5Q7	
3073	2	102.3323	97720.90	B4-5P4 B24-4P2	
3074	42	102.3365	97716.87	B'0-3P5 C10-5R1	
3075	59	102.3396	97713.92	D0-4R1	
3076	69	102.3437	97710.00	B7-0P4	
3077		102.3447	97709.03	D0-4R2	sh
3078	3	102.3473	97706.57	B19-3P2	
3079	19	102.3699	97684.98	D0-4R0	
3080	1	102.3748	97680.28	D11-11Q1	
3081	6	102.3786	97676.64	C11-5Q5 D'2-7Q3	
3082	6	102.3798	97675.55	D2-5P6	
3083	13	102.3856	97669.98	B0-1R1	
3084	11	102.3879	97667.77	D0-4R3 C0-0R8	
3085	11	102.4021	97654.27	B30-5R3 C10-5Q1	
3086	0	102.4239	97633.49	D11-11Q2 B20-3R5	
3087	75	102.4317	97626.04	D3-6Q1 C10-5R2	
3088	1	102.4357	97622.23	C6-3Q7	
3089	0	102.4377	97620.29	B6-0R0	
3090	1	102.4486	97609.90		
3091	6	102.4566	97602.31	B14-2R0	
3092	3	102.4614	97597.76	C8-4P5	
3093	0	102.4633	97595.88	B29-5R0	
3094	100	102.4679	97591.58	B10-1P1 D0-4R4	sh
3095	64	102.4698	97589.73	C3-2R1	
3096	1	102.4737	97586.06	D0-4Q1 C0-0P6 B19-3R3	
3097	37	102.4780	97581.90	B24-4R3	
3098	109	102.4808	97579.24	B15-0P14 C5-3R0 D'2-7Q4	
3099	101	102.4867	97573.66	B8-0R7 C3-2R0	
3100	5	102.4888	97571.66	C5-3R1	
3101		102.4917	97568.89	B30-5P3 C4-2P7	
3102		102.4968	97563.99	B6-0R1 B7-0R5 D11-11Q3 C11-5P5	r 97562.25
3103		102.5005	97560.51	B6-0R1 B7-0R5	r 97562.25
3104	15	102.5026	97558.52	D3-6Q2	
3105	23	102.5092	97552.18	C0-0Q7 D1-4Q11	
3106	3	102.5136	97548.05	B11-0R11 B12-0R12	
3107	1	102.5215	97540.50	B14-1P9	
3108					HeII 102.5246 Ke P
3109					HeII 102.5302 Ke P
3110	38	102.5305	97531.92	C3-2R2 C2-1P7	
3111	2	102.5359	97526.83	B9-0P8 C3-0P16	
3112	64	102.5394	97523.51	B14-2R1 B10-1R2 B25-4R5	
3113	3	102.5417	97521.28	C2-1R9	
3114	3	102.5443	97518.79	D0-4Q2	
3115	17	102.5499	97513.51	C5-3R2	
3116	17	102.5558	97507.87		HI 102.57223 Ly β
3117	135	102.5886	97476.68	C5-3Q1 C10-5P2 D11-11Q4 B13-0R13 B12-1R7	
3118					
3119	460	102.5910	97474.46	C3-2Q1 D2-5Q8	
3120	11	102.5956	97470.07	C4-2Q8 B24-4P3	
3121					MgII 102.59681
3122	3	102.5999	97466.04	B'5-5R6	
3123	5	102.6019	97464.11	B14-2P1	
3124	31	102.6078	97458.44	B17-0R16 D3-6Q3 D2-7Q5	
3125					MgII 102.61133
3126	2	102.6172	97449.56	B10-0R10 B17-1P12 C2-1Q8	
3127	8	102.6207	97446.22	B19-3P3	
3128	1	102.6318	97435.68	B16-2P6	
3129	5	102.6390	97428.87	D6-8Q2	
3130	0	102.6411	97426.90	B'0-3R8	
3131		102.6441	97423.99	C8-4Q6	
3132	43	102.6468	97421.42	C3-2R3	
3133	1	102.6492	97419.19	B4-5P5	
3134		102.6513	97417.17	B6-0R2	r 97415.90
3135		102.6540	97414.64	B6-0R2	r 97415.90
3136	26	102.6628	97406.31	D0-4Q3	
3137	6	102.6698	97399.59	D0-4P2	
3138	5	102.6764	97393.37	B10-1P2 B13-1P8	
3139	9	102.6795	97390.44	B'0-3P6	
3140	0	102.6809	97389.11	B29-5R2	
3141	18	102.6829	97387.24	C5-3R3	
3142	9	102.6897	97380.80	B14-2R2	
3143		102.6998	97371.15	C6-3P7	
3144	48	102.7048	97366.40	C5-3Q2	
3145	131	102.7065	97364.87	C3-2Q2 B23-4R0	
3146	3	102.7180	97353.91	C10-5Q3	
3147	6	102.7333	97339.47	D6-8Q3	
3148	1	102.7397	97333.36	D5-7Q8 B5-5P7	CI I 102.73386
3149					
3150					OI 102.74307
3151		102.7431	97330.15		OI 102.74307
3152	3	102.7461	97327.28	B29-5P2 D3-6Q4	
3153	4	102.7658	97308.64	C3-0Q17	
3154		102.7727	97302.07	C1-1R1	r 97301.22
3155		102.7745	97300.36	C1-1R1	r 97301.22
3156		102.7780	97297.09	C1-1R0 B10-1R3	r 97296.46
3157		102.7793	97295.84	C1-1R0	r 97296.46
3158	8	102.7826	97292.70	B11-1P5	
3159					CuII 102.78311
3160	13	102.7878	97287.81	D9-10Q1 C1-0P11	
3161	11	102.7929	97283.02	B14-2P2	

97300
97400
97500
97600
97700
97800
102.3
102.4
102.5
102.6
102.7
102.8
3060
3065
3070
3075
3080
3085
3090
3095
3100
3105
3110
3115
3120
3125
3130
3135
3140
3145
3150
3155
3160
3165
3170
3175

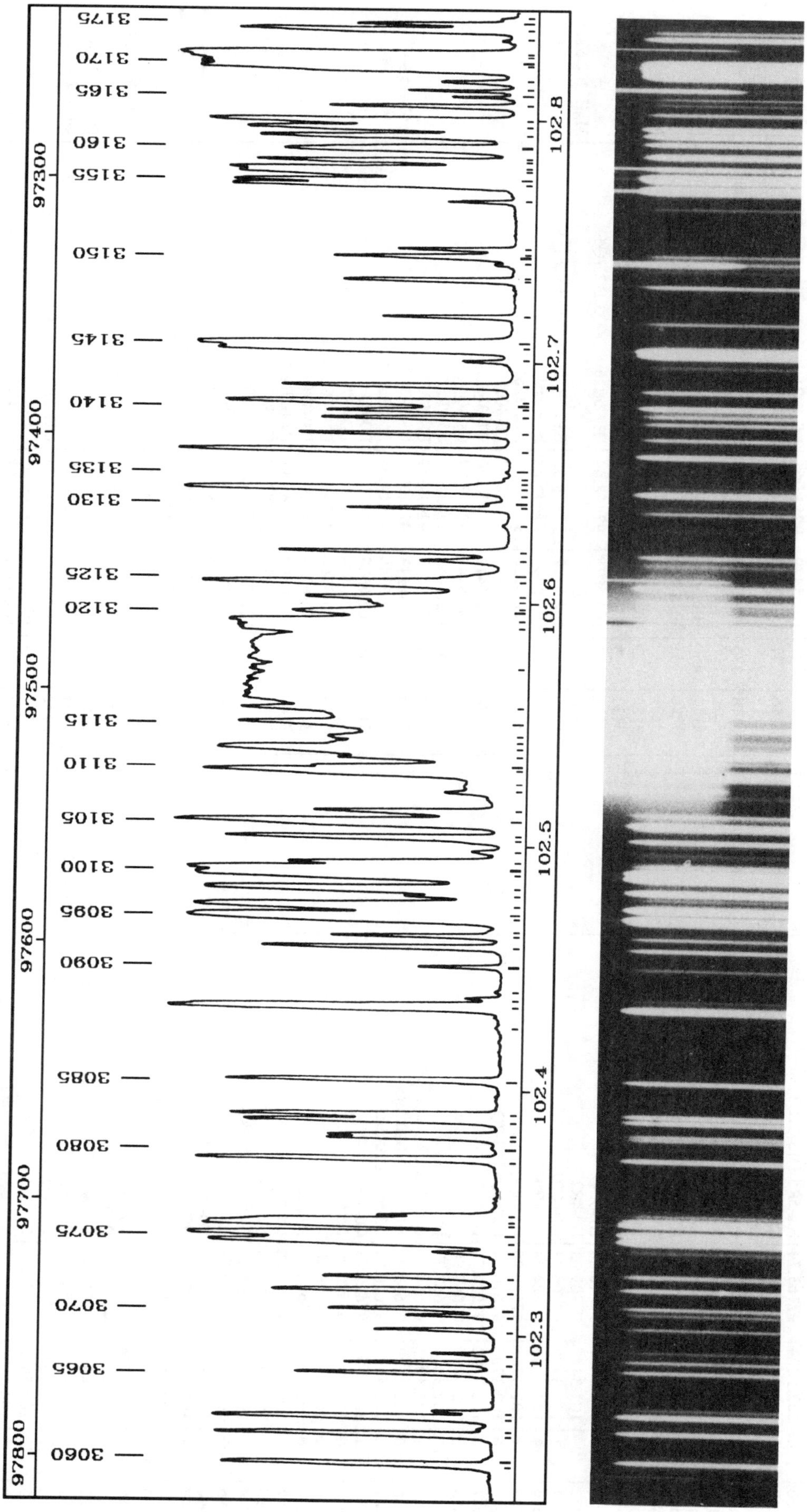

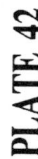

PLATE 42

TABLE 42 *continued*

Nb	I	λ (nm)	σ (cm⁻¹)	Assignment	Comment
3162	21	102.7964	97279.64	C7-4R0 C1-0R13 C2-0P14	
3163	26	102.7991	97277.09	C5-3P2	
3164	4	102.8054	97271.16	C10-5P3	
3165		102.8093	97267.47	B6-0P2	r 97266.27
3166		102.8118	97265.08	B6-0P2 D0-4Q4	r 97266.27
3167					OI 102.81571
3168		102.8157	97261.40		OI 102.81571
3169	43	102.8214	97255.99	C7-4R1 C3-2R4 C10-5R4	
3170	49	102.8232	97254.29	B30-5P4 C3-2P2	
3171	78	102.8249	97252.71	B7-0P5	
3172	87	102.8259	97251.76	C1-1R2 B23-4P1	
3173					CuII 102.83279
3174	16	102.8369	97241.39	D0-4P3	
3175	2	102.8395	97238.91	B18-3R0	
3176	1	102.8414	97237.07	D9-10Q2	

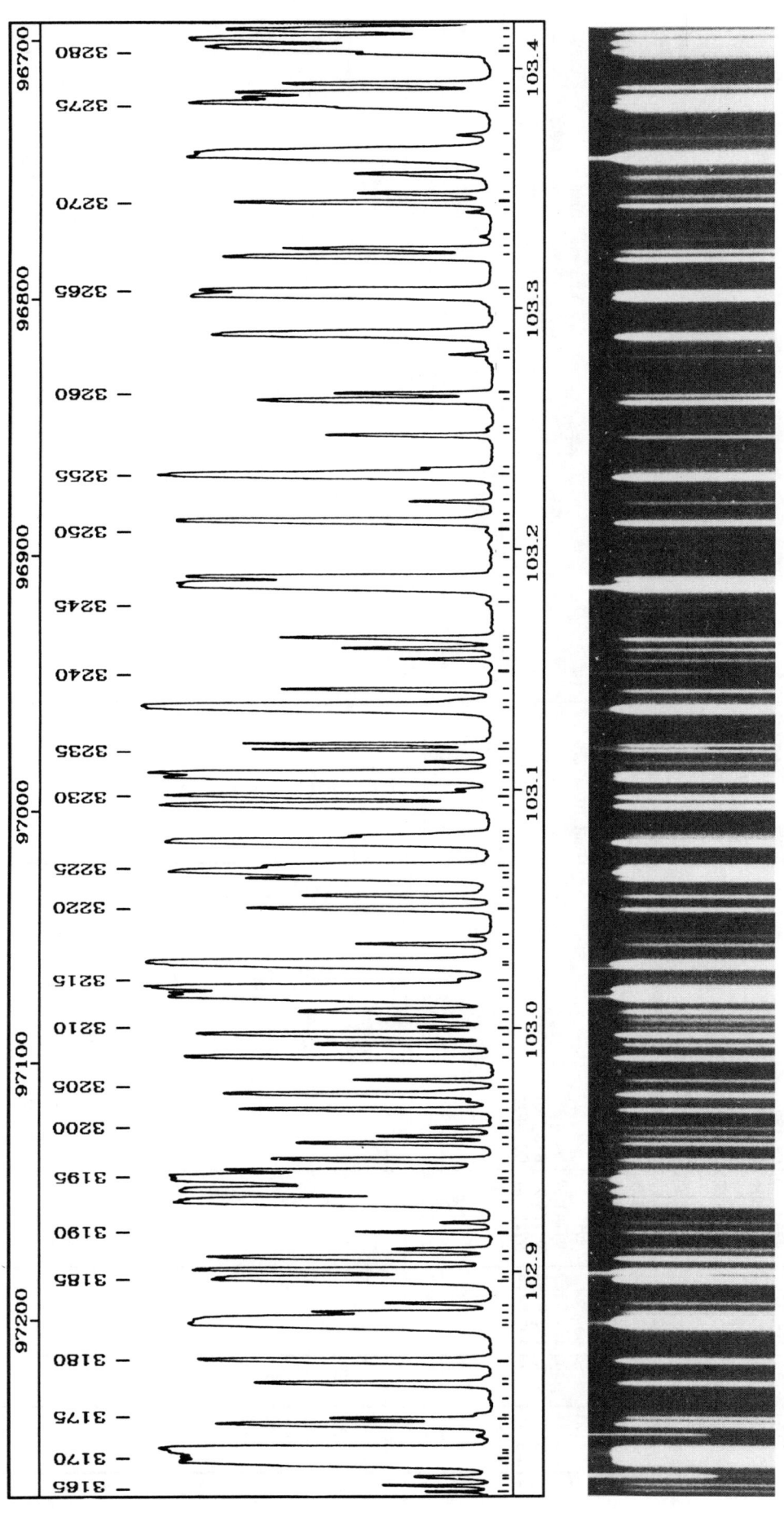

PLATE 43

TABLE 43

Nb	I	λ (nm)	σ (cm⁻¹)	Assignment	Comment
3177	0	102.8478	97231.02	C8-4P6	
3178	7	102.8539	97225.31	B9-0R9 C9-4Q9	
3179	1	102.8564	97222.95	D6-8Q4	
3180	32	102.8630	97216.72	D3-8Q1	
3181	75	102.8778	97202.68	C3-2Q3 C5-3Q3	
3182	75	102.8799	97200.69	B8-0P7 C1-0Q12 C6-3Q8	
3183	3	102.8829	97197.93	C5-3R4	
3184	3	102.8868	97194.24	D1-4Q12 D2-5Q9	
3185		102.8965	97185.02	C1-1Q1 B6-0R3 C7-4Q1	r 97183.32
3186		102.8994			
3187		102.9001	97181.62	C1-1Q1 B6-0R3	r 97183.32
3188	16	102.9053	97176.74	B18-3R1	
3189	2	102.9089	97173.30	D3-8Q2 B29-5R3	
3190	2	102.9161	97166.57	D3-6Q5 C4-2P8	
3191	1	102.9203	97162.60	D9-1Q03	
3192	38	102.9288	97154.55	B14-2R3 B'1-4R0 B'3-5R0	
3193	100	102.9332	97150.36	B'1-4R1	
3194	128	102.9376	97146.22	B'3-5R1	
3195		102.9384	97145.44	B24-4P4 C1-1R3	sh
3196	9	102.9413	97142.79	B23-4R2	
3197	4	102.9462	97138.16	C0-0Q8	
3198	9	102.9531	97131.59	C2-0Q15	
3199	2	102.9561	97128.78	B12-1P7	
3200	1	102.9597	97125.36	B16-1R12	
3201	0	102.9627	97122.55	C5-1P16	
3202	7	102.9670	97118.46	B18-3P1 C10-5Q4	
3203	0	102.9709	97114.84		
3204	12	102.9735	97112.37	B10-1P3 C4-0Q19	sh
3205		102.9750	97110.94	B19-3P4	
3206	6	102.9795	97106.75	D3-8Q3	
3207	26	102.9888	97097.94	B'1-4R2	
3208	4	102.9943	97092.80	D0-4Q5 B11-1R6	
3209	17	102.9982	97089.02	B3-5R2	
3210	0	103.0012	97086.21	B29-5P3	
3211	2	103.0045	97083.12	B28-5R0 C7-4R3 D5-7Q9	
3212	5	103.0080	97079.82	D6-8Q5 B7-0R6 B23-4P2	
3213	180	103.0145	97073.72	C1-1Q2 C7-4Q2	
3214	101	103.0175	97070.91	C5-3P3 C2-1P8	
3215	0	103.0211	97067.52	B13-1R9	
3216					CuII 103.02633
3217	163	103.0280	97061.03	C3-2P3	
3218	3	103.0360	97053.47	D0-4P4	
3219	2	103.0399	97049.72	B15-0R15 C2-1R10	
3220	5	103.0511	97039.28	C3-2R5 C8-4Q7	
3221	5	103.0565	97034.11	B18-3R2 B12-0P12	
3222	1	103.0590	97031.80	B11-0P11	
3223	8	103.0638	97027.27	B'0-3P7 C5-3R5	
3224	75	103.0665	97024.76	B'3-5P1	
3225	6	103.0686	97022.72	D3-8Q4 B28-5R1 C1-1R4	
3226	113	103.0793	97012.74	B'1-4P1	
3227	3	103.0817	97010.46	C4-2Q9 B17-2P8	
3228	2	103.0830	97009.20	C10-5P4	
3229	35	103.0947	96998.21	B'1-4R3 C9-4P9	
3230	21	103.0985	96994.58	C7-4P2 B14-2P3	
3231	1	103.1015	96991.85	C2-1Q9	
3232	30	103.1064	96987.21	C3-2Q4 C5-3Q4	
3233	31	103.1082	96985.54	B'3-5R3	
3234	2	103.1130	96980.95	B28-5P1	
3235		103.1180	96976.29	B6-0P3	r 96975.14
3236		103.1204	96973.99	B6-0P3	r 96975.14
3237		103.1356	96959.76	B30-5R5 C1-1P2	r
3238		103.1384	96957.12	B10-0P10	sh
3239	4	103.1431	96952.72	B10-1R4	Cl I 103.15070
3240					
3241	2	103.1558	96940.77	B8-0R8	
3242	4	103.1602	96936.64	B18-3P2	sh
3243		103.1633	96933.70	B13-0P13	
3244	7	103.1645	96932.56	B23-4R3	
3245	0	103.1791	96918.85	D3-8Q5	
3246	280	103.1866	96911.84	C1-1Q3 D6-8Q6 C6-3P8	
3247	23	103.1896	96908.99	C7-4Q3	
3248	0	103.1981	96900.98	C7-4R4	b
3249	0	103.2037	96895.80	D2-5Q10 C4-1Q15	
3250	0	103.2093	96890.49	D0-4Q6	
3251	38	103.2133	96886.70	B'3-5P2	
3252	2	103.2156	96884.56	B28-5R2 B29-5R4	
3253	2	103.2217	96878.85	B19-3R5	
3254	1	103.2282	96872.79	B30-5P5	
3255	70	103.2326	96868.67	B'1-4P2	
3256	9	103.2354	96865.97	B6-0R4	
3257	4	103.2493	96852.99	B'1-4R4	
3258	0	103.2520	96850.41	B14-2R4	
3259	12	103.2637	96839.42	B'3-5R4 B23-4P3	
3260	3	103.2666	96836.70	D0-4P5	
3261	0	103.2806	96823.64	C10-5Q5	
3262	1	103.2830	96821.38	D1-4Q13	
3263	24	103.2912	96813.63	B18-3R3 C5-3P4	
3264	38	103.3072	96798.68	C3-2P4	
3265	17	103.3090	96797.02	C1-1R5	
3266	13	103.3233	96783.60	C7-4P3	
3267	5	103.3266	96780.49	B9-0P9	
3268	0	103.3319	96775.56	B11-1P6	
3269	1	103.3420	96766.04	D3-6Q7	
3270	16	103.3458	96762.48	C13-6Q1 D4-9Q1	
3271	5	103.3497	96758.90	C3-2R6 C11-5P7 C3-0P17	
3272	3	103.3578	96751.25	B10-1P4 B24-4P5	
3273	340	103.3661	96743.54	C1-1P3 C5-3R6	
3274	1	103.3743	96735.89	C6-3Q9	

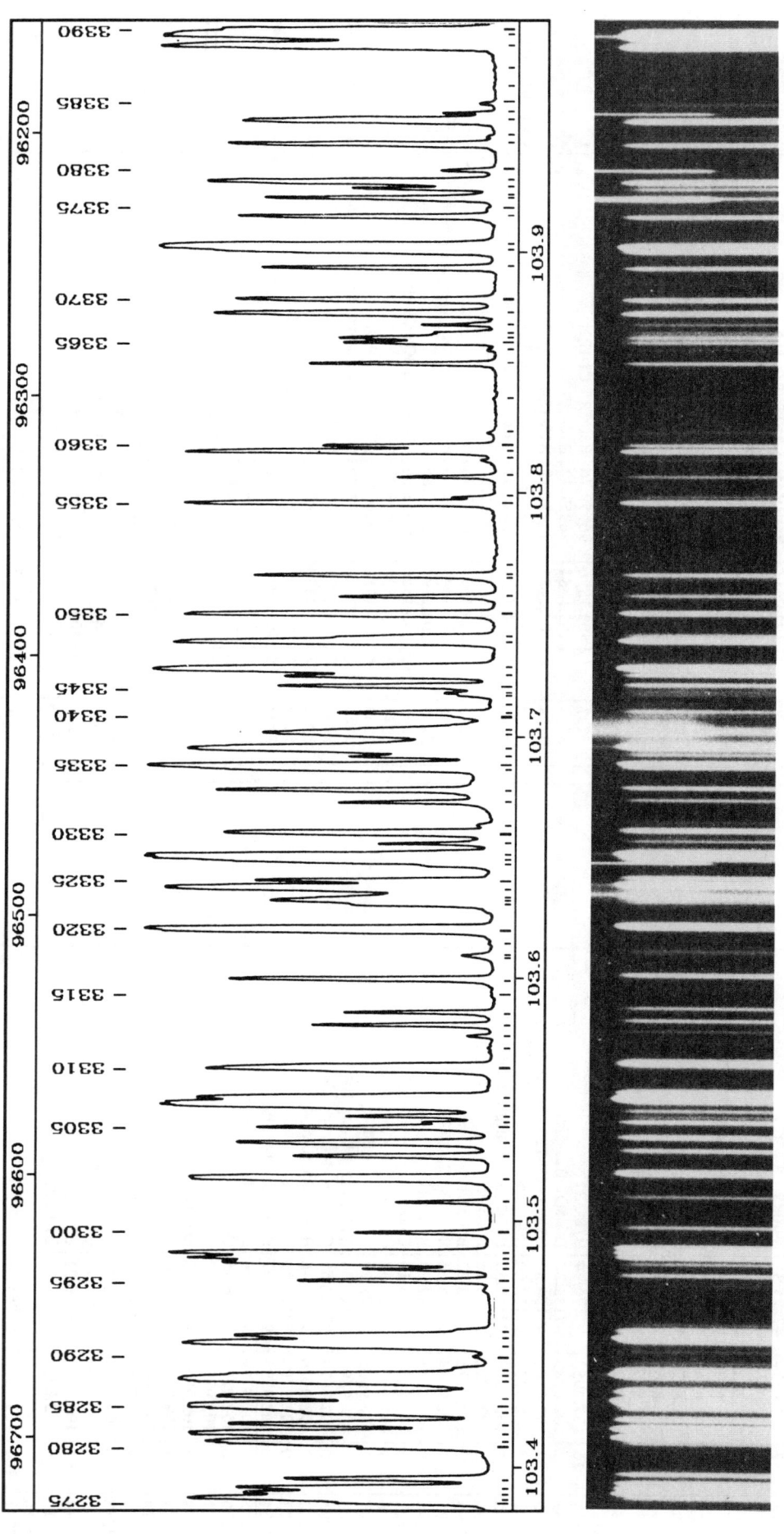

PLATE 44

TABLE 44

Nb	I	λ (nm)	σ (cm⁻¹)	Assignment	Comment
3275	28	103.3856	96725.24	B16-0R16 D6-8Q7 D4-9Q2	
3276	8	103.3876	96723.42	C3-2Q5	sh
3277	8	103.3895	96721.61	C5-3Q5	
3278	5	103.3915	96719.71	B7-0P6	
3279	3	103.3953	96716.16	C0-0P8	
3280	59	103.4079	96704.44	B19-3P5	
3281	52	103.4104	96702.11	B3-5P3	
3282	16	103.4140	96698.65	C1-1Q4	
3283	11	103.4178	96695.17	B22-4R0 D4-7Q1	
3284	122	103.4224	96690.87	C7-4Q4 C10-5P5 B13-2R0	
3285	20	103.4255	96687.98	D1-5R1 B13-1P9 D7-9Q1	
3286	210	103.4291	96684.61	C0-0Q9	sh
3287	0	103.4357	96678.44	D1-5R2 B1-4P3	
3288	1	103.4376	96676.61	B18-3P3 B20-3R7 C9-4Q10	
3289	73	103.4406	96673.88	B28-5R3	
3290	11	103.4455	96669.29	D4-9Q3	
3291	1	103.4512	96663.98	B1-4R5 C7-4R5 D1-5R0	
3292	1	103.4538	96661.49	C9-5R0 D0-4Q7 B11-0R12	
3293	5	103.4562	96659.26	C4-2P9	
3294					
3295	2	103.4725	96644.00	B23-4R4	
3296	14	103.4766	96640.19	C9-5R1	
3297	21	103.4811	96635.97	D7-9Q2	
3298	24	103.4842	96633.16	B22-4R1 D4-7Q2	
3299	5	103.4860	96631.44	B0-3P8 D1-5R3	
3300	5	103.4876	96629.93	B13-2R1	
3301	11	103.4961	96622.04	B10-0R11	r
3302	14	103.5085	96610.45	C8-4Q8 B12-0R13	
3303	8	103.5182	96601.42	B6-0P4	
3304	1	103.5327	96592.86	C2-0P15 D0-4P6	
3305	2	103.5391	96587.83	C1-0Q13 C2-1R11	
3306	240	103.5410	96581.89	B22-4P1	
3307	22	103.5437	96580.10	D2-5Q11	
3308	21	103.5490	96577.55	B28-5P3	
3309	0	103.5512	96572.68	C9-5Q1	
3310	1	103.5634	96570.59	C9-5Q1 C4-0P19	d
3311	6	103.5719	96559.21	B13-2P1 C7-4P4 C2-1P9 D7-9Q3	
3312	3	103.5766	96551.31	B11-1R7 B30-5R6	
3313	0	103.5811	96546.93	D1-5R4	d
3314	1	103.5863	96542.72	D4-7Q3 B27-5R0	
3315	0	103.5953	96537.86	C1-1R6	
3316		103.6001	96529.50	B29-5R5 D3-6Q8	
3317		103.6099	96525.01	B7-0R7	
3318		103.6117	96515.85	B18-3R4 C3-0Q18	
3319		103.6151	96514.20	C4-2Q10	
3320		103.6208	96511.01	B17-1P13	sh
3321	73	103.6309	96505.73	C5-3P5 D1-5Q2	
3322		103.6328	96496.34	B8-0P8	+CII 103.63367
3323	2	103.6348	96494.58	B22-4R2 C3-2R7	CII 103.63367
3324	51	103.6378	96489.94	C3-2P5	
3325	7	103.6405	96487.39	B13-2R2	CuII 103.64695
3326					sh
3327	129	103.6489	96479.54	B27-5R1 B9-0R10	
3328		103.6513	96477.35	C1-1P4 C10-5Q6	
3329	5	103.6560	96472.98	B3-5P4 B13-0R14 B12-1P8	
3330	16	103.6604	96468.85	B6-0R5 B14-2R5	
3331	1	103.6637	96465.81	B34-6P1	
3332	5	103.6729	96457.19	C9-5Q2 D7-9Q4	
3333	21	103.6781	96452.43	B17-3R0 B19-3R6	
3334	8	103.6867	96444.35	B1-4P4	+CII 103.70182
3335	48	103.6882	96442.99	B27-5P1	CII 103.70182
3336	3	103.6918	96439.60	C1-1Q5 D1-4Q14	sh
3337	80	103.6954	96436.31	B30-5P6	r 96418.23
3338	0	103.7016	96430.55		r 96418.23
3339					
3340		103.7084	96424.18	C9-5R3	
3341	2	103.7098	96422.93	D4-7Q4 C7-4Q5	
3342		103.7130	96419.93	B5-0R1 C9-4P10	
3343		103.7167	96416.52	B5-0R1 C8-4P8 D1-5R5	
3344	3	103.7179	96415.36	C2-0Q16	
3345	8	103.7207	96412.76	C3-2Q6 C5-3R7	
3346	6	103.7249	96408.88	C5-3Q6 B22-4P2	
3347	85	103.7278	96406.20	D1-5Q3 D0-4Q8	
3348	39	103.7393	96395.53	D1-5P2 C6-3P9	
3349		103.7410	96393.91	C9-5P2	
3350	23	103.7507	96384.92	B17-3R1	sh
3351	5	103.7579	96378.24	B'5-6R0	
3352		103.7650	96371.59	B9-1P2	sh
3353	8	103.7665	96370.25	B13-2P2	
3354	0	103.7702	96366.82	C7-4R6	
3355	25	103.7963	96342.58	B'5-6R1 B27-5R2	
3356	2	103.7987	96340.36	B18-3P4	
3357	4	103.8072	96332.47	D7-9Q5 C0-0R11	r 96324.37
3358		103.8141	96326.06	B5-0P1 B13-1R0 B17-2P9	r 96324.37
3359		103.8177	96322.68	B5-0P1 B17-3P1 B15-2R7 D0-4P7	
3360	5	103.8201	96320.46	C12-6R1	
3361	2	103.8255	96315.42	B10-1P5	
3362	0	103.8393	96302.64	C10-5R7 D0-6Q1	
3363	0	103.8542	96288.84	C9-5Q3	
3364	0	103.8599	96283.58	B23-4R5	
3365	5	103.8629	96280.74	B22-4R3	
3366	6	103.8649	96278.93	C12-6Q1	
3367		103.8667	96277.25	B5-0R2 D4-7Q5	r 96275.53
3368		103.8704	96273.80	B5-0R2 B27-5P2 D11-12Q1	r 96275.53
3369	16	103.8751	96269.49	B28-5P4 B14-0R15 B'5-6P1 D13-14Q4	
3370	11	103.8807	96264.28	B13-2R3	
3371	9	103.8940	96251.98	C7-4P5	
3372	16	103.9005	96245.92	B17-3R2 B'5-6R2 D11-12Q2 C4-0Q20	

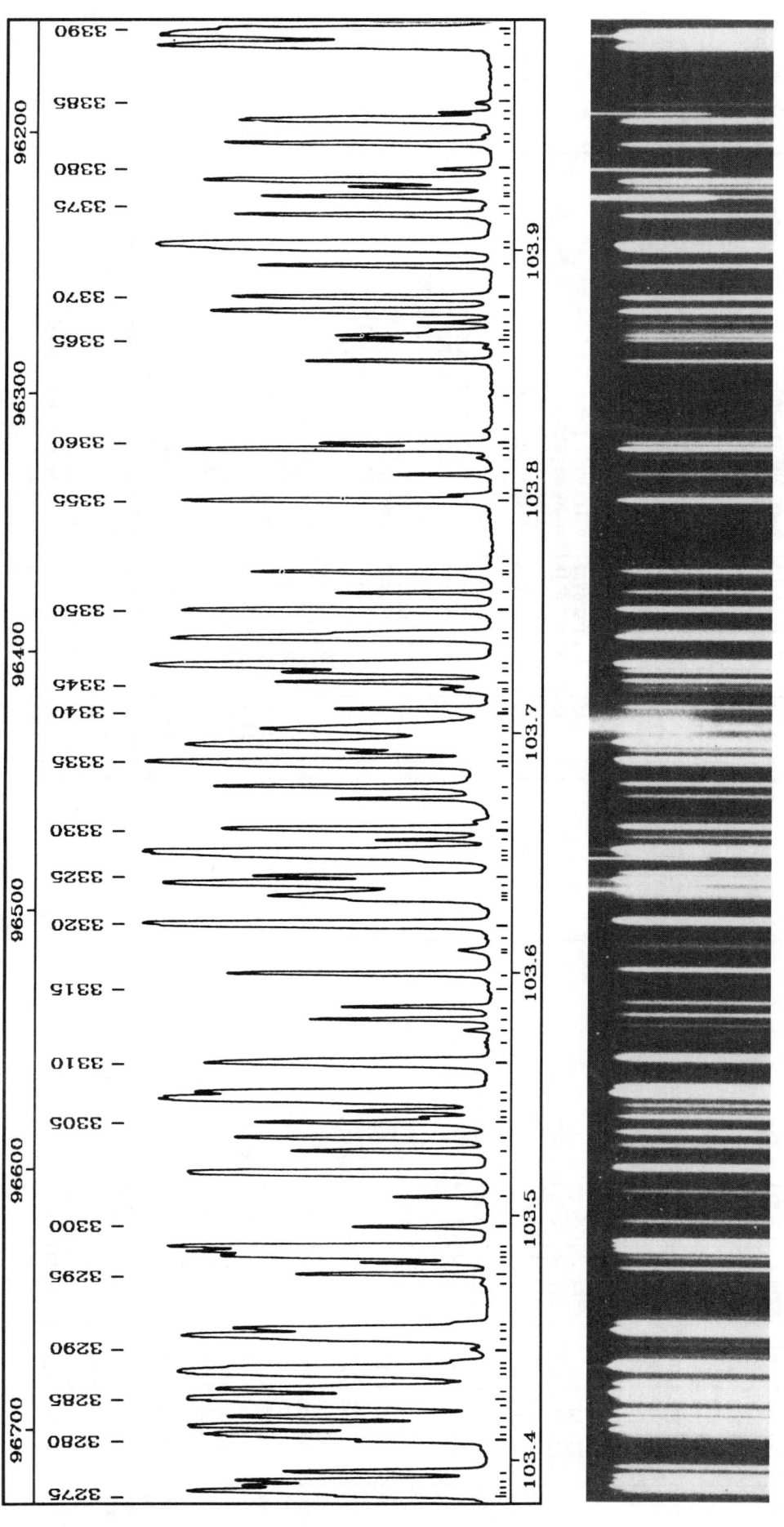

PLATE 44

TABLE 44 *continued*

Nb	I	λ (nm)	σ (cm⁻¹)	Assignment	Comment
3373	129	103.9030	96243.62	B8-0R9 D1-5P3 B19-3P6 D13-14Q3	
3374	9	103.9152	96232.36	C1-1R7	
3375	2	103.9183	96229.46	C9-5R4 D13-14Q2	
3376		103.9229	96225.23		OI 103.92304
3377					OI 103.92304
3378	5	103.9268	96221.56	D13-14Q1	
3379	17	103.9297	96218.91	B16-1R13 C0-0P9	
3380					CuII 103.93477
3381	9	103.9451	96204.64	D11-12Q3 B3-5P5 B0-3P9	
3382	0	103.9480	96201.96	B33-6R1	
3383	13	103.9550	96195.52	C9-5P3 C0-0Q10	
3384					CuII 103.95821
3385	1	103.9618	96189.22	B11-1P7	
3386	1	103.9676	96183.80	B12-1R9	
3387	0	103.9743	96177.66	B'1-4P5 B'1-4R7	
3388	54	103.9857	96167.06	C1-1P5	sh
3389	175	103.9902	96162.91	B22-4P3	
3390		103.9920	96161.21		

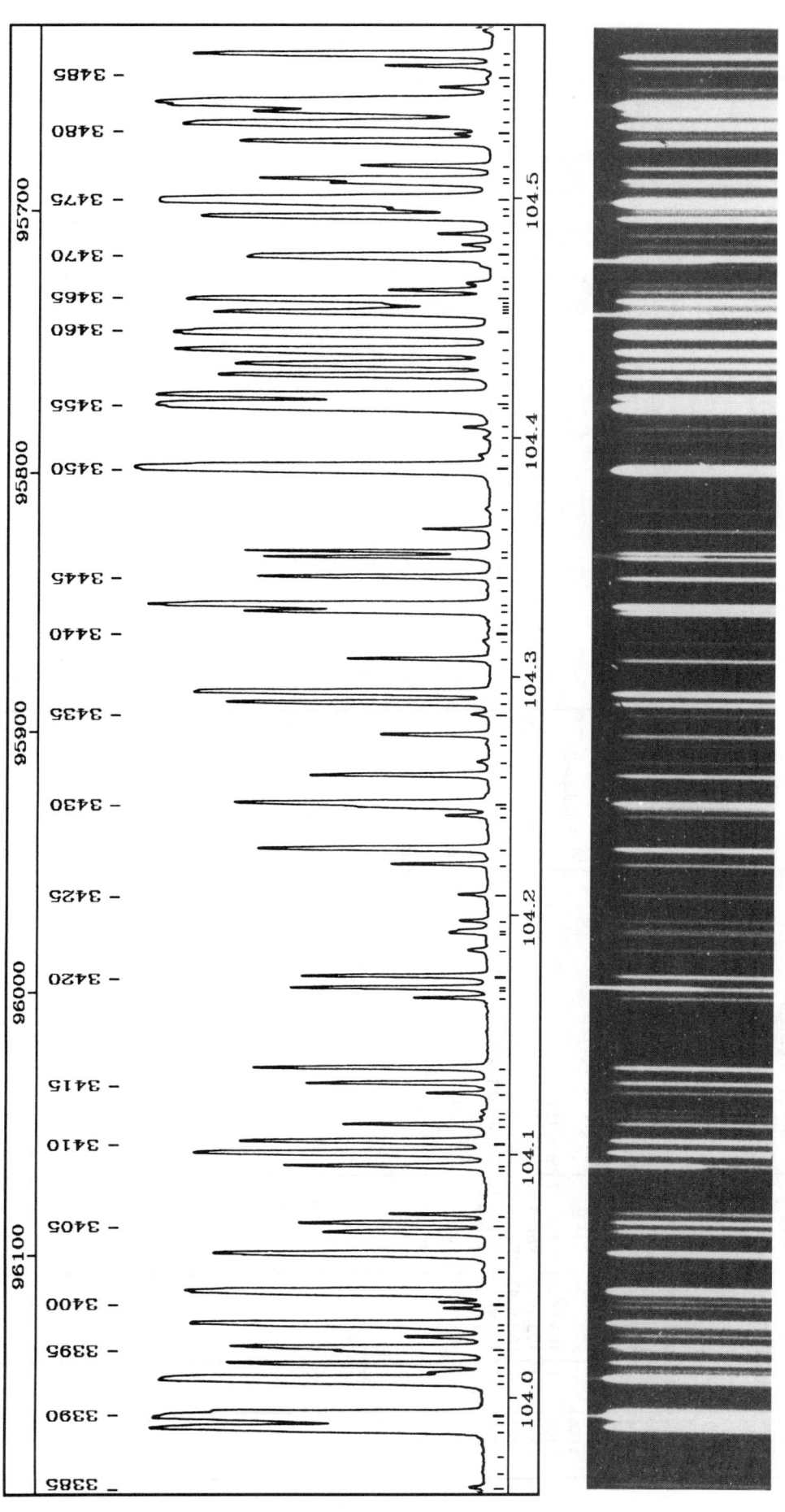

PLATE 45

TABLE 45

Nb	I	λ (nm)	σ (cm⁻¹)	Assignment	Comment
3391	141	104.0059	96148.42	B6-0P5	
3392	2	104.0090	96145.57	B18-3R5	
3393	12	104.0125	96142.34	B17-3P2 D0-6P2 ?	
3394		104.0178	96137.37	C5-3P6 C8-4Q9	sh
3395	7	104.0191	96136.19	C3-2P6	
3396	3	104.0235	96132.13	B27-5R3	
3397		104.0269	96128.95	D0-4Q9	sh
3398	24	104.0287	96127.34	C1-1Q6 B'5-6P2	
3399		104.0354	96121.17	B5-0P2	r 96119.99
3400		104.0379	96118.80	B5-0P2	r 96119.99
3401	26	104.0418	96115.20	D1-5Q5 B7-0P7	
3402	0	104.0510	96106.63	C7-4Q6 D4-7Q6	
3403	21	104.0574	96100.78	B13-2P3 B23-4P5	
3404	5	104.0666	96092.31	B10-0P11 C7-4R7 B9-1P3	
3405	10	104.0705	96088.74	B12-0P13 C9-4Q11	
3406	4	104.0743	96085.25	B'5-6R3	
3407	0	104.0942	96066.81	B30-5R7 C9-5Q4	+OI 104.09425
3408					OI 104.09425
3409	23	104.0994	96062.01	D1-5P4	
3410	12	104.1044	96057.46	C3-2Q7	
3411	4	104.1113	96051.02	C5-3Q7 C2-1R12	
3412		104.1144	96048.18	B5-0R3	
3413	2	104.1171	96045.73	B5-0R3	r 96046.96
3414	7	104.1244	96038.99	D8-10Q1 B15-2P7	r 96046.96
3415	5	104.1285	96035.17	B27-5P3 C1-0R15	
3416	3	104.1349	96029.28	B17-3R3	
3417		104.1640	96002.44	B21-4R0	OI 104.16876
3418					+OI 104.16876
3419	3	104.1688	95998.01	C1-0P13	
3420	4	104.1732	95994.00	B6-0R6 D8-10Q2 B22-4R4	
3421	1	104.1840	95983.99	C4-2Q11	
3422	2	104.1914	95977.25	B9-0P10 B15-0R16	
3423	2	104.1923	95976.42	B26-5R0	
3424	2	104.1965	95972.48	C2-1Q11	
3425	1	104.2073	95962.55	B13-2R4 B11-1R8	
3426	5	104.2199	95950.97	C1-0Q14	
3427	8	104.2267	95944.75	B21-4R1 C9-5P4	
3428	1	104.2407	95931.78	B18-3P5	
3429		104.2445	95928.29	D1-5Q6 D8-10Q3	sh
3430	8	104.2460	95926.93	C6-3P10 B'5-6P3	
3431	3	104.2579	95916.02	B26-5R1 D4-7Q7	
3432	1	104.2635	95910.85	D2-5Q13	
3433	0	104.2707	95904.19	C12-6P3	
3434	2	104.2746	95900.68	B7-0R8 C7-4P6 B'3-5P6	
3435	0	104.2832	95892.70	C8-4P9 B28-5P5	
3436	8	104.2880	95888.36	B21-4P1	
3437	22	104.2924	95884.30	B17-3P3	
3438	3	104.3063	95871.46	B26-5P1	
3439	0	104.3139	95864.46	C1-1R8 C9-4P11 D'0-7Q1	
3440	0	104.3173	95861.37	B'1-4R8	
3441	0	104.3208	95858.23		
3442	5	104.3263	95853.15	B'1-4P6	
3443	23	104.3288	95850.80	B27-5R4 D1-5P5	
3444	0	104.3348	95845.33	B33-6R3	
3445	5	104.3408	95839.79	C1-1P6 B22-4P4	
3446		104.3489	95832.37	B5-0P3 C2-0P16 D0-4Q10	r 95831.23
3447		104.3514	95830.10	B5-0P3	r 95831.23
3448	1	104.3608	95821.45	B32-6P1	
3449	0	104.3692	95813.67	B21-4R2	
3450	85	104.3863	95798.00	C4-3R0 C4-3R1	
3451	0	104.3918	95792.99	B17-0P17 C9-5Q5	
3452	0	104.3993	95786.11	B33-6P3	
3453	0	104.4037	95782.06	B26-5R2	
3454		104.4115	95773.89	C1-1Q7	sh
3455	103	104.4133	95773.22	D2-6R1	
3456	24	104.4168	95770.01	D2-6R2 B10-1P6 D'0-7Q3 ?	
3457	10	104.4253	95762.22	C6-4R0	
3458	9	104.4302	95757.74	B12-2R0 B12-1P9 B8-0P9	
3459	22	104.4357	95752.71	C6-4R1 B13-2P4 B'0-3P10	
3460	46	104.4433	95745.72	D2-6R0 C4-3R2 C7-4Q7	
3461					CuII 104.45188
3462	6	104.4519	95737.82	C3-2P7	
3463		104.4531	95736.71	B5-0R4	r 95735.72 sh
3464		104.4553	95734.73	B5-0R4	r 95735.72
3465	16	104.4571	95733.08	D2-6R3 B10-0R12	
3466	2	104.4612	95729.37	B11-0R13	
3467	1	104.4641	95726.67	B27-5P4 D1-7Q1 B16-1P13	
3468					CuII 104.47435
3469	0	104.4714	95719.94	D3-6Q11	
3470	6	104.4751	95716.58	B21-4P2 D1-5Q7 C3-2R9	
3471	2	104.4800	95712.06	C3-0Q19 C4-0P20	
3472	1	104.4848	95707.71	B26-5P2	
3473	10	104.4917	95701.34	B12-2R1	
3474	1	104.4949	95698.50	B14-0P15 C6-3Q11	
3475	143	104.4983	95695.33	C4-3Q1 D'0-7Q4 C7-4R8	
3476	3	104.5057	95688.58	C0-0P10	
3477	5	104.5074	95687.03	C6-4R2	
3478	5	104.5128	95682.07	D1-7Q2 C2-0Q17	
3479	11	104.5231	95672.68	C0-0Q11 C5-3R9	
3480	0	104.5263	95669.69	B'5-6P4	
3481	37	104.5305	95665.88	C6-4Q1	
3482	8	104.5356	95661.16	C3-2Q8 B9-0R11	
3483	188	104.5388	95658.28	D2-6Q1 D2-6R4	
3484	1	104.5457	95651.91	C5-3Q8 D1-4Q16	
3485	0	104.5491	95648.88	B19-3P7	
3486	0	104.5546	95643.77	C9-5P5	
3487	12	104.5595	95639.31	C4-3R3	
3488	2	104.5696	95630.11	B12-0R14	

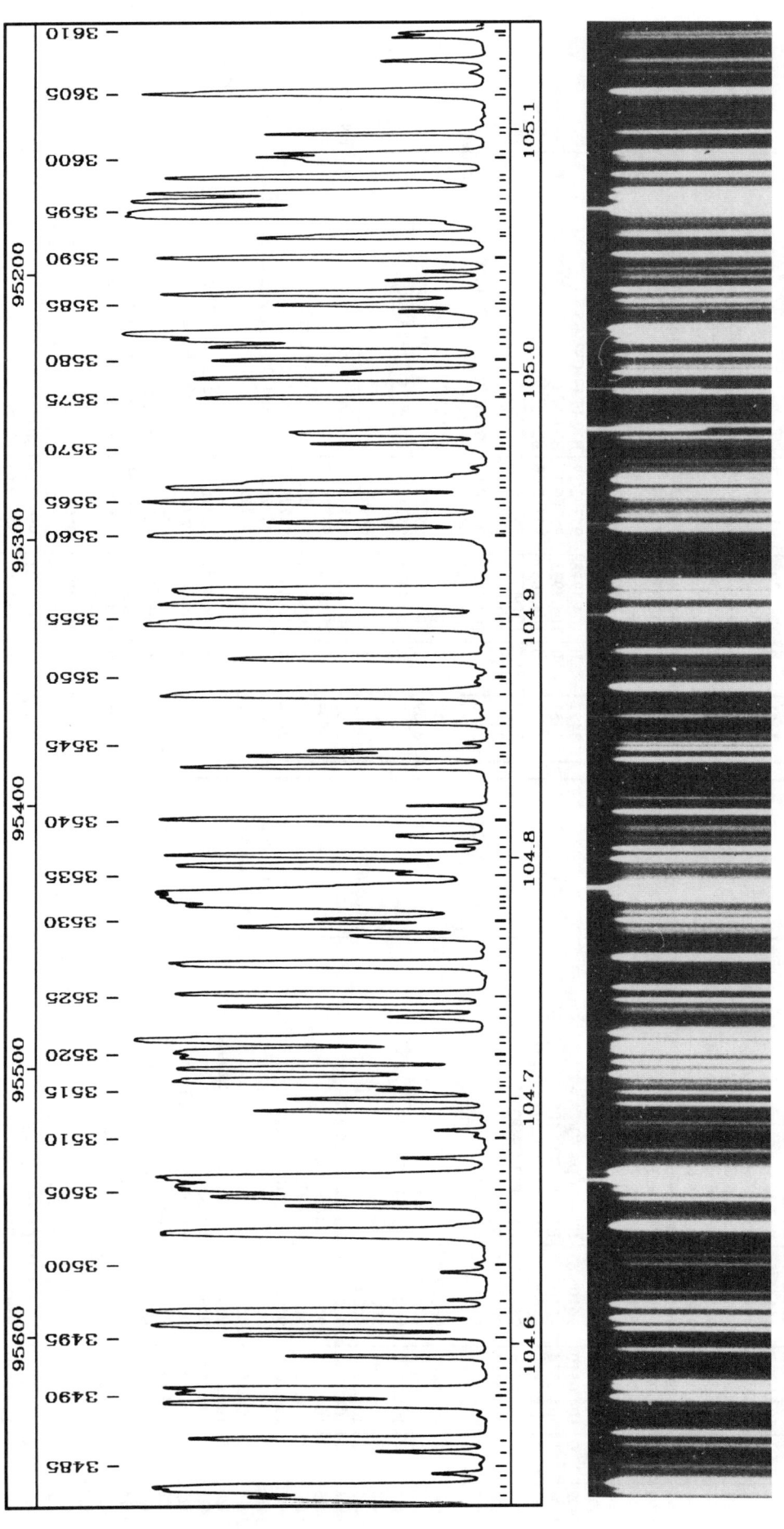

PLATE 46

TABLE 46

Nb	I	λ (nm)	σ (cm⁻¹)	Assignment	Comment
3489	29	104.5742	95625.88	B12-2P1 C8-4Q10	
3490	13	104.5785	95622.00	B16-3R0	
3491	11	104.5803	95620.31	B6-0P6	
3492	0	104.5826	95618.21	B16-0R17 D1-7Q3	
3493	0	104.5900	95611.47	C11-6R1	
3494	4	104.5936	95608.14	D1-5P6	
3495	6	104.6022	95600.27	B8-1R1	
3496	40	104.6065	95596.37	D2-6Q2	
3497	34	104.6123	95591.05	C4-3Q2	
3498	1	104.6172	95586.57	B13-2R5	
3499	1	104.6287	95576.11	B26-5R3	
3500	0	104.6323	95572.80	B3-5P7	
3501	37	104.6445	95561.68	B16-3R1 B17-1P14 C6-4Q2 C11-6Q1	
3502	0	104.6479	95558.58	D9-11Q1	
3503	3	104.6555	95551.64	B17-3P4	
3504	11	104.6592	95548.25	C6-4R3	
3505	240	104.6631	95544.70	C2-2R1	
3506	138	104.6659	95542.09	C2-2R0	
3507		104.6680	95540.17		sh
3508	2	104.6751	95533.72	D12-14Q3	
3509	0	104.6779	95531.19	D1-7Q4 B13-1R11	
3510	0	104.6841	95525.50	C2-1R13	
3511	1	104.6864	95523.42	B1-4R9 D9-11Q2	
3512	0	104.6905	95519.67	B30-5R8 C11-6R2 D0-4Q11	
3513	3	104.6942	95516.30	B8-1P1	sh
3514	3	104.6991	95511.81	C1-1R9	
3515	2	104.7029	95508.34	B31-6R0 C5-1P18 D12-14Q2	
3516	62	104.7065	95505.04	D2-6Q3 C7-4P7 B1-4P7	
3517	0	104.7094	95502.45	B27-5R5 C2-1P11	
3518	17	104.7114	95500.57	B16-3P1	
3519	29	104.7162	95496.25	C4-3P2	
3520	31	104.7179	95494.68	B14-2R7 D2-6P2 B'2-5P1 D12-14Q1	
3521	64	104.7233	95489.73	C2-2R2	
3522		104.7253	95487.94	B8-0R10	sh
3523	1	104.7334	95480.54	C4-3R4 D1-5Q8	b
3524	7	104.7374	95476.92	D4-7Q9 C6-4P2	
3525	14	104.7427	95472.09	B21-4P3 B26-5P3 D9-11Q3	
3526	38	104.7552	95460.64	B5-0P4 B8-1R2	
3527	0	104.7609	95455.44	B18-3P6	
3528	3	104.7668	95450.11	B22-4P5 B12-1R10 B13-0R15	b
3529	6	104.7710	95446.31	B6-0R7 B31-6R1	
3530	2	104.7742	95443.32	B7-0P8 C11-6Q2	
3531	14	104.7799	95438.17	B12-2P2 C6-4R4	
3532	44	104.7821	95436.09	C4-3Q3	
3533	460	104.7843	95434.14	C2-2Q1	
3534	9	104.7881	95430.70	B20-3R9 C4-1Q17	
3535	1	104.7931	95426.11	B15-2P8 C8-5R0 D'1-7Q5	
3536	25	104.7964	95423.14	B16-3R2 C1-1P7 C4-2Q12	
3537	14	104.8007	95419.17	C2-2R3	
3538	1	104.8052	95415.09	B31-6P1	b
3539	2	104.8094	95411.33	C8-5R1	
3540	16	104.8161	95405.18	C6-4Q3	ArI 104.82199
3541		104.8218	95400.02		
3542	7	104.8379	95385.36	C1-0R16 D2-6Q4	
3543	5	104.8424	95381.25	C1-1Q8	
3544	4	104.8445	95379.34	B25-5R0	
3545	1	104.8478	95376.39		
3546	5	104.8560	95368.86	C1-0P14 C4-0Q21	
3547	0	104.8604	95364.87	B23-4R7 C11-6R3	
3548	79	104.8681	95357.87	D2-6P3 B'2-5P2	
3549	0	104.8736	95352.92	C5-3P8	
3550	0	104.8755	95351.19	B27-5P5	
3551	0	104.8797	95347.30	C6-3P11 C8-4P10	
3552	11	104.8829	95344.41	B5-0R5 C8-5R2	
3553	0	104.8869	95340.76	C7-4Q8	
3554	138	104.8975	95331.17	C2-2Q2 C8-5Q1	
3555		104.9003	95328.61	D1-5P7	sh
3556	50	104.9063	95323.18	B12-2R3	
3557	22	104.9100	95319.77	B16-3P2 B8-1P2	
3558	22	104.9121	95317.92	B25-5R1 B14-6P1	
3559	0	104.9182	95312.39	B31-6R2	
3560	94	104.9347	95297.37	C4-3P3	CuII 104.93640
3561		104.9394	95293.12	C1-0Q15	
3562	12	104.9409	95291.70	B21-4R4	
3563	3	104.9455	95287.55	C3-2P8	
3564	4	104.9486	95284.76	B20-4R0 B'0-4R1	b
3565	49	104.9541	95279.76	C6-4P3 B'0-4R0	
3566	27	104.9564	95277.70	B25-5P1	
3567	5	104.9586	95275.70	B'0-3P11	
3568	2	104.9619	95272.69	C0-0R13	
3569	1	104.9693	95265.99	C11-6Q3 D2-8Q1	
3570	1	104.9717	95263.77	C2-2R4	
3571	2	104.9767	95259.25	B31-6P2 C4-3R5 C7-4R9	CuII 104.97554
3572		104.9837	95252.93	D1-4Q17	
3573	5	104.9908	95246.45	B'0-4R2	
3574	0	104.9922	95243.68	B4-0R1	Al II 104.9923 Ke
3575	14	104.9938	95239.08	B4-0R1 D2-6Q5	r 95241.38
3576		104.9989	95236.93	B8-1R3 D11-13Q1	r 95241.38
3577		105.0013	95232.46	C3-4Q4	
3578		105.0062	95227.19	C3-2Q9 D2-8Q2 D1-5Q9	
3579	3	105.0120	95224.76	B20-4R1 C8-5Q2	
3580	8	105.0147	95222.53	C2-2P2 C8-5R3	
3581	8				
3582	18				
3583	103				
3584	2	105.0259	95214.60	C5-3Q9	
3585	6	105.0290	95211.81	B7-0R9	
3586		105.0303	95210.61	C5-3R10	sh

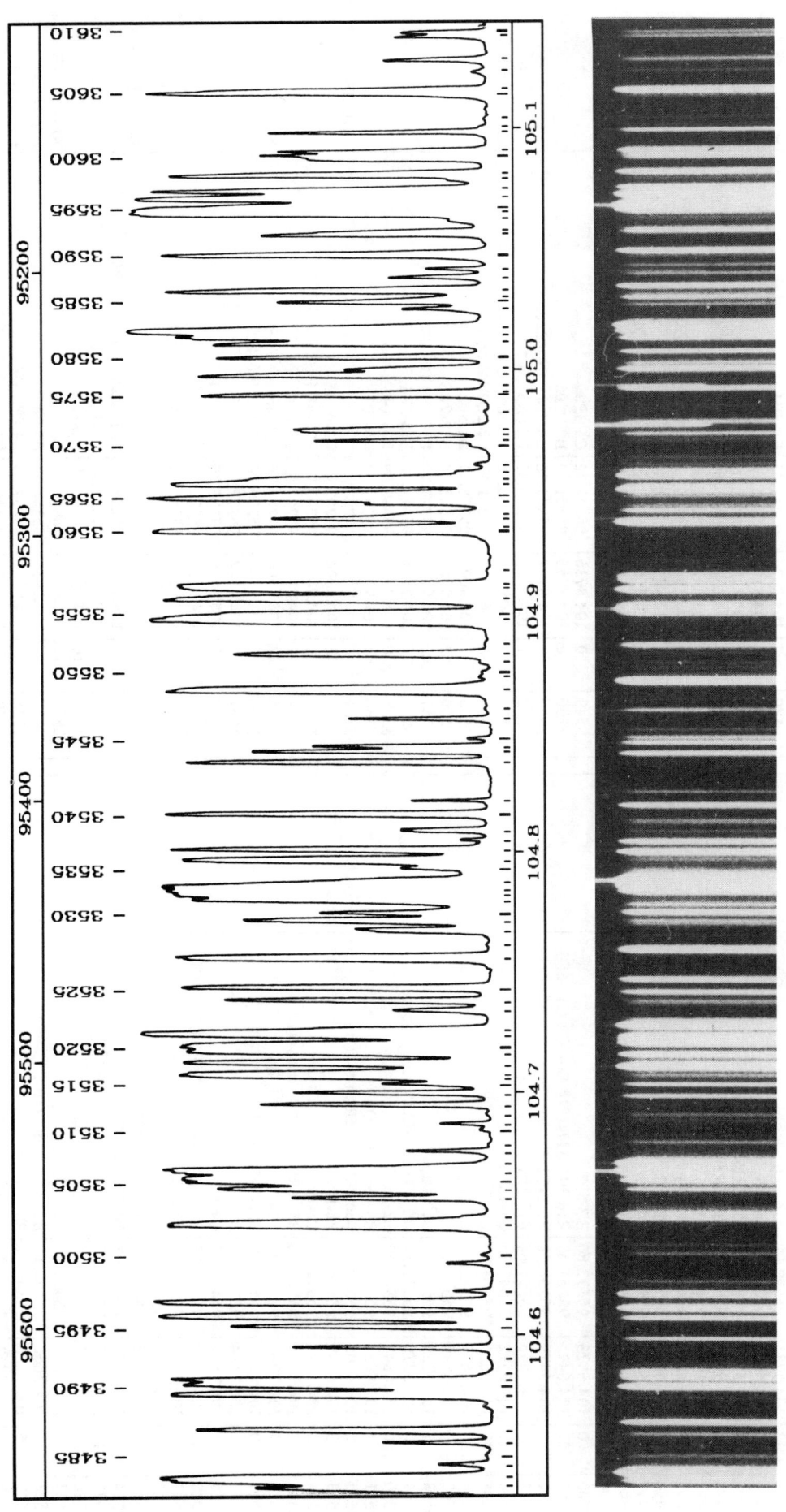

PLATE 46

TABLE 46 *continued*

Nb	I	λ (nm)	σ (cm⁻¹)	Assignment	Comment
3587	14	105.0336	95207.62	B16-3R3 B10-1P7	
3588	2	105.0388	95202.88	B18-3R7 D6-9Q1	
3589	1	105.0424	95199.64	C6-4Q4	
3590	16	105.0485	95194.13	B14-0R16 D2-6P4	
3591	6	105.0573	95186.16		
3592	5	105.0584	95185.13	B25-5R2	
3593	0	105.0623	95181.63	B10-0P12	
3594		105.0660	95178.23	B'2-5P3	sh
3595	190	105.0676	95176.79	C2-2Q3	
3596	46	105.0724	95172.46	B12-2P3	
3597	13	105.0752	95169.96	B20-4P1 D2-8Q3	
3598	0	105.0793	95166.25	B26-5P4	
3599	20	105.0818	95163.97	B'0-4R3	
3600	4	105.0887	95157.72	B21-4P4 C8-5P2	
3601	15	105.0906	95155.95	B11-0P13 C0-0P11 D6-9Q2	
3602	4	105.0994	95148.06	B17-3P5	r 95144.70
3603		105.1014	95146.17	B4-0P1	r 95144.70
3604		105.1047	95143.22	B4-0P1	sh
3605		105.1141	95134.67	C3-0P19	b
3606	36	105.1163	95132.76	C6-3Q12 B'0-4P1	
3607	1	105.1242	95125.56	B9-1R6 B'1-4P8	
3608	4	105.1290	95121.21	C0-0Q12	
3609	4	105.1388	95112.40	C2-0P17	
3610	3	105.1405	95110.86	B25-5P2	

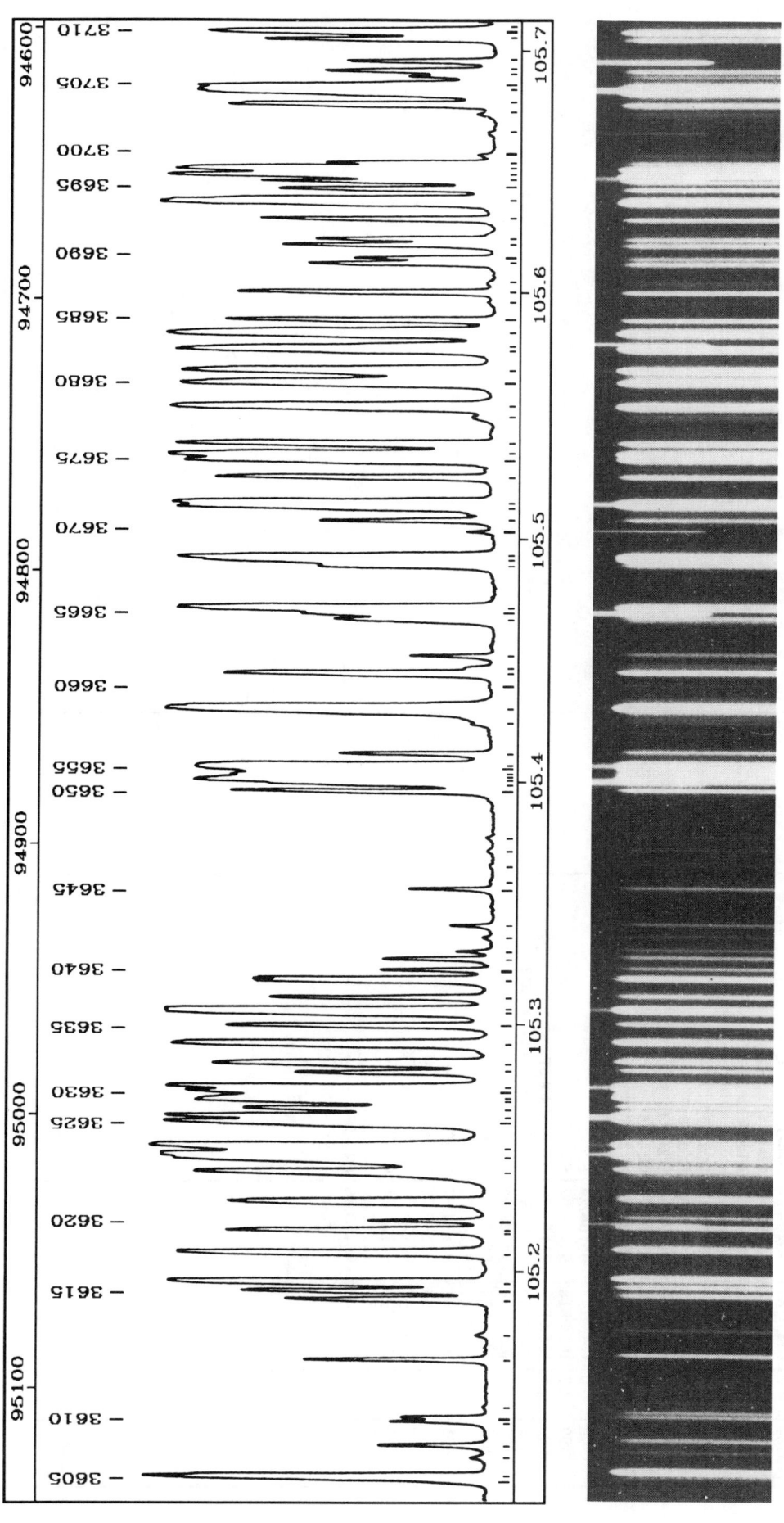

PLATE 47

TABLE 47

Nb	I	λ (nm)	σ (cm⁻¹)	Assignment	Comment
3611	0	105.1445	95107.17	B31-6R3	
3612	3	105.1640	95089.57	B20-4R2 B27-5R6	
3613	1	105.1740	95080.50	C1-1R10	
3614	3	105.1883	95067.58	C8-5Q3 D2-6Q6	
3615	4	105.1917	95064.53	C2-2R5	
3616	24	105.1953	95061.29	B16-3P3	
3617	15	105.2070	95050.73	C4-3P4	
3618	6	105.2160	95042.55	B8-1P3	
3619					CuII 105.21747
3620	2	105.2199	95039.03	B'0-4R4	
3621	5	105.2277	95032.06	B12-2R4 B31-6P3 C6-4P4	
3622	11	105.2395	95021.40	B6-0P7	
3623	225	105.2457	95015.75	C2-2P3	
3624	61	105.2497	95012.16	B5-0P5	r 95002.16
3625		105.2596	95003.18	C0-1R1	r 95002.16
3626		105.2619	95001.14	C0-1R1	
3627	7	105.2653	94998.09	B20-4P2	
3628	31	105.2687	94995.04	B'4-6P3	sh
3629		105.2696	94994.21	B'0-4P2 B22-4P6	
3630		105.2723	94991.76	C0-1R0	r 94990.93
3631		105.2741	94990.10	C0-1R0	r 94990.93
3632	4	105.2802	94984.66	B12-1P10 C1-1P8	
3633	10	105.2840	94981.21	B25-5R3 C4-3Q5	
3634	23	105.2917	94974.28	C2-2Q4	
3635	9	105.2992	94967.51	B30-6P1 C8-5P3	
3636		105.3044	94962.81	C0-1R2	r 94962.20
3637		105.3057	94961.58	C0-1R2	r 94962.20
3638	5	105.3109	94956.90	B'2-5P4	
3639	13	105.3182	94950.35	C1-1Q9	
3640	2	105.3222	94946.72	C6-4Q5	
3641		105.3266	94942.77	B4-0P2 B21-4R5	r 94941.23
3642		105.3298	94939.90	B4-0P2	r 94941.23
3643	2	105.3354	94934.86	B8-1R4 C2-0Q18	
3644	1	105.3405	94930.29	D3-9Q1	
3645	2	105.3553	94916.92	B16-3R4	
3646	0	105.3585	94914.02	B18-3P7	
3647	0	105.3648	94908.35	B14-2R8	
3648	0	105.3711	94902.65	B17-3R6 C2-1P12	
3649	0	105.3766	94897.72	C7-4Q9 D3-9Q2 ?	
3650		105.3959	94880.36	B4-0R3 B20-4R3 B15-0R17	r 94878.73
3651		105.3994			NI 105.3988 Ke
3652		105.3995	94877.10	B4-0R3 C0-1Q1 B5-0R6	r 94878.73 +r 94876.59
3653	14	105.4006	94876.09	C0-1Q1 C5-3P9	r 94876.59
3654		105.4032	94873.80	D2-6Q7 B'0-4R5 B25-5P3	
3655		105.4055	94871.72	C0-1R3	r 94871.50
3656		105.4060	94871.28	C0-1R3	r 94871.50
3657	3	105.4117	94866.16	C3-2P9	
3658	0	105.4243	94854.78	D3-9Q3 ?	d
3659	198	105.4299	94849.77	D3-7Q1	
3660	0	105.4394	94841.20	B13-2P6	
3661	10	105.4446	94836.58	B12-2P4 C4-2Q13	
3662	2	105.4473	94834.13	B31-6R4 C2-1Q13 C3-2R11	
3663	2	105.4520	94829.87	B6-0R8	
3664	5	105.4666	94816.74	B10-0R13 C2-2R6	
3665					b CuII 105.46901
3666	49	105.4712	94812.58	B30-6P2 C8-5R5 B'0-4P3	
3667	6	105.4886	94796.94	B21-4P5	
3668	26	105.4908	94794.97	B11-2R0 B26-5P5	
3669	50	105.4922	94793.79	D3-7Q2 B9-0R12	GeII 105.50261
3670					
3671	4	105.5070	94780.44	D2-6P6	
3672		105.5128	94775.22	C0-1Q2	r 94774.54
3673		105.5143	94773.86	C0-1Q2	r 94774.54
3674	11	105.5248	94764.48	B'4-6P4 B24-5R0 B11-0R14	
3675	25	105.5320	94758.03	C4-3P5 C3-2Q10	
3676	54	105.5342	94756.05	C2-2P4	
3677	17	105.5385	94752.15	B20-4P3	
3678	1	105.5496	94742.21	C5-3Q10 C8-4P11	
3679	68	105.5538	94738.39	B11-2R1 D11-14Q4	
3680	36	105.5635	94729.71	C0-1R4 B16-3P4	
3681	38	105.5687	94725.04	C2-2Q5	
3682	46	105.5775	94717.17	B15-3R1 C6-4P5 B'1-4P9	
3683					CuII 105.57968
3684	63	105.5838	94711.48	B7-0P9 D3-7Q3	
3685	7	105.5893	94706.59	B24-5R1 B25-5R4 C10-6R1	
3686	1	105.5966	94700.04	C1-0R17 C5-1P19	
3687	7	105.6008	94696.29	B'2-5P5	
3688	1	105.6043	94693.09	B15-2P9 C0-0R14	
3689	5	105.6124	94685.87	B8-1P4 C4-3Q6 B12-1R11	
3690	3	105.6146	94683.84	D11-14Q3	
3691	10	105.6201	94678.99	B17-3P6 C1-0P15	
3692	5	105.6225	94676.79	B8-0R11	
3693	6	105.6307	94669.46	D7-10Q1 B'0-4R6 D1-5Q11	
3694	48	105.6381	94662.83	B11-2P1 B12-2R5 D2-6Q8	
3695	5	105.6430	94658.47	B24-5P1	
3696		105.6460	94655.74	B4-0P3	r 94654.60
3697		105.6486	94653.44	B4-0P3 C0-1P2	r 94654.60
3698	1	105.6511	94651.17	B15-3P1	
3699	1	105.6536	94648.89	C6-4Q6 D11-14Q2	
3700	1	105.6569	94646.00	C10-6Q1 C7-4P9	Al II 105.6661 Ke
3701		105.6660			
3702	2	105.6737	94630.91	D7-10Q2	
3703	12	105.6775	94627.53	C10-6R2 D11-14Q1	
3704		105.6829	94622.66	C0-1Q3	r 94622.02
3705		105.6844	94621.38	C0-1Q3	r 94622.02
3706	4	105.6891	94617.15	B14-0P16 C1-0Q16 B'3-5P9	
3707	6	105.6914	94615.10	B12-0R15	

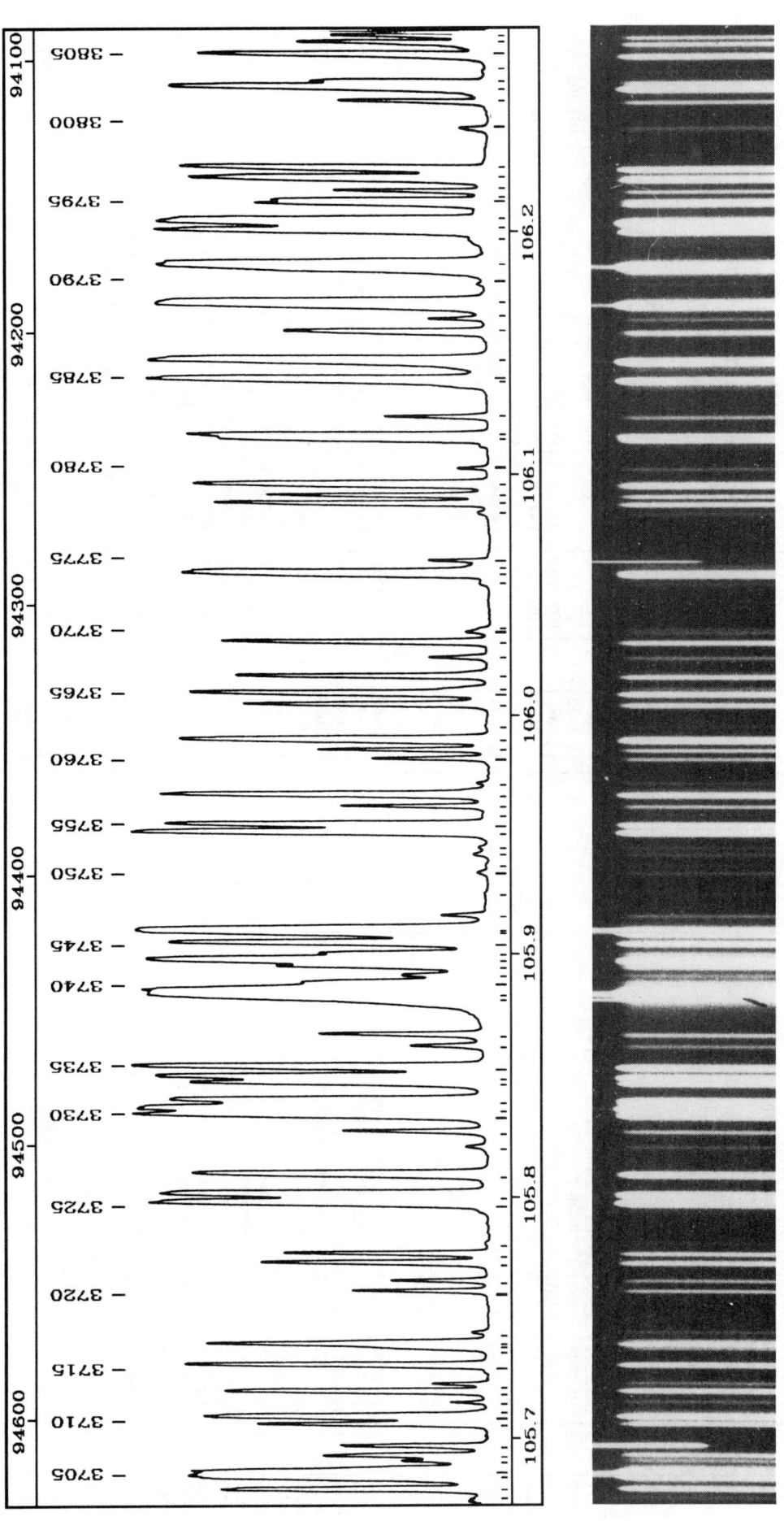

PLATE 48

TABLE 48

Nb	I	λ (nm)	σ (cm⁻¹)	Assignment	Comment
3708	7	105.7042	94603.66	D3-7Q4 C8-5Q5	CuII 105.69546
3709	17	105.7073	94600.87	B11-2R2	
3710		105.7094	94598.97	B20-4R4	sh
3711	3	105.7134	94595.41	B9-1R7 C1-1R11	
3712	11	105.7179	94591.40	B'0-4P4	
3713	2	105.7209	94588.67	B30-6P3	
3714	10	105.7285	94581.88	B15-3R2 B29-6R0	
3715	2	105.7356	94575.54	B24-5R2	
3716		105.7371	94574.19	B4-0R4 B7-1R0 B8-1R5 D7-10Q3	r 94573.71
3717		105.7382	94573.23	B4-0R4 B7-1R0	r 94573.71
3718		105.7420	94569.77	B25-5P4	
3719	0	105.7591	94554.55	B16-3R5	
3720	3	105.7634	94550.71	C0-0P12	
3721	3	105.7708	94544.09	C0-0Q13	
3722	11	105.7745	94540.79	B19-4R0	
3723	5	105.7789	94536.78	C10-6Q2 C6-3Q13	
3724	1	105.7954	94522.03	C0-1R5 B29-6R1	
3725	40	105.7989	94518.91	B7-1R1	
3726	43	105.8073	94511.48	C1-1P9 C2-2R7	
3727	10	105.8186	94501.32	D7-10Q4	
3728	2	105.8248	94495.83	B24-5P2 B31-6R5	
3729	3	105.8314	94489.93	B5-0P6 B4-6P5	
3730	18	105.8341	94487.47	C2-2P5	
3731	53	105.8373	94484.65	B19-4R1 C1-1Q10 B29-6P1 D0-5R2	
3732	19	105.8450	94477.76	B22-4P7 D0-5R1	
3733	11	105.8473	94475.77	B11-2P2	
3734	26	105.8514	94472.09	B15-3P2 D3-7Q5	
3735	24	105.8603	94464.15	B7-0R10	
3736	2	105.8649	94460.02	D0-5R3	
3737					
3738					
3739	680	105.8824	94444.40	C0-1P3 D0-5R0	CuII 105.87988
3740	4	105.8875	94439.85	C8-5P5	
3741	1	105.8899	94437.70	C6-4P6	
3742	5	105.8937	94432.30	B20-4P4	
3743	48	105.8961	94432.15	B7-1P1 C2-2Q6 D2-6Q9	
3744	4	105.8988	94429.81	B'0-4R7	
3745	19	105.9032	94425.88	B19-4P1	
3746	148	105.9083	94421.30	C0-1Q4 C4-3P6	CuII 105.90960
3747					
3748	1	105.9154	94415.02	B17-3R7 B27-5P7	
3749	0	105.9246	94406.82	D0-5R4	
3750	1	105.9329	94399.64	B'2-5P6	
3751	0	105.9361	94396.54	B13-0R16	
3752	1	105.9409	94392.29	B29-6R2	
3753	2	105.9440	94389.52		
3754	35	105.9502	94383.94	B11-2R3	
3755	18	105.9535	94381.03	B7-1R2	
3756	0	105.9579	94377.10	B31-6P5	

Nb	I	λ (nm)	σ (cm⁻¹)	Assignment	Comment
3757	4	105.9611	94374.22	B24-5R3 C10-6Q3	
3758	17	105.9659	94369.97	B15-3R3	
3759	0	105.9707	94365.68	B25-5R5	
3760	4	105.9807	94356.80	B6-0P8	
3761	3	105.9845	94353.45	B19-4R2	
3762	14	105.9887	94349.68	D0-5Q1 C4-3Q7 C3-2P10	
3763	1	105.9945	94344.50	C2-1R15	
3764	8	106.0032	94336.74	B5-0R7	
3765	20	106.0079	94332.61	C7-4R11 B'0-4P5	
3766		106.0096	94331.11	B29-6P2	sh
3767	9	106.0150	94326.27	B16-3P5 D10-13Q7	
3768	4	106.0229	94319.22	C2-0P18 D3-7Q6	
3769	8	106.0293	94313.56	C0-1R6	
3770	2	106.0335	94309.80	D8-11Q1	
3771	1	106.0346	94308.85	C6-4Q7	
3772	0	106.0540	94291.56	C10-6P3	
3773	43	106.0580	94288.00	B4-0P4 D0-5Q2	
3774	0	106.0608	94285.51	B13-2P7	CuII 106.06343
3775					
3776	1	106.0832	94265.68	B14-2R9	
3777	8	106.0871	94262.15	B24-5P3	
3778	5	106.0904	94259.24	C3-2Q11	
3779	9	106.0951	94255.06	B19-4P2 D10-13Q5 B8-1P5	
3780	0	106.1018	94249.07	B20-4R5	
3781	12	106.1138	94238.44	C5-3Q11 C5-3R12 D10-13Q1 D10-13Q3	
3782	17	106.1152	94237.25	B7-1P2 D10-13Q2	
3783	4	106.1232	94230.11	B10-0P13 C2-1Q14	
3784	0	106.1359	94218.86	B12-2R6	
3785	46	106.1384	94216.59	B15-3P3	
3786	70	106.1464	94209.54	B11-2P3	
3787	5	106.1584	94198.84	B25-5P5 D0-5Q3	
3788	0	106.1640	94193.86	B29-6R3	
3789	190	106.1701	94188.50	C0-1P4 B4-0R5	
3790	170	106.1784	94181.11		
3791	170	106.1860	94174.39	C0-1Q5 D0-5P2 B'4-0P6	d
3792	30	106.1962	94165.36	B12-1P11 C6-3P13	
3793	170	106.2004	94161.62	B7-1R3	
3794	6	106.2041	94158.31	D4-8Q1 B17-3P7	
3795	8	106.2116	94151.67	B19-4R3	
3796	5	106.2140	94149.55	B6-0R9	
3797	12	106.2174	94146.58	D3-7Q7	
3798	24	106.2226	94141.92	C2-2P6 B'3-6R0	
3799	0	106.2267	94138.31	B'3-6R1 C1-1R12	
3800	5	106.2437	94123.20	B16-3R6 B23-5R0	
3801	38	106.2545	94113.69	B8-0P11	
3802	5	106.2598	94108.93	C5-4R1 B14-0R17 D4-8Q2	
3803	3	106.2623	94106.76	B29-6P3 C8-5P6 C0-0R15	
3804	16	106.2671	94102.52	B14-1P13	
3805		106.2733	94096.99	C2-2Q7	

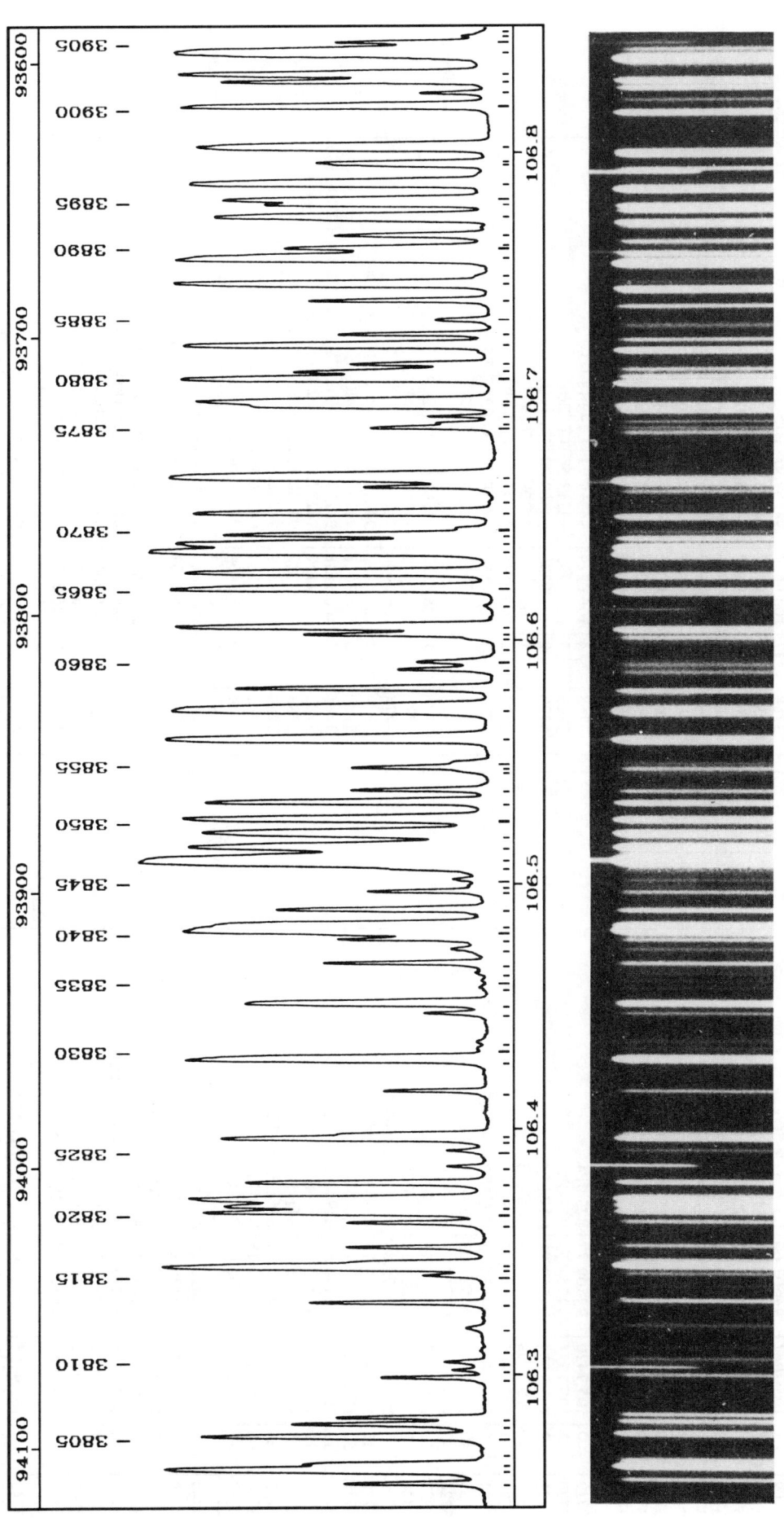

PLATE 49

TABLE 49

Nb	I	λ (nm)	σ (cm⁻¹)	Assignment	Comment
3806	6	106.2784	94092.49	B'3-6R2	
3807	3	106.2811	94090.08	B11-2R4	
3808	2	106.2973	94075.74	B23-5R1	
3809					CuII 106.30052
3810	1	106.3039	94069.92	B28-6R0 B'2-5P7	
3811	1	106.3087	94065.65	D1-5Q13	
3812	1	106.3176	94057.80	C6-4P7	
3813	4	106.3276	94048.99	B20-4P5	
3814	1	106.3335	94043.72	B21-4R7	
3815	3	106.3387	94039.19	B'0-4P6	
3816	55	106.3420	94036.24	D4-8Q3	
3817	7	106.3440	94034.47	C0-1R7 D0-5P3	
3818	3	106.3500	94029.15	C4-3P7	
3819	5	106.3601	94020.25	B13-2R8 C7-5R1	
3820	13	106.3643	94016.53	B'3-6P1	
3821	15	106.3667	94014.35	B23-5P1 B28-6R1 C5-4Q1 C1-0R18	
3822	24	106.3693	94012.11	B19-4P3 B12-0P15	
3823	9	106.3762	94006.02	B'3-6R3	
3824					CI II 106.38311
3825	3	106.3894	93994.31	C1-0P16	
3826	16	106.3943	93989.96	C1-1Q11 C7-5R2	
3827		106.3959	93988.62	C1-1P10	sh
3828	3	106.4138	93972.77	B28-6P1 C4-3Q8	
3829	40	106.4270	93961.10	B7-1P3 B24-5P4	
3830	0	106.4313	93957.30	D3-7Q8	
3831	0	106.4340	93954.92	B18-3R9 C8-5Q7	
3832	3	106.4456	93944.72	C0-0Q14	
3833	11	106.4498	93941.01	D4-8Q4 C0-0P13	
3834	1	106.4562	93935.38	D0-5Q5	
3835		106.4590	93932.87	B3-0P1	r 93931.50
3836		106.4621	93930.13	B3-0P1 B29-6R4 C6-4Q8 D2-6Q11	r 93931.50
3837	10	106.4663	93926.44	C1-0Q17	
3838	1	106.4720	93921.41	B7-0P10	
3839	2	106.4760	93917.85	B23-5R2	
3840	115	106.4798	93914.49	B1-5R1 C5-4Q2	
3841	25	106.4819	93912.66	B'1-5R0 B12-2P6	
3842	5	106.4885	93906.89	B14-3R0 C7-4Q11	
3843	2	106.4959	93900.29	C3-3R1	r 93897.29
3844		106.4978	93898.69	B3-0R2	r 93897.29
3845		106.5009	93895.88	B3-0R2	
3846	0	106.5050	93892.27	B19-4R4	
3847	300	106.5091	93888.66	C0-1P5 B15-3P4	
3848	39	106.5147	93883.78	C0-1Q6	
3849	32	106.5207	93878.45	C3-3R0 B9-0R13	
3850	31	106.5264	93873.48	C5-3P11 B'1-5R2	
3851	13	106.5334	93867.24	B11-2P4	
3852	3	106.5383	93863.00	B7-1R4	
3853	1	106.5459	93856.23	B16-3P6 B10-0R14	
3854	4	106.5476	93854.76	B23-5P2	

Nb	I	λ (nm)	σ (cm⁻¹)	Assignment	Comment
3855	2	106.5497	93852.88	B'0-4R9	
3856	84	106.5595	93844.28	B4-0P5	
3857	120	106.5714	93833.83	B14-3R1 B20-4R6	
3858	11	106.5798	93826.39	D4-8Q5 C5-4P2	
3859	2	106.5875	93819.65	B28-6P2 B'4-6P7	
3860	3	106.5905	93816.97	B8-0R12 B29-6P4 C3-2R13	
3861		106.5999	93808.73	B13-1R13	
3862	5	106.6017	93807.11	C3-2P11	sh
3863	31	106.6049	93804.33	B10-2R0	
3864					CuII 106.61343
3865	43	106.6206	93790.54	B'1-5R3	
3866	20	106.6272	93784.72	C3-3Q1	
3867	111	106.6360	93776.93	B'1-5P1	
3868	55	106.6389	93774.38	B14-3P1	
3869	7	106.6427	93771.09	B18-4R0 B24-5R5	
3870	2	106.6454	93768.71	C5-4Q3 D10-14Q4	
3871	21	106.6517	93763.20	C2-2P7	
3872	0	106.6560	93759.37	B16-1P15	
3873	4	106.6622	93753.92	B15-0R18 D3-7Q9	
3874	115	106.6665	93750.13	B10-2R1 B11-0R15	+ArI 106.66599
3875	3	106.6867	93732.39	C3-2Q12 B'3-6R5	
3876		106.6885	93730.79	B3-0P2	r 93729.56
3877		106.6913	93728.34	B3-0P2	r 93729.56
3878	7	106.6960	93724.26	B23-5R3 C2-2Q8 C8-5P7	
3879	14	106.6978	93722.68	B11-2R5 B'3-6P3	
3880	34	106.7070	93714.56	B18-4R1 B'0-4P7	
3881	5	106.7099	93712.06	C0-1R8	
3882	4	106.7129	93709.38	B12-2R7 B'3-5P11	
3883	23	106.7206	93702.63	B14-3R2 D10-14Q3	
3884	3	106.7251	93698.65	B19-4P4	
3885	1	106.7309	93693.54	D4-8Q6 B28-6R3	
3886	4	106.7389	93686.53	C3-3Q2	
3887		106.7459	93680.47	B3-0R3 D1-6R1 D1-6R2	
3888		106.7489	93677.75	B3-0R3	
3889	93	106.7563	93671.28	B10-2P1	
3890	6	106.7603	93667.82	B'1-5R4 C4-3P8	
3891		106.7612			NI 106.7616 Ke
3892	9	106.7656	93663.15	B7-0R11	
3893	23	106.7730	93656.63	B18-4P1 D10-14Q2	b
3894	7	106.7785	93651.86	C5-3R13 D1-6R3	
3895	14	106.7802	93650.31	D1-6R0	
3896	59	106.7871	93644.28	B'1-5P2	CI II 106.79442
3897					
3898	4	106.7956	93636.81	C5-4P3	
3899	27	106.8022	93631.03	D10-14Q1 B6-0P9 B23-5P3 C9-6R1 B16-3R7	
3900	25	106.8189	93616.38	B10-2R2	
3901	3	106.8241	93611.81	C1-1R13	
3902	8	106.8291	93607.47	B7-1P4	

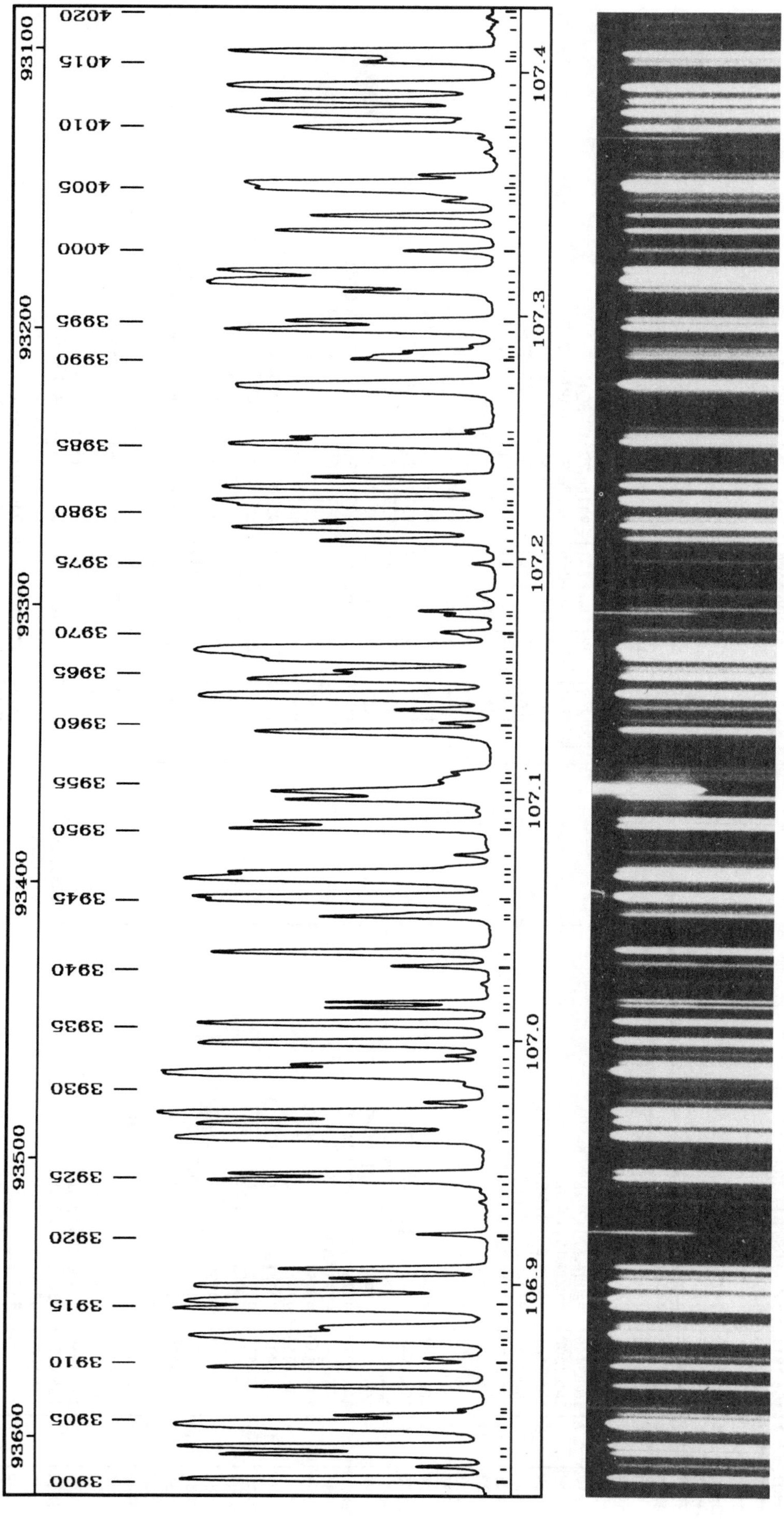

PLATE 50

TABLE 50

Nb	I	λ (nm)	σ (cm⁻¹)	Assignment	Comment
3903	19	106.8321	93604.83	B14-3P2 C2-1Q15	
3904	69	106.8412	93596.90	B28-6P3 C4-2Q15 B'5-7R0 D5-9Q1	
3905	3	106.8450	93593.51	B24-5P5	
3906		106.8477			NI 106.8477 Ke
3907	0	106.8488	93590.19	D1-6R4	
3908	7	106.8568	93583.19	B18-4R2	
3909	13	106.8650	93576.05	C3-3P2 C5-4Q4 C7-5P3	
3910	0	106.8686	93572.82	B8-1R7 B12-0R16	
3911		106.8765	93565.93	B'5-7R1	sh
3912	48	106.8779	93564.71	D1-6Q1 C9-6R2	
3913	5	106.8808	93562.17	C9-6Q1 C4-3Q9 C2-0P19	
3914	30	106.8893	93554.72	D5-9Q2	
3915	74	106.8921	93552.30	C0-1Q7	
3916	58	106.8982	93546.94	C0-1P6	
3917	4	106.9015	93544.09	C3-3Q3	
3918	6	106.9054	93540.64	C4-3Q7	
3919	0	106.9191	93528.62	B27-6R0	
3920					CuII 106.91954
3921	0	106.9273	93521.50	B17-2R13	
3922	1	106.9326	93516.86	B'3-6P4	
3923	1	106.9369	93513.10	C6-4Q9	
3924	10	106.9416	93509.00	D1-6Q2	
3925	8	106.9439	93506.94	C7-5R5 B'1-5R5	
3926	63	106.9591	93493.70	B14-3R3 B19-4R5 C1-1P11 B'5-7P1 D5-9Q3	
3927	15	106.9648	93488.68	B7-1R5 B15-3P5 C0-0R16 D1-6R5	
3928	58	106.9687	93485.30	B10-2P2 B18-4P2	
3929	1	106.9735	93481.11	B'5-7R2	
3930	2	106.9808	93474.75	B17-1P16	
3931	163	106.9859	93470.31	B27-6R1 B'1-5P3	
3932	5	106.9890	93467.53	C1-1Q12	
3933	0	106.9929	93464.19	B29-6P5	
3934	35	106.9981	93459.65	B6-1R0 B23-5R4 C9-6Q2	
3935	21	107.0062	93452.57	B11-2P5 B22-5R0	
3936		107.0126	93446.91	B3-0P3	r 93445.87
3937		107.0151	93444.84	B3-0P3 C9-6R3	r 93445.87
3938	1	107.0198	93440.70	B28-6R4	
3939	1	107.0231	93437.75	B27-6P1	
3940	2	107.0302	93431.55	D1-6Q3	
3941	18	107.0359	93426.56	D5-9Q4	
3942	7	107.0509	93413.51	B6-0R10	
3943		107.0521	93412.48	C2-0Q20	sh
3944		107.0563	93408.78	C2-0Q20	sh
3945	92	107.0584	93406.95	B6-1R1 B10-2R3	
3946	45	107.0664	93399.97	B33-7P1 C3-3P3 D1-6P2	
3947	11	107.0690	93397.75	B22-5R1	
3948	2	107.0711	93395.85		
3949	1	107.0764	93391.23	B13-2R9	
3950	1	107.0878	93381.30	B3-0R4 D4-8Q8 C1-2R1	r 93379.48
3951	0	107.0905	93378.96	B3-0R4 B18-4R3 C0-1R9	r 93379.48
3952	0	107.0953	93374.76	C7-5P4	
3953	4	107.0998	93370.83	C1-2R0	
3954					Cl II 107.10358
3955	1	107.1077	93363.94	B20-4R7 C7-4Q12	
3956	1	107.1093	93362.57	B'5-7P2	
3957	1	107.1113	93360.82	B12-2P7 B'0-4P8	
3958	9	107.1263	93347.75	C3-3Q4	sh
3959	1	107.1279	93346.38	B22-5P1 B27-6R2 C2-2P8	
3960	3	107.1317	93343.08	C1-2R2	
3961	90	107.1373	93338.19	B'5-7R3 C5-4Q5	
3962	16	107.1429	93333.33	B14-3P3	
3963	14	107.1497	93327.39	B4-0P6 D'0-8Q1	
3964	6	107.1510	93326.25	C0-0Q15	
3965	6	107.1534	93324.19	B16-3P7 C0-0P14	
3966	11	107.1584	93319.81	B19-4P5 B15-3R6 B9-1R9	
3967	106	107.1603	93318.11	D1-6Q4 B13-0R17	
3968	0	107.1623	93316.40	B6-1P1 B18-3R10 C2-2Q9 D5-9Q5	
3969	1	107.1680	93311.48	D3-7Q11	
3970	0	107.1695	93310.12	B28-6P4	
3971	2	107.1738	93306.39	C9-6Q3 B'1-5R6	
3972					Cl II 107.17667
3973	2	107.1783	93302.45	B22-4P9	
3974	2	107.1857	93296.05	B23-5P4 D'0-8Q2	d
3975	1	107.1982	93285.15	C3-2P12 C1-0R19	
3976	1	107.2037	93280.34	B11-2R6	
3977	10	107.2075	93277.08	D9-13Q7	
3978	28	107.2130	93272.32	B27-6P2 C1-0P17	
3979	5	107.2155	93270.11	B6-1R2	
3980	0	107.2189	93267.16	B22-5R2 B9-0P13	
3981	17	107.2226	93263.97	B13-1P13	
3982	33	107.2241	93262.64	D1-6P3	
3983	32	107.2296	93257.83	C1-2Q1 D"0-8Q3	
3984	4	107.2339	93254.12	C7-4R13 B'1-5P4	
3985	37	107.2477	93242.13	C1-2R3	
3986	5	107.2501	93239.99	B18-4P3	
3987	1	107.2532	93237.31	B5-0P8 C3-2R14	
3988	128	107.2718	93221.17	D"1-9Q1 B11-1P11	
3989	0	107.2788	93215.03	B10-2P3 B8-0P12	
3990	2	107.2831	93211.34	D9-13Q6	
3991	3	107.2845	93210.11	B14-3R4 B29-6R6 D"0-8Q4	
3992	2	107.2863	93208.55	C4-3P9	
3993	2	107.2888	93206.42	D4-8Q9	
3994	22	107.2955	93200.56	D5-9Q6	
3995	8	107.2988	93197.70	B8-1P7	
3996	1	107.3076	93190.07	B4-0R7	
3997	2	107.3111	93186.97	D1-6Q5	
3998	91	107.3151	93183.52	B22-5P2 D6-10Q1 C0-1Q8	

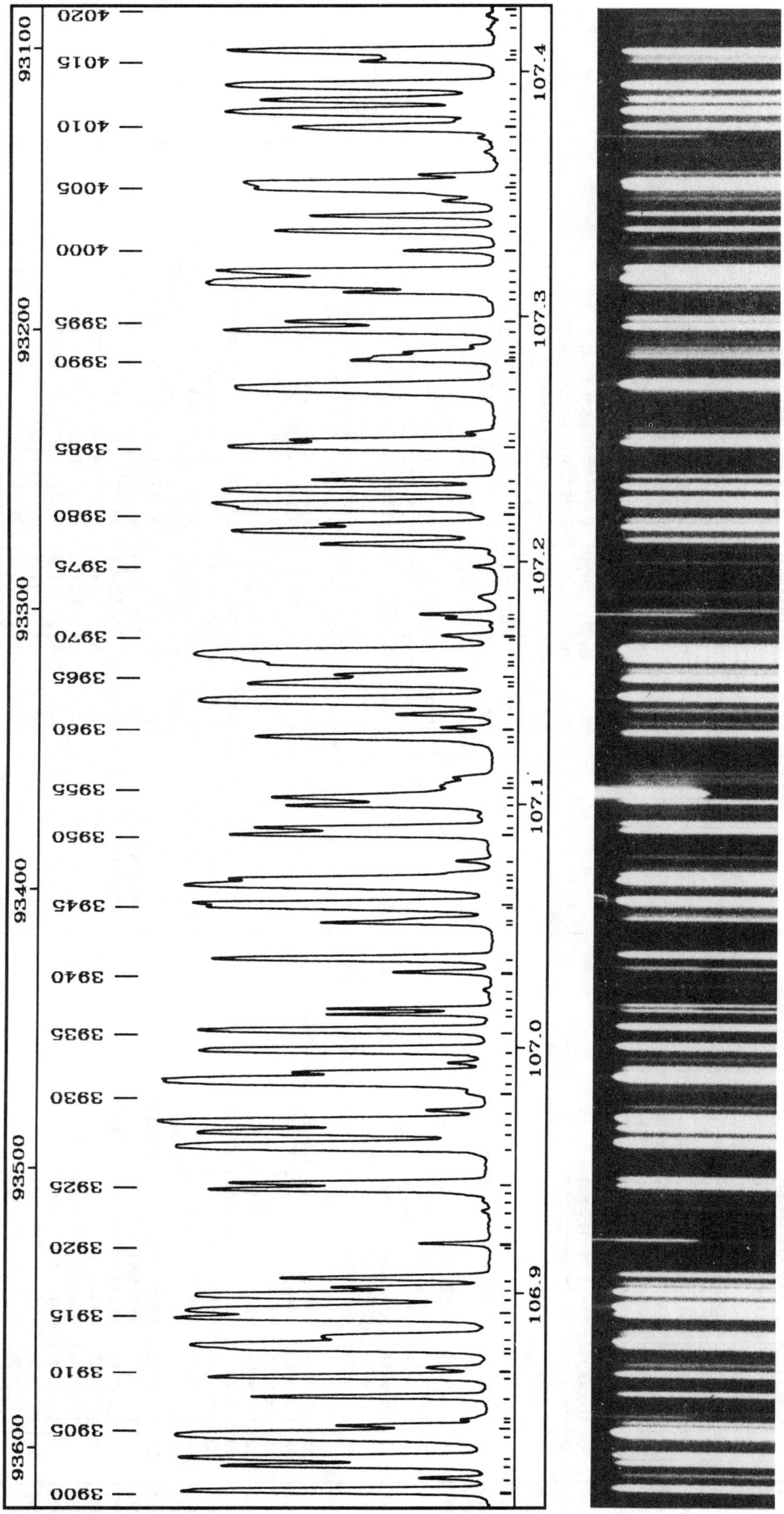

PLATE 50

TABLE 50 *continued*

Nb	I	λ (nm)	σ (cm⁻¹)	Assignment	Comment
3999	13	107.3199	93179.40	C3-2Q13 B7-1P5 B'5-7P3	
4000	3	107.3280	93172.37	C6-4P9 D9-13Q5	
4001	9	107.3360	93165.38	B24-5P6 C1-2Q2	
4002	5	107.3425	93159.79	C3-3P4	
4003	2	107.3486	93154.47	C1-2R4 D''0-8Q5	
4004	0	107.3509	93152.42	B27-6R3	
4005	23	107.3535	93150.21	D6-10Q2	
4006	42	107.3555	93148.51	C0-1P7 C5-3Q13	
4007	3	107.3592	93145.22	D9-13Q4	
4008	0	107.3687	93137.05	B12-2R8	
4009					CuII 107.37454
4010	20	107.3786	93128.40	D9-13Q3	
4011	0	107.3813	93126.08	B23-5R5	
4012	54	107.3850	93122.88	B6-1P2 C5-4P5	
4013	13	107.3898	93118.75	D9-13Q2	
4014	63	107.3955	93113.77	C3-3Q5 D9-13Q1	
4015	4	107.4055	93105.07	D1-6P4	
4016	3	107.4071	93103.76	B18-4R4	
4017	27	107.4095	93101.65	D6-10Q3	b
4018	0	107.4182	93094.12	B19-4R6 C7-5P5	
4019	0	107.4229	93090.06	B20-4P7	
4020	0	107.4249	93088.28	D''0-8Q6	

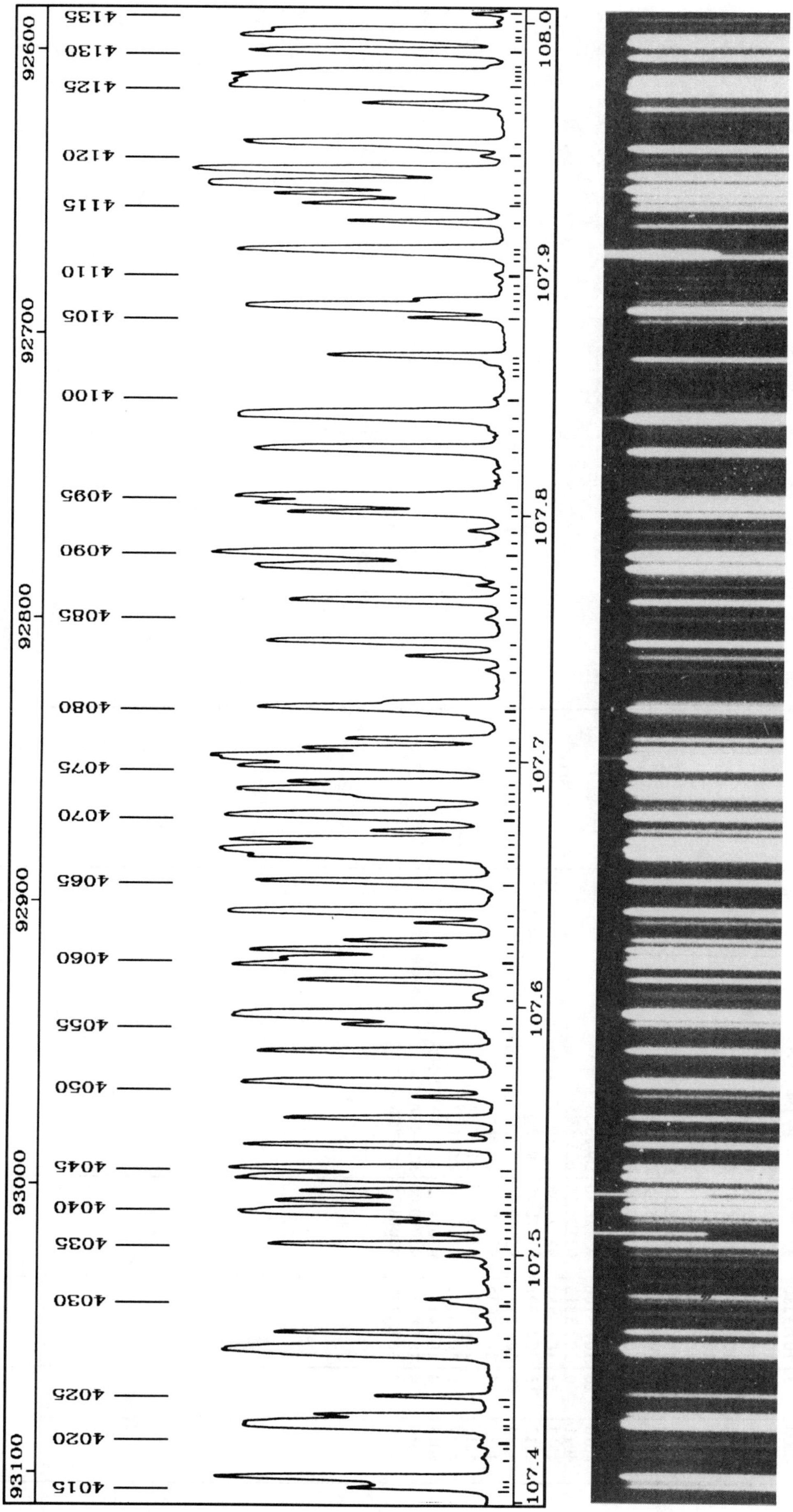
92600
92700
92800
92900
93000
93100
4015
4020
4025
4030
4035
4040
4045
4050
4055
4060
4065
4070
4075
4080
4085
4090
4095
4100
4105
4110
4115
4120
4125
4130
4135
107.4
107.5
107.6
107.7
107.8
107.9
108.0

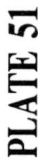

PLATE 51

TABLE 51

Nb	I	λ (nm)	σ (cm⁻¹)	Assignment	Comment
4021	7	107.4294	93084.40	B10-2R4	
4022	42	107.4314	93082.66	B3-0P4 B7-0P11 C1-1R14 D5-9Q7	
4023	5	107.4344	93080.03	B1-5R7	
4024	0	107.4394	93075.73	B32-7R1	
4025	4	107.4423	93073.20	B22-5R3	
4026	8	107.4585	93059.15	B5-0R9	
4027	55	107.4608	93057.22	B6-1R3 C5-4Q6	
4028	7	107.4674	93051.44	B29-6P6 C1-2P2	
4029	0	107.4747	93045.16	B32-7P1	
4030	1	107.4797	93040.79	B7-1R6	
4031	4	107.4816	93039.16	D6-10Q4	
4032	0	107.4880	93033.63	D1-6Q6	
4033	0	107.4947	93027.87	D4-8Q10	
4034	1	107.4986	93024.46	B15-3P6	
4035	8	107.5030	93020.68	C1-2Q3	
4036					GeII 107.50720
4037	1	107.5098	93014.79	B13-2P9	
4038	2	107.5122	93012.69	B1-5P5	
4039	43	107.5159	93009.47	B14-3P4	
4040		107.5170	93008.56	D2-7R2 C7-5R7	sh
4041	9	107.5205	93005.51		ClII 107.52293
4042					
4043	10	107.5243	93002.26	B3-0R5	
4044	60	107.5300	92997.33	D2-7R1	
4045	18	107.5336	92994.21	B13-3R0	
4046	10	107.5428	92986.22	B8-1R8 D2-7R3	
4047	1	107.5475	92982.15	B0-4P9	
4048	7	107.5537	92976.81	B17-4R0	
4049	2	107.5625	92969.20	B11-2P6	
4050		107.5664	92965.87	C0-1R10	sh
4051	26	107.5686	92963.96	B28-6P5 D6-10Q5 D2-7R0	
4052	0	107.5777	92956.09	C1-2R5 C3-3R7	
4053	12	107.5810	92953.17	B22-5P3	
4054	0	107.5872	92947.85	C4-2Q16	
4055	4	107.5918	92943.90	B23-5P5	
4056	122	107.5957	92940.54	D7-11Q1 B13-3R1	
4057	3	107.6037	92933.57	D2-7R4 C1-1P12	
4058	8	107.6101	92928.10	B18-4P4 B9-0R14	
4059	26	107.6166	92922.47	B17-4R1 D1-6P5	
4060	14	107.6189	92920.48	C1-1Q13	
4061	19	107.6223	92917.56	D7-11Q2	
4062	7	107.6262	92914.22	B8-0R13 B21-4R9	
4063	3	107.6336	92907.79	B26-6R1	
4064	72	107.6381	92903.94	D8-12Q1	
4065	18	107.6506	92893.10	D8-12Q2 B19-4P6 B27-6R4	
4066	28	107.6609	92884.20	D7-11Q3 C2-2P9	
4067	101	107.6635	92881.98	D2-7Q1 B10-2P4	
4068	28	107.6674	92878.64	D8-12Q3 D6-10Q6 C3-3P5	
4069	3	107.6715	92875.05	C2-2Q10	
4070	55	107.6779	92869.59	B13-3P1 B24-5R7	
4071	1	107.6811	92866.82	B20-4R8	
4072	2	107.6855	92863.00	B26-6P1 C11-7R0	
4073	28	107.6885	92860.45	B17-4P1 B10-0R15 C0-0R17 D8-12Q4	
4074	9	107.6915	92857.84	B14-3R5	
4075	27	107.6979	92852.31	C1-2P3 B6-0P10	
4076	141	107.7013	92849.35	B6-1P3	
4077	6	107.7052	92846.06	D8-12Q5	
4078	5	107.7092	92842.62	D7-11Q4 D4-8Q11	
4079	1	107.7177	92835.30	C11-7R1 C3-3Q6	
4080	24	107.7215	92831.98	D2-7Q2 D8-12Q6	
4081	2	107.7238	92829.97	C1-2Q4 D8-12Q7	
4082	1	107.7372	92818.49	B1-5R8	
4083	3	107.7427	92813.74	B7-0R12 D5-9Q9	
4084	17	107.7483	92808.88	B13-3R2 B22-5R4	
4085	0	107.7583	92800.26	B14-2R11	
4086	9	107.7651	92794.46	B17-4R2 C7-4Q13 D7-11Q5	
4087		107.7680	92791.92	B2-0R1	r 92790.49
4088		107.7713	92789.05	B2-0R1	r 92790.49
4089	54	107.7786	92782.78	B9-2R0 B11-2R7 B15-3R7 C11-7Q1	
4090	48	107.7840	92778.13	C0-1Q9	
4091	1	107.7890	92773.88	C5-3P13	
4092	1	107.7940	92769.53	B29-6R7 B27-6P4	
4093	6	107.8008	92763.68	B6-1R4	
4094	22	107.8044	92760.58	B18-4R5 B21-5R0 B32-7R3 C0-1P8	
4095	31	107.8073	92758.12	D2-7Q3	
4096		107.8187	92748.28	B12-2P8	
4097	28	107.8267	92741.44	B4-0P7 D7-11Q6 B10-2R5	
4098	0	107.8356	92733.72	B16-3P8	
4099	138	107.8404	92729.61	B9-2R1 D2-7P2 B1-5P6	
4100	0	107.8471	92723.86	C4-3P10	
4101	0	107.8575	92714.92	B11-0R16	
4102	0	107.8596	92713.09	D1-6P6	
4103	0	107.8621	92710.94	B26-6P2	
4104	5	107.8647	92708.70	B21-5R1	
4105	1	107.8802	92695.42	C3-2P13	
4106	38	107.8846	92691.66	B13-3P2 B17-4P2 C3-2R15 D6-10Q8 C0-0Q16	
4107	1	107.8877	92688.95	B32-7P3 D7-11Q7	
4108		107.8906	92686.50	B2-0P1	r 92685.01
4109		107.8941	92683.51	B24-5P7	r 92685.01
4110	0	107.8979	92680.17	B7-1P6	
4111	8	107.9068	92672.57	C0-0P15 D5-9Q10	
4112					ClII 107.90796
4113	0	107.9089	92670.76	B31-7R1 D9-14Q4	
4114	1	107.9194	92661.75	D2-7Q4	
4115	2	107.9264	92655.71	D4-8Q12	
4116	1	107.9282	92654.17	B22-5P4	

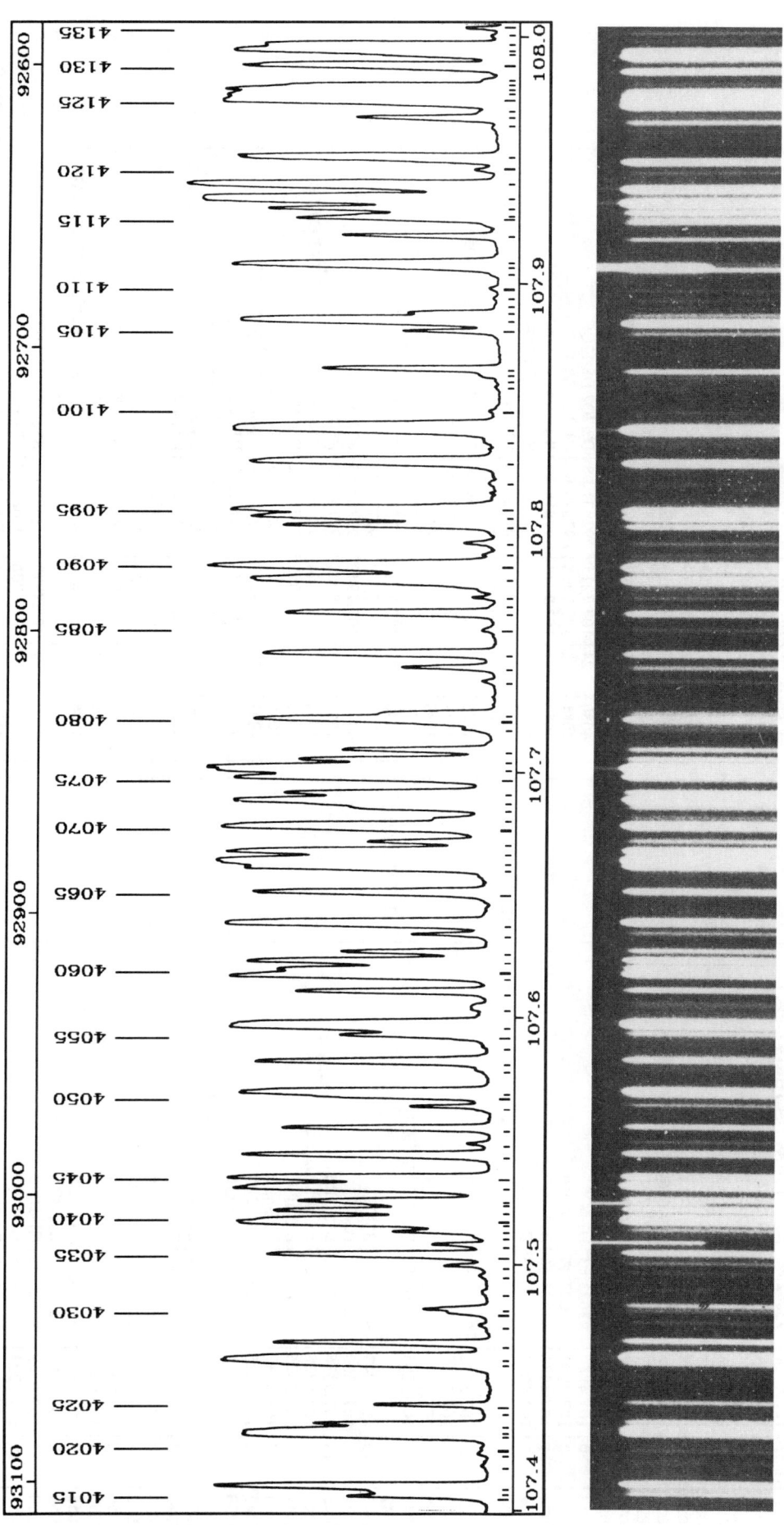

PLATE 51

TABLE 51 *continued*

Nb	I	λ (nm)	σ (cm⁻¹)	Assignment	Comment
4117	4	107.9303	92652.36	B21-5P1	
4118	118	107.9342	92649.03	B9-2P1	
4119	54	107.9399	92644.14	B3-0P5 B'4-7R0 C4-3Q11	
4120	1	107.9467	92638.28	B31-7P1 D7-11Q8	
4121	26	107.9511	92634.51	B'4-7R1 C2-0Q21	
4122	0	107.9636	92623.80	B19-4R7	
4123	2	107.9673	92620.65	B6-0R11	
4124	0	107.9697	92618.54	B15-0R19 C11-7R3	
4125	68	107.9743	92614.65	B'2-6R1	
4126	31	107.9766	92612.66	B'2-6R0 B14-3P5	
4127	19	107.9784	92611.09	D2-7P3	
4128	2	107.9800	92609.74	C1-2P4	
4129	0	107.9827	92607.43	C3-2Q14	
4130	13	107.9883	92602.61	B13-3R3 B18-3R11	sh
4131	0	107.9933	92598.36	B4-0R8	
4132	32	107.9946	92597.21	B9-2R2 C1-2Q5 D6-10Q9	
4133	9	107.9969	92595.26	B17-4R3 B26-6R3 D7-11Q9	
4134	0	108.0007	92591.98	C1-0R20 D9-14Q3	
4135	0	108.0041	92589.07	B21-5R2	

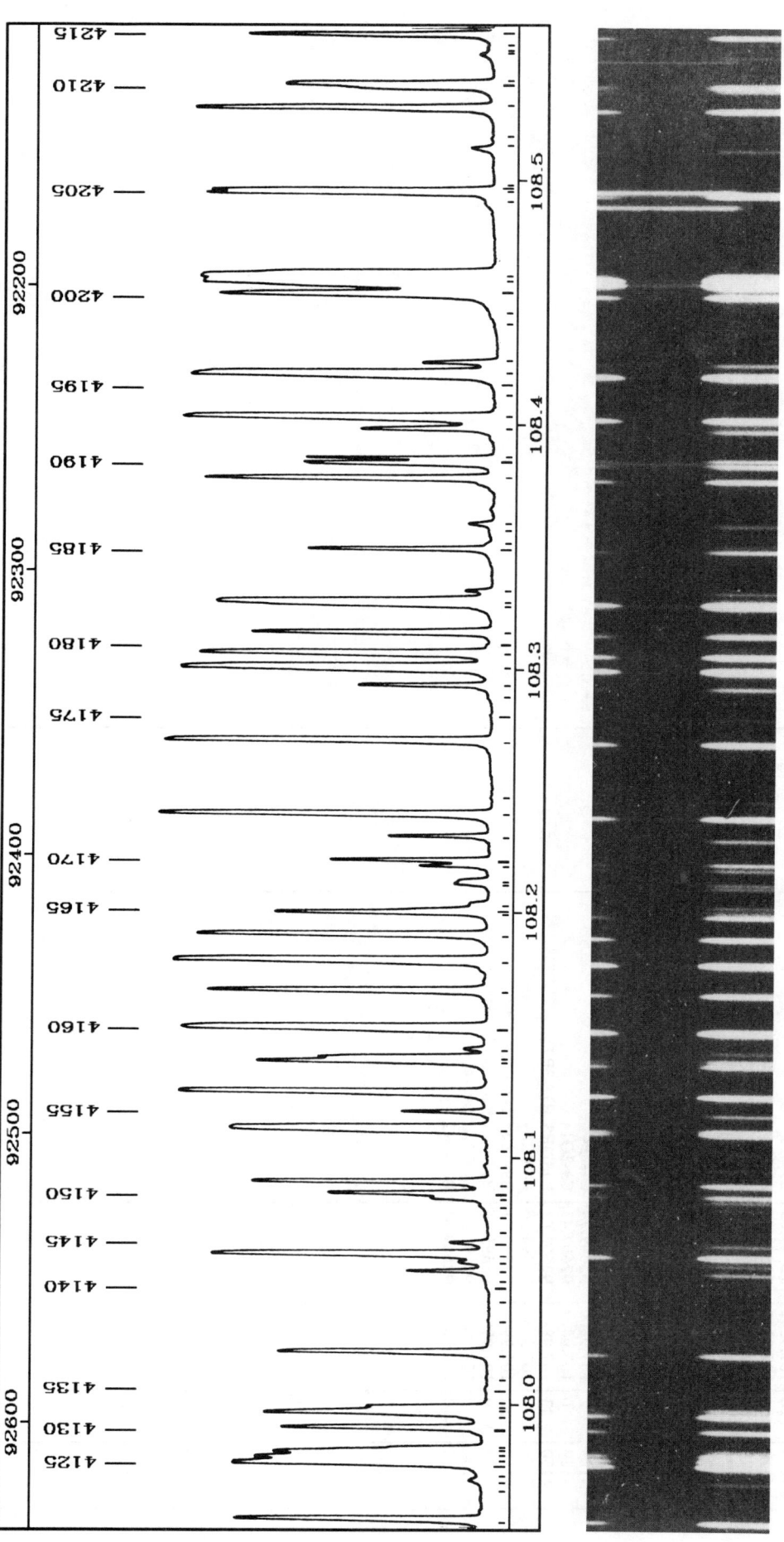

PLATE 52

TABLE 52

Nb	I	λ (nm)	σ (cm⁻¹)	Assignment	Comment
4136	0	108.0091	92584.77	B29-6P7 C6-4Q11	
4137	26	108.0193	92576.04	B'2-6R2 B4-7R2	
4138	1	108.0322	92565.01	C1-0P18 D7-11Q10	
4139	0	108.0409	92557.58	C3-3P6	
4140	0	108.0462	92553.02	D7-11Q11	
4141	0	108.0494	92550.26	B3-0R6	
4142	5	108.0526	92547.52	B18-4P5	
4143	2	108.0561	92544.50	D2-7Q5	
4144	41	108.0595	92541.65	B'4-7P1	
4145	5	108.0641	92537.66	C0-1R11 D9-14Q2	
4146	0	108.0693	92533.24	D5-9Q11	
4147	0	108.0746	92528.66	B'1-5R9	
4148	0	108.0789	92524.98	B7-1R7	
4149	4	108.0822	92522.16	B5-0P9	
4150	7	108.0841	92520.57	B23-5P6 C6-5R0 C11-7Q3	
4151	14	108.0887	92516.61	C6-5R1 C3-3Q7	
4152	1	108.0950	92511.25	C1-0Q19	
4153	0	108.1016	92505.57	D9-14Q1	
4154	26	108.1100	92498.37	B6-1P4 B12-0R17 B'2-6R3	
4155	2	108.1170	92492.36	B21-5P2 B26-6P3	
4156	67	108.1249	92485.62	B'2-6P1	
4157	20	108.1372	92475.09	B10-2P5	
4158	8	108.1390	92473.58	B'4-7R3 D4-8Q13 D1-6P7	
4159	1	108.1425	92470.63	D2-7P4	
4160	72	108.1510	92463.34	B9-2P2 C6-5R2	
4161	24	108.1665	92450.09	B17-4P3	
4162	72	108.1790	92439.33	B13-3P3	
4163	32	108.1898	92430.12	C6-5Q1	
4164	13	108.1991	92422.22	B27-6P5 B12-1P13	
4165		108.2012	92420.42	B11-2P7 B21-5R3 B17-1P17	sh
4166	4	108.2029	92419.00	B'1-5P7	
4167	2	108.2111	92411.99	C8-6R0	
4168	2	108.2124	92410.88	B'4-7P2	
4169	5	108.2184	92405.72	C2-2Q11	
4170	0	108.2208	92403.71	C8-6R1	
4171	7	108.2310	92395.01	B6-1R5	
4172	41	108.2401	92387.23	B9-2R3	
4173	2	108.2469	92381.43	B'2-6R4	
4174	32	108.2703	92361.40	B'2-6P2	
4175	1	108.2806	92352.67	C1-1Q14	
4176	8	108.2875	92346.80	C8-6R2	
4177	0	108.2931	92342.02	C0-1Q10	
4178	65	108.3004	92335.80	B19-4P7 C4-4R1 C6-5Q2	
4179	34	108.3064	92330.68	B5-0R10 C4-4R0 C1-1P13	
4180	0	108.3093	92328.22	B'4-7R4 B17-4R4	
4181	16	108.3147	92323.55	C8-6Q1 B13-3R4 C1-2P5	
4182		108.3266	92313.46	B5-1R0	sh
4183	115	108.3277	92312.52	D3-8Q1 B10-2R6 C0-1P9	
4184	2	108.3324	92308.49	D2-7P5	
4185	8	108.3496	92293.88	C4-4R2	
4186	0	108.3516	92292.16	B22-5P5	
4187	0	108.3561	92288.30		
4188	5	108.3600	92284.95	B8-0P13 C4-3P11 C4-2Q17	
4189	17	108.3785	92269.26	D3-8Q2 B25-6P1	
4190	0	108.3848	92263.89	B5-1R1 B21-5P3	r 92263.07
4191	0	108.3867	92262.25	B5-1R1	r 92263.07
4192	20	108.3988	92251.98	C6-5R4 C6-5P2	
4193	43	108.4033	92248.12	B'4-7P3	
4194	0	108.4108	92241.74	C8-6R3	
4195	0	108.4152	92238.00	B11-2R8	
4196	98	108.4211	92232.96	C4-4Q1 C6-4P11	
4197	6	108.4258	92228.99	C8-6Q2 B2-6R5	
4198	0	108.4398	92217.04	B10-0P15	
4199	0	108.4451	92212.53	B'1-5R10	
4200	28	108.4539	92205.09	D3-8Q3 C4-4R3	
4201	176	108.4591	92200.63	B9-2P3 C0-0R18	
4202	107	108.4609	92199.14	C3-3P7 B'2-6P3	
4203		108.4912			HeII 108.4913 Ke P
4204		108.4954	92169.85	B5-1P1	r 92169.29
4205		108.4968	92168.61	B5-1P1	r 92169.29
4206		108.4977			HeII 108.4975 Ke P
4207	4	108.5141	92153.91	B2-0R4 B16-4R0	
4208	0	108.5176	92150.98	B14-3P6	
4209	30	108.5305	92140.04	B17-4P4 C4-4Q2 C1-2R8 B'4-7R5	
4210	7	108.5388	92132.93	B3-0P6	
4211	37	108.5402	92131.77	B5-1R2	
4212		108.5509			GeII 108.5513 Ke
4213	14	108.5522	92121.56	D3-8Q4	
4214	0	108.5549	92119.27	D2-7P6	
4215	12	108.5603	92114.68	B13-3P4 B25-6P2 B7-1P7	

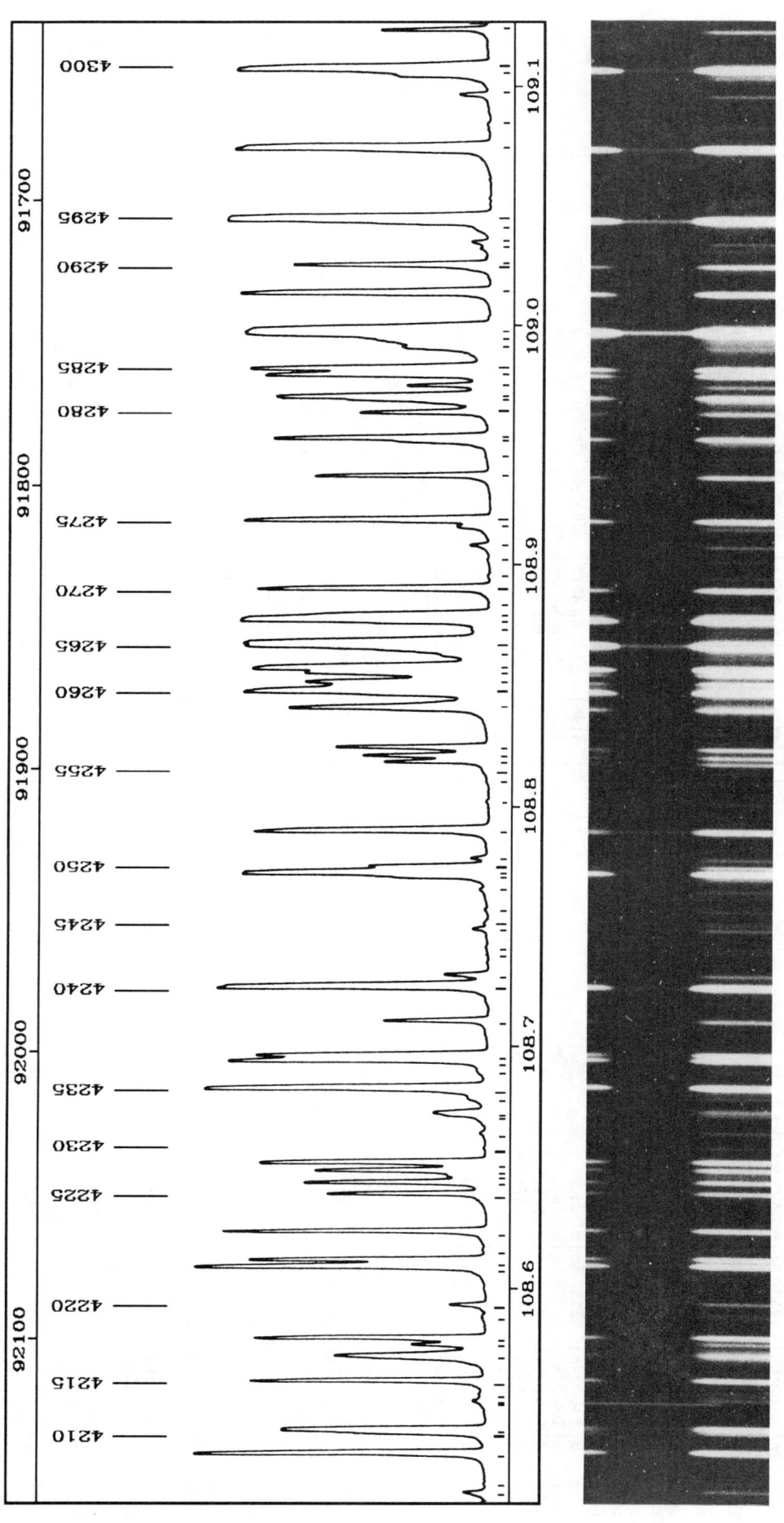

PLATE 53

TABLE 53

Nb	I	λ (nm)	σ (cm⁻¹)	Assignment	Comment
4216	94	108.5705	92106.09	B21-5R4	+NII 108.5701 Ke
4217	6	108.5754	92101.89	B9-2R4	
4218	13	108.5780	92099.65	B16-4R1	
4219	2	108.5866	92092.36	B4-0P8 B16-3P9 C0-1R12	
4220	5	108.5920	92087.82	C8-6Q3 C8-6R4	
4221	36	108.6075	92074.68	B6-1P5	
4222	19	108.6103	92072.31	C3-2R16 C6-5P3	
4223	0	108.6141	92069.12	C4-4R4	
4224	11	108.6221	92062.25	B12-3R0	
4225	7	108.6376	92049.18	B'0-5R1	
4226	7	108.6422	92045.26	C4-4P2	
4227	2	108.6446	92043.23	C0-0Q17 B'2-6R6	
4228	8	108.6473	92040.98	B20-5R0 B4-7P4	
4229	16	108.6507	92038.11	B16-4P1 B23-5P7 B'0-5R0	
4230	1	108.6564	92033.21	C1-2P6	
4231	1	108.6628	92027.83	B3-0R7	
4232		108.6708	92021.03	B'0-5R2	sh
4233	9	108.6718	92020.23	D3-8Q5 B6-0P11	
4234	4	108.6773	92015.55	C3-2Q15 C6-5Q4	
4235	60	108.6818	92011.69	B12-3R1 B15-0P19	
4236	0	108.6888	92005.79	B17-4R5 C1-2Q7	
4237	24	108.6933	92002.00	C4-4Q3	
4238	19	108.6954	92000.23	B10-1R12 B'2-6P4	
4239	4	108.7093	91988.45	B20-5R1 C8-6P3	
4240	73	108.7235	91976.45	B5-1P2 B13-3R5	
4241	2	108.7283	91972.36	B16-4R2 B21-5P4 B8-0R14	
4242	1	108.7380	91964.18	B13-2R11	
4243	0	108.7394	91962.96	D8-13Q5 B10-2P6	
4244	2	108.7474	91956.20	B'0-5R3	
4245	1	108.7498	91954.23	B6-1R6 B14-3R7	
4246	0	108.7553	91949.58	C10-7R0	
4247	1	108.7637	91942.42	B9-0R15	
4248	2	108.7691	91937.89	B4-0R9	
4249	37	108.7712	91936.13	B12-3P1 C2-2P11	
4250	5	108.7735	91934.15	B20-5P1 C1-1R16	
4251	1	108.7769	91931.26	B15-3P8 C10-7R1	
4252	50	108.7890	91921.03	B5-1R3 D8-13Q4 B7-0R13	
4253	1	108.8006	91911.25	C2-2Q12	
4254	0	108.8093	91903.92	D3-8Q6	
4255	0	108.8136	91900.27		
4256	5	108.8177	91896.86	B'0-5P1	b
4257	5	108.8201	91894.79	D8-13Q3 B25-6P3	
4258	6	108.8236	91891.82	B12-3R2	
4259	26	108.8404	91877.61	C0-1Q11	
4260	72	108.8475	91871.64	C10-7Q1 C1-0R21 D4-9Q1	
4261	7	108.8512	91868.57	B16-4P2	
4262	8	108.8551	91865.22	D8-13Q1 B20-5R2 B26-6P5	
4263	38	108.8572	91863.47	B9-2P4 C4-4P3 C10-7R2	
4264	5	108.8621	91859.30	B12-0P17	d
4265	194	108.8671	91855.13	C2-3R1 B'0-5R4	
4266	0	108.8726	91850.50	D4-8Q17	
4267	112	108.8775	91846.36	C2-3R0 C6-5P4	
4268		108.8791	91844.97	B2-0P4 B18-3R12	sh
4269	0	108.8830	91841.68	B30-5P13	
4270	24	108.8903	91835.55	B10-0R16 C0-1P10 C1-2R9 D4-9Q2	
4271	2	108.8951	91831.46	C1-0P19	
4272	0	108.9020	91825.67	C6-5R6 B12-2R10 B'2-6R7	
4273	4	108.9082	91820.47	C4-4Q4	
4274	4	108.9158	91814.01	B10-2R7 B11-2P8	
4275	25	108.9189	91811.40	C2-3R2	
4276	9	108.9374	91795.83	B4-7P5	
4277	0	108.9453	91789.22	C6-5Q5	
4278	4	108.9517	91783.77	B2-0R5	
4279	24	108.9534	91782.35	D4-9Q3 B6-0R12	
4280	4	108.9641	91773.34	B16-4R3 B20-5P2 C10-7Q2 D3-8Q7	
4281	2	108.9690	91769.21	B'0-5P2	
4282	25	108.9707	91767.73	B2-6P5 C1-1Q15	
4283	4	108.9754	91763.83	B17-4P5	
4284	16	108.9801	91759.87	B12-3P2	
4285	30	108.9827	91757.68	C2-3R3	
4286	1	108.9916	91750.19	B5-0P10 B24-6R0	
4287	4	108.9946	91747.63		d
4288	348	108.9984	91744.49	C2-3Q1 B9-2R5 C10-7R3	
4289	31	109.0144	91731.02	B8-2R0	d
4290	1	109.0237	91723.19	B1-5P9	
4291	14	109.0258	91721.45	B13-3P5 C4-3P12	
4292	0	109.0323	91715.95	C10-7P2 B'0-5R5	
4293	1	109.0351	91713.55	D4-9Q4	
4294	0	109.0399	91709.56	B29-7P1	
4295	184	109.0450	91705.29	B5-1P3 C8-6R6	
4296	121	109.0747	91680.27	B8-2R1	
4297	1	109.0838	91672.65	B20-5R3 C8-6Q5	
4298	4	109.0967	91661.83	B12-3R3	
4299	4	109.1048	91654.99	C1-2P7 B11-0R17	
4300	98	109.1079	91652.43	C2-3Q2	
4301	4	109.1238	91639.01	C4-4P4	

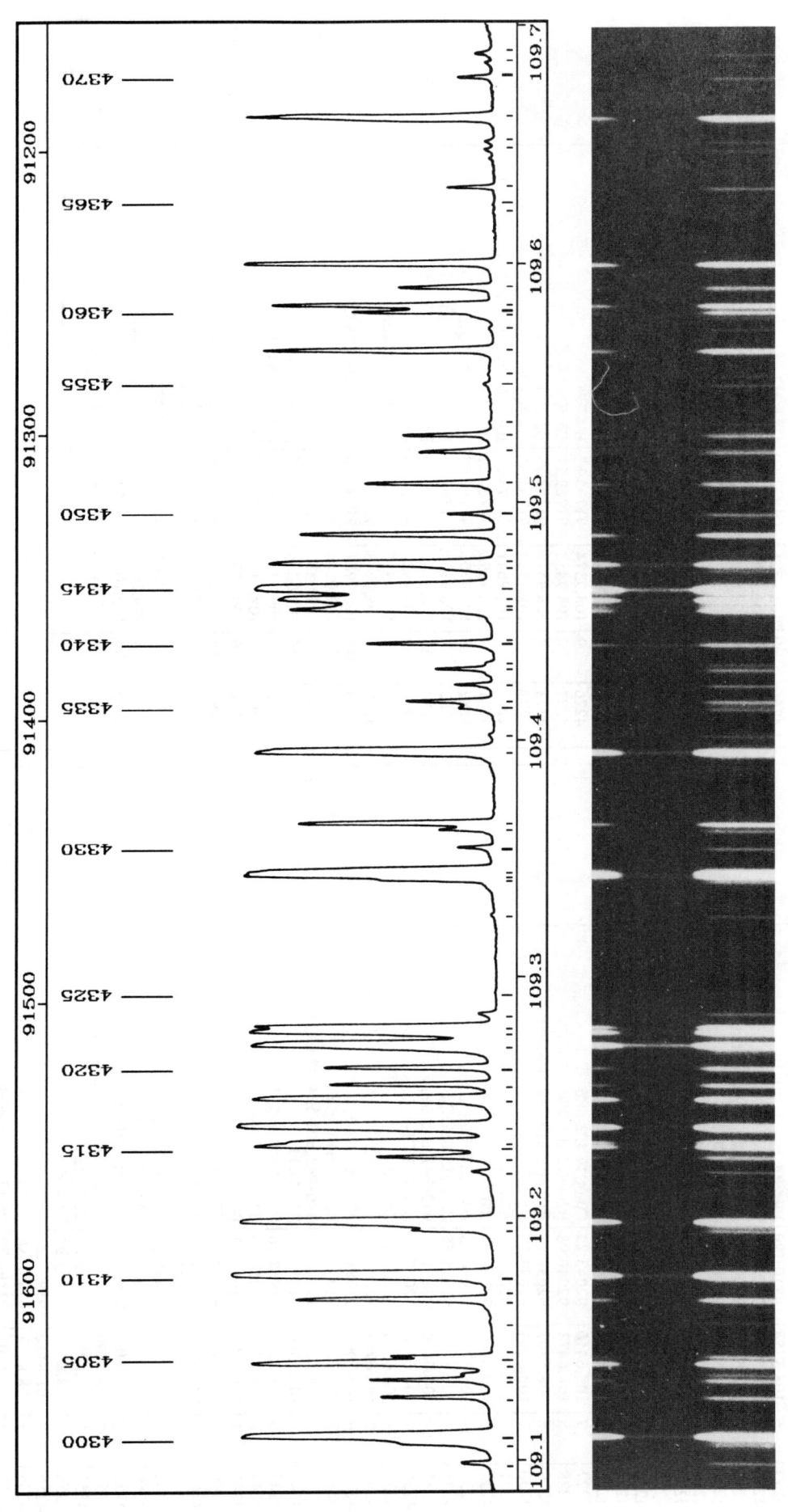

PLATE 54

TABLE 54

Nb	I	λ (nm)	σ (cm⁻¹)	Assignment	Comment
4302	7	109.1309	91633.06	B5-1R4 D3-8Q8	
4303	4	109.1332	91631.13	D4-9Q5	
4304	40	109.1379	91627.19	B16-4P3 B14-3P7 C10-7Q3	
4305	2	109.1407	91624.87	C2-3R4	
4306	0	109.1437	91622.32	C4-3Q13	
4307	0	109.1547	91613.10	B25-6P4	
4308	16	109.1644	91604.94	C0-1R13 B'0-5P3	
4309	0	109.1679	91602.04	B18-4P7	
4310	112	109.1746	91596.36	B8-2P1 C4-4Q5	
4311	4	109.1933	91580.69	B6-1P6 B11-2R9	
4312	76	109.1972	91577.40	B13-0P18 D5-10Q1	
4313	2	109.2180	91560.00	C6-5P5	
4314	7	109.2241	91554.86	B3-0P7	
4315	28	109.2286	91551.15	B8-2R2	
4316	25	109.2301	91549.83	D5-10Q2 B5-0R11	
4317	74	109.2369	91544.16	C2-3P2 C10-7P3 B20-5P3	
4318	40	109.2484	91534.56	D7-12Q1 D7-12Q7	
4319	10	109.2543	91529.57	D7-12Q2	
4320	8	109.2615	91523.59	C6-5Q6 D7-12Q3	
4321	130	109.2711	91515.54	C2-3Q3 B4-7P6 D7-12Q5	
4322	37	109.2763	91511.16	B12-3P3 C8-6P5	
4323	23	109.2784	91509.43	D5-10Q3	
4324	2	109.2839	91504.78	B16-4R4 B'2-6P6	
4325	1	109.2925	91497.59	C3-2P15	
4326		109.3254	91470.07	B15-0R20	sh
4327		109.3401	91457.76	D5-10Q4	
4328	32	109.3427	91455.61	B9-2P5	
4329	32	109.3437	91454.71	D6-11Q1 C2-3R5	
4330	3	109.3540	91446.10	B6-1R7	
4331	4	109.3619	91439.56	B3-0R8 B10-2P7	
4332	11	109.3647	91437.18	D6-11Q2 C10-7Q4	
4333	97	109.3951	91411.81	B2-0P5 B8-2P2 B14-3R8 D6-11Q3 C10-7R5	
4334	4	109.4017	91406.25	B'0-5P4	
4335	2	109.4131	91396.75	C2-2P12 D5-10Q5	
4336	5	109.4162	91394.17	B12-3R4 B24-6R3 C2-2Q13	
4337	7	109.4229	91388.58	C0-1Q12	
4338	4	109.4296	91382.97	B4-0P9	
4339	0	109.4317	91381.23	D6-11Q4	
4340		109.4403	91374.04	C4-4P5	CuII 109.44025
4341	5	109.4547	91361.98	B12-2P10 B'3-7R1	
4342	11	109.4565	91360.51	B29-7P3 C0-1P11	
4343	8	109.4592	91358.24	B5-1P4 B'3-7R0 D8-14Q3	
4344	38	109.4636	91354.57	C2-3P3	
4345	232	109.4716	91347.90	C0-0P17	
4346	4	109.4743	91345.67	B8-2R3 D6-11Q5	
4347	37	109.4790	91341.74	B2-0R6	
4348	0				
4349	14	109.4866	91335.41	C2-3Q4 C4-4Q6	

Nb	I	λ (nm)	σ (cm⁻¹)	Assignment	Comment
4350	6	109.4952	91328.20	B17-4P6 D5-10Q6 C10-7P4 C1-1R17 B'3-7R2	
4351	6	109.5082	91317.36	B9-2R6 B16-4P4	
4352	2	109.5216	91306.23	B7-1R9 B15-4R0	
4353	6	109.5285	91300.44	B19-5R0 B9-1P11	
4354	0	109.5334	91296.38	D8-14Q2	
4355	1	109.5497	91282.77	B24-6P3	
4356	2	109.5537	91279.44	B9-0P15	
4357	13	109.5640	91270.83	B5-1R5 B7-0P13 B25-6P5	
4358	0	109.5730	91263.37	B13-3P6	
4359		109.5782	91259.02	B'3-7R3 C1-2P8	sh
4360	5	109.5802	91257.35	D8-14Q1	
4361	7	109.5830	91255.02	B15-4R1 B10-2R8 D5-10Q7	
4362	2	109.5905	91248.80	B19-5R1 B20-5P4	
4363	12	109.6005	91240.50	C2-3R6 B'3-7P1	
4364	0	109.6219	91222.67	C6-5Q7	
4365	0	109.6249	91220.12	B4-0R10	
4366	2	109.6325	91213.83	B2-6P7	
4367	0	109.6485	91200.52	C10-7Q5 B'4-7P7	
4368	0	109.6519	91197.66	B12-3P4	
4369	6	109.6624	91189.00	B15-4P1 B19-5P1	
4370	1	109.6789	91175.26	B'0-5P5	
4371	1	109.6847	91170.38	B16-4R5 B10-0P16	
4372	1	109.6888	91167.05	B21-5P6 C1-1Q16	

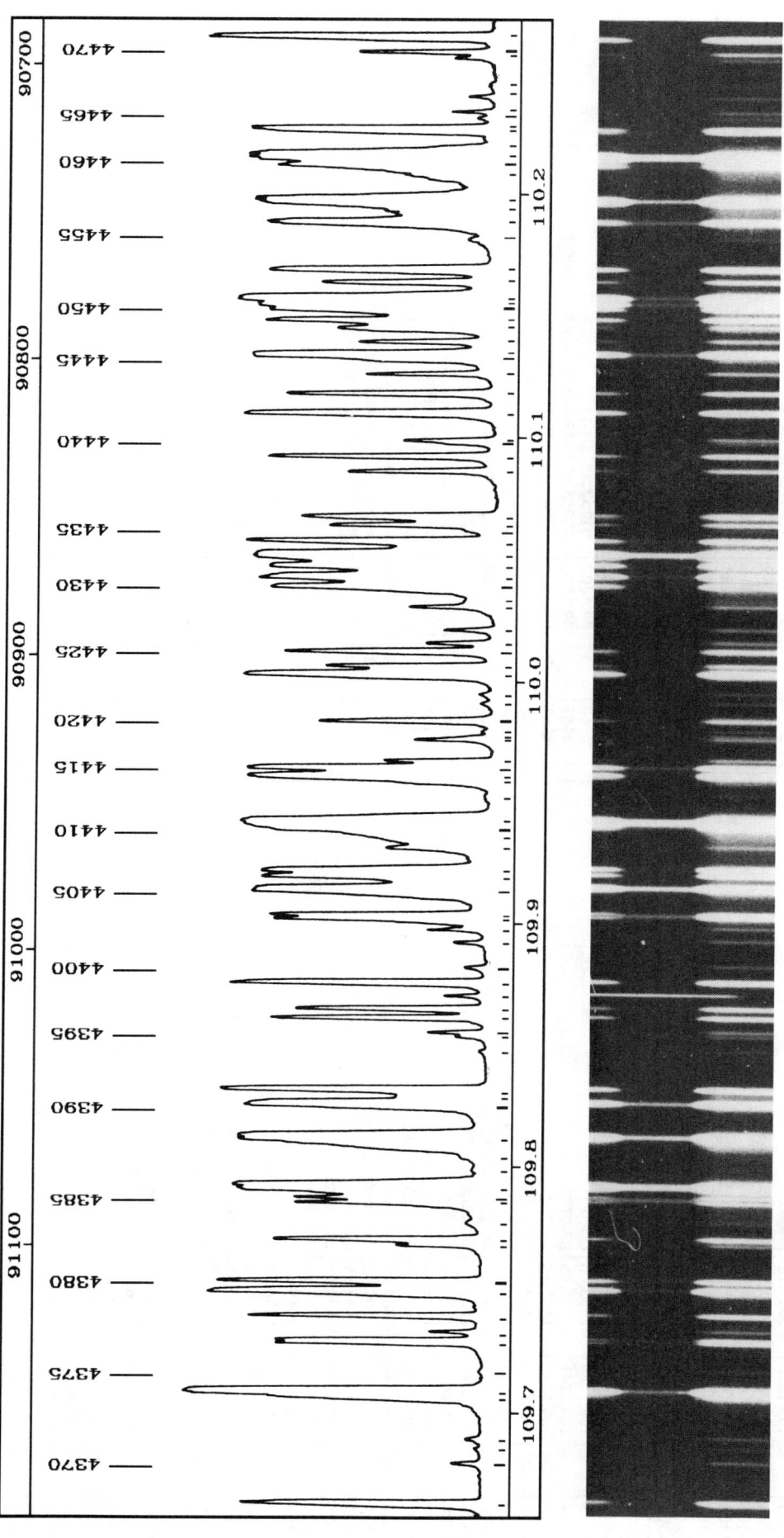

PLATE 55

TABLE 55

Nb	I	λ (nm)	σ (cm⁻¹)	Assignment	Comment
4373		109.7059	91152.79	B11-2P9 C1-1P15 C4-3P13	sh
4374	112	109.7078	91151.25	B8-2P3	
4375	0	109.7162	91144.28	B6-0P12	
4376		109.7280	91134.43	B4-1R0	r
4377	0	109.7323	91130.86	B15-4R2	
4378	5	109.7385	91125.69	B3-7P2	
4379	34	109.7488	91117.22	C2-3P4	
4380	14	109.7525	91114.10	C2-3Q5	
4381	2	109.7673	91101.80	B13-3R7	
4382	4	109.7697	91099.81	B11-3R0 B13-0R19	
4383		109.7767	91093.98	B29-7P4 C0-1R14	d
4384		109.7850	91087.09	B4-1R1	r 91086.52
4385		109.7864	91085.94	B4-1R1	r 91086.52
4386	249	109.7909	91082.23	B16-3R11 C0-2R1	
4387	0	109.7983	91076.10	C2-0Q23	
4388	230	109.8075	91068.46	B23-6R1 C4-4P6 C10-7P5	
4389	135	109.8114	91065.25	C0-2R0 B8-2R4 C1-0Q21	sh
4390		109.8253	91053.69	C0-2R2	
4391	1	109.8287	91050.85	B12-3R5	
4392	25	109.8311	91048.89	B11-3R1 B14-3P8 B18-3R13	
4393	0	109.8470	91035.68	C4-4Q7	
4394	0	109.8531	91030.69	B19-5P2	
4395	2	109.8548	91029.22	B5-8R1 C1-2R11	
4396	9	109.8607	91024.33	C7-6R0	
4397	8	109.8646	91021.16	B6-1P7 B15-4P2 B26-6P7	
4398		109.8703	91012.13		GeII 109.8710 Ke
4399	17	109.8755	91006.64	C7-6R1	
4400	0	109.8821	90998.03	B23-6P1 C12-8R1	
4401	1	109.8925	90998.03	B8-0R15	
4402	2	109.8979	90993.55	B15-3R10 C7-6R2	
4403		109.9022	90989.70	B4-1P1 B7-0R14	r 90989.34
4404		109.9038	90988.70	B4-1P1	r 90989.34
4405	101	109.9143	90980.01	C0-2R3 B9-2P6	
4406	13	109.9196	90975.63	B'3-7P3	
4407	18	109.9220	90973.62	B11-3P1 C2-3R7	
4408	0	109.9319	90965.45	C12-8Q1	
4409		109.9351	90962.79		sh
4410		109.9382	90960.17	B4-1R2	r 90958.81
4411	1	109.9415	90957.45	B4-1R2 C0-2Q1 B'5-8P1 B21-5R7	r 90958.81
4412		109.9530	90947.98	B'1-5P11	
4413		109.9584	90943.45	B19-5R3 B16-4P5	sh
4414	52	109.9612	90941.18	C7-6Q1	
4415	53	109.9644	90938.54	B5-1P5	
4416	2	109.9675	90935.96	B15-4R3	
4417	4	109.9763	90928.72	B5-0P11 B9-0R16	
4418		109.9776	90927.64	B1-0P3	r 90926.77
4419		109.9797	90925.90	B1-0P3	r 90926.77
4420	6	109.9837	90922.54	B11-3R2 B23-6R2	
4421	0	109.9910	90916.53	D1-7R2	
4422	1	109.9951	90913.17	B3-0P8	
4423	33	110.0032	90906.46	B2-0P6 B'1-6R1	
4424	3	110.0064	90903.84	D1-7R1 D1-7R3 D'1-9Q1	
4425	11	110.0125	90898.81	B'1-6R0 B6-0R13	
4426	3	110.0159	90895.98	B1-0R4 C7-6R3	
4427	2	110.0214	90891.45	B20-5P5 C0-0R20	
4428	5	110.0309	90883.59	D'1-9Q2 C2-2P13	
4429	1	110.0329	90881.89	B8-1R11 C3-2R18	
4430	26	110.0393	90876.61	C2-3P5 B'1-6R2 C0-1Q13	
4431	60	110.0434	90873.27	B6-1R8 C5-5R1	
4432	28	110.0477	90869.68	C5-5R0 C8-6P7	
4433	304	110.0520	90866.12	C0-2Q2 C9-7R0 D1-7R4	
4434	26	110.0577	90861.47	C0-2R4 C12-8Q2 C4-4R9	
4435	1	110.0622	90857.74	B23-6P2 C2-2Q14	
4436	8	110.0645	90855.86	C9-7R1 D'1-9Q3	
4437	11	110.0681	90852.87	C7-6Q2 C2-3Q6	
4438	6	110.0867	90837.50	B5-1R6 B'5-8P2	
4439	11	110.0928	90832.49	C5-5R2	
4440		110.0984	90827.83	B2-0R7	
4441	4	110.0995	90826.96	B9-2R7	sh
4442	33	110.1108	90817.64	B8-2P4 B14-3R9 C0-1P12	
4443	10	110.1189	90810.96	B'1-6R3	
4444	2	110.1271	90804.20	B7-1P9 B19-5P3	
4445		110.1327	90799.58	C9-7R2	
4446	92	110.1349	90797.71	B4-1P2 B11-3P2	sh
4447	4	110.1403	90793.32	B'3-7P4 D1-7R5	
4448	5	110.1463	90788.38	B3-0R9 C3-2Q17 D1-7Q1	
4449	15	110.1492	90785.97	C9-7Q1 B10-0R17	
4450	15	110.1537	90782.28	B15-4P3 C12-8R3	
4451	37	110.1558	90780.48	B12-3P5	
4452	121	110.1582	90778.57	C5-5Q1	
4453	8	110.1647	90773.15	B16-4R6 C7-6P2	
4454	25	110.1699	90768.89	B'1-6P1 B29-7P5	
4455	0	110.1823	90758.65	C7-6R4	d
4456		110.1893	90752.96	B4-1R3	r
4457	0	110.1932	90749.74		
4458	192	110.1980	90745.78	B23-6R3 C0-2P2 C5-5R3 B13-3P7	d
4459	1	110.2083	90737.31	B8-2R5	
4460	24	110.2128	90733.58	C0-2Q3	
4461	472	110.2162	90730.79	C2-3R8	
4462		110.2202	90727.46	B11-3R3 C1-1R18	
4463	23	110.2258	90722.87	C7-6Q3 B5-0R12	
4464		110.2275	90721.44	C0-0Q19	
4465	3	110.2318	90717.89	C4-4P7	
4466	2	110.2344	90715.82	B'1-6R4 B20-5R6	
4467	2	110.2405	90710.76	B19-5R4 C12-8Q3	
4468	1	110.2455	90706.61	B21-5P7 C9-7R3	sh
4469	2	110.2562	90697.88	B27-7R1 C9-7Q2	sh
4470	5	110.2586	90695.91		

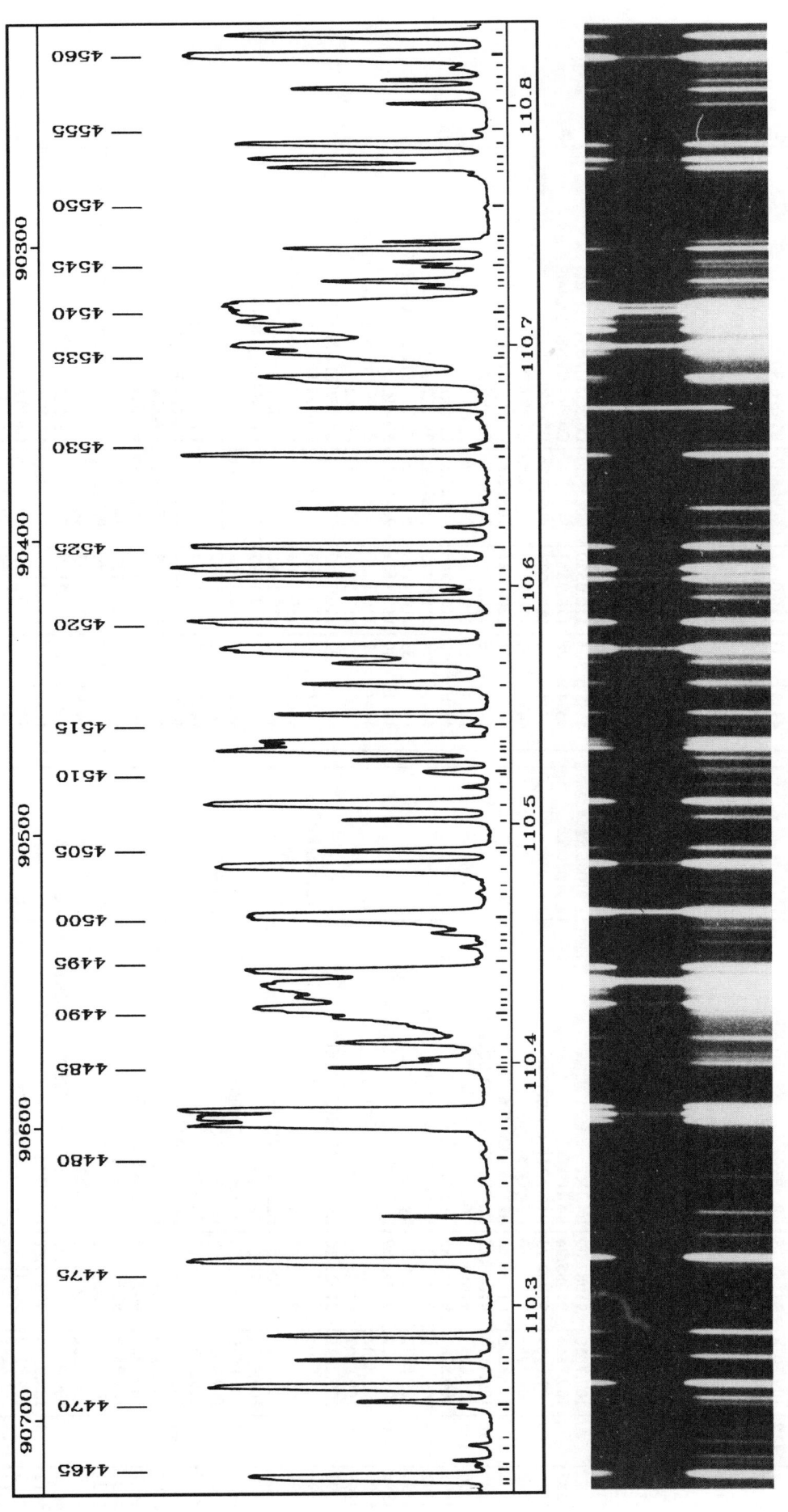

PLATE 56

TABLE 56

Nb	I	λ (nm)	σ (cm-1)	Assignment	Comment
4471	43	110.2646	90690.95	C5-5Q2	
4472	10	110.2758	90681.74	C0-2R5	
4473	0	110.2782	90679.75	D1-7Q3	
4474	8	110.2860	90673.32	B5-8P3 B15-4R4	
4475	0	110.3134	90650.81	B23-6P3 C12-8P3	
4476	42	110.3171	90647.78	B7-2R0 B'1-6P2	
4477	1	110.3266	90640.01	B10-2R9	
4478	1	110.3360	90632.26	C9-7P2	
4479	1	110.3510	90619.93	B4-0P10	
4480	0	110.3615	90611.33	B7-1R10	
4481	18	110.3738	90601.20	C5-5R4 C5-5P2	
4482	69	110.3767	90598.91	B7-2R1 B12-2P11	
4483	26	110.3797	90596.36	C7-6P3	
4484	4	110.3978	90581.49	C0-1R15 B'3-7P5	
4485	3	110.3993	90580.27	B27-7R2	
4486	1	110.4016	90578.37	B'1-6R5 C7-6R5	
4487	0	110.4038	90576.60	B11-0R18	
4488	6	110.4084	90572.84	B1-0P4	
4489	1	110.4188	90564.26	C2-3P6	
4490	28	110.4228	90560.90	C5-5Q3	
4491	0	110.4258	90558.58		b
4492	5	110.4279	90556.85	C2-3Q7	
4493	432	110.4322	90553.29	C0-2P3 C0-2Q4 C1-1Q17	
4494	62	110.4384	90548.25	B11-3P3 C7-6Q4 C9-7R4	
4495	0	110.4432	90544.29	C1-1P16	
4496	0	110.4489	90539.59	D7-13Q3	
4497	0	110.4523	90536.81	B'1-5P12	
4498	1	110.4550	90534.63	B1-0R5	
4499	270	110.4585	90531.77		sh
4500	0	110.4620	90528.88	B4-1P3	
4501	0	110.4713	90521.23	C4-3Q15	
4502	0	110.4748	90518.34	C6-5Q9	
4503	118	110.4822	90512.34	B7-2P1 B19-5P4 D7-13Q2 B27-7P2	
4504	11	110.4888	90506.91	C0-2R6	
4505	11	110.4901	90505.81	B16-4P6	sh
4506	3	110.5021	90496.04	D7-13Q1	
4507	52	110.5086	90490.72	B'1-6P3	
4508	1	110.5156	90484.94	C5-5R5 B12-1P15	
4509	1	110.5204	90481.02	D2-8R2	
4510	1	110.5212	90480.38	D2-8R3	
4511	2	110.5268	90475.81	B15-4P4 B20-5P6	
4512	17	110.5309	90472.41	B7-2R2	
4513	11	110.5330	90470.76	B4-1R4	
4514	9	110.5343	90469.69	C9-7P3	
4515	1	110.5413	90463.92	B11-2P10 C4-4R10 B'5-8P4	
4516	6	110.5458	90460.21	D2-8R1	
4517	5	110.5583	90450.04	B5-1P6 B11-3R4 B4-0R11 B13-0P19	
4518	7	110.5671	90442.85	B9-2P7	
4519	163	110.5727	90438.24	B14-4R0 C3-4R1	

Nb	I	λ (nm)	σ (cm-1)	Assignment	Comment
4520	80	110.5839	90429.10	C5-5P3 C12-8P4	
4521	3	110.5944	90420.47	B18-5P1 D2-8R0	
4522	1	110.5978	90417.69	C7-6P4	
4523	33	110.6023	90414.05	B8-2P5 B'1-6R6 B14-3P9	
4524	105	110.6067	90410.45	C3-4R0	
4525	36	110.6159	90402.93	C3-4R2	
4526	4	110.6241	90396.21	C2-3R9	
4527	6	110.6319	90389.87	C5-5Q4	
4528	1	110.6361	90386.38	C9-7Q4	
4529	15	110.6543	90371.55	B14-4R1	
4530	1	110.6582	90368.38	B13-2R13	
4531	1	110.6698	90358.92	C1-0P21	
4532		110.6740			GeII 110.6737 Ke
4533	9	110.6855	90346.09	B23-6P4 C0-1Q14 B15-4R5	
4534	10	110.6868	90345.04	B'3-7P6 D2-8Q1	sh
4535		110.6949	90338.37	C4-4Q9	
4536	14	110.6971	90336.62	B2-0P7 B5-1R7 B13-1P16 C7-6Q5	
4537	110	110.7001	90334.13	B12-3P6 C0-2Q5	
4538	49	110.7070	90328.55	B7-2P2	
4539	37	110.7104	90325.78	C3-4R3	
4540	410	110.7146	90322.34	C3-4Q1	
4541	238	110.7172	90320.25	C0-2P4 C11-8R0	
4542	5	110.7247	90314.06	B8-0P15 B15-0R21	
4543	6	110.7273	90311.98	B14-4P1 B7-0P14 B12-2R12	
4544	2	110.7301	90309.66	B27-7P3 C2-2P14	
4545	4	110.7334	90307.03	D2-8Q2	
4546	8	110.7354	90305.34	B12-0R19 C2-2Q15	
4547	9	110.7409	90300.91	B'1-6P4	
4548	2	110.7438	90298.52	C11-8R1	
4549	1	110.7459	90296.80	B15-3R11	
4550	1	110.7578	90287.08	B8-2R6	
4551	2	110.7713	90276.12	B9-2R8 B20-5R7	
4552	29	110.7748	90273.25	C0-1P13	
4553	35	110.7784	90270.32	B7-2R3	
4554	31	110.7845	90265.31	C0-2R7 C9-7P4	
4555	1	110.7898	90261.06	B18-5P2	
4556	1	110.8013	90251.65	B14-4R2 D2-8Q3	
4557	8	110.8075	90246.63	C11-8Q1 C12-8Q5	
4558	4	110.8111	90243.66	B6-1R9 B17-4P8 B9-0P16	
4559	3	110.8159	90239.78	C3-2P17	
4560	160	110.8212	90235.41	B18-3R14 C3-4Q2	
4561	16	110.8299	90228.40	B11-3P4 B6-0P13 C2-3Q8	

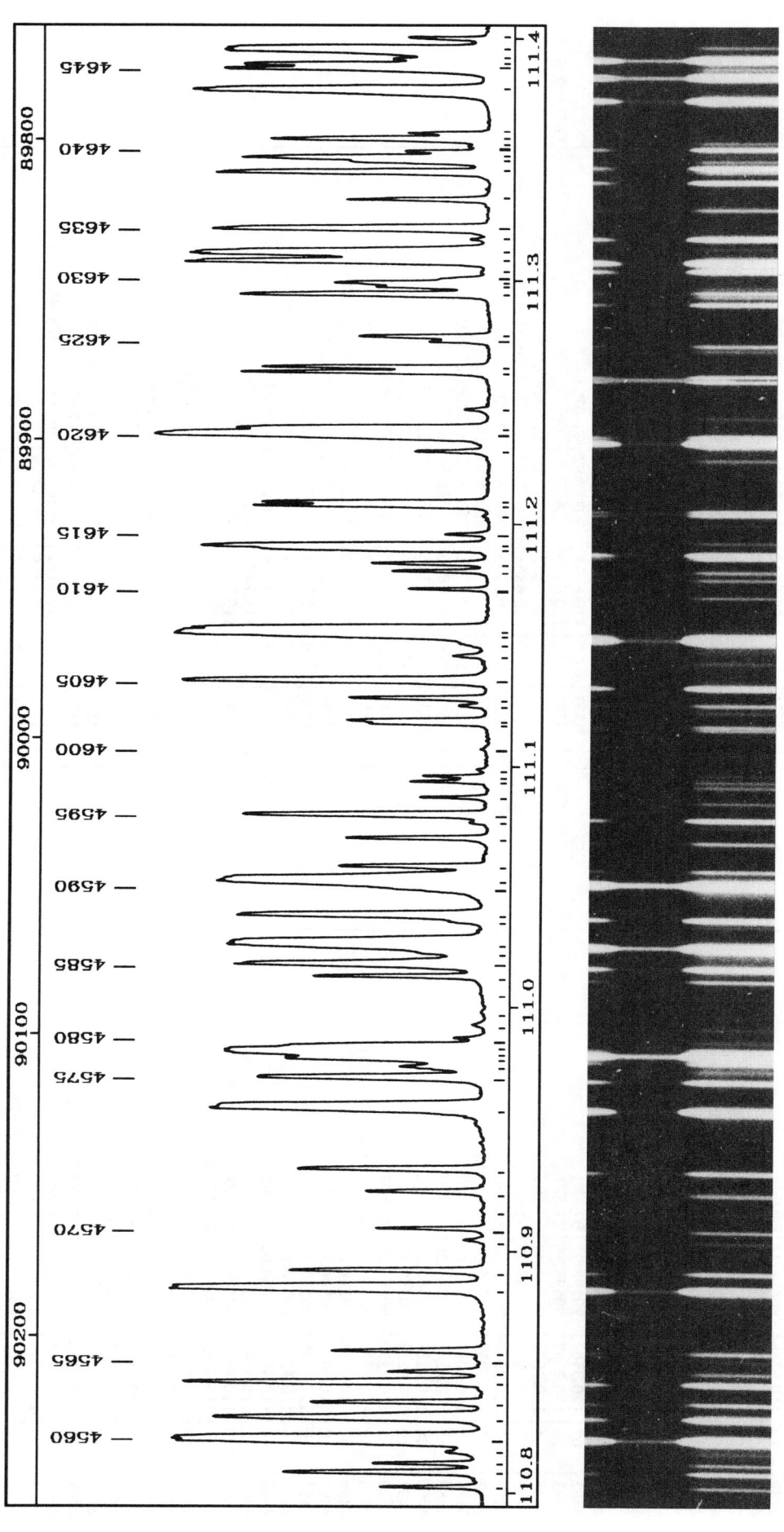

PLATE 57

TABLE 57

Nb	I	λ (nm)	σ (cm⁻¹)	Assignment	Comment
4552	8	110.8360	90223.39	C3-2R19 C2-3P7 B'1-6R7	
4553	16	110.8441	90216.77	C5-5P4 C0-0R21	
4554	4	110.8485	90213.23	B3-0P9	
4555	0	110.8536	90209.08	B'5-8P5	
4556	7	110.8569	90206.37	C3-4R4 D2-8P2	
4567	82	110.8827	90185.42	B4-1P4 B10-2P9 B14-3R10 B11-2R11	
4568	12	110.8897	90179.72	C5-5Q5 D2-8Q4	
4569	1	110.9021	90169.65	B12-3R7 C9-7Q5	
4570	3	110.9068	90165.80	C7-6P5	
4571	0	110.9142	90159.76	C12-8P5 C3-2Q18	
4572	3	110.9220	90153.43	B14-4P2 C11-8Q2	
4573	17	110.9312	90145.94	B1-0P5	
4574	104	110.9570	90125.00	C3-4P2	
4575	23	110.9692	90115.10	B4-1R5	
4576	1	110.9736	90111.50	B11-3R5 C11-8R3	d
4577	7	110.9771	90108.65	B10-3R0	
4578	174	110.9800	90106.29	C3-4Q3	
4579	3	110.9817	90104.92	B15-4P5 C11-8P2 D2-8P3	sh
4580	3	110.9854	90101.96	B1-0R6 B10-0P17	
4581	1	110.9910	90097.41	B'3-7P7	
4582	0	110.9960	90093.32	D2-8Q5	
4583	0	111.0036	90087.20	C7-6Q6	
4584	9	111.0109	90081.28	B'1-6P5 B3-0R10	
4585	39	111.0161	90077.00	C0-2Q6	sh
4586		111.0213	90072.82		
4587	193	111.0248	90070.00	B7-2P3 B7-1P10	
4588	2	111.0338	90062.66	B5-0P12	
4589	22	111.0366	90060.41	B10-3R1 B14-4R3	sh
4590		111.0477	90051.42	D6-12Q3	
4591	303	111.0512	90048.55	C0-2P5 D6-12Q1 D6-12Q2	
4592	9	111.0568	90044.05	C3-4R5 C0-0Q20	
4593	5	111.0687	90034.33	B18-5P3	
4594	2	111.0755	90028.88	C2-3R10 C4-3P15	
4595	18	111.0786	90026.31	B7-0R15 D3-9Q1	
4596	3	111.0857	90020.56	B17-4R9 C9-7P5	
4597	3	111.0924	90015.19	B23-6P5	
4598	2	111.0945	90013.44	C11-8Q3 C5-5R7	
4599	1	111.0973	90011.19	B16-4P7	
4600		111.1057	90004.42	B20-5P7 B'4-8P1	
4601	4	111.1166	89995.57	B8-0R16 D3-9Q2	
4602	5	111.1177	89994.65	B'7-2R4	
4603	4	111.1235	89989.93	C0-0P19	
4604	7	111.1266	89987.44	C0-2R8 C4-4P9	
4605	23	111.1340	89981.51	B10-3P1 B6-0R14	
4606	1	111.1442	89973.20	B10-2R10 B13-0R20	
4607		111.1492	89969.12	B15-4R6	sh
4608	139	111.1536	89965.57	B29-7P7 C3-4P3	
4609		111.1551	89964.38	C5-5P5	sh
4610	1	111.1714	89951.19	D3-9Q3	
4611	4	111.1788	89945.23	B8-2P6 C12-8Q6 C11-8R4	
4612	4	111.1819	89942.67	C11-8P3 B13-3R9	sh
4613		111.1880	89937.79	B10-3R2	
4614	30	111.1892	89936.76	C3-4Q4	
4615	3	111.1941	89932.86	C5-5Q6 C1-1Q18	
4616	0	111.2017	89926.68	C1-2P11	
4617		111.2057	89923.43	B3-1R0	r 89922.74
4618		111.2074	89922.04	B3-1R0	r 89922.74
4619	5	111.2285	89904.99	C1-1P17	
4620	104	111.2359	89899.04	B14-4P3	
4621	14	111.2381	89897.25	B5-1P7	
4622	4	111.2461	89890.80	B9-0R17	
4623		111.2618	89878.12	B3-1R1	r 89877.49
4624		111.2634	89876.86	B3-1R1	r 89877.49
4625	5	111.2743	89868.05	B7-1R11	
4626	7	111.2762	89866.45	C2-3Q9	
4627	8	111.2936	89852.42	D4-10Q1	
4628	3	111.2969	89849.81	B5-0R13 C2-3P8	
4629	5	111.2983	89848.62	D5-11Q1	
4630	2	111.2999	89847.32	B9-2P8	
4631	0	111.3031	89844.76	B15-3P11	
4632	19	111.3074	89841.29	B11-3P5	
4633	65	111.3109	89838.45	B'2-7R1	
4634	1	111.3165	89833.94	B'1-6P6	
4635	17	111.3208	89830.49	B'2-7R0 C3-4R6 C11-8Q4 D4-10Q2	
4636	3	111.3327	89820.89	B12-3P7	
4637	11	111.3443	89811.51	B'2-7R2	
4638	3	111.3487	89807.98	B4-0P11	
4639	9	111.3503	89806.64	B10-3P2	
4640	1	111.3529	89804.62	B8-2R7 B17-2R17	
4641	1	111.3553	89802.67	C7-6Q7	
4642	21	111.3580	89800.47	C0-1Q15	
4643	2	111.3605	89798.48	D4-10Q3	
4644	61	111.3783	89784.06	C0-2Q7	
4645		111.3870	89777.12	B3-1P1	r 89776.32
4646		111.3889	89775.52	B3-1P1	r 89776.32
4647	0	111.3912	89773.71	B5-1R8	
4648	75	111.3952	89770.49	B4-1P5 C11-8R5	
4649	2	111.4002	89766.46	C2-2P15	

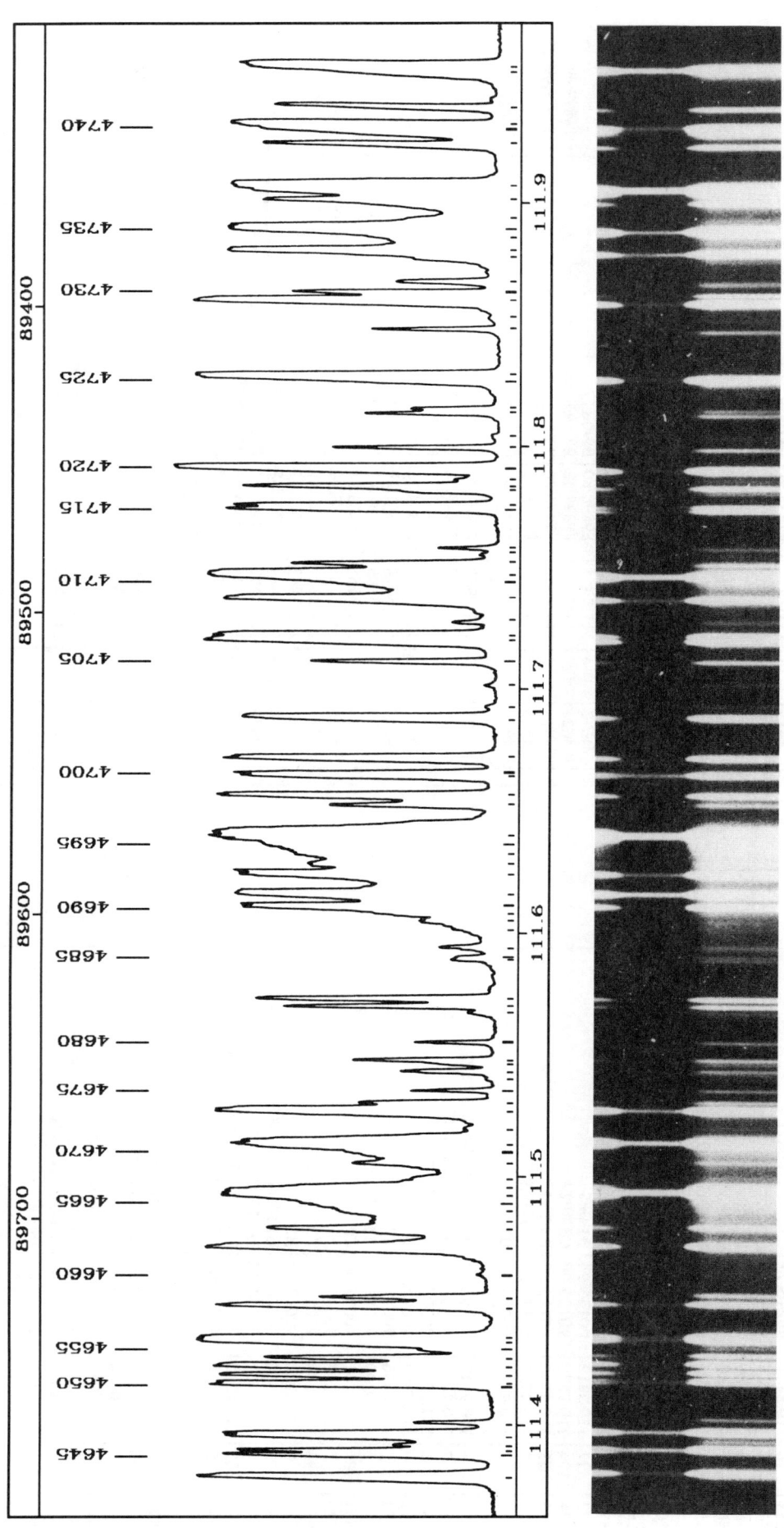

PLATE 58

TABLE 58

Nb	I	λ (nm)	σ (cm⁻¹)	Assignment	Comment
4650		111.4154	89754.20	B3-1R2	r 89753.70
4651	17	111.4166	89753.20	B3-1R2	r 89753.70
4652	26	111.4195	89750.88	B'2-7R3	
4653	6	111.4233	89747.86	C3-4P4	
4654		111.4265	89745.25	B10-3R3	
4655	1	111.4298	89742.63	B18-5P4	
4656	39	111.4332	89739.83	C0-2P6	
4657	39	111.4343	89738.98	B7-2P4 B30-8R0 C2-2Q16 C11-8P4	
4658	22	111.4476	89728.23	C3-4Q5	
4659	4	111.4511	89725.47	B6-1P9 C0-1P14	
4660	0	111.4603	89718.05	B10-0R18	
4661	32	111.4713	89709.17	B'2-7P1 D4-10Q5	
4662	12	111.4789	89703.07	B2-0P8 C0-2R9	
4663	1	111.4821	89700.52	C6-5Q11	
4664	0	111.4858	89697.54	C5-5R8	
4665		111.4887	89695.20		sh
4666	321	111.4931	89691.62	B30-8R1 C1-3R1	
4667		111.4958	89689.46	B4-1R6	sh
4668	0	111.4993	89686.70	B17-5R1	
4669	2	111.5056	89681.60	C2-3R11	
4670		111.5092	89678.70	C4-4R12	sh
4671	269	111.5134	89675.33	C1-3R0	
4672	0	111.5195	89670.44	B11-2P11	
4673	127	111.5268	89664.55	C1-3R2	
4674	6	111.5300	89661.96	C5-5P6	
4675	1	111.5350	89657.98	B'2-7R4 D4-10Q6	
4676	0	111.5399	89654.05	B15-2P15 B17-4P9 B30-8P1	
4677	2	111.5430	89651.55	C5-5Q7	
4678	1	111.5458	89649.24	B1-0P6	
4679	4	111.5473	89648.04	B7-2R5	
4680	1	111.5546	89642.18	B9-2R9 C3-4R7	
4681	0	111.5672	89632.10	B4-0R12 B10-1P14	
4682	5	111.5698	89630.00	B12-0P19 C8-7R0	
4683	9	111.5729	89627.50	C8-7R1	
4684		111.5889	89614.65	B0-0P3	r 89614.17
4685		111.5901	89613.68	B0-0P3	r 89614.17
4686	1	111.5943	89610.35	B2-0R9	
4687	0	111.6017	89604.39	D4-10Q7	
4688	0	111.6050	89601.70	B21-5P9 C11-8Q5	
4689	0	111.6081	89599.23	B1-0R7	
4690	25	111.6112	89596.76	B14-4P4 C12-8Q7 B'2-7P2	
4691	79	111.6164	89592.55	C1-3R3	
4692	70	111.6246	89585.99	B3-1P2	
4693	0	111.6290	89582.48	C8-7R2 B25-7R0	
4694		111.6328	89579.43		sh
4695		111.6370	89576.01		sh
4696	571	111.6407	89573.11	C1-3Q1 C3-2P18	
4697	8	111.6534	89562.85	B15-3R12 B16-3R13 C7-6P7 B'1-6P7	
4698	29	111.6573	89559.79	B10-3P3 B6-1R10	
4699		111.6653	89553.36	B3-1R3	r 89552.87
4700		111.6665	89552.38	B3-1R3	r 89552.87
4701	44	111.6728	89547.32	C8-7Q1	
4702	25	111.6896	89533.85	B6-2R0 B25-7R1 B'2-7R5	
4703	0	111.6942	89530.16	C4-4Q11	
4704	0	111.7031	89523.03	C3-2Q19	
4705	6	111.7126	89515.43	C1-3R4	
4706	56	111.7214	89508.34	C6-6R1	
4707	30	111.7235	89506.71	C6-6R0	
4708	0	111.7290	89502.28	B12-2R13 B13-2R14	
4709	54	111.7386	89494.61	C3-4P5 C8-7R3	
4710		111.7441	89490.23	C11-8P5	
4711	369	111.7481	89487.01	B6-2R1 C1-3Q2 C7-6Q8 C0-1R17	
4712	7	111.7526	89483.37	B11-0R19 C3-4Q6	
4713	1	111.7563	89480.45	B10-2P10	
4714	2	111.7592	89478.10	C2-3Q10	
4715	12	111.7750	89465.42	C6-6R2	
4716	2	111.7763	89464.39	B16-4P8 C8-7Q2	
4717	2	111.7821	89459.73	B3-0P10 B14-3R11	
4718	20	111.7843	89458.03	C0-2Q8	
4719	0	111.7871	89455.73	C5-5R9	
4720	88	111.7918	89452.02	B'2-7P3	
4721	3	111.8001	89445.34	B10-3R4	
4722	0	111.8047	89441.68	B21-6R3	
4723	4	111.8141	89434.14	C2-3P9	
4724	3	111.8161	89432.58	B8-2P7	
4725		111.8269	89423.93	B25-7R2	sh
4726	113	111.8294	89421.93	C6-6Q1 B11-2R12	
4727	2	111.8489	89406.32	B17-4R10 C10-8R0	
4728	2	111.8536	89402.59	C9-7P7	
4729	72	111.8608	89396.81	B6-2P1 C5-5P7 C8-7P2	
4730	4	111.8641	89394.20	C10-8R1	d
4731	2	111.8686	89390.62	B11-3P6 B18-5P5	sh
4732		111.8781	89382.98	B'2-7R6	
4733	57	111.8814	89380.33	C0-2P7	
4734		111.8870	89375.86		
4735	201	111.8907	89372.96	C1-3P2	
4736	1	111.8947	89369.76	C0-0Q21	
4737	13	111.9022	89363.73	B6-2R2	
4738	410	111.9078	89359.32	C1-3Q3 C3-4R8	
4739	23	111.9252	89345.35	C1-3R5	
4740		111.9316	89340.30	C8-7Q3 C6-6Q2	sh
4741	69	111.9331	89339.05	B7-2P5 C10-8R2 C5-5Q8	
4742	14	111.9407	89333.05	C10-8Q1 C11-8Q6	
4743		111.9545	89321.97	B3-0R11 B7-0P15 B13-0P20 B13-1P17	sh
4744	404	111.9571	89319.91	B3-1P3	

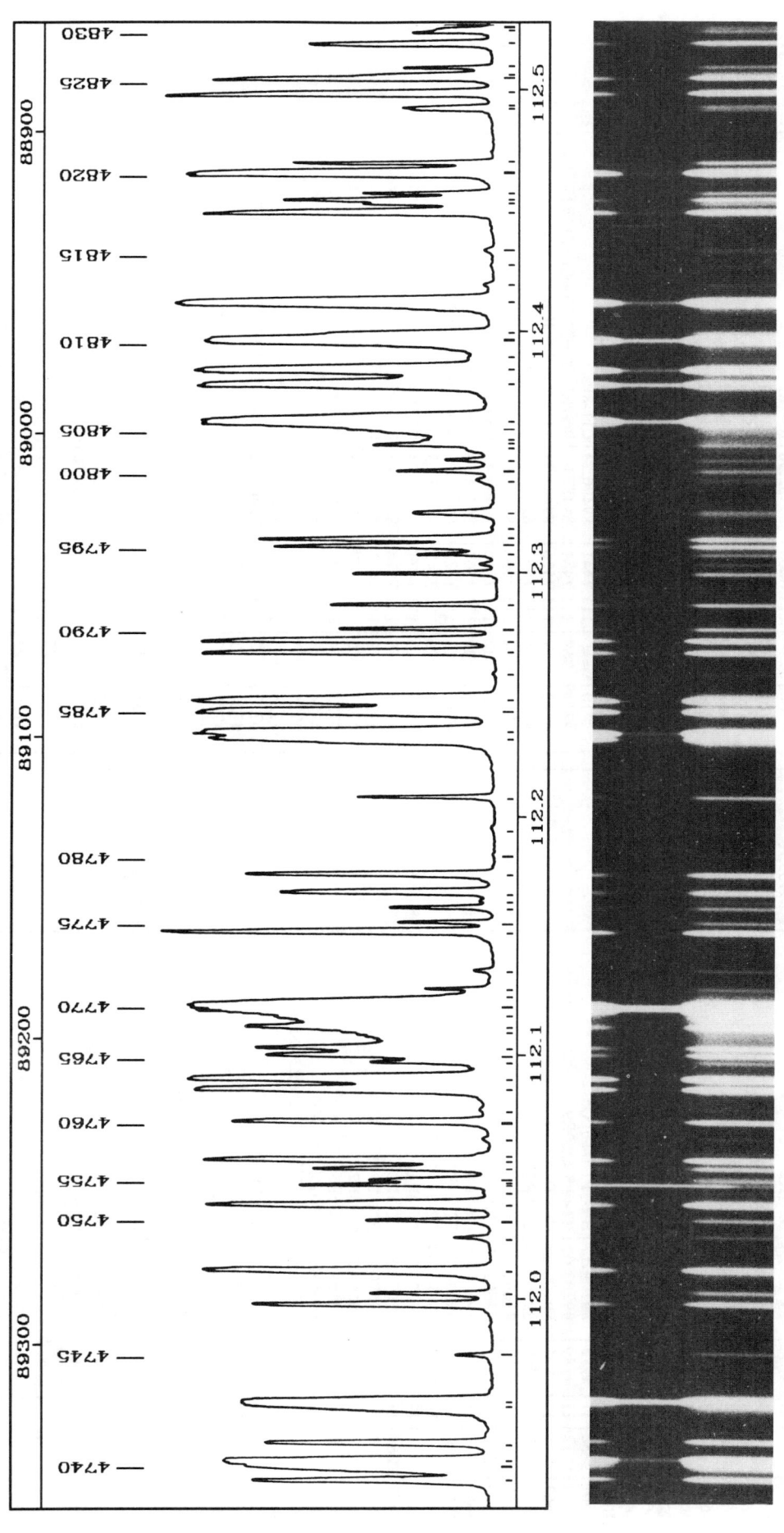

PLATE 59

TABLE 59

Nb	I	λ (nm)	σ (cm⁻¹)	Assignment	Comment
4745	3	111.9768	89304.25	C1-1Q19	
4746	20	111.9973	89287.87	B4-1P6 B7-1P11 B8-0P16	
4747	5	112.0018	89284.26	B5-1P8 B21-6P3	
4748	38	112.0112	89276.81	B3-1R4 B'2-7P4	
4749	3	112.0247	89266.04	B0-0P4 C2-3R12	
4750	5	112.0315	89260.59	B8-2R8 B10-2R11	
4751	23	112.0378	89255.56	C6-6P2 C6-5Q12	
4752	0	112.0428	89251.65	B12-3P8	
4753	0	112.0458	89249.19	B19-5P7	
4754		112.0460			GeII 112.0458 Ke
4755	4	112.0477	89247.68	C10-8Q2 B8-1R13	
4756	4	112.0525	89243.92	B11-3R7 B10-3P4	
4757	29	112.0559	89241.15	C8-7P3 B13-4P2 C0-1Q16	
4758	0	112.0582	89239.31	C10-8R3	
4759	1	112.0647	89234.18	B7-2R6	
4760	10	112.0716	89228.66	B14-4P5	
4761	0	112.0766	89224.69		
4762	37	112.0851	89217.94	C6-6Q3 C4-4R13	
4763	49	112.0895	89214.44	B6-2P2	
4764	1	112.0961	89209.19	C8-6P11	
4765	12	112.0992	89206.67	C3-4P6 B'2-7R7	
4766	12	112.1022	89204.36	C3-4Q7	
4767	0	112.1086	89199.20	C11-8P6	
4768	13	112.1108	89197.46	B4-1R7 B9-2P9	
4769	6	112.1145	89194.52		
4770	711	112.1188	89191.13	C1-3Q4 C1-3P3 C8-7R5 C10-8P2	
4771	2	112.1239	89187.06	B12-0R20 B15-4P7	
4772	4	112.1265	89184.97	B9-0P17	
4773	2	112.1343	89178.76	C8-7Q4	
4774	30	112.1504	89165.98	B6-2R3	
4775	8	112.1544	89162.83	C2-2Q17 C13-9Q1	
4776	7	112.1606	89157.86	B5-0P13	
4777	2	112.1632	89155.80	C2-2P16	
4778	14	112.1670	89152.76	B5-1R9 C1-3R6	
4779	29	112.1744	89146.92	C0-1P15	
4780	0	112.1822	89140.67	C7-6Q9	
4781	0	112.1939	89131.39	C6-6R5	
4782	5	112.2068	89121.14	C10-8Q3	
4783	62	112.2311	89101.85	C0-2Q9	
4784	149	112.2336	89099.91	C4-5R1	
4785	85	112.2425	89092.80	C6-6P3 C10-8R4 C4-4Q12	
4786	92	112.2471	89089.17	C4-5R0	
4787	0	112.2581	89080.40	B7-1R12 B15-3P12	
4788	22	112.2670	89073.41	B'2-7P5	
4789	21	112.2720	89069.44	C4-5R2	
4790	8	112.2767	89065.69	C2-3Q11	
4791	7	112.2868	89057.66	B12-3R9 C6-6Q4 C13-9Q2	
4792	3	112.2997	89047.45	C8-7P4	
4793	1	112.3036	89044.36	C3-4R9	
4794	3	112.3076	89041.20	B'0-6R1	
4795	9	112.3111	89038.39	C10-8P3 C5-5P8	
4796	17	112.3141	89036.01	B7-0R16 C0-2P8	
4797	0	112.3185	89032.57	B1-0R8	
4798	5	112.3249	89027.48	B20-6R0 B6-0R15 C11-8Q7	d
4799	0	112.3384	89016.75	B14-3P11 B21-6P4 B10-0P18	
4800	3	112.3422	89013.78	B2-0P9 B17-4P10	
4801	0	112.3467	89010.22	D6-13Q1	
4802	1	112.3506	89007.11	B'2-7R8	
4803	0	112.3531	89005.13	B13-4P3	
4804	0	112.3547	89003.81	B9-2R10	sh
4805		112.3588	89000.63		
4806	371	112.3632	88997.13	C4-5Q1 C4-5R3 C5-5Q9 B6-1P10	
4807	126	112.3784	88985.09	C1-3Q5	
4808	134	112.3843	88980.38	B3-1P4 B20-6R1 B12-2P13 B18-5P6	
4809	0	112.3888	88976.84	C8-7Q5	
4810	267	112.3967	88970.60	B'0-6R3	
4811		112.3990	88968.75	C1-3P4 C0-2R11	sh
4812	216	112.4123	88958.24	B24-7R0 B8-0R17	
4813	1	112.4192	88952.75	B6-2P3	
4814	1	112.4273	88946.39	B4-0P12	
4815	2	112.4339	88941.19	C3-2P19	
4816	42	112.4499	88928.53	B5-0R14 B26-7R6	
4817	3	112.4533	88925.84	B3-1R5	
4818	12	112.4550	88924.44	B20-6P1	
4819	2	112.4577	88922.36	C1-3R7	
4820	105	112.4663	88915.52	B24-7R1	
4821	5	112.4703	88912.40	C4-5Q2 B2-0R10	
4822	3	112.4924	88894.87	B16-5R0	
4823	6	112.4935	88894.04	B6-2R4	
4824	21	112.4989	88889.79	C3-4Q8	
4825	26	112.5052	88884.76	C6-6P4 B'0-6P1	sh
4826		112.5066	88883.71	C3-4P7	
4827	5	112.5095	88881.37	C4-5R4	
4828	11	112.5196	88873.42	B11-3P7	
4829	5	112.5243	88869.71	B7-2P6 B24-7P1	
4830	3	112.5255	88868.72	B16-4P9 C2-3R13 B20-6R2	

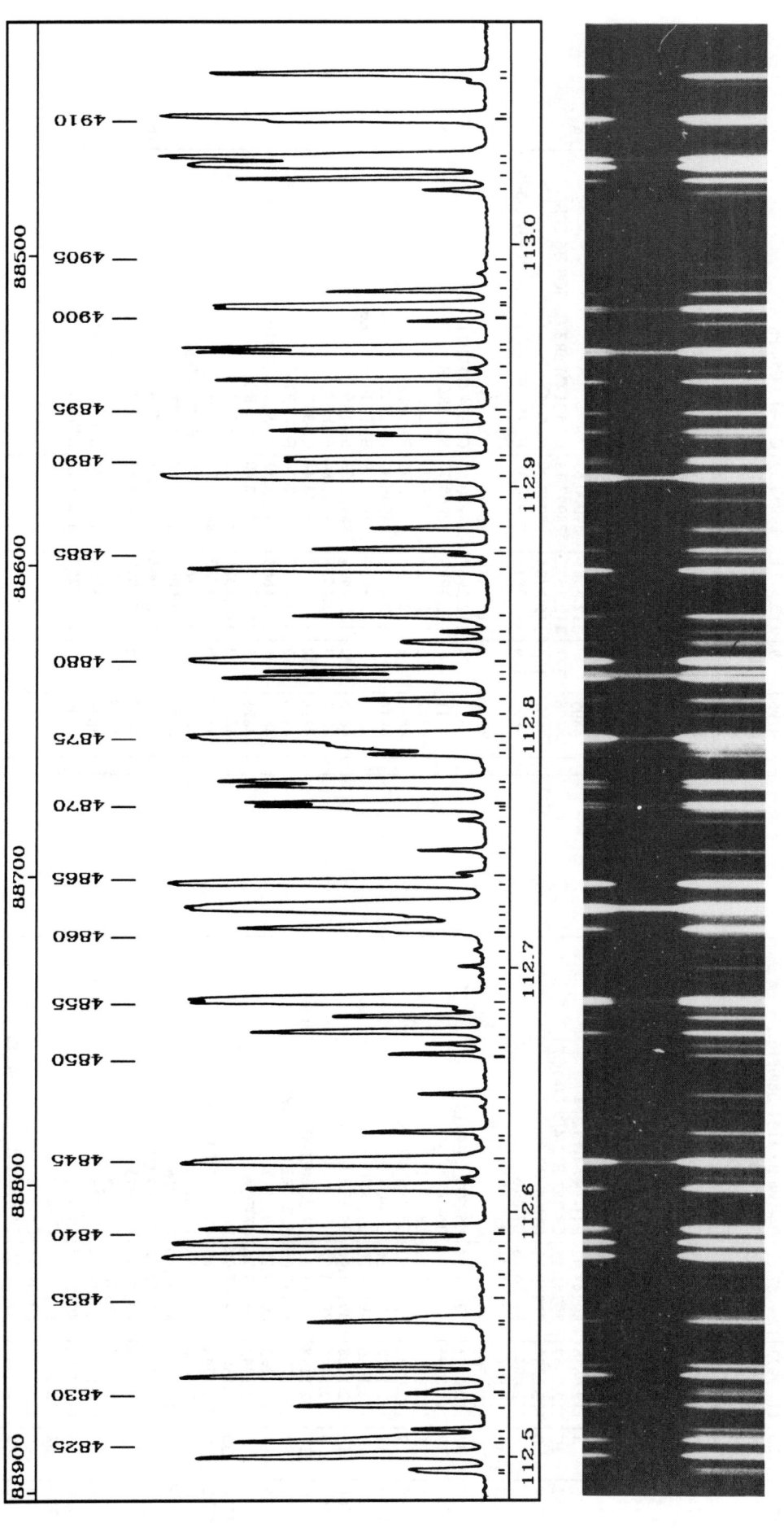

PLATE 60

TABLE 60

Nb	I	λ (nm)	σ (cm⁻¹)	Assignment	Comment
4831	23	112.5317	88863.85	B10-3P5 B16-5R1	
4832	11	112.5355	88860.82	C6-6Q5	
4833		112.5538	88846.43	B0-0P5 C10-8P4	r 88845.87
4834		112.5552	88845.31	B0-0P5 B'2-7P6	r 88845.87
4835	1	112.5645	88837.94	B13-0R21 C1-1R21	
4836	0	112.5699	88833.72	B9-0R18	
4837	0	112.5745	88830.09		
4838	71	112.5806	88825.25	B6-1R11 B'3-8R1	
4839	69	112.5862	88820.83	C4-5P2	
4840	1	112.5911	88816.98	B8-2P8 C8-7P5	sh
4841	24	112.5919	88816.33	B'3-8R0	
4842	15	112.6085	88803.25	B'3-8R2 B16-5P1	
4843	1	112.6127	88799.93	B14-4P6	
4844	130	112.6196	88794.50	C4-5Q3	
4845	0	112.6223	88792.36	C6-5Q13 B'2-7R9	
4846	4	112.6289	88787.17	C5-5R11 B'0-6R5	
4847	5	112.6315	88785.09	B9-3P2	
4848	0	112.6419	88776.93	B20-6P2 B'0-6P2	
4849	3	112.6472	88772.76	B4-0R13 B'0-6P2	
4850	2	112.6634	88759.99	B'5-9R0	
4851	1	112.6674	88756.86	B7-2R7 B15-3R13 C3-4R10	
4852	14	112.6725	88752.77	B16-3R14 B'3-8R3	
4853	4	112.6790	88747.71	B16-5R2	
4854		112.6821	88745.27	B28-8R0 C8-7Q6 C12-9R0	
4855	48	112.6854	88742.66	C1-3Q6	b
4856	0	112.6866	88741.73	B4-1P7 B'5-9R1 B19-5P8 B11-3R8	
4857	0	112.6952	88734.91	B24-7P2	
4858	1	112.6998	88731.33	B10-2P11	
4859	1	112.7066	88725.92	B9-3R3	
4860	20	112.7141	88720.03	C12-9R1 C9-7P9	sh
4861		112.7161	88718.46	C4-5R5 C0-2Q10	sh
4862		112.7213	88714.35		
4863	334	112.7250	88711.51	B21-6P5 C1-3P5	
4864	82	112.7350	88703.56	B'3-8P1 B13-4P4	
4865	1	112.7389	88700.55	B28-8R1	
4866	3	112.7484	88693.04	B20-6R3	
4867	0	112.7615	88682.78	B'5-9R2	sh
4868	0	112.7661	88679.12	C12-9Q1	r 88677.54
4869		112.7672	88678.29	B2-1R0	r 88677.54
4870		112.7691	88676.80	B2-1R0	
4871	23	112.7757	88671.57	D0-9Q3 C0-1Q17	
4872	13	112.7778	88669.96	C1-1Q20 B'5-9P1	
4873	2	112.7889	88661.23	B8-2R9	
4874	7	112.7927	88658.19	B3-0P11 B28-8P1 B12-2R14 C5-5P9	
4875	221	112.7961	88655.52	C4-5P3	
4876	2	112.8056	88648.12	B24-7R3 D'0-9Q4	
4877		112.8115	88643.47	B4-1R8 B16-5P2 C1-3R8	r 88635.33
4878	8	112.8206	88636.33	B2-1R1 C0-2P9	r 88635.33
4879		112.8231	88634.33	B2-1R1 C4-4Q13	
4880	53	112.8277	88630.70	B6-2P4 C6-6P5 C2-3Q12 B12-3P9 / C5-5Q10 C6-6Q6	b
4881	7	112.8352	88624.85	B13-3R11 B13-2R15 C1-1P19 B'0-6P3	
4882	3	112.8398	88621.24	B11-2R13	b
4883	11	112.8464	88616.03	B5-1P9 C10-8P5	
4884	41	112.8656	88600.92	B'3-8P2	
4885	2	112.8715	88596.33	C2-3P11 B28-8R2	
4886	7	112.8741	88594.28	B12-4R0 C2-2P17 B'2-7P7	
4887	7	112.8826	88587.62	C12-9Q2 C0-2R12	
4888	1	112.8949	88577.97	B26-7R7 B'5-9R3	
4889	185	112.9046	88570.38	B3-1P5	
4890	8	112.9108	88565.51	B16-5R3	
4891	7	112.9120	88564.54	B20-6P3 B'5-9P2	
4892	7	112.9213	88557.28	C0-1P16	
4893	11	112.9236	88555.44	C3-4Q9 B6-2R5	
4894	0	112.9277	88552.20	C8-7P6	
4895	8	112.9313	88549.42	B12-4R1 C12-9P2	
4896	14	112.9442	88539.32	B9-3P3	
4897	1	112.9485	88535.90	B24-7P3	
4898		112.9555	88530.43	B2-1P1	r 88529.69
4899		112.9574	88528.94	B2-1P1	r 88529.69
4900	4	112.9680	88520.61	C3-4P8	
4901		112.9736	88516.28	B2-1R2	r 88515.76
4902		112.9779	88515.24	B2-1R2	r 88515.76
4903	6	112.9806	88510.81	B3-1R6	
4904	1	112.9874	88505.42	B21-6R6	
4905	1	112.9933	88500.84	B9-2P10	
4906	4	113.0225	88477.94	B5-1R10 C8-7Q7	
4907	9	113.0277	88473.90	B12-4P1	
4908	96	113.0336	88469.27	B'3-8P3	sh
4909	48	113.0367	88466.87	C1-3Q7	
4910		113.0519	88454.97	C1-3P6	
4911	46	113.0534	88453.74	C4-5P4 C12-9Q3	
4912	1	113.0678	88442.52	D5-12Q1	
4913	20	113.0711	88439.93	C4-5Q5 B28-8R3 B12-4R2	

88000
88100
88200
88300
88400
4910
4915
4920
4925
4930
4935
4940
4945
4950
4955
4960
4965
4970
4975
4980
4985
4990
4995
5000
5005
5010
5015
113.1
113.2
113.3
113.4
113.5
113.6

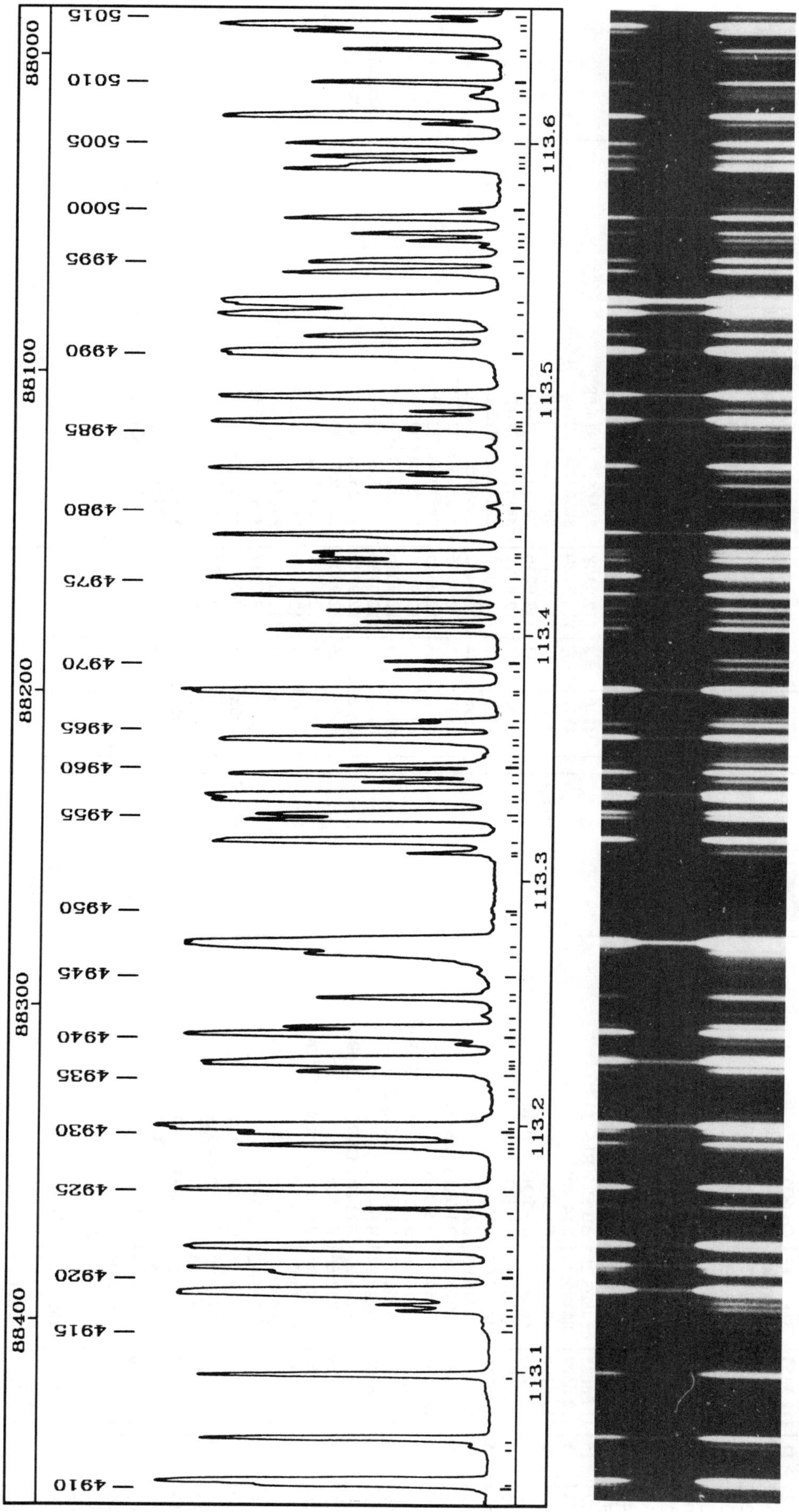

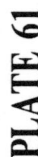

PLATE 61

TABLE 61

Nb	I	λ (nm)	σ (cm⁻¹)	Assignment	Comment
4914	12	113.0972	88419.51	B'5-9P3 B16-5P3 C6-6P6	
4915	0	113.1169	88404.16	B1-0R9	
4916	0	113.1201	88401.66	D6-14Q1	
4917	3	113.1237	88398.82	C2-3R14 D1-8R3	
4918	3	113.1261	88396.97	C12-9P3	
4919	214	113.1313	88392.87	C2-4R1	
4920	9	113.1390	88386.88	B5-2R0 C1-3R9	
4921	32	113.1414	88384.95	B10-3P6 D1-8R4	
4922	107	113.1497	88378.51	C2-4R0 C3-4R11	
4923	0	113.1569	88372.88	C7-6Q11	
4924	6	113.1648	88366.69	C6-6Q7	
4925	39	113.1728	88360.44	C2-4R2	
4926	3	113.1888	88347.93	C10-8P6 C4-5R7 C0-1R19	
4927	15	113.1905	88346.65	B7-2P7	
4928		113.1932	88344.52	B28-8P3	sh
4929	15	113.1955	88342.70	B5-2R1	
4930		113.1972	88341.41	B2-1P2 D1-8R5	r 88340.98
4931		113.1983	88340.55	B2-1P2	r 88340.98
4932	0	113.2020	88337.69	B13-4P5	
4933	0	113.2129	88329.19	B17-4P11	
4934	0	113.2157	88326.95	B23-7R0	
4935	6	113.2207	88323.09	C9-8R0 C2-4R3	
4936		113.2236	88320.80	B2-1R3	r 88320.32
4937		113.2248	88319.84	B2-1R3 B16-5R4	r 88320.32
4938		113.2259	88319.04	C9-8R1	sh
4939	1	113.2319	88314.36	B14-4P7 C6-5Q14	
4940	71	113.2359	88311.24	C0-2Q11 B'3-8P4	
4941	4	113.2387	88309.05	B12-4P2	
4942	2	113.2437	88305.15	B7-0P16 C9-7P10	
4943	10	113.2508	88299.60	B6-0P15 B9-2R11	
4944	0	113.2538	88297.26	B32-9R0	
4945	1	113.2611	88291.54	B23-7R1	
4946	5	113.2694	88285.11	B19-6R0 B15-3P13	
4947	475	113.2732	88282.15	C2-4Q1	
4948	0	113.2819	88275.35	C9-8R2	
4949	0	113.2858	88272.35	B21-6P6 C12-9Q4	
4950	0	113.2875	88270.95	B2-0P10	
4951	4	113.3100	88253.45	B7-1R13	
4952		113.3113	88252.44	B32-9R1	
4953	41	113.3155	88249.19	B5-2P1 C9-8Q1 C8-7P7	
4954	14	113.3245	88242.16	C7-7R0	
4955	22	113.3266	88240.54	B19-6R1 B8-0P17 C5-5Q11	
4956	48	113.3327	88235.76	C7-7R1 B'5-9P4	
4957	69	113.3345	88234.42	B6-2P5	
4958	3	113.3397	88230.34	B23-7P1	
4959	33	113.3431	88227.69	B12-4R3	
4960	5	113.3466	88224.97	B9-3P4 B6-1P11	
4961	0	113.3486	88223.39	B5-2R2 C5-5P10	
4962	0	113.3507	88221.79	B7-2R8 B32-9P1	
4963	55	113.3553	88218.20	B5-0P14	sh
4964	11	113.3575	88216.48	C4-5P5	
4965	11	113.3628	88212.38	C0-2P10 C2-4R4	
4966	4	113.3651	88210.61	C4-5Q6	
4967		113.3749	88202.92	B27-8R0	sh
4968	105	113.3770	88201.34	C2-4Q2 C12-9P4	b
4969	4	113.3856	88194.59	C3-4P9	
4970	4	113.3893	88191.77	C9-8R3 C3-4Q10	
4971	11	113.4020	88181.87	B19-6P1 C8-7Q8 B19-5P9	
4972	8	113.4053	88179.30	C2-3Q13	
4973	8	113.4102	88175.52	B8-2P9	
4974	6	113.4164	88170.66	C9-8Q2	+NI 113.41653
4975	62	113.4239	88164.83	C7-7Q1 C0-2R13 C4-4Q13	
4976	13	113.4302	88159.94	C1-3Q8	
4977	8	113.4325	88158.13	B27-8R1	
4978	8	113.4340	88156.99	B23-7R2	+NI 113.44149
4979	0	113.4414	88151.21	B26-7R8	
4980	1	113.4531	88142.13	B13-3P11	
4981	7	113.4607	88136.27	B4-1P8	
4982	3	113.4657	88132.38	B19-6R2 B9-3R5	
4983	26	113.4685	88130.19	B'1-7R1 B3-8P5	
4984	1	113.4772	88123.45	D4-11Q1	
4985	1	113.4840	88118.18	B27-8P1	
4986		113.4856	88116.92	B'1-7R0 C2-3P12	sh
4987	49	113.4876	88115.33	C1-3P7 C6-6P7	
4988	2	113.4914	88112.44	B'1-7R2	
4989	5	113.4977	88107.50	C9-8P2	+NI 113.49803
4990	102	113.5158	88093.46	B3-1P6 C2-4P2 B23-7P2 C0-1Q18	
4991	12	113.5224	88088.38	C7-7Q2	
4992	126	113.5315	88081.28	C2-4Q3	
4993	560	113.5362	88077.60	B2-1P3 B12-4P3	
4994	18	113.5485	88068.09	B5-2P2 C9-8R4 C2-4R5	
4995	14	113.5529	88064.68	B15-5R0 B'1-7R3	
4996	3	113.5587	88060.22	C1-3R10	
4997	7	113.5612	88058.27	B4-0P13	
4998	9	113.5641	88056.02	C9-8Q3 B27-8R2	
4999	32	113.5704	88051.11	B2-1R4 C12-9Q5	
5000	3	113.5739	88048.40	B6-1R12 B6-0R16 B11-2P13	
5001	0	113.5839	88040.66	B'2-7P9	
5002	15	113.5904	88035.61	B8-3R0	
5003	12	113.5920	88034.39	B19-6P2 C1-1Q21	
5004	21	113.5952	88031.89	B4-1R9	
5005	28	113.6006	88027.73	B3-1R7 C11-9R0	
5006	4	113.6082	88021.78	B15-4P9	
5007	52	113.6117	88019.14	B15-5R1 B7-0R17	
5008	2	113.6201	88012.62	B16-5R5 C11-9R1 B'5-9P5	
5009	2	113.6219	88011.23	B8-2R10	
5010	9	113.6252	88008.66	C7-7P2	
5011	6	113.6356	88000.59	B5-0R15	

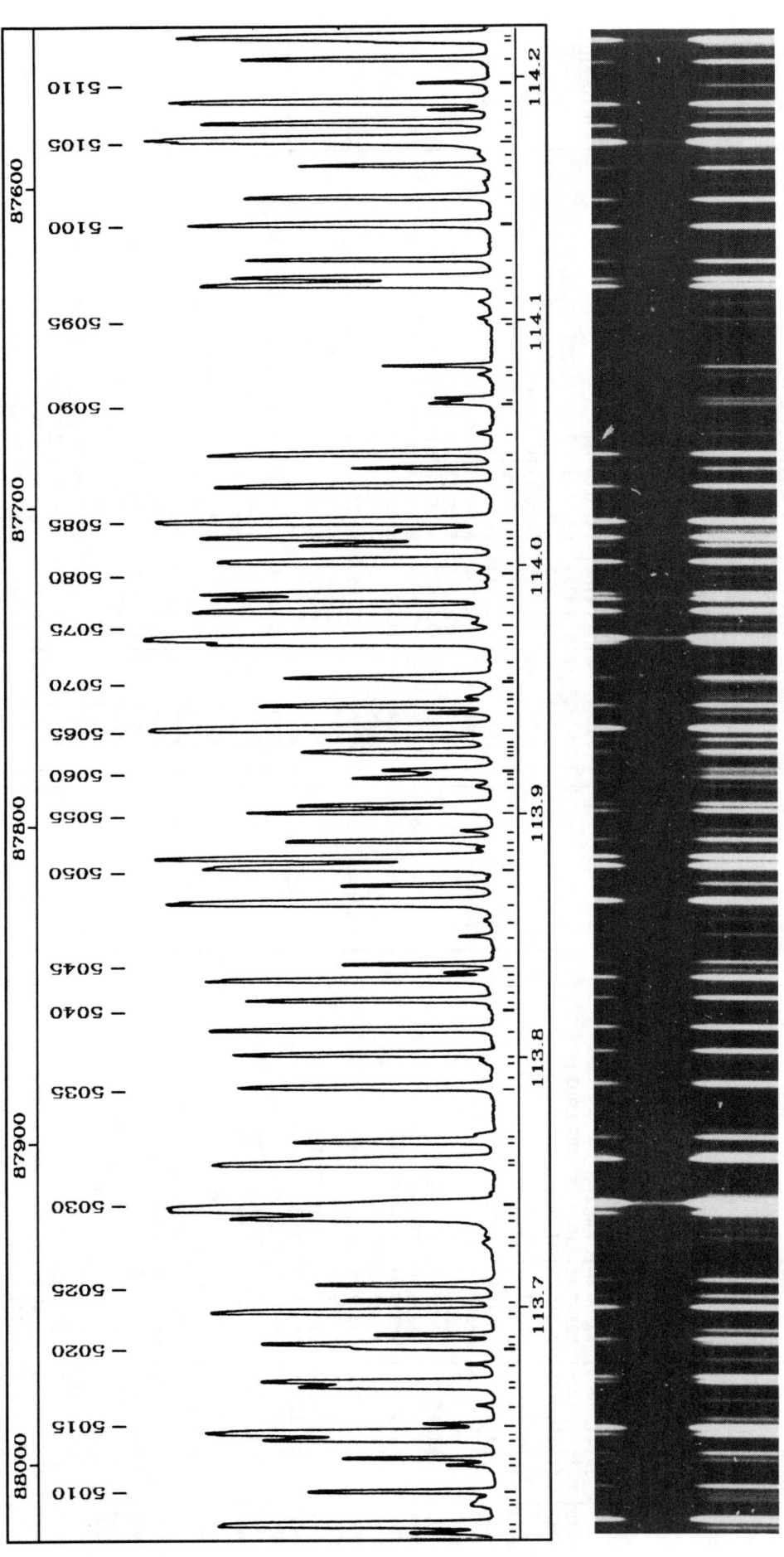

PLATE 62

TABLE 62

Nb	I	λ (nm)	σ (cm⁻¹)	Assignment	Comment
5012	8	113.6384	87998.44	B23-7R3	
5013	12	113.6459	87992.60	B'1-7P1	
5014	62	113.6485	87990.58	B8-3R1	
5015	7	113.6520	87987.91	B27-8P2 B'1-7R4 C8-7P8	
5016	1	113.6595	87982.05	B12-4R4	
5017	22	113.6670	87976.30	B14-3R13 C0-1P17	
5018	22	113.6690	87974.71	B15-3R14 C7-7Q3	
5019	4	113.6759	87969.42	B24-7P5 C2-3R15 C3-4R12	
5020	7	113.6825	87964.27	B19-6R3 B12-3P10 C12-9P5	
5021	17	113.6840	87963.11	C9-8P3	
5022	5	113.6876	87960.32	C11-9Q1	
5023	25	113.6970	87953.05	B15-5P1	
5024	9	113.7016	87949.54	C4-5Q7	
5025	9	113.7081	87944.52	C4-5P6 B10-2P12	
5026	0	113.7250	87931.42	B'3-8P6	
5027	0	113.7283	87928.90	C9-7P11	
5028	21	113.7353	87923.48	B8-0R18 C2-4Q4 C10-8Q8	
5029	344	113.7394	87920.32	C2-4P3	
5030		113.7417	87918.50	B10-0P19	sh
5031	40	113.7573	87906.46	B8-3P1 B15-5R2	
5032		113.7589	87905.21	B23-7P3 B13-3R12 C9-8Q4	sh
5033	14	113.7660	87899.72	B10-3P7	
5034	2	113.7686	87897.75	B5-1P10 B27-8R3	
5035	31	113.7877	87882.95	B26-7P8 C0-2Q12 B'1-7P2	
5036	0	113.7921	87879.56	C11-9Q2	
5037	1	113.7979	87875.08	C5-5P11 B4-0R14	
5038	16	113.8008	87872.88	B8-3R2	
5039	18	113.8105	87865.33	C5-6R1	
5040	0	113.8196	87858.33	C8-7Q9 C11-9R3	
5041	11	113.8227	87855.97	C5-6R0	
5042	0	113.8304	87850.02	B17-2R19	
5043	23	113.8310	87849.56	C7-7P3	
5044	2	113.8338	87847.38	B31-9R1	
5045	5	113.8371	87844.81	B9-3P5	
5046	2	113.8486	87835.93	C5-6R2	
5047	2	113.8550	87830.98	B24-7R6 C11-9P2 C5-5Q12	
5048	60	113.8626	87825.14	C1-3Q9 C7-7Q4 B19-6P3	
5049	7	113.8695	87819.80	C6-5Q15	
5050	54	113.8770	87814.02	B5-2P3 B3-0P12 B31-9P1	
5051	34	113.8809	87811.07	B'4-9R1	
5052	1	113.8844	87808.34	B'4-9R0	
5053	18	113.8880	87805.53	C3-4Q11 B0-0P7	
5054	2	113.8919	87802.52	B27-8P3	
5055	15	113.8999	87796.41	B15-5P2	
5056	10	113.9026	87794.30	C0-2P11 B11-2R14	
5057	2	113.9099	87788.68	C12-9Q6 B12-2R15	
5058	7	113.9136	87785.81	B12-4P4 C6-6P8	
5059	2	113.9152	87784.61	B1-0P9	
5060	6	113.9167	87783.42	C9-8P4 C3-4P10	
5061	5	113.9226	87778.86	B12-3R11	
5062	8	113.9237	87778.08	B14-4P8 B23-7R4 B4-9R2	
5063	0	113.9265	87775.89		
5064	7	113.9288	87774.10	B6-2P6	
5065	82	113.9335	87770.51	C5-6Q1 C0-1R20	
5066	2	113.9396	87765.82	C5-6R3	
5067	17	113.9428	87763.37	B7-2P8 C1-3P8	
5068	0	113.9458	87761.04	B9-0R19	
5069		113.9475	87759.73	C11-9Q3	sh
5070		113.9528	87755.66	C6-6Q9	sh
5071	9	113.9539	87754.77	B5-1R11	
5072	0	113.9604	87749.75	B19-6R4	
5073	25	113.9683	87743.73	B'1-7P3	
5074	172	113.9702	87742.22	B2-1P4	
5075	1	113.9755	87738.14	B18-5R9	
5076	30	113.9812	87733.79	B9-2P11 B8-3P2	
5077	22	113.9862	87729.94	C2-4Q5	
5078	20	113.9883	87728.32	B15-5R3	
5079	0	113.9921	87725.36	B'3-8P7	
5080	1	113.9964	87722.06	B13-2R16 C0-2R14	
5081	45	114.0014	87718.21	C9-8Q5 B'4-9P1	
5082	9	114.0081	87713.05	B'4-9R3	
5083	38	114.0115	87710.47	B2-1R5 C2-3Q14	
5084	1	114.0139	87708.63	B10-2R13	
5085	75	114.0180	87705.42	C2-4P4 B11-3P9	
5086	20	114.0323	87694.46	C5-6Q2 C7-7P4	
5087	8	114.0396	87688.88	C11-9P3 B27-8R4 C1-3R11	
5088	26	114.0455	87684.30	B8-3R3	
5089	1	114.0538	87677.93	B6-2R7	
5090	4	114.0660	87668.52	B3-0R13 C2-3P13	
5091	4	114.0680	87667.04	B12-4R5	
5092	2	114.0775	87659.68	C2-4R7 C4-5Q8	
5093	4	114.0810	87657.04	B22-7R0	
5094	0	114.0980	87643.93	C5-6R4	
5095	0	114.1001	87642.35	C7-7Q5	
5096	0	114.1069	87637.13	B26-8R0	
5097	40	114.1137	87631.93	B11-4R0	
5098	18	114.1165	87629.74	C4-5P7 B7-2R9	
5099	13	114.1238	87624.13	B22-7R1 B17-4P12 B4-9P2 C8-7P9 C3-4R13	
5100	25	114.1383	87613.04	C5-6P2 B7-1P13	
5101	17	114.1492	87604.69	B31-9R3	
5102	0	114.1554	87599.92	B26-8R1	
5103	6	114.1621	87594.76	B18-3R17	
5104	1	114.1668	87591.12	C10-8Q9	
5105		114.1716	87587.46	B11-4R1	sh
5106	135	114.1728	87586.55	C5-6Q3	
5107	20	114.1795	87581.43		
5108	1	114.1847	87577.42	B'1-7P4	

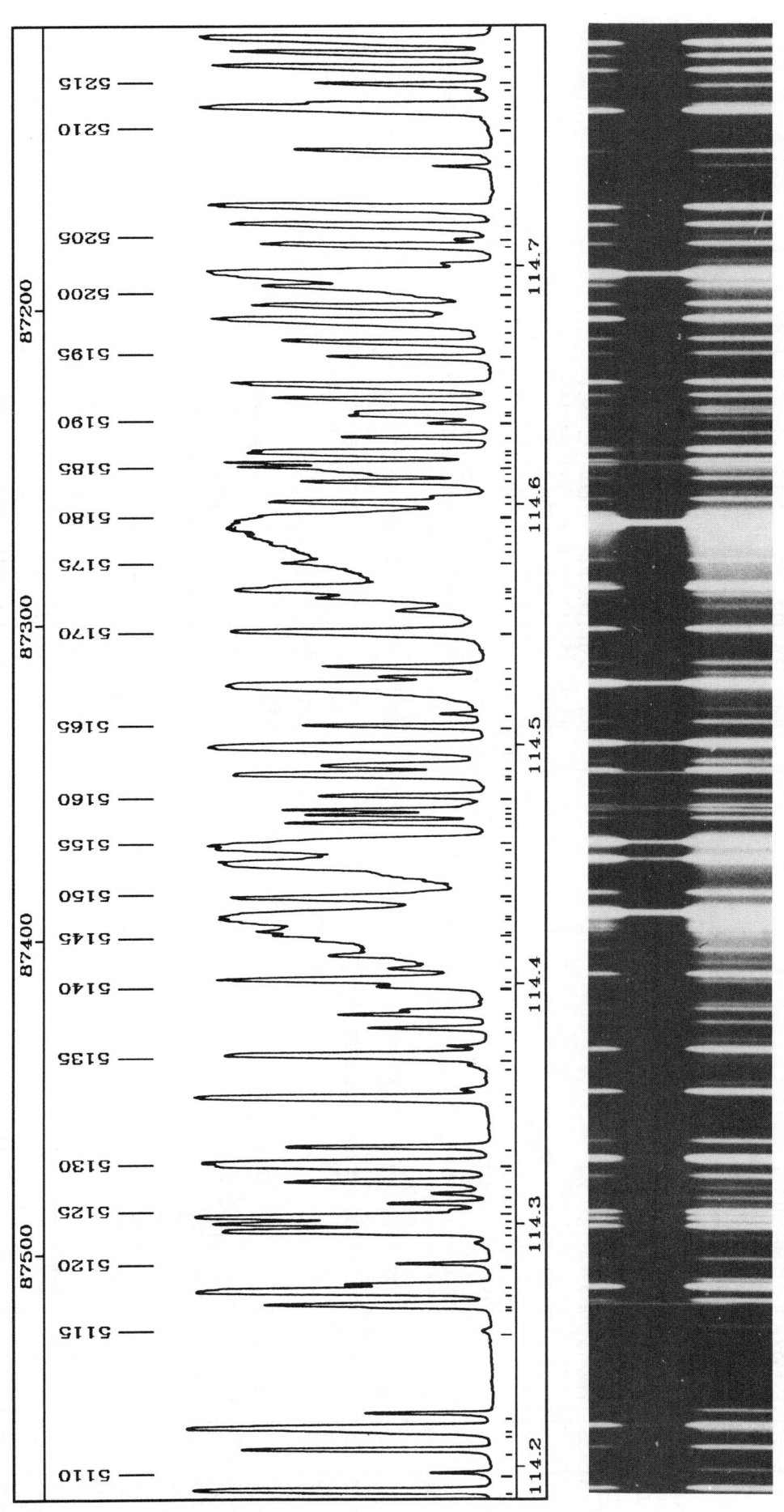

PLATE 63

TABLE 63

Nb	I	λ (nm)	σ (cm^{-1})	Assignment	Comment
5109	32	114.1881	87574.78	B15-5P3	
5110	3	114.1956	87569.04	C9-8P5	
5111	12	114.2052	87561.68	B22-7P1	
5112		114.2124	87556.17		sh
5113	40	114.2142	87554.83	B19-6P4 C5-6R5	
5114	4	114.2202	87550.17	B3-1P7 B9-2R12	
5115	0	114.2545	87523.94	B26-8P1	
5116	15	114.2652	87515.75	C7-6Q13	CuII 114.26405
5117	80	114.2709	87511.35	B18-6R0	
5118	6	114.2735	87509.32	B11-4P1 C0-1Q19 C8-7Q10	
5119	6	114.2822	87502.72	B22-7R2 C11-9P4	
5120	0	114.2907	87496.20	B23-7R5 C2-4Q6	
5121	36	114.2958	87492.25	B26-8R2	
5122	26	114.2990	87489.84	C2-4P5	
5123	50	114.3018	87487.70	B5-2P4 B8-3P3	
5124	1	114.3043	87485.77	B15-5R4 B4-9P3	
5125	4	114.3073	87483.51	B8-2P10 C12-9Q7	
5126	1	114.3115	87480.29	B3-1R8	
5127	12	114.3163	87476.58	B2-0P11	
5128	37	114.3230	87471.49	B4-1P9	
5129	55	114.3243	87470.47	B11-4R2 B8-1R15 C7-7P5	
5130	8	114.3310	87465.33	B18-6R1	
5131	47	114.3517	87449.54	C1-3Q10 C2-3R16	
5132	0	114.3552	87446.87	C5-6P3	NI 114.36458 Ke
5133	37	114.3650	87439.34	B14-3P13	+NI 114.36508 Ke
5134	0	114.3691	87436.21	C0-2Q13	
5135	2	114.3730	87433.25	B13-4P7 C5-6Q4	
5136	3	114.3803	87427.68	B8-3R4 C7-7Q6 B27-8R5	
5137	2	114.3859	87423.39	C5-5P12 B22-7P2	
5138	1	114.3877	87421.98	B26-8P2	
5139	20	114.3975	87414.47	C6-6P9 C6-6Q10	
5140	4	114.4005	87412.22	B18-6P1 B6-1P12 B26-7P9	
5141	5	114.4051	87408.66	B19-6R5	
5142	0	114.4108	87404.34	B30-9R1 B9-3P6 C5-5Q13	
5143		114.4170	87399.62	C3-4Q12 B15-4P10	
5144		114.4187	87398.30	B1-1R0	r 87397.54
5145		114.4207	87396.79	B1-1R0 B12-4P5 B7-1R14 B24-7R7	r 87397.54
5146	1300	114.4262	87392.60	C0-3R1 B13-3P12	
5147		114.4298	87389.79	C10-8P9	sh
5148	25	114.4346	87386.19	C1-3P9 B1-7P5	
5149		114.4366	87384.65	C0-1P18	sh
5150	245	114.4438	87379.10	C0-3R2	
5151		114.4488	87375.34	B2-0R12	sh
5152	730	114.4518	87373.06	C0-3R0	
5153		114.4560	87369.86	B4-1R10 B30-9P1	sh
5154	10	114.4591	87367.42	B18-6R2	
5155		114.4663	87361.98	B18-6R2	

Nb	I	λ (nm)	σ (cm^{-1})	Assignment	Comment
5157		114.4697	87359.35	B1-1R1	r 87358.56
5158		114.4718	87357.78	B1-1R1	r 87358.56
5159	0	114.4736	87356.36	C4-5P8	
5160	9	114.4776	87353.34	C3-4P11 C1-3R12	CuII 114.48556
5161					
5162	37	114.4865	87346.56	B11-4P2 B22-7R3	
5163	6	114.4902	87343.73	B10-3P8 B26-8R3 C4-5Q9	b
5164	170	114.4978	87337.94	B2-1P5	
5165	6	114.5068	87331.05	B'4-9P4	
5166	1	114.5121	87327.04	C1-1P21	
5167	270	114.5230	87318.68	C0-3R3	
5168	2	114.5272	87315.48	B8-2R11	
5169	6	114.5314	87312.27	C0-2P12	
5170	35	114.5456	87301.52	B2-1R6 B'2-8R1	
5171	1	114.5548	87294.46	C11-9P5 B6-0P16	
5172	2	114.5561	87290.43	B12-4R6 B15-5P4	
5173	32	114.5629	87288.32	B11-4R3	
5174	12	114.5643	87287.20	B9-3R7 B'2-8R0 B'2-8R2	
5175	11	114.5742	87279.66	C0-2R15	
5176		114.5788	87276.20	B27-8P5	sh
5177		114.5823	87273.55	B28-8R7	sh
5178		114.5883	87268.93	B7-0P17	sh
5179	3320	114.5900	87267.63	C0-3Q1	
5180		114.5940	87264.58	B18-6P2	sh
5181	13	114.5999	87260.11	C5-6P4	
5182	0	114.6025	87258.12	B12-3P11	
5183	7	114.6087	87253.41	B6-2P7	
5184	4	114.6119	87251.01	C5-6Q5 C9-8Q7 B5-0P15	
5185		114.6147	87248.87	B1-1P1	r 87248.23
5186		114.6163	87247.59	B1-1P1	r 87248.23
5187		114.6204	87244.51	B1-1R2	r 87244.11
5188		114.6215	87243.71	B1-1R2	r 87244.11
5189	8	114.6277	87238.96	B26-8P3	
5190	1	114.6338	87234.33	B19-6P5	
5191	10	114.6370	87231.90	B6-1R13 C8-7P10	
5192	9	114.6384	87230.78	B16-5R7 C2-3Q15	
5193	14	114.6438	87226.72	B22-7P3	
5194	45	114.6493	87222.52	C0-3R4 C7-7P6	
5195	10	114.6615	87213.27	C2-4P6	
5196	15	114.6675	87208.67	B4-2R0	
5197	0	114.6709	87206.12	B24-7P7	
5198	93	114.6760	87202.23	C3-5R1	
5199	25	114.6822	87197.47	B14-5R0	
5200		114.6868	87194.00	B0-0P8 B14-4P9	
5201	15	114.6900	87191.54	B18-6R3	
5202	700	114.6949	87187.83	B15-5R5 B11-2P14 C0-3Q2	
5203	2	114.7001	87183.92	B14-3R14 C7-7Q7 C0-1R21	
5204	17	114.7074	87178.36	C3-5R2 B8-3P4	
5205	0	114.7102	87176.23	B23-7R6 B8-0P18 C3-4R14 B'2-8R4	

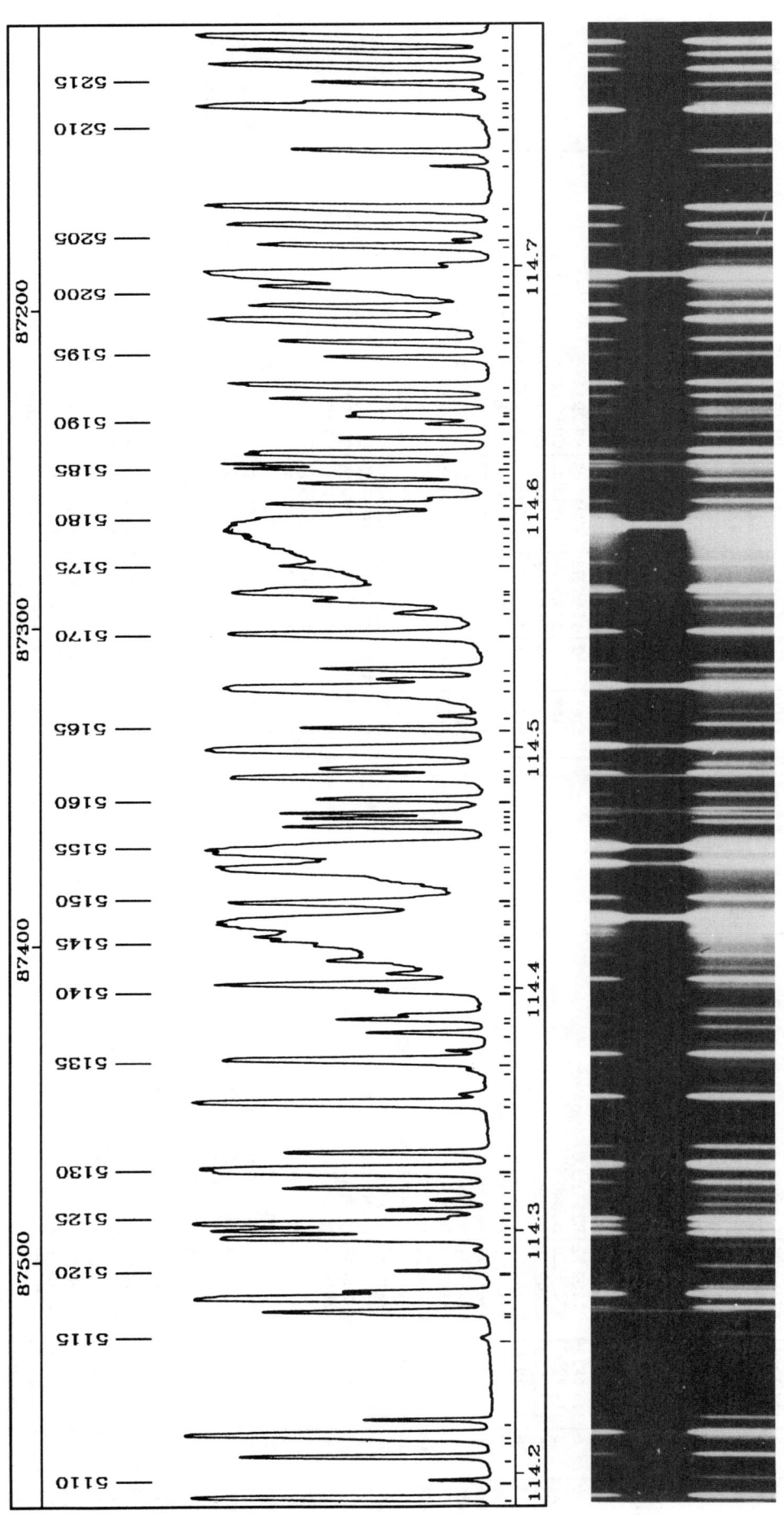

PLATE 63

TABLE 63 *continued*

Nb	I	λ (nm)	σ (cm⁻¹)	Assignment	Comment
5206	32	114.7155	87172.20	B'2-8P1	
5207	65	114.7228	87166.62	B4-2R1 B30-9R3	
5208	1	114.7402	87153.45	B13-3R13	
5209	9	114.7462	87148.83	B15-3R15 B'4-9P5	
5210	0	114.7549	87142.28	C8-7Q11	
5211	0	114.7597	87138.60	B26-8R4	
5212	110	114.7636	87135.67	B14-5R1	
5213	7	114.7656	87134.14	B5-1P11	
5214	1	114.7711	87129.98	B4-0P14	
5215	11	114.7740	87127.76	B7-2P9 B22-7R4	
5216	30	114.7805	87122.81	B8-3R5 B10-2P13	
5217	18	114.7866	87118.19	C3-5R3	
5218	90	114.7920	87114.09	B11-4P3	

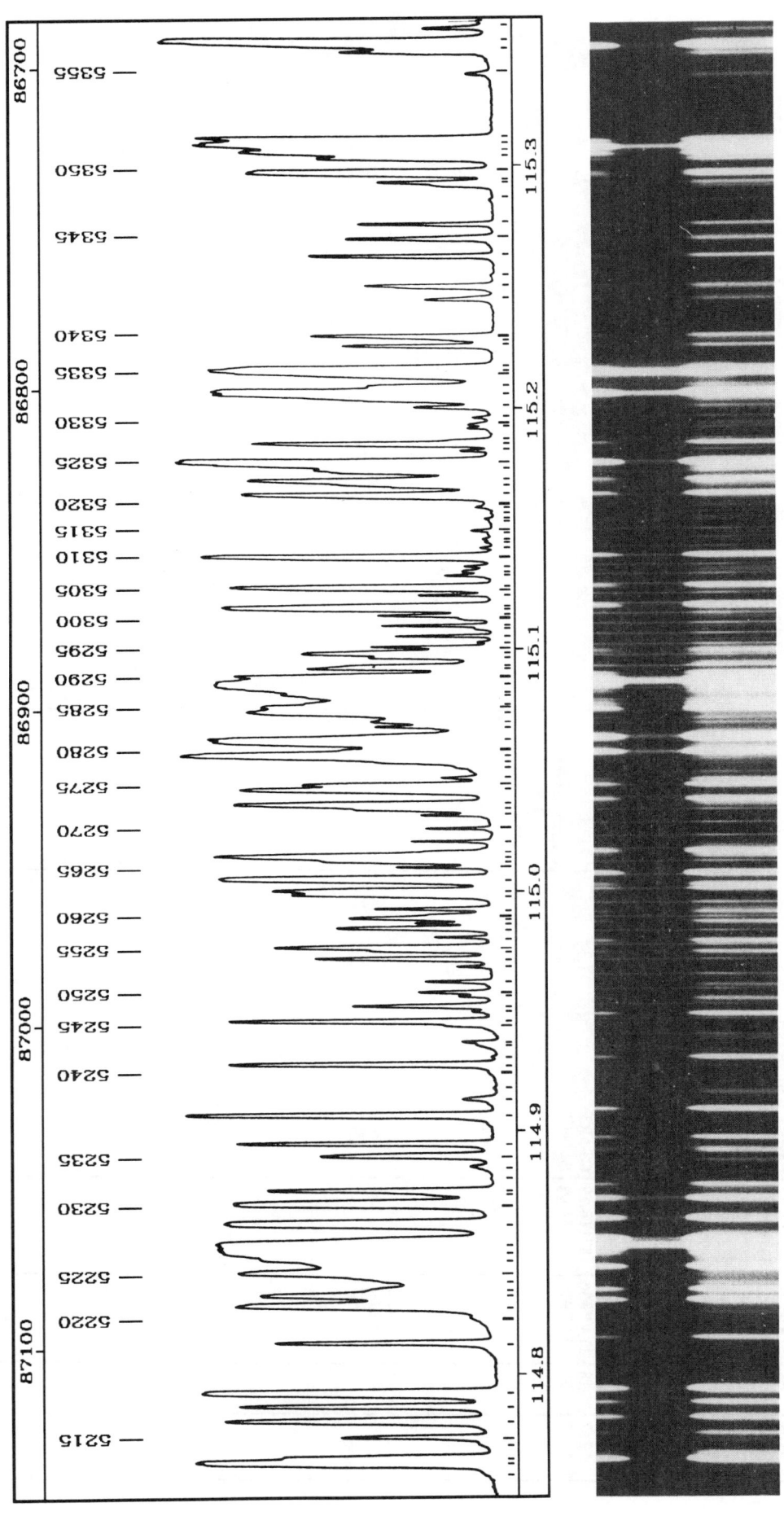

PLATE 64

TABLE 64

Nb	I	λ (nm)	σ (cm⁻¹)	Assignment	Comment
5219	12	114.8129	87098.25	B5-2P5	
5220	0	114.8232	87090.39	C8-8R1	
5221	65	114.8284	87086.46	C3-5Q1 C8-8R0	
5222	25	114.8326	87083.26	C1-3Q11 B'2-8R5	
5223	0	114.8352	87081.30	B30-9P3	
5224	0	114.8382	87079.00		
5225	65	114.8420	87076.15	B14-5P1	
5226		114.8478	87071.75	C0-3R5 B'2-8P2	
5227	1080	114.8509	87069.44	C0-3Q3 B4-2P1	
5228	420	114.8532	87067.67	C0-3P2	sh
5229		114.8620	87061.00	B1-1P2	
5230		114.8702	87054.78	B1-1R3 B18-6P3	r
5231		114.8744	87051.57	C6-6Q11 B1-0P10	r
5232	17	114.8758	87050.52	B4-2R2 B16-3R16	sh
5233	0	114.8810	87046.64	C10-9Q1	
5234	1	114.8859	87042.87	B6-0R17	
5235	8	114.8903	87039.54	B11-4R4 B25-8R0	
5236	14	114.8952	87035.83	C5-6P5	
5237	30	114.9069	87026.97	B14-5R2	
5238	3	114.9140	87021.61	B9-0P19 C3-5R4	
5239		114.9182	87018.44		CO B-X(0-0)R22
5240		114.9247	87013.47		CO R21
5241	15	114.9282	87010.85	C3-5Q2 C8-8Q1	
5242		114.9311	87008.64		CO R20
5243	1	114.9361	87004.84	C4-5Q10	
5244		114.9374	87003.86		CO R19
5245		114.9443	86998.70		CO R18
5246	15	114.9461	86997.30	B25-8R1	
5247		114.9504	86994.08		CO R17
5248	5	114.9526	86992.39	C4-5P9 C2-3R17	
5249		114.9566	86989.36		CO R16
5250	3	114.9583	86988.04	B5-1R12 B7-2R10 C8-8R3 B24-7R8	
5251	1	114.9627	86984.72	B1-0R11 B7-0R18	+CO R15
5252	15	114.9689	86980.06		CO R14
5253	12	114.9720	86977.74	C3-4Q13	+CO R13 sh
5254	5	114.9750	86975.41	C9-8Q8	
5255	15	114.9765	86974.28	C10-9Q2 B22-7P4 C0-2Q14 B9-2P12	
5256	3	114.9810	86970.91	C9-7P13 C1-3P10 B'2-8R6	CO R12
5257	8	114.9848	86968.05		CO R11
5258		114.9869	86966.42		
5259	7	114.9891	86964.76	B21-7R0 C5-5Q14	
5260	0	114.9907	86963.58	B29-9R0 C3-4P12	sh
5261	0	114.9929	86961.91	B18-6R4	+CO R10
5262	10	114.9986	86957.56	B3-1P8 C2-4Q8	+CO R9
5263	13	115.0003	86956.32	B25-8P1 C1-3R13	
5264	56	115.0051	86952.64	B7-3R0	+CO R8
5265		115.0101	86948.89		CO R7
5266		115.0125	86947.05	B4-0R15 B15-5P5	sh
5267	35	115.0144	86945.62	B'2-8P3	+CO R6
5268	0	115.0168	86943.83	C7-7P7 B'1-7P7 B'4-9P6	CO R5
5269		115.0212	86940.50		CO R4
5270		115.0267	86936.35		CO R3
5271		115.0322	86932.18		sh
5272		115.0346	86930.38	B3-0P13	
5273	30	115.0364	86929.03	C0-3R6 B14-5P2	
5274	22	115.0425	86924.41	C0-1Q20 B21-7R1	
5275	7	115.0444	86922.98	B29-9R1 B11-2R15	
5276		115.0476	86920.51		CO R0
5277	0	115.0517	86917.47	C10-9P2	
5278		115.0534	86916.19		CO B-X(1-1)R14
5279	175	115.0564	86913.90	C0-3Q4	
5280	260	115.0595	86911.56	C7-7Q8	
5281		115.0623	86909.03	B7-3R1 C2-4P7	CO (0-0)P3
5282		115.0683	86904.94	C10-9P3	
5283	0	115.0703	86903.43	B9-3P7	
5284	55	115.0742	86900.46	C3-5P2	
5285	21	115.0762	86898.97	B25-8R2 C3-5Q3	
5286	0	115.0774	86898.00	B12-2R16	+CO P5
5287	0	115.0809	86895.38	B21-5P13	
5288	1820	115.0827	86894.05		CO P6
5289	28	115.0854	86892.01	C0-3P3	
5290	7	115.0886	86889.56	B4-2P2 B26-8R5	
5291	5	115.0920	86887.02		+CO P8
5292	2	115.0929	86886.33	B29-9P1 C3-5R5	
5293	15	115.0967	86883.46	B10-2R14	+CO P9
5294		115.0984	86882.15	B3-1R9	
5295		115.1010	86880.23		CO P10
5296		115.1028	86878.83		CO (1-1)R5
5297		115.1054	86876.92		CO (0-0)P11
5298		115.1083	86874.75		CO (1-1)R4
5299		115.1097	86873.63		CO (0-0)P12
5300	0	115.1125	86871.54	C8-8P2	
5301		115.1139	86870.47		CO P13
5302	63	115.1172	86868.00	B2-1P6 B21-7P1 B17-4P13 C10-9Q3	
5303		115.1182	86867.22		CO P14
5304		115.1223	86864.12		CO P15
5305	30	115.1256	86861.67	B4-2R3 B8-0R19	
5306		115.1266	86860.90		CO P16
5307		115.1305	86857.95		CO P17
5308	1	115.1323	86856.59	B12-4R7 B22-7R5	
5309		115.1343	86855.08		CO P18
5310	37	115.1384	86852.04	B14-5R3	+CO P19
5311		115.1420	86849.30		CO P20
5312		115.1444	86847.45		CO (1-1)P2
5313		115.1457	86846.48		CO (0-0)P21
5314	0	115.1478	86844.93	B15-5R6	
5315		115.1492	86843.83		CO(0-0)P22 +(1-1)P3
5316		115.1530	86840.99		CO (0-0)P23

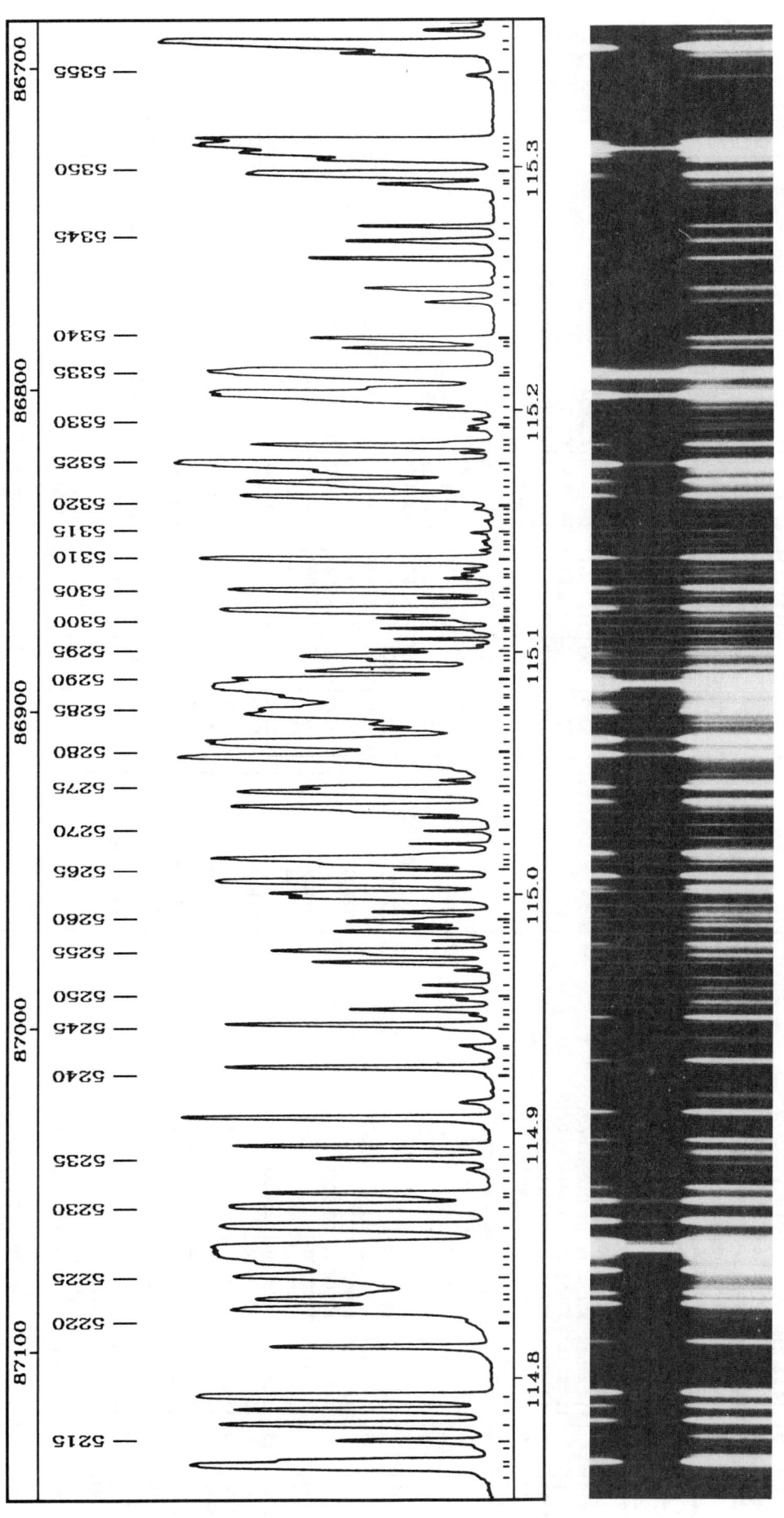

PLATE 64

TABLE 64 *continued*

Nb	I	λ (nm)	σ (cm-1)	Assignment	Comment
5317		115.1540	86840.25		CO (1-1)P4
5318		115.1566	86838.31		CO (0-0)P24
5319		115.1591	86836.41		CO (1-1)P5
5320		115.1598	86835.83		CO (0-0)P25
5321	29	115.1642	86832.54	C0-2P13 B'2-8R7	+CO P26
5322	0	115.1686	86829.24	B29-9R2	+CO (1-1)P7
5323	21	115.1701	86828.08	B2-1R7 B21-7R2	+CO (0-0)P28
5324	8	115.1748	86824.56	B25-8P2	
5325	230	115.1773	86822.63	B7-3P1	
5326		115.1826	86818.67		CO P32
5327	17	115.1853	86816.63	B11-4P4	
5328		115.1868	86815.48		CO (1-1)P11
5329	0	115.1919	86811.67	C0-2R16	+CO P12
5330	0	115.1935	86810.42	B10-0P20	+CO (0-0)P36
5331		115.1963	86808.37		CO P37
5332	1	115.2005	86805.18	B16-4P12 C8-7P11	
5333	400	115.2061	86800.99	B1-1P3	
5334	0	115.2092	86798.65	B13-2R17	
5335					OI 115.21512
5336	65	115.2156	86793.84	B7-3R2 B'2-8P4	+ OI 115.21512
5337		115.2164	86793.22	B1-1R4	sh
5338	6	115.2258	86786.11	B18-6P4	
5339		115.2288	86783.87	B3-0R14 C10-9P3 C0-1P19	sh
5340	13	115.2299	86783.06	C9-8P8	
5341	5	115.2454	86771.40	B27-8R7 B9-2R13 B24-7P8 C3-4R15	b
5342	7	115.2507	86767.36	B4-1P10 B29-9P2 C5-6P6	
5343	0	115.2550	86764.16	B16-5R8 C10-9R5	
5344	8	115.2633	86757.88	C3-5P3	
5345	6	115.2706	86752.36	B8-2P11 C3-5Q4	
5346	5	115.2767	86747.83	B25-8R3	
5347	1	115.2866	86740.38	C2-3Q16	sh
5348		115.2926	86735.81	B10-3P9	
5349	7	115.2940	86734.77	C8-8P3	
5350	17	115.2980	86731.78	B21-7P2 B15-4P11 C6-7R1	
5351	8	115.3043	86727.05	B11-4R5	
5352	28	115.3065	86725.40	C0-3R7 C6-7R0	
5353	142	115.3096	86723.12	C0-3Q5	
5354	25	115.3117	86721.50	B17-6R0	
5355	3	115.3390	86700.94	C6-7R2 C6-6P11	
5356	4	115.3480	86694.21	B21-7R3	
5357	240	115.3517	86691.40	B14-5P3 C2-3P15	
5358	1	115.3578	86686.81	B29-9R3	

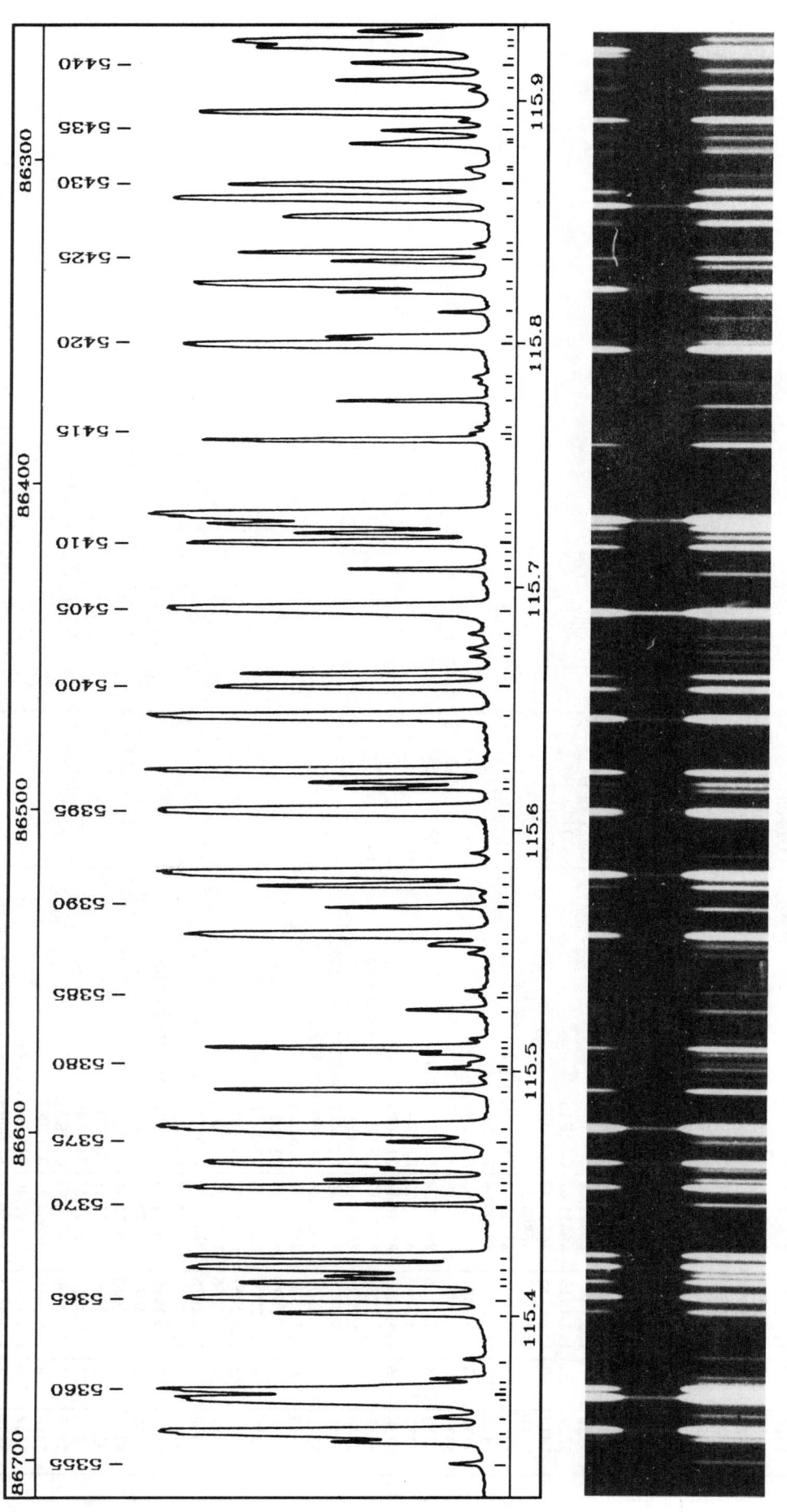

PLATE 65

TABLE 65

Nb	I	λ (nm)	σ (cm⁻¹)	Assignment	Comment
5359	240	115.3651	86681.32	C0-3P4 B19-6R7 C1-3Q12	
5360	70	115.3687	86678.66	B17-6R1	
5361	2	115.3709	86677.01	B6-2P8 B9-0R20	
5362	3	115.3734	86675.06	B18-6R5 C9-8Q9	
5363	2	115.3815	86669.03	B22-7P5	
5364	11	115.4001	86655.01	B4-1R11	
5365	122	115.4064	86650.31	B7-3P2	
5366	21	115.4124	86645.78	B2-0P12 C6-7Q1 C4-5Q11	
5367	5	115.4152	86643.73	C2-4Q9 B5-2P6	
5368	70	115.4187	86641.05	B10-4R0 B25-8P3	
5369	41	115.4232	86637.70	B4-2P3 B14-3P14	
5370	5	115.4445	86621.75	B'2-8P5	
5371	55	115.4515	86616.45	B17-6P1	
5372	3	115.4545	86614.21	B14-5R4 C6-7R3	
5373	2	115.4589	86610.94	C4-5P10 B13-3P13	
5374	60	115.4614	86609.01	B7-3R3	
5375	1	115.4703	86602.36	B4-2R4	
5376	360	115.4759	86598.16	B10-4R1 B29-9P3	
5377	20	115.4914	86586.55	C1-3P11	
5378	0	115.4963	86582.86	B33-10R1	
5379	3	115.5005	86579.73	C2-4P8	
5380	2	115.5026	86578.15	B8-2R12 B15-3P15	
5381	2	115.5064	86575.27	C6-7Q2	
5382	11	115.5088	86573.50	B17-6R2	
5383	1	115.5123	86570.91	C6-7R4 C3-5Q5	
5384	4	115.5243	86561.86	C3-5P4 B6-1P13 C1-3R14	
5385	0	115.5298	86557.75	B33-10P1	
5386	1	115.5320	86556.08	C3-5R7	
5387	1	115.5476	86544.42	B25-8R4	
5388	2	115.5513	86541.65	C5-6P7 C3-4Q14	
5389	35	115.5558	86538.27	B22-7R6 B2-0R13 B21-7P3 B'3-9R1	
5390	4	115.5671	86529.82	B24-7R9 B'3-9R2	
5391	0	115.5713	86526.71	B0-0P9 C5-6Q8	
5392	10	115.5761	86523.07	B'3-9R0	
5393	260	115.5812	86519.28	B10-4P1 B9-1R17	
5394	1	115.5891	86513.37	C5-5Q15	
5395	43	115.6075	86499.56	B7-1R15 C0-3Q6 C0-2Q15 B'3-9R3	
5396	5	115.6166	86492.79	C3-4P13	
5397	6	115.6193	86490.73	B33-10R2 C6-7P2 C0-3R8	
5398	60	115.6243	86487.05	B10-4R2	
5399	160	115.6465	86470.41	B1-1P4 B17-6P2 C6-7Q3	
5400	26	115.6589	86461.16	B1-1R5 B18-6P5	
5401	15	115.6642	86457.21	B11-4P5	
5402	1	115.6707	86452.31	B28-9R0	
5403	1	115.6742	86449.69	B33-10P2 C9-8P9 B'3-9R4	
5404	280	115.6802	86445.22	B7-2P10	
5405		115.6905	86437.55	B10-0R21 C0-3P5 C6-7R5	
5406	0	115.7038	86427.62		
5407	2	115.7069	86425.26	B24-8R0 B21-7R4	

Nb	I	λ (nm)	σ (cm⁻¹)	Assignment	Comment
5408	0	115.7103	86422.76	B12-1P19	
5409	0	115.7142	86419.84	C10-9P5	
5410	32	115.7180	86416.99	B'3-9P1	
5411	7	115.7220	86413.97	B28-9R1	
5412	22	115.7257	86411.21	B14-5P4	
5413	410	115.7293	86408.53	B7-3P3 B17-6R3	
5414	18	115.7605	86385.22	B24-8R1	
5415	0	115.7630	86383.38	B15-5R7 B29-9P4	
5416	0	115.7655	86381.55	B6-1R14	
5417	4	115.7768	86373.13	B28-9P1	
5418	0	115.7839	86367.81	B14-3R15	
5419	0	115.7866	86365.80	C7-7P9 C8-8P5	
5420	150	115.8003	86355.59	B7-3R4 B10-4P2 B11-4R6	
5421	3	115.8033	86353.32	B33-10R3 C0-2P14	
5422	2	115.8133	86345.90	C6-7P3	
5423	5	115.8218	86339.54	B13-4P9 C0-2R17	
5424	50	115.8253	86336.96	B2-1P7 B24-8P1 C0-1Q21 C3-5P5	
5425	5	115.8344	86330.15	B5-1P12 B11-3P11	
5426	13	115.8381	86327.42	B'3-9P2	
5427	0	115.8414	86324.94	B28-9R2	
5428	10	115.8526	86316.55	B4-2P4 B14-5R5 C3-5R8	
5429	78	115.8600	86311.06	B10-4R3	
5430	26	115.8656	86306.89	B3-1P9 C2-4Q10 C3-4R16	
5431	0	115.8712	86302.71	B11-2P15 C7-7Q10	
5432	1	115.8725	86301.73	B7-2R11	
5433	6	115.8828	86294.11	B2-1R8 B33-10P3	
5434	4	115.8840	86293.18	B21-7P4	
5435	4	115.8881	86290.10	B25-8R5 B24-8R2	
5436	2	115.8919	86287.28	C4-5P11	
5437	25	115.8956	86284.59	B13-5R0	
5438	2	115.9046	86277.87	C6-6Q13	
5439	10	115.9087	86274.80	B4-2R5 B1-0P11	sh
5440		115.9146	86270.43	B6-0P17 B32-10R0	
5441	10	115.9160	86269.34	C4-5Q12	
5442	42	115.9225	86264.57	B8-3R7 C1-3Q13	
5443	41	115.9248	86262.83	B17-6P3 C6-6P12	
5444	7	115.9289	86259.74	B10-2P14 B16-5R9 B29-9R5	

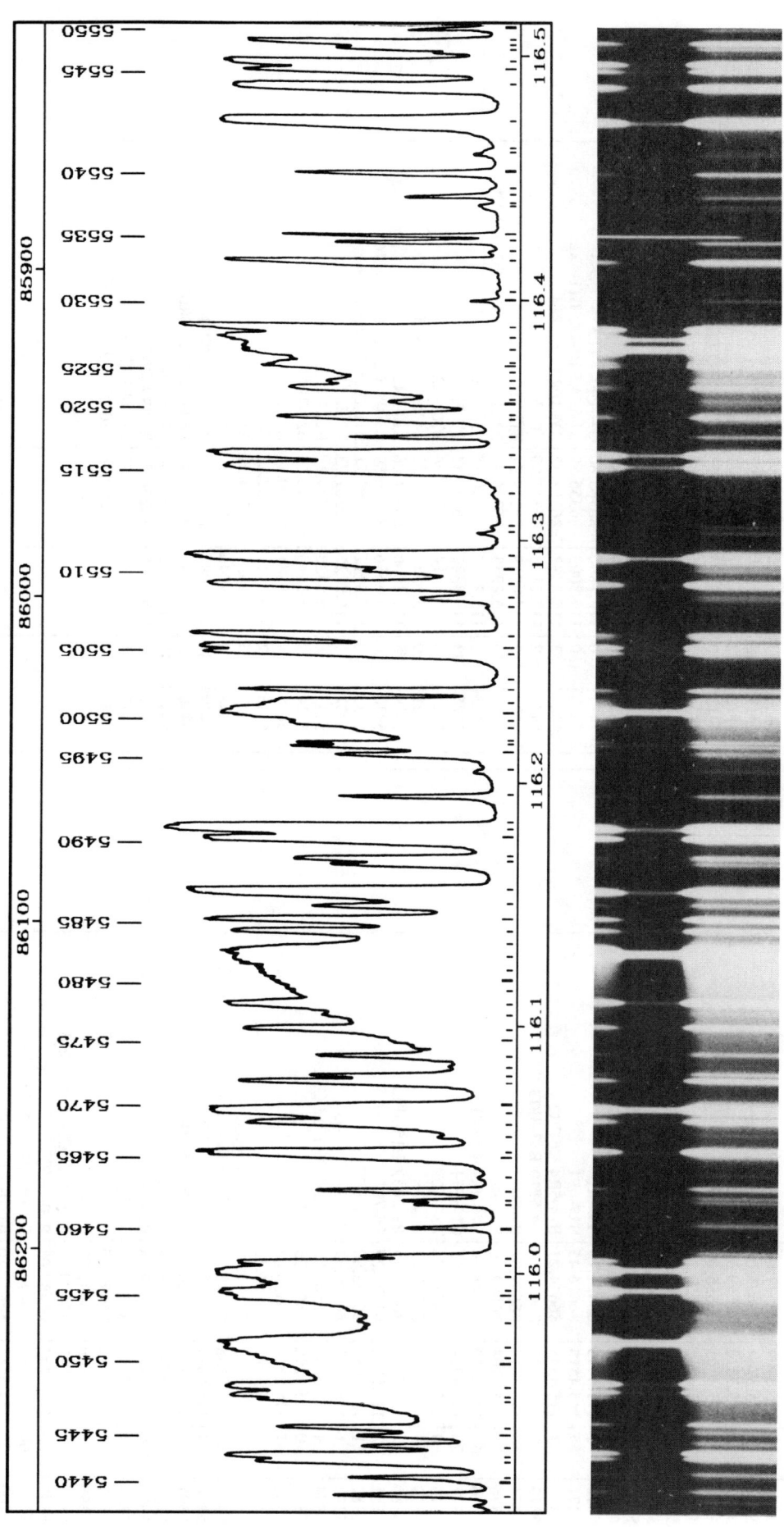

PLATE 66

TABLE 66

Nb	I	λ (nm)	σ (cm⁻¹)	Assignment	Comment
5445	5	115.9334	86256.41	B28-9P2 B5-0P16	
5446	17	115.9365	86254.13	C0-3R9	
5447	13	115.9477	86245.75	B20-7R0	
5448	80	115.9498	86244.23	C0-3Q7	
5449	260	115.9531	86241.78	B13-5R1 B19-6R8 C2-3Q17	
5450	5	115.9631	86234.36	C5-6Q9	
5451	10	115.9663	86231.98	B32-10R1	
5452	5800	115.9698	86229.32	C1-4R1 B3-1R10 C5-6P8	
5453	0	115.9807	86221.21	B2-8P7	
5454		115.9889	86215.12	B7-0P18 B'3-9P3	
5455	780	115.9911	86213.53	C1-4R2	sh
5456		115.9938	86211.50	C2-4P9	sh
5457	2500	115.9994	86207.37	C1-4R0	
5458	66	116.0032	86204.50	B20-7R1 B26-8R7	
5459	2	116.0063	86202.21	B32-10P1 B13-1P20	
5460	5	116.0178	86193.68	B9-3R9	
5461	3	116.0273	86186.58	B28-9R3	
5462	3	116.0287	86185.56	B17-6R4	
5463	13	116.0328	86182.50	B5-1R13	
5464	1	116.0385	86178.33	B16-3R17	
5465		116.0469	86172.05	B4-0P15	sh
5466	87	116.0479	86171.35	B13-5P1 C0-1P20	
5467	0	116.0550	86166.06	C6-7Q5 C6-7P4	
5468	49	116.0604	86162.02	C0-3P6 B21-7R5 B9-2P13	
5469	440	116.0652	86158.45	C1-4R3	
5470		116.0674	86156.84	B18-5P11 C2-3P16 D1-9R4	sh
5471	21	116.0773	86149.51	B20-7P1	
5472	10	116.0799	86147.58	B32-10R2 C1-3P12 D1-9R5	
5473	0	116.0836	86144.82	D1-9R3	
5474	8	116.0882	86141.40	B24-8R3	
5475	0	116.0944	86136.77	C8-8P6	
5476	32	116.0991	86133.31	B13-5R2 C1-3R15	
5477	5	116.1053	86128.74	B5-2P7	
5478	135	116.1096	86125.50	B10-4P3	
5479	0	116.1140	86122.24	B29-9P5	
5480	0	116.1207	86117.30	C3-5Q7	
5481		116.1250	86114.08		sh
5482	4230	116.1296	86110.73	C1-4Q1	
5483		116.1332	86108.03		sh
5484	50	116.1391	86103.64	C1-4R4 B20-7R2	
5485	39	116.1440	86100.01	B7-3P4 B8-0P19	
5486	11	116.1498	86095.71	B32-10P2 C3-4Q15 C9-8P10	
5487	224	116.1552	86091.72	C4-6R1	
5488	4	116.1672	86082.82	B16-4P13 B10-3P10 B'3-9P4	
5489	17	116.1692	86081.31	B0-1R0	
5490	93	116.1774	86075.25	C4-6R0	
5491	173	116.1816	86072.12	B1-1P5 C4-6R2	
5492	93	116.1833	86070.91	B14-5P5	sh
5493	10	116.1954	86061.92	B1-1R6	
5494	2	116.2059	86054.16	C5-5Q16	
5495	7	116.2130	86048.88	B6-2P9 C3-5R9	
5496	13	116.2162	86046.52	B0-1R1	r 86045.98
5497		116.2177	86045.44	B0-1R1	r 86045.98
5498	2	116.2220	86042.22		
5499	10	116.2259	86039.31	B11-4P6 B5-0R17 B11-2R16 B24-7R10	
5500	960	116.2304	86036.01	C1-4Q2 B7-3R5	
5501	13	116.2304	86033.22	B10-4R4 B10-2R15 C9-9R1	
5502	14	116.2392	86029.50	B24-8P3 B15-4P12 C9-9R0	b
5503	0	116.2439	86026.04	C8-7P13	
5504	50	116.2549	86017.87	C4-6R3	
5505	55	116.2571	86016.22	B13-5P2 B21-7P5 B32-10R3 C0-2Q16 C0-1R23	
5506	21	116.2615	86013.01	B20-7P2 B4-1P11 B3-0P14	
5507	0	116.2744	86003.45	B28-9R4	
5508	3	116.2744	86000.95	C9-9R2 C11-10R0 C3-4P14	b
5509	63	116.2832	85996.98	B3-2R0 B17-6P4	
5510	2	116.2893	85992.45	C11-10R1 B25-8R6	
5511	360	116.2940	85988.96	C4-6Q1 C2-3R19	
5512	3	116.3043	85981.31	B8-2P12 B29-9R6	
5513	0	116.3070	85979.33	B12-2R17	
5514	0	116.3206	85969.30	C6-7Q6 C7-7Q11	
5515	171	116.3318	85961.05	C1-4R5 B13-5R3 C0-3Q8 B14-5R6	
5516	195	116.3365	85957.52	B3-2R1 B9-2R14	
5517	6	116.3442	85951.84	C2-4Q11 C11-10R2	
5518		116.3500	85947.57	B18-6R7	sh
5519	18	116.3524	85945.78	B20-7R3 C0-3R10	
5520	2	116.3573	85942.17	B24-8R4	
5521	2	116.3590	85940.95	C11-10Q1	
5522	22	116.3646	85936.77	B0-1R2 C6-7P5 C9-9R3	
5523	8	116.3673	85934.79	B7-0R19 B'3-9P5	
5524	8	116.3703	85932.59	B15-5R8	
5525		116.3733	85930.34	B0-1P1 B4-2P5 B11-4R7	
5526		116.3747	85929.30	B0-1P1	r 85929.82
5527	1700	116.3805	85925.03	B9-0P20 C1-4Q3	r 85929.82
5528	560	116.3847	85921.95	C1-4P2 C5-6Q10	
5529	98	116.3886	85919.11	C4-6Q2 B17-6R5	
5530	1	116.4005	85910.28	B27-9R0	
5531	0	116.4040	85907.68	C6-6P13	
5532	16	116.4164	85898.55	B16-6R0 B4-1R12 C9-9Q2	
5533	1	116.4206	85895.49	C5-6P9	
5534	9	116.4244	85892.65	B8-3P7	
5535	11	116.4273			GeII 116.4273 Ke
5536	2	116.4383	85882.37	B4-2R6	
5537	3	116.4391	85881.80	C2-4P10	
5538	6	116.4431	85878.88	C4-5Q13	
5539	2	116.4455	85877.07	B28-9P4 C3-4R17 C8-8P7	
5540	7	116.4527	85871.78	B27-9R1 C6-6Q14	
5541	2	116.4599	85866.46	B3-0R15	
5542	1	116.4615	85865.32	B31-10R0	

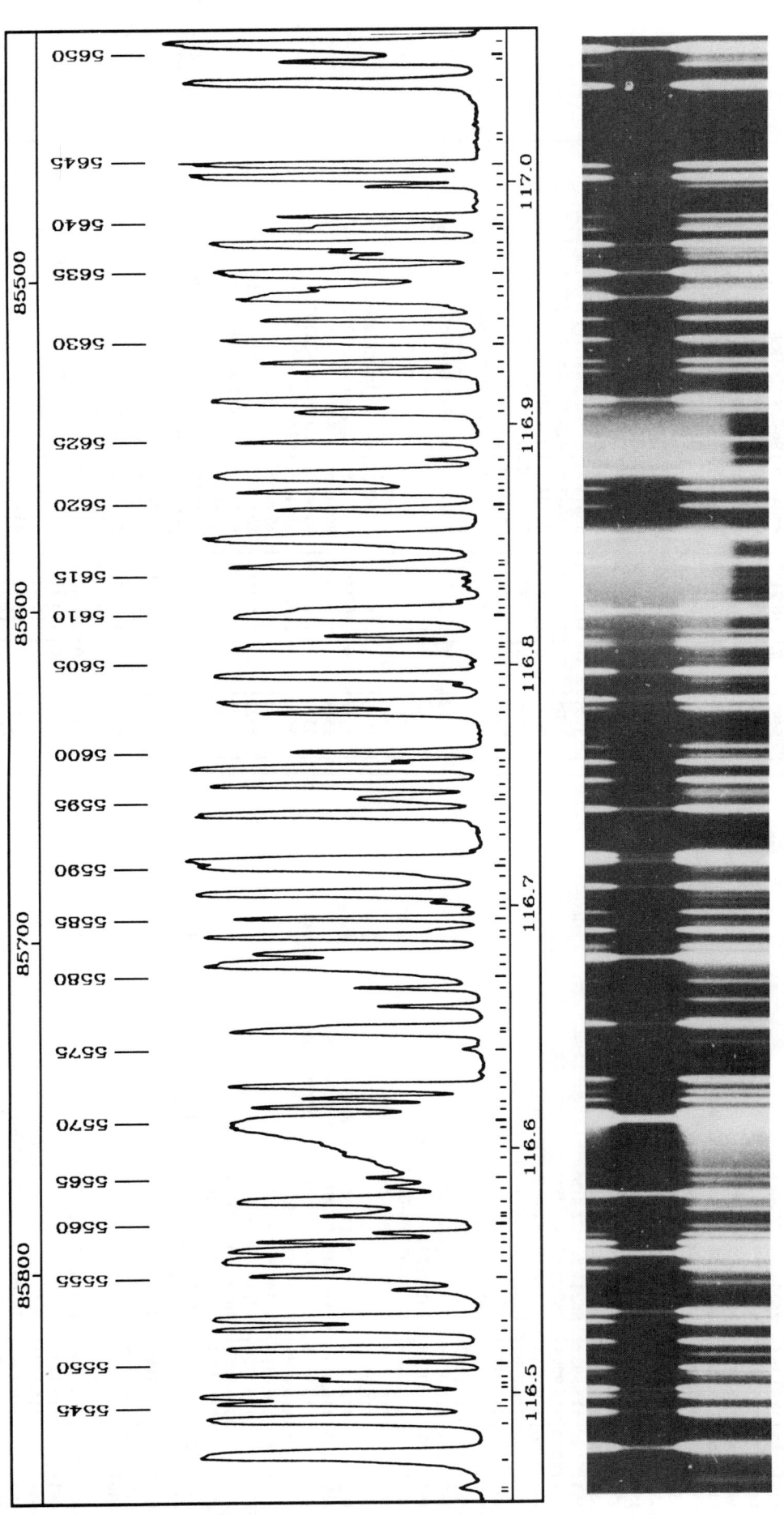

PLATE 67

TABLE 67

Nb	I	λ (nm)	σ (cm⁻¹)	Assignment	Comment
5543	230	116.4739	85856.14	B3-2P1 B16-6R1	
5544	73	116.4887	85845.24	B3-2R2 B21-7R6 B32-10R4 C4-5P12	
5545	49	116.4952	85840.42	C0-3P7	
5546	85	116.4981	85838.29	B6-3R0	
5547	2	116.5023	85835.25	B12-4R9	
5548	9	116.5043	85833.72	C9-9P2 C1-3Q14	
5549	26	116.5067	85831.96	B10-4P4 B27-9P1	
5550	2	116.5123	85827.84	B31-10R1 B26-8R8	
5551	33	116.5176	85823.94	C4-6P2	
5552	33	116.5258	85817.90	B20-7P3	
5553	122	116.5301	85814.74	C4-6Q3	
5554	4	116.5419	85806.05	B8-2R13	
5555	28	116.5477	85801.79	C1-4R6 C9-9Q3	
5556	460	116.5539	85797.19	B6-3R1 B13-5P3 B24-8P4 B14-3P15	
5557	56	116.5579	85794.25	B16-6P1 B31-10P1 B7-1P15	
5558	7	116.5614	85791.73	B4-10R1	
5559	6	116.5649	85789.09	C4-6R5 B28-9R5 B8-0R20	
5560	0	116.5685	85786.47	C9-8P11	
5561	5	116.5720	85783.86	B23-8R0	
5562	0	116.5737	85782.65	B27-9R2	
5563	230	116.5781	85779.42	C1-4Q4	
5564	2	116.5838	85775.23	B2-0P13 C11-10Q3	
5565	1	116.5876	85772.44	B4-10R2	
5566	0	116.5969	85765.57	B12-3P13	
5567	0	116.5996	85763.52	B8-3R8	
5568	1920	116.6044	85760.07	B19-6R9	sh
5569		116.6088	85756.80	C1-4P3	sh
5570		116.6111	85755.16	B0-1R3 B23-8R1	
5571	15	116.6168	85750.98	B16-6R2 C0-1Q22	
5572	10	116.6203	85748.34	B2-1P8 B9-3P9	b
5573	37	116.6254	85744.58	B0-1P2 B10-4R5 B31-10R2	
5574	1	116.6315	85740.12	C2-3Q18	
5575	2	116.6409	85733.23	B20-7R4	
5576	54	116.6484	85727.68	B7-3P5	
5577		116.6501	85726.43	B13-5R4	sh
5578	5	116.6587	85720.18	B7-2P11	
5579	5	116.6663	85714.53	B27-9P2	sh
5580		116.6744	85708.64	C9-9P3 C1-3R16	
5581	300	116.6765	85707.10	B6-3P1	
5582	22	116.6806	85704.04	B2-1R9 C7-8R2	
5583	38	116.6877	85698.81	C7-8R0 C7-8R1 B4-10P1	
5584	4	116.6904	85696.83	B10-0P21 B24-8R5	
5585	12	116.6953	85693.26	B23-8P1	
5586	0	116.6993	85690.32	B31-10P2	
5587	3	116.7021	85688.23	C4-6R6	
5588	85	116.7060	85685.41	B6-3R2	
5589	0	116.7128	85680.41	B6-1P14	
5590	133	116.7167	85677.57	B3-2P2 B17-6P5 B13-4P10 C4-6Q4	
5591	126	116.7189	85675.94	C4-6P3 B14-5P6 C9-9Q4 C1-3P13	

Nb	I	λ (nm)	σ (cm⁻¹)	Assignment	Comment
5592	0	116.7304	85667.51	B29-9R7	
5593	0	116.7345	85664.45	B2-0R14	+ NI 116.74485
5594	110	116.7385	85661.57	B3-2R3	
5595	0	116.7444	85657.18		
5596	30	116.7457	85656.27	B7-3R6 C11-10R5	
5597	60	116.7504	85652.79	B25-8R7 C0-3Q9	
5598	34	116.7581	85647.14	B16-6P2	
5599	3	116.7606	85645.36	B27-9R3 C11-10Q4	b
5600	12	116.7646	85642.43	C7-8R3 C3-4Q16	
5601	25	116.7806	85630.67	C7-7P11 C0-3R11	
5602	61	116.7846	85627.76	C7-8Q1	
5603	0	116.7921	85622.26	B28-9P5	
5604	79	116.7962	85619.25	B9-4R0	
5605	0	116.8010	85615.68	B31-10R3	
5606	0	116.8035	85613.91	B21-7P6	
5607	50	116.8074	85611.04	C1-4R7	
5608	31	116.8092	85609.71	B1-1P6	
5609	5	116.8126	85607.22	B3-1P10 B9-3R10 B'4-10P2	
5610	340	116.8213	85600.86	C1-4Q5	
5611	20	116.8234	85599.32	B1-1R7 B11-3P12	
5612	2	116.8271	85596.61	B8-1R17	
5613	0	116.8334	85592.94	C9-9R6	
5614	1	116.8368	85591.99	B32-10R5 C5-6Q11	
5615	0	116.8368	85589.47	C5-5Q17	
5616	30	116.8418	85585.83	B16-6R3 B9-0R21 C0-1P21	sh
5617		116.8433	85584.69	B7-1R16	
5618	250	116.8530	85577.65	B9-4R1 C2-4Q12 C3-4P15	
5619	11	116.8651	85568.73	B20-7P4 B23-8P2 B11-4P7	
5620					HeI 2x58.43340
5621	18	116.8724	85563.39	C7-8Q2	
5622		116.8752	85561.38	C3-5Q9	sh
5623	400	116.8790	85558.57	B5-2P8 C1-4P4	
5624	4	116.8852	85553.99	B14-5R7 C7-8R4 C9-9P4	
5625	10	116.8928	85548.46	B27-9P3	
5626	17	116.9054	85539.24	B31-10P3 C0-3P8	
5627	182	116.9099	85535.99	B6-3P2	
5628	20	116.9219	85527.19	B3-1R11	
5629	31	116.9259	85524.25	C0-2Q17 C9-9Q5	
5630		116.9337	85518.55	B18-6R8 B24-8P5	sh
5631	26	116.9351	85517.54	B13-5P4 B24-7R11	
5632	28	116.9437	85511.23	B13-4R11 C4-6Q5	
5633	173	116.9532	85504.33	B6-3R3 B19-7R0 B'4-10P3	
5634	12	116.9566	85501.82	B0-1R4 B6-1R15	
5635	250	116.9633	85496.94	B9-4P1 C4-6P4 C6-7P7	
5636	9	116.9696	85492.28	B23-8R3 C4-6R7	
5637	12	116.9724	85490.25	B5-1P13	
5638	116	116.9750	85488.33	B0-1P3	
5639	12	116.9815	85483.63	C7-8P2	
5640	9	116.9828	85482.68	B21-7R7 C2-4P11	

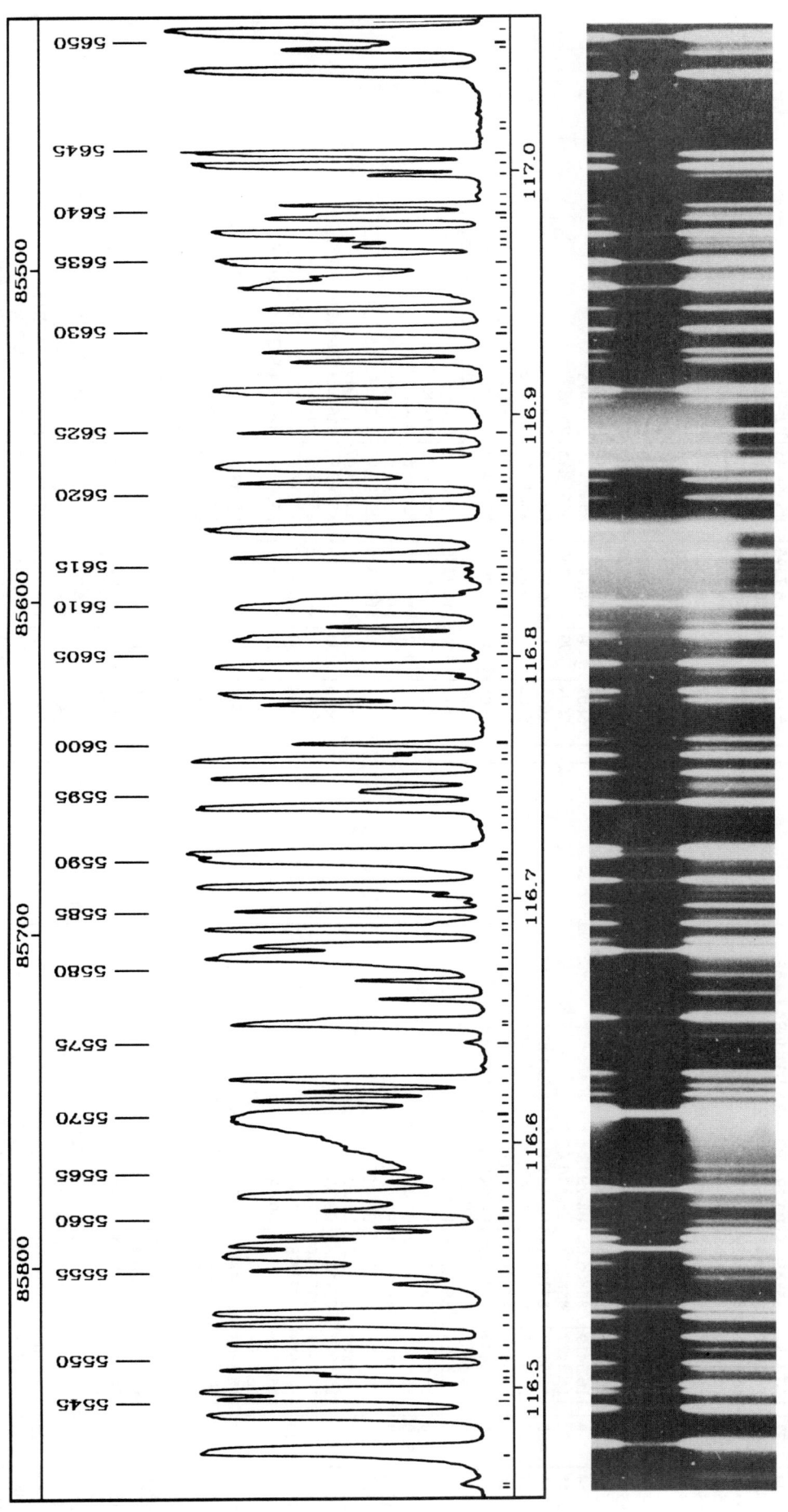

PLATE 67

TABLE 67 *continued*

Nb	I	λ (nm)	σ (cm⁻¹)	Assignment	Comment
5641	10	116.9865	85479.98	B10-4P5 B4-2P6	
5642	1	116.9904	85477.08	C4-5Q14	
5643	7	116.9992	85470.70	B11-4R8	
5644	75	117.0031	85467.84	B9-4R2 B20-7R5 C7-8Q3	
5645	39	117.0080	85464.22	B19-7R1	
5646	0	117.0174	85457.39	C6-6Q15	
5647	0	117.0196	85455.80	B1-8P1	
5648	123	117.0413	85439.91	B16-6P3 C3-4R18	
5649	16	117.0501	85433.52	B13-5R5 C0-1R24	sh
5650		117.0513	85432.60	C7-8R5	
5651	340	117.0569	85428.53	B3-2P3 B4-2R7 B15-5R9	

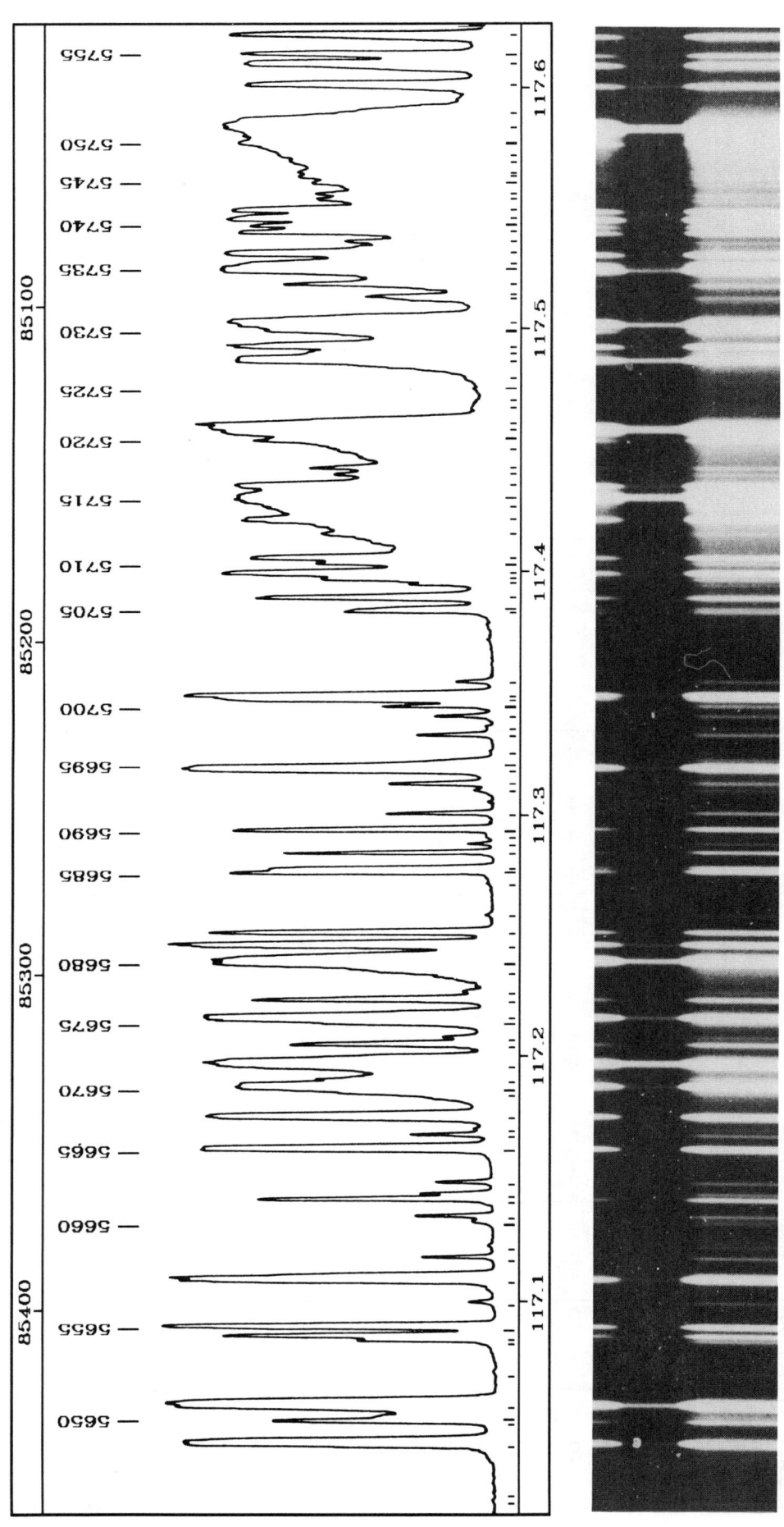

PLATE 68

TABLE 68

Nb	I	λ (nm)	σ (cm⁻¹)	Assignment	Comment
5652	0	117.0696	85419.29	B30-10R0	
5653	6	117.0827	85409.70	B26-8R9 C4-5P13	
5654	24	117.0845	85408.44	B3-2R4	
5655	29	117.0884	85405.54	B19-7P1	
5656	3	117.0986	85398.12	B23-8P3	
5657	100	117.1077	85391.47	C1-4Q6 C1-3Q15 B10-2P15 B10-3P11	
5658	3	117.1163	85385.21	B30-10R1	
5659	2	117.1209	85381.89	B10-4R6	
5660	7	117.1311	85374.42	B6-2P10	
5661	5	117.1333	85372.86	C1-4R8 C9-9P5	
5662	8	117.1402	85367.79	B19-7R2	
5663	6	117.1423	85366.26	C0-2R19	
5664	3	117.1469	85362.90	B16-6R4	
5665	49	117.1610	85352.61	B12-5R0	
5666	2	117.1664	85348.68	B30-10P1 B32-10R6	
5667	0	117.1687	85347.01	B31-10P4	
5668	63	117.1742	85343.03	B26-9R0 B5-1R14 C7-8Q4 C7-8P3	
5669	0	117.1843	85335.66	B27-9P4	
5670	182	117.1867	85333.89	B9-4P2	
5671	12	117.1893	85332.02	C0-2P16	
5672	620	117.1958	85327.33	C1-4P5	
5673	18	117.2035	85321.71	C0-3Q10	
5674	2	117.2063	85319.67	B8-3P8 B9-2P14	sh
5675		117.2123	85315.30	C4-6Q6	
5676	203	117.2147	85313.53	B12-5R1	
5677	13	117.2221	85308.15	B26-9R1 B17-6P6 C0-3R12	
5678	1	117.2252	85305.91	B30-10R2	
5679	0	117.2333	85300.02	C11-10Q6	d
5680	510	117.2381	85296.52	B6-3P3 B7-3P6 B23-8R4 B15-4P13	
5681	80	117.2447	85291.69	B9-4R3	
5682	27	117.2501	85287.77	C4-6P5	
5683	2	117.2571	85282.75	B17-4P15	
5684	0	117.2700	85273.33	C4-6R8 C7-8R6	
5685	17	117.2752	85269.53	B19-7P2	
5686	10	117.2763	85268.70	B20-7P5	
5687	6	117.2833	85263.62	B26-9P1	
5688	3	117.2871	85260.85	C1-3R17	
5689	0	117.2900	85258.76	B12-4R10	
5690	25	117.2933	85256.36	B6-3R4	
5691	7	117.2998	85251.62	C1-3P14	
5692	2	117.3093	85244.70	B30-10P2 C5-6Q12	
5693	7	117.3127	85242.27	B5-0P17 B19-6R10	
5694	138	117.3187	85237.90	B12-5P1	
5695	0	117.3216	85235.80	C2-3Q19	
5696	5	117.3244	85233.76	B31-10R5	
5697	5	117.3323	85228.05	B14-5P7 B6-0P18	
5698	2	117.3350	85226.08	C5-6P11	
5699	3	117.3399	85222.51	B26-9R2 B21-7P7	
5700	4	117.3441	85219.45	B4-1P12	
5701		117.3472	85217.22	B19-7R3 B7-3R7	sh
5702	60	117.3487	85216.14	B12-5R2	
5703	3	117.3542	85212.12	C7-8P4 B8-3R9	
5704	9	117.3826	85191.52	B4-0P16 C2-4Q13	
5705	12	117.3836	85190.77	C7-8Q5	
5706	32	117.3885	85187.22	C0-3P9	
5707	3	117.3941	85183.13	B30-10R3	
5708	6	117.3963	85181.56	C3-4Q17	
5709	45	117.3989	85179.69	B0-1R5 B13-5P5	
5710	7	117.4026	85176.97	B8-2P13 C8-7P15	
5711	25	117.4053	85175.05	B16-6P4	
5712	1	117.4151	85167.91	C0-1Q23	
5713	71	117.4212	85163.51	B0-1P4 C1-4R9	
5714	2170	117.4259	85160.06	B17-6R7 B10-2R16	sh
5715	143	117.4303	85156.93	C2-5R1 B20-7R6 C7-8R7	
5716	5	117.4349	85153.60	C1-4Q7	
5717	5	117.4392	85150.43	B7-0P19	
5718	1	117.4421	85148.37	B26-9P2	
5719	1	117.4489	85143.43		
5720	21	117.4539	85139.78	B23-8P4 C9-9Q7 D0-11Q6 ?	sh
5721	1000	117.4577	85137.05	B11-2R17 C2-5R0	
5722		117.4598	85135.49	C2-5R2	
5723	0	117.4673	85130.09	B13-1P21	
5724	0	117.4692	85128.71	C6-7R11	
5725	0	117.4741	85125.13	C2-3P18	
5726	0	117.4786	85121.92	C5-5Q18	
5727	310	117.4873	85115.61	C2-5R3 B9-2R15	
5728	12	117.4895	85114.00	B22-8R0 B31-10P5	
5729	78	117.4927	85111.71	B3-2P4	
5730	41	117.4990	85107.13	B2-1P9	
5731	440	117.5017	85105.18	B4-1R13 B9-4P3 C1-4P6 C4-6R9	
5732	2	117.5113	85098.18	B14-5R8	
5733	4	117.5125	85097.35	B30-10P3	
5734	14	117.5178	85093.44	C4-6Q7	
5735	340	117.5242	85088.85	C5-7R1 B26-9R3 C10-10R1	
5736		117.5257	85087.77	B1-1P7 B3-2R5 B13-5R6 C10-10R0	sh
5737	71	117.5305	85084.31	B12-5P2 B16-6R5	
5738	1	117.5351	85080.95	B27-9P5 D0-11Q5 ?	
5739	37	117.5395	85077.77	B19-7P3	
5740	24	117.5421	85075.88	B1-1R8 B22-8R1 B21-7R8	
5741	138	117.5451	85073.73	C5-7R0	
5742	52	117.5485	85071.26	C5-7R2	
5743	5	117.5526	85068.28	B3-0P15 C6-6P15	
5744	5	117.5548	85066.72	C4-5Q15	
5745	7	117.5595	85063.31	C2-4P12	
5746	8	117.5621	85061.46	B2-1R10 B32-10R7	
5747	7	117.5645	85059.69	C10-10R2	

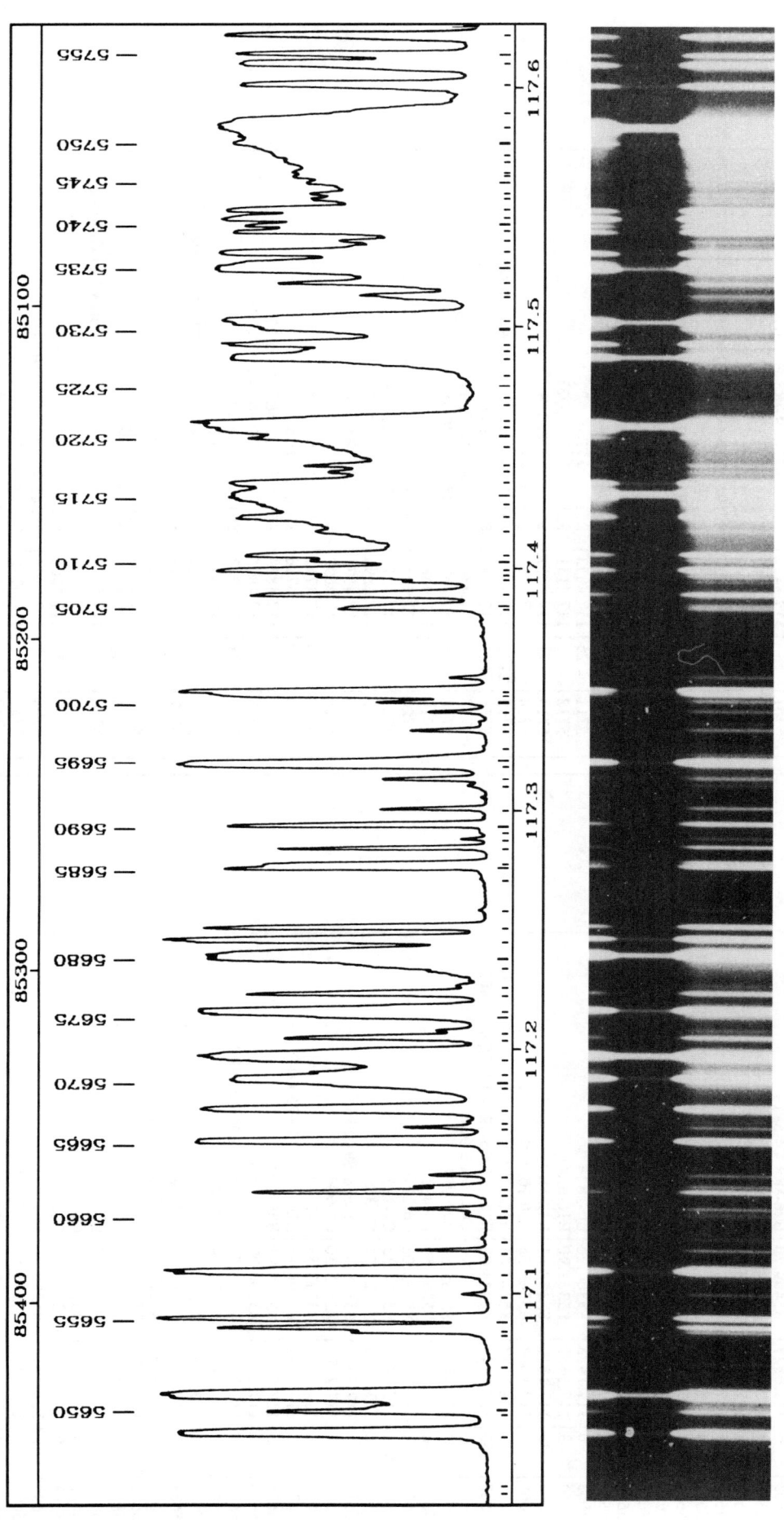

PLATE 68

TABLE 68 *continued*

Nb	I	λ (nm)	σ (cm-1)	Assignment	Comment
5748	10	117.5681	85057.10		d
5749	13	117.5712	85054.87		d
5750	55	117.5769	85050.70	B9-4R4 B15-6R0 B23-8R5 C4-6P6	
5751	4400	117.5831	85046.27	B11-4P8 C2-5Q1	
5752		117.5949	85037.70	C6-6Q16	sh
5753	92	117.6004	85033.72	B10-4P6 D0-11Q4 ?	
5754	85	117.6089	85027.58	B19-7R4 B13-4P11 C2-5R4 C10-10Q1	
				C4-5P14 C0-2Q18	
5755	15	117.6129	85024.71	B22-8P1	
5756	65	117.6208	85018.99	C5-7R3 C7-8P5	

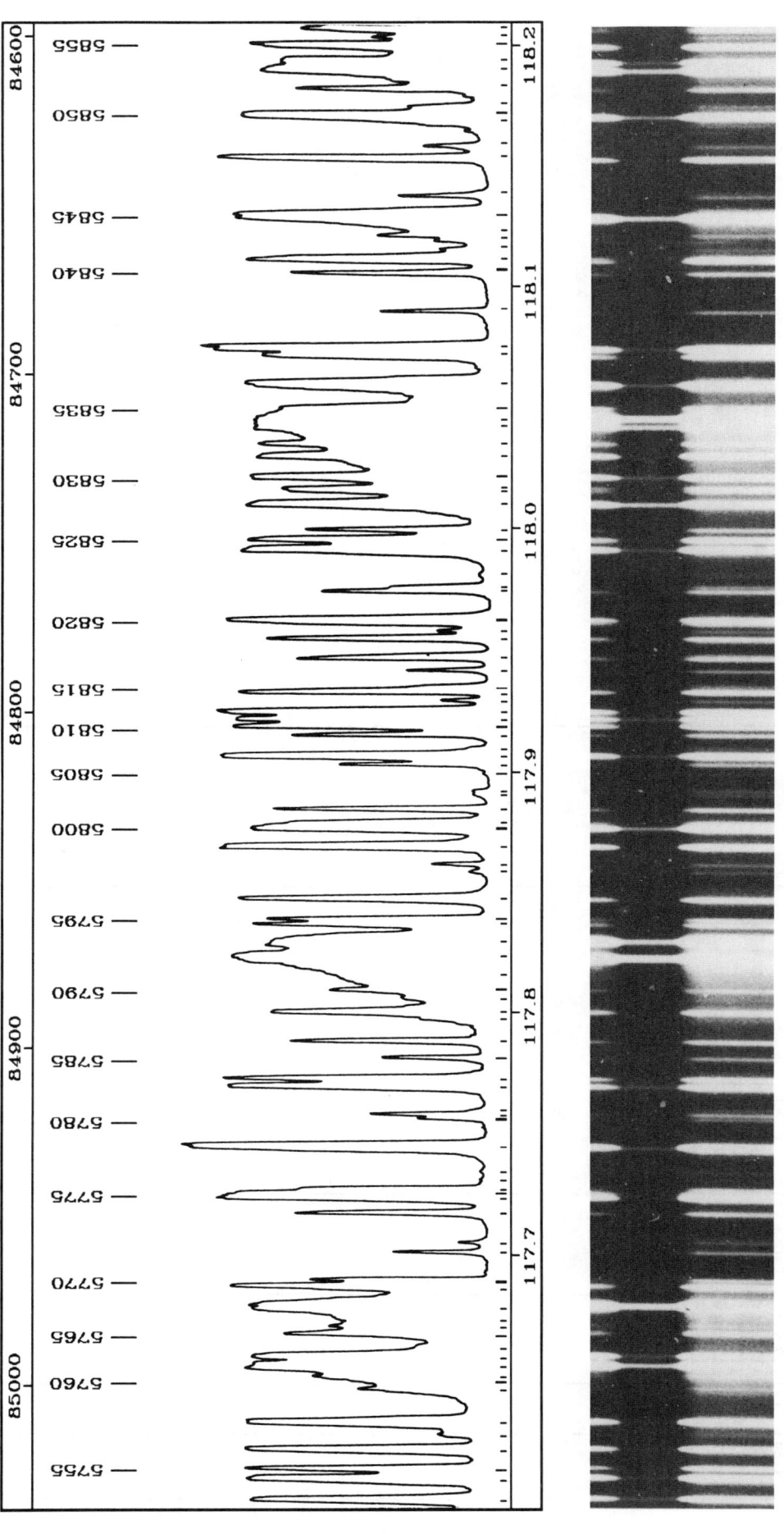

PLATE 69

TABLE 69

Nb	I	λ (nm)	σ (cm⁻¹)	Assignment	Comment
5757	1	117.6273	85014.31	B8-0P20 B'2-9R1	
5758	81	117.6319	85010.95	B15-6R1 C7-8Q6 B4-0R17	
5759	3	117.6459	85000.86	B8-2R14 C10-10R3	
5760		117.6502	84997.72		sh
5761	25	117.6512	84997.01	D0-11Q3	
5762	540	117.6552	84994.11	C5-7Q1	
5763	116	117.6588	84991.53	B6-3P4 C0-1P22	
5764	3	117.6628	84988.62	B14-4R13	
5765	12	117.6682	84984.72	B22-8R2 B26-9P3	
5766	6	117.6713	84982.50	B6-0R19 C12-11R0 C2-3R21	
5767	6	117.6756	84979.38	B24-7R12	
5768	1030	117.6793	84976.75	C2-5Q2	
5769	56	117.6872	84970.98	C0-3Q11 C12-11R1	
5770	5	117.6897	84969.22	C10-10Q2	
5771	7	117.7010	84961.05	B10-4R7 B9-3R11	
5772	2	117.7052	84957.99	B7-2P12 C3-4R19	
5773	19	117.7170	84949.49	C0-3R13	
5774	82	117.7233	84944.92	B15-6P1	
5775	0	117.7249	84943.76	B6-3R5	sh
5776	0	117.7264	84942.71	C1-3Q16	
5777	0	117.7309	84939.49	B5-2P9 B14-3P16	
5778	0	117.7340	84937.26	B29-10R0	
5779	131	117.7444	84929.89	C5-7Q2 C12-11Q1 C12-11R2 B12-3P14	
5780	2	117.7560	84921.33	B3-0R16 B'5-11P1	
5781	6	117.7577	84920.17	B20-7P6 C5-7R4	
5782	0	117.7595	84918.85	B4-2R8 C9-9P7	
5783	111	117.7694	84911.73	B13-2R19 C2-5R5	
5784	20	117.7720	84909.80	B15-6R2 B26-9R4 C10-10R4	
5785	4	117.7809	84903.43	B29-10R1 B17-6P7	
5786	8	117.7876	84898.57	B22-8P2	
5787	0	117.7971	84891.71	B24-6P15	
5788	58	117.8000	84889.65	C1-4Q8 C1-4R10 C5-6Q13	
5789	0	117.8057	84885.53	B2-9P1	
5790	6	117.8091	84883.06	C1-10Q3	
5791	1640	117.8175	84877.05	B25-8R9 C0-2R20	
5792	640	117.8222	84873.67	B7-0R20 C2-5Q3	sh
5793	29	117.8290	84868.77	C2-5P2 B12-5P3 B23-8P5 B8-1P17 C12-11Q2	
5794	18	117.8364	84863.46	B3-1P11	
5795	37	117.8382	84862.13	C5-7R5	
5796	2	117.8466	84856.06	B16-6P5 B7-1P16 C0-1R25 C12-11R3	
5797	3	117.8578	84848.00	C4-6Q8	
5798	76	117.8609	84845.76	B18-5P13 C10-10R5 B'5-11P2	
5799	200	117.8688	84840.08	C5-7P2 B22-8R3	
5800	23	117.8765	84834.55	B24-8P7 C5-7Q3	
5801	1	117.8784	84833.20	B19-7P4 B11-3P13	sh
5802	1	117.8841	84829.08	C0-2P17	
5803	0	117.8907	84824.33	B29-10R2	
5804	1	117.8916	84823.71	B9-0P21	
5805	0	117.8996	84817.95	C1-3R18	

Nb	I	λ (nm)	σ (cm⁻¹)	Assignment	Comment
5806	10	117.9031	84815.45	C0-3P10	
5807	94	117.9066	84812.90	B9-4P4 C4-6R10 B7-2R13	
5808	2	117.9098	84810.61	C7-8Q7	
5809	18	117.9152	84806.72	B7-3P7 C7-8P6	b
5810	49	117.9192	84803.82	C8-9R1	
5811	114	117.9219	84801.90	B20-7R7 B22-7R10 C1-4P7	
5812	48	117.9246	84799.94	B15-6P2 C10-10P3	
5813	2	117.9291	84796.75	B12-5R4	
5814	23	117.9331	84793.87	C8-9R0	
5815	5	117.9349	84792.58	C2-4Q14 B0-1R6	
5816	6	117.9416	84787.72	B13-5P6	
5817	12	117.9471	84783.79	B3-1R12 C8-9R2	
5818	28	117.9549	84778.16	C12-11Q3 C1-3P15	
5819	1	117.9581	84775.90	B26-9P4	
5820	73	117.9622	84772.92	B0-1P5 B6-1P15 C4-6P7	
5821	18	117.9745	84764.07	C2-5R6	
5822	1	117.9763	84762.79	B29-10P2	
5823	0	117.9809	84759.46	B2-0R15	
5824	134	117.9918	84751.64	B2-2R0	
5825	54	117.9956	84748.91	B9-4R5 B15-6R3	
5826	4	117.9997	84745.96	B'5-11P3 B25-9R0	
5827	240	118.0106	84738.15	C2-5Q4	
5828	11	118.0159	84734.32	B14-5P8 C8-9R3	
5829	12	118.0169	84733.64	B18-7R0 B17-6R8	
5830	132	118.0218	84730.15	B3-2P5 C2-3Q20	
5831	95	118.0306	84723.77	C8-9Q1 B7-3R8 C12-11P3	
5832	30	118.0357	84720.10	B22-8P3 C3-4Q18	
5833	300	118.0428	84715.06	B2-2R1	
5834	2300	118.0461	84712.64	C2-5P3 C5-7R6	
5835	40	118.0503	84709.62	B25-9R1 B8-0R21 C5-7Q4	
5836	240	118.0600	84702.70	C5-7P3 B13-5R7 B3-2R6 B29-10R3 B19-6R11	
5837	32	118.0718	84694.21	B18-7R1 B26-9R5	
5838	145	118.0750	84691.97	B5-3R0	
5839	6	118.0901	84681.14	C2-4P13	
5840	8	118.1059	84669.74	B25-9P1	
5841	27	118.1118	84665.57	C8-9Q2 B12-4R11	
5842	0	118.1157	84662.75	B10-3P12 B24-8R8 C10-10P4 C8-8P11	
5843		118.1192			
5844	3	118.1217	84658.45	B6-2P11	
5845	440	118.1289	84653.26	B5-3R1	
5846	4	118.1374	84647.21	B7-1R17 B22-8R4	
5847	30	118.1531	84635.94	B18-7P1 B21-7R9	
5848	3	118.1583	84632.26	B18-5R14 C10-10Q5 C3-4P17	
5849	0	118.1672	84625.87	B25-9R2	
5850	198	118.1708	84623.30	B6-3P5	
5851	4	118.1746	84620.58	B5-1P14 C2-3P19	
5852	10	118.1821	84615.22	B29-10P3 B8-3R10 C6-6Q17	
5853	400	118.1899	84609.63	B2-2P1 C6-6P16	

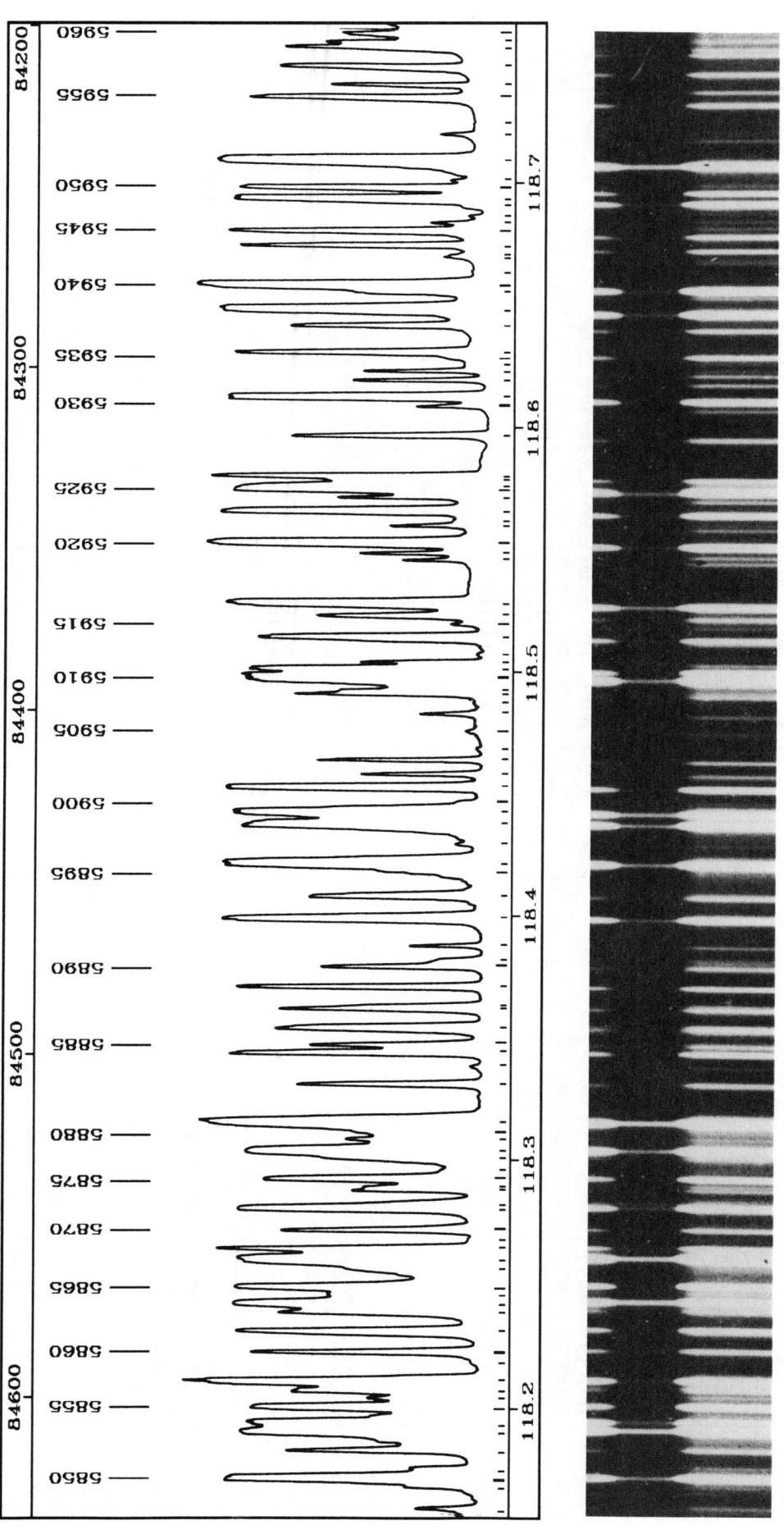

PLATE 70

TABLE 70

Nb	I	λ (nm)	σ (cm⁻¹)	Assignment	Comment
5854	136	118.1933	84607.19	B2-2R2	
5855	118	118.1996	84602.67	C1-4Q9 C0-3Q12	
5856	6	118.2032	84600.12	B14-5R9 B12-5P4	
5857	9	118.2066	84597.62	B18-7R2 C8-9P2 B6-1R16	
5858	117	118.2099	84595.28	B15-6P3	
5859	0	118.2175	84589.84	C0-1Q24	
5860	37	118.2221	84586.54	B10-4P7	
5861	43	118.2303	84580.67	C4-6Q9 C8-9Q3	
5862	24	118.2386	84574.77	C1-4R11 C4-5P15	
5863	400	118.2421	84572.28	C2-5Q5 C2-5R7	
5864	9	118.2457	84569.67	B6-3R6 C7-8P7	
5865	78	118.2484	84567.74	B8-4R0	
5866		118.2557	84562.50		sh
5867	490	118.2598	84559.57	B5-3P1	
5868	48	118.2637	84556.78	C5-7Q5	
5869	0	118.2667	84554.65	B9-1R19	
5870	11	118.2713	84551.37	B25-9P2	
5871	123	118.2804	84544.84	B5-3R2 C4-6P8	
5872		118.2827	84543.22	B15-4P14 C8-9R5	sh
5873	7	118.2878	84539.57	B19-7P5 B29-10R4	
5874	7	118.2891	84538.66	C5-7R7	
5875	45	118.2922	84536.44	C5-7P4	
5876	0	118.2953	84534.20	B32-11R0	
5877	300	118.3006	84530.42	B15-6R4	sh
5878	2	118.3036	84528.26	B8-4R1 B20-7P7 C0-2Q19	
5879	2	118.3084	84524.83	B26-9P5	
5880		118.3114	84522.68	C5-6Q14	sh
5881	570	118.3150	84520.12	C2-5P4 C4-6R11	
5882	25	118.3307	84508.92	B1-1P8 B12-5R5	
5883	2	118.3377	84503.89	B32-11R1	
5884	27	118.3435	84499.81	B18-7P2 C10-10P5	
5885	15	118.3465	84497.68	B1-1R9	
5886	30	118.3531	84492.92	B22-8P4 B25-9R3 B9-0R22 C1-4P8	b
5887	27	118.3612	84487.16	B10-4R8 C1-3Q17	
5888		118.3621	84486.49	B16-6P6	sh
5889	19	118.3702	84480.73	B17-4P16 C8-9P3	
5890	16	118.3786	84474.73	B5-1R15 C8-9R6	d
5891	4	118.3813	84472.82	B32-11P1 B11-2P17	
5892	6	118.3870	84468.72	C8-9Q4	
5893	135	118.3987	84460.37	B9-4P5 B14-4P13	
5894	21	118.4078	84453.87	C0-3P11 B9-2P15	sh
5895		118.4174	84447.06	B24-8P8	
5896	330	118.4213	84444.26	B8-4P1 B18-7R3	
5897	0	118.4291	84438.74	B20-7R8	
5898	300	118.4381	84432.26	B2-2P2 B32-11R2	
5899	222	118.4426	84429.06	B2-2R3 B23-8R7	
5900	1	118.4470	84425.91	B28-10R0 B29-10P4	
5901	69	118.4528	84421.83	B8-4R2 B21-8R0	
5902	12	118.4582	84417.99	B2-1P10	

Nb	I	λ (nm)	σ (cm⁻¹)	Assignment	Comment
5903	15	118.4639	84413.93	C2-5R8	
5904	1	118.4685	84410.62	B4-2P8 B16-3R19	
5905	4	118.4758	84405.45	B22-8R5	
5906	7	118.4828	84400.46	C0-1P23	
5907	0	118.4872	84397.29	B32-10R9	
5908	8	118.4908	84394.71	B28-10R1 C8-8P12	
5909		118.4924	84393.57	B9-3P11	
5910	320	118.4978	84389.76	B5-3P2 B11-5R0 B9-4R6 B4-1P13	
5911	57	118.5010	84387.50	B21-8R1 B25-9P3 B17-6P8	
5912	6	118.5038	84385.51	C2-4Q15	
5913	0	118.5070	84383.20	B32-11P2	
5914	85	118.5140	84378.23	C2-5Q6 C5-7Q6	
5915	3	118.5193	84374.42	B16-6R7	
5916	18	118.5226	84372.10	B2-1R11	
5917	176	118.5276	84368.52	B5-3R3	
5918	10	118.5458	84355.58	B4-2R9 C1-3R19	
5919	6	118.5482	84353.87	B28-10P1	
5920	120	118.5527	84350.69	B11-5R1	
5921	0	118.5597	84345.69	B13-5P7	
5922	0	118.5659	84344.12	B8-2P14	
5923	80	118.5659	84341.31	C5-7P5 B0-1R7 B11-4R10 B13-4P12	
5924	4	118.5716	84337.26	B29-10R5 C8-9P4	
5925	161	118.5749	84334.90	C2-5P5	
5926		118.5763	84333.88	B15-6P4	sh
5927		118.5784	84332.38	C8-9Q5	sh
5928	15	118.5800	84331.25	B21-8P1	
5929	20	118.5968	84319.30	B0-1P6 B28-10R2 C1-3P16	
5930	8	118.6089	84310.69	C5-7R8 C0-2P18	
5931	78	118.6128	84307.92	B18-7P3	
5932	7	118.6196	84303.10	B21-8R2	
5933	10	118.6234	84300.44	C1-4R12	
5934	0	118.6251	84299.16	B15-5R11	
5935	0	118.6287	84296.61	C4-6Q10	
5936	34	118.6310	84295.00	C1-4Q10 C8-9R7	
5937	22	118.6420	84287.18	B3-2P6	
5938	178	118.6488	84282.39	B8-4P2	
5939	5	118.6555	84277.63	B4-1R14	
5940	162	118.6584	84275.53	B11-5P1 B9-3R12	
5941	1	118.6636	84271.82	C4-6R12	
5942	3	118.6699	84267.36	B10-2R17	
5943	3	118.6710	84266.61	B7-3P8	
5944	13	118.6748	84263.87	B17-6R9 C11-11R0 C11-11R1	
5945	38	118.6807	84259.74	B3-2R7 B15-6R5	
5946	4	118.6839	84257.41	C3-4Q19	
5947	0	118.6870	84255.20	B32-11P3	
5948	0	118.6907	84252.58	B9-2R16	
5949	80	118.6941	84250.16	B8-4R3 B28-10P2	
5950	25	118.6985	84247.04	B11-5R2	
5951	4	118.7017	84244.80	C2-4P14	

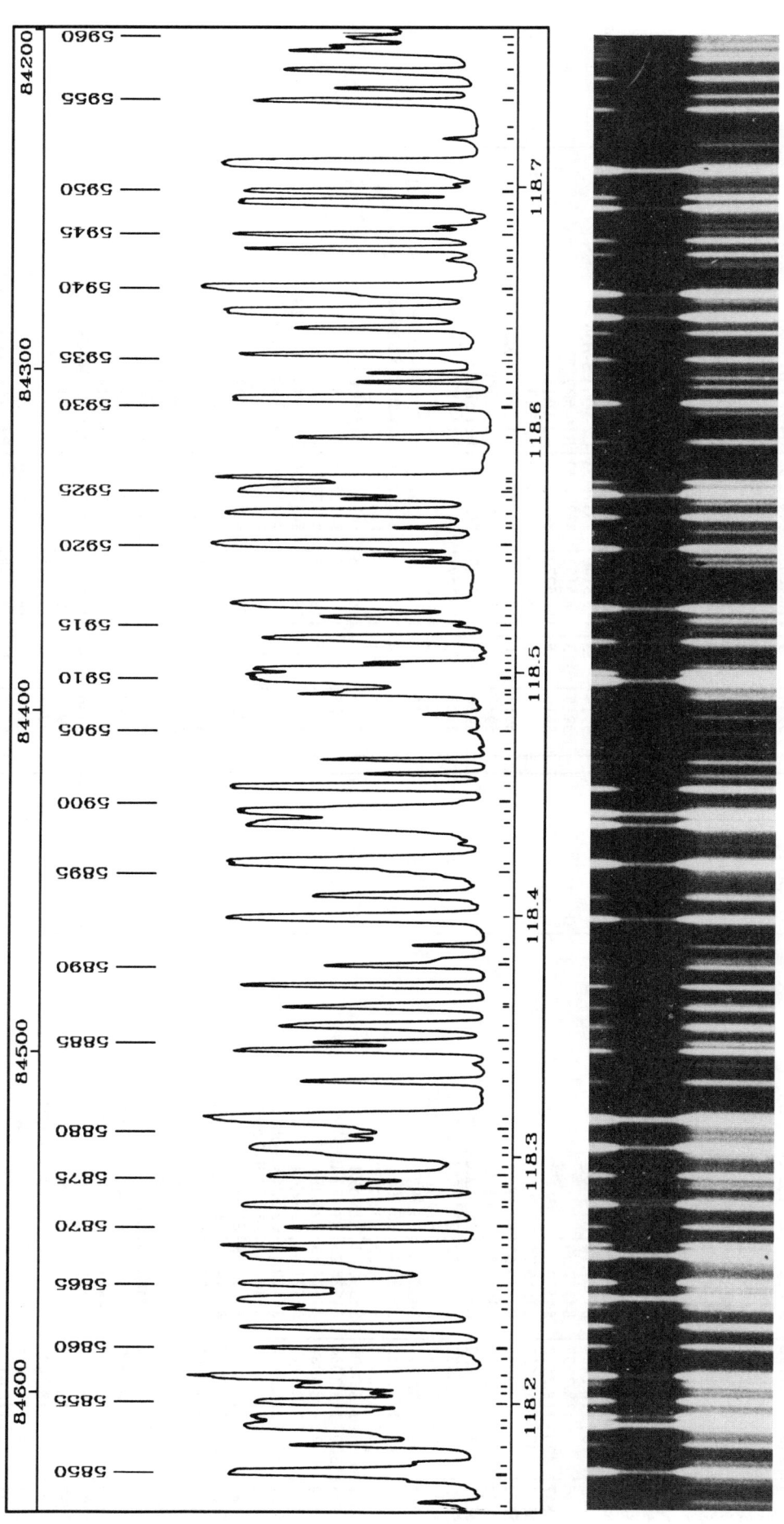

PLATE 70

TABLE 70 *continued*

Nb	I	λ (nm)	σ (cm⁻¹)	Assignment	Comment
5952	360	118.7098	84239.03	B12-5P5 B18-7R4 C11-11R2	
5953	4	118.7202	84231.67	C4-5Q17	
5954	0	118.7240	84228.98	C4-6P9 C2-3Q21	
5955	45	118.7349	84221.24	B11-2R18 C0-3Q13	
5956	8	118.7402	84217.49	B22-8P5	
5957	15	118.7477	84212.18	B13-5R8 B5-0P18 C11-11Q1	
5958	10	118.7558	84206.45	B21-8P2	
5959	14	118.7578	84204.98	C0-3R15	
5960	4	118.7614	84202.43	B28-10R3 B29-10P5 C6-6P17	

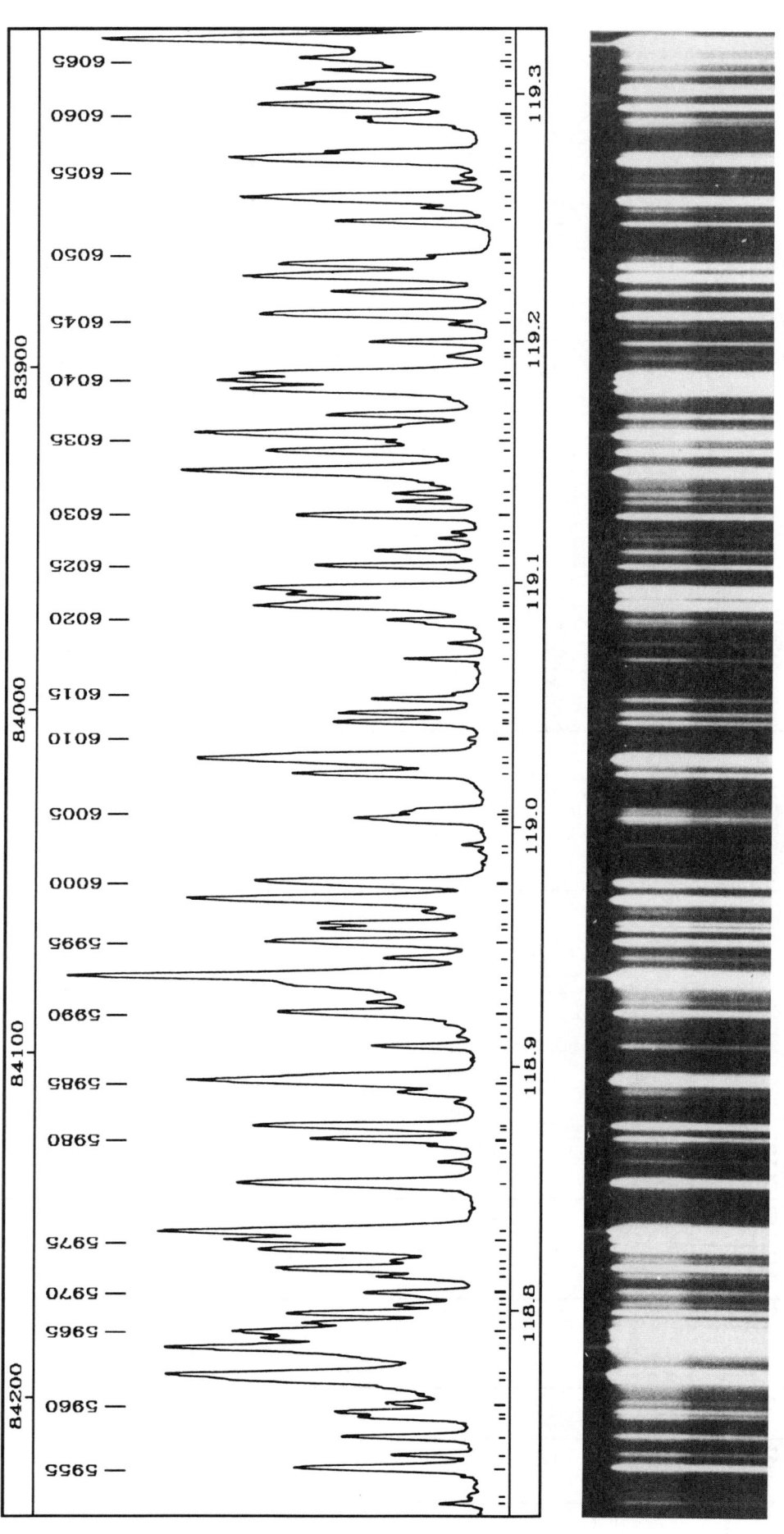

PLATE 71

TABLE 71

Nb	I	λ (nm)	σ (cm⁻¹)	Assignment	Comment
5961	790	118.7712	84195.50	C6-8R1 B6-3P6	
5962	800	118.7735	84193.89	C3-6R1	
5963	110	118.7841	84186.39	B2-2P3 B21-8R3 B32-11R4 C11-11R3	
5964	250	118.7884	84183.32	B2-2R4 B14-6R0	
5965		118.7910	84181.48	C3-6R2 C6-8R0	sh
5966	34	118.7941	84179.28	B25-9P4 B7-3R9 C2-5R9	
5967	49	118.7982	84176.35	C6-8R2 C5-7Q7	
5968	4	118.8014	84174.12	B6-0P19 C8-9Q6	
5969		118.8050	84171.55	B8-2R15 C4-5P16	sh
5970	9	118.8065	84170.50	C5-7R9 C8-9P5	
5971	8	118.8133	84165.66	B12-5R6 B13-4R13 B21-7R10	
5972	76	118.8166	84163.35	C1-4P9 B7-2P13	
5973	9	118.8197	84161.10	C11-11Q2	
5974	140	118.8246	84157.63	C2-5Q7	
5975	250	118.8283	84155.07	C3-6R0	
5976	780	118.8315	84152.74	B5-3P3	
5977	207	118.8516	84138.57	C3-6R3 B6-3R7	
5978	3	118.8605	84132.25	B24-9R0 C3-4P18	
5979	3	118.8673	84127.44	B4-6R1	
5980	34	118.8698	84125.64	B5-3R4	
5981		118.8732			GeII 118.8732 Ke
5982	94	118.8751	84121.93	B11-5P2	
5983	3	118.8853	84114.71	B12-2R19	
5984	6	118.8895	84111.72	B13-3P16 C8-9R8 C11-11P2	
5985	400	118.8935	84108.89	C6-8Q1 B12-3P15	d
5986	10	118.8956	84107.38	B28-10P3 C6-8R3 C5-7P6	sh
5987	2	118.9079	84098.74	B24-9R1 B3-0P16	
5988	2	118.9124	84095.49	B25-9R5	
5989	58	118.9175	84091.90	B32-11P4 C5-6P14 C2-3P20	
5990	70	118.9218	84088.88	B8-3P10 C2-5P6 C6-8R4	
5991	2	118.9263	84085.68	C11-11Q3	
5992	70	118.9344	84079.98	B11-5R3 B3-1P12	
5993	2240	118.9365	84078.51	C3-6Q1	
5994	8	118.9442	84073.02	B10-4P8 B14-5R10 B14-3P17	
5995	69	118.9511	84068.16	B14-6P1 B16-6P7 B31-11P1	
5996	26	118.9567	84064.20	C3-6R4	
5997	26	118.9590	84062.55	B18-7P4 B19-7R7 B4-11R1	
5998	4	118.9638	84059.20	B4-11R2	
5999	470	118.9686	84055.79	B8-4P3	
6000	109	118.9757	84050.79	B9-4P6 B24-9P1 C6-8Q2	
6001	0	118.9891	84041.31	B11-3P14	
6002	2	118.9917	84039.45	B4-11R3	
6003	0	119.0018	84032.34	C0-3P12	
6004	16	119.0027	84031.67	B21-8P3	
6005	6	119.0052	84029.89	B14-6R2 C0-2Q20	
6006	6	119.0065	84029.01	C1-3Q18	
6007	38	119.0214	84018.54	B15-6P5 C0-1Q25	
6008	500	119.0265	84014.93	B8-4R4 B24-9R2 C11-11P3	
6009		119.0283	84013.62	C3-6Q2 C9-10R1	sh
6010	0	119.0360	84008.22	C13-12Q1	SiII 119.04160
6011					
6012	18	119.0427	84003.49	B5-0R19 C9-10R0	
6013	19	119.0462	84000.97	B3-1R13	
6014	12	119.0521	83996.86	B27-9R9 C9-10R2 C4-6Q11	
6015	3	119.0544	83995.22	C8-9Q7	
6016	8	119.0688	83985.08	C6-8R5	
6017	5	119.0753	83980.48	C8-9P6 B18-7R5	
6018	0	119.0803	83976.95	B31-11P2	
6019	5	119.0832	83974.90	B8-3R11	
6020	11	119.0848	83973.76	B9-4R7	
6021	100	119.0907	83969.60	C1-4Q11 B4-11P1	
6022		119.0924	83968.39	B10-4R9 C1-4R13	sh
6023	60	119.0956	83966.13	C6-8P2	
6024	150	119.0980	83964.46	C6-8Q3	
6025	38	119.1073	83957.90	C3-6R5 B24-8P9	
6026	1	119.1109	83955.38	B18-6R11 B4-11R5	
6027	11	119.1134	83953.61	B3-0R17 B16-6R8 C5-7Q8 C9-10R3	
6028	4	119.1186	83949.97	B15-6R6 C13-12Q2	
6029	2	119.1212	83948.08	C4-6R13	
6030	65	119.1284	83943.07	B29-10P6 C9-10Q1	
6031	8	119.1341	83939.02	B17-7R0	
6032	6	119.1375	83936.60	B2-0P15 B24-9P2	
6033	760	119.1468	83930.11	B25-9P5 B25-8R11 C0-4R1	
6034	117	119.1551	83924.23	B8-0P21 C0-4R2 C5-7R10	
6035	2	119.1582	83922.02	B28-10P4 C5-7P7	
6036	780	119.1622	83919.24	C3-6Q3	
6037	2	119.1654	83916.97	C2-5R10	
6038	32	119.1698	83913.87	C2-5Q8 C1-3R20	
6039	290	119.1810	83905.97	B11-5P3 B6-2P12 B10-3P13	
6040	440	119.1843	83903.64	C3-6P2 C4-6P10	
6041	260	119.1870	83901.79	C0-4R0 B17-7R1	
6042	5	119.1939	83896.93	B17-6P9 B7-1P17 B32-11P5	
6043		119.1948	83896.27	C11-11P4 B11-4P10	sh
6044	17	119.2002	83892.50	C9-10Q2	
6045	4	119.2074	83887.42	B24-9R3	
6046	202	119.2116	83884.43	C0-4R3 C9-10R4 C8-9R9	
6047	34	119.2207	83878.06	B1-1P9	
6048	202	119.2272	83873.46	B2-2P4 B14-6R3	
6049	100	119.2319	83870.14	B2-2R5 B28-10R5	
6050	5	119.2352	83867.86	B1-1R10 C11-11Q5	
6051	26	119.2495	83857.77	B12-5P6	
6052	4	119.2552	83853.80	B11-5R4	
6053	168	119.2593	83850.93	B5-3P4 B31-11P3 C6-8Q4	
6054	3	119.2657	83846.43	C2-4P15	
6055	2	119.2688	83844.21	B6-1P16	
6056	228	119.2756	83839.42	B17-7P1 B29-10R7 C6-8P3	
6057	20	119.2782	83837.62	C1-3P17	
6058	2	119.2882	83830.57	B0-1R8	

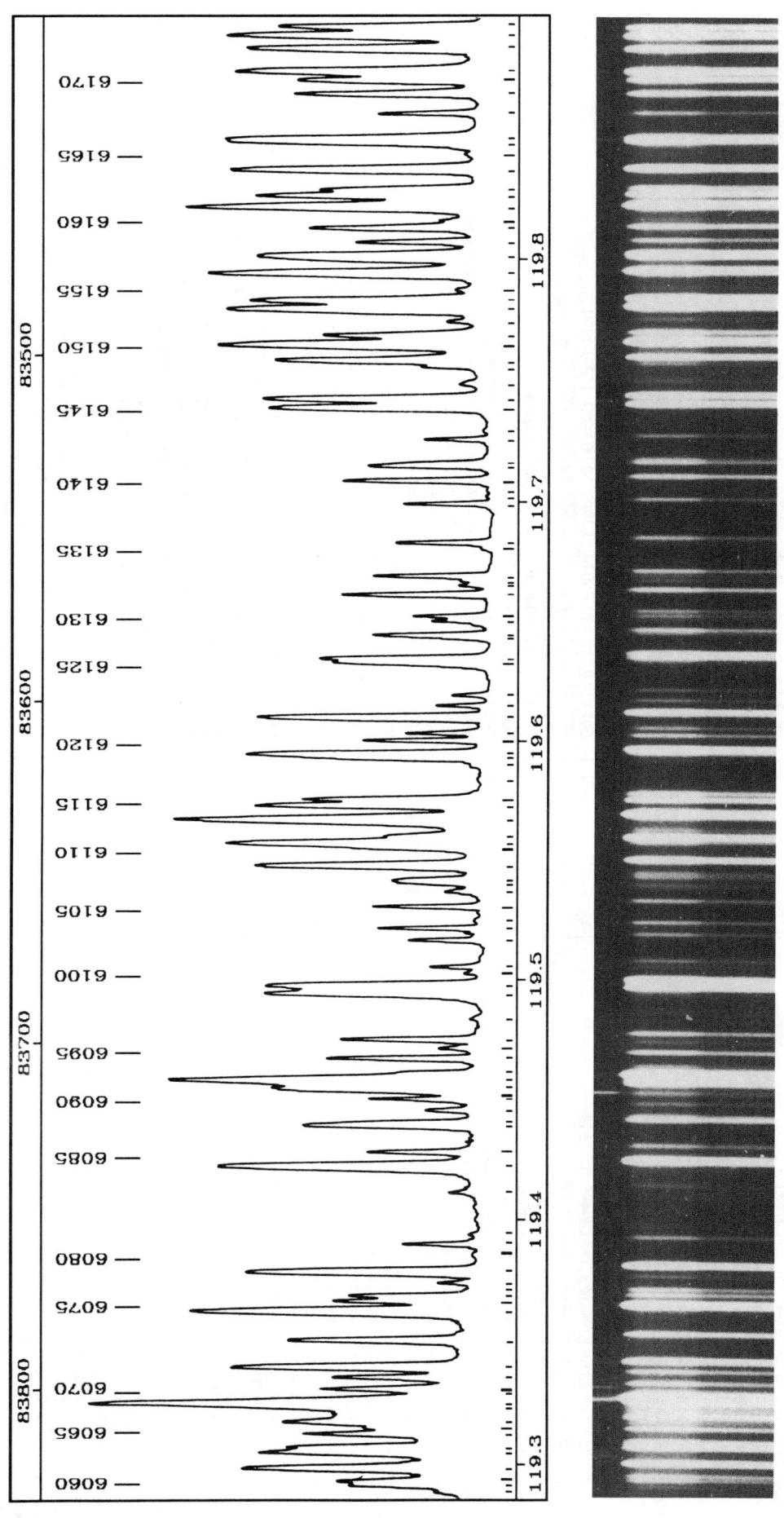

PLATE 72

TABLE 72

Nb	I	λ (nm)	σ (cm⁻¹)	Assignment	Comment
6059	9	119.2909	83828.72	C9-10P2	
6060	10	119.2925	83827.60	C0-3Q14 B2-0R16	
6061	105	119.2978	83823.84	C2-5P7	
6062	55	119.3044	83819.20	B5-3R5 B16-5R13 C0-1P24	
6063	29	119.3062	83817.96	C9-10Q3	
6064	15	119.3118	83813.98	B4-11P3	
6065	2	119.3144	83812.20	B27-10P1 C0-3R16 C4-5Q18	
6066	42	119.3169	83810.47	C0-4R4 C3-6R6 B21-8P4	
6067		119.3199	83808.31	B17-7R2 B7-0R21	sh
6068	1480	119.3241	83805.39	C0-4Q1 B0-1P7	
6069					SiII 119.32898
6070	17	119.3301	83801.12	B4-2P9 C0-2P19	
6071	16	119.3351	83797.66	C8-9Q8 C1-4P10	
6072	111	119.3395	83794.57	C3-6Q4	
6073	42	119.3506	83786.73	B3-2P7	
6074	360	119.3627	83778.25	C3-6P3	
6075	17	119.3667	83775.49	B24-9P3 C9-10R5 C5-6Q16	
6076	13	119.3688	83773.98	B19-7P7	
6077	0	119.3715	83772.07	C6-6Q19	
6078	4	119.3744	83770.05	B12-5R7	
6079	132	119.3790	83766.83	B8-4P4 B18-7P5 C8-9P7	
6080	0	119.3857	83762.15	B22-7R12	
6081	7	119.3905	83758.78	B3-2R8 B17-6R10	
6082	1	119.3950	83755.62	C11-11P5	
6083	3	119.4119	83743.73	B4-2R10 C8-9R10 C6-6P18	
6084	260	119.4220	83736.69	C4-0Q2 C5-7R11	
6085	10	119.4279	83732.50	B11-4R11 B15-4P15	
6086	29	119.4390	83724.76	C9-10P3 B13-5R9	
6087		119.4399	83724.10	B5-1P15	sh
6088	4	119.4452	83720.40	B9-0P22 C9-10Q4	
6089	5	119.4498	83717.18	B4-11P4	
6090					SiII 119.45004
6091	50	119.4544	83713.94	B6-3P7 B24-9R4 C6-8Q5 C9-10R6 C4-5P17	
6092	640	119.4572	83712.01	B14-6P3 C5-7Q9	
6093		119.4602	83709.90	B27-10P2 B21-8R5	sh
6094	16	119.4665	83705.49	B17-7P2	
6095	4	119.4706	83702.57	B20-8R0 B1-0P14 B28-10P5	
6096	15	119.4742	83700.08	C3-6R7	
6097	1	119.4837	83693.42	B7-1R18	
6098	87	119.4939	83686.25	C0-4R5 C6-8R7	
6099	78	119.4968	83684.23	C6-8P4 C2-5R11 B2-1P11 C4-6Q12	
6100	1	119.5022	83680.45	C3-4P19	
6101	6	119.5047	83678.73	B7-3P9	
6102	7	119.5158	83670.94	B6-1R17	
6103	10	119.5210	83667.32	B20-8R1 B19-7R8 C4-6R14	
6104	0	119.5243	83665.02	B17-4P17	
6105	10	119.5302	83660.86	B17-7R3	
6106	3	119.5365	83656.45	B14-6R4 B30-11R1 C5-6P15	
6107	6	119.5401	83653.96	B15-6P6 B8-1R19 B29-10P7 C5-7P8	
6108	6	119.5416	83652.86	B6-3R8	
6109	70	119.5473	83648.93	C2-5Q9	
6110	160	119.5552	83643.40	C1-4R14	
6111	5	119.5572	83641.98	C3-6Q5	
6112		119.5600	83639.99	B2-1R12 B25-9P6	sh
6113	430	119.5668	83635.26	C0-4Q3	
6114	57	119.5727	83631.10	B11-5P4	
6115	27	119.5751	83629.46	C1-4Q12	
6116	0	119.5831	83623.87	B13-4P13	
6117	0	119.5891	83619.68	B30-11P1	
6118	107	119.5921	83617.54	C0-3P13	
6119	12	119.5936	83616.51	C0-4P2	
6120	6	119.5994	83612.43	B20-8P1	
6121	107	119.6025	83610.30	B16-6P8 B'4-11P5	
6122	5	119.6090	83605.74	C3-6P4	
6123	5	119.6138	83602.40	C9-10Q5	
6124	33	119.6181	83599.38	C9-10P4	
6125	39	119.6321	83589.62	B9-4P7	
6126	11	119.6337	83588.51	C8-9Q9 B'5-12R1	
6127	6	119.6424	83582.42	B10-2P17	
6128	7	119.6433	83581.80	B5-1R16 B'5-12R0	
6129	24	119.6493	83577.62	B20-8R2 C0-4R6 C6-8R8	
6130	5	119.6512	83576.28	B'5-12R2 C11-11Q7	
6131	4	119.6602	83569.95	B11-5R5 B24-9P4 C1-3Q19	
6132	12	119.6641	83567.22	C12-12R0 B9-2P16	
6133	0	119.6659	83566.00	C12-12R1	
6134	10	119.6681	83564.48	B27-10P3 B21-8P5	
6135	8	119.6788	83556.96	B9-1R20 B9-3R13 B29-10R8	
6136		119.6819	83554.80	C6-8Q6 C2-4Q17	sh
6137		119.6981	83543.49	B'5-12R3	
6138		119.7008	83541.63	C12-12R2	sh
6139	0	119.7040	83539.38	B11-2P18 B25-9R7	
6140	22	119.7078	83536.73	C2-5P8	
6141	17	119.7138	83532.53	C0-2Q21	
6142	8	119.7149	83531.78	B4-1P14 B30-11P2	
6143	5	119.7246	83525.00	C12-12Q1	
6144	0	119.7283	83522.42	B14-4R15	
6145	84	119.7382	83515.53	B17-7P3 B'5-12P1	
6146	102	119.7416	83513.18	B4-3R0 B10-4P9	
6147	2	119.7475	83509.08	B9-4R8 B18-6P11	
6148	3	119.7550	83503.80	C3-6R8	
6149	60	119.7577	83501.94	C0-4Q4	
6150	270	119.7642	83497.43	B2-2P5 B16-6R9 B31-11P5	
6151	21	119.7677	83494.94	B2-2R6 C12-12R3	
6152	2	119.7735	83490.90	B23-9R0	
6153	225	119.7791	83487.04	B5-3P5 B20-8P2 B8-2P15	
6154	125	119.7823	83484.79	C6-8P5 B7-4R0	
6155	1	119.7866	83481.79	C12-12Q2	

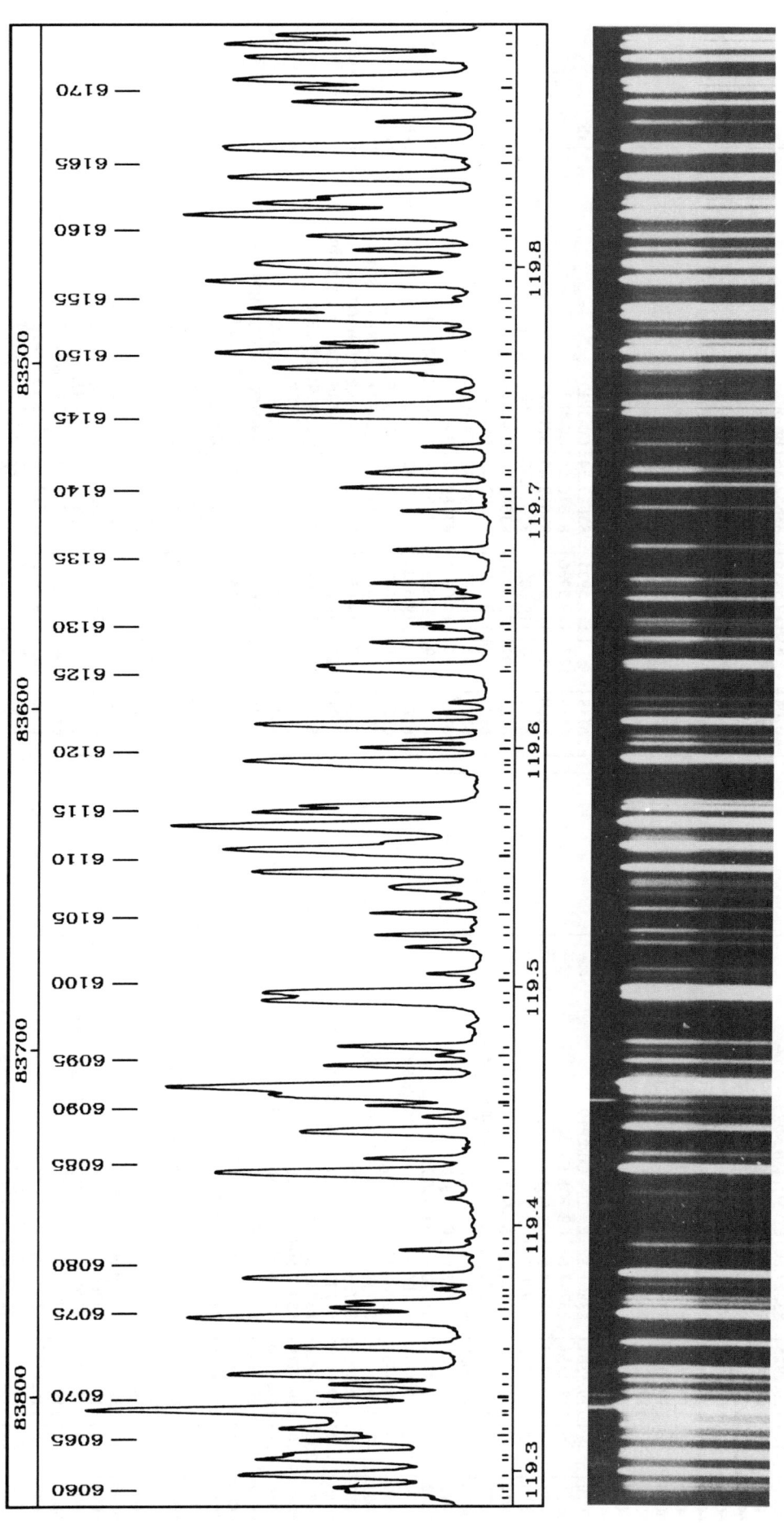

PLATE 72

TABLE 72 *continued*

Nb	I	λ (nm)	σ (cm⁻¹)	Assignment	Comment
6156	340	119.7939	83476.74	B4-3R1	
6157	150	119.8006	83472.05	B1-2R0 C1-4P11	b
6158	8	119.8065	83467.91	B23-9R1	
6159	33	119.8125	83463.75	C3-6Q6 C1-3R21	
6160	0	119.8157	83461.51	B17-7R4	
6161	400	119.8215	83457.50	C0-4P3 B'5-12P2 C5-7Q10	
6162	73	119.8261	83454.29	B14-6P4 C0-1Q26	
6163	20	119.8283	83452.73	B5-3R6 C9-10P5	
6164	180	119.8368	83446.79	B7-4R1	
6165	0	119.8425	83442.86	C12-12P2 C5-7R12	
6166		119.8477	83439.23	B20-8R3	sh
6167	230	119.8489	83438.38	B1-2R1	
6168	10	119.8597	83430.88	C7-9R2 B21-8R6	
6169	39	119.8680	83425.09	C0-3Q15 B18-7P6	
6170	41	119.8738	83421.03	B12-5P7 B4-1R15 C0-3R17	
6171	150	119.8772	83418.73	B8-4P5 C12-12Q3 C12-12R4	
6172	105	119.8865	83412.22	B8-3P11 C7-9R1 C0-4R7	
6173	207	119.8919	83408.46	B30-11P3 C3-6P5 C11-11Q8	
6174	52	119.8953	83406.11	B23-9P1 B10-4R10 C7-9R0	

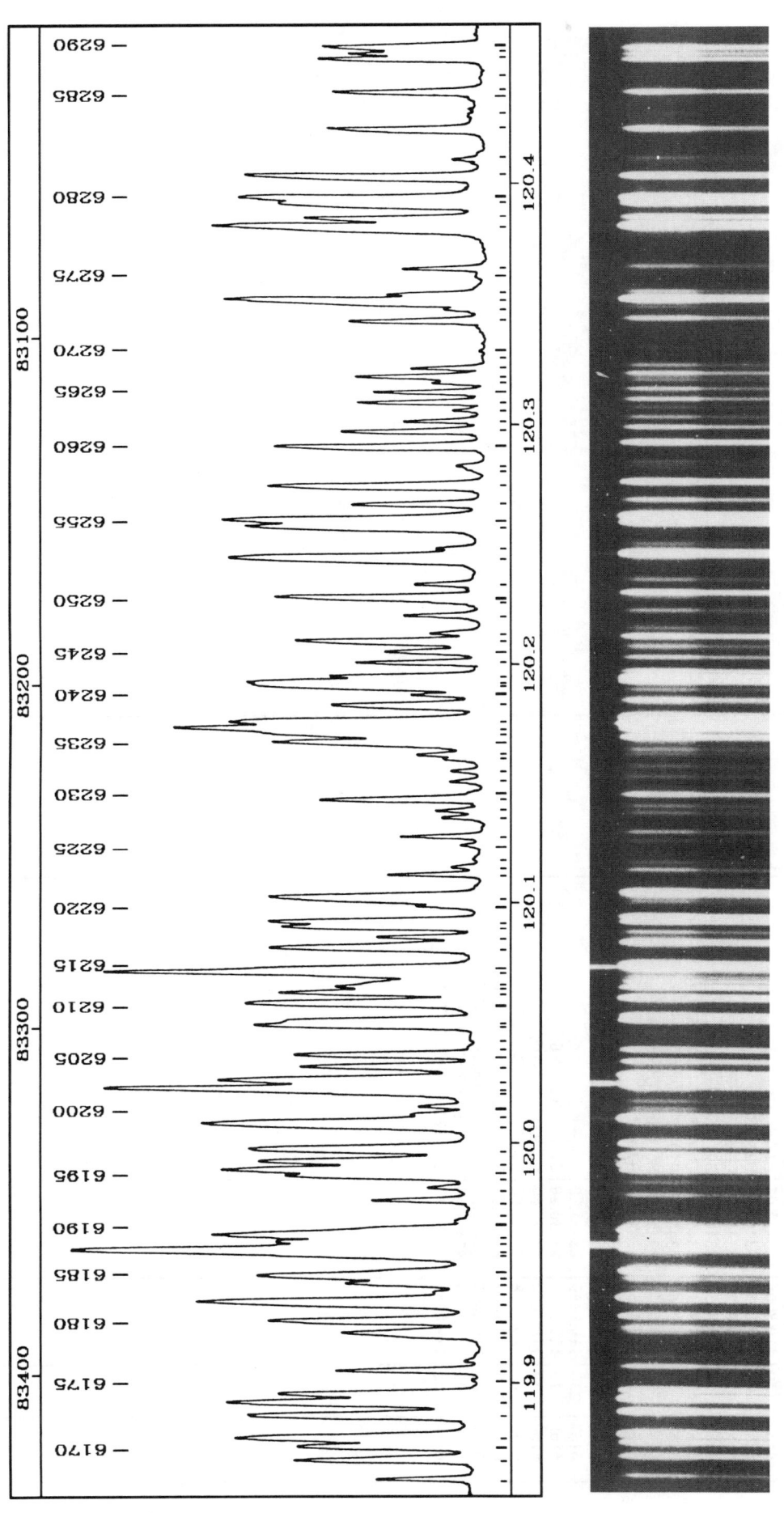

PLATE 73

TABLE 73

Nb	I	λ (nm)	σ (cm-1)	Assignment	Comment
6175	0	119.9016	83401.75	C5-6Q17 C0-2R23	
6176	17	119.9051	83399.31	B10-5R0 B9-0R23	
6177	2	119.9090	83396.56	C4-5Q19	
6178	8	119.9194	83389.31	C2-5R12	
6179	23	119.9207	83388.41	B14-6R5 C7-9R3 C2-4P16	
6180	69	119.9258	83384.91	B18-6R12 B5-12P3	
6181	370	119.9335	83379.52	B4-3P1 B27-10P4	
6182	1	119.9373	83376.93	B4-11P7	
6183	15	119.9413	83374.14	C6-8Q7 C5-7P9 C6-8R9	
6184	82	119.9444	83371.98	B4-3R2 B9-2R17	
6185	0	119.9465	83370.50	C8-8P15 C1-3P18	
6186	740	119.9549	83364.70	C2-5Q10	+ NI 119.95496
6187					NI 119.95496
6188	50	119.9585	83362.17	B10-5R1 B17-6P10 B10-2R18 C12-12P3 C4-6Q13	
6189	197	119.9613	83360.23	B7-4P1 C10-11R1	
6190	0	119.9650	83357.64	B8-4R6	
6191	0	119.9667	83356.48	B19-7P8	
6192	7	119.9758	83350.15	C10-11R0 C12-12R5	
6193	4	119.9812	83346.41	B13-5P9 C10-11R2	
6194	49	119.9865	83342.74	B7-4R2 C6-8P6 C6-6P19 C8-9P9	
6195	200	119.9889	83341.07	C7-9Q1 B7-2P14	
6196	102	119.9921	83338.80	C0-4Q5	
6197	137	119.9972	83335.27	B1-2R2 C12-12Q4	
6198	380	120.0078	83327.95	B1-2P1 B12-5R8 B18-7R7	
6199	5	120.0113	83325.51	C7-9R4 C4-6R15	
6200	4	120.0148	83323.05	B24-9P5 C5-6P16	
6201	540	120.0224	83317.76	B26-10R0 B25-9P7	+NI 120.02233
6202					NI 120.02233
6203	300	120.0257	83315.50	C4-7R1	
6204	38	120.0315	83311.45	B20-8P3 C10-11R3	
6205	41	120.0364	83308.05	C4-7R2	
6206	0	120.0393	83306.08	B13-2P20	
6207	0	120.0427	83303.71	C9-10R9 C4-5P18	
6208	95	120.0488	83299.46	B11-5P5 B8-3R12 B12-4R13	
6209	48	120.0504	83298.36	C10-11Q1 B5-12P4	
6210	138	120.0582	83292.95	B11-2R19 C4-7R0	
6211	47	120.0627	83289.81	C7-9Q2	
6212	18	120.0652	83288.05	B26-10R1 C1-4R15	
6213	5	120.0667	83287.04	C9-10P6 C3-6R9	
6214					NI 120.07098
6215	100	120.0715	83283.69	B10-5P1	
6216	75	120.0813	83276.89	C1-4Q13 C0-2P20	+NI 120.07098
6217	11	120.0855	83274.01	B17-7P4	
6218	59	120.0903	83270.68	C4-7R3 B13-6R0 B12-3P16	
6219	65	120.0923	83269.27	C0-4P4	
6220	7	120.0989	83264.72	B0-1R9 C12-12P4	
6221	77	120.1020	83262.53	C3-6Q7 B10-5R2 B3-1P13	
6222	7	120.1117	83255.82	C10-11Q2	

Nb	I	λ (nm)	σ (cm-1)	Assignment	Comment
6223	3	120.1148	83253.71	B13-3P17	
6224	1	120.1173	83251.95	B20-8R4 C10-11R4	
6225	2	120.1243	83247.08	B22-7R13 C0-1P25	
6226	7	120.1282	83244.41	B15-6P7 B26-10P1	
6227	5	120.1364	83238.72	B0-1P8	
6228	5	120.1392	83236.77	B19-7R9 C7-9R5 C12-12Q5	
6229	23	120.1441	83233.36	B3-2P8 B13-6R1	
6230	0	120.1467	83231.56	B11-5R6	
6231	4	120.1517	83228.13	B11-3P15 C10-11R5 C11-11Q9	
6232	4	120.1562	83225.03	B17-7R5	
6233	5	120.1613	83221.46	B17-6R11	
6234	7	120.1632	83220.14	C0-4R8	
6235	60	120.1685	83216.47	B26-10R2 B11-4P11 B25-9R8 C2-5P9	
6236	100	120.1723	83213.82	C7-9Q3	
6237	660	120.1748	83212.16	C4-7Q1	
6238	240	120.1770	83210.58	B4-3P2 C7-9P2	
6239	25	120.1838	83205.87	B3-2R9 C0-3P14	
6240	6	120.1884	83202.71	C4-7R4	
6241	77	120.1921	83200.12	B4-3R3 B1-1P10 C10-11P2	
6242	77	120.1938	83198.97	B7-4P2	
6243	22	120.1959	83197.50	B21-8P6 B'5-12P5	
6244	18	120.2019	83193.40	C10-11Q3 B13-5R10	
6245	10	120.2064	83190.25	B1-1R11	
6246	40	120.2107	83187.30	C3-6P6 C5-7Q11	
6247	6	120.2139	83185.08	B3-1R14	
6248	8	120.2213	83179.92	B6-3P8 B14-3P18	
6249		120.2272	83175.83	C6-8Q8	sh
6250	50	120.2290	83174.59	B7-4R3	
6251	7	120.2344	83170.88	B4-0P18 B5-0P19	
6252	250	120.2454	83163.24	B1-2R3 B13-6P1	
6253	5	120.2491	83160.70	B27-10P5	
6254	162	120.2586	83154.14	C4-7Q2	
6255	300	120.2613	83152.27	B1-2P2	
6256	19	120.2756	83147.74	C0-4Q6 C9-10O8 C12-12P5	
6257	84	120.2828	83142.35	B14-6P5 B26-10P2 B4-2P10 B26-9P9	
6258	3	120.2905	83137.38	B29-11P1 C8-9Q11 C6-8R10 C2-4Q18	
6259	3	120.2923	83136.13	B13-6R2	
6260	52	120.2999	83130.84	B10-5P2 C5-7R13	
6261	18	120.3063	83126.41	C2-5R13	
6262	8	120.3105	83123.54	C6-8P7	
6263	2	120.3147	83120.62	C12-12Q6 B6-2P13	
6264	14	120.3182	83118.20	B6-3R9	
6265	12	120.3224	83115.22	C7-9Q4 C7-9R6 C10-11R6 B16-7R0	
6266	6	120.3263	83112.52	B16-6P9 B6-0P20 B10-3P14 C10-11Q4	
6267	15	120.3215	83110.70	C10-11P3	
6268	7	120.3249	83108.32	B21-8R7 C3-6R10 B3-0P17	
6269	0	120.3301	83104.70	B26-10R3	
6270	0	120.3324	83103.16	B10-5R3	

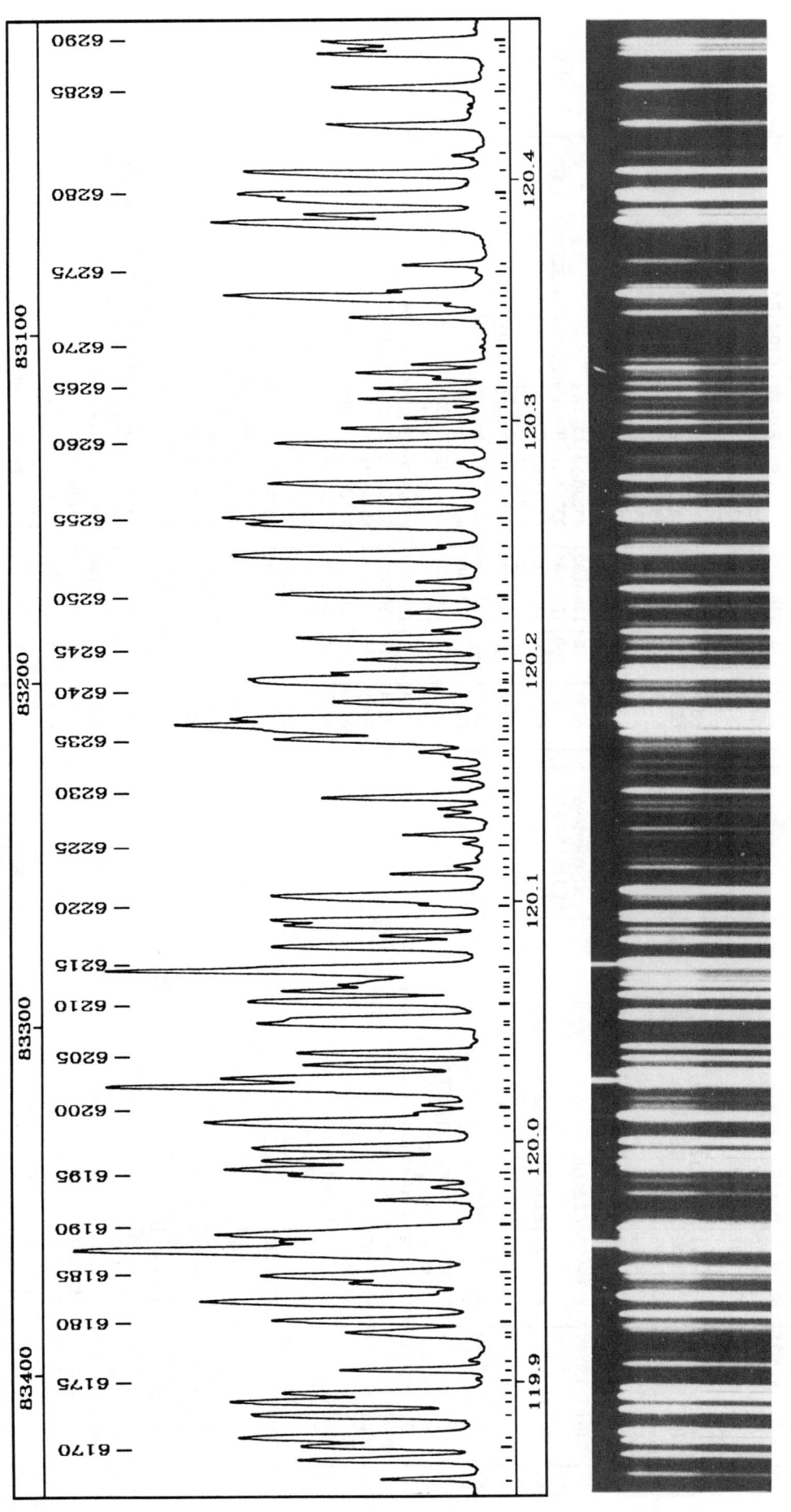

PLATE 73

TABLE 73 *continued*

Nb	I	λ (nm)	σ (cm⁻¹)	Assignment	Comment
6271	16	120.3448	83094.55	C4-7R5 C1-4P12	
6272	5	120.3497	83091.22	C9-10P7 B11-4R12	
6273	152	120.3530	83088.89	C7-9P3 B4-2R11	
6274	8	120.3558	83086.97	B20-8P4	
6275	0	120.3641	83081.25	B'5-12P6	
6276	8	120.3666	83079.54	B9-4P8 B16-7R1	
6277	300	120.3838	83067.64	C4-7Q3 C2-5Q11	
6278	38	120.3873	83065.24	B5-3P6	
6279	83	120.3934	83061.01	B2-2P6	
6280	118	120.3957	83059.43	B2-2R7 C4-7P2	
6281	31	120.4049	83053.13	C0-4P5	
6282	1	120.4119	83048.25	B7-3P10 C5-7P10 B29-11P2 C7-9R7	
6283	8	120.4246	83039.53	C3-6Q8 B18-7P7 C0-2Q22	
6284	0	120.4298	83035.93	C4-6Q14	
6285	0	120.4360	83031.62	B26-9R10 C0-4R9	
6286	5	120.4402	83028.75	B5-3R7 C4-7R6	
6287	0	120.4447	83025.65	C4-6R16	
6288	22	120.4540	83019.24	B13-6P2	
6289	17	120.4563	83017.64	B16-7P1 B20-8R5	
6290	23	120.4590	83015.81	B8-4P6 C10-11Q5 C0-3Q16	

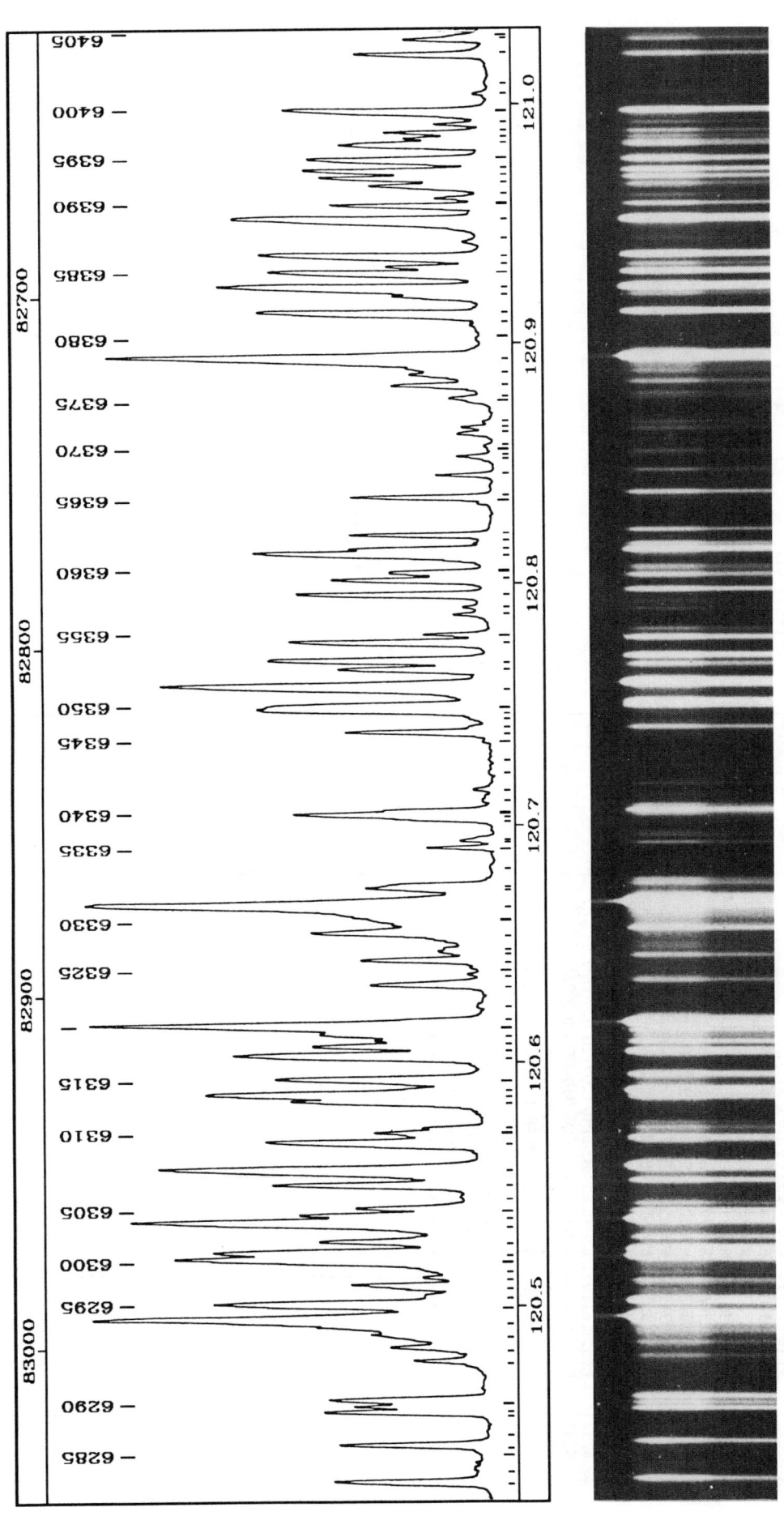

PLATE 74

TABLE 74

Nb	I	λ (nm)	σ (cm⁻¹)	Assignment	Comment
6291	7	120.4755	83004.45	C10-11P4 C0-3R18	
6292	8	120.4810	83000.63	B26-10P3 B13-2R21	
6293	15	120.4860	82997.18	B4-0R19 B7-0P21	
6294	1700	120.4917	82993.27	C1-5R1 C12-12Q7	
6295	280	120.4982	82988.79	C1-5R2	
6296	2	120.5022	82986.03	C6-8R11 B16-7R2 C2-4P17 C4-5Q20	
6297	9	120.5060	82983.43	B17-7P5 C7-9P4	
6298	3	120.5097	82980.89	B13-6R3	
6299	0	120.5126	82978.90	B2-0P16	
6300	600	120.5166	82976.14	B4-3P3	
6301	320	120.5194	82974.19	B7-4P3	
6302	15	120.5241	82970.95	B9-4R9 C9-10Q9	
6303	640	120.5319	82965.58	B25-9P8 C1-5R0	
6304	24	120.5352	82963.32	B4-3R4 C6-8Q9	
6305	7	120.5381	82961.31	B7-3R11 B19-8R0	
6306	0	120.5435	82957.58	B25-8R13	
6307	27	120.5479	82954.60	C4-7Q4 C9-10P8	
6308	340	120.5537	82950.60	C1-5R3 B8-4R7	
6309	43	120.5654	82942.53	C3-6P7 B7-4R4 C1-4R16	
6310	7	120.5695	82939.75	C2-5P10	
6311	4	120.5713	82938.45	B12-5P8	
6312	23	120.5828	82930.57	C0-4Q7	
6313	300	120.5853	82928.88	C4-7P3	
6314		120.5879	82927.06	B19-8R1	sh
6315	48	120.5918	82924.36	B1-2R4 B7-1P18 B29-11P3	
6316	168	120.6017	82917.60	B10-5P3 C1-5R4 C0-2R24	
6317	30	120.6054	82915.00	B11-5P6 C1-4Q14	
6318	4	120.6080	82913.27	B2-1P12 B10-4P10	
6319	33	120.6105	82911.53	B23-9P4 C5-7R14	
6320	1100	120.6134	82909.51	B1-2P3 C5-7Q12	
6321		120.6169	82907.09	C8-9Q12	sh
6322	0	120.6225	82903.27	C10-11Q6 C0-1Q27	sh
6323	15	120.6311	82897.35	B19-7P9 B6-1P17	b
6324	1	120.6359	82894.04	C4-6P13	
6325	0	120.6386	82892.24	B8-1P19	
6326	20	120.6414	82890.26	C1-3P19 B7-13P1	
6327	1	120.6449	82887.90	C6-8P8	
6328	1	120.6461	82887.08	C5-6P17	
6329	24	120.6530	82882.31	B16-7P2 C10-11P5 B5-12P7	sh
6330		120.6578	82879.03	B6-0R21	
6331	2740	120.6641	82874.71	B25-9R9 C1-5Q1	
6332	12	120.6714	82869.66	B2-1R13 C10-11R8	
6333	6	120.6728	82868.69	B19-8P1 C7-9R14	
6334	0	120.6799	82863.85	B14-5R12	
6335	8	120.6888	82857.73	C11-12R1	
6336	3	120.6917	82855.72	C7-9Q6	
6337	2	120.6929	82854.94	C4-5P19	
6338		120.7012	82849.24	C11-12R0 C11-12R2 B21-8P7	sh
6339	31	120.7030	82847.96	B10-5R4	
6340	0	120.7050	82846.64	C3-6R11	sh
6341	0	120.7094	82843.58	B19-8R2 B11-5R7	
6342	1	120.7137	82840.68	B12-5R9	
6343	0	120.7265	82831.88	B8-0P22	
6344	0	120.7306	82829.04		
6345	0	120.7358	82825.46	B'3-11R3 B'6-13R0	
6346	17	120.7378	82824.08	C7-9P5 C11-12R3	
6347	0	120.7403	82822.36	C5-7P11	
6348	0	120.7436	82820.11	B22-9R0 B26-10P4	
6349	95	120.7469	82817.84	B13-6P3	
6350	50	120.7487	82816.64	C4-7Q5 B20-8P5	
6351	560	120.7566	82811.18	C1-5Q2 C0-4P6 B9-3R14	
6352	20	120.7642	82805.99	B10-4R11 B5-1P16 C11-12Q1	
6353	45	120.7678	82803.51	C1-5R5	
6354	28	120.7754	82798.29	C3-6Q9	
6355	7	120.7787	82796.05	B17-6P11 C2-5R14	
6356	3	120.7870	82790.35	B22-9R1 B18-6R13 B'3-11R1	
6357	1	120.7903	82788.10	B'6-13P1 C9-10Q10	
6358	33	120.7955	82784.55	C8-10R1 C10-11R9	
6359	21	120.8016	82780.38	B14-6P6 B1-0P15 C10-11Q7	
6360	8	120.8045	82778.36	C8-10R2 C11-12R4 B12-4P13	
6361	61	120.8122	82773.09	B26-10R5 C11-12Q2 C4-7P4	
6362	16	120.8143	82771.68	C0-3P15	
6363	0	120.8177	82769.34	B29-11P4 B13-6R4 B'6-13R2	
6364	16	120.8203	82767.53	C8-10R0	
6365	0	120.8337	82758.40	C11-12R5	
6366	17	120.8362	82756.63	C2-5Q12 B7-13P3	
6367	6	120.8455	82750.29	C8-10R3	
6368		120.8526	82745.43	C9-10P9	
6369	4	120.8537	82744.69	B19-8P2	
6370	0	120.8569	82742.47	C10-11P6	
6371	1	120.8589	82741.15	B20-8R6 C6-8R12	
6372	4	120.8631	82738.26	C6-8Q10 B22-9P1	
6373	2	120.8656	82736.50	B13-5P10 B7-0R22	
6374	0	120.8681	82734.80	B'6-13P2	
6375	6	120.8764	82729.13	B6-1R18	
6376	5	120.8777	82728.20	B7-1R19	
6377	8	120.8830	82724.63	C11-12Q3 C11-12P2	
6378	6	120.8876	82721.47	B25-10R0 C2-4Q19	
6379	840	120.8938	82717.26	C1-5Q3	
6380	0	120.9038	82710.40	B19-8R3 B22-9R2 C7-9R9	
6381	90	120.9082	82707.35		
6382	90	120.9127	82704.32	C8-10Q1 C4-6Q15	
6383	8	120.9201	82699.21	C8-10R4 C7-9Q7 C4-7R8	
6384	280	120.9232	82697.13	C1-5P2	
6385	67	120.9295	82692.79	B16-7P3	
6386	5	120.9324	82690.83	C0-4Q8 B25-10R1	
6387	80	120.9366	82687.98	B7-4P4 C1-4P13	
6388	2	120.9435	82683.26	B17-6R12 C0-1P26	

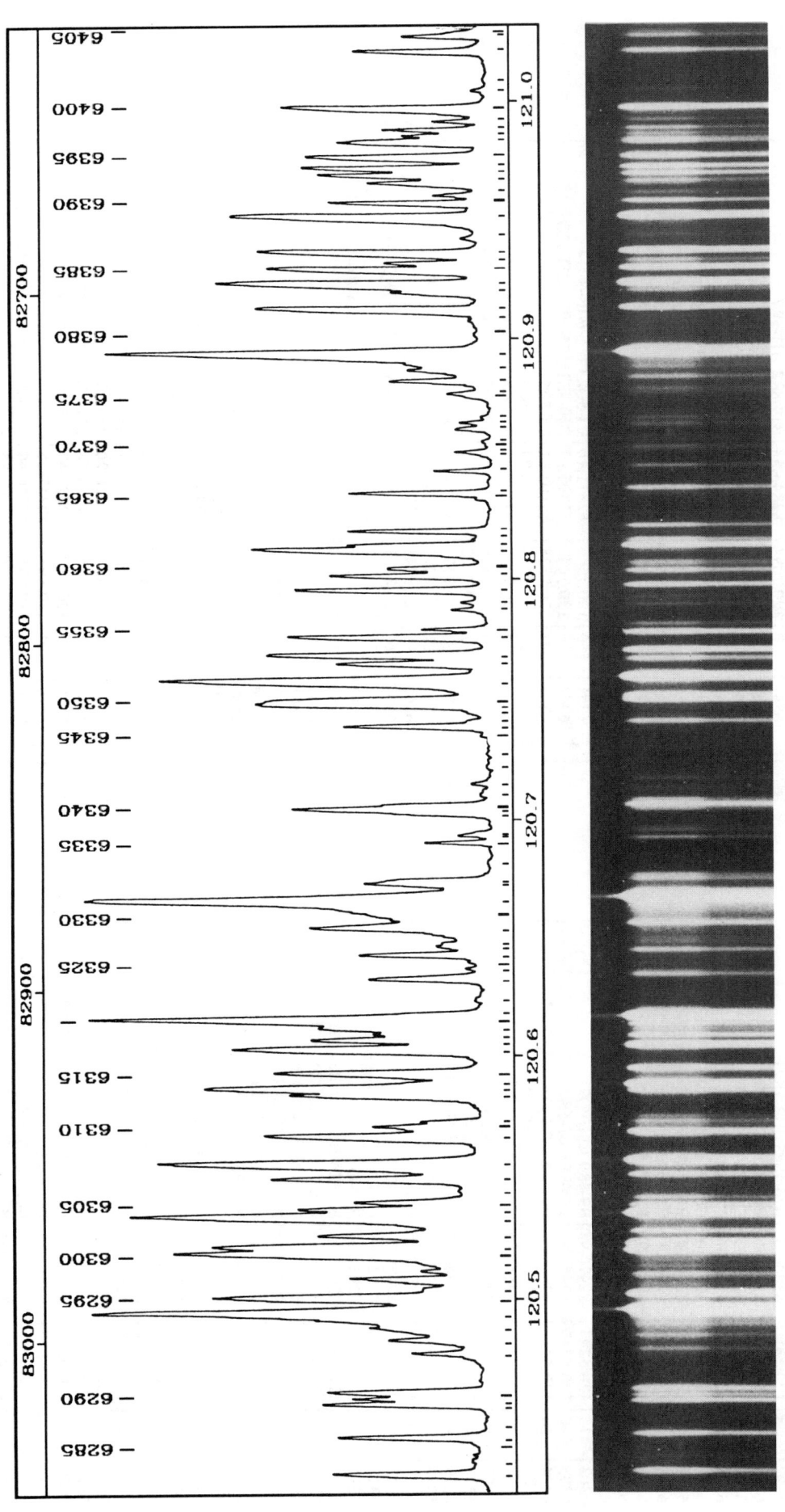

PLATE 74

TABLE 74 *continued*

Nb	I	λ (nm)	σ (cm⁻¹)	Assignment	Comment
6389	121	120.9515	82677.77	B4-3P4 C1-5R6 B'4-12R3 B'4-12R5	
6390	15	120.9577	82673.52	B23-9P5	
6391	4	120.9611	82671.23	C8-9Q13 B'4-12R2	b
6392	12	120.9661	82667.78	B5-1R17 B'3-11P1	
6393	19	120.9693	82665.59	C3-6P8 B9-2P17	
6394	20	120.9721	82663.66	B4-3R5 C11-12Q4	
6395	30	120.9767	82660.53	B10-2P18 C8-10Q2 C1-3Q21 B'6-13P3	
6396	16	120.9832	82656.11	C4-7Q6 B'4-12R1	b
6397	6	120.9858	82654.34	C11-12P3	
6398	7	120.9882	82652.67	B7-4R5	
6399	3	120.9917	82650.27	B25-10P1 C10-11Q8 C7-9P6	
6400	50	120.9971	82646.64	B10-5P4 B4-1P15	
6401	2	121.0046	82641.47		
6402	0	121.0087	82638.69	B26-9R11 C10-11R10	
6403	13	121.0208	82630.39	B3-2P9 B'4-12R0	
6404	8	121.0273	82625.98	C5-7Q13 B13-5R11	
6405	8	121.0288	82624.93	B22-9P2	sh

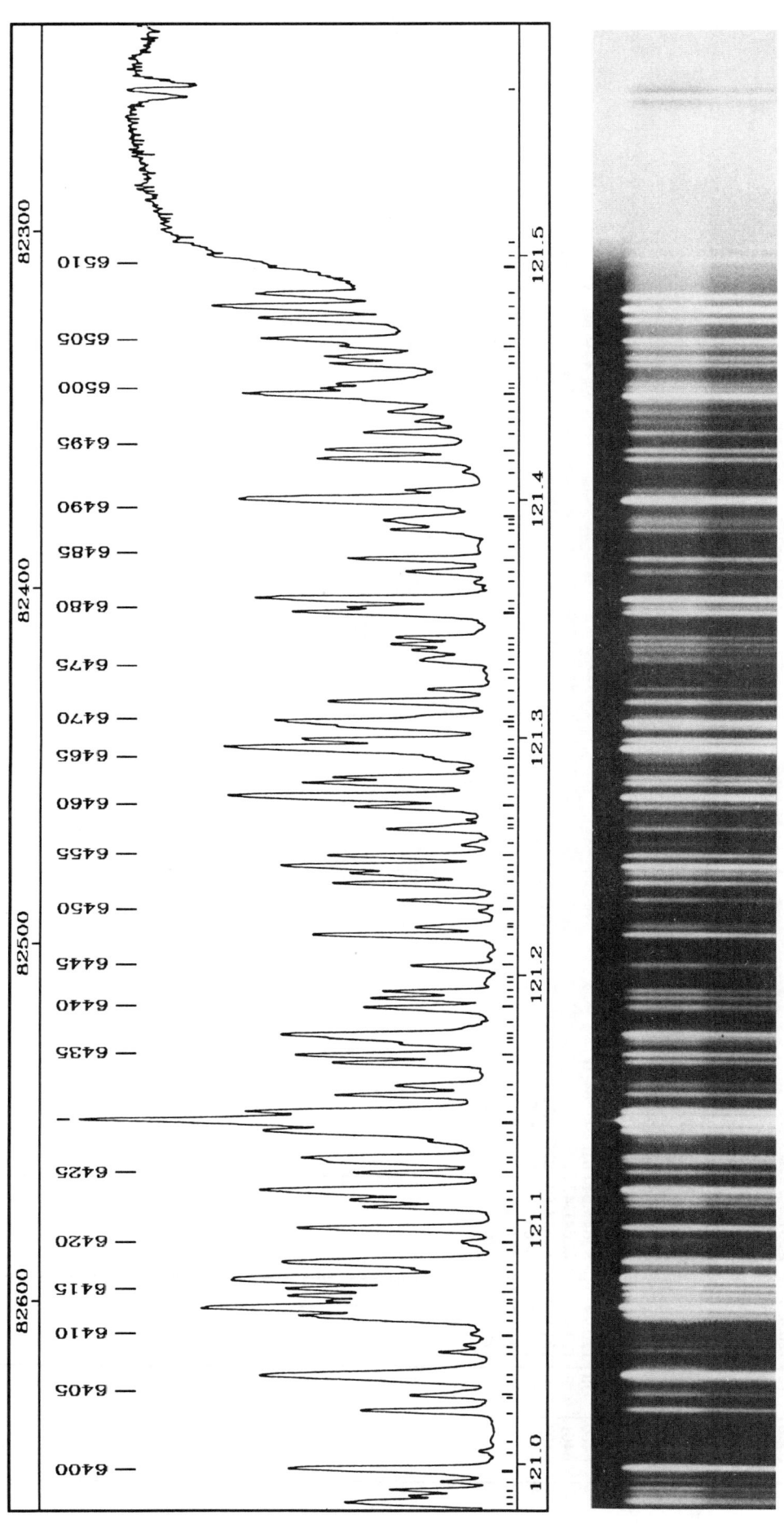

PLATE 75

TABLE 75

Nb	I	λ (nm)	σ (cm⁻¹)	Assignment	Comment
6406	92	121.0352	82620.61	B1-2R5 B0-1P9	
6407		121.0390	82617.99	B18-7P8 B9-0P23	sh
6408	6	121.0449	82613.95	C0-3R19	
6409	2	121.0479	82611.90	B8-2P16	
6410	2	121.0524	82608.82	B'3-11P2	
6411	30	121.0597	82603.84	B26-10P5 B3-2R10 C0-3Q17 C9-10Q11	
6412	250	121.0631	82601.55	B1-2P4	
6413	1	121.0661	82599.51	B6-3P9	
6414	30	121.0682	82598.09	B11-2P19 C2-5P11 C11-12P7	
6415	30	121.0709	82596.22	C8-10Q3	
6416	157	121.0748	82593.56	B12-4R14 C1-5Q4 C4-7P5 C8-10P2 C7-9R10	
6417	5	121.0787	82590.91	B8-3R13 C11-12Q5	
6418	45	121.0817	82588.87	B5-3P7 B10-5R5 B25-9P9 C5-7R15	
6419	0	121.0890	82583.87	B29-11P5 C4-5Q21	
6420	2	121.0903	82582.97	C4-7R9	
6421	32	121.0958	82579.22	B8-4P7	
6422	15	121.1048	82573.10	C1-4R17 C1-3R23	
6423	13	121.1077	82571.15	B19-8P3 C11-12P4	
6424	78	121.1114	82568.64	B2-2R8 B2-2P7 B6-13P4	
6425	13	121.1186	82563.70	B4-12P1	
6426	20	121.1236	82560.32	B13-6P4 C3-6R12	
6427	30	121.1251	82559.26	B'5-13R2 B'5-13R3	
6428	5	121.1319	82554.63	B9-1R21	
6429	66	121.1362	82551.74	B5-3R8 C5-8R1 C6-8R13 C0-2Q23	
6430	900	121.1402	82549.00	C1-5P3 C11-12R8 B'5-13R1	
6431	75	121.1439	82546.47	B25-10P2 B11-4P12 B8-0R23 C5-8R2	
6432	13	121.1509	82541.71	C1-4Q15 B'5-13R4	
6433	7	121.1546	82539.17	C3-6Q10 B'3-11P3	
6434	18	121.1644	82532.47	B6-3R10 B4-1R16	
6435	25	121.1675	82530.37	C9-10P10 C10-11R11 B'5-13R0	
6436	7	121.1721	82527.25	C5-8R0 C7-9Q8	
6437	29	121.1741	82525.90	C12-13R1 C0-4P7	
6438	0	121.1758	82524.74	B9-4P9	
6439	0	121.1787	82522.73	C12-13R2 C1-5R7	
6440	12	121.1869	82517.14	C2-4P18	
6441	15	121.1910	82514.39	B6-13R5 C0-4R11	
6442	8	121.1936	82512.58	C10-11Q9 C2-5R15	
6443	0	121.1962	82510.82	C8-10Q4 C5-8R3	
6444	0	121.1994	82508.66	C12-13R3	
6445	7	121.2045	82505.17	C11-12Q6	
6446	0	121.2084	82502.52	B20-8P6 C6-8Q11 B'4-12P2	b
6447	30	121.2173	82496.50	C5-7P12	
6448	6	121.2206	82494.23	C8-10P3	
6449		121.2215	82493.59	B7-2P15	sh
6450	0	121.2277	82489.41	B8-4R8	
6451	9	121.2320	82486.46	C12-13Q1	
6452	17	121.2391	82481.63	B11-5P7	
6453	13	121.2433	82478.80	B1-1P11 C11-12P5 C12-13R4	
6454	67	121.2460	82476.92	C10-11P8 B'5-13P1 B9-2R18	
6455	16	121.2503	82474.05	C4-7Q7	
6456	2	121.2553	82470.65	B1-1R12	
6457	10	121.2620	82466.05	C12-13R5 B22-9P3	
6458	0	121.2635	82465.07	C12-13Q2 B'6-13P5	sh
6459	0	121.2662	82463.19	B21-8P8	
6460	7	121.2713	82459.74	C7-9P7	
6461	148	121.2754	82456.93	C5-8Q1	
6462	22	121.2810	82453.12	C4-5P20 B'4-12P3 B'5-13R5	
6463	15	121.2833	82451.57	B16-7P4	
6464	2	121.2870	82449.05	B8-2R17	sh
6465	0	121.2913	82446.14	B4-2P11	
6466	155	121.2930	82444.99	B10-2R19	sh
6467		121.2958	82443.10	C1-5Q5 C0-2R25 C11-12R9	
6468	37	121.2991	82440.87	B'5-13P2	
6469		121.3047	82437.04	C3-6P9 B9-4R10	
6470	62	121.3066	82435.72	C2-5Q13	
6471	5	121.3088	82434.28	C12-13Q3	
6472	15	121.3149	82430.13	B20-8R7 C0-4Q9 C6-8P10	
6473	7	121.3204	82426.40	B16-5R15 C12-13P2 C1-3P20	
6474	0	121.3252	82423.12	C11-12Q7	
6475	0	121.3285	82420.87	C9-10Q12	
6476	5	121.3329	82417.90	B11-4R13 B12-3P17	d
6477	5	121.3366	82415.38	B3-1P14	
6478	6	121.3390	82413.74	B12-5P9	
6479	6	121.3416	82411.96	C8-10Q5	
6480	38	121.3522	82404.76	B11-3P16 B19-7P10 C5-8Q2	
6481	15	121.3544	82403.30	B25-10P3	
6482	97	121.3576	82401.07	B'5-13P3	
6483	0	121.3652	82395.92	C12-13Q4	
6484	6	121.3692	82393.22	B23-9P6 B4-2R12 C7-9R11 B'4-12P4	
6485	16	121.3745	82389.65	C4-7P6	
6486	5	121.3803	82385.71	B'6-13R6	
6487	5	121.3867	82381.37	C8-10P4	
6488	5	121.3885	82380.15	B7-3P11 C9-10P11 C10-11Q10	
6489	6	121.3904	82378.83	C12-13P3	
6490	0	121.3922	82377.60	B9-5R0 C4-6R18	
6491	150	121.3992	82372.86	C1-5P4 B14-6P7 C11-12P6 C4-6Q16	b
6492	4	121.4029	82370.34	B6-4R0	
6493	0	121.4114	82364.61	B21-8R9	
6494	26	121.4162	82361.36	B11-2R20 C5-7R16 C13-14Q1	
6495	20	121.4198	82358.88	B26-10P6 B'5-13P4	
6496	9	121.4271	82353.95	C12-13Q5 C13-14Q2	
6497	2	121.4320	82350.63	B19-8P4	
6498	5	121.4360	82347.90	C7-9Q9 C4-7R10	
6499	12	121.4401	82345.11		
6500	87	121.4428	82343.31	B7-4P5	
6501	17	121.4453	82341.63	B9-5R1 C3-6R13	

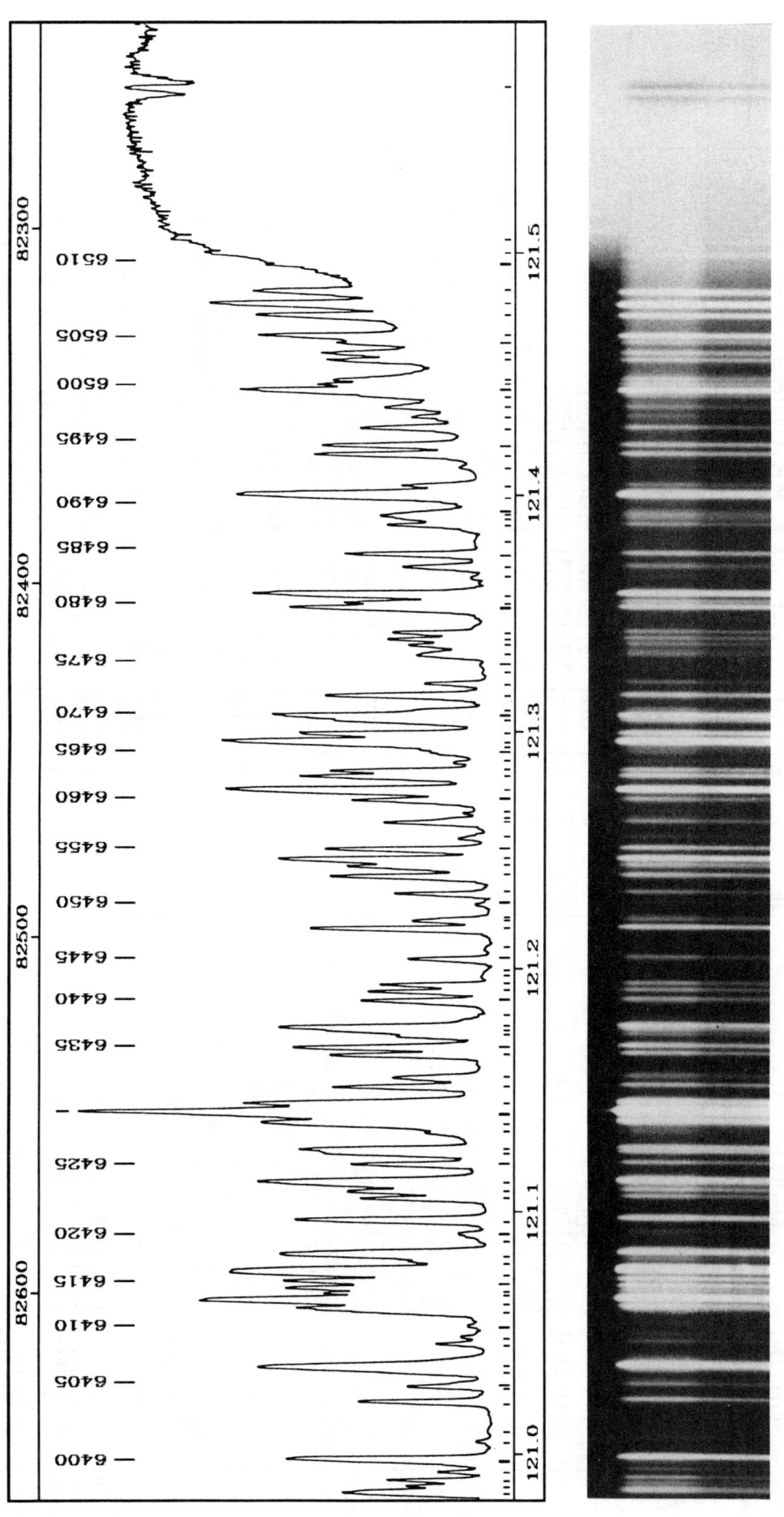

PLATE 75

TABLE 75 *continued*

Nb	I	λ (nm)	σ (cm-1)	Assignment	Comment
6502	12	121.4475	82340.08	B3-1R15 C5-7Q14	
6503	13	121.4556	82334.63	B6-4R1 C11-12Q8 C1-4P14	
6504	16	121.4585	82332.68	C0-3P16 B4-12P5	
6505	10	121.4630	82329.62	C1-5R8	
6506	52	121.4656	82327.85	C5-8Q3	
6507	51	121.4741	82322.10	B10-5P5	
6508	165	121.4789	82318.84	B4-3P5 C10-11P9	
6509	48	121.4840	82315.37	B10-3P15 C5-8P2 B'5-13P5	
6510	15	121.4965	82306.92	C11-12P7	
6511	16	121.4998	82304.65	B7-4R6 B4-3R6 B12-6R1 B7-14R0	
6512	7	121.5047	82301.31	B3-3R0	HI 121.56701 Ly α
6513					

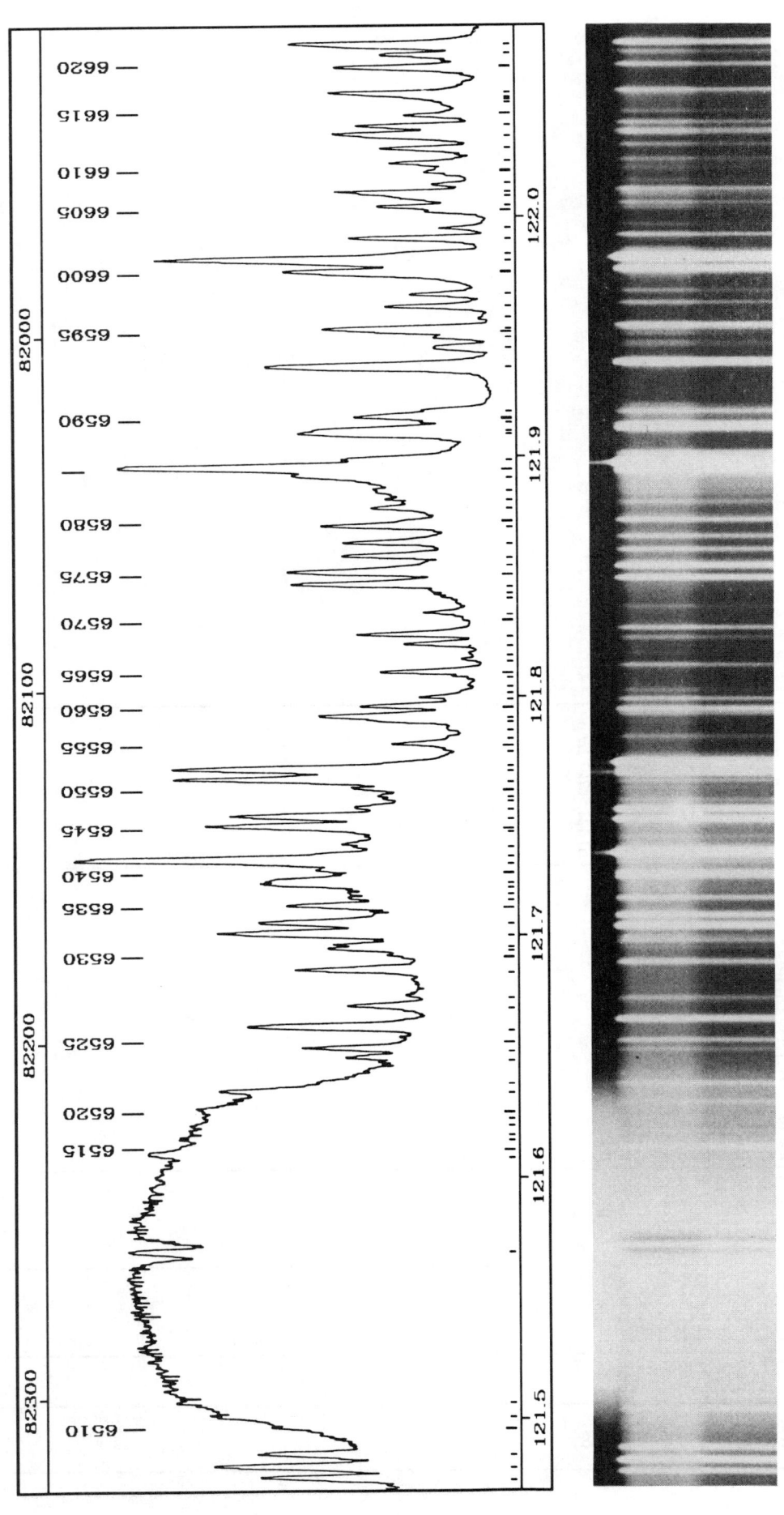

PLATE 76

TABLE 76

Nb	I	λ (nm)	σ (cm-1)	Assignment	Comment
6514	360	121.6071	82232.01	B1-2P5 B14-5R13	
6515	0	121.6103	82229.87	B12-6P1	
6516	3	121.6141	82227.30	C5-8Q4	
6517	2	121.6158	82226.18	B'6-13P7	
6518	3	121.6168	82225.49	C9-11R4	
6519	2	121.6182	82224.53	B25-10P4	
6520	12	121.6258	82219.37	B12-6R2	
6521	28	121.6336	82214.15	C9-11Q1 C1-3Q22	
6522	0	121.6377	82211.35	C0-3R20	
6523	8	121.6481	82204.32	C12-13P6 B'6-14R0	
6524	22	121.6517	82201.91	B'6-14R1	
6525	0	121.6555	82199.30		
6526	103	121.6604	82196.01	C5-8P3 C4-5Q22	
6527	10	121.6691	82190.15	B18-6R14 C11-12P8 B6-14R2 C0-3Q18	
6528	0	121.6731	82187.46		
6529	26	121.6847	82179.58	C9-11Q2 B'6-14P1	
6530	0	121.6909	82175.43	C12-13P7	
6531	15	121.6937	82173.52	C1-4Q16	
6532	7	121.6952	82172.52	B14-4P16 B'6-14R3	
6533	165	121.7000	82169.30	B15-7P1 C1-5P5 C1-5R9	
6534	70	121.7042	82166.43	B3-3P1 B3-3R2 C8-10Q7 C2-5R16	
6535	25	121.7117	82161.37	B16-7P5 B18-7P9 B10-4R12 C9-10P12 B'5-13P7|	
6536	0	121.7153	82158.91	C7-9Q10	
6537	0	121.7166	82158.04	B18-8R1 C9-11R5 B'6-14R4	
6538	60	121.7208	82155.21	B0-2R0 B20-8P7	
6539	60	121.7226	82154.01	C6-8P11 C4-7P7 B'6-14P2	
6540	15	121.7262	82151.61	C0-4Q10 C10-11P10	
6541	2200	121.7304	82148.73	C2-6R1	
6542		121.7343	82146.09	B15-7R2	sh
6543	3	121.7372	82144.15	B13-5P11	
6544		121.7417	82141.12	B4-0P19	sh
6545	360	121.7450	82138.86	C2-6R2	
6546	200	121.7493	82136.01	C2-6R3	
6547	7	121.7529	82133.60	C3-6P10	
6548	2	121.7546	82132.42	C10-11Q12	
6549	2	121.7556	82131.75	C0-1P27	
6550	15	121.7617	82127.62	C9-11Q3	
6551	280	121.7646	82125.69	B0-2R1 B'6-14P3	
6552	860	121.7689	82122.81	C2-6R0	
6553	0	121.7724	82120.44	B5-0P20	
6554	0	121.7777	82116.85	C4-7R11	
6555	6	121.7793	82115.73	C9-11P2	
6556	0	121.7817	82114.18	B20-8R8	
6557	0	121.7859	82111.33	C5-7R17	
6558	9	121.7879	82109.95	B9-5P2	
6559	38	121.7911	82107.80	B2-1P13 C2-5Q14 C2-4P19	
6560	13	121.7952	82105.06	C5-8Q5	
6561	4	121.7987	82102.66	C8-10P6 B3-0P18	
6562	0	121.8011	82101.07		

Nb	I	λ (nm)	σ (cm-1)	Assignment	Comment
6563	0	121.8043	82098.91	B18-8P1	
6564	0	121.8067	82097.32	C6-8R15	
6565	12	121.8093	82095.53	C7-9P9 B'6-14P4	
6566	0	121.8146	82091.96	B24-6P19	
6567	0	121.8176	82089.96	C11-12P9	
6568	6	121.8211	82087.58	B12-6P2 B19-8P5	
6569	15	121.8252	82084.81	B6-4P2	
6570	0	121.8317	82080.41	C4-6R19	
6571	2	121.8341	82078.83	B23-9P7	
6572	0	121.8418	82073.62	C0-2Q24	
6573	0	121.8441	82072.09	B18-8R2 B5-14R4	
6574	46	121.8464	82070.52	B6-4R3 C2-6R4	
6575	48	121.8515	82067.13	C1-5Q7	
6576		121.8536	82065.71	B2-1R14	sh
6577	12	121.8580	82062.73	B5-3P8 C9-11Q4 C4-5P21	
6578	13	121.8637	82058.92	C4-7Q9	
6579	27	121.8708	82054.13	C5-8P4	
6580	7	121.8727	82052.80	C5-7Q15	
6581	9	121.8783	82049.09	B8-4P8 C3-6R14	
6582	10	121.8817	82046.80	B21-8P9 B17-6R13	
6583	11	121.8853	82044.36	B6-0P21 C7-9R13 C4-6Q17	
6584	110	121.8909	82040.58	B12-6R3 B9-3R15	
6585	3720	121.8946	82038.12	C2-6Q1 C10-11Q13	
6586	35	121.8980	82035.78	B15-7P2 C9-11P3	
6587	110	121.9091	82028.36	B0-2R2 B13-5R12 C8-10Q8	
6588	56	121.9109	82027.10	B2-2R9 B5-3R9	
6589	8	121.9134	82025.46	C6-8Q13	
6590	20	121.9161	82023.62	B2-2P8	
6591	5	121.9185	82022.01	B22-9P5 B5-14R3	
6592	310	121.9369	82009.63	B0-2P1 B25-10P5	
6593	5	121.9446	82004.48	C12-14R4 B11-5P8	
6594	4	121.9493	82001.31	B15-7R3 B2-0P17	
6595		121.9516	81999.72	B3-3R3 B17-4P19	sh
6596	42	121.9528	81998.93	B3-3P2	
6597	0	121.9571	81996.06	B27-11P2	
6598	10	121.9619	81992.81	C3-6Q12	
6599	6	121.9671	81989.34	B8-4R9	
6600	82	121.9765	81983.03	B0-1R11 B3-2P10 C2-6R5 C9-11Q5	
6601	76	121.9810	81980.00	C2-6Q2 C1-5P6	
6602		121.9844	81977.73	B6-3P10	sh
6603	19	121.9905	81973.58	C12-14R3 B4-0R20	
6604	2	121.9949	81970.67	C4-7P8	
6605	4	122.0024	81965.61	C7-9Q11 B3-0R19	
6606	8	122.0042	81964.37	C0-4P9	
6607	7	122.0078	81961.94	B8-3P13 C5-8Q6 C0-4R13	
6608	16	122.0098	81960.66	B3-2R11	
6609	1	122.0134	81958.20	C10-12R2 B24-9P9	
6610	2	122.0192	81954.31	B0-1P10 C10-12R1	
6611	9	122.0224	81952.15	C1-3P21	

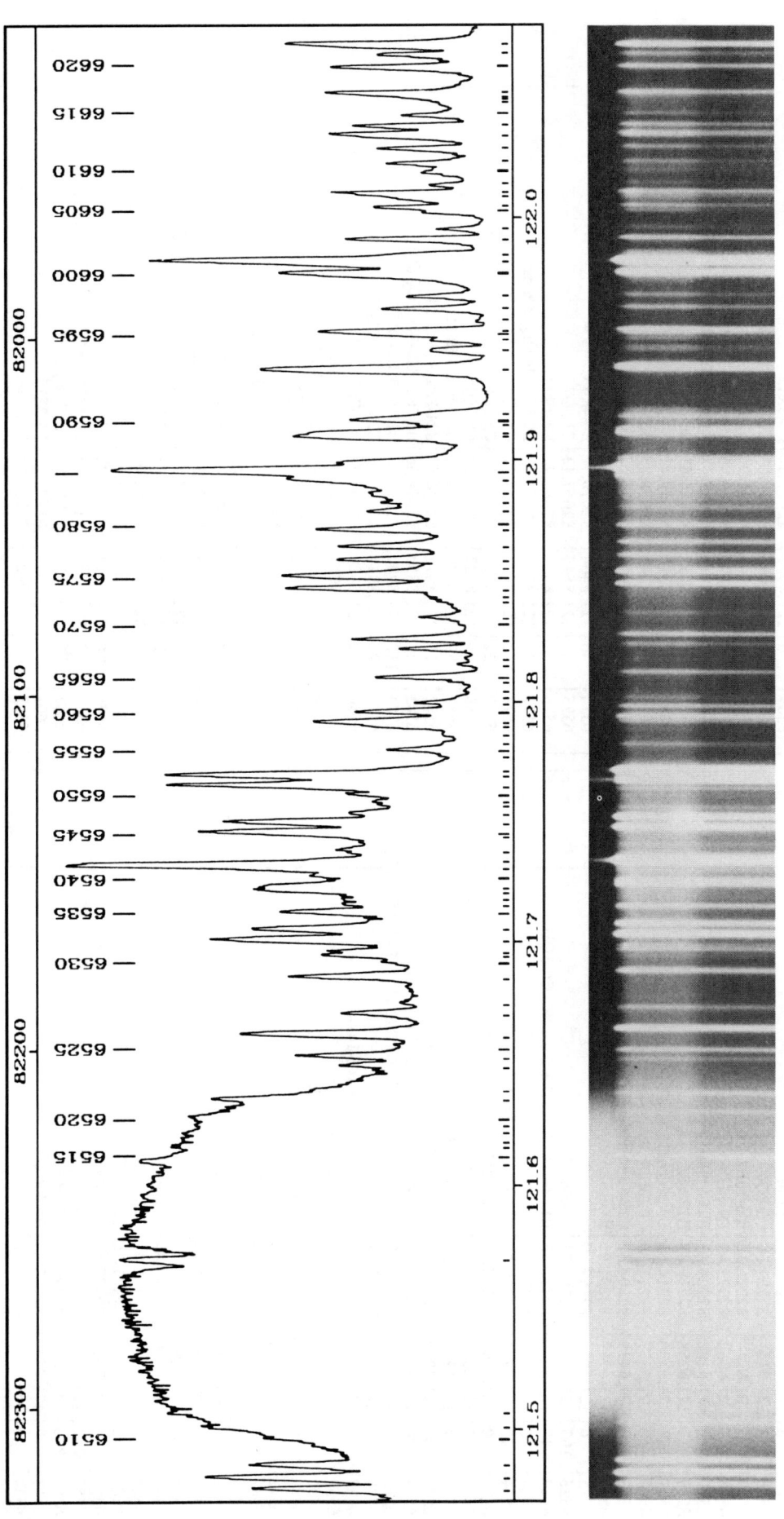

PLATE 76

TABLE 76 *continued*

Nb	I	λ (nm)	σ (cm⁻¹)	Assignment	Comment
6612	7	122.0289	81947.82	C1-5R10	
6613	16	122.0343	81944.15	B7-4P6 B7-1P19 C12-14R2 C9-10P13	
6614	12	122.0380	81941.72	C9-11P4 B'5-14R1	
6615	5	122.0426	81938.63	C8-10P7 B18-8R3	
6616	0	122.0454	81936.72	B6-1P18	
6617	0	122.0477	81935.21	C10-12R0 C4-7R12	
6618		122.0501	81933.56	B9-4P10 C1-4P15	sh
6619	25	122.0516	81932.57	B24-10P2	
6620	24	122.0624	81925.31	B5-0R21 C2-5P13	
6621	6	122.0675	81921.86	B14-6P8 B18-5P17 C12-14Q4	
6622	58	122.0716	81919.14	C12-14R1 B6-3R11	

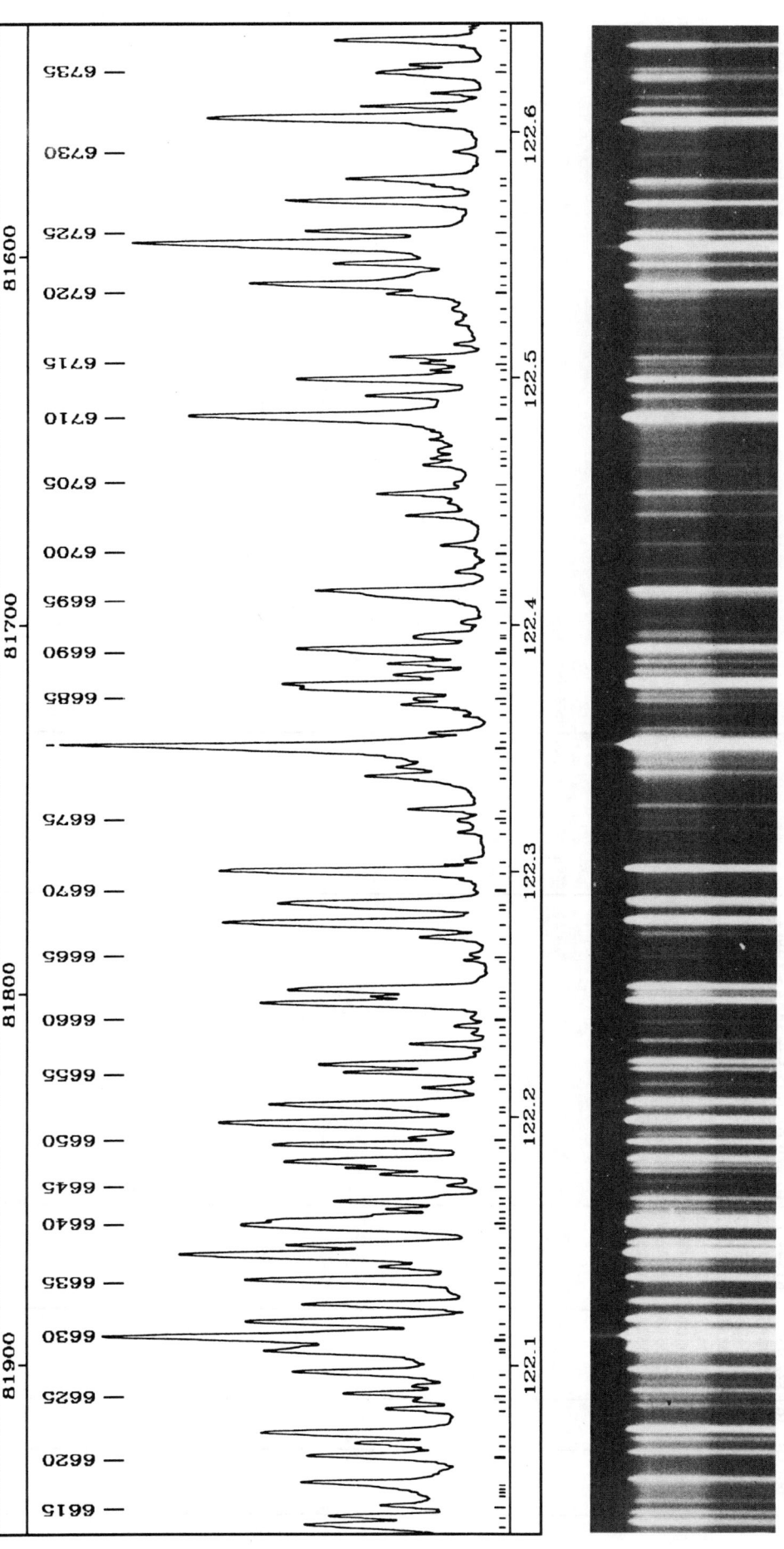

PLATE 77

TABLE 77

Nb	I	λ (nm)	σ (cm⁻¹)	Assignment	Comment
6623	4	122.0811	81912.77	B11-5R9	
6624	3	122.0846	81910.44	B10-5P6	
6625	15	122.0872	81908.64	C6-9R1 C0-3P17 C5-7P14	
6626	3	122.0905	81906.48	C2-4Q21	
6627	25	122.0958	81902.91	B4-3P6 B7-4R7	
6628	37	122.1046	81896.99	C12-14Q3 C12-14R0 B9-5P3	
6629	1080	122.1064	81895.82	C9-11Q6	
6630		122.1090	81894.01	C2-6Q3	
6631		122.1116	81892.33	B13-6P6 C10-12Q1	sh
6632	85	122.1157	81889.53	B4-3R7 B12-6P3 B19-7P11 C5-8P5 C6-9R0	
6633	30	122.1230	81884.66	C8-10Q9 C12-14Q2 B'5-14P4	
6634	0	122.1279	81881.35	B'5-14P2	
6635	83	122.1326	81878.20	C12-14Q1 C11-13R5 C11-13R7	
6636	4	122.1380	81874.58	B27-11P3 B10-5R7	
6637	380	122.1427	81871.47	C2-6P2 B5-1P17	
6638	28	122.1464	81868.99	C2-6R6 C11-13R6 C10-12Q2 B'5-14P1 B12-4R15	
6639	175	122.1541	81863.84	B0-2R3 B'5-14P3	
6640	63	122.1562	81862.38	B6-4P3 C7-9P10	
6641	7	122.1582	81861.04	B'5-14P2	
6642	8	122.1609	81859.25	C2-5R17	
6643	11	122.1640	81857.21	C0-4Q11 C6-8P12	
6644	1	122.1661	81855.78	B8-3R14	
6645	7	122.1709	81852.59	B12-5P10	
6646	10	122.1752	81849.67	C12-14P4	
6647	32	122.1778	81847.94	B15-7P3 C6-8R16	
6648	45	122.1802	81846.34	C1-5Q8 B13-6R7 B9-4R11 C12-14P2	
6649	1	122.1870	81841.79	B11-4P13	
6650	215	122.1900	81839.79	C12-14P3 C1-4R19	
6651	48	122.1958	81835.85	C11-13R3 C11-13R4 B12-6R4	
6652	3	122.2015	81832.07	B0-2P2 B1-0P16 C9-11P5	
6653	18	122.2033	81830.86	C10-12Q3 C11-13R2	
6654	22	122.2103	81826.15	B1-2R7 C4-7Q10	
6655	5	122.2165	81822.00	B16-7P6 C4-5Q23	
6656		122.2195	81819.98	C6-9Q1 C5-7R18	sh
6657	0	122.2279	81814.35	B6-0R22 C3-6P11 C11-13R1	
6658		122.2328	81811.13	C10-12P2 C0-3R21	sh
6659	4	122.2357	81809.17	C10-12R9	
6660	0	122.2395	81806.59	B15-7R4	
6661	70	122.2448	81803.07	B1-2P6 C5-8Q7	
6662	7	122.2474	81801.31	B18-8P3 C9-11Q7	
6663	52	122.2503	81799.42	C11-13R0 C1-4Q17 C3-6R15	
6664	0	122.2627	81791.10	B24-10P3	
6665	0	122.2650	81789.56	C10-12Q4	
6666	9	122.2720	81784.87	C6-8Q14	
6667	147	122.2775	81781.18	B12-5R11 C2-6Q4	
6668		122.2837	81777.02	C6-9Q2 C0-3Q19 C10-12R10	sh
6669	68	122.2854	81775.94	C2-5Q15 B6-1R19 C1-3Q23	
6670	0	122.2917	81771.67	C7-9Q12 C5-7Q16	
6671	128	122.2986	81767.07	B3-3P3 C11-13Q1	
6672	1	122.3032	81764.00	B25-10P6	
6673	2	122.3152	81755.98	C11-13Q2 B18-8R4	
6674	0	122.3178	81754.24	B7-1R20	
6675	5	122.3199	81752.87	B9-2P18 B16-5R16	
6676	5	122.3244	81749.81	C10-12P3 B15-6P10	
6677	10	122.3381	81740.67	C11-13Q3 B4-1P16 B19-7R12	
6678	5	122.3422	81737.95	C10-12Q5 C8-10Q10 B5-1R18	
6679		122.3475	81734.40	B23-9P8	sh
6680	1390	122.3497	81732.93	C2-6P3	
6681		122.3527	81730.94	B26-10P8	sh
6682	5	122.3557	81728.90	B10-2P19 B11-4R14	
6683	1	122.3633	81723.85	C11-13Q4	
6684	6	122.3674	81721.09	B27-11P4 B8-2P17 C4-6Q18	
6685	3	122.3699	81719.46	B1-1P12	
6686	30	122.3739	81716.78	C9-11P6 C2-6R7	
6687	40	122.3757	81715.58	C1-5P7	
6688	4	122.3795	81713.06	B1-1R13 B4-2P12	
6689	7	122.3834	81710.40	C11-13Q5 C6-9Q3	
6690		122.3879	81707.40	C4-7P9 C11-13P2	sh
6691	34	122.3895	81706.35	C3-6Q13 C7-9R15	
6692	5	122.3938	81703.49	B14-5P13	
6693	5	122.3950	81702.70	C11-13Q7	
6694	1	122.4001	81699.32	C11-13Q6	
6695	5	122.4080	81694.02	B15-5R15	
6696	12	122.4109	81692.06	C6-9P2 C5-8P6	
6697	22	122.4129	81690.76	C1-5R11	
6698	2	122.4157	81688.90		
6699	0	122.4187	81686.88		
6700	0	122.4207	81685.56	C4-7R13	
6701	3	122.4313	81678.43	B7-3P12 B8-1R21	
6702	7	122.4435	81670.35	C11-13P3 C2-4P20	
6703	2	122.4489	81666.73	B7-0R23	
6704	8	122.4522	81664.52	B4-2R13	
6705	0	122.4566	81661.60	C6-8R17	
6706	3	122.4638	81656.78	C0-4R14	
6707	1	122.4667	81654.87	C6-8P13	
6708	0	122.4702	81652.50	B11-2P20	
6709	0	122.4747	81649.50	C4-5P22	
6710	216	122.4836	81643.59	C2-6Q5 B12-6P4	
6711	0	122.4889	81640.04	C8-10P9	
6712	12	122.4921	81637.92	B4-1R17	
6713	38	122.4990	81633.34	B0-2R4 C11-13P4	
6714	10	122.5026	81630.90		
6715	6	122.5054	81629.07	C5-8Q8 B7-2P16	
6716	6	122.5078	81627.48	C7-9P11 B9-5P4	
6717	1	122.5131	81623.92	C6-9Q4	

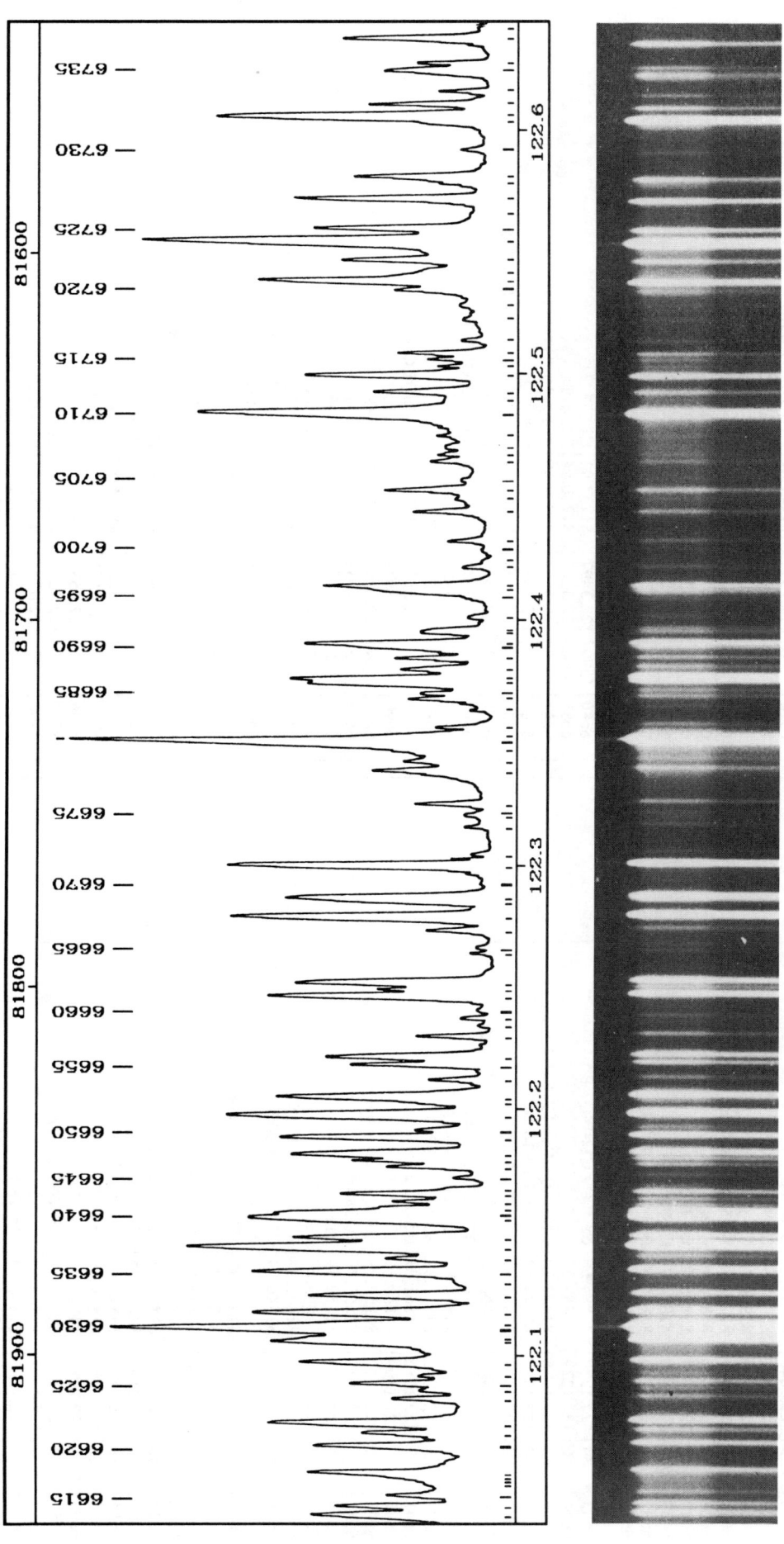

PLATE 77

TABLE 77 *continued*

Nb	I	λ (nm)	σ (cm⁻¹)	Assignment	Comment
6718	0	122.5218	81618.16	B24-10P4	
6719	0	122.5282	81613.85	B15-7P4 B10-4P12	
6720	6	122.5340	81610.00	C1-5Q9	
6721	46	122.5376	81607.62	B21-8P10 C10-12P5 C0-2Q25	d sh
6722		122.5402	81605.87	C2-6R8 C9-11Q9	
6723	10	122.5461	81601.92	B0-2P3 C11-13P5 B7-3R13	
6724	670	122.5537	81596.90	C4-7Q11 C8-10Q11	
6725	15	122.5589	81593.45		
6726	0	122.5654	81589.10	C6-9P3 C11-13P7 C5-7R19	
6727	30	122.5712	81585.23	B18-8P4 B12-6R5 B14-5R14	sh
6728		122.5779	81580.75	B26-11R0 C7-9Q13	
6729	12	122.5799	81579.46	B6-4P4 C9-11P7	
6730	3	122.5910	81572.07	B9-2R19	
6731	3	122.6019	81564.79	B15-7R5 C11-13P6	
6732	210	122.6045	81563.09	C2-6P4	
6733	11	122.6093	81559.90	B26-11R1 B12-3P18 C2-5P14 C5-7P15	
6734	4	122.6144	81556.47	C3-6P12	
6735	12	122.6226	81551.02	B11-3P17 C0-4Q12 C6-8Q15 C1-4P16	b
6736	4	122.6257	81548.95	C5-8P7	
6737	20	122.6357	81542.31	B3-1P15 B27-11P5	
6738	0	122.6411	81538.73	C10-12Q9	

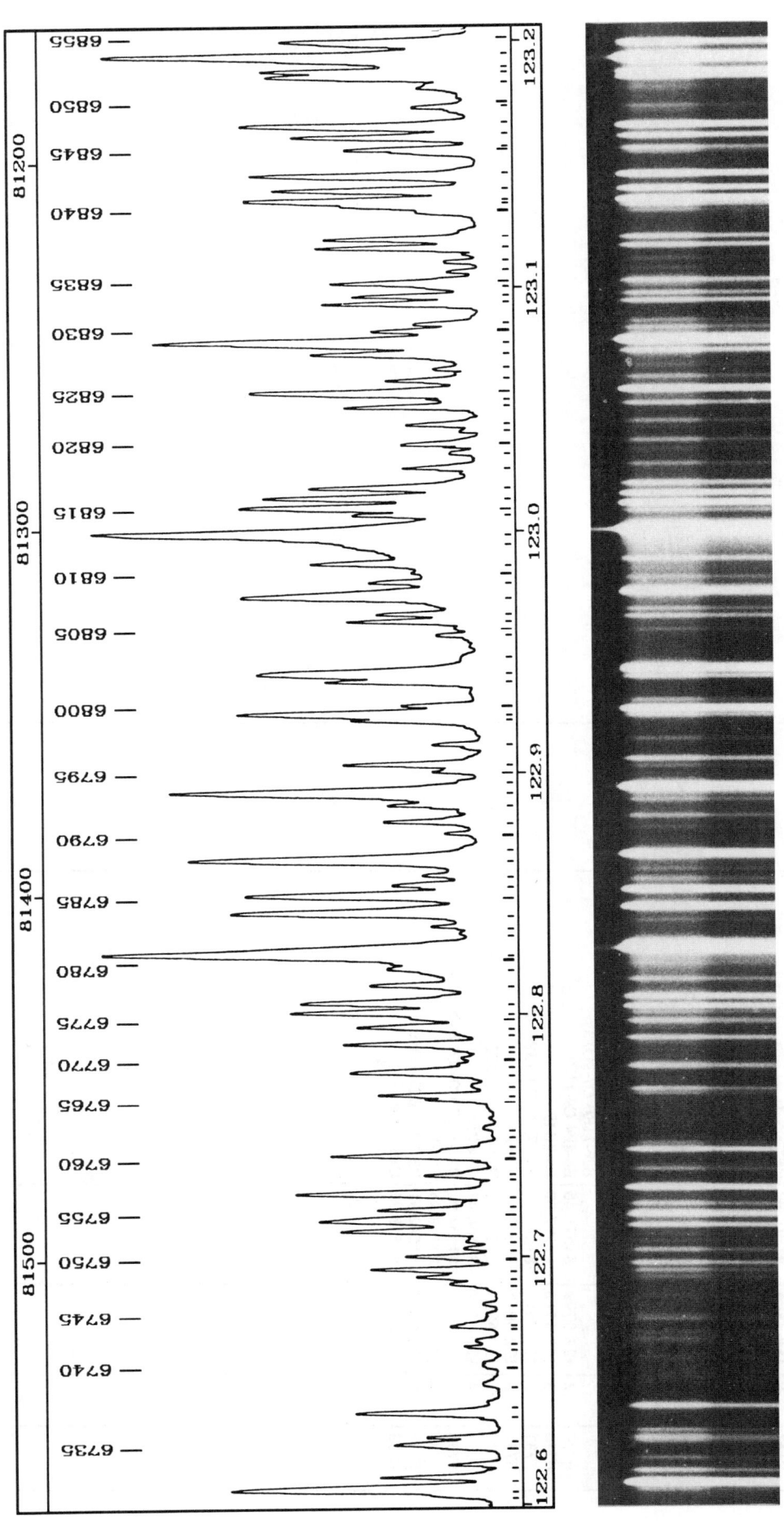

PLATE 78

TABLE 78

Nb	I	λ (nm)	σ (cm⁻¹)	Assignment	Comment
6739	0	122.6473	81534.62	B18-6R15	
6740	0	122.6552	81529.36	B18-8R5	
6741	3	122.6627	81524.39	B10-2R20 C2-5R18	
6742	1	122.6659	81522.27	C10-12P6	
6743	3	122.6710	81518.83	C6-9Q5 C10-12Q10	sh
6744		122.6723	81518.02	B13-5P12	
6745	1	122.6765	81515.19	B26-11P1 B21-9R5	
6746	2	122.6802	81512.71	C2-4Q22 C2-4R24	
6747	5	122.6881	81507.47	B13-3P19 C9-11P8 C9-11Q10 C1-3P22	
6748	6	122.6908	81505.69	B10-4R13	
6749	8	122.6938	81503.73	B8-4P9 B26-11R2	
6750	5	122.6990	81500.24	B10-5P7	
6751	1	122.7027	81497.81	C3-6R16	
6752	1	122.7054	81495.99	B10-3P16 C5-7Q17	
6753	18	122.7095	81493.25	B7-4P7 C1-4R20	
6754	20	122.7140	81490.27	B5-3P9 B13-6P7 B25-10P7	
6755	7	122.7181	81487.57	B11-5P9	
6756	0	122.7212	81485.49	C4-7R14	
6757	40	122.7249	81483.05	C2-6Q6	
6758	6	122.7323	81478.14	C1-5R12	
6759		122.7336	81477.25	B3-3R5	
6760	28	122.7405	81472.72	B3-3P4 C0-3P18	
6761	6	122.7428	81471.19	B3-1R16	
6762	0	122.7452	81469.60		
6763	0	122.7494	81466.80	C7-9P12	
6764	0	122.7532	81464.27	B23-10R0	
6765	5	122.7639	81457.14	B5-3R10	
6766	10	122.7654	81456.15	C6-9P4 C8-10Q12	
6767	0	122.7696	81453.40	B7-4R8	
6768	11	122.7751	81449.73	B16-7P7 C1-5P8	
6769	1	122.7788	81447.27	B23-10R1	
6770	1	122.7813	81445.64	B10-5R8	
6771	0	122.7844	81443.56	B8-4R10	
6772	23	122.7865	81442.18	C5-8Q9 C2-5Q16	
6773	0	122.7902	81439.70	B26-10P9	
6774	16	122.7934	81437.60	B2-2R10 C4-7P10	
6775	6	122.7971	81435.11	B14-6P9	
6776	33	122.7998	81433.41	B4-3P7 B22-9P7	
6777	33	122.8037	81430.79	B2-2P9	
6778	14	122.8109	81426.03	B11-2R21 C1-4Q18	
6779	6	122.8177	81421.48	B4-3R8	
6780	1440	122.8237	81417.52	C3-7R1 C3-7R2	
6781		122.8264	81415.75	C3-6Q14 C2-6R9 C0-3R22	sh
6782	0	122.8335	81410.99	B14-3P20	
6783	1	122.8354	81409.78	B20-9R0 B13-5R13 C4-6Q19	
6784	166	122.8411	81405.98	C2-6P5	
6785	107	122.8481	81401.33	B14-7R0 B19-8P7 B24-10P5	
6786	4	122.8523	81398.57	B13-6R8 B11-5R10 B16-6P12 C6-9Q6 C7-9Q14	
6787	3	122.8566	81395.70	B17-8R0	
6788	332	122.8628	81391.63	C3-7R3	
6789	0	122.8703	81386.62	B23-10P1	
6790	2	122.8744	81383.95	B14-4P17	
6791	13	122.8792	81380.77	B20-9R1 B28-11P8	
6792	7	122.8860	81376.26	B11-6R0 B17-7P9 C9-11P9	
6793	600	122.8911	81372.85	C3-7R0 C6-8P14	
6794	1	122.8969	81369.03	B21-9P5 C7-10R4	
6795	4	122.8999	81367.03	B20-8P9 B23-9P9 C0-3Q20	
6796	17	122.9027	81365.15	C11-14R4 B17-8R1	
6797	5	122.9111	81359.59	C9-11Q12 C0-4R15	
6798	14	122.9213	81352.88	C1-5Q10	
6799	146	122.9238	81351.20	B1-2R8 B14-7R1	
6800	5	122.9270	81349.08	C4-7Q12	
6801	24	122.9374	81342.21	B11-6R1 C10-12P9	
6802	88	122.9405	81340.14	C3-7R4 B0-2R5 C0-4P11	
6803	0	122.9484	81334.96	B23-9R10	
6804	2	122.9565	81329.56	C5-8P8	
6805	0	122.9591	81327.88	C8-10Q13 C6-8Q16	
6806	14	122.9620	81325.96	B8-5R0 B20-9P1	
6807	9	122.9649	81324.01	B15-7P5	
6808	97	122.9724	81319.05	B1-2P7 B6-3P11	
6809	6	122.9781	81315.31	B18-8P5 B25-9P12	
6810	0	122.9822	81312.61		
6811	14	122.9859	81310.11	B12-6P5	
6812	15	122.9927	81305.65	B20-9R2 C0-2P24	
6813	3350	122.9981	81302.08	B9-5P5 B17-8P1 C3-7Q1 C2-6Q7	
6814	11	123.0062	81296.70	B3-2P11	
6815	144	123.0092	81294.70	B0-2P4	
6816	56	123.0131	81292.18	B8-5R1 B14-7P1 C10-12P10	
6817	20	123.0173	81289.41	B15-7R6 C6-9P5	
6818	6	123.0252	81284.13	B17-8R2 C7-10Q2 C4-5P23	
6819	1	123.0315	81280.02	B9-4P11	
6820	6	123.0353	81277.49	B3-2R12 B0-1R12	
6821	1	123.0407	81273.91	B21-9R6	
6822	5	123.0433	81272.18	B2-1P14	
6823	18	123.0529	81267.15	B11-6P1	
6824	1	123.0549	81265.85	B14-7R2	
6825	80	123.0568	81263.30	C3-7R5 C6-9Q7 B6-3R12 C2-4P21	
6826	6	123.0617	81260.04	B12-5P11 B18-8R6	
6827	1	123.0667	81256.74	C8-10P11	
6828	27	123.0731	81252.54	B9-3R16 C11-14R2	
6829	780	123.0775	81249.64	B11-6R2 C3-7Q2 C11-14Q4	
6830	8	123.0823	81246.47	B0-1P11 C5-8Q10	
6831	7	123.0851	81244.58	C3-6R17	
6832	19	123.0936	81239.02	B6-4P5	
6833	18	123.0966	81237.00	C2-5P15	
6834		123.1002	81234.61	C4-7P11	sh
6835	19	123.1018	81233.58	B2-1R15 C0-4Q13 C7-9P13 C5-7Q18	

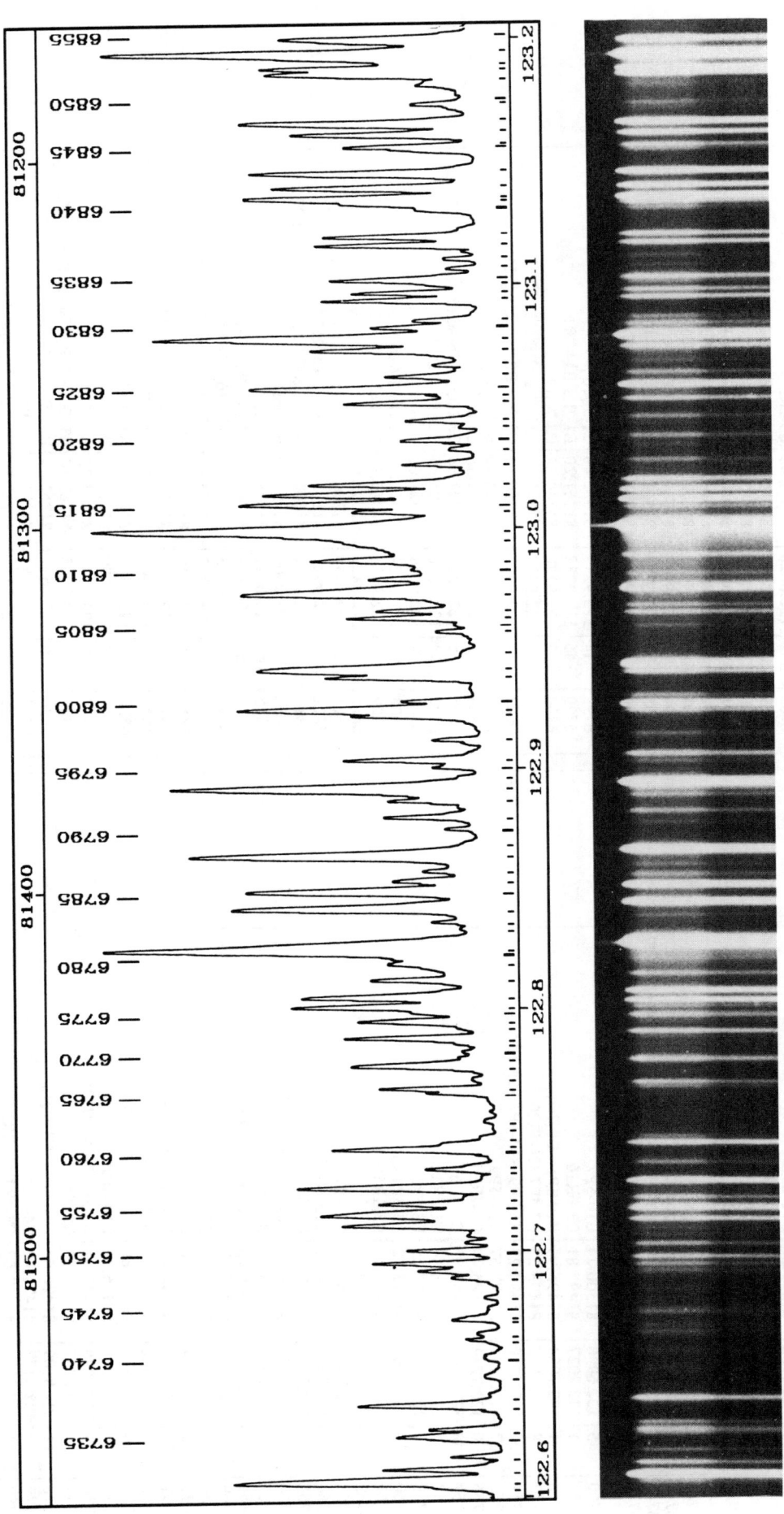

PLATE 78

TABLE 78 *continued*

Nb	I	λ (nm)	σ (cm⁻¹)	Assignment	Comment
6836	1	123.1067	81230.32	C7-10Q3 C7-9Q15	
6837	2	123.1108	81227.63	C4-7R15	
6838	28	123.1165	81223.89	B5-4R0 B19-7R13	
6839	20	123.1201	81221.48	C3-6P13 B9-4R12	
6840	9	123.1312	81214.18	B20-9P2 C2-5R19	
6841	25	123.1339	81212.43	C1-5R13	
6842	141	123.1360	81211.02	C11-14R1	
6843	47	123.1403	81208.18	B8-5P1 C2-6R10	
6844	85	123.1463	81204.27	C11-14Q3	
6845		123.1556	81198.09	B15-6P11 B8-3P14	sh
6846	15	123.1571	81197.08	B8-5R2 B25-10P8	
6847	32	123.1623	81193.65	C2-6P6	
6848	108	123.1668	81190.74	B5-4R1	
6849	0	123.1726	81186.90	C6-9P6	
6850	3	123.1745	81185.61	B20-9R3	
6851	2	123.1827	81180.20	B17-8P2 B24-9P11	
6852	66	123.1872	81177.29	C11-14R0	
6853	60	123.1893	81175.89	C11-14Q2	
6854	1085	123.1949	81172.20	C3-7Q3	
6855	32	123.2011	81168.09	B14-7P2 C1-5P9	

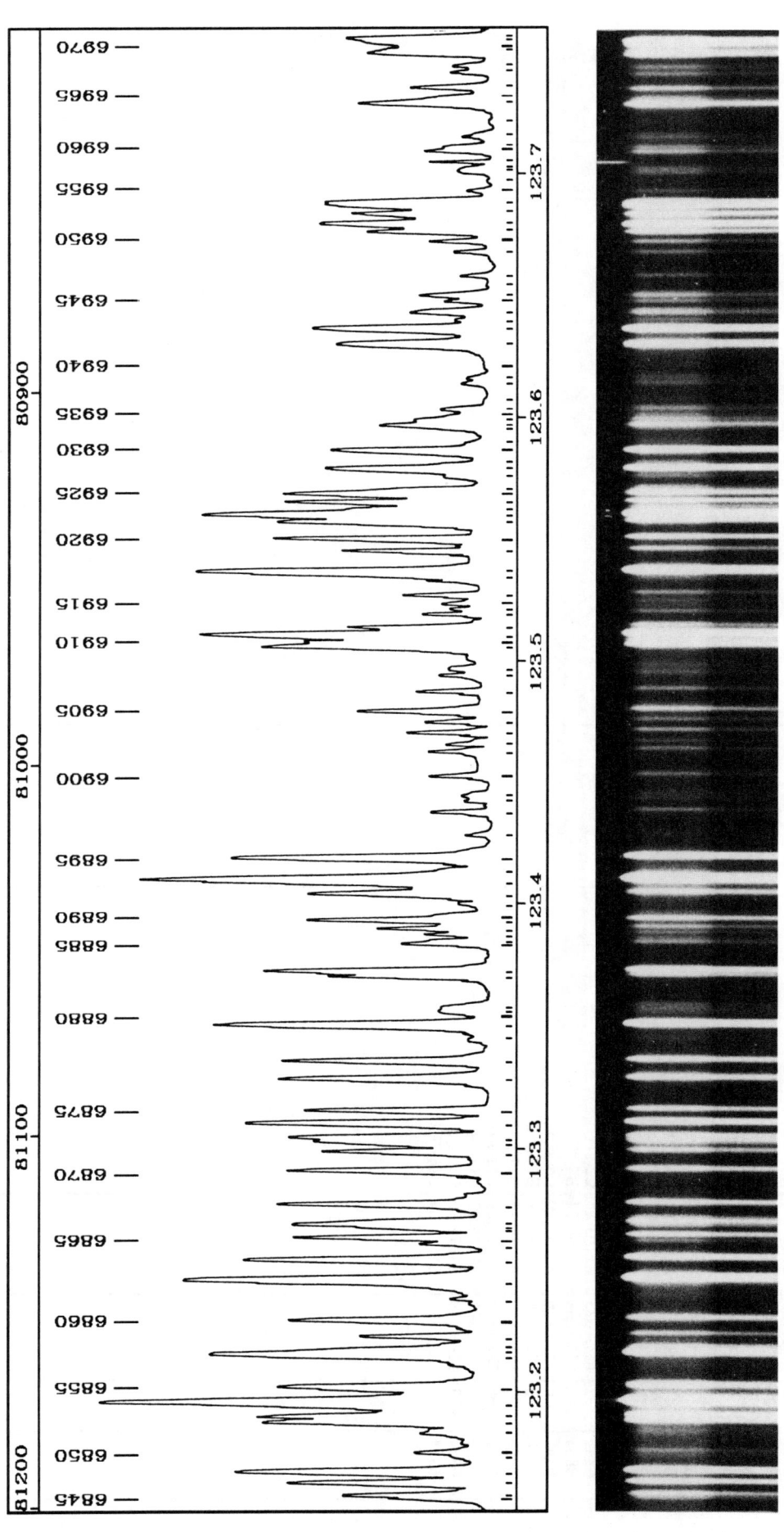

PLATE 79

TABLE 79

Nb	I	λ (nm)	σ (cm⁻¹)	Assignment	Comment
6856	233	123.2144	81159.35	C11-14Q1 C7-10Q4	
6857	0	123.2165	81157.98	C11-14P4	
6858		123.2190	81156.31	B12-5R12	sh
6859	11	123.2219	81154.40	B17-8R3	
6860	33	123.2285	81150.03	C3-7R6 C1-4P17 C2-4R25	
6861	0	123.2376	81144.06	B21-8P11 C8-10Q15	
6862	395	123.2451	81139.16	B28-11P9 C3-7P2	
6863	86	123.2533	81133.73	C11-14P3 C2-4Q23	
6864	4	123.2602	81129.18	B12-4R16 C10-13R7	
6865	29	123.2627	81127.54	C11-14P2	
6866		123.2667	81124.89	B11-6P2	sh
6867	34	123.2681	81124.01	C3-6Q15 B15-5P15	
6868		123.2694	81123.13	B11-4P14	sh
6869	32	123.2758	81118.93	B3-3P5 C6-9Q8 C6-8Q17	
6870	34	123.2896	81109.86	C2-5Q17	
6871	16	123.2974	81104.68	B18-7P11 C7-10P3	
6872	18	123.3016	81101.94	B4-0P20 C2-6Q8 C5-8P9	
6873	27	123.3033	81100.83	C4-7Q13 B11-6R3	
6874	77	123.3088	81097.22	B5-4P1 B8-3R15 B22-9P8	
6875	26	123.3141	81093.74	B5-4R2	
6876	32	123.3271	81085.18	C10-13R6 C1-5Q11 B3-0P19	
6877	25	123.3342	81080.50	C3-7R7	
6878	0	123.3432	81074.61	C7-10Q5	
6879	120	123.3488	81070.92	C3-7Q4	
6880		123.3534	81067.89	C1-3P23	
6881	5	123.3547	81067.04	C10-13R5	
6882	4	123.3562	81066.07	B5-0P21	
6883	22	123.3694	81057.36	B8-5P2	
6884	62	123.3714	81056.04	B2-3R0 C1-4Q19	
6885	9	123.3825	81048.76	C0-3P19	
6886	3	123.3836	81048.03	B13-6P8	
6887	3	123.3888	81044.62	B16-5R17	
6888	6	123.3888	81044.62	C5-8Q11 C0-4R16	
6889	18	123.3924	81042.24	B8-5R3	
6890	0	123.3941	81041.15	B21-9P6	
6891	0	123.3979	81038.65	B29-12R8	
6892	24	123.4037	81034.87	B16-7P8 B14-5P14 C2-6R11	d
6893	700	123.4091	81031.27	B10-5P8 B19-8P8 C3-7P3	
6894	0	123.4132	81028.62	C7-10P4 C0-3R23	
6895	142	123.4183	81025.28	B2-3R1	
6896	0	123.4279	81018.96	C4-7R16	
6897	4	123.4376	81012.56	C6-9P7	
6898	0	123.4417	81009.92	B18-8P6	
6899	5	123.4436	81008.68	B2-0P18	
6900	5	123.4524	81002.88	B11-4R15	
6901	3	123.4659	80995.96	B7-4P8	
6902	2	123.4659	80994.02	B15-7P6 C3-6R18	
6903	5	123.4707	80990.89	B30-13R5	
6904	4	123.4750	80988.07	C10-13R4 C5-7Q19	
6905	10	123.4792	80985.27	B0-2R6	
6906	5	123.4875	80979.85	B17-8R4 C0-4P12	
6907	2	123.4944	80975.30	B16-7R9 B6-0P22 C8-11Q1	
6908	2	123.4973	80973.40	B28-12R5 B10-5R9	
6909	43	123.5060	80967.75	C2-6P7 C9-12R10	
6910	6	123.5078	80966.54	C10-13R3 B6-1P19	
6911	258	123.5106	80964.70	B14-7P3	
6912	12	123.5136	80962.73	C6-9Q9 B25-11R1 C0-3Q21 B13-6R9	
6913	6	123.5193	80959.01	C9-12R9 C1-5R14	
6914	1	123.5216	80957.49	B7-4R9	
6915	2	123.5238	80956.08	B23-10P4 B18-8R7 B7-1P20	
6916	3	123.5274	80953.73	B11-5P10 B3-0R20	
6917		123.5338	80949.48	B4-2P13	sh
6918	131	123.5368	80947.55	C3-7Q5 C8-11Q2 B30-13R6 B7-3P13	
6919	13	123.5457	80941.74	B4-0R21 C10-13R2	
6920	35	123.5506	80938.48	B5-4P2	
6921	25	123.5577	80933.85	B5-4R3	
6922	224	123.5602	80932.21	B5-2P5	
6923		123.5633	80930.16	C3-7R8	sh
6924	23	123.5656	80928.69	B2-3R2	
6925	22	123.5689	80926.48	B11-6P3 B1-1P13 B9-5P6	
6926		123.5710	80925.15	C4-7P12 C9-12R6	sh
6927	4	123.5761	80921.81	B1-1R14 B25-11P1 B5-1P18	
6928	110	123.5793	80919.70	B2-3P1 B10-4P13	
6929		123.5812	80918.45	B8-4P10	sh
6930	80	123.5867	80914.85	B4-3P8 C10-13R1	
6931	0	123.5895	80913.03	B14-5R15	
6932		123.5962	80908.61	B25-9P13 C9-12R7 C0-4Q14 C10-13Q7	sh
6933	10	123.5972	80907.97	B25-11R2 B15-7R7	
6934	7	123.5994	80906.57	B4-3R9 B18-6R16 C8-11Q3	
6935	4	123.6011	80905.41	B4-2R14	
6936	4	123.6035	80903.86	C9-12R3 C7-10P5	
6937	0	123.6077	80901.11	C9-12R4	
6938	2	123.6144	80896.75	B11-6R4 B17-7P10 C9-12R8	
6939	2	123.6166	80895.30	B30-13R7 C9-12R2	
6940	0	123.6223	80891.54	B8-1P21	
6941	63	123.6304	80886.23	B25-10P9 C10-13R0 C2-6Q9 C2-5R20	
6942	182	123.6367	80882.10	C3-7P4	
6943	3	123.6403	80879.80	C8-11P2 C9-12R5	
6944	11	123.6435	80877.68	C9-12R1 B5-3P10 B5-0R22	
6945	6	123.6481	80874.71	B25-9R14 C3-6P14	
6946	7	123.6504	80873.14	B7-3R14 C10-13Q6 C7-10Q7	
6947	1	123.6553	80869.96	B1-0P17	
6948	4	123.6586	80867.78	B13-5P13 C7-9P15 C4-6P19	
6949	4	123.6685	80861.33	B8-4R11 C6-9R13	
6950	9	123.6729	80858.43	C2-5P16	
6951	10	123.6773	80855.61	C8-11Q4 C10-13Q5 C1-5P10	
6952	160	123.6802	80853.66	C10-13Q1	
6953	34	123.6844	80850.96	C10-13Q2 C4-7Q14 C9-12R0	b

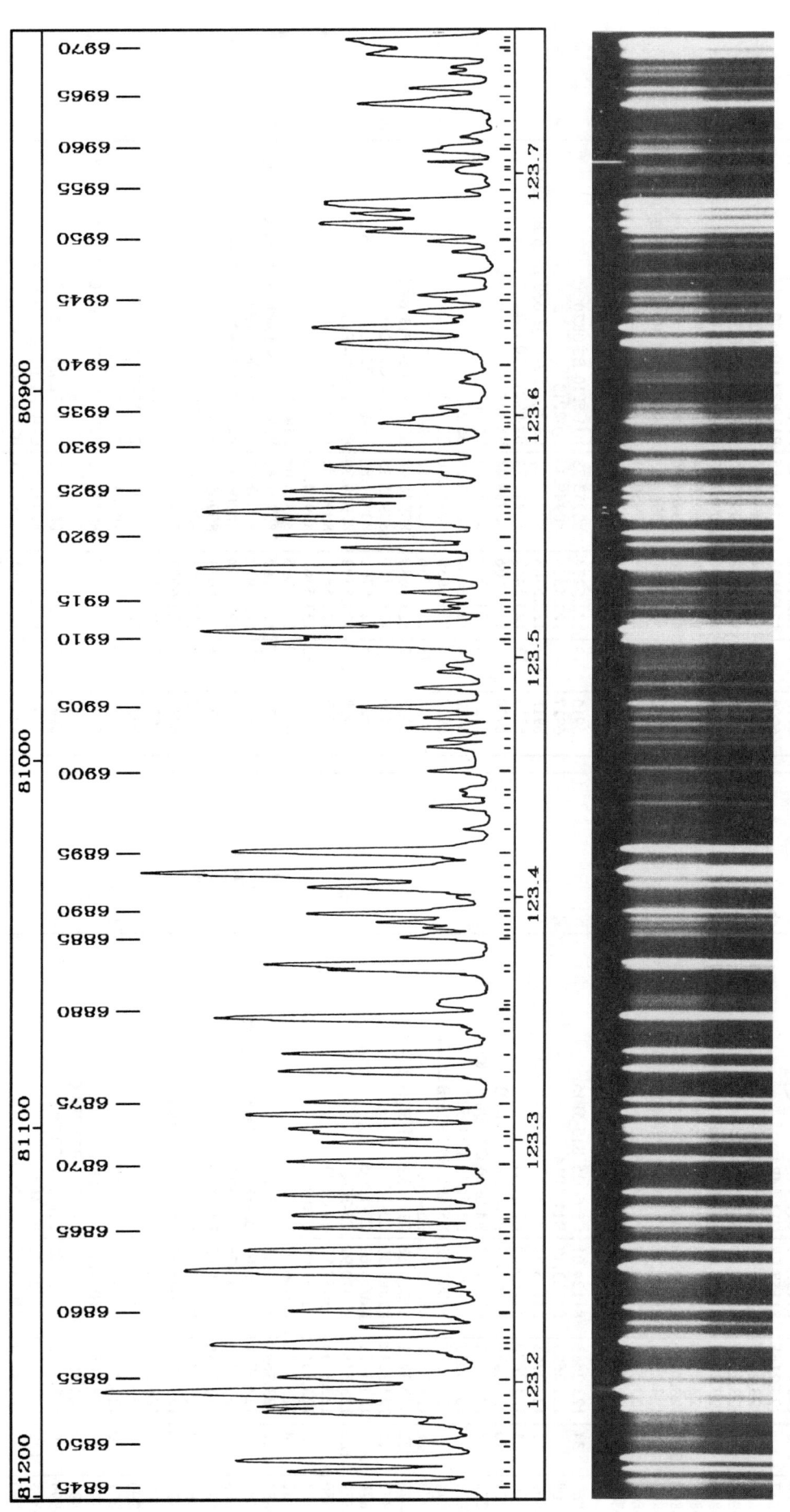

PLATE 79

TABLE 79 *continued*

Nb	I	λ (nm)	σ (cm⁻¹)	Assignment	Comment
6954	93	123.6883	80848.38	C10-13Q3 C10-13Q4 B5-3R11 B8-5P3 B11-5R11	
6955	2	123.6934	80845.03	B6-4P6 B29-12R9	
6956	1	123.6972	80842.57	C2-4P22	
6957	5	123.7013	80839.92	B7-0P23 C5-8Q12	
6958	5	123.7027	80838.96	C5-8P10	
6959					GeII 123.70589
6960	14	123.7102	80834.08	C8-11R10 C3-6Q16 C0-2P25	
6961	10	123.7118	80833.01	B9-2P19	
6962	2	123.7160	80830.31	B8-5R4 C6-9P8	b
6963	0	123.7230	80825.72	B6-4R7 B30-13P6	
6964	30	123.7300	80821.14	B1-2R9 B20-9R5 C2-4R26	
6965	15	123.7320	80819.85	B25-11R3 B8-2P18 B10-4R14	
6966	15	123.7362	80817.04	B4-1P17	
6967	5	123.7422	80813.20	B14-6R11 B6-1R20 C8-10P15	
6968	4	123.7450	80811.35	B16-6P13 C8-11R12 C8-11R9	
6969	38	123.7504	80807.86	C9-12Q1	sh
6970		123.7519	80806.87	C8-11P3 C1-5Q12 C8-11R13	
6971	34	123.7544	80805.20	B2-2R11 C2-6R12 C6-9Q10	
6972	34	123.7559	80804.25	C3-7Q6	

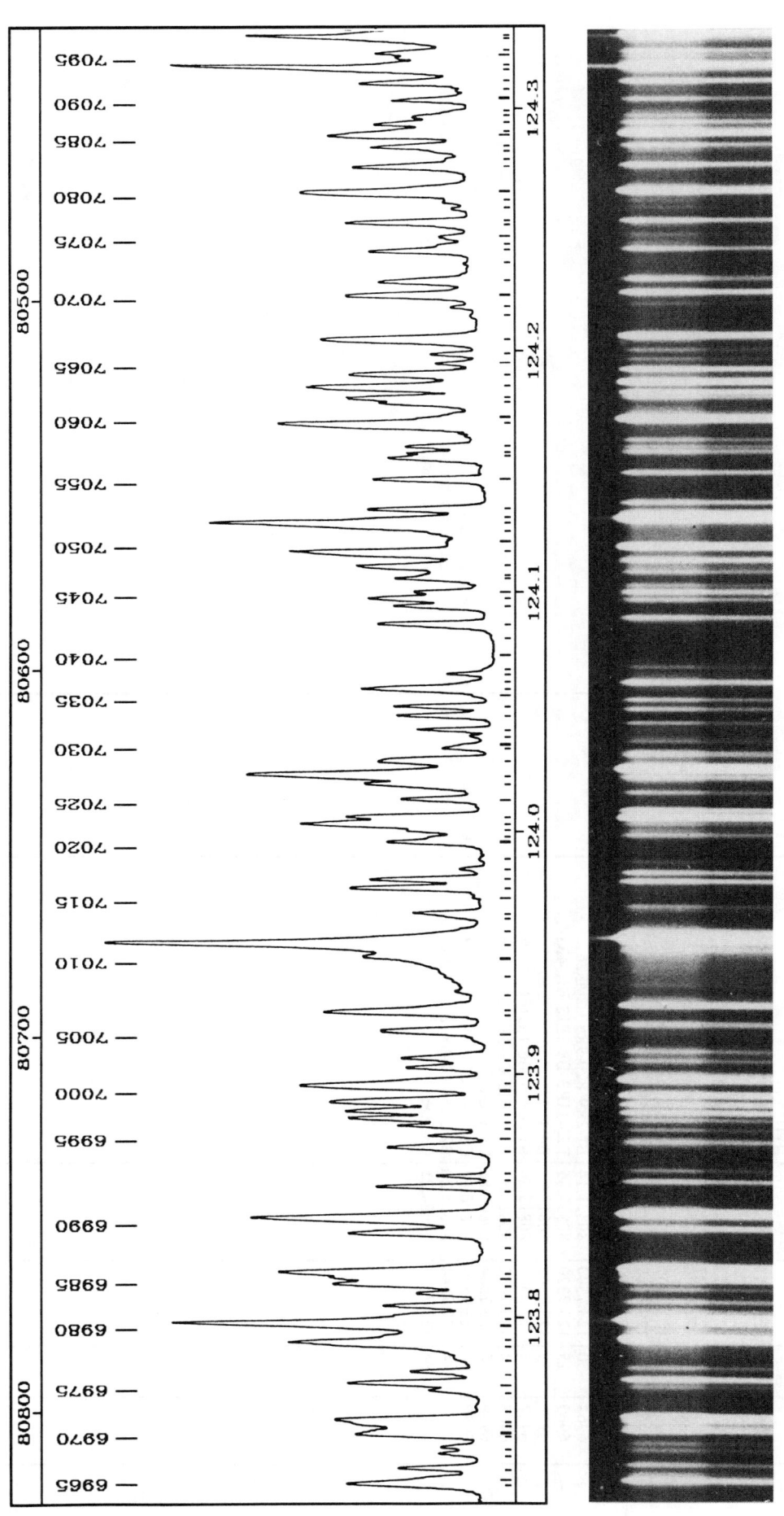

PLATE 80

TABLE 80

Nb	I	λ (nm)	σ (cm⁻¹)	Assignment	Comment
6973	0	123.7618	80800.39	B28-12R6 B5-1R19	
6974	8	123.7682	80796.20	B2-2P10 B22-10R0 B10-2P20	
6975	40	123.7712	80794.25	C10-13P2 C8-11Q5	
6976	11	123.7759	80791.15	C9-12Q2	
6977	0	123.7812	80787.74	C1-4R22	
6978	300	123.7880	80783.28	C4-8R2 B1-2P8	
6979		123.7910	80781.29	C2-5Q18 C8-11R11	sh
6980	1980	123.7955	80778.40	C4-8R1	
6981		123.7991	80776.06	B7-1R21 C10-13P7	sh
6982	10	123.8029	80773.56	B17-8R5 C7-10P6 C4-7R17	
6983	9	123.8082	80770.12	B22-10R1	
6984	57	123.8123	80767.43	B2-3R3 C9-12Q3 C1-4P18 C2-4Q24	
6985	50	123.8148	80765.80	C10-13P3	
6986	320	123.8172	80764.24	C4-8R3 C3-7R9	
6987	0	123.8241	80759.73	C7-10Q8	
6988	0	123.8284	80756.94	B22-9P9	
6989	64	123.8333	80753.73	B2-3P2 B23-10P5	
6990	920	123.8392	80749.87	C4-8R0 B7-2P17	
6991	24	123.8523	80741.34	C10-13P4	
6992	5	123.8567	80738.44	B21-9P7 C9-12Q4	
6993	0	123.8601	80736.26	B29-13R0 B29-12R10	
6994	18	123.8686	80730.72	B14-7P4 B27-12R0 B29-13R1	
6995	7	123.8734	80727.59	C9-12P2 C8-11Q6	
6996	5	123.8774	80725.01	C2-6P8	
6997	38	123.8806	80722.87	C10-13P5 C8-11P4	
6998	31	123.8839	80720.73	C4-8R4 B4-1R18	
6999	108	123.8877	80718.29	B5-4P3 B22-10P1 B27-12R1	
7000	225	123.8943	80713.98	C3-7P5 B5-4R4	
7001		123.8965	80712.56	C10-13P6	sh
7002	9	123.9019	80709.01	B3-3P6 B11-2P21 C3-6R19	
7003	9	123.9056	80706.61	C9-12Q5 B22-13R2 B29-13R2 C6-9R14	
7004	3	123.9136	80701.40	B25-11R4 B11-3P18 C7-10R13	
7005	14	123.9167	80699.37	B30-13P7 C0-5R2	
7006	83	123.9248	80694.12	C0-5R1	
7007		123.9274	80692.42	B29-13P1 B14-7R5 C1-4Q20	sh
7008	0	123.9332	80688.66	B12-3P19	
7009	0	123.9415	80683.24	B19-7R14	
7010	18	123.9486	80678.64	C5-8P11 C1-5R15	
7011	3820	123.9532	80675.64	B27-12P1 B27-12R2 B11-6P4 C4-8Q1 C0-5R3 C9-12P3 C9-12Q6	
7012		123.9577	80672.67	B8-2R19	sh
7013	4	123.9645	80668.25	B29-13R3	
7014	8	123.9659	80667.37	B19-9R0 B18-8P7	
7015	0	123.9733	80662.55	B9-2R20 B10-3P17	
7016	30	123.9769	80660.22	C0-5R0	
7017	12	123.9804	80657.94	C2-6Q10	
7018	1	123.9845	80655.24	C8-11Q7	
7019	0	123.9891	80652.26	C0-3R24	
7020	10	123.9959	80647.82	C9-12Q7 B3-1P16	
7021	0	123.9981	80646.43	C9-12Q10	
7022	1	124.0007	80644.68	C7-10Q9 C6-9Q11	
7023	80	124.0037	80642.75	B9-4P12 B13-3P20 C3-7Q7 C3-7R10	
7024	26	124.0067	80640.82	C4-8R5	
7025	10	124.0141	80635.98	B19-9R1 B11-6R5 B12-5P12	
7026	12	124.0207	80631.69	C5-8Q13	
7027	720	124.0241	80629.50	C8-11P5 C7-10P7 C4-7P13	
7028	10	124.0298	80625.79	C4-8Q2 C9-12Q8 C0-4P13 C7-10R15 B19-8P9 B6-3P12 B15-7P7 C9-12Q9	
7029	1	124.0353	80622.22	C0-5R4	
7030	0	124.0365	80621.41	C6-9P9	
7031	0	124.0401	80619.06	B15-6P12 B18-8R8 B20-9P5	
7032	9	124.0429	80617.30	B22-10P2 C6-9R15	
7033	6	124.0488	80613.44	B29-13R4 C9-12P4	
7034	15	124.0527	80610.90	C4-8R6 C0-3P20	
7035	0	124.0571	80608.02	B27-12R3 C2-6R13	
7036	36	124.0600	80606.17	B18-7P12	
7037	0	124.0622	80604.71	C4-7Q15 C3-6P15	
7038	2	124.0661	80602.20	C2-5R21	
7039	0	124.0678	80601.09	B30-14R1 B22-10R3	
7040	0	124.0765	80595.46	B10-2R21 B7-0R24	
7041	0	124.0855	80589.59	B30-14R0	
7042	14	124.0871	80588.54	B16-7P9	
7043	8	124.0948	80583.59	C1-5P11	
7044	25	124.0978	80581.61	C8-11Q8 B13-6P9 B20-9R6	
7045	6	124.1005	80579.87	B8-5R5 B8-5P4 B19-9P1 B3-1R17	
7046	10	124.1065	80575.94	B29-13P3 C0-4Q15	
7047		124.1084	80574.70	B6-3R13	sh
7048	23	124.1113	80572.84	B3-2P12	
7049	182	124.1170	80569.15	B0-2R7 B16-8R0	
7050	0	124.1208	80566.68	B19-9R2 C0-5Q1 B9-4R13	
7051	0	124.1248	80564.08	C0-3Q22 B30-14P1	
7052	920	124.1291	80561.29	B25-10P10	
7053	17	124.1313	80559.86	C4-8Q3 B29-13R5	
7054	35	124.1352	80557.33	B3-2R13	
7055		124.1477	80549.20	B25-11R5 C9-12P5	sh
7056	9	124.1563	80543.64	C3-6Q17	
7057	7	124.1583	80542.34	B2-3R4 B12-5R13	
7058	15	124.1612	80540.45	B16-8R1	
7059	420	124.1703	80534.59	C2-5P17	
7060		124.1727	80533.01	C4-8P2 B28-12R8	sh
7061	5	124.1791	80528.85	B0-1R13 B29-12P10 C8-11P6	
7062	27	124.1811	80527.53	C0-5R5 C7-10Q10 C2-4R27	
7063	157	124.1855	80524.68	B14-3P21 C3-7P6	
7064	26	124.1907	80521.31	B2-3P3	
7065	2	124.1949	80518.60	B27-12R4 B28-12P7 C1-5Q13 C5-8R17	
7066	3	124.1987	80516.11	B10-5P9 C8-11Q9	

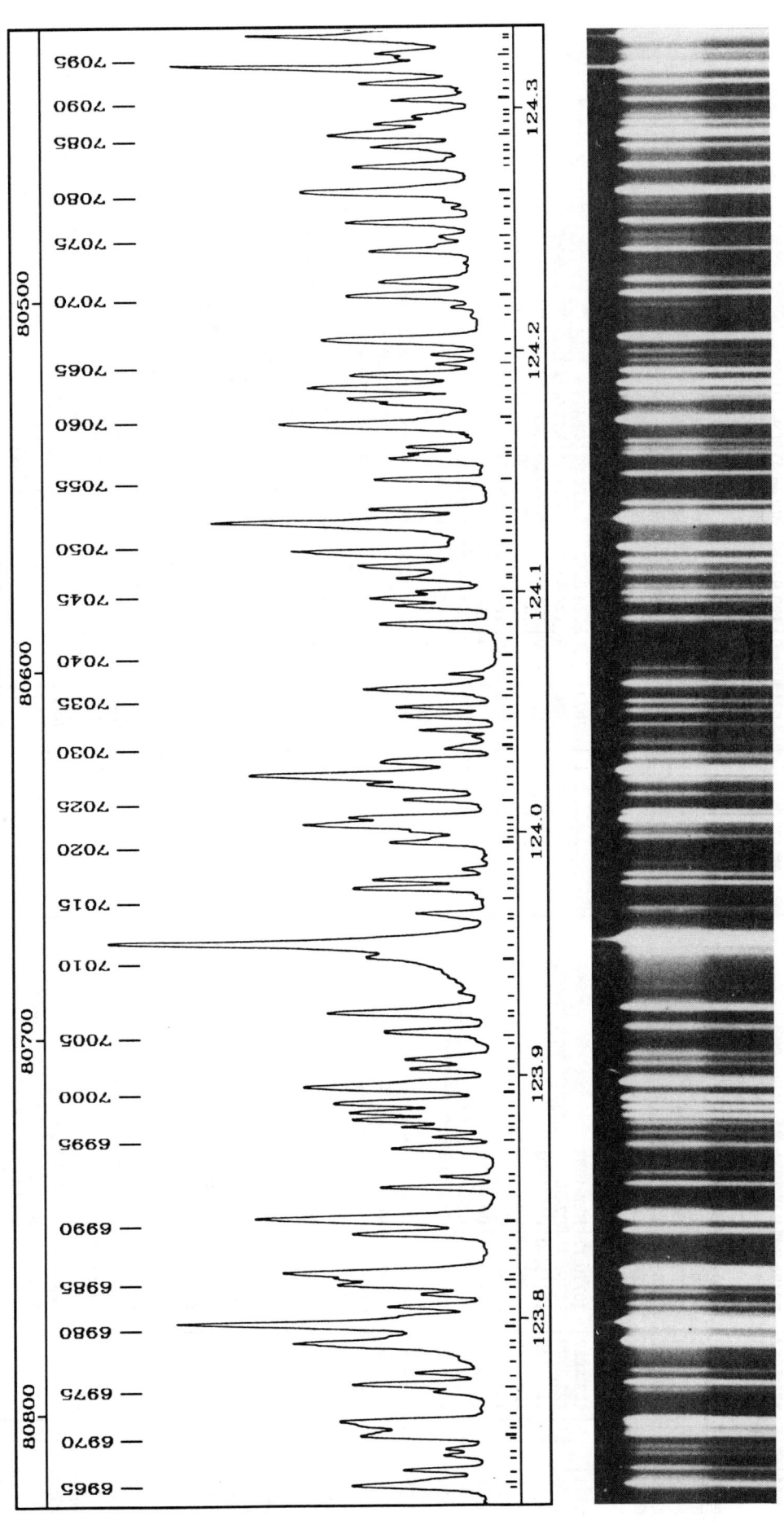

PLATE 80

TABLE 80 *continued*

Nb	I	λ (nm)	σ (cm-1)	Assignment	Comment
7067	71	124.2050	80512.09	B0-2P6 C0-5Q2	
7068	0	124.2143	80506.05	B29-13P4	
7069	1	124.2182	80503.53	B9-5P7 B29-13R6	
7070	22	124.2230	80500.39	C4-8R7 B0-1P12	
7071	12	124.2287	80496.67	B15-6R13 C9-12P6 C6-9P10 C6-9R16	
7072	0	124.2357	80492.16	B23-10R7	
7073	17	124.2415	80488.43	B13-7R0	
7074	2	124.2436	80487.04	C6-9Q12	
7075	2	124.2472	80484.74	B13-6R10 B16-7R10	
7076		124.2518	80481.75	B22-10P3	sh
7077	24	124.2532	80480.81	B16-8P1	
7078	1	124.2589	80477.15	C7-10P8	
7079	2	124.2618	80475.25	C2-4P23	
7080	112	124.2658	80472.64	C4-8Q4	
7081	14	124.2765	80465.74	C3-7Q8	
7082	2	124.2793	80463.90	C1-4R23	
7083	0	124.2831	80461.49	B16-8R2 B22-10R4	
7084	30	124.2848	80460.39	C2-5Q19	
7085	92	124.2898	80457.07	B13-7R1 B19-9R3 C8-11Q10 B17-8R6	
7086		124.2913	80456.16	B7-4P9 C6-9R18	sh
7087	10	124.2942	80454.28	C0-5R6 B29-13R7 C2-6P9 C6-9R17	
7088	4	124.2971	80452.40	B16-5P17	
7089	10	124.2986	80451.40	B9-3R17	sh
7090	10	124.3040	80447.91	B14-7P5	
7091	0	124.3076	80445.58	B18-5P19	
7092	25	124.3115	80443.07	C3-7R11	sh
7093		124.3158	80440.31		
7094	280	124.3178	80439.01	B5-4P4	
7095	6	124.3220	80436.27	C9-12P8 C5-8Q14	
7096	9	124.3241	80434.95	B5-4R5	
7097		124.3294	80431.53	C5-8P12	sh
7098	225	124.3313	80430.25	C10-14R3	

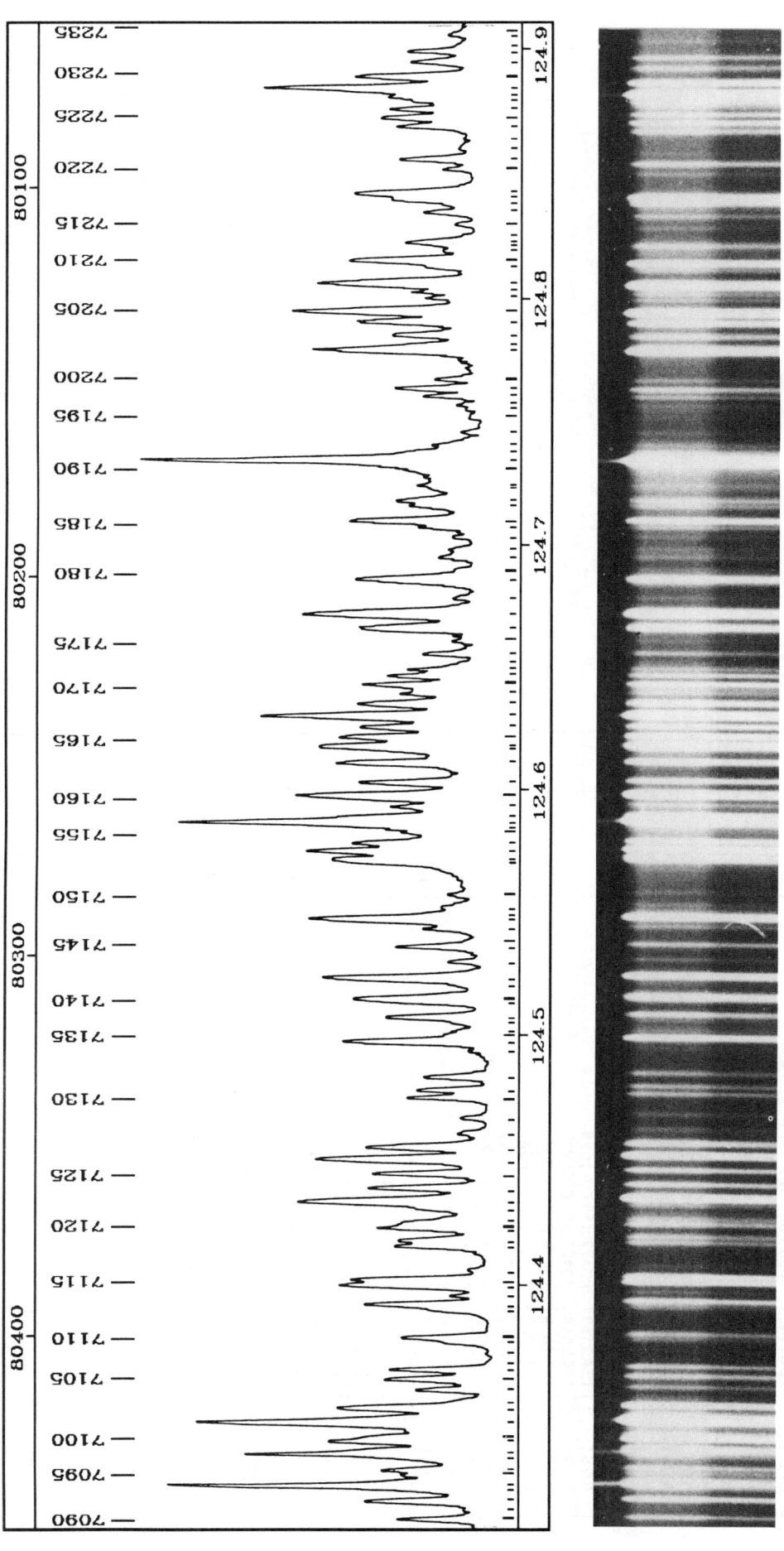

PLATE 81

TABLE 81

Nb	I	λ (nm)	σ (cm⁻¹)	Assignment	Comment
7099	40	124.3359	80427.33	C0-5Q3	
7100	1180	124.3371	80426.50	C9-12P7 C8-11P7 C2-4Q25	
7101		124.3433	80422.52	C4-8P3	sh
7102	39	124.3492	80418.71	C2-6Q11	
7103	2	124.3519	80416.96	B27-12R5	
7104	6	124.3565	80413.97	B8-3P15 C8-11Q11	
7105	16	124.3608	80411.16	B2-1P15 C1-5R16	
7106	12	124.3646	80408.74	C8-11Q13	
7107	0	124.3677	80406.71	B14-7R6	
7108	0	124.3714	80404.32	B21-9P8 C0-4R18	
7109	2	124.3760	80401.40	B28-12R9 C4-7P14 C4-7R19	
7110	8	124.3775	80400.37	C9-12P9 B6-4P7	
7111	4	124.3890	80392.99	B17-7P11 C8-11Q12	
7112	22	124.3910	80391.67	C0-5P2 C7-10P9 B25-11P5	
7113	2	124.3945	80389.43	C8-11P8 C9-12P10	
7114	73	124.3986	80386.77	B13-7P1 B25-11R6	
7115	46	124.4007	80385.39	B10-6R0	
7116	1	124.4045	80382.95	B6-4R8 B11-4P15	
7117	18	124.4144	80376.55	B2-1R16 C1-4P19	
7118	10	124.4166	80375.14	C10-14Q4 C4-8R8	
7119	17	124.4217	80371.81	B13-7R2 B12-4R17	
7120	9	124.4238	80370.49	B14-6P11 B11-6P5	
7121	0	124.4261	80369.01	B24-11R0	
7122	8	124.4296	80366.75	C4-7Q16	
7123	140	124.4323	80365.01	B21-9R9 C4-8Q5	
7124	25	124.4376	80361.60	C10-14R2 C5-8R18	
7125	27	124.4434	80357.80	B16-8P2	
7126	157	124.4494	80353.94	B10-6R1 C2-6R14	
7127	17	124.4541	80350.89	B4-3P9 B24-11R1	
7128	3	124.4596	80347.35	B4-3R10 B20-9P6 B17-7R12	
7129	2	124.4660	80343.24	B29-13P6 B14-5P15	
7130	19	124.4744	80337.80	C1-4Q21	
7131	8	124.4773	80335.91	B11-6R6 C6-9Q13	
7132	6	124.4826	80332.49	B16-8R3 B11-5P11	
7133	0	124.4930	80325.81	C0-5R7 C2-5R22	
7134	40	124.4973	80323.03	C3-7P7	
7135	0	124.5000	80321.31	C7-10Q12	
7136	0	124.5018	80320.11	B8-3R16	
7137	0	124.5048	80318.20	B20-9R7	
7138	12	124.5068	80316.90	C0-5Q4 B19-9P3	
7139		124.5134	80312.63	C4-8R9 B22-9P10	sh
7140	46	124.5144	80312.02	C10-14Q3	
7141	0	124.5205	80308.07	B28-12P8	
7142	167	124.5228	80306.61	C10-14R1	
7143	3	124.5291	80302.50	B24-11P1	
7144	0	124.5333	80299.79		
7145	9	124.5356	80298.30	B24-11R2 B8-4P11	
7146	5	124.5431	80293.53	B18-8P8 C6-9P11	
7147	180	124.5469	80291.05	C4-8P4	
7148	1	124.5491	80289.64	B22-10R5 C0-3R25	sh
7149	1	124.5516	80287.98	B28-12R10 C0-4P14	
7150		124.5566	80284.78	B26-11P8	sh
7151		124.5669	80278.16	B14-6R12	
7152	150	124.5713	80275.29	B10-6P1 C3-7Q9 C3-6P16	
7153	240	124.5751	80272.86	B11-4R16 C5-9R2 C3-6Q18 C1-5P12 C5-8R19	
7154	45	124.5784	80270.74	B18-6R17 C10-14Q2 C8-11P9	sh
7155		124.5830	80267.76	B11-5R12 B29-13P7	sh
7156		124.5844	80266.88	C10-14P4	
7157	1470	124.5871	80265.11	C5-9R1 B10-6R2	sh
7158		124.5897	80263.46	C10-14R0	
7159	7	124.5931	80261.27	C5-8P13 B8-5P5	
7160	260	124.5975	80258.42	B2-3R5 C5-9R3	
7161	38	124.6030	80254.87	B13-7P2 B24-10P9	
7162	70	124.6110	80249.75	C0-5P3	
7163		124.6159	80246.58	B3-3P7 B8-4R12	sh
7164	248	124.6177	80245.42	C10-14Q1 C5-8Q15 C3-6R21	
7165	90	124.6215	80243.00	B7-5R0 B1-2R10	
7166	29	124.6255	80240.41	C4-8Q6	
7167	664	124.6299	80237.55	C5-9R0 C7-10Q13	
7168	36	124.6349	80234.37	B2-3P4 B13-7R3 B8-5R6	
7169	5	124.6392	80231.59	B15-7P8 C2-6P10 C1-5Q14 C7-10P10	
7170	34	124.6428	80229.29	C10-14P3	
7171	10	124.6464	80226.94	B5-3P11	
7172	3	124.6489	80225.32	C3-7R12	
7173	0	124.6507	80224.18	B28-13R0	
7174	8	124.6553	80221.21	B28-13R1 B14-5R16	
7175	8	124.6603	80218.02	B24-11P2	
7176	45	124.6658	80214.45	C5-9R5 C10-14P2 B24-11R3	
7177	348	124.6716	80210.74	B7-5R1 C5-9R4 C0-3P21	
7178	2	124.6777	80206.81	B19-7P14 B23-9P12 C4-7R20	
7179	47	124.6856	80201.75	B1-2P9 B28-13R2 B5-3R12	
7180	1	124.6889	80199.58	B25-11R7 B13-5P14	
7181	2	124.6946	80195.96	B10-4P14 B19-8P10 B16-6P14 C6-9Q14	
7182	1	124.6975	80194.06	B25-11P6 B7-3P14	
7183	1	124.7029	80190.59	B27-11P11	
7184	6	124.7067	80188.13	C9-13R7	
7185	84	124.7095	80186.38	B16-8P3	
7186	7	124.7155	80182.50	B18-8R9	
7187	8	124.7176	80181.13	C0-5Q5 C0-3Q23	
7188	0	124.7223	80178.15	C8-11P10 C0-5R8	
7189	3	124.7239	80177.11	C7-10Q14	
7190		124.7304	80172.91	C2-6Q12	sh
7191	2480	124.7340	80170.62	C5-9Q1	
7192		124.7372	80168.54	B28-13R3	sh
7193	6	124.7401	80166.68	B26-12R0 B27-12R7 C2-5P18	
7194	1	124.7453	80163.36	B28-12P9	
7195	0	124.7517	80159.21	B4-2P14	

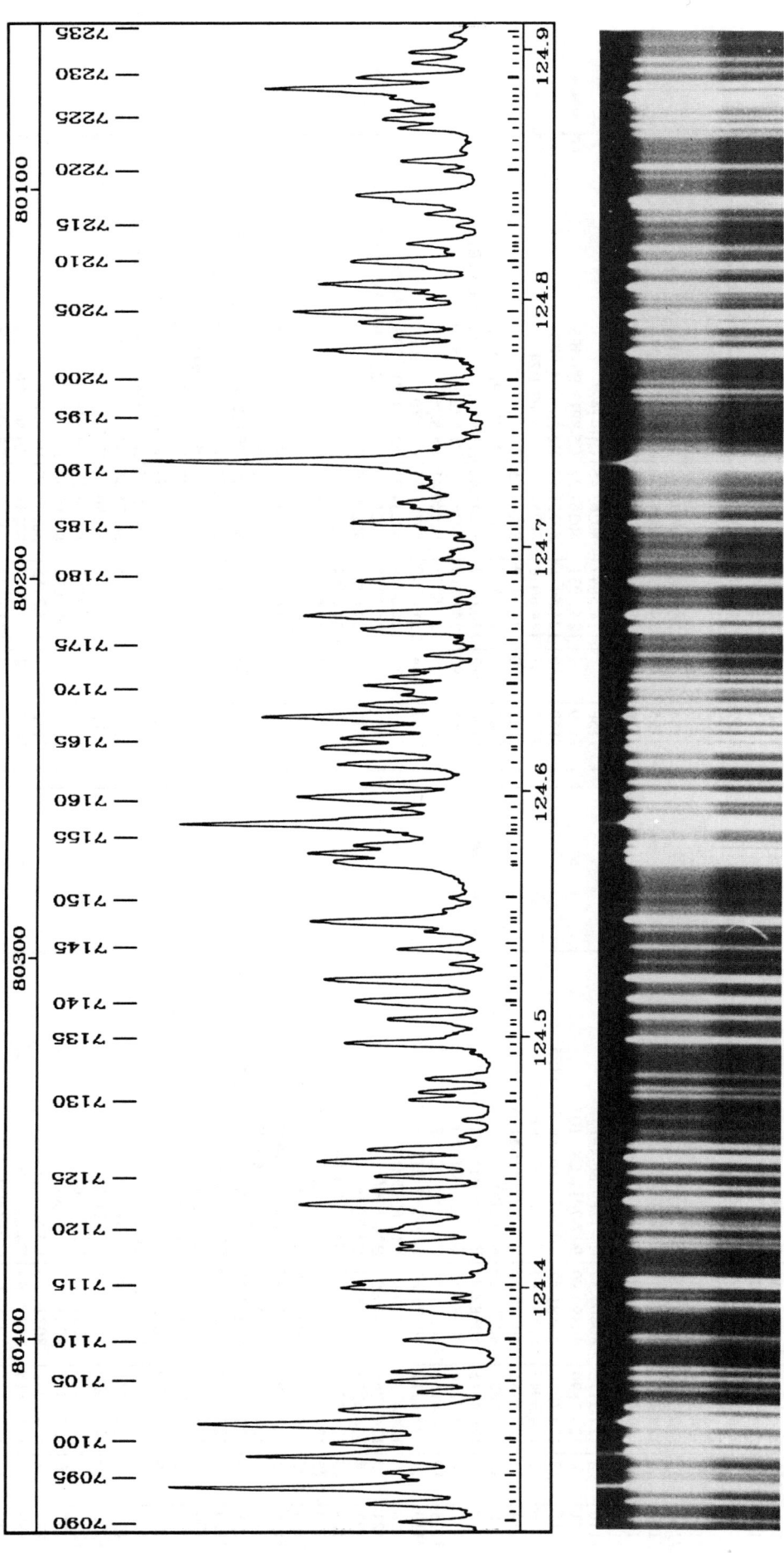

PLATE 81

TABLE 81 *continued*

Nb	I	λ (nm)	σ (cm⁻¹)	Assignment	Comment
7196	0	124.7548	80157.21	B16-8R4	
7197	0	124.7569	80155.89	B19-8R11	
7198	5	124.7603	80153.72	B26-12R1	
7199	17	124.7632	80151.85	C7-10Q15 C2-6R15	
7200	8	124.7670	80149.43	C2-5Q20 C1-4R24	
7201	240	124.7792	80141.56	C4-8P5	
7202		124.7807	80140.58	C5-9R6	sh
7203	24	124.7851	80137.77	C4-8R10 C4-7Q17	
7204	78	124.7908	80134.13	B10-6P2 B2-2R12	
7205	540	124.7949	80131.50	C5-9Q2	
7206	6	124.7998	80128.36	C1-5R17	
7207	4	124.8027	80126.51	B28-13R4 B7-3R15 B21-8P13	
7208	310	124.8062	80124.20	B7-5P1 B10-6R3	
7209	6	124.8127	80120.04	B4-2R15 B14-7P6 B26-11R10	
7210	105	124.8152	80118.44	B7-5R2 B2-2P11	
7211	2	124.8191	80115.98	B26-12R2 B16-7P10	
7212		124.8207	80114.90	C9-13R6	sh
7213	13	124.8224	80113.84	B10-4R15 C4-7P15	
7214		124.8250	80112.15	C2-4P24 C2-4Q26	sh
7215	2	124.8297	80109.17	B26-12P1	
7216	6	124.8344	80106.10	B0-2R8	
7217		124.8376	80104.07	C8-11P12	sh
7218	30	124.8399	80102.62	B5-4P5 B1-1P14 B1-1R15 B21-10R0	
7219	65	124.8421	80101.20	B5-4R6 B24-11P3 B24-11R4 C4-8Q7	
				C0-4R19	
7220	4	124.8517	80095.01	B13-5R15	
7221	11	124.8561	80092.23	B28-13R5 C3-7P8	
7222	0	124.8603	80089.52	B22-10R6	
7223	0	124.8644	80086.89		
7224	13	124.8695	80083.62	B29-14P1 C0-5P4 C8-11P13	
7225	35	124.8729	80081.43	B21-10R1 C7-10P11	
7226	20	124.8763	80079.23	B14-7R7 C3-7R13	
7227		124.8800	80076.89	C6-9P12 C8-11P11	sh
7228	23	124.8817	80075.76	C3-7Q10 C4-7R21	
7229	750	124.8846	80073.92	C5-9Q3 C6-9Q15	
7230	90	124.8894	80070.84	B13-7P3 B26-11P9	
7231	9	124.8953	80067.04	B19-9R5 B18-7P13 C5-8Q16	
7232	14	124.8997	80064.28	B29-14P2 B16-6R15	
7233	0	124.9025	80062.45	C2-5R23	
7234	2	124.9064	80059.97	B4-0P21	
7235	0	124.9084	80058.65	B3-0P20	

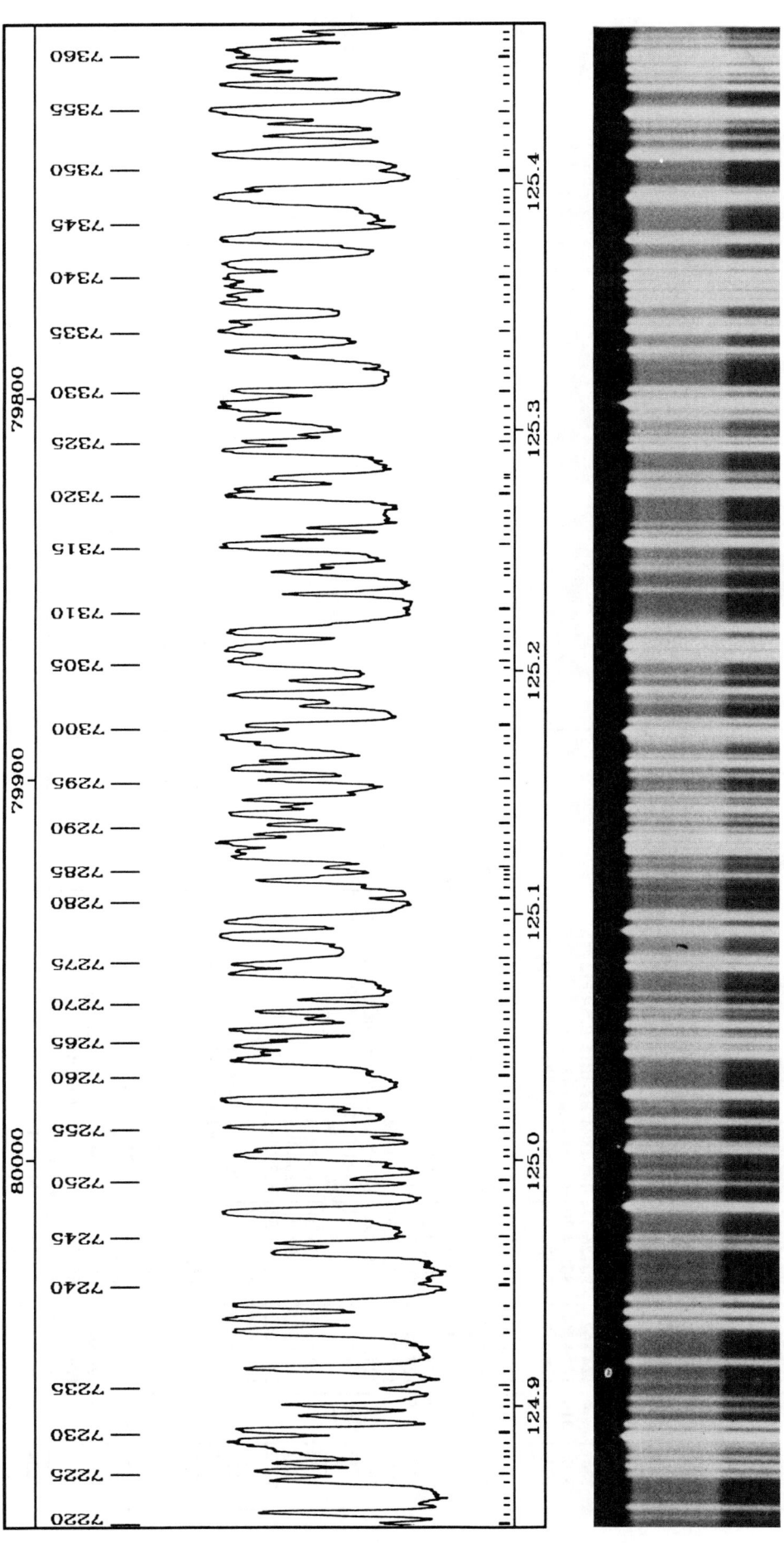

PLATE 82

TABLE 82

Nb	I	λ (nm)	σ (cm⁻¹)	Assignment	Comment
7236	27	124.9145	80054.77	C5-9R7 B20-9R8 B26-12R3	
7237	158	124.9296	80045.07	B4-4R0 B13-7R4 B20-9P7 B29-14P3	
7238	280	124.9347	80041.83	B26-12P2 B27-12R8 C5-9P2	
7239	72	124.9401	80038.33	B0-2P7 B9-5P8	
7240	0	124.9485	80032.95	B16-7R11	
7241	0	124.9510	80031.38	B29-14P4	
7242	0	124.9558	80028.27	B28-12P10 C3-6R22	
7243	20	124.9615	80024.62	B21-10P1 B15-6P13 B21-10R2	
7244	13	124.9652	80022.28	B12-5P13 B21-9R10 C0-5Q6	
7245	0	124.9678	80020.62	B11-6P6	
7246		124.9742	80016.50	C7-10P12	sh
7247	610	124.9775	80014.43	B4-4R1	
7248		124.9801	80012.73		sh
7249	23	124.9869	80008.41	B5-0P22 C3-6Q19	
7250	5	124.9911	80005.67	C5-8P14 C1-4P20	
7251	1	124.9967	80002.12	B2-0P19 B25-11R8	
7252	115	125.0011	79999.32	C5-9Q4 C3-6P17	
7253	17	125.0031	79998.02	C9-13R5	
7254	4	125.0081	79994.83	C1-4Q22	
7255	63	125.0124	79992.08	C1-6R2 C4-7R22	
7256		125.0141	79990.96	B6-1P20	sh
7257	3	125.0192	79987.69	B28-13R6 B11-6R7	
7258	340	125.0232	79985.14	C1-6R1	
7259		125.0258	79983.51		sh
7260	0	125.0338	79978.35	B25-11P7	
7261		125.0358	79977.12	B9-5R9 B13-6R11	sh
7262	180	125.0399	79974.48	B7-5P2 C4-8P6 C6-9Q16	
7263	15	125.0426	79972.77	B26-12R4 C4-8R11 C8-12R10	
7264	86	125.0453	79971.02	B10-5P10 C1-6R3	
7265	6	125.0478	79969.43	B16-8P4	
7266	107	125.0517	79966.93	B7-5R3 B24-10R11	sh
7267		125.0567	79963.75	C6-9P13 B7-1P21	
7268	7	125.0594	79962.02	B23-10R9 B5-1P19	
7269	18	125.0642	79958.91	C9-13R4 B24-11R5	
7270	8	125.0686	79956.12	B9-4P13	
7271	0	125.0708	79954.72	B24-11P4	
7272	136	125.0756	79951.66	B24-10P10	
7273	55	125.0788	79949.62	C1-6R0 B26-12P3 B21-10R3 B28-13R7 C4-8Q8	
7274				C2-6P11	sh
7275	0	125.0809	79948.26	B27-12R9	
7276		125.0860	79944.98	C0-3R26	
7277	820	125.0905	79942.10	C5-9P3	
7278	186	125.0969	79938.06	B10-6P3 B16-8R5 C1-5Q15	
7279	0	125.1022	79934.65	B3-0R21	
7280	2	125.1058	79932.32	C1-5P13	
7281	1	125.1107	79929.19	B15-6R14 B22-9P11	
7282	15	125.1136	79927.37	B21-10P2 C9-13Q7 C5-9R8	
7283	1	125.1154	79926.23	C0-4P15	
7284	3	125.1179	79924.62	B28-13P5	sh
7285		125.1190	79923.89	C4-7Q18	
7286	142	125.1237	79920.89	B4-4R2	
7287	91	125.1261	79919.40	C9-13R3 C2-6Q13	
7288	480	125.1294	79917.26	B4-4P1 B2-3R6 C0-4Q17	
7289	10	125.1332	79914.86	C3-7P9	
7290	14	125.1373	79912.26	B17-8P7 B26-11R11 C6-9Q17	
7291	129	125.1416	79909.50	B6-4P8 B9-2P20 B8-2P19 B6-0P23	
7292	5	125.1448	79907.44	B12-5R14 C5-9Q5	
7293	23	125.1474	79905.78	B4-0R22 C5-8Q17	
7294	0	125.1501	79904.02	B6-3P13 C6-10R4	
7295	12	125.1555	79900.60	C6-9Q18	
7296	168	125.1604	79897.49	B27-11P12 C5-9R9	
7297	10	125.1633	79895.62	C6-10R2 B18-9R0	
7298		125.1707	79890.93	B6-4R9 C0-5P5	
7299	980	125.1734	79889.16	B10-6R4 B8-5P6 B18-8P9 B1-0P18 C6-10R1 C7-10P13 C2-6R16	sh
7300	35	125.1778	79886.36	C4-7P16	
7301	4	125.1856	79881.42	B2-3P5	
7302	86	125.1902	79878.43	B22-10R7 B4-1P18	sh
7303	4	125.1959	79874.85	C9-13R2 B26-12R5 B7-4P10	
7304		125.2016	79871.17	C1-6R5 C7-10P14	sh
7305	143	125.2046	79869.29	B18-9R1	
7306	580	125.2084	79866.87	B17-7P12 C6-10R3 C6-10R5 C3-7Q11	
7307		125.2115	79864.38	C1-6Q1 C4-8R12 C3-7R14	
7308	420	125.2159	79862.09	B25-10P12 B8-5R7 C9-13Q6	sh
7309		125.2186	79860.37	B6-3R14 C6-10R0 C3-6R23	
7310	0	125.2251	79856.22	B10-2P21 B7-2P18 C1-5R18	
7311	11	125.2316	79852.06	C1-4R25	
7312		125.2390	79847.35	B28-13P6 C2-5Q21	sh
7313	11	125.2406	79846.32	B6-1R21	
7314	2	125.2429	79844.85	B5-1R20 B11-3P19 C2-5P19	
7315	460	125.2511	79839.60	C0-5Q7	
7316	12	125.2549	79837.20	C9-13R1	
7317	5	125.2587	79834.75	B13-7P4	
7318		125.2621	79832.62	B26-11P10 C8-12R9	
7319	0	125.2648	79830.91	C0-5R10 C2-4Q27	
7320	167	125.2721	79826.25	B5-0R23	
7321	40	125.2742	79824.88	C5-9P4	
7322	10	125.2782	79822.36	C9-13Q5 C8-12R8	
7323	10	125.2795	79821.52	B3-2P13 B12-3P20	
7324	138	125.2905	79814.49	B10-3P18 C6-10R6	
7325	27	125.2938	79812.39	C1-6Q2	
7326	2	125.2967	79810.55	B18-9P1 B3-2R14 B14-6P12	
7327		125.3018	79807.30	B13-7R5 B24-11R6	
7328	29	125.3035	79806.24	B27-12P8 C0-3P22 C0-3Q24	
7329	2020	125.3088	79802.83	C5-9Q6 C8-12R6	
7330	42	125.3147	79799.12	C6-10Q1 C9-13R0 C8-12R7 B25-11R9 B18-13P7 B18-9R2 C9-13Q4 B28-13P7 B18-9R2	sh

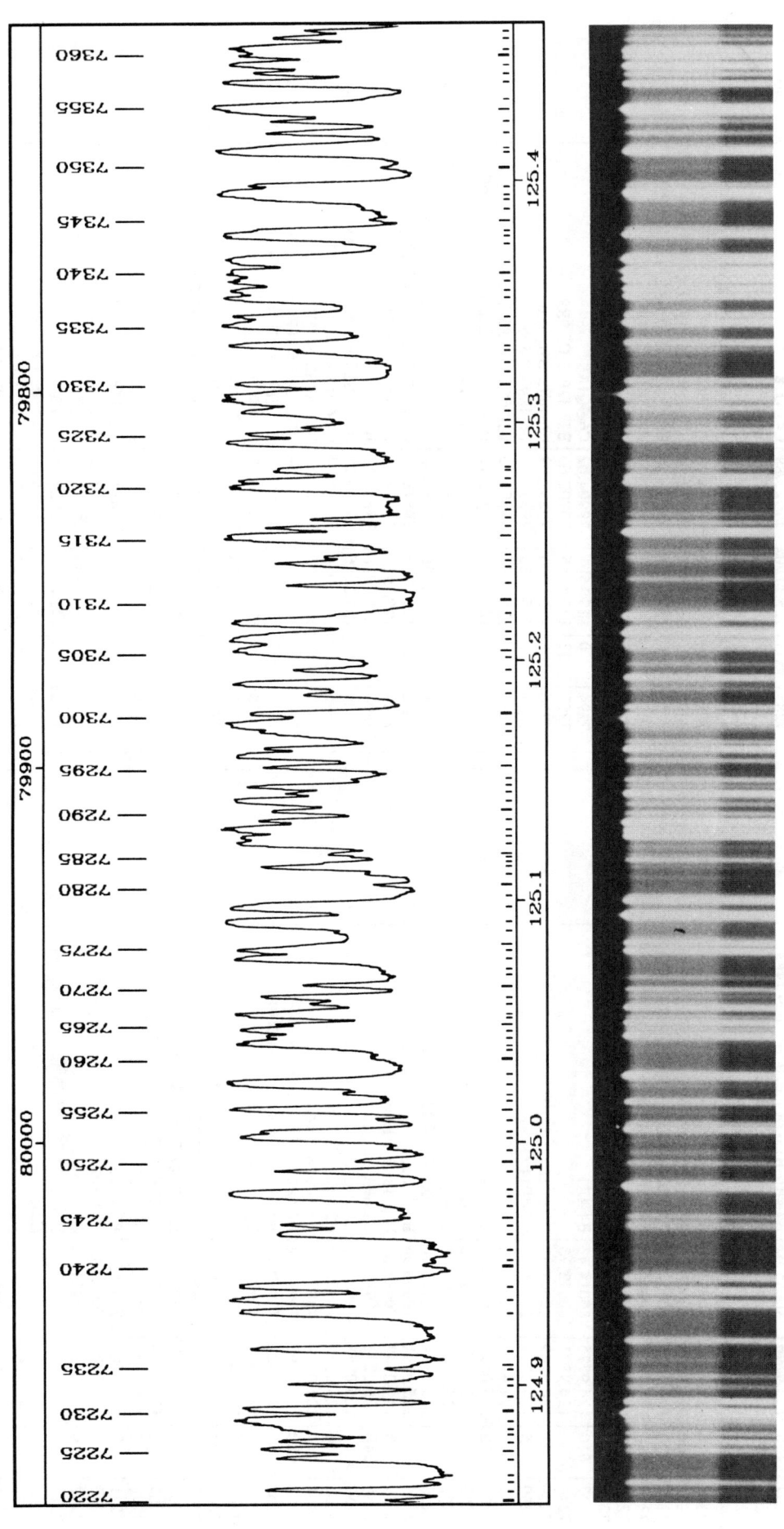

PLATE 82

TABLE 82 *continued*

Nb	I	λ (nm)	σ (cm-1)	Assignment	Comment
7331	0	125.3237	79793.38	B21-10P3	
7332	1	125.3262	79791.80	C0-4R20	
7333	3	125.3282	79790.49	B4-1R19	
7334	45	125.3309	79788.78	C4-8Q9 C7-10P15	
7335	300	125.3394	79783.35	C9-13Q3 C1-6R6 C6-9P14 C7-11R13	
7336	33	125.3424	79781.44	C4-8P7 C5-9R10	
7337	200	125.3506	79776.27	B1-3R0	
7338	220	125.3538	79774.23	C9-13Q2 C5-8Q18	
7339	380	125.3576	79771.77	C6-10Q2 C2-4P25	
7340	880	125.3612	79769.53	C9-13Q1	
7341	500	125.3659	79766.50	B4-4R3 B7-5P3 B11-2P22 B27-12R10	
7342		125.3735	79761.67	C7-11R12	sh
7343	260	125.3765	79759.80	B4-4P2 B21-10R4 B26-10P14 C3-6Q20	
7344		125.3784	79758.57	B7-5R4	sh
7345	0	125.3825	79755.92	B0-1R14 B13-3P21	
7346	0	125.3878	79752.55	B14-7P7	
7347	700	125.3903	79750.98	B25-11P8 B9-2R21 C5-8P15	sh
7348		125.3937	79748.82	B1-3R1 C6-10R8 B4-3R11	
7349	22	125.3970	79746.70	B4-3P10 B19-8P11 C6-10R7	
7350	0	125.4043	79742.08	B15-7P9 B23-9R14	
7351	200	125.4113	79737.62	C1-6Q3	
7352		125.4129	79736.63	B3-3P8 B3-1P17 B18-6P17	sh
7353	11	125.4187	79732.94	C5-9R11 B23-9P13	
7354	14	125.4238	79729.68	B12-4P17 C4-7Q19	
7355	580	125.4288	79726.53	C6-10Q3 C8-12R5 B23-11R0 B15-8R0	
7356	2	125.4327	79724.03	B11-5P12	
7357	73	125.4396	79719.64	B23-11R1 B0-1P13	
7358	22	125.4436	79717.13	B14-6R13 B27-11P13 C8-12R4	
7359	135	125.4471	79714.86	C7-11R3 B5-4R7 B5-4P6 B20-9P8 B25-9P16	
7360	65	125.4511	79712.36	C7-11R2	
7361	120	125.4536	79710.77	C7-11R4 C9-13P2 B16-8P5	
7362	7	125.4581	79707.91	B26-12P5 B6-0R24 C7-11R7	
7363	3	125.4609	79706.08	C9-13P7 B23-10R10	

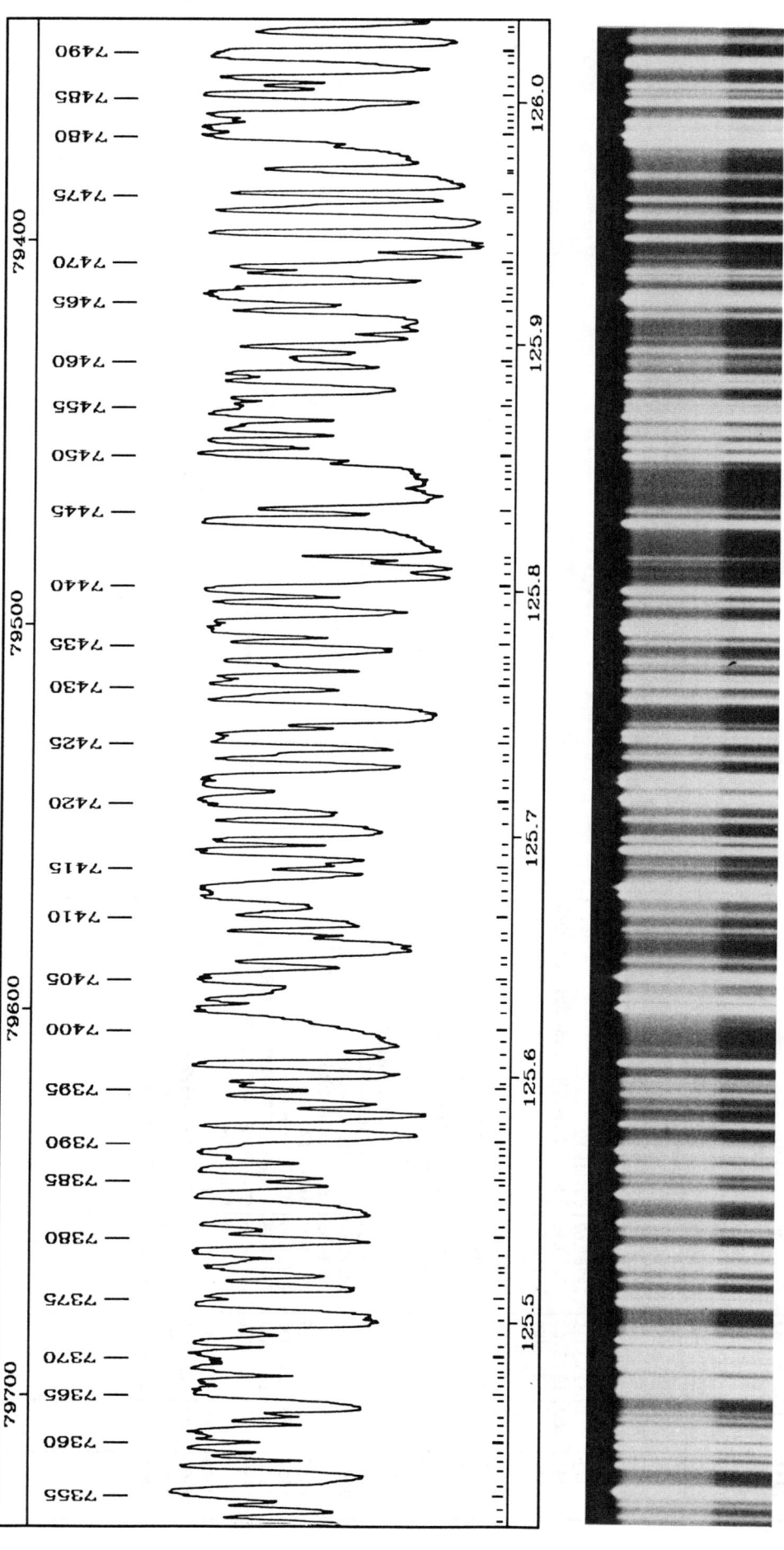

PLATE 83

TABLE 83

Nb	I	λ (nm)	σ (cm⁻¹)	Assignment	Comment
7364	155	125.4681	79701.53	B18-9P2 B26-12R6 C3-7R15	sh
7365	70	125.4701	79700.26	C8-12R3 C1-6P2	
7366		125.4738	79697.93	B15-8R1 C7-11R5	sh
7367		125.4779	79695.30	C4-8R13	
7368	205	125.4797	79694.15	C5-9P5	
7369	20	125.4825	79692.36	C5-9Q7	
7370	280	125.4854	79690.52	C9-13P3 B10-6P4	
7371	240	125.4902	79687.48	C6-10P2 C9-13P6 C0-5P6 C3-6P18	
7372	10	125.4948	79684.59	C7-11R10 C6-9P15 B18-9R3	
7373	0	125.5001	79681.22	B24-10R12 B10-2R22	
7374	130	125.5059	79677.53	B16-8R6 B27-13R1 C8-12R2 C9-13P4	
7375	70	125.5087	79675.75	C9-13P5 B3-1R18 C5-8Q19	
7376	20	125.5150	79671.77	B19-8R12 B21-9P10 C6-10R9 C2-6R17	
7377	80	125.5209	79668.01	C6-10Q4 C7-11R6 C1-6R7 C7-11R8 C7-11R11 C3-7P10	
7378		125.5231	79666.62	B21-8P14 C2-6Q14 C1-4Q23	sh
7379	740	125.5270	79664.12	B11-5R13 B23-10P9 C7-11R1	
7380	10	125.5344	79659.47	B10-6R5 B27-12P9 B23-11P1	
7381	200	125.5383	79656.95	B1-3R2 B20-9R9 C3-7Q12 C2-6P12	
7382		125.5455	79652.41	B19-7P15 B25-10R14	sh
7383	535	125.5500	79649.58	B8-4P12 B14-3P22 C8-12R1 C1-5P14	
7384	4	125.5553	79646.19	C4-8P8	
7385		125.5585	79644.14	C1-5Q16	sh
7386	360	125.5604	79642.95	B19-9P6 B14-5P16 C7-11R0	
7387	30	125.5650	79640.01	B23-11R2	
7388	700	125.5678	79638.25	B1-3P1 B9-3R18 C1-6Q4 B27-13P1	
7389		125.5706	79636.45	C7-11R9 B11-4P16	sh
7390		125.5736	79634.55	B24-10P11 C0-5R11	sh
7391	72	125.5784	79631.51	B15-8P1	
7392	7	125.5810	79629.89	C1-4P21	
7393	5	125.5855	79627.05	B27-13R3 B16-7P11 B21-10P4 B11-6P7	
7394	22	125.5907	79623.69	B1-2R11	
7395	15	125.5947	79621.22	C4-8Q10 C0-3R27	
7396	25	125.5963	79620.16	B15-8R2 C4-7P17	
7397	250	125.6034	79615.69	C8-12R0	
7398	2	125.6088	79612.27	B12-4R18 B8-3P16 C5-8P16	
7399	0	125.6132	79609.50		d
7400	0	125.6207	79604.70	B22-10P7 B26-11P11 B8-4R13	
7401	760	125.6262	79601.22	B26-12R7 C6-10P3	
7402	94	125.6302	79598.71	B21-9R11 C6-10Q5	
7403	0	125.6336	79596.52	C8-12Q10	
7404	1360	125.6389	79593.19	C7-11Q1 C5-9R12	sh
7405		125.6420	79591.22	B27-13P2 B24-11P6	
7406	15	125.6460	79588.70	B0-2R9	
7407		125.6479	79587.48	C0-4Q18	sh
7408	5	125.6560	79582.34	B27-13R4 C4-8R14	
7409	18	125.6589	79580.50	C6-10R11 C1-5R19	
7410	11	125.6655	79576.33	B1-2P10 C6-10R10	
7411	340	125.6730	79571.59	C7-11Q2 C2-5Q22	
7412	1440	125.6765	79569.40	B25-12R0 B24-11R7 C8-12Q1 C1-6P3 C5-9Q8	
7413		125.6790	79567.76	B22-9R13 B26-12P6 B16-6P15 B11-2R23	sh
7414	7	125.6845	79564.33	B23-11R3 C0-4P16	
7415	2	125.6872	79562.63	C4-7Q20	
7416	250	125.6922	79559.44	C8-12Q2	
7417	65	125.6967	79556.60	B12-7R0 B13-7P5 B25-12R1	
7418	25	125.7048	79551.40	B4-4R4	
7419	80	125.7114	79547.27	B18-9P3 C6-9P16	
7420	400	125.7137	79545.82	C8-12Q3 B5-3P12	
7421	730	125.7195	79542.17	B4-4P3	
7422	420	125.7222	79540.46	B7-0R25 C7-11Q3 C8-12Q9 C6-10R15	
7423	30	125.7302	79535.36	B14-5R17 C5-9P6	
7424	9	125.7322	79534.15	B25-11R10 B27-12P10 C3-6Q21	
7425	65	125.7378	79530.61	B9-5P9 B22-9P12 C8-12Q4	
7426	103	125.7401	79529.11	B12-7R1 B27-13P3 B2-1P16 B18-9R4 B11-4R17 B8-3R17	
7427	9	125.7438	79526.81	B5-3R13 B26-10P15	
7428	38	125.7540	79520.34	C6-10Q6 B2-3R7 B13-5P15 B5-2R17	
7429	12	125.7555	79519.37	B25-12R2 C6-10R13 C1-6R8	
7430	77	125.7606	79516.11	C8-12Q5 B25-11P9	
7431	25	125.7630	79514.63	C1-6Q5 B0-2P8 B25-12P1 B18-7P14	
7432	5	125.7676	79511.74	C8-12Q8	
7433	45	125.7699	79510.30	B15-8P2 C6-10R12 C2-5P20	
7434		125.7718	79509.06	B24-9P15	
7435	15	125.7776	79505.39	C8-12Q6	
7436	380	125.7826	79502.23	B1-3R3 B7-5P4 C8-12Q7 C7-11Q4 C6-9P17	
7437	173	125.7857	79500.33	C6-10P4 B2-1R17	
7438	39	125.7939	79495.13	B15-8R3 B7-5R5 B13-6P11	
7439	300	125.7991	79491.83	B8-5P7 C7-11P2 C8-12P2 C3-7R16	
7440	5	125.8017	79490.17	B17-8P8	
7441	4	125.8062	79487.33	C9-14R4 C0-4R21	
7442	7	125.8101	79484.86	C6-10R14	
7443		125.8126	79483.31	B2-3P6	
7444	400	125.8272	79474.09	B1-3P2 B18-8P10	
7445	18	125.8322	79470.91	B23-11P3 B10-4P15	
7446	0	125.8408	79465.47	C2-4P26	
7447	0	125.8437	79463.67		
7448	2	125.8459	79462.27		
7449	87	125.8512	79458.93	B27-13P4 B25-12R3	
7450		125.8548	79456.63	C7-11Q5 C6-9P18	
7451	152	125.8598	79453.51	B12-7P1 B12-7R2	
7452	42	125.8648	79450.31	B8-5R8 C4-8Q11 C5-9R13 C0-3Q25	
7453	17	125.8666	79449.20	B21-10P5	
7454	340	125.8715	79446.10	C8-12P3	
7455	26	125.8747	79444.06	B25-12P2 C3-7Q13	
7456	29	125.8773	79442.47	C5-9Q9	

7490
7485
7480
7475
7470
7465
7460
7455
7450
7445
7440
7435
7430
7425
7420
7415
7410
7405
7400
7395
7390
7385
7380
7375
7370
7365
7360
7355
79400
79500
79600
79700
126.0
125.9
125.8
125.7
125.6
125.5

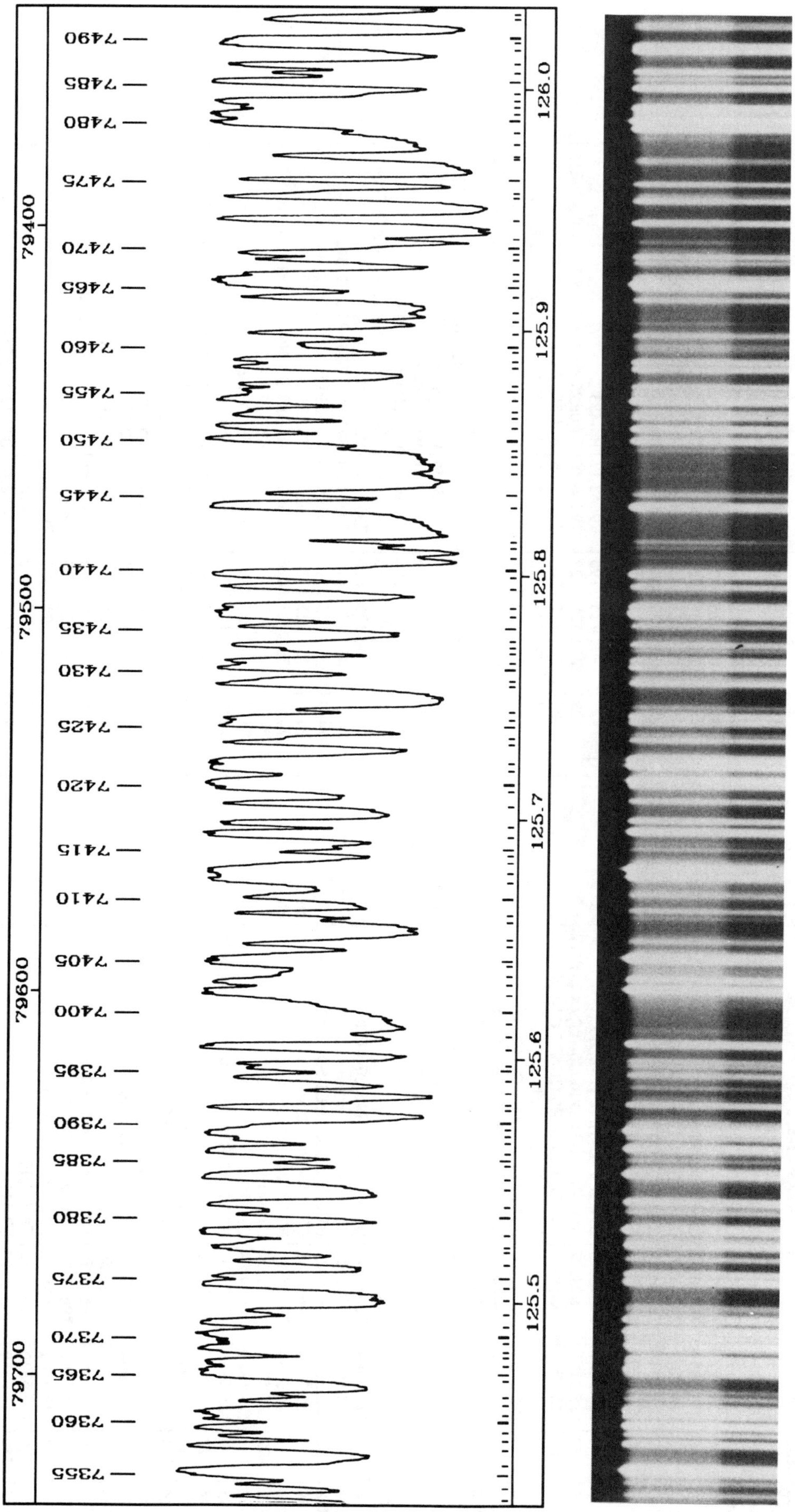

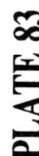

PLATE 83

TABLE 83 *continued*

Nb	I	λ (nm)	σ (cm⁻¹)	Assignment	Comment
7457	33	125.8848	79437.71	C4-8P9 C5-9P7	
7458	37	125.8876	79435.94	C6-10Q7 C0-5Q9	
7459	14	125.8932	79432.39	C0-5R12	
7460	14	125.8942	79431.80	C3-6P19 C4-7Q21	
7461	17	125.8988	79428.88	B2-2R13	
7462	1	125.9036	79425.84	B26-12P7	
7463	0	125.9071	79423.65	B13-5R16	
7464	25	125.9136	79419.55	C5-9R14 B7-3P15 C4-8R15	
7465		125.9188	79416.28	C1-6P4 C3-7P11 B15-6P14	sh
7466	700	125.9206	79415.10	C7-11P3	
7467	35	125.9226	79413.86	C2-6Q15 C4-7P18 C0-3P23	
7468	17	125.9290	79409.85	C2-6P13	
7469	22	125.9315	79408.24	B2-2P12 C7-11Q6	
7470	5	125.9338	79406.78	C1-6R9	
7471	5	125.9374	79404.51	B23-9R15	
7472	70	125.9455	79399.40	C8-12P4	
7473	37	125.9547	79393.65	B10-6P5 B10-5P11	
7474	8	125.9567	79392.38	B10-4R16 C5-8P17	
7475	23	125.9616	79389.25	B24-11R8 C2-7R3	
7476	0	125.9670	79385.90		
7477	12	125.9715	79383.05	C9-14R3 B20-10R0 B27-13P5 B16-8R7 B24-11P7	
7478	0	125.9769	79379.63		
7479	1	125.9810	79377.05	B6-4P9	
7480	185	125.9851	79374.49	C2-7R2	
7481	700	125.9885	79372.36	C2-7R1 C1-6Q6	
7482	40	125.9913	79370.54	C7-11P4	
7483	170	125.9936	79369.14	B6-4R10 C6-10P5	
7484	1	125.9965	79367.31	B20-9P9 B25-10R15	
7485	112	126.0019	79363.87	B9-6R0	
7486	13	126.0056	79361.56	B20-10R1 B10-6R6	
7487	33	126.0087	79359.59	B7-3R16 C7-11Q7	
7488		126.0150	79355.61	C4-7Q22	sh
7489	120	126.0174	79354.16	C8-12P5 C1-5Q17	
7490	80	126.0191	79353.03	B25-12P3 B18-9P4 B23-10P10	
7491	14	126.0273	79347.88	C6-10Q8	
7492	11	126.0282	79347.35	B4-2P15 B23-10R11 C2-7R4	

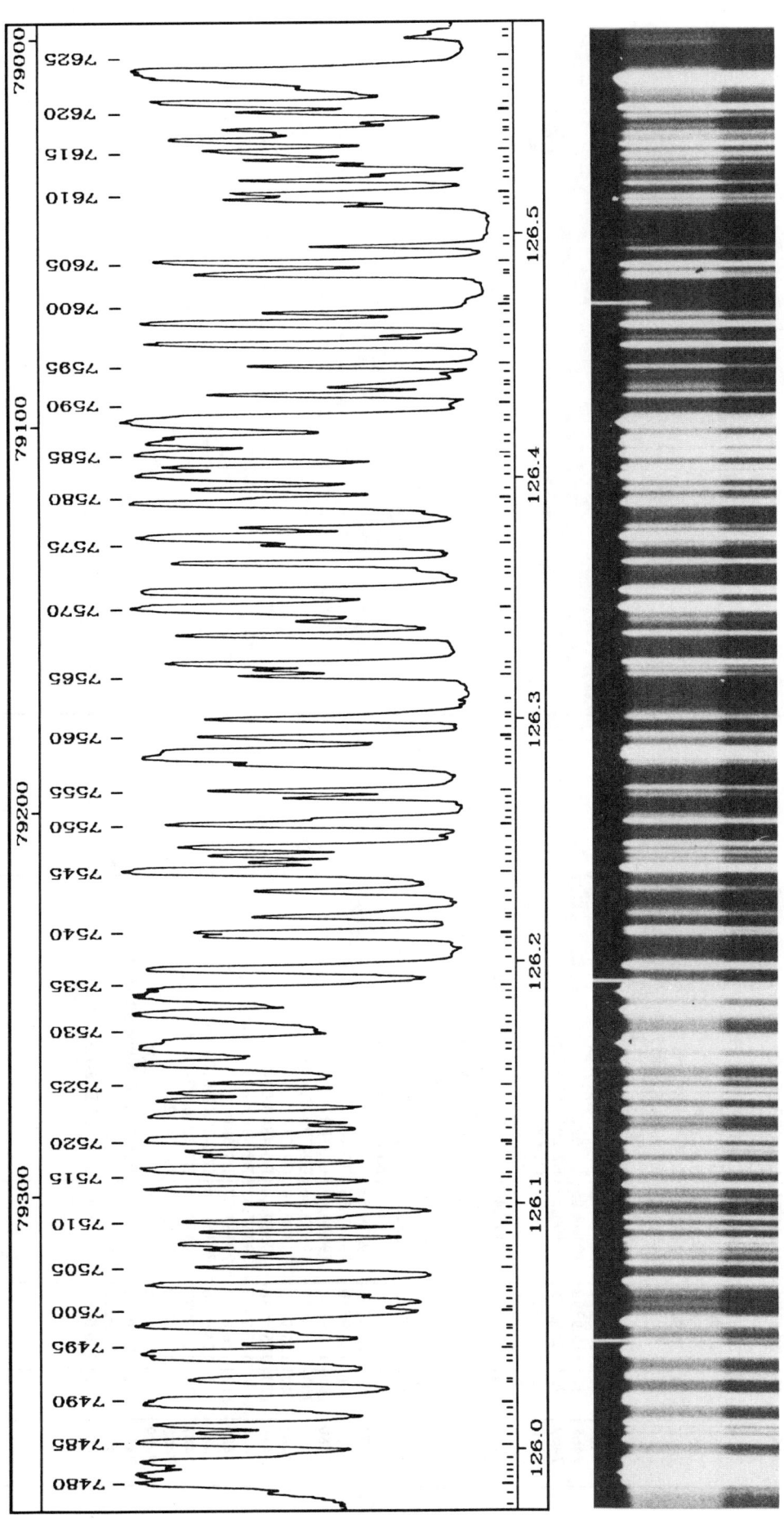

PLATE 84

TABLE 84

Nb	I	λ (nm)	σ (cm⁻¹)	Assignment	Comment
7493		126.0351	79342.96	B24-10R13	sh
7494	290	126.0380	79341.18	C2-7R0 B15-8P3	
7495					SiII 126.04223
7496	12	126.0423	79338.45	C3-7R17	
7497	4	126.0467	79335.65	C3-6Q22 B23-9P14	
7498	360	126.0501	79333.53	B9-6R1 B18-9R5 B17-7P13	
7499		126.0536	79331.33	B15-6R15 C1-5R20	sh
7500	1	126.0572	79329.09	B25-9P17	
7501	0	126.0584	79328.32	B22-10P8	
7502	1	126.0619	79326.13	B12-5P14	
7503	5	126.0644	79324.57	C9-14Q4 B15-8R4 C1-5P15	
7504	93	126.0667	79323.11	B12-7P2	
7505	12	126.0742	79318.39	C8-12P8 C8-12P10 C7-11Q13	
7506	12	126.0785	79315.66	B24-10P12 C5-9R17 C4-8R16 C2-5Q23	
7507	7	126.0814	79313.87	B4-2R16 C7-11Q8	
7508	18	126.0838	79312.36	C8-12P6 C5-9Q10	
7509	13	126.0886	79309.30	C6-10P6	
7510	25	126.0928	79306.66	B20-10P1	
7511	4	126.0948	79305.43	C5-9R16	
7512	8	126.1000	79302.12	C5-9R15	
7513	1	126.1028	79300.40	B20-10R2	
7514	68	126.1062	79298.22	B23-11P4 C9-14R2 C5-8P18	
7515		126.1117	79294.78	C8-12P9	sh
7516	165	126.1139	79293.40	B12-7R3	
7517	8	126.1193	79290.00	B19-9P7 C1-4P22	
7518	17	126.1213	79288.72	C2-7R5	
7519	140	126.1261	79285.76	B1-3R4 C7-11P5	
7520		126.1284	79284.32	B25-11P10 B19-8P12 B21-9P11	sh
7521	1	126.1320	79282.00	B5-4R8	
7522	1	126.1361	79279.43	B4-4R5 C4-8Q12	
7523	27	126.1423	79275.52	B5-4P7 C7-11Q9	
7524	21	126.1454	79273.60	C8-12P7	
7525	6	126.1495	79271.01	C5-9P8	sh
7526		126.1553	79267.40	B7-4P11	
7527	147	126.1573	79266.10	B4-4P4	
7528	1140	126.1641	79261.88	C2-7Q1 C0-4Q19	
7529	45	126.1666	79260.31	B12-5R15 C6-10Q9 C7-11Q12	
7530	0	126.1714	79257.26	B1-1R16	
7531	17	126.1760	79254.36	B1-1P15 B9-4P14 B21-9R12	
7532	320	126.1782	79252.97	B9-6P1	
7533	1160	126.1858	79248.20	B1-3P3 C7-11Q10	
7534	100	126.1885	79246.55	B9-6R2 C9-14Q3	
7535					GeII 126.19053
7536	57	126.1969	79241.26	C7-11Q11 B25-12P4 C1-6P5 C1-6R10	
7537	0	126.2031	79237.33		
7538	0	126.2066	79235.17		
7539	17	126.2107	79232.57	B13-7P6 B19-8R13 C3-7Q14	
7540	20	126.2124	79231.51	C9-14R1	
7541	8	126.2190	79227.42	C4-8P10 C2-6R19	
7542	0	126.2210	79226.15	C0-5P8	
7543		126.2263	79222.77	C0-4P17	sh
7544	6	126.2297	79220.69	C3-7P12	
7545	220	126.2380	79215.48	C2-7Q2	
7546	9	126.2417	79213.15	B26-12P8	
7547	8	126.2446	79211.34	C1-6Q7 C5-8P19	
7548	25	126.2481	79209.14	B20-10P2 C2-7R6 C2-5P21 C0-5Q10	
7549	0	126.2527	79206.25	C0-5R13	
7550	23	126.2578	79202.99	C7-11P6 C9-14P4	
7551	4	126.2598	79201.77	B20-10R3	
7552	0	126.2636	79199.36		
7553	0	126.2658	79198.02	C0-4R22	
7554	8	126.2687	79196.20	B9-4R15 C3-7R18	
7555	15	126.2716	79194.35	C9-14Q2 C3-6P20	
7556	12	126.2823	79187.63	B14-6P13 B24-11R9 C4-7P19	
7557	108	126.2853	79185.80	B7-5P5 C6-10P7 C4-8R17	
7558	35	126.2874	79184.48	B21-8P15 C5-9Q11	
7559	15	126.2925	79181.29	B3-3P9 C2-4P27	sh
7560	15	126.2934	79180.69	C9-14R0 B7-5R6	
7561	0	126.2974	79178.17		
7562	15	126.3012	79175.85	C6-10Q10	
7563		126.3034	79174.46	B24-11P8 C3-6Q23	sh
7564	0	126.3096	79170.56	B21-10R7	
7565	10	126.3190	79164.69	C2-6Q16	
7566	5	126.3214	79163.16	B21-10P6	
7567	45	126.3240	79161.53	C9-14Q1 B6-3P14	
7568	25	126.3359	79154.08	C9-14P3	
7569	4	126.3419	79150.33	B11-6R9 B21-8R16	
7570	300	126.3471	79147.06	C2-7Q3	
7571	220	126.3540	79142.75	B12-7P3	
7572		126.3593	79139.40	B14-6R14	
7573	2	126.3626	79137.37		
7574	30	126.3659	79135.26	B23-11P5	
7575	8	126.3735	79130.51	C9-14P2	
7576	182	126.3765	79128.62	B6-5R0 B15-8P4 B22-9P13	
7577	10	126.3804	79126.20	B6-3R15 B22-9R14	
7578		126.3862	79122.54		
7579	52	126.3905	79119.88	C7-11P7 B18-9P5	
7580	9	126.3924	79118.71	C2-6P14 C4-8R18	
7581	12	126.3964	79116.19	B12-7R4 B26-13R0 B26-13R1 C0-3Q26	
7582	200	126.4020	79112.68	B9-6P2 B25-12P5 B15-8R5	
7583	34	126.4052	79110.65	C4-8Q13	
7584	110	126.4108	79107.15	C2-7P2	
7585	110	126.4122	79106.29	B4-3P11	
7586	110	126.4154	79104.29	B9-6R3 B17-8P9	
7587	27	126.4173	79103.09	C5-9P9 B17-9R0	
7588	620	126.4238	79099.03	B6-5R1 B26-13R2 C6-10Q11	

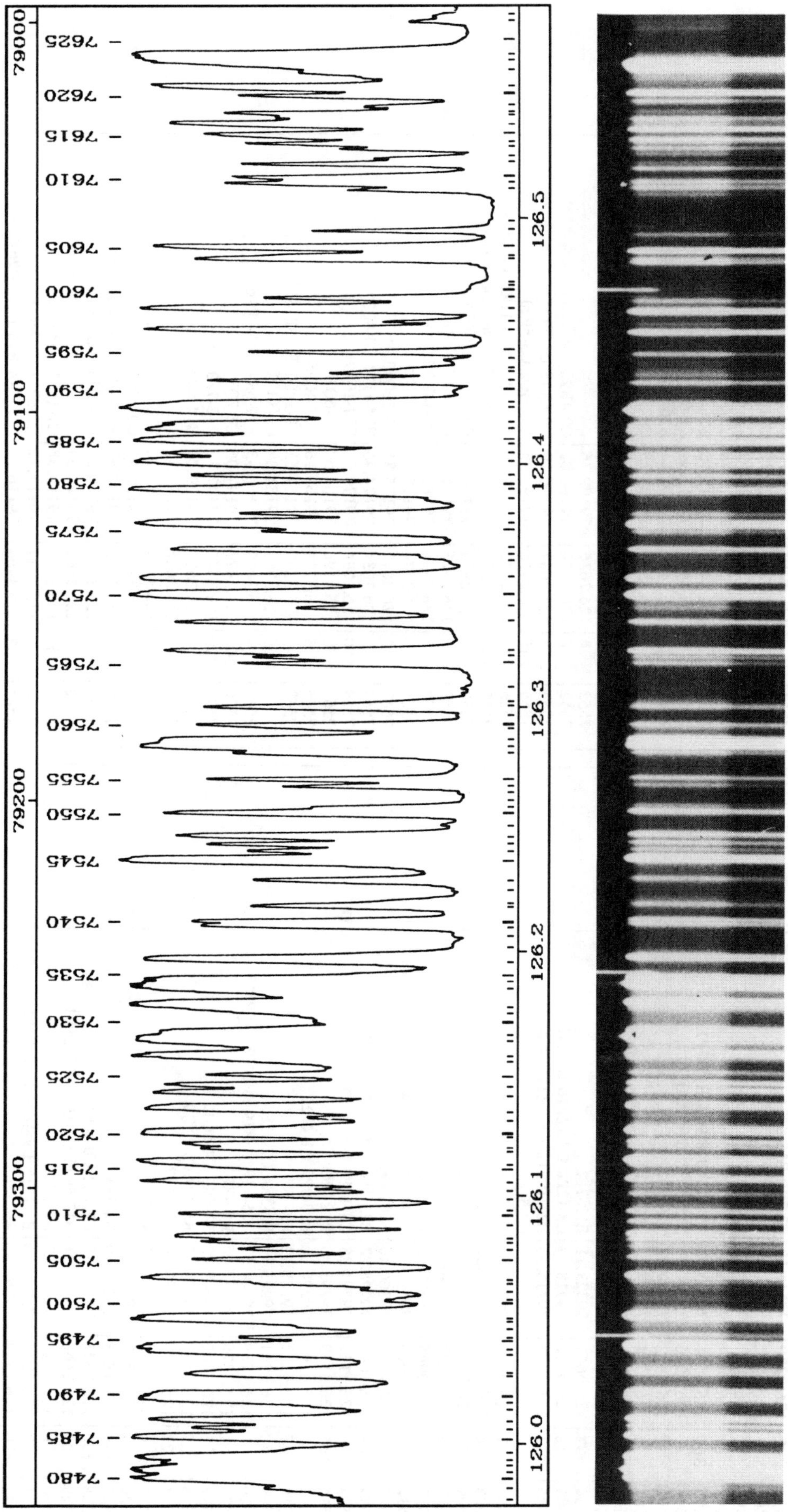
79000
79100
79200
79300
7480
7485
7490
7495
7500
7505
7510
7515
7520
7525
7530
7535
7540
7545
7550
7555
7560
7565
7570
7575
7580
7585
7590
7595
7600
7605
7610
7615
7620
7625
126.0
126.1
126.2
126.3
126.4
126.5

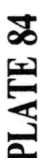

PLATE 84

TABLE 84 *continued*

Nb	I	λ (nm)	σ (cm-1)	Assignment	Comment
7589		126.4259	79097.75	C2-7R7	sh
7590	0	126.4308	79094.65	B26-12P9	
7591	12	126.4352	79091.92	C4-8P11	
7592	2	126.4385	79089.83	B11-5P13 B27-14R3	
7593	2	126.4398	79089.01	C4-8R19	
7594	2	126.4434	79086.76	B19-7P16 C1-5R21	
7595	7	126.4470	79084.54	C1-6P6	
7596	45	126.4562	79078.78	B17-9R1	
7597	2	126.4593	79076.85	B16-8P7	
7598	57	126.4647	79073.44	B20-10P3 B2-3R8 C1-4Q25	
7599	8	126.4690	79070.78	B26-13P1 B27-14R1 C1-5Q18	
7600	0	126.4716	79069.15	B25-11P11	
7601					SiII 126.47379
7602	0	126.4736	79067.86	B20-10R4	
7603	11	126.4850	79060.73	C5-9Q12	
7604	8	126.4859	79060.21	C6-10P8	
7605	45	126.4898	79057.77	C2-7Q4 B22-11R0	
7606	5	126.4964	79053.64	B16-8R8 B22-10P9	
7607					SiII 126.50022
7608	5	126.5133	79043.09	B3-2P14 B11-5R14 C1-6R11	
7609	14	126.5161	79041.30	B22-11R1	
7610	16	126.5184	79039.88	B3-2R15 B23-10P11 C0-3P24	
7611	10	126.5240	79036.40	C6-10Q12	
7612	2	126.5263	79034.95	C1-6Q8	
7613	5	126.5302	79032.49	B20-9P10	
7614	7	126.5325	79031.08	B27-14P1 C7-11P8 C3-7R19	
7615	15	126.5360	79028.87	B2-3P7 B3-0P21 B19-7R17 C2-7R8	
7616	40	126.5404	79026.12	C4-7P20	
7617	4	126.5430	79024.52	C3-7Q15	
7618	6	126.5447	79023.45	B0-2R10 B26-13P2	
7619	5	126.5477	79021.62	C7-11P9	
7620	8	126.5519	79018.93	C1-5P16	
7621	58	126.5556	79016.69	B10-6R7	
7622	1	126.5620	79012.69	B17-9P1 B4-0P22	
7623		126.5652	79010.66	B6-1P21	sh
7624	620	126.5673	79009.35	B17-9R2 B27-14P2 B25-10P15	sh
				B1-3R5 B6-5R2 B6-5P1 B24-10P13	
				C2-6R20	
7625		126.5712	79006.92	B8-5P8 B18-6R19	
7626	2	126.5823	78999.99	B5-1P20	
7627	1	126.5849	78998.39	B8-2P20	

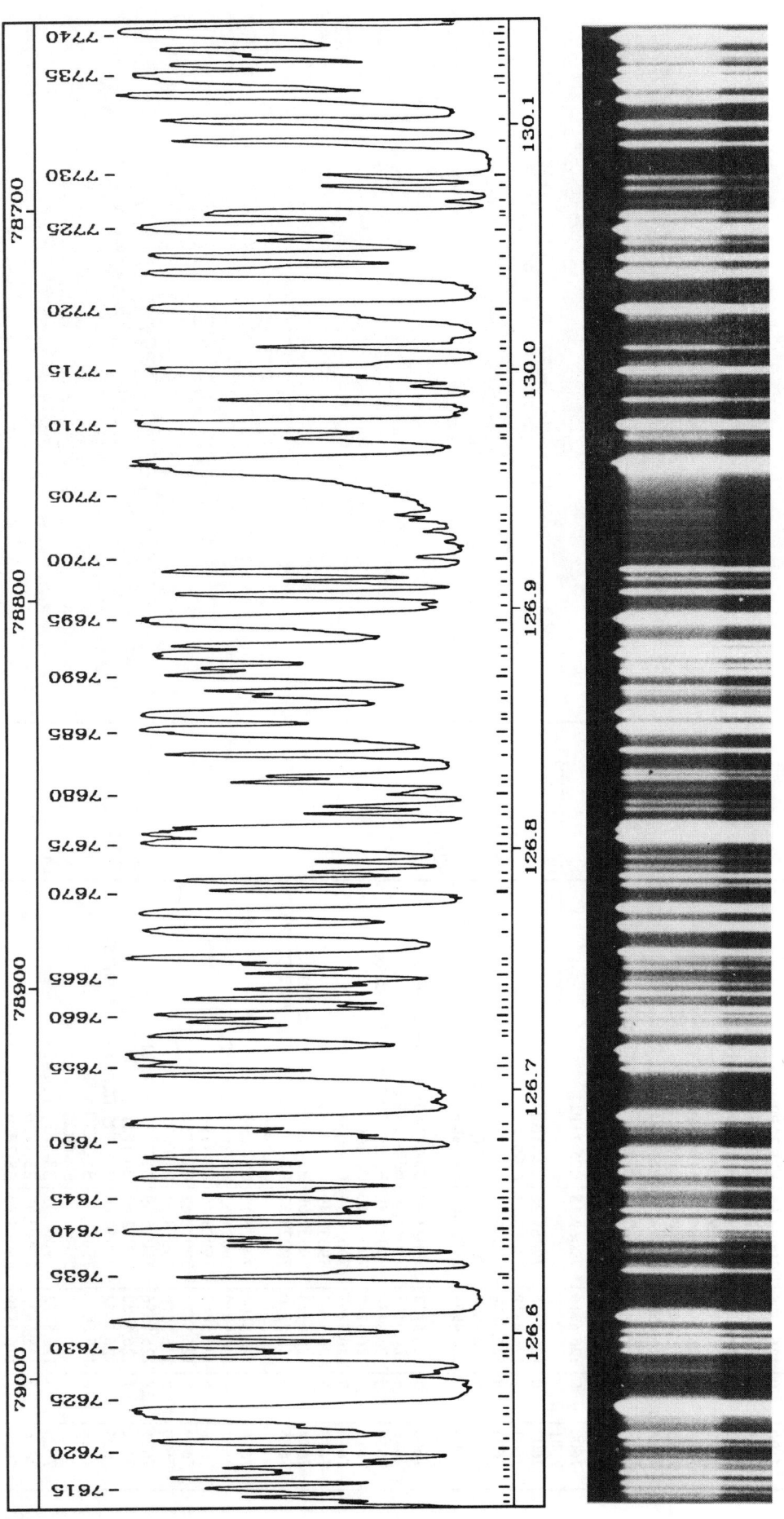

PLATE 85

TABLE 85

Nb	I	λ (nm)	σ (cm⁻¹)	Assignment	Comment
7628	20	126.5904	78994.90	B8-5R9 C6-10Q15	
7629	5	126.5928	78993.45	B27-14P3	
7630	27	126.5950	78992.06	B22-11R2 B9-5P10 C6-10Q13 B23-9R16	
7631	15	126.5984	78989.94	B22-11P1 B2-0P20 B11-3P20	
7632	340	126.6046	78986.05	C2-7P3 B9-2P21	
7633	0	126.6196	78976.74	B7-1P22 B10-3P19	
7634	15	126.6234	78974.37	B8-4P13 C6-10Q14	
7635	3	126.6253	78973.15	C0-5Q11	
7636	4	126.6316	78969.21	C0-5P9 C7-11P13	
7637	7	126.6361	78966.46	B26-13P3	
7638	9	126.6383	78965.03	B1-2R12	
7639	250	126.6418	78962.85	B1-3P4	sh
7640	14	126.6435	78961.79	B9-5R11 B7-2P19 B18-8P11	
7641	1	126.6478	78959.11	C3-7P13 B12-3P21 C1-4P23	
7642	1	126.6508	78957.25	B12-4P18	
7643	1	126.6519	78956.59	B18-7P15 B5-0P23	
7644	1	126.6550	78954.62	C7-11P12	
7645	8	126.6570	78953.38	B4-4R6 B24-12R0	
7646	2	126.6592	78952.01	C7-11P10	
7647	65	126.6637	78949.19	C2-7Q5 C4-8Q14	
7648	35	126.6677	78946.74	B23-11P6 B0-1R15 B23-9P15 C5-9Q13 C0-4Q20	
7649	40	126.6723	78943.85	B0-2P9 B24-12R1	
7650	3	126.6806	78938.71	B14-5P17	
7651	12	126.6832	78937.08	B8-4R14 C3-6P21	
7652	180	126.6867	78934.87	B4-4P5 B4-1P19 B13-6P12	
7653	0	126.6950	78929.74	B10-2P22	
7654	30	126.7064	78922.59	C2-6Q17	
7655	40	126.7107	78919.95	B12-7P4 C6-10P9 C0-4R23 C3-7R20	
7656	500	126.7141	78917.83	B9-6P3	
7657	38	126.7226	78912.50	B1-2P11 B22-11R3 B3-0R22	
7658	8	126.7257	78910.61	B21-10P7 B0-1P14 B24-12R2 C4-7P21	
7659	15	126.7284	78908.92	B9-6R4	
7660	37	126.7313	78907.11	B17-9P2 B22-11P2	
7661	2	126.7348	78904.90	C5-9P10	
7662	15	126.7379	78903.01	B20-10P4 B15-5R19 C7-11P11 C2-5P22	
7663	7	126.7419	78900.52	B17-9R3 B20-10R5	
7664	3	126.7444	78898.94	B26-13P4	
7665	8	126.7486	78896.32	B24-12P1 B1-0P19 B21-9P12	
7666	4	126.7524	78893.96	C2-7R9 C1-6R12	
7667	83	126.7551	78892.30	B5-1R21 B26-13R6 C3-8R2	
7668	125	126.7660	78885.53	B12-7R5 B24-11P9 C3-8R3	
7669	300	126.7737	78880.72	C3-8R1 B6-1R22	
7670	19	126.7831	78874.86	C2-6P15 B4-0R23	
7671	29	126.7870	78872.50	B15-8P5	
7672	5	126.7905	78870.28	B13-7P7 C0-4P18	
7673	4	126.7947	78867.67	C6-10P10	sh
7674		126.7996	78864.62	B11-4P17	
7675	220	126.8030	78862.54	B6-5R3	
7676	320	126.8065	78860.31	B6-5P2 C1-6P7	
7677	24	126.8090	78858.76	C3-8R4 B14-8R0	
7678	5	126.8150	78855.05	C4-8P12	b
7679	4	126.8180	78853.13	B18-9P6 B4-1R20 B24-12R3	
7680	3	126.8235	78849.73	B12-4R19 B6-0P24 C2-6R21 C1-5R22	
7681	12	126.8282	78846.82	C5-9Q14 C3-7R21	
7682	10	126.8306	78845.32	C1-6Q9 C8-13R5	
7683	27	126.8398	78839.58	B9-2R22 C2-7P4	
7684		126.8480	78834.50	B5-3P13	sh
7685	235	126.8496	78833.49	B3-4R0 B11-2P23	
7686	520	126.8557	78829.71	C3-8R0 B24-12P2	
7687		126.8578	78828.45	C3-7Q16	sh
7688	5	126.8638	78824.71	B13-7R8 B5-3R14 B19-8P13	
7689	10	126.8659	78823.39	C2-7Q6	
7690	27	126.8725	78819.28	B7-5P6 B7-1R23 B9-3R19 C5-9P11	
7691	12	126.8753	78817.53	B7-5R7	
7692	620	126.8807	78814.23	B14-8R1 B17-7P14 B13-5P16	
7693	35	126.8846	78811.80	B3-1P18 C3-8R5	
7694		126.8915	78807.51	B6-4P10 C4-7P22	sh
7695	820	126.8949	78805.39	B3-4R1 B5-4R9 B6-4R11	
7696	37	126.9027	78800.56	B8-3P17	
7697		126.9060	78798.49	C4-8Q15	
7698	10	126.9116	78795.02	B15-7P11 C1-5Q19	
7699	35	126.9155	78792.61	B22-11P3 B5-4P8	
7700	0	126.9216	78788.78	B10-5P12 B5-0R24	
7701	0	126.9280	78784.82	B14-3P23	
7702	0	126.9324	78782.11	C8-13Q7	
7703	0	126.9369	78779.31	B11-4R18	
7704	0	126.9391	78777.91	C8-13R4	
7705		126.9548	78768.19		sh
7706	50	126.9573	78766.63	C5-9Q15	
7707	1530	126.9600	78764.97	B10-2R23 C3-8Q1	
7708	7	126.9704	78758.54	B3-1R19	
7709		126.9713	78757.93	B26-13P6 B23-10P12	
7710	165	126.9760	78755.07	B17-9P3 B14-8P1	
7711	0	126.9806	78752.20	B17-9R4	
7712	10	126.9864	78748.60	C6-10P11	
7713	2	126.9921	78745.08	B13-5R17	
7714	2	126.9956	78742.87	C2-7R10	
7715	97	126.9988	78740.92	B14-8R2 B24-12P3 C3-6P22	
7716		127.0017	78739.13	B23-11P7	
7717	6	127.0082	78735.07	C3-8R6 B22-9P14	
7718	1	127.0115	78733.02	B24-10P14	
7719	2	127.0208	78727.24	B10-4P16 B8-3R18 B15-6R16 C0-5Q12	b
7720	300	127.0246	78724.89	B11-6P9 C3-8Q2	
7721	180	127.0392	78715.86	B3-4R2 C8-13R3	
7722		127.0413	78714.59	C5-9Q16	sh
7723	87	127.0460	78711.65	C2-7P5	
7724	6	127.0524	78707.68	B21-8P16 C3-8R7	

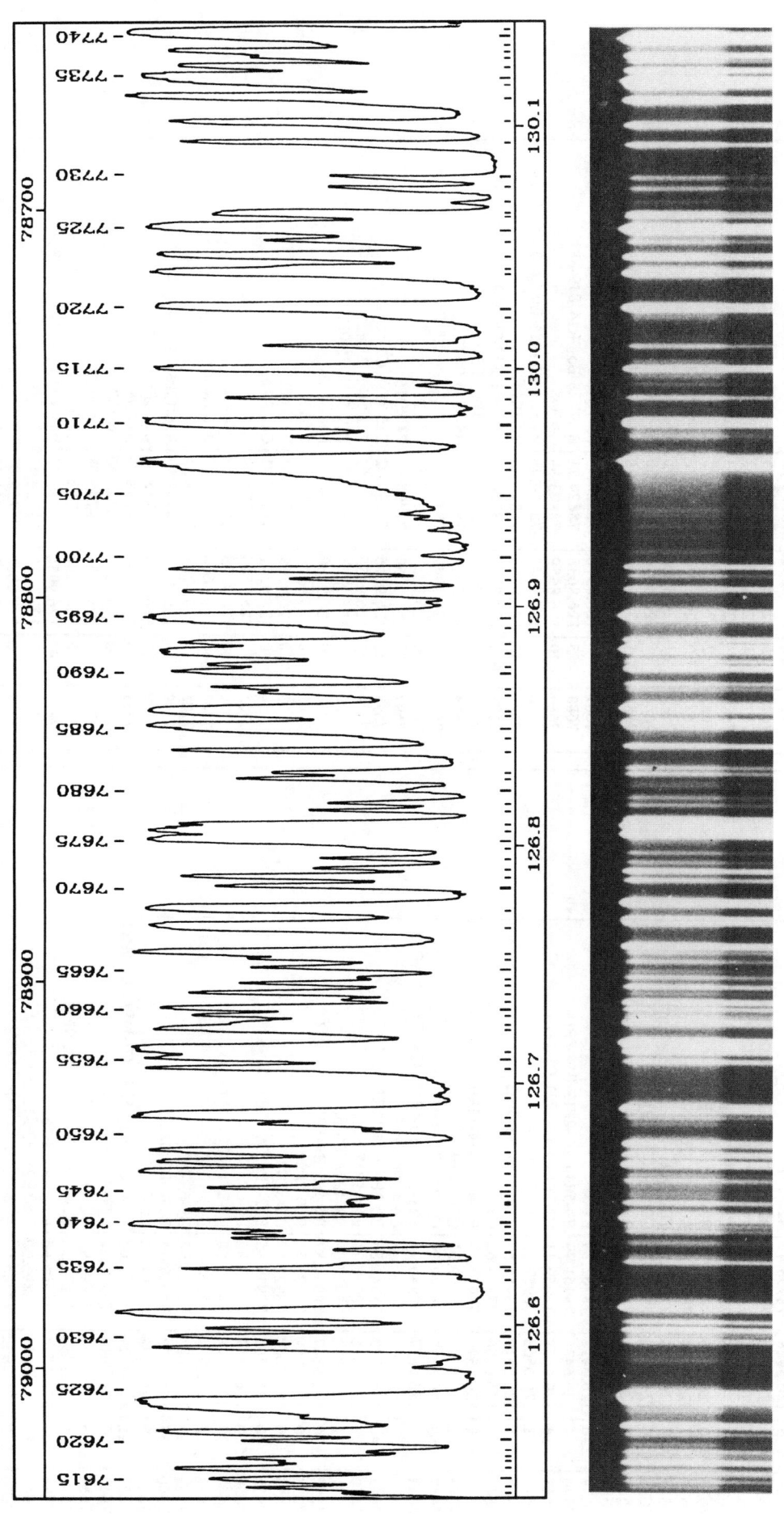

PLATE 85

TABLE 85 *continued*

Nb	I	λ (nm)	σ (cm⁻¹)	Assignment	Comment
7725	840	127.0578	78704.32	B3-4P1 C0-5P10 C1-5P17	
7726	15	127.0634	78700.93	B20-10P5 B7-0P25	
7727	15	127.0642	78700.37	C5-9Q17	
7728	2	127.0690	78697.38	C1-6R13	
7729	7	127.0748	78693.83	B2-2R14	
7730	8	127.0796	78690.84	B11-6R10 C2-6Q18 C3-7P14	
7731	13	127.0933	78682.34	B22-10P10 C2-7Q7	
7732	21	127.1016	78677.22	B1-3R6	
7733	90	127.1122	78670.66	B9-6P4 B22-11R5	
7734		127.1183	78666.90	B2-2P13	s h
7735	400	127.1202	78665.74	C3-8Q3	
7736	26	127.1245	78663.03	B9-6R5 C4-8Q16	
7737	7	127.1285	78660.57	C8-13R2 C2-5P23 C7-12R9	
7738	30	127.1309	78659.08	B6-5R4 B10-4R17 B6-0R25	
7739	0	127.1338	78657.27		
7740	940	127.1381	78654.64	B6-5P3 B12-5P15	s h
7741		127.1404	78653.20	C5-9P12	

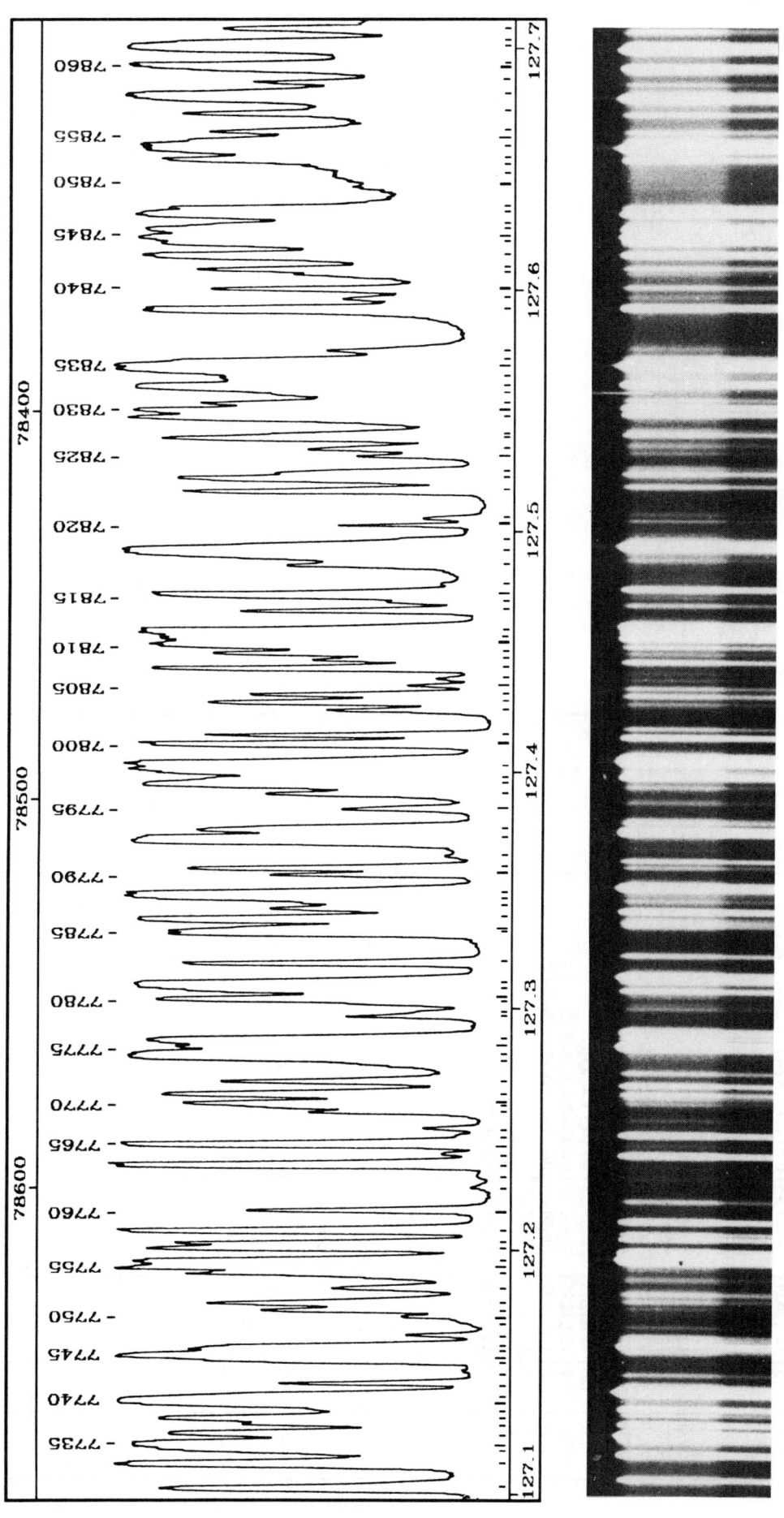

PLATE 86

TABLE 86

Nb	I	λ (nm)	σ (cm⁻¹)	Assignment	Comment
7742	6	127.1448	78650.47	B22-11P4	
7743	0	127.1488	78648.01	B11-2R24 C1-5R23	
7744	121	127.1547	78644.33	B10-6P7 C1-6Q10	sh
7745	14	127.1568	78643.06	B14-8P2 C3-7Q17	
7746	8	127.1590	78641.67	C4-8P13	
7747	0	127.1599	78641.13	B19-10R0 C0-4Q21	
7748	1	127.1644	78638.39	C1-6P8	
7749	5	127.1695	78635.19	B21-10P8 C8-13Q5 C1-4P24	
7750	10	127.1723	78633.50	B10-6R8	
7751	3	127.1751	78631.73	B24-12P4 C2-7R11	
7752	4	127.1781	78629.90	B2-1P17 B7-3P16	
7753	360	127.1843	78626.04	B7-4P12 C6-10P12	
7754	80	127.1912	78621.81	C6-10P13 B19-10R1	
7755	30	127.1929	78620.73	B1-3P5	
7756	17	127.1953	78619.24	B14-8R3	
7757	45	127.2008	78615.81	B12-7P5	
7758	8	127.2082	78611.29	C3-8P2	
7759	1	127.2166	78606.10	B12-7R6 C8-13R1	
7760	70	127.2260	78600.31	C3-8R8 C6-10P15 B2-1R18	
7761	48	127.2301	78597.74	B12-5R16	
7762	0	127.2358	78594.26	C7-12R8	
7763	3	127.2399	78591.68	B11-7R0	
7764	5	127.2444	78588.94	C8-13Q4 B14-6P14	
7765	10	127.2477	78586.91	C3-8Q4	
7766	0	127.2504	78585.20	B3-3P10	
7767	3	127.2574	78580.88	C3-6P23	
7768	5	127.2599	78579.38	B23-9P16 C2-6P16	
7769	10	127.2613	78576.52	B7-3R17	
7770	12	127.2651	78576.17	B2-3R9 B15-8P6 B17-9R5	
7771	18	127.2702	78572.96	B4-4R7	
7772	10	127.2771	78568.73	C5-9P13	
7773		127.2814	78566.08		sh
7774	500	127.2846	78564.10	B3-4R3 B11-7R1 C8-13R0	
7775	25	127.2871	78562.56	B17-9P4 B19-10R2	
7776	41	127.2971	78556.39	B19-10P1 C8-13Q3 C6-10P14	
7777	2	127.2983	78555.67	B18-9P7	
7778	10	127.3020	78553.36	C7-12R7	sh
7779	23	127.3047	78551.72	C8-13P7	
7780	6	127.3084	78549.41	B4-4P6 B14-6R15	sh
7781	440	127.3105	78548.12	C4-8Q17	sh
7782	20	127.3192	78542.73	B3-4P2	
7783	12	127.3321	78534.77	C8-13Q2	
7784	10	127.3333	78534.08	B15-8R7 C0-4P19	
7785	92	127.3381	78531.06	B9-4P15 C2-7P6	
7786	6	127.3426	78528.30	C8-13Q1 C1-5Q20	
7787	4	127.3452	78526.71	B19-7P17 C2-7Q8	sh
7788	6	127.3481	78524.91	B2-3P8	
7789	480	127.3565	78519.78	C3-8P3	
7790	8	127.3565	78519.78	B25-13R0	
7791	22	127.3592	78518.08	B25-13R1 C1-6R14	
7792	0	127.3641	78515.08	B4-2P16	
7793	215	127.3720	78510.21	C4-9R2 C4-9R3 B8-5P9	
7794	20	127.3758	78507.87	B24-12P5 B26-14R1 C3-7P15	
7795	4	127.3843	78502.60	B25-13R2 B8-5R10 C8-13P6	
7796	6	127.3909	78498.57	C4-8P14	
7797	63	127.3952	78495.92	C3-8Q5	
7798	12	127.3969	78494.86	B26-14R0 C3-8R9	
7799	820	127.4021	78491.63	B11-7P1 B4-2R17 C4-9R1 C4-9R4	
7800	57	127.4118	78485.64	B11-7R2 B9-4R16	
7801	9	127.4153	78483.52	B22-11P5	
7802	7	127.4259	78476.98	B25-13P1 C0-5Q13	
7803	15	127.4292	78474.94	B25-13R3 B13-7P8 C8-13P2 C3-7Q18	
7804	12	127.4322	78473.10	B19-10R3 C2-6Q19	
7805	1	127.4358	78470.88	B20-10P6 C1-5R24	
7806	1	127.4389	78468.96	C2-7R12	
7807	23	127.4435	78466.14	B19-10P2	
7808	4	127.4468	78464.11	C4-8Q18	
7809	13	127.4498	78462.25	C8-13P3 B26-14P1	
7810	240	127.4537	78459.85	B0-3R0 C8-13P4	
7811	110	127.4561	78458.39	B14-8P3	
7812	320	127.4586	78456.83	C4-9R0 C4-9R6	
7813	7	127.4671	78451.63	B14-8R4	
7814	1	127.4706	78449.46	B4-3R13	
7815	38	127.4740	78447.36	C4-9R5 C7-12R6	
7816	0	127.4779	78444.98	B26-14P2	
7817	4	127.4867	78439.56	B13-7R9 B11-5P14	sh
7818	880	127.4922	78436.20	B0-3R1 C1-6Q11	
7819		127.4964	78433.58	B4-3P12 C3-8R10	
7820	20	127.5026	78429.79	B26-14P3	
7821	2	127.5055	78428.02	B25-13P2	
7822	10	127.5174	78420.69	C7-12R5	
7823	20	127.5227	78417.42	B0-2R11 B24-12R7	
7824	8	127.5248	78416.11	C5-9P14 C4-8Q19	
7825	2	127.5315	78411.98	B18-7P16 C1-5P18	
7826	3	127.5346	78410.08	B7-5R8 B14-7R11	
7827	18	127.5393	78407.21	B7-5P7	
7828	7	127.5409	78406.24	C1-6P9	
7829	40	127.5480	78401.83	B6-5R5	
7830	92	127.5508	78400.14	C3-8P4 C6-11R11	
7831	10	127.5539	78398.21	B6-3P15 B25-13R5 B18-6R20	
7832	167	127.5607	78394.07	B6-5P4 B11-5R15	
7833	1	127.5639	78392.08	B1-1R17	
7834	1500	127.5687	78389.17	C4-9Q1 C3-8Q6 C4-9R7 B22-9P15	CuII 127.55717
7835	17	127.5715	78387.40	C7-12R4	
7836	1	127.5753	78385.08	B1-1P16 B22-10P11	
7837	110	127.5928	78374.30	B9-6P5	
7838	5	127.5972	78371.65	B24-12P6 B6-3R16	

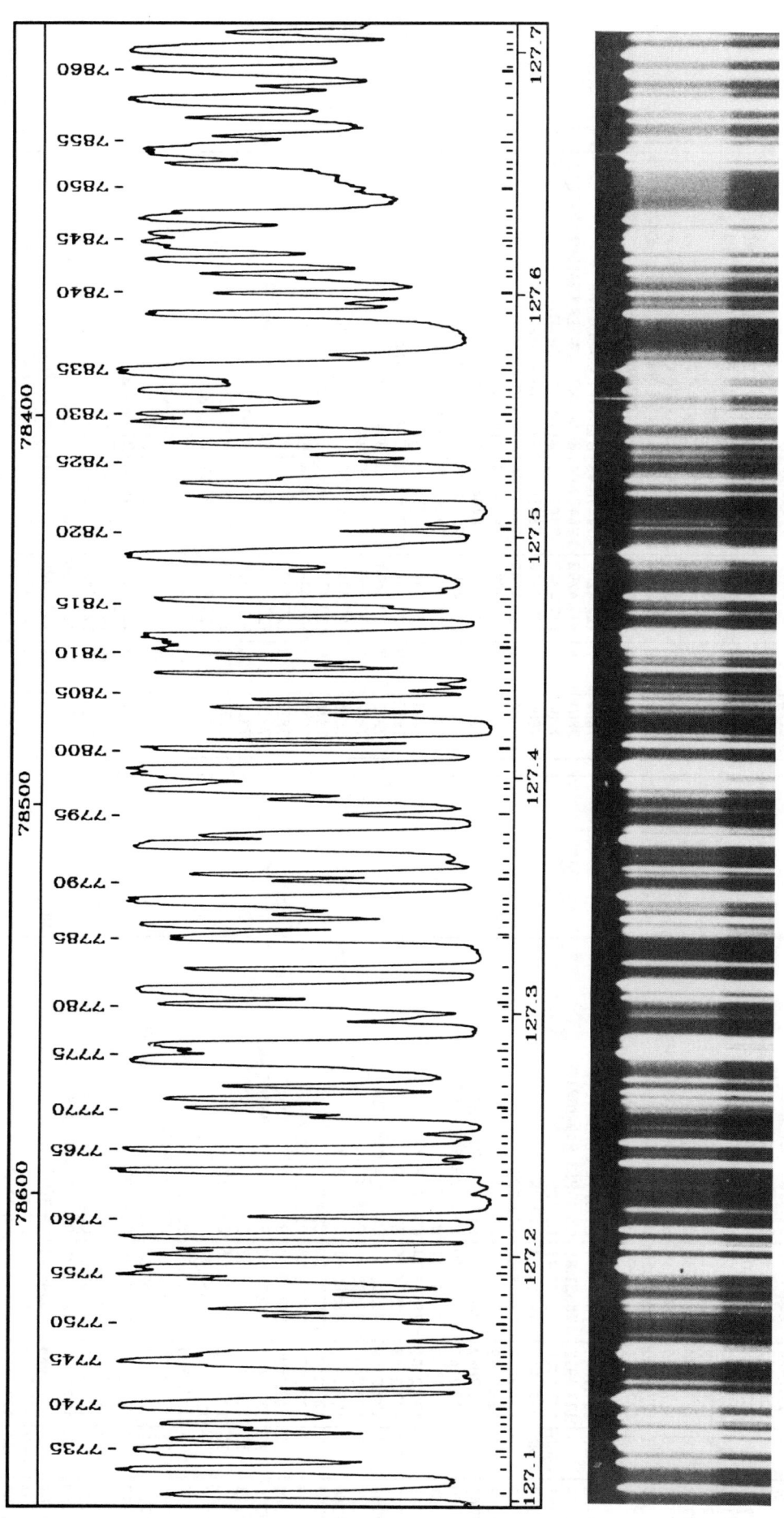

PLATE 86

TABLE 86 *continued*

Nb	I	λ (nm)	σ (cm⁻¹)	Assignment	Comment
7840	10	127.6012	78369.18	B9-6R6 B25-13P3 B22-11R7	
7841	5	127.6072	78365.50	B21-11R0	
7842	10	127.6094	78364.12	C2-7Q9	
7843	134	127.6152	78360.58	B11-7P2 B19-10R4	
7844	50	127.6204	78357.38	B3-4R4 C1-4P25	
7845	300	127.6226	78356.00	C4-9Q2	
7846	115	127.6258	78354.07	B11-7R3 C7-12R3	
7847		127.6276	78352.96	B21-11R1 B13-6P13 C2-6P17	sh
7848	250	127.6325	78349.96	B0-3R2	
7849	20	127.6348	78348.54	C2-7P7 C2-7R13 C0-4Q22	
7850	0	127.6434	78343.24	B21-10P9 C0-3P26	
7851	0	127.6483	78340.28		
7852	0	127.6516	78338.23		
7853	30	127.6552	78336.01	B17-9P5	
7854	1200	127.6603	78332.92	B3-4P3 B19-10P3	
7855	17	127.6631	78331.16	B0-2P10	
7856	13	127.6667	78328.98	C3-7Q19 C1-5R25	
7857	18	127.6742	78324.36	C7-12R2	
7858	920	127.6812	78320.06	B0-3P1	
7859	5	127.6868	78316.61	C4-9R8 C1-6R15	
7860		127.6921	78313.35	B23-12R1 B23-12R0	sh
7861	113	127.6940	78312.20	B8-6R0 C4-9R9	
7862	400	127.7018	78307.43	C4-9Q3	
7863		127.7037	78306.25	C4-8P15	sh
7864	8	127.7091	78302.93	C6-11R8 C6-11R9 B12-7P6 B25-13P4 C8-14R4	

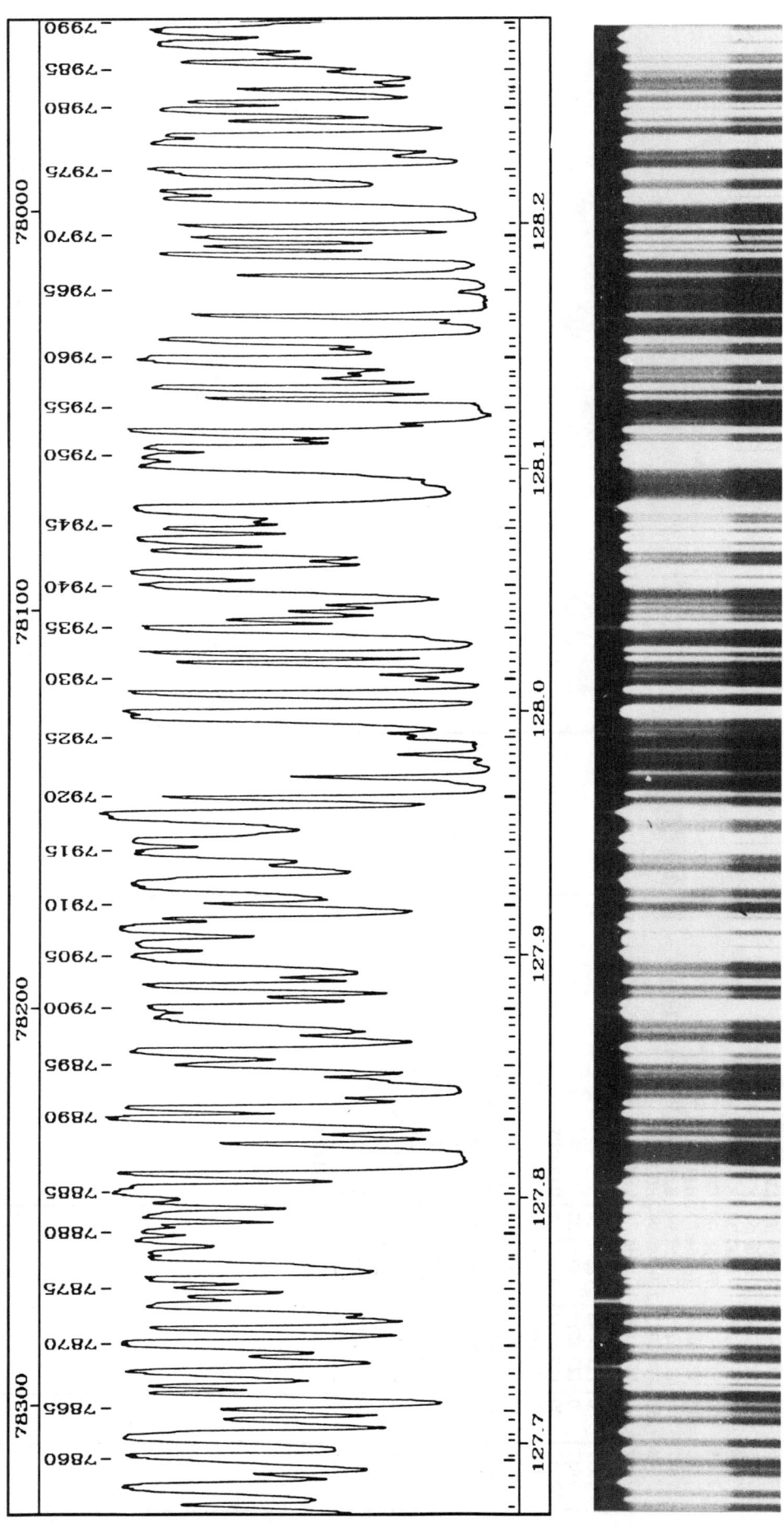

PLATE 87

TABLE 87

Nb	I	λ (nm)	σ (cm⁻¹)	Assignment	Comment
7865	9	127.7134	78300.34	B17-9R6 C3-8R11 C7-12Q9	
7866	25	127.7200	78296.30	B21-11P1 B22-11P6	
7867	17	127.7229	78294.52	B12-7R7 C5-10R5 C6-11R10	
7868	52	127.7286	78291.03	B1-3R7	
7869	7	127.7350	78287.12	B5-4R10 B23-9P17	
7870	400	127.7400	78284.05	B8-6R1 B20-10R8 B15-7P12 C1-5Q21	
7871	39	127.7464	78280.09	B16-9R0 B8-4P14	
7872	1	127.7514	78277.05	B17-8R12	
7873	175	127.7557	78274.39	B16-8R10 C5-10R3 C2-6Q20	
7874	20	127.7591	78272.29	B1-2R13 C5-9P15	
7875	15	127.7632	78269.82	C3-8Q7 C3-7P16	
7876	135	127.7665	78267.77	C5-10R2 B5-4P9	
7877	102	127.7753	78262.38	C3-8P5 C5-10R6	
7878	175	127.7774	78261.13	B21-11R3 C4-9P2 C5-9P16	
7879	400	127.7825	78257.98	C7-12R1	
7880	140	127.7860	78255.83	B16-9R1	
7881	20	127.7883	78254.43	C5-10R4	
7882	52	127.7921	78252.10	B23-12R2 B23-12P1 B15-8P7 B8-4R15 C6-11R6	
7883		127.7934	78251.29	B21-8P17 C6-11R5	sh
7884	25	127.7978	78248.58	B14-8P4 C6-11R4	
7885	740	127.8021	78245.99	C5-10R1 B3-2R16 B11-6P10	
7886	58	127.8040	78244.81	C4-9Q4	
7887	18	127.8092	78241.64	B14-8R5 C7-12Q8 B3-2P15	
7888	10	127.8215	78234.08	B18-9P8 B15-7R13 C6-11R7	
7889	4	127.8255	78231.63	B25-13P5 B24-12P7	
7890	150	127.8318	78227.82	C7-12R0 C5-10R7	
7891	57	127.8360	78225.20	B1-3P6	
7892	3	127.8405	78222.51	B15-8R8 B22-10R13 C1-6Q12	
7893	2	127.8472	78218.35	B10-6P8	
7894	3	127.8494	78217.03	B20-10P7	
7895	22	127.8539	78214.26	B1-2P12 B21-11P2	
7896	310	127.8589	78211.20	B10-6R9 B6-4R12 C5-10R0 C3-7Q20	
7897	2	127.8659	78206.93	C7-12Q7	
7898		127.8712	78203.69	B23-12R3 B6-4P11 C4-9R10	sh
7899	420	127.8734	78202.36	B0-3R3	
7900	440	127.8769	78200.18	B8-6R2 B8-6P1 B11-6R11	
7901	5	127.8818	78197.25	C5-10R9	
7902	80	127.8864	78194.37	B16-9P1 C0-4P20	
7903	2	127.8899	78192.26	C2-7Q10	
7904		127.8969	78187.99	B16-9R2 B23-12P2	sh
7905	500	127.8984	78187.04	C7-12Q1 C7-12Q6 C5-10R11 C4-8P16	
7906	135	127.9027	78184.46	C6-11R2 C5-9P17	
7907	105	127.9042	78183.50	C7-12Q2 B18-9R9	
7908	610	127.9103	78179.75	B11-7P3 C7-12Q3 C7-12Q5 C6-11R3 C8-14R3	
7909	18	127.9143	78177.35	C7-12Q4	
7910	13	127.9206	78173.50	B11-7R4	
7911		127.9261	78170.13	C4-9Q5 C2-7R14	sh
7912	490	127.9284	78168.76	C4-9P3	
7913		127.9305	78167.43	B19-10P4	sh
7914	2	127.9364	78163.84	B25-13P6 B10-5P13	
7915	610	127.9421	78160.38	C6-11R1	
7916	540	127.9465	78157.70	B0-3P2 C5-10R8 C0-5P12	
7917		127.9488	78156.26	B15-6P16 C3-8R12	sh
7918		127.9555	78152.17	B4-4R8 C5-10R10	sh
7919	1200	127.9568	78151.38	C5-10Q1 C2-7P8	
7920	15	127.9644	78146.71	B19-10R5	
7921	5	127.9731	78141.40	C3-8Q8 C5-10R12	
7922	0	127.9792	78137.73	B11-3P21	
7923	4	127.9825	78135.66	C3-7Q21	
7924	0	127.9859	78133.58	C1-6R16	
7925	1	127.9890	78131.71	B10-3P20	
7926	1	127.9913	78130.34	B23-12R4	
7927	280	127.9973	78126.67	C5-10Q2	
7928	280	127.9998	78125.10	C6-11R0	
7929	40	128.0077	78120.33	B4-4P7 B22-11R8 B13-5P17 B15-6R17	
7930	2	128.0135	78116.79	C8-14Q4 C1-5P19	
7931	5	128.0158	78115.38	B0-1R16	
7932	17	128.0207	78112.36	C3-8P6 C4-9R11	
7933	28	128.0245	78110.09	B23-12P3	
7934	0	128.0306	78106.33	B25-13P7 C5-10R13	
7935	92	128.0346	78103.88	B21-11P3 B12-3P22 C7-12P2	
7936	5	128.0382	78101.71	B21-11R4 B5-3P14	
7937	5	128.0418	78099.52	B5-3R15 B11-4P18	
7938	3	128.0445	78097.84	C2-6Q21	
7939		128.0516	78093.53	B6-5R6 C3-7P17	
7940	55	128.0527	78092.89	B3-4R5 B22-11P7	
7941	320	128.0575	78089.94	C5-10Q3 C3-8R13	
7942	2	128.0626	78086.85	C4-9R12 B8-2P21	
7943	50	128.0668	78084.28	C4-9Q6 B16-9P2	
7944	240	128.0713	78081.49	B6-5P5 C8-14R2	
7945	35	128.0760	78078.64	B16-9R3	
7946	1	128.0788	78076.93	B17-9P6 B13-5R18 C2-6P18	
7947	870	128.0836	78074.00	C6-11Q1 B0-1P15	
7948	0	128.0944	78067.44	B9-2P22	
7949	150	128.1020	78062.80	B8-6R3 C4-9P4	
7950	420	128.1050	78060.94	B3-4P4 B8-6P2 B13-7P9 B7-2P20	
7951	192	128.1089	78058.60	C6-11Q2	
7952	0	128.1123	78056.52	B17-9R7	
7953	85	128.1155	78054.54	C7-12P3 C1-5Q22 C2-7R15	
7954	1	128.1189	78052.53	B21-10R11	
7955	0	128.1248	78048.90		
7956	13	128.1291	78046.32	C7-12P4	
7957	45	128.1334	78043.66	C5-10Q4	
7958	5	128.1376	78041.11	B2-3R10	
7959	2	128.1399	78039.68	B23-12R5	
7960	390	128.1448	78036.75	C6-11Q3 C5-10P2 B6-1P22	

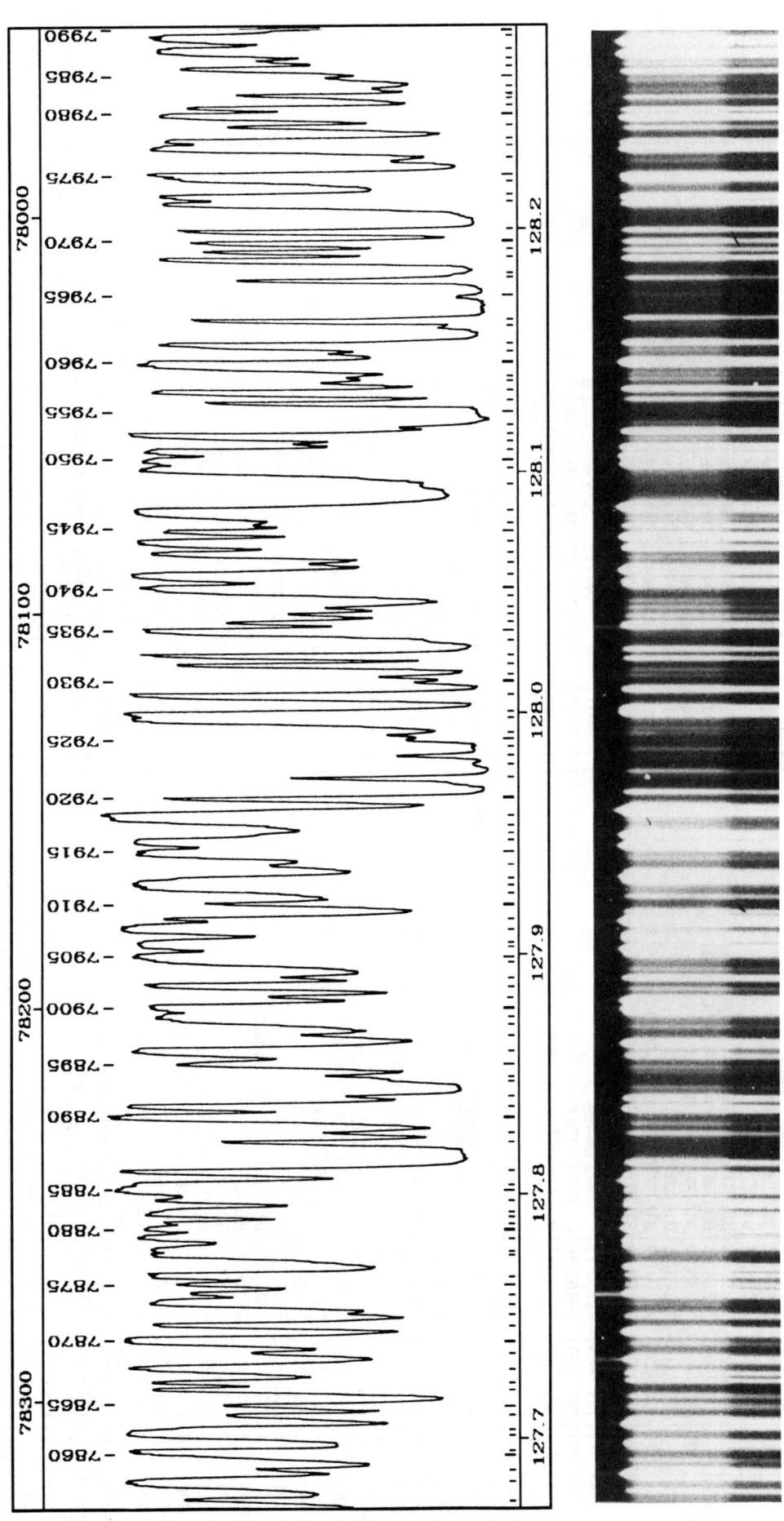

PLATE 87

TABLE 87 *continued*

Nb	I	λ (nm)	σ (cm⁻¹)	Assignment	Comment
7961	4	128.1496	78033.81	B5-1P21	
7962	26	128.1528	78031.87	B9-6R7 B9-6P6 C4-8P17 C0-3P27	
7963	1	128.1602	78027.34	B11-4R19	
7964	13	128.1633	78025.47	C8-14Q3	
7965	1	128.1730	78019.53	B13-7R10	
7966	7	128.1801	78015.25	C2-7Q11	
7967	0	128.1826	78013.73		
7968	32	128.1885	78010.17	C6-11Q4	
7969	15	128.1917	78008.16	C7-12P5	
7970	19	128.1953	78005.97	C3-8Q9 C1-6Q13 C4-9R13 B10-2P23	
7971	25	128.2000	78003.10	C8-14R1	
7972	40	128.2112	77996.28	B14-8P5 B3-0P22 B19-7P18	
7973	72	128.2141	77994.57	B0-3R4	sh
7974		128.2212	77990.24	C4-9Q7	
7975	60	128.2227	77989.32	C5-10Q5	
7976	2	128.2291	77985.43	C8-14P4	
7977	120	128.2343	77982.28	B5-5R0 B2-3P9 B4-1P20 B19-7R19	
				C7-12P6	
7978	40	128.2369	77980.68	B8-3P18 B14-6P15 C6-11Q5	
7979	9	128.2430	77976.95	B21-11R5 B8-5R11 B8-5P10	
7980	97	128.2474	77974.32	B4-0P23 B10-4P17 C6-11P2	
7981	22	128.2506	77972.38	B19-10P5 B12-5P16	
7982	4	128.2544	77970.02	B2-0P21	
7983	10	128.2557	77969.24	B21-11P4	
7984	1	128.2588	77967.38	C7-12P7	
7985	2	128.2633	77964.61	B19-10R6 B25-14R4	
7986	22	128.2667	77962.53	C8-14Q2 B7-5R9 C3-8R14	
7987	1	128.2699	77960.59	B7-4P13	
7988	390	128.2741	77958.07	C5-10P3 B3-3P11	
7989	410	128.2793	77954.92	B5-5R1 B20-10R9 C4-8P18	sh
7990		128.2819	77953.31	B7-5P8	

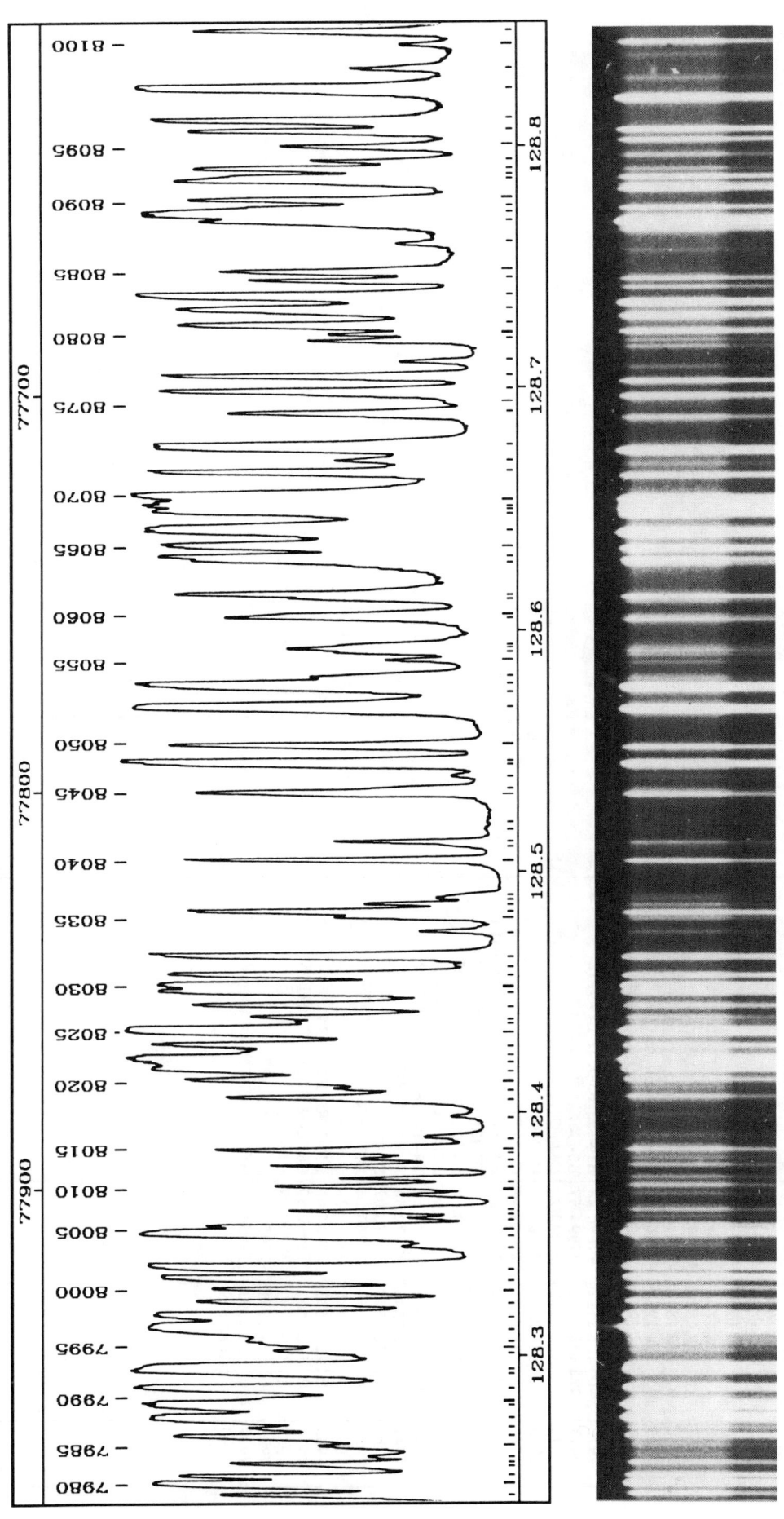

PLATE 88

TABLE 88

Nb	I	λ (nm)	σ (cm⁻¹)	Assignment	Comment
7991	80	128.2863	77950.66	B11-7P4 C6-11Q6 C3-8P7	
7992	117	128.2937	77946.17	C4-9P5 B11-7R5 B12-7P7	
7993	62	128.2958	77944.87	B13-8R0 C8-14R0 B20-10P8	
7994	5	128.3019	77941.15	B12-7R8 C4-9R15 C1-6R17	
7995	5	128.3041	77939.83	C1-6P11	
7996	6	128.3062	77938.56	B25-14R3 B22-11R9	
7997	1540	128.3111	77935.56	B0-3P3 B5-1R22 B9-2R23	
7998	128	128.3160	77932.62	B16-9P3 C2-7P9 B2-2R15	
7999	15	128.3219	77929.00	C5-10Q6 B16-9R4	
8000	16	128.3274	77925.70	C8-14P3 B12-5R17	
8001	70	128.3320	77922.92	C6-11Q7 C8-14Q1 B25-14R2	
8002	146	128.3364	77920.22	B13-8R1 B10-4R18 B8-3R19 C6-11Q11	
8003	4	128.3451	77914.93	B6-1R23	
8004	260	128.3504	77911.70	C6-11P3	
8005	18	128.3532	77910.05	B25-14R1 B4-1R21	
8006	3	128.3570	77907.74	B15-8P8 B11-2P24	
8007	8	128.3598	77906.03	B24-13R1 B5-0P24	
8008	1	128.3612	77905.18	B24-13R0	
8009	2	128.3665	77901.98	C6-11Q8	
8010	7	128.3700	77899.81	B2-2P14	
8011	5	128.3732	77897.88	B25-14R0	
8012	3	128.3783	77894.78	C8-14P2 C6-11Q10	
8013	4	128.3813	77892.99	B24-13R2 B18-9P9 C3-8R15	
8014	8	128.3846	77890.94	C4-9Q8 C6-11Q9 B3-0R23	
8015		128.3864	77889.88	B16-8R11 C0-4P21	sh
8016	1	128.3905	77887.41	B18-9R10 C3-7P18	
8017	0	128.3989	77882.27	B22-11P8 C2-7R16	
8018	10	128.4065	77877.70	B3-1P19	
8019	3	128.4104	77875.29	C0-5P13 C4-8P19	
8020	15	128.4136	77873.35	B8-6R4 B15-8R9	
8021	84	128.4184	77870.49	C5-10P4 B18-10R0	
8022	600	128.4216	77868.52	B5-5R2 B8-6P3 B25-14P1 B24-13R3	
8023		128.4249	77866.50	C3-8Q10	sh
8024	13	128.4274	77864.99	C5-10Q7	
8025	440	128.4334	77861.40	B5-5P1	
8026	1	128.4371	77859.12	C2-6P19	
8027	4	128.4390	77857.97	B24-13P1 B23-12P5	
8028	12	128.4441	77854.89	B1-3R8	
8029	70	128.4499	77851.35	B18-10R1	
8030	138	128.4524	77849.83	B13-8P1 B7-1R24 C1-5Q23	
8031	33	128.4568	77847.21	B13-8R2 B25-14P2	
8032	3	128.4606	77844.87	C1-5P20	
8033	42	128.4646	77842.46	C6-11P4 B4-0R24	
8034	2	128.4751	77836.09	B24-13R4 C2-7Q12 C7-13R7	
8035	7	128.4812	77832.41	B25-14P3 B3-1R20	
8036	21	128.4834	77831.08	B21-11P5	
8037	4	128.4865	77829.18	B7-3P17	
8038	2	128.4890	77827.71	B25-14P4	
8039	1	128.4900	77827.05	B21-11R6 B23-12R7	
8040	12	128.5049	77818.04	C4-9P6	
8041	1	128.5113	77814.15	B5-2R19	
8042	4	128.5128	77813.24	B24-13P2 B9-5R13	
8043	0	128.5180	77810.11	B18-6R21	
8044	0	128.5261	77805.23		
8045	10	128.5331	77800.95	B17-9P7 C5-10Q8	
8046		128.5350	77799.82	B9-4P16	sh
8047	1	128.5401	77796.77	B6-0P25	
8048		128.5460	77793.75	B18-10R2 B17-8R13	sh
8049	50	128.5460	77793.18	B18-10P1	
8050	14	128.5527	77789.12	B7-3R18 C4-9Q9 C1-6Q14 C3-8R16	
8051	100	128.5685	77779.53	B1-3P7	
8052	4	128.5767	77774.61	B3-4R6	
8053	78	128.5779	77773.85	C5-10P5 B11-5P15	
8054	3	128.5809	77772.07	B0-2R12	
8055	2	128.5881	77767.70	C3-8P8	
8056	2	128.5915	77765.65	C2-7P10 C2-7R17	
8057	5	128.5929	77764.79	B19-10P6	
8058	2	128.5944	77763.86	B9-4R17 B15-7P13	
8059	10	128.6059	77756.93	B24-13P3 B13-6P14	
8060		128.6070	77756.30	B10-6R10 B10-6P9	sh
8061		128.6141	77751.98	B5-0R25 C3-7P19	sh
8062	20	128.6156	77751.04	B19-10R7 C6-11P5 C3-8R17	
8063	10	128.6300	77742.37	B16-9R5	
8064	22	128.6313	77741.55	B16-9P4	
8065	20	128.6359	77738.80	C5-10Q9 B6-5R7 C6-11P6	
8066	310	128.6425	77734.80	B3-4P5 B5-4R11 B4-3P13	
8067	85	128.6503	77730.09	B13-8P2	
8068	138	128.6525	77728.75	B0-3R5	
8069		128.6540	77727.89	B13-8R3	sh
8070	118	128.6564	77726.40	B5-5R3 C3-8Q11	
8071	27	128.6666	77720.28	B6-5P6	
8072	3	128.6714	77717.37	B2-1P18	
8073	260	128.6770	77713.99	B5-5P2	
8074	8	128.6909	77705.59	B14-8R7 B14-8P6 B5-4P10	
8075	0	128.6952	77702.99	C0-6R2	
8076	23	128.7000	77700.09	B18-10R3 B2-1R19	
8077	24	128.7062	77696.36	B18-10P2 B24-13R7	
8078	2	128.7120	77692.85	B24-13P4 B20-10R10	
8079	4	128.7206	77687.62	C4-9Q10 C1-6P12	
8080	3	128.7233	77686.02	C0-6R1	
8081	20	128.7270	77683.76	B4-4R9 C5-10Q10 B22-11P9 B11-6R12	
8082	20	128.7336	77679.81	B4-0P11	
8083	90	128.7391	77676.49	B11-7R6 B11-7P5	
8084	8	128.7456	77672.58	C6-11P7	
8085	10	128.7492	77670.37	C4-9P7 B4-2P17	
8086	1	128.7612	77663.17	B20-10P9	
8087	15	128.7704	77657.62	C5-10P6 B21-11R7 C2-7Q13	
8088	320	128.7732	77655.90	B0-3P4	

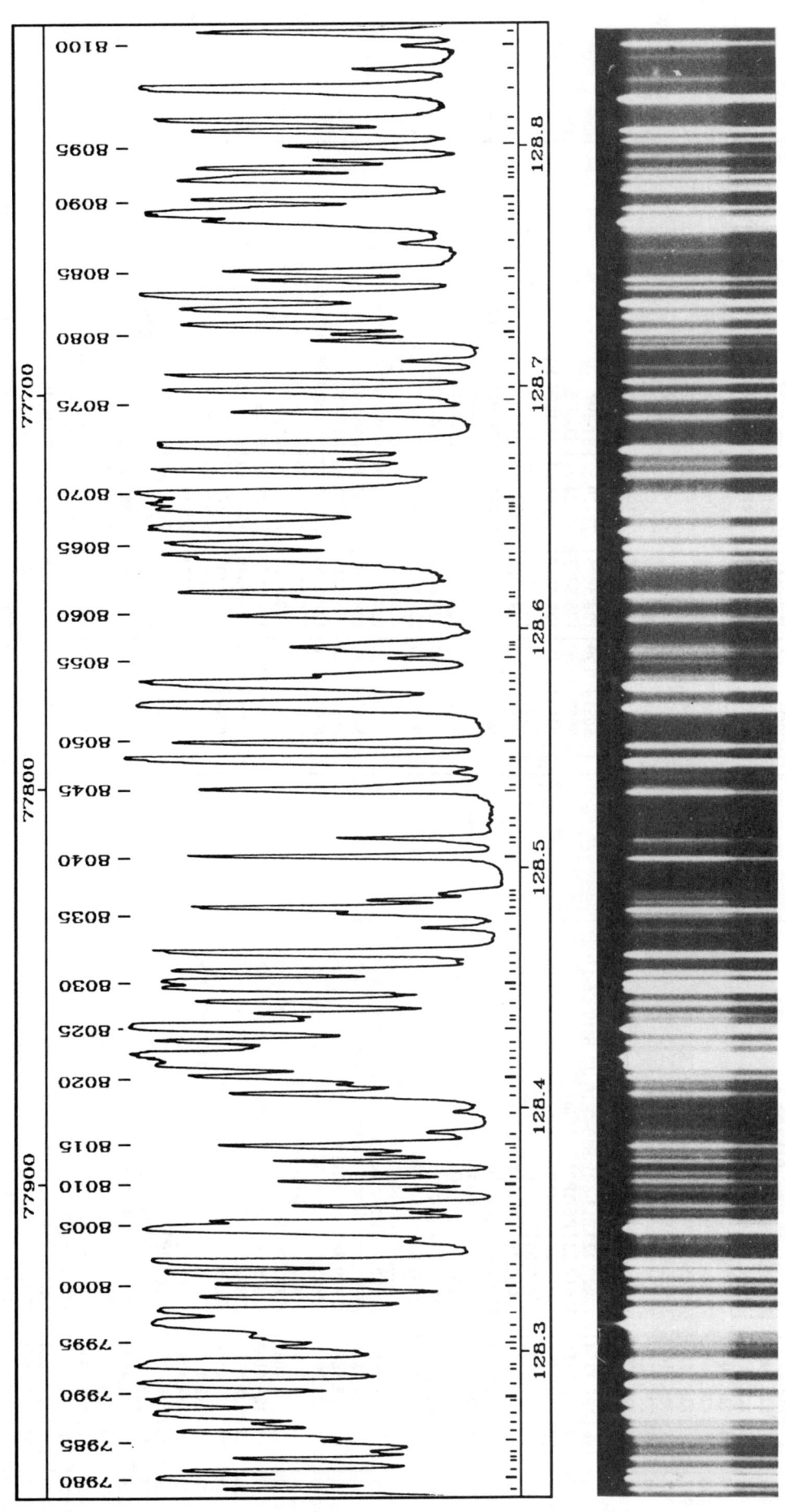

PLATE 88

TABLE 88 *continued*

Nb	I	λ (nm)	σ (cm-1)	Assignment	Comment
8089	11	128.7755	77654.50	B9-6R8 B4-2R18	sh
8090	20	128.7790	77652.41	B8-6R5	
8091	10	128.7871	77647.52	B9-6P7	
8092	12	128.7884	77646.73	B20-11R0 C0-6R0 C3-7P20	
8093	4	128.7920	77644.56	B4-4P8 C3-8P9	
8094	5	128.7954	77642.55	B22-12R0	
8095	18	128.8013	77638.99	C7-13R6 C5-10Q11	
8096	33	128.8077	77635.14	B22-12R1 C2-6P20	
8097	113	128.8119	77632.57	B20-11R1	
8098	3	128.8252	77624.57	B8-6P4 B24-13P5	
8099	1	128.8332	77619.72	B6-3P16 B6-0R26	
8100	14	128.8431	77613.79	C6-11P8	
8101		128.8485	77610.52	C5-10P7 C5-10Q12	

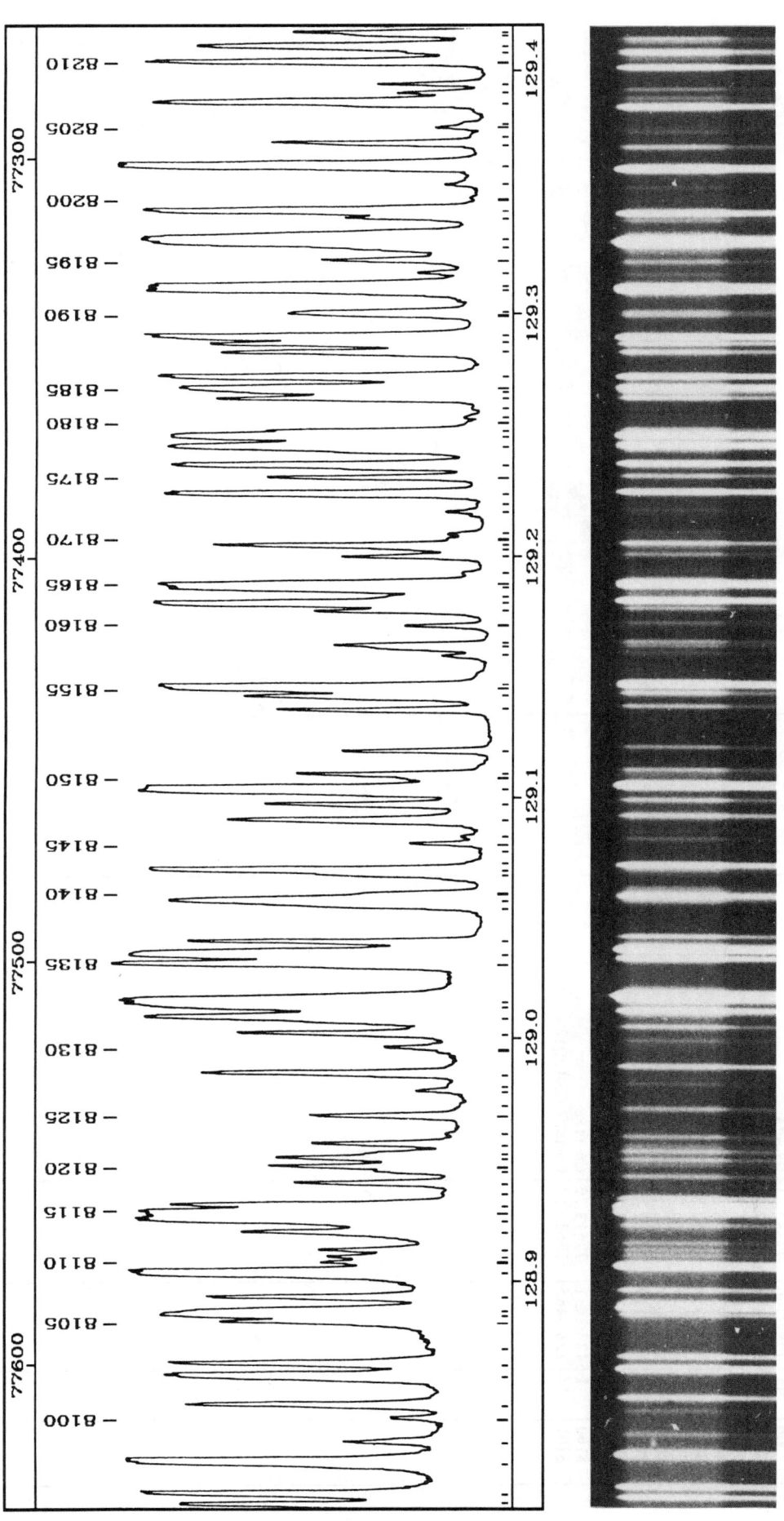

PLATE 89

TABLE 89

Nb	I	λ (nm)	σ (cm⁻¹)	Assignment	Comment
8102	58	128.8604	77603.38	B10-7R0 B22-12R2 C5-10Q13 B6-3R17	
8103	16	128.8655	77600.30	C0-6R5	
8104	0	128.8715	77596.66		
8105	8	128.8829	77589.79	C4-9Q11 C0-4P22	
8106	62	128.8856	77588.20	B2-4R0 B6-4R13	
8107		128.8871	77587.29	B20-11R2 C4-9P8 C3-8Q12 B23-12P7	sh
8108	14	128.8930	77583.77	B22-12P1	
8109	198	128.9030	77577.72	B10-7R1 B20-11P1	
8110	3	128.9072	77575.18	C1-6Q15 C1-5P21	
8111	2	128.9096	77573.77	B6-4P12 B13-7R11	
8112	2	128.9122	77572.16	B18-10R4	
8113	7	128.9197	77567.66	B8-4P15	
8114	220	128.9252	77564.36	B13-8R4 B13-8P3	
8115	260	128.9274	77563.05	B2-4R1 B18-10P3	
8116	14	128.9307	77561.07	B24-13P6 C0-6Q1	
8117		128.9352	77558.35	C0-6R6	
8118	6	128.9397	77555.63	B12-7R9	
8119	2	128.9446	77552.68	B12-7P8	
8120	6	128.9465	77551.55	B22-12R3 B8-4R16	
8121	6	128.9499	77549.51	B1-2R14 C6-11P9	
8122	1	128.9516	77548.46	B18-9R11	
8123	3	128.9559	77545.88	C2-7P11	
8124	5	128.9597	77543.61	C1-5Q25	
8125	5	128.9674	77539.00	B18-9P10 C7-13R5 C3-7P21	
8126	0	128.9714	77536.59		
8127	1	128.9780	77532.60	B15-6P17	
8128	0	128.9805	77531.09	B15-5R21	
8129	16	128.9852	77528.30	B5-5R4	
8130	2	128.9957	77521.98	B16-9R6	
8131	9	129.0015	77518.50	B22-12P2	
8132	28	129.0100	77514.36	B16-9P5 B19-10R8 C0-6Q2	
8133		129.0130	77511.58	B20-11R3 B1-1R18	sh
8134	720	129.0146	77510.59	B5-5P3 B10-5P14 B24-13P7	
8135	38	129.0304	77501.12	B10-7R2 C5-10P8	
8136	27	129.0343	77498.83	B10-7P1 B1-1P17 C4-9Q12 B15-8R10	b
8137	13	129.0396	77495.59	B20-11P2	
8138		129.0549	77486.38	C7-13Q7	sh
8139	25	129.0567	77485.31	B1-2P13	
8140		129.0593	77483.74	C2-7Q14	sh
8141		129.0675	77478.86	B7-5R10 B22-12R4	sh
8142	46	129.0695	77477.67	B2-4R2 B21-11R8	
8143	0	129.0722	77476.04	B16-8R12	
8144		129.0772	77473.00	B15-8P9	
8145	2	129.0807	77470.93	B17-9R9 C0-6R7	
8146	1	129.0835	77469.23	B16-7P15	
8147	12	129.0904	77465.12	B2-3R11	
8148	6	129.0969	77461.23	B7-5P9	
8149	260	129.1031	77457.49	B2-4P1 C3-8P10	
8150		129.1076	77454.75	B14-7R13 C2-6P21	sh
8151	9	129.1094	77453.69	C3-8Q13	
8152	5	129.1191	77447.88	C7-13R4 C0-6Q3	
8153	7	129.1362	77437.64	C4-9P9	
8154	11	129.1418	77434.25	C6-12R10 B3-2R17	
8155	24	129.1449	77432.39	B15-9R0	
8156	24	129.1460	77431.77	B22-12P3	
8157	2	129.1586	77424.21	B8-5R12 B13-5P18	
8158	3	129.1617	77422.32	B3-2P16 B20-10R11	
8159	7	129.1632	77421.44	C4-9Q13	
8160	3	129.1711	77416.67	C1-6P13	
8161	5	129.1774	77412.90	B18-10R5	
8162	82	129.1805	77411.07	B15-9R1	sh
8163	0	129.1834	77409.30	B20-11R4	
8164	30	129.1872	77407.06	B0-3R6	
8165	30	129.1878	77406.66	B3-4R7	
8166	0	129.1931	77403.54	B20-10P10	
8167	5	129.1999	77399.47	B21-11P7 C5-10P9	
8168		129.2032	77397.46	B2-3P10	
8169	11	129.2046	77396.64	B18-10P4	
8170	0	129.2070	77395.17	C0-6P2	
8171	1	129.2091	77393.94	B22-12R5	
8172	1	129.2183	77388.40	B14-8R8	
8173	0	129.2215	77386.48		
8174	45	129.2260	77383.82	B20-11P3 B13-5R19	
8175	7	129.2326	77379.86	B14-8P7	
8176	39	129.2378	77376.75	B10-7R3 C7-13Q6	
8177	40	129.2454	77372.19	B1-3R9	
8178	115	129.2495	77369.72	B10-7P2 C7-13R3	
8179		129.2514	77368.63	B11-7R7	sh
8180	0	129.2560	77365.83	C1-6Q16	
8181	0	129.2574	77365.03	B14-6P16	
8182	0	129.2613	77362.67		
8183	12	129.2653	77360.30	B11-7P6 C0-6Q4 C4-9Q14	
8184		129.2682	77358.52	B13-8R5	sh
8185	50	129.2695	77357.75	B3-4P6 B5-3R16	
8186	42	129.2742	77354.98	B13-8P4	
8187	13	129.2845	77348.79	B5-3P15 C1-5P22	
8188	16	129.2880	77346.68	B15-9R2	
8189	75	129.2909	77345.00	B15-9P1	
8190	6	129.3001	77339.48	B6-5R8	
8191	6	129.3009	77338.97	B8-6R6	
8192	170	129.3105	77333.27	B2-4R3	
8193	170	129.3116	77332.55	B8-6P5 B24-14R3 B11-4P19	
8194	2	129.3171	77329.33	C3-8Q14	
8195	5	129.3224	77326.10	B22-12P4 C6-12R9	
8196		129.3270	77323.37		
8197	600	129.3308	77321.11	B0-3P5 C2-7P12 C4-9Q15	
8198	4	129.3400	77315.63	C2-7Q15	
8199	46	129.3428	77313.93	B6-5P7 B24-14R2	sh

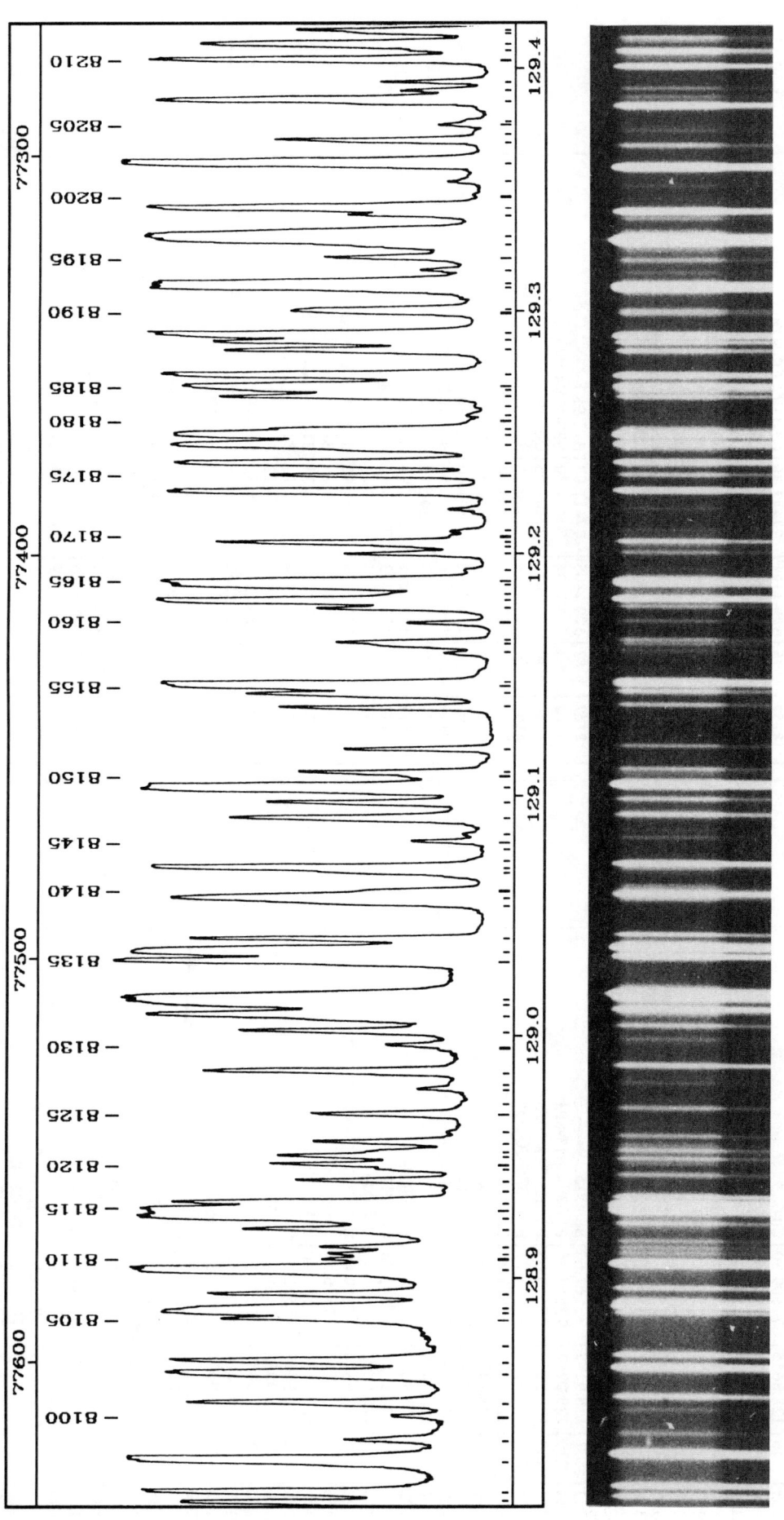

PLATE 89

TABLE 89 *continued*

Nb	I	λ (nm)	σ (cm-1)	Assignment	Comment
8200	0	129.3480	77310.85	B22-10P15	
8201	1	129.3536	77307.49	C7-13R2	
8202	146	129.3615	77302.77	B2-4P2	
8203	7	129.3705	77297.35	C7-13Q5 B24-14R1	
8204	0	129.3736	77295.50	B3-3P12	
8205	1	129.3768	77293.59	B11-3P22 C4-9P10	
8206	0	129.3781	77292.83	B21-11R9	
8207	26	129.3874	77287.26	B1-3P8	
8208	2	129.3911	77285.05	B12-5P17	
8209	2	129.3947	77282.94	B24-14R0 B20-11R5	
8210	23	129.4039	77277.42	B5-5R5 B11-4R20	
8211	0	129.4075	77275.29	B7-4P14 C5-10P10	
8212	13	129.4103	77273.63	B10-6R11 C5-10P11 C3-8P11 C0-6R9	
8213		129.4149	77270.86	B16-9R7	sh
8214	7	129.4159	77270.28	C0-6P3	

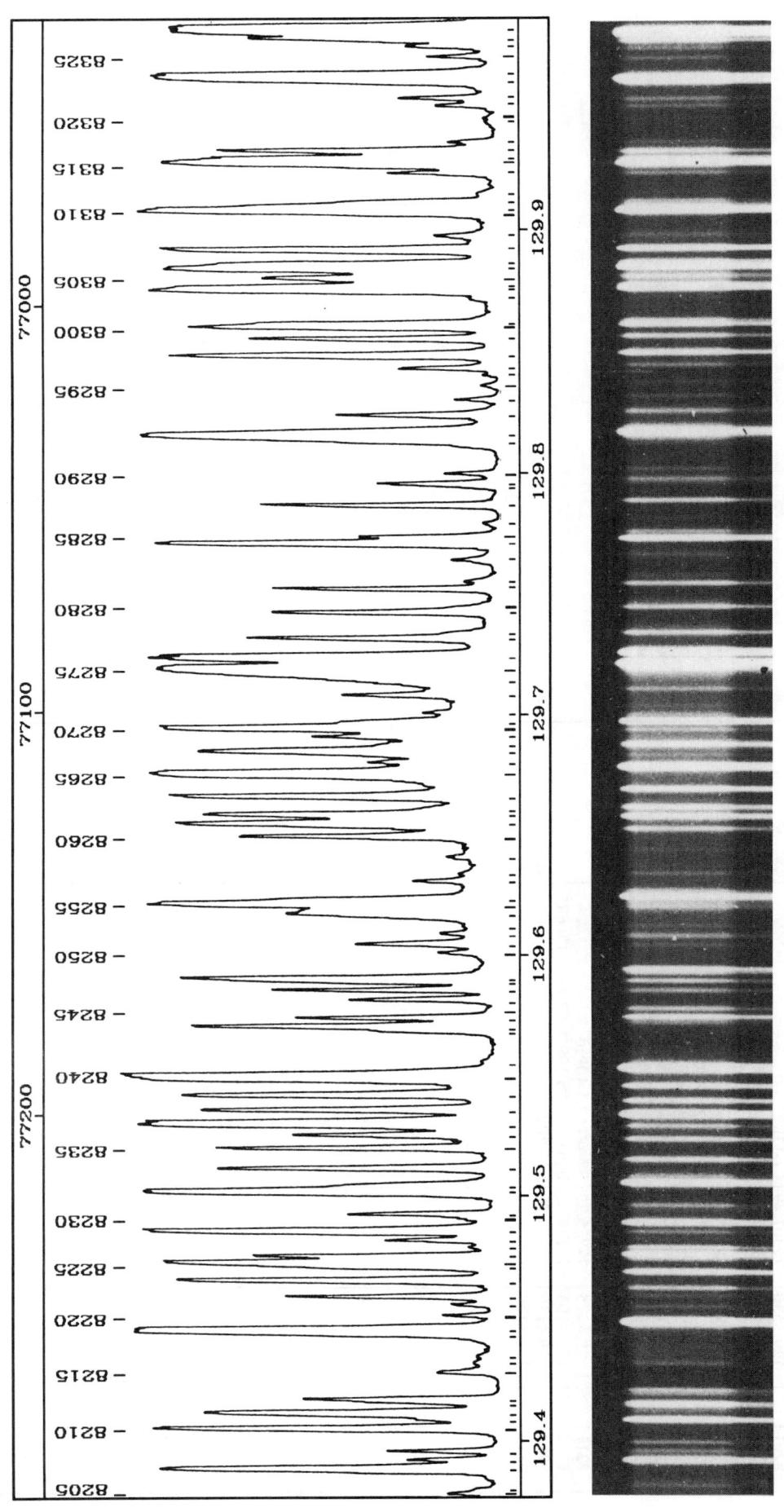

PLATE 90

TABLE 90

Nb	I	λ (nm)	σ (cm⁻¹)	Assignment	Comment
8215	1	129.4271	77263.60	B23-13R0 B19-10R9 B10-6P10	
8216	0	129.4310	77261.22	B0-1R17	
8217	0	129.4341	77259.42	B12-3P23 B18-6R22	
8218		129.4432	77253.96	B16-9P6	sh
8219	147	129.4442	77253.39	B5-5P4 C0-6Q5	
8220	1	129.4504	77249.69	B24-14P1	
8221	1	129.4549	77246.97	C7-13P7	
8222	9	129.4581	77245.10	B20-11P4	
8223	23	129.4647	77241.15	B15-9R3	
8224		129.4704	77237.72	C7-13Q4 C1-7R4	sh
8225	45	129.4719	77236.88	B15-9P2	
8226	10	129.4745	77235.31	B13-13R2	
8227	1	129.4782	77233.10	B24-14P2 B15-7P14	
8228	1	129.4811	77231.36	C6-12R8 C4-9P11	
8229	45	129.4849	77229.09	B7-6R0 C5-10P13	
8230		129.4900	77226.03	B18-10R6	sh
8231	3	129.4917	77225.03	B9-6P8	
8232	74	129.5009	77219.52	C7-13R1 B23-13R3 B24-14P3 B10-4P18	
8233		129.5031	77218.25	C3-8Q15 B0-1P16	sh
8234	14	129.5104	77213.90	B23-13P1 B9-6R9	
8235	12	129.5189	77208.82	B9-5R14 C1-7R1	
8236	6	129.5242	77205.64	B22-12P5	
8237	160	129.5291	77202.74	B7-6R1 B9-5P13 C5-10P12	
8238	20	129.5346	77199.45	B18-10P5	
8239	27	129.5409	77195.69	C7-13Q3	
8240	310	129.5483	77191.28	B10-7P3 B23-13R4	
8241	0	129.5547	77187.48	B21-11P8	
8242	2	129.5677	77179.72	B8-2P22 B10-4R19 C1-7R5	
8243	20	129.5695	77178.68	C7-13R0	
8244	7	129.5730	77176.59	B4-4R10	
8245	0	129.5765	77174.46		
8246	6	129.5806	77172.04	C7-13P6 B23-13P2	
8247	7	129.5847	77169.61	C1-7R0	
8248		129.5887	77167.24	B10-7R4	sh
8249	20	129.5895	77166.75	C7-13Q2 C1-6Q17	
8250	2	129.5997	77160.70	B7-2P21	
8251	3	129.6037	77158.32	B23-13R5 B13-6P15 B8-3P19 C2-7Q16	
8252	1	129.6078	77155.87	B9-2P23	
8253	0	129.6113	77153.79	B20-10R12	
8254	5	129.6162	77150.84	C3-8P12 B5-4R12 B2-2R16	
8255	67	129.6201	77148.53	B17-9R10 C7-13Q1	
8256		129.6216	77147.63	C2-7P13 B11-6R13	sh
8257	1	129.6295	77142.96	B12-7R10	
8258	0	129.6326	77141.09	C1-5P23	
8259	0	129.6397	77136.86	B20-11R6	
8260	10	129.6486	77131.57	B2-4R4	
8261	21	129.6539	77128.44	B4-4P9 B12-7P9 C0-6Q6	
8262	18	129.6551	77127.68	C0-6P4	
8263	0	129.6576	77126.22	B23-13P3 B23-13R6 B17-7P17 C7-13P5	
8264	30	129.6651	77121.74	C3-8Q16	
8265	190	129.6745	77116.15	B7-6R2	
8266	1	129.6794	77113.26	B13-8R6 B7-6P1	
8267	20	129.6839	77110.60	B5-4P11 B21-11R10	
8268	0	129.6868	77108.88	B2-2P15 B8-3R20	
8269	4	129.6901	77106.89	B13-7R12	
8270	57	129.6935	77104.89	B23-13R7 B11-5P16 C7-13P4	
8271	1	129.6958	77103.49	B13-8P5 B17-9P9	
8272	1	129.6998	77101.13	B15-8R11	
8273	3	129.7070	77096.85	B18-9P11	
8274	440	129.7162	77091.38	B15-9R4	
8275	123	129.7178	77090.42	B0-2R13	
8276	11	129.7224	77087.70	B2-4P3 C1-7Q1	
8277		129.7306	77082.84	B15-9P3 C7-13P2	sh
8278	0	129.7326	77081.60	B20-11P5	
8279	12	129.7413	77076.46	C6-12R7	
8280		129.7431	77075.36	B17-10R0 C4-9P12	sh
8281	9	129.7512	77070.57	B22-12P6 B15-8P10	
8282	0	129.7545	77068.64	C7-13P3	
8283	1	129.7632	77063.43	B5-1P22	
8284	40	129.7701	77059.37	B6-1P23	
8285	6	129.7726	77057.88	B17-10R1	
8286	0	129.7782	77054.56	B9-4P17	
8287	10	129.7858	77050.04	C3-8Q17	
8288		129.7937	77045.32	B16-8R13 C1-7Q2	sh
8289	2	129.7947	77044.77	B4-3R15	
8290	1	129.7947	77042.39	B14-8R9	
8291		129.8100	77035.65	B11-7R8 B9-2R24	sh
8292	61	129.8145	77033.00	B9-4R18	
8293	6	129.8229	77027.99	B0-3R7 C6-12Q10	
8294	1	129.8294	77024.17	B4-1P21	
8295	0	129.8352	77020.70	B14-8P8	CuII 129.83949
8296				B7-3P18	
8297	2	129.8424	77016.42	B18-10R7 B7-1P24 C2-7Q17	
8298	24	129.8477	77013.32	B12-8R0 B4-3P14	
8299	7	129.8547	77009.15	B8-6R7	
8300	16	129.8596	77006.22	B11-7P7	
8301	0	129.8608	77005.52	B17-10R2	
8302		129.8717	76999.09	B11-2P25	sh
8303	45	129.8749	76997.18	B17-10P1	
8304		129.8758	76996.64	B8-6P6 B16-9R8	sh
8305	5	129.8798	76994.27	B0-2P12 B7-3R19	
8306	66	129.8838	76991.92	B3-4R8 B12-8R1	
8307		129.8853	76991.03	C1-7Q3	sh
8308	34	129.8915	76987.32	B7-6R3 C0-6Q7	
8309	2	129.8970	76984.10	B5-1R23	
8310		129.9075	76977.83	B7-6P2 B20-11R7	sh
8311	113	129.9093	76976.78	B5-5R6	

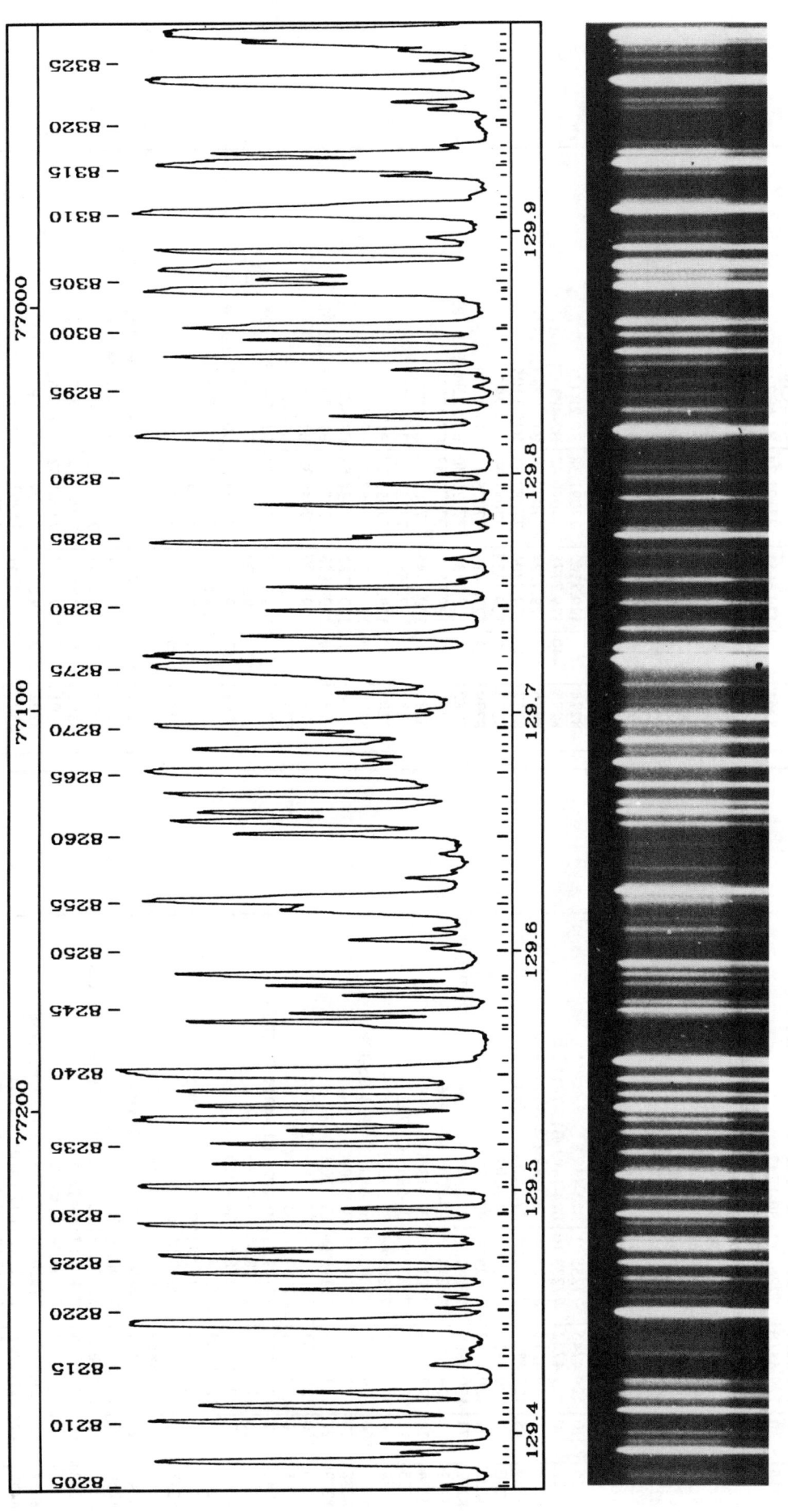

PLATE 90

TABLE 90 *continued*

Nb	I	λ (nm)	σ (cm⁻¹)	Assignment	Comment
8312	5	129.9108	76975.90	B18-10P6 B23-13P5	
8313	0	129.9148	76973.51	C6-12R5 B21-11P9	
8314	2	129.9230	76968.67	C3-8P13 B4-1R22	
8315	58	129.9273	76966.14	B10-7P4 B16-9P7 B3-0P23 C0-6P5	
8316		129.9286	76965.35	B10-7R5	sh
8317	10	129.9320	76963.34	B7-5R11	
8318	1	129.9356	76961.21	C1-6P15	
8319	0	129.9441	76956.19	B6-1R24 C1-5P24	d
8320	0	129.9459	76955.10	B14-7R14	d
8321	1	129.9508	76952.21	B5-2R20 C4-9P13	
8322	3	129.9538	76950.45	B21-12R0 B2-0P22	
8323		129.9614	76945.90	B21-12R1	sh
8324	194	129.9625	76945.26	B5-5P5 B6-4R14 C1-7R8	
8325	1	129.9708	76940.37	B22-12P7	
8326	2	129.9752	76937.75	B3-1P20 B4-0P24	
8327	10	129.9784	76935.88	B7-5P10 C1-7P2 C2-7P14	
8328	140	129.9818	76933.87	B0-3P6 B3-4P7	

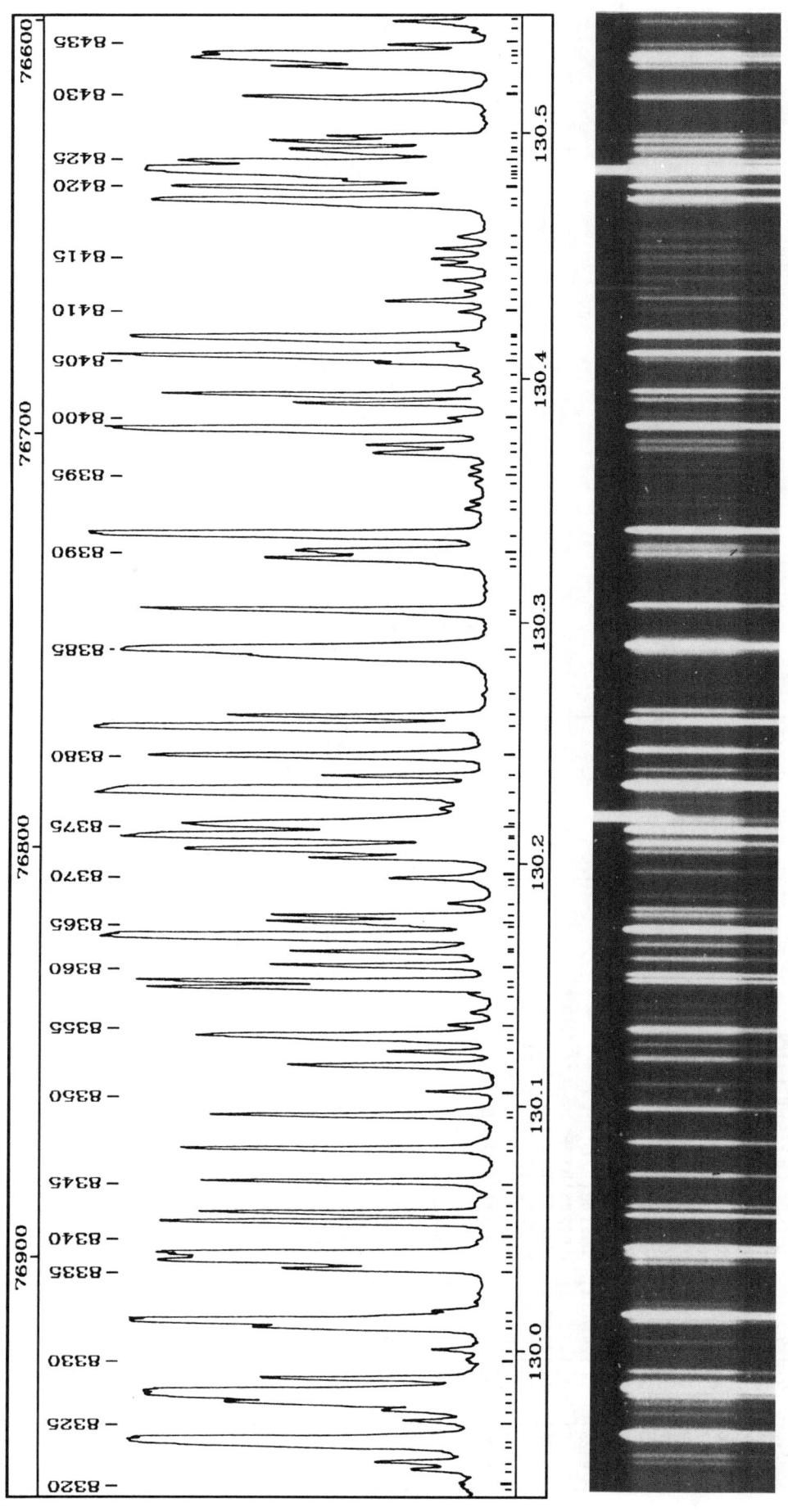

PLATE 91

TABLE 91

Nb	I	λ (nm)	σ (cm⁻¹)	Assignment	Comment
8329	9	129.9882	76930.05	B12-8R2 B16-7P16	
8330	0	129.9960	76925.41	B14-7P13 B17-8R15	
8331	2	130.0001	76923.03	B21-12R2 C6-12R4	
8332	6	130.0089	76917.80	B15-9R5 B6-4P13	
8333	108	130.0114	76916.35	B12-8P1 B17-10R3	
8334	2	130.0158	76913.74	C1-7Q4 C6-12Q9	
8335	8	130.0326	76903.82	B19-11R0	
8336	25	130.0359	76901.81	B17-10P2 B3-1R21	
8337		130.0380	76900.59	B20-11P6 B6-5R9 B15-9P4	sh
8338	28	130.0389	76900.07	B21-12R3	
8339	0	130.0442	76896.93	B20-10R13	
8340	0	130.0480	76894.67		
8341	20	130.0519	76892.40	B19-11R1	
8342	12	130.0557	76890.15	B21-12P1	
8343	0	130.0602	76887.49	B7-1R25	
8344	0	130.0655	76884.31	B12-6P14 C1-7R9	
8345	12	130.0683	76882.68	C2-8R3	
8346	11	130.0818	76874.72	B2-4R5	
8347	0	130.0847	76873.00	B3-0R24 B21-11R11	
8348	10	130.0957	76866.52	B6-5P8 B5-0P25 B23-13P7	
8349	0	130.0980	76865.14	C2-8R4	
8350	4	130.1058	76860.54	B10-5P15	
8351	7	130.1163	76854.34	B13-8R7 B2-3R12	
8352	4	130.1219	76851.04	B19-11R2	
8353		130.1280	76847.39	B8-5R13	sh
8354	12	130.1287	76847.00	B1-3R10	
8355	2	130.1330	76844.45	B8-4P16	
8356	1	130.1384	76841.25	B8-4R17	
8357	0	130.1454	76837.15	C2-8R5	
8358	21	130.1486	76835.28	C2-8R1	
8359	23	130.1512	76833.74	B19-11P1	
8360	9	130.1577	76829.86	B6-3P17 C6-12Q8 C1-5P25	
8361	5	130.1637	76826.37	B21-12P2 B8-5P12	
8362		130.1656	76825.22	B6-3R18	
8363	110	130.1701	76822.59	B2-4P4 C1-7P3	
8364	0	130.1746	76819.90	C1-7Q5 B4-0R25	
8365	8	130.1760	76819.07	C6-12R3	
8366	8	130.1784	76817.64	B13-8P6	
8367	2	130.1837	76814.56	B17-9R11 B4-2P18	
8368	2	130.1911	76810.14	B22-12P8	
8369	4	130.1938	76808.56	B4-2R19	
8370	0	130.1962	76807.16	B4-5R0	
8371	6	130.2024	76803.48	B7-6R4 B1-2R15	
8372	17	130.2061	76801.33	B12-8P2	
8373	55	130.2116	76798.09	C2-8R0	sh
8374	12	130.2122	76797.73	B17-10R4 C6-12R2 B2-1P19	
8375	12	130.2158	76795.61		
8376					OI 130.21685
8377	0	130.2231	76791.30	C0-6P6 B18-10R8	

Nb	I	λ (nm)	σ (cm⁻¹)	Assignment	Comment
8378	360	130.2303	76787.07	B7-6P3 B12-8R3 B2-1R20 B9-6R10	
8379	6	130.2367	76783.27	B19-11R3	
8380	22	130.2450	76778.41	B4-5R1 B2-3P11 B21-12R4	
8381	70	130.2570	76771.33	B17-10P3 C6-12Q7	
8382	9	130.2613	76768.80	B9-6P9	
8383	0	130.2752	76760.58	B21-11P10	
8384		130.2865	76753.95	B10-6R12 B17-9P10 C6-12R1	
8385	60	130.2886	76752.70	B1-3P9 B19-11P2 B14-6P17	
8386		130.3039	76743.65	B10-6P11	
8387	22	130.3052	76742.93	B21-12P3	
8388		130.3240	76731.86	B13-5P19 B18-6R23	sh
8389	7	130.3262	76730.55	B1-2P14 C6-12Q6	
8390	7	130.3295	76728.62	B18-10P7 B12-7R11	
8391	77	130.3361	76724.70	C2-8Q1 C2-8R7	
8392	0	130.3467	76718.47	B15-9R6	
8393	0	130.3497	76716.71	B23-14R4	
8394	0	130.3575	76712.12	B13-5R20 C1-7Q6 C5-11R8	
8395	0	130.3606	76710.27	C6-12R0	
8396	0	130.3641	76708.24	B22-12P9	
8397	3	130.3700	76704.79	B16-9R9 B20-11P7 C2-8R8	
8398	2	130.3732	76702.86	B10-7R6 C6-12Q5 B19-11R4	
8399	70	130.3796	76699.14	B10-7P5 B21-12R5 B15-8R12 C3-8P15	
8400	1	130.3842	76696.42	B4-5R2	
8401	7	130.3905	76692.71	C1-7P4	
8402	17	130.3940	76690.62	C2-8Q2	
8403	0	130.3965	76689.16	B14-8R10	
8404	0	130.4020	76685.95	C6-12Q4	
8405	2	130.4074	76682.75	B23-14R3	
8406	37	130.4100	76681.25	B4-5P1	
8407	0	130.4142	76678.78	B12-7P10	
8408	37	130.4175	76676.80	B15-9P5 C5-11R5	
8409		130.4205	76675.06	C6-12Q3	sh
8410	1	130.4284	76670.44	C6-12Q2	
8411	5	130.4330	76667.73	B23-14R1 C6-12Q1	
8412					OI 130.43711
8413	2	130.4413	76662.83	B15-7P15	
8414	3	130.4477	76659.09	C7-14Q3	
8415	3	130.4501	76657.65	B17-10R5	
8416	3	130.4544	76655.16	B23-14R2	
8417	1	130.4594	76652.18	B16-9P8 B15-8P11	
8418		130.4724	76644.58	B21-12P4	sh
8419	38	130.4740	76643.65	B19-11P3 B14-8P9	
8420	20	130.4794	76640.48	C2-8Q3 B23-14R0	
8421	2	130.4827	76638.50	B8-6R8 C1-7R11	
8422					OI 130.48576
8423	122	130.4858	76636.69	B8-6P7	
8424	122	130.4867	76636.16	B12-8P3	
8425	20	130.4897	76634.42	B12-8R4 B4-4R11	sh
8426	0	130.4944	76631.66	B13-7R13 C7-14P4	

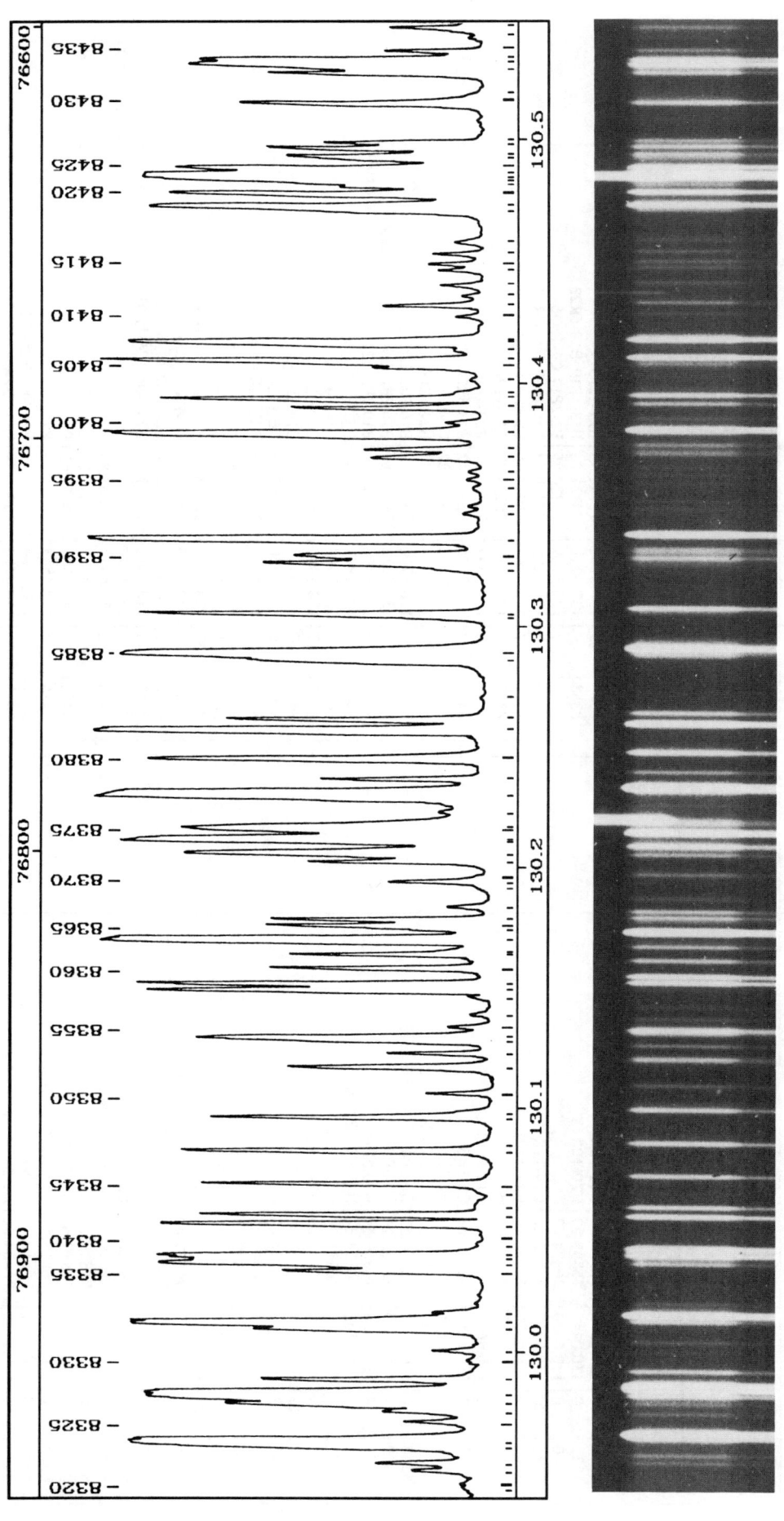

PLATE 91

TABLE 91 *continued*

Nb	I	λ (nm)	σ (cm⁻¹)	Assignment	Comment
8427	6	130.4950	76631.27	C3-9R3 C3-9R4	
8428	6	130.4984	76629.32	B5-5R7	sh
8429	5	130.5007	76627.94	B11-7R9	
8430	10	130.5165	76618.65	B11-7P8 C3-9R2	
8431		130.5181	76617.72	B1-1R19	
8432	9	130.5290	76611.35	C7-14R1	
8433	23	130.5324	76609.31	B0-3R8 B3-2R18 C4-10R9	
8434	16	130.5342	76608.25	B17-10P4 B3-3P13	
8435	5	130.5382	76605.92	B23-14P1 B21-12R6	
8436	0	130.5439	76602.56	B12-5P18 B11-6R14	
8437	4	130.5479	76600.23	B5-3R17	

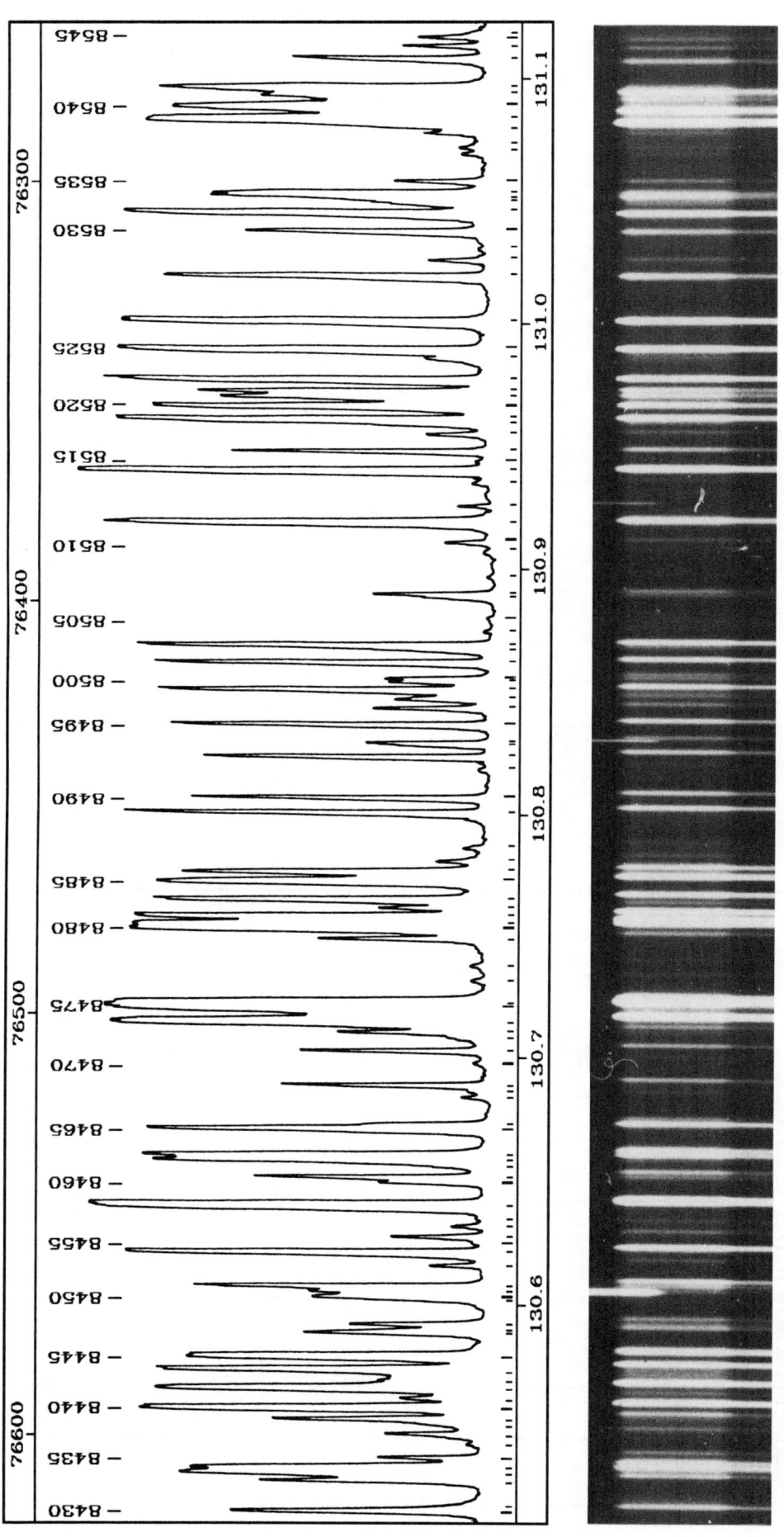

PLATE 92

TABLE 92

Nb	I	λ (nm)	σ (cm⁻¹)	Assignment	Comment
8438		130.5522	76597.68	B1-1P18 B17-7P18	sh
8439	6	130.5538	76596.75	C3-9R7 B22-13R1 C5-11R4	
8440	50	130.5583	76594.13	B22-13R0 B6-0R27 C3-9R1	
8441	1	130.5622	76591.85	B23-14P2 C1-7Q7	
8442	43	130.5663	76589.44	B5-5P6 B23-14P3 B3-2P17	
8443	0	130.5696	76587.53	B9-5R15 B22-13R2 B12-5R19	
				C7-14Q2 C6-12P2	
8444	27	130.5739	76585.01	B20-11R9 C0-6P7 C5-11R3	
8445	23	130.5790	76581.98	B9-7R0 C2-8P2 B5-3P16	
8446	5	130.5887	76576.29	B4-4P10	
8447		130.5896	76575.79	C2-7P16 C2-8Q4	sh
8448	5	130.5920	76574.35	B7-4P15	
8449					OI 130.60286
8450	3	130.6037	76567.51	B22-13R3	
8451	6	130.6056	76566.39	B7-6R5 C3-8P16	
8452	13	130.6075	76565.27	B2-4R6 B14-9R0	
8453	4	130.6162	76560.16	B23-14P4 B21-11P11	
8454	37	130.6214	76557.15	B9-7R1 C7-14R0	
8455	0	130.6259	76554.49	C5-11R2	
8456	4	130.6279	76553.35	B21-12P5	
8457	1	130.6322	76550.81	C6-12P3	
8458	0	130.6354	76548.91	C5-11Q12	
8459	77	130.6403	76546.03	B7-6P4 C1-7P5 B22-13P1	
8460	3	130.6501	76540.30	C7-14Q1 B22-13R4	
8461	7	130.6522	76539.11	B5-4R13	
8462		130.6570	76536.27	B13-6P16 C3-9R0 C4-10R4	sh
8463	25	130.6590	76535.09	B4-5P2	
8464	18	130.6607	76534.10	B3-4R9 B11-4R21	
8465	26	130.6717	76527.68	B14-9R1 B19-11R5	
8466	3	130.6735	76526.59	C4-10R3	
8467	0	130.6850	76519.89	C6-12P7	
8468	0	130.6874	76518.47	C6-12P4 C6-12P6 C1-6P17	
8469	8	130.6895	76517.24	C5-11R1	
8470	0	130.6986	76511.92	B22-13R5	
8471	6	130.7033	76509.14	B19-11P4	
8472	0	130.7089	76505.88	B21-12R7	
8473	4	130.7111	76504.59	C4-10R2 C3-8P17	
8474	124	130.7153	76502.14	B2-4P5 B13-8R8 B22-13P2	
8475	166	130.7215	76498.51	B0-3P7 B17-9R12 B20-11P8	
8476		130.7226	76497.85	B13-8P7 C2-8Q5	sh
8477	0	130.7326	76492.00	B5-4P12	
8478	0	130.7388	76488.39	B22-13R6	
8479	5	130.7499	76481.92	B9-7R2	
8480	73	130.7540	76479.51	C2-8P3 B18-10R9	
8481	79	130.7556	76478.57	C3-9Q1	
8482	51	130.7589	76476.65	B9-7P1	
8483	2	130.7625	76474.52	C5-11R0 C6-12P5	
8484	28	130.7659	76472.52	C4-10R1	
8485	39	130.7734	76468.16	B14-9P1 B14-9R2	
8486	15	130.7773	76465.84	B3-4P8 B18-10P8	b
8487	2	130.7823	76462.94	B10-4P19	
8488	1	130.7878	76459.73	B11-3P23 C5-11Q11	
8489	16	130.8021	76451.38	C3-9Q2	
8490	11	130.8082	76447.80	B22-13P3	
8491	0	130.8200	76440.90	B10-4R20	
8492	9	130.8251	76437.92	B12-8P4	
8493					CuII 130.82971
8494	4	130.8309	76434.54	B12-8R5	
8495	15	130.8378	76430.51	C4-10R0	
8496	4	130.8445	76426.59	B6-5R10	
8497	3	130.8483	76424.38	C1-7P6 B17-10R6 B15-9R7	b
8498	15	130.8518	76422.33	C5-11Q1 B15-9P6	
8499	3	130.8554	76420.24	B7-5R12	
8500	4	130.8566	76419.54	B11-5P17	
8501	15	130.8630	76415.79	B17-10P5	
8502		130.8682	76412.75	C5-11Q2 B20-11R10	sh
8503	22	130.8699	76411.74	C3-9Q3	
8504	0	130.8760	76408.21	C2-8Q6	
8505	0	130.8815	76404.99	B21-12R8 B17-8P15	
8506		130.8899	76400.08	C5-11Q3	sh
8507	3	130.8911	76399.38	B10-7R7	
8508	0	130.8987	76394.98	B19-11R6	
8509	0	130.9089	76388.98	B17-9P11	
8510	2	130.9124	76386.94	B22-13P4	
8511	25	130.9204	76382.27	B0-2R14 B6-5P9 B7-5P11	
8512					SiII 130.92766
8513	0	130.9371	76372.55	B21-12P6	
8514	60	130.9419	76369.77	C4-10Q1 B14-9P2 C5-11Q9	
8515					CuII 130.94633
8516	7	130.9499	76365.08	B14-9R3	
8517	2	130.9572	76360.82	C3-9Q4 C6-13R7	
8518		130.9612	76358.50	B16-7P17	sh
8519	37	130.9629	76357.53	B9-7R3 C2-8P4	
8520	23	130.9682	76354.45	B10-7P6 B19-11P5	
8521	12	130.9721	76352.13	B2-2R17	
8522	11	130.9744	76350.82	C4-10Q2	
8523	36	130.9790	76348.11	B9-7P2	
8524	1	130.9880	76342.87	B0-1P17	
8525	55	130.9913	76340.98	B16-9R10 C3-9P2	
8526	71	131.0028	76334.23	B4-5P3	
8527	17	131.0213	76323.47	B22-13P5 C4-10Q3 C5-11P2 C1-7Q9	
8528	2	131.0283	76319.41	B16-9P9	
8529	0	131.0335	76316.34	B21-12R9	
8530	8	131.0396	76312.81	B9-6R11 B9-4P18	
8531	49	131.0473	76308.35	B1-4R0 C2-8Q7	
8532	1	131.0519	76305.65	B9-4R19	
8533	45	131.0543	76304.23	B8-3R21 B2-2P16	
8534		131.0550	76303.83		sh

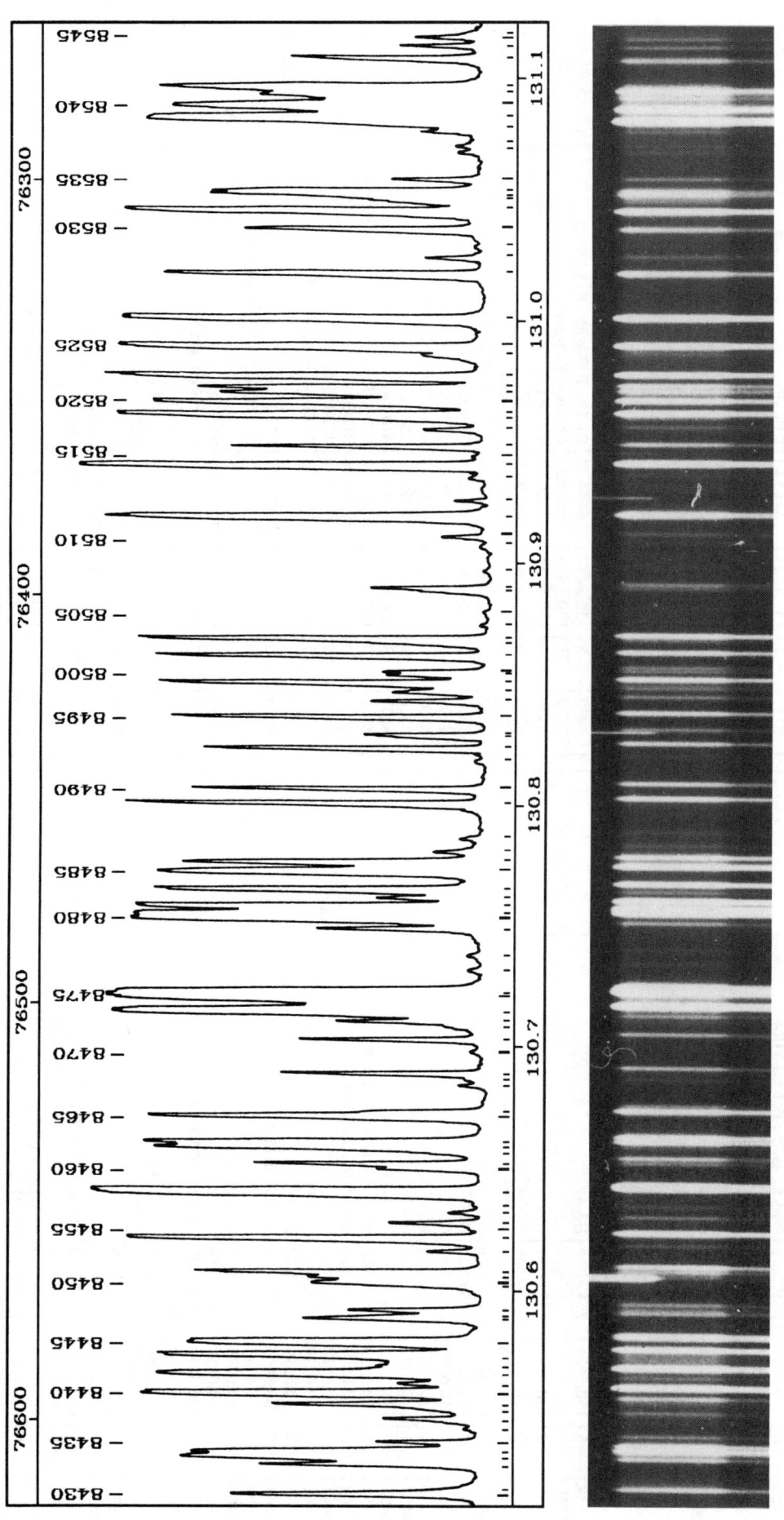

PLATE 92

TABLE 92 *continued*

Nb	I	λ (nm)	σ (cm⁻¹)	Assignment	Comment
8535	2	131.0608	76300.46	C3-9Q5	
8536	1	131.0721	76293.91	B20-11P9	
8537	1	131.0745	76292.50	B12-6P15	
8538	1	131.0809	76288.80	C4-10Q4	
8539	200	131.0850	76286.37	B1-4R1 B7-6R6	
8540	31	131.0900	76283.48	B1-3R11 B9-6P10	
8541	8	131.0947	76280.76	B6-4R15	
8542	31	131.0975	76279.09	B0-2P13 B8-2P23	b
8543	10	131.1102	76271.71	B4-3P15	
8544	2	131.1157	76268.50	C5-11P3	
8545	2	131.1193	76266.41	B14-8R11	
8546	0	131.1214	76265.23	B22-13P6 B7-2P22	

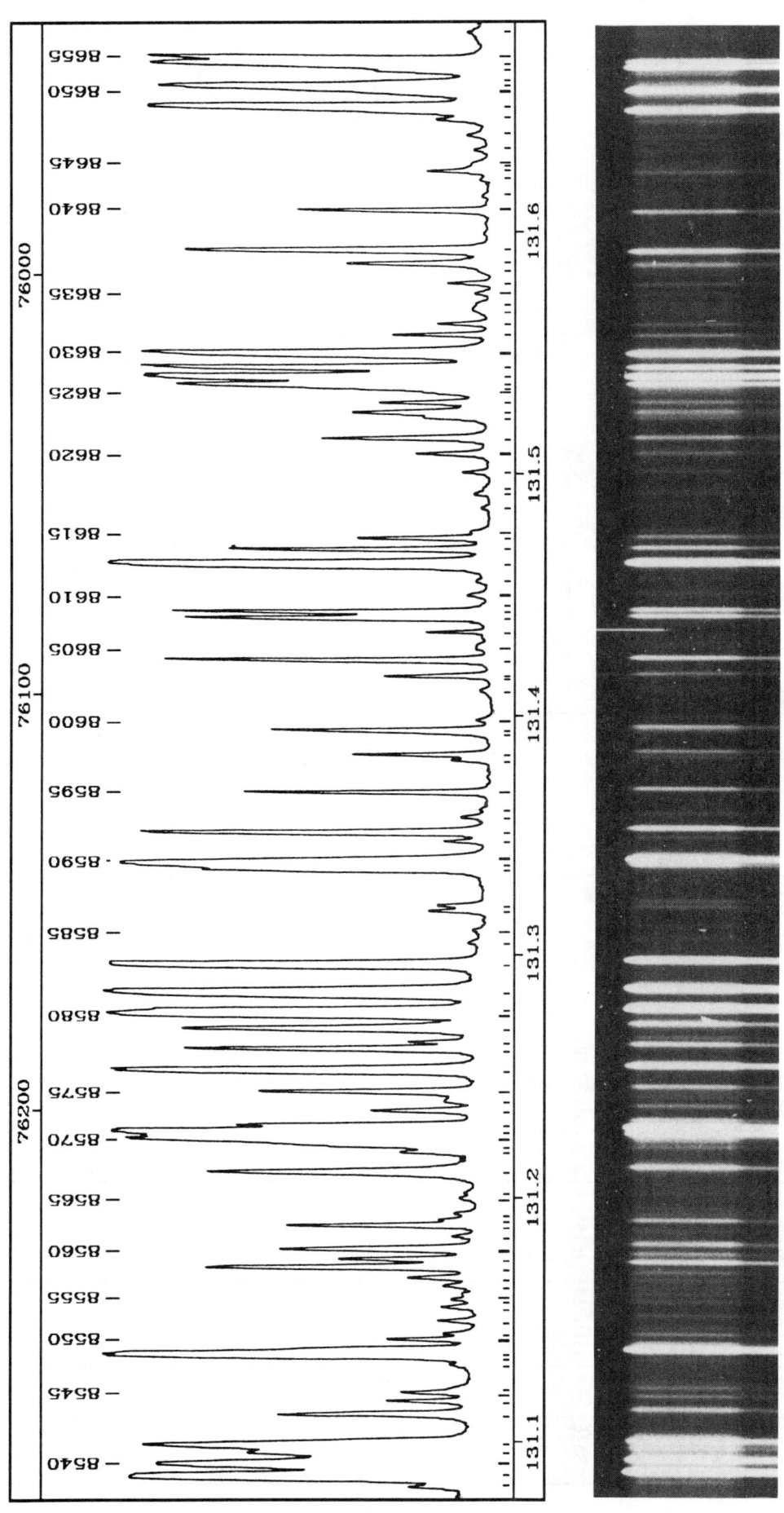

PLATE 93

TABLE 93

Nb	I	λ (nm)	σ (cm⁻¹)	Assignment	Comment
8547	0	131.1306	76259.86	C2-8P5	
8548	86	131.1349	76257.38	B7-6P5 B18-10R10	
8549		131.1369	76256.18	C4-10P2 B9-2P24	sh
8550	2	131.1411	76253.74	B16-10R0	
8551	0	131.1435	76252.35	B8-5R14 B21-12R10	
8552	1	131.1487	76249.33	C4-10Q5	
8553	1	131.1545	76245.99	B21-12P7	
8554	0	131.1577	76244.12	B6-4P14 B15-8R13	
8555	0	131.1601	76242.72	B20-11R11	
8556	0	131.1633	76240.82	B14-8P10 C4-10Q15	
8557	1	131.1666	76238.92	B5-5R8 B19-11R7	
8558	15	131.1708	76236.49	B16-10R1 B12-7R12	
8559	4	131.1740	76234.59	B10-6R13	
8560	6	131.1782	76232.18	B8-6R9 B20-12R0 C3-9Q6	
8561	0	131.1836	76229.04	B17-10R7	
8562	7	131.1880	76226.52	B20-12R1	
8563	0	131.1902	76225.22	B11-7R10	
8564	0	131.1928	76223.67	B14-9R4	
8565	0	131.1991	76220.03	B22-13P7	
8566	0	131.2013	76218.78	B8-5P13	
8567	10	131.2097	76213.89	B2-3R13	
8568	2	131.2182	76208.95	B7-3P19	
8569	55	131.2213	76207.12	B2-4R7 B12-7P11 C4-10Q6	
8570	165	131.2236	76205.81	B1-4R2	
8571		131.2260	76204.41	B14-9P3 B10-6P12 C2-8Q8	sh
8572	7	131.2286	76202.91	B11-7P9	
8573	4	131.2348	76199.31	B17-10P6 B20-12R2	
8574	1	131.2388	76197.00	B12-8R6 B10-5P16 B7-3R20	
8575	5	131.2427	76194.73	B8-6P8 B19-11P6	
8576	39	131.2515	76189.62	B5-5P7 B18-10P9	
8577	13	131.2603	76184.52	C4-10P3 B9-7R4	
8578	3	131.2629	76183.00	B16-10R2	
8579	13	131.2686	76179.68	B1-3P10	
8580	14	131.2749	76176.05	B1-4P1	
8581		131.2761	76175.35	B16-10P1	sh
8582	108	131.2838	76170.85	B9-7P3 B20-12P1 B13-8R9 C3-9P4	
8583	84	131.2954	76164.14	B12-8P5 B15-9R8 C4-10Q7	
8584	0	131.3016	76158.99	C3-9Q7	
8585	0	131.3096	76155.87	C5-11P5	
8586	4	131.3180	76151.04	B20-12R3 B15-7P16	
8587	2	131.3202	76149.77	B13-8P8	
8588		131.3356	76140.82	B15-9P7	sh
8589	42	131.3377	76139.62	B0-3R9	
8590	0	131.3398	76138.38	C4-10Q14	
8591	2	131.3470	76134.20	B18-11R0	
8592	19	131.3509	76131.94	B2-4P6 B14-6P18	
8593	1	131.3569	76128.45	B2-3P12	
8594	0	131.3601	76126.65	B13-7R14	
8595	9	131.3673	76122.47	B18-11R1 B20-11P10	

Nb	I	λ (nm)	σ (cm⁻¹)	Assignment	Comment
8596	2	131.3805	76114.81	B6-6R0	
8597	7	131.3829	76113.42	B8-4P17	
8598		131.3919	76108.17	B5-1P23	sh
8599	8	131.3930	76107.53	B20-12P2 C4-10P4	
8600	0	131.3971	76105.20	B17-9R13	
8601	0	131.4093	76098.11	C5-11P7 B6-1P24	
8602					CuII 131.41495
8603	5	131.4154	76094.58	B16-10R3 C2-8Q9 C5-11P11	
8604	13	131.4219	76090.81	B6-6R1 B20-11R12	
8605	0	131.4276	76087.50	B5-2R21 C5-11P6	b
8606					CuII 131.43366
8607	16	131.4394	76080.71	B4-5P4 B18-11R2	
8608	14	131.4420	76079.21	B16-10P2	
8609	0	131.4448	76077.56	C4-10Q13	
8610	1	131.4488	76075.25	B4-1P22	
8611	0	131.4547	76071.81	B17-7P19	
8612	95	131.4618	76067.76	B1-4R3 C3-9P5	
8613	11	131.4679	76064.20	B18-11P1	
8614	6	131.4725	76061.55	B4-4R12	
8615	0	131.4750	76060.08	B10-7R8	
8616	0	131.4855	76054.02	B13-5R21	
8617	0	131.4907	76051.03	B11-8R0 C6-13Q7	
8618	0	131.4926	76049.93	B7-1P25	
8619	0	131.4996	76045.84	B14-9R5	
8620	6	131.5072	76041.49	B6-3R19	
8621	7	131.5137	76037.73	B3-4R10 B5-1R24	
8622	1	131.5222	76032.79	B6-3P18	
8623	4	131.5242	76031.66	B1-2R16 B11-6R15	
8624	3	131.5283	76029.28	B11-8R1 B16-9R11	
8625	0	131.5326	76026.79	B4-1R23	
8626		131.5360	76025.66	C4-10P5	
8627	21	131.5391	76024.83	B10-7P7	sh
8628	95	131.5429	76023.01	B1-4P2 B20-12P3	
8629	27	131.5487	76020.80	B14-9P4	
8630	36	131.5562	76017.50	B0-3P8	
8631	5	131.5608	76013.13	B6-6R2	
8632	2	131.5663	76010.49	B18-11R3	
8633	0	131.5676	76007.32	B6-1R25	
8634	0	131.5693	76006.54	B17-10R8	
8635	0	131.5734	76005.60	C3-9Q9 B20-12R5	
8636	3	131.5776	76003.23	B18-10R11	
8637	0	131.5858	76000.76	B6-6P1	
8638	7	131.5915	75996.05	B3-1P21	
8639	15	131.5992	75992.79	B4-4P11	
8640	9	131.6080	75983.23	B18-11P2	
8641	0	131.6140	75979.74	B20-11R13	
8642	0	131.6209	75975.76	C2-8P7	
8643	2	131.6245	75973.73	B16-10R4 B17-10P7	
8644	0	131.6263	75972.64	B16-9P10	

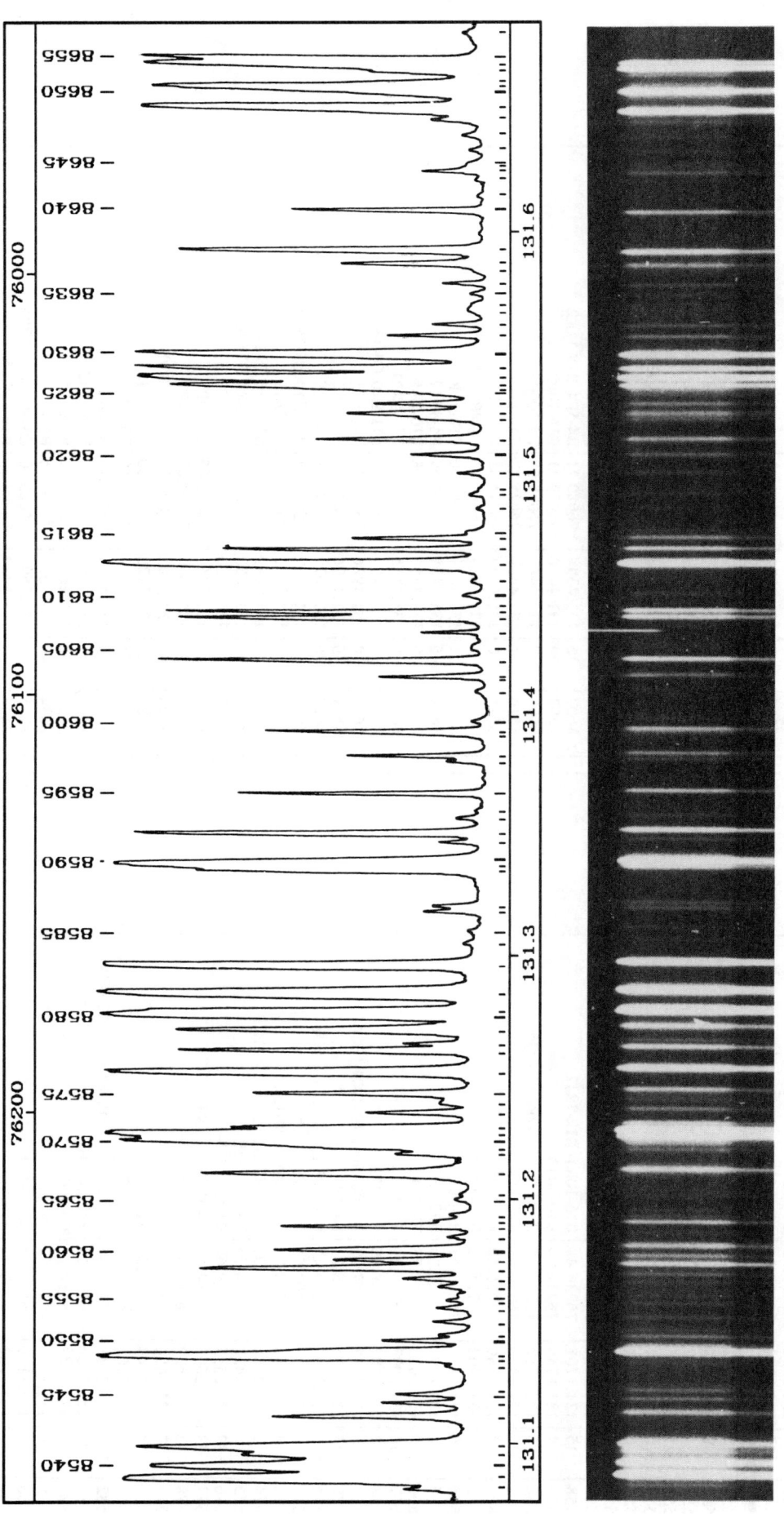

PLATE 93

TABLE 93 *continued*

Nb	I	λ (nm)	σ (cm-1)	Assignment	Comment
8645	0	131.6277	75971.83	B3-1R22	
8646	0	131.6334	75968.59	B9-7R5	
8647	0	131.6393	75965.18	B3-3R15 B17-8P16	
8648	1	131.6458	75961.41	B7-6R7 B11-8R2	
8649	33	131.6503	75958.83	B3-4P9 B4-2R20	
8650		131.6563	75955.33	B9-5R16 C3-9P6	sh
8651	30	131.6586	75954.02	B1-2P15 B11-8P1	
8652		131.6609	75952.68	B4-2P19	sh
8653	3	131.6656	75949.96	B19-11P7	
8654	52	131.6682	75948.50	B16-10P3	
8655	24	131.6704	75947.20	B9-7P4	
8656	0	131.6822	75940.44	C4-10P6 B3-0P24	

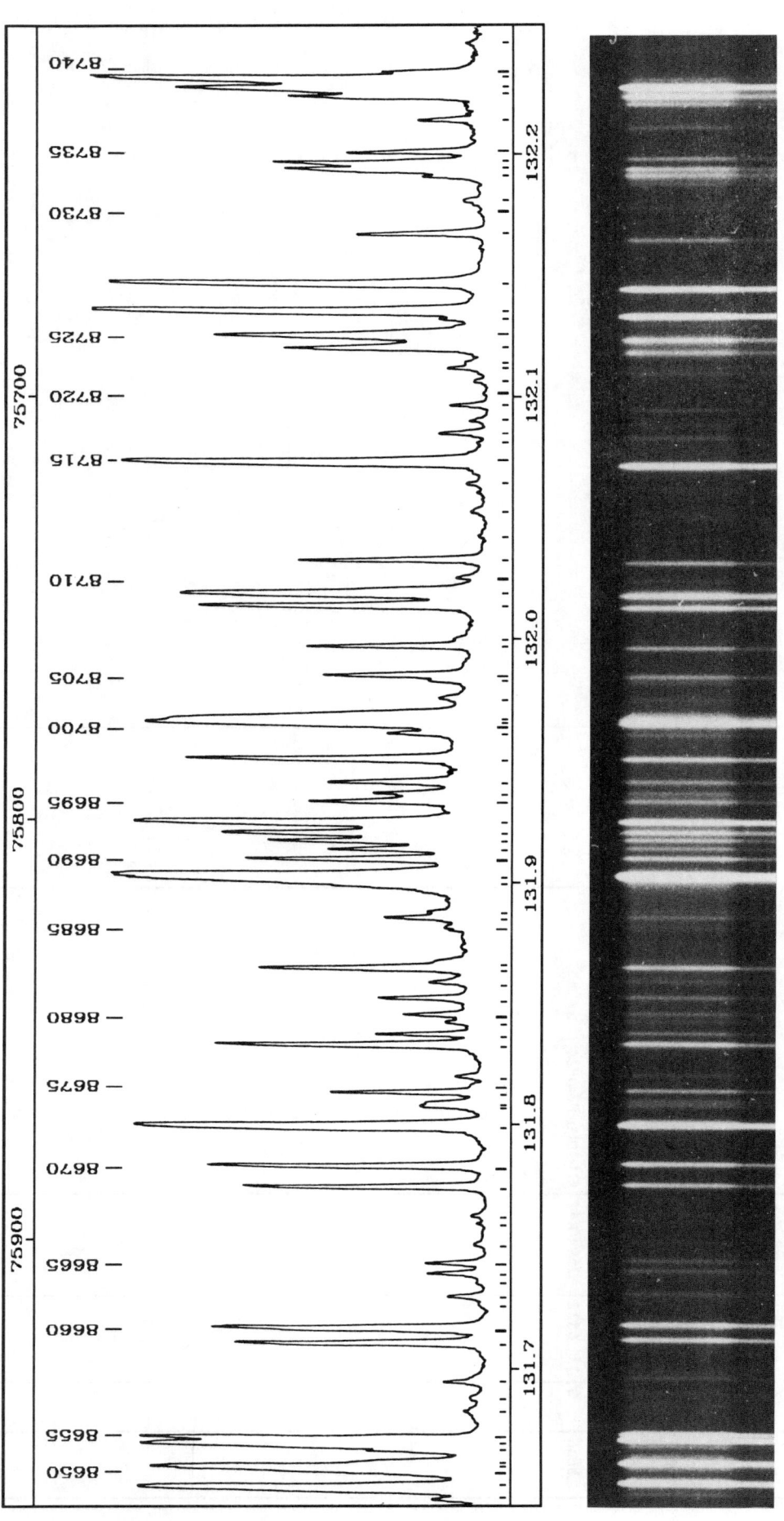

PLATE 94

TABLE 94

Nb	I	λ (nm)	σ (cm⁻¹)	Assignment	Comment
8657	0	131.6873	75937.45	B22-14P1	
8658	1	131.6938	75933.71	B2-0P23 B13-6P17	
8659	12	131.7095	75924.67	B7-6P6 B9-5P15	
8660	15	131.7157	75921.08	B12-8R7 B6-5R11 B20-12P4	
8661	0	131.7244	75916.07	B20-12R6 B21-13R3	
8662	2	131.7284	75913.78	B18-11R4 B12-5P19	
8663	0	131.7338	75910.63	B18-10P10	
8664	3	131.7380	75908.24	B21-13R1 B4-0P25	
8665	3	131.7421	75905.90	B5-4R14 B21-13R2	
8666	0	131.7497	75901.52	B21-13R0	
8667	0	131.7582	75896.60	B3-3P14	
8668	0	131.7612	75894.89	B19-11R9	
8669	10	131.7732	75887.97	B12-8P6	
8670	10	131.7817	75883.09	B6-6R3	
8671	45	131.7979	75873.75	B1-4R4 B18-11P3 B15-9R9 B14-8R12	
8672	2	131.8062	75868.96	B2-1R21	
8673	2	131.8076	75868.18	B2-1P20	
8674	4	131.8121	75865.60	B6-5P10	
8675	0	131.8148	75864.02	B6-6P2 B16-7P18	
8676	1	131.8187	75861.79	B7-4P16 B3-0R25	
8677	11	131.8314	75854.45	B7-5R13	
8678	3	131.8358	75851.93	B21-13P1	
8679	2	131.8408	75849.08	B15-9P8	
8680	2	131.8441	75847.17	B5-4P13	
8681	3	131.8510	75843.19	B15-8R14 C4-10P7	
8682	1	131.8576	75839.39	C3-9P7 B21-13R4	
8683	8	131.8637	75835.91	B11-8P2 B14-9R6 C6-13R3	
8684	0	131.8660	75834.53	B5-3R18	
8685	1	131.8805	75826.24	B20-12R7 C2-8P8	
8686	1	131.8847	75823.81	B14-8P11	
8687	0	131.8871	75822.40	B16-10R5	
8688	213	131.9001	75814.95	B9-6R12 B11-4P21	sh
8689	7	131.9017	75814.03	B1-4P3	
8690	7	131.9091	75809.76	B5-5R9 B13-8R10 B21-13P2	
8691	2	131.9133	75807.35	C0-6P11	
8692	6	131.9170	75805.27	B20-12P5 B4-0R26	
8693	9	131.9201	75803.44	B7-5P12 B2-4R8 B5-3P17	
8694	32	131.9246	75800.87	B14-9P5 B11-4R22	
8695	4	131.9331	75796.02	B18-11R5 B11-7R11 B21-13R6	
8696	1	131.9364	75794.10	B8-6R10	
8697	3	131.9412	75791.33	B13-8P9	
8698	11	131.9508	75785.84	B16-10P4 B21-13R7	
8699	1	131.9610	75779.96	C6-13R2 B11-7P10	
8700	43	131.9657	75777.28	B4-5P5	
8701	53	131.9669	75776.58	B3-2R19	
8702		131.9691	75775.34		sh
8703	1	131.9754	75771.71	B17-9R14	
8704	0	131.9831	75767.31	B17-10R9 B20-12R8	
8705	3	131.9850	75766.17	B12-7R13	
8706	6	131.9972	75759.16	B21-13P3	
8707	0	132.0003	75757.42	B19-11P8	
8708	10	132.0141	75749.49	B5-5P8 B18-10R12 C4-10P15	
8709	17	132.0191	75746.63	B8-6P9 B3-2P18 B9-6P11	
8710	2	132.0253	75743.09	B11-5P18	
8711	8	132.0327	75738.80	B18-11P4	
8712	0	132.0425	75733.22		
8713	0	132.0527	75727.34	B6-0P27	
8714	0	132.0635	75721.14	B12-7P12	
8715	31	132.0731	75715.64	B2-4P7 C6-13R1 B1-1R20	
8716	0	132.0812	75711.01	B10-4P20	
8717	2	132.0847	75708.99	B9-7R6 B10-4R21 C3-9P8	
8718	0	132.0901	75705.90	B21-13P4	
8719	1	132.0963	75702.34	B6-6R4	
8720	0	132.1017	75699.25		
8721	0	132.1063	75696.61	B16-9R12	
8722	1	132.1117	75693.54	B10-6R14 B12-6P16	
8723	0	132.1139	75692.25	C6-13Q3	
8724	6	132.1204	75688.55	B10-7R9 B1-1P19	
8725	10	132.1256	75685.57	B1-3R12 C4-10P11 C4-10P14	
8726	0	132.1318	75681.99	C1-7P10	
8727	32	132.1355	75679.90	B9-7P5 B20-12P6	
8728	28	132.1469	75673.37	B11-8P3	
8729	3	132.1673	75661.66	B17-10P8 C6-13R0 C2-8P9	
8730	0	132.1766	75656.37	B18-11R6	
8731	0	132.1806	75654.08	C6-13Q2	
8732	1	132.1908	75648.23	B10-6P13	
8733	4	132.1941	75646.33	B0-2R15	
8734	4	132.1966	75644.93	B10-7P8 B16-10R6 C3-9P9	
8735	5	132.2010	75642.40	B8-5R15 B11-3P24	
8736	3	132.2139	75634.99	B13-7R15	
8737	5	132.2235	75629.51	C6-13Q1	
8738	12	132.2268	75627.63	B0-3R10 B15-7P17	
8739	33	132.2304	75625.56	B1-4R5 C4-10P12	
8740	0	132.2333	75623.94	B16-9P11	
8741	0	132.2461	75616.60	B12-8R8	

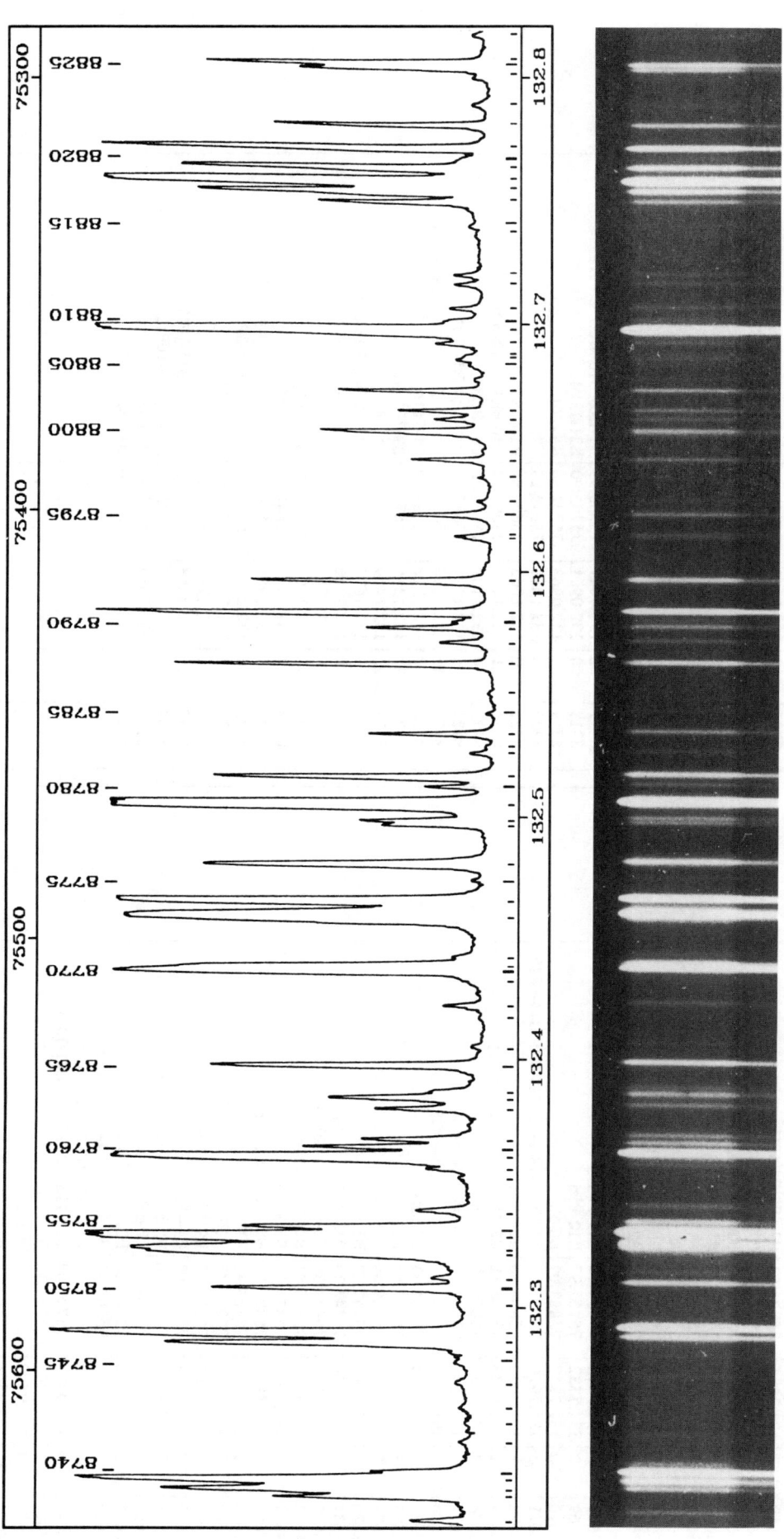

PLATE 95

TABLE 95

Nb	I	λ (nm)	σ (cm⁻¹)	Assignment	Comment
8742	0	132.2529	75612.72		
8743	0	132.2602	75608.54	B20-12R9 C4-10P13	
8744	0	132.2696	75603.16	B6-4R16	
8745	0	132.2777	75598.53	B7-6R8	
8746	0	132.2818	75596.21		
8747	18	132.2853	75594.20	B16-10P5 B14-9R7	
8748	99	132.2892	75591.98	B3-5R0 C6-13P5	
8749	0	132.3002	75585.69		d
8750	11	132.3075	75581.50	B18-11P5	
8751	0	132.3115	75579.20	C6-13P4 B9-4R20	sh
8752		132.3225	75572.94	B12-8P7	
8753	39	132.3236	75572.30	B1-3P11	
8754	379	132.3285	75569.53	B3-5R1 B19-11R11 C6-13P2	
8755	7	132.3325	75567.24	C6-13P3 B9-4P19	
8756	1	132.3390	75563.51	B13-9P1 B21-13P6 B15-9R10	
8757	0	132.3520	75556.07	B18-10P11	
8758	0	132.3560	75553.80	B19-11P9 B6-4P15	
8759	60	132.3607	75551.12	B1-4P4 B7-6P7 B20-12P7 B20-12R10	
8760	4	132.3645	75548.98	B14-9P6	
8761	3	132.3675	75547.25	B2-3R14	
8762	3	132.3800	75540.12	B2-2R18	
8763	3	132.3845	75537.57	B0-2P14 B17-8P17	
8764	1	132.3871	75536.06	B14-6P19	
8765	16	132.3975	75530.15	B8-7R0 B10-5P17	
8766	0	132.4051	75525.77	B21-13P7	
8767	0	132.4161	75519.51	B17-10R10	
8768	2	132.4223	75516.01	B4-3P16	
8769	0	132.4327	75510.07	B0-1R19 C6-14R4	
8770	54	132.4345	75507.91	B8-7R1	
8771	15	132.4379	75507.09	B3-4R11	
8772	1	132.4420	75504.77	B18-11R7 B18-10R13	
8773	80	132.4593	75494.88	B0-3P9	
8774	92	132.4656	75491.32	B3-5R2	
8775	0	132.4740	75486.52	B19-12R1	
8776	17	132.4808	75482.63	B2-2P17	
8777	5	132.4972	75473.30	B6-6R5 B13-9R3	
8778	5	132.4989	75472.34	B15-9P9	
8779	296	132.5057	75468.45	B3-5P1 B11-8P4	
8780	3	132.5128	75464.40	B14-8R13 B11-6R16	
8781	17	132.5169	75462.10	B4-4R13	
8782	3	132.5261	75456.83	B13-9P2	
8783	3	132.5292	75455.04	B0-1P18	
8784	5	132.5343	75452.14	B2-3P13	
8785	1	132.5419	75447.86	B17-9R15	
8786	0	132.5497	75443.43	B16-10R7	
8787	13	132.5626	75436.04	B8-7R2	
8788	3	132.5717	75430.91	B13-8R11	
8789	5	132.5776	75427.54	B19-12P1 B4-5P6 B20-12P8	
8790	1	132.5802	75426.05	B15-8R15	
8791	28	132.5839	75423.96	B8-7P1	
8792	8	132.5969	75416.54	B3-4P10	
8793	1	132.6146	75406.47	B14-8P12 B8-4R19	
8794	2	132.6154	75406.03	B18-11P6	
8795	4	132.6239	75401.19	B17-10P9 B7-3R21	
8796	0	132.6296	75397.95	C5-12R5 B7-3P20	CuII 132.63954
8797					
8798	3	132.6464	75388.40	B6-5R12	
8799	0	132.6487	75387.09		
8800	10	132.6580	75381.81	B4-4P12	
8801	3	132.6630	75378.96	B8-4P18 B21-14R3 B13-5P21	
8802	3	132.6664	75377.04	B16-10P6 B7-2P23	
8803	5	132.6747	75372.35	B9-7P6	
8804	1	132.6794	75369.67	B9-2P25	
8805	0	132.6848	75366.61	B16-7P19	
8806	0	132.6876	75364.99	B19-12P2	
8807	0	132.6895	75363.89	B21-14R4	
8808	1	132.6941	75361.30	B16-9R13	
8809	139	132.6994	75358.32	B3-5R3	
8810	0	132.7032	75356.16	B2-4R9	
8811	3	132.7085	75353.10	B13-8P10 B19-11P10	
8812	2	132.7185	75347.44	B11-7R12 B18-11R8	
8813	1	132.7223	75345.30	B5-5R10	
8814	0	132.7384	75336.14	B13-9R4	
8815	0	132.7425	75333.81	B13-6P18	
8816	6	132.7522	75328.30	B8-6R11	
8817	11	132.7565	75325.88	B1-4R6 B15-10P1 B21-14R2	
8818	158	132.7606	75323.56	B3-5P2	
8819	12	132.7662	75320.39	B6-5P11	
8820	0	132.7692	75318.66	B9-5R17 B20-12P9	
8821	21	132.7743	75315.80	B8-7R3	
8822	9	132.7836	75310.52	B13-9P3	
8823	0	132.7915	75306.01		
8824	0	132.8025	75299.79	B21-14R1 C6-14R3	
8825	8	132.8063	75297.63	B9-6R13	
8826	12	132.8083	75296.51	B8-7P2	
8827	0	132.8189	75290.52	B10-7R10	+NI 132.79170 Ke

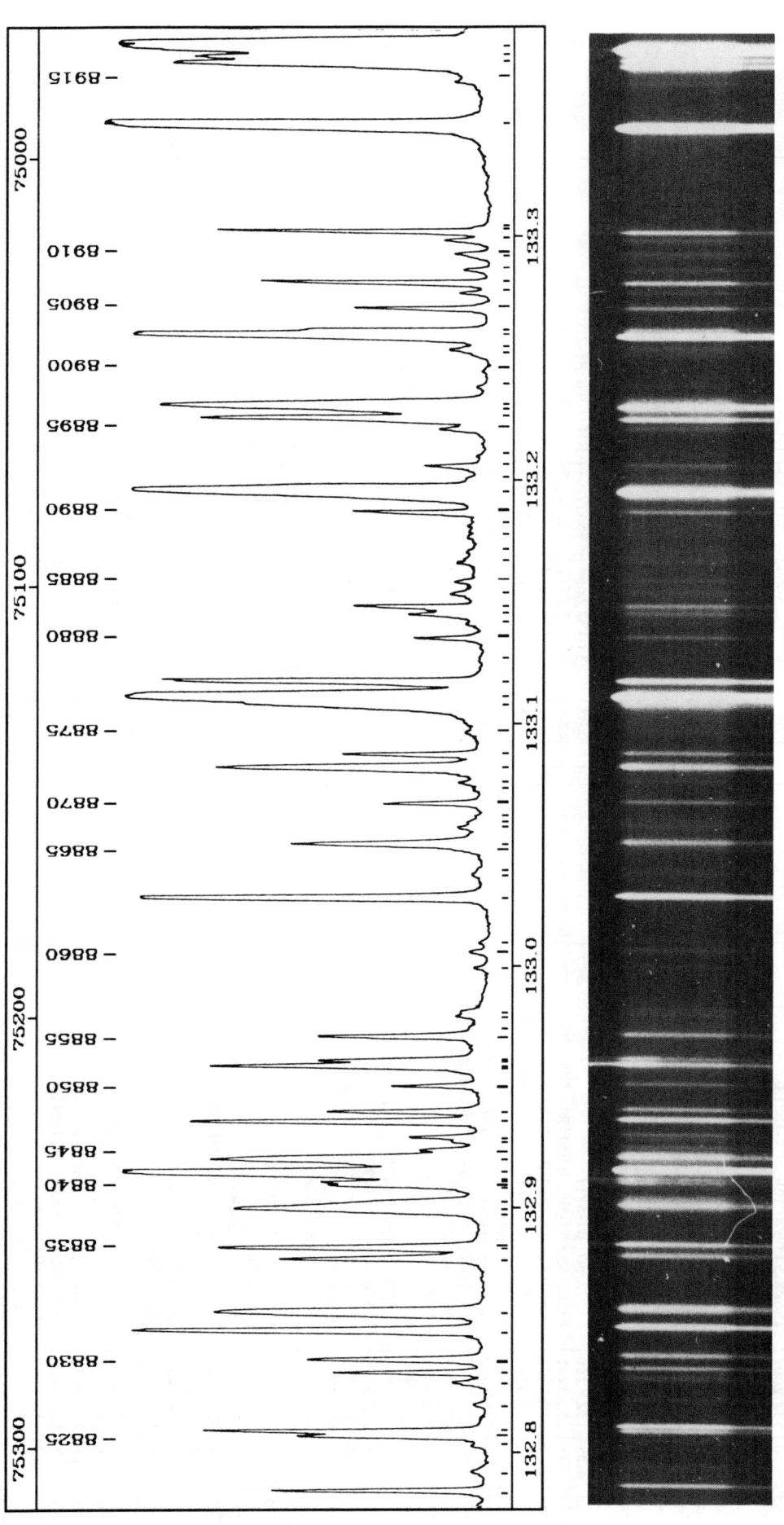

PLATE 96

TABLE 96

Nb	I	λ (nm)	σ (cm⁻¹)	Assignment	Comment
8828	2	132.8280	75285.36	B12-8R9 B12-7R14	
8829	6	132.8322	75282.95	B19-12P3	
8830	7	132.8374	75280.03	B11-7P11 B21-14R0	
8831	18	132.8494	75273.21	B5-5P9	
8832	11	132.8569	75268.96	B17-11P1 B14-9P7 B8-6P10 B7-5R14 B17-10R11 B9-5P16	
8833	10	132.8786	75256.67	B2-4P8 B6-3R20	+ CI 132.88333
8834	13	132.8832	75254.06	B5-4R15	CI 132.88333
8835					sh
8836		132.8994	75244.91	B1-2R17	
8837	12	132.9005	75244.28	B12-7P13	
8838	2	132.9019	75243.47	B15-9R11 B21-14P1	
8839	4	132.9085	75239.71	C4-11R9	+ CI 132.90853
8840					CI 132.91004
8841		132.9098	75239.00		CI 132.91004
8842					
8843	77	132.9140	75236.65	B1-4P5	
8844	12	132.9193	75233.65	B10-7P9 B9-6P12 B6-3P19	
8845	6	132.9230	75231.50	B15-10P2 B21-14P4	
8846	0	132.9265	75229.56	B21-14P2	
8847	1	132.9282	75228.59	B12-8P8	
8848	14	132.9348	75224.88	B11-8P5 B5-2R22 C5-12R3	
8849	6	132.9389	75222.55	B21-14P3	
8850	4	132.9495	75216.51	B18-11P7	
8851					CI 132.95775
8852					CI 132.95775
8853		132.9577	75211.90		CI 132.96005
8854					CI 132.96005
8855	4	132.9598	75210.73	B7-5P13	
8856	0	132.9699	75205.01		
8857	0	132.9735	75202.97	B7-6R9 C5-12Q7	
8858	0	132.9783	75200.25	B6-6R6	
8859	0	132.9799	75199.32	B17-11P2	
8860	0	132.9982	75188.97	B19-12P4	
8861	0	133.0050	75185.16	B5-4P14	
8862	26	133.0090	75182.89	B3-5R4	
8863	0	133.0279	75172.20	B3-3P15	
8864	0	133.0360	75167.62	C5-12R2	
8865	0	133.0389	75161.00	B19-11P11	
8866	6	133.0477	75159.48	B1-2P16	
8867	0	133.0504	75155.66	C6-14Q3 B5-1P24	
8868	0	133.0572	75154.49	C1-7P13	
8869	0	133.0592	75152.56		
8870	5	133.0627	75149.96	B8-7R4	
8871	0	133.0672	75146.55	B17-9R16	
8872	1	133.0733	75145.09	B6-1P25	
8873	11	133.0759	75141.61	B7-6P8 B15-9P10 B7-4P17 B10-6R15	
8874	3	133.0821	75138.43	B16-10P7 C5-12Q6	
8875	0	133.0877	75133.39	B20-13P1	
8876	416	133.1083	75126.80	B13-9P4 B4-1P23	sh
8877	26	133.1110	75125.27	B3-5P3 B17-10P10	
8878	0	133.1177	75121.51	B8-7P3	
8879	2	133.1276	75115.93	B15-7P18	
8880	0	133.1359	75111.21	B18-11R9 C5-12R1	
8881	3	133.1418	75107.91	C6-14R1	
8882	3	133.1457	75105.69	B20-13R7 B4-2R21	
8883	2	133.1488	75103.94	B15-10P3 C6-14P4	
8884	0	133.1546	75100.65	B5-1R25	
8885	0	133.1595	75097.91	B12-6P17	
8886	0	133.1676	75093.33	C5-12Q5 B4-1R24	
8887	0	133.1731	75090.24	B20-13P2	
8888	0	133.1782	75087.35	B4-2P20	
8889	4	133.1835	75084.38	B11-4R23	
8890	32	133.1880	75081.87	B17-11P3	
8891	0	133.1963	75077.19	B0-3R11 B19-12P5	
8892	4	133.2017	75074.09	C6-14Q2	
8893	0	133.2072	75071.02	B11-5P19 B6-1R26	
8894	2	133.2127	75067.90	B10-6P14	
8895	14	133.2226	75062.37	C5-12Q4 B5-3R19	
8896	26	133.2265	75060.16	B10-8R0 C5-12R0	
8897	0	133.2316	75057.28	B1-3R13	
8898	0	133.2342	75055.80	B3-1P22	
8899	0	133.2403	75052.35	B14-8R14	sh
8900	0	133.2482	75047.90	B17-10R12	
8901	2	133.2550	75044.06	B14-9R9 B18-10P13	
8902	62	133.2575	75042.71	B3-1R23	
8903	5	133.2610	75040.69	B10-8R1	
8904	4	133.2633	75039.43	C5-12Q3 B13-8R12 B20-13P3 C6-14R0	
8905	2	133.2727	75034.11	B4-5P7 B16-9R14	
8906	7	133.2793	75030.42	C6-14P3	
8907	2	133.2831	75028.26	B9-7P7	
8908	0	133.2887	75025.09	C5-12Q2	
8909	2	133.2941	75022.10	B8-5R16	sh
8910	3	133.2954	75021.34	C6-14Q1	
8911	8	133.3007	75018.35	B18-11P8 B5-3P18	
8912		133.3042	75016.40	C5-12Q1 B15-8R16	
8913					CuII 133.30452 Ke
8914	368	133.3475	74992.02	B0-4R0	
8915	0	133.3658	74981.72	B20-13P4	
8916	50	133.3729	74977.75	B1-4R7	
8917	48	133.3754	74976.34	B10-8R2 B18-11R10	
8918	1408	133.3797	74973.93	B0-4R1	

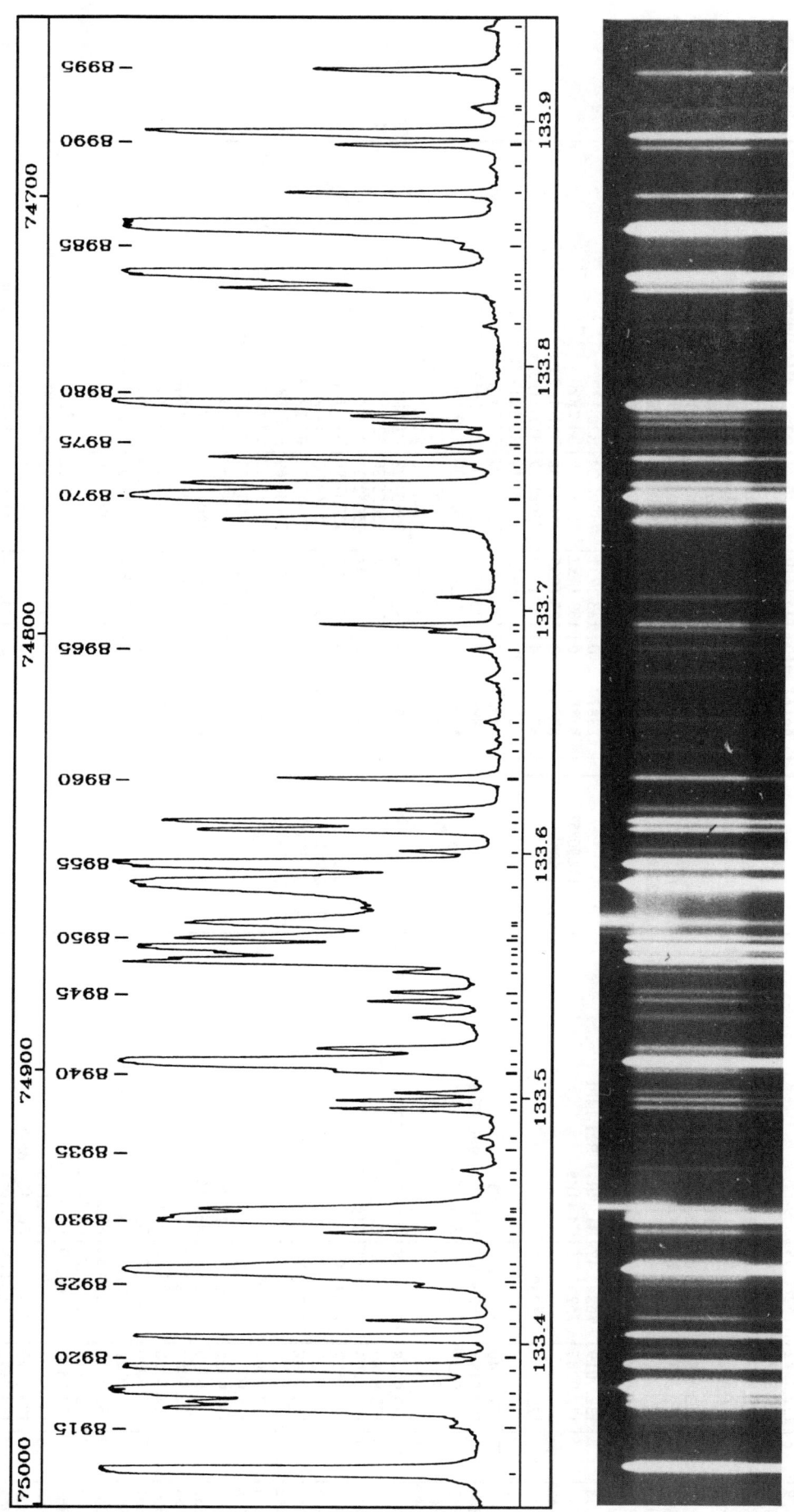

PLATE 97

TABLE 97

Nb	I	λ (nm)	σ (cm-1)	Assignment	Comment
8919	124	133.3893	74968.54	B5-6R0 B10-4P21	
8920	0	133.3945	74965.61	B14-9P8	
8921	43	133.4012	74961.84	B10-8P1	
8922	5	133.4081	74957.99	B8-7R5	
8923	0	133.4146	74954.31	B14-6P20	
8924	1	133.4220	74950.17	B17-11P4 B2-1R22	
8925		133.4250	74948.45	B13-8P11	sh
8926	448	133.4280	74946.81	B5-6R1	
8927		133.4301	74945.59	B3-4R12 B11-8P6 B15-10P4	sh
8928	10	133.4436	74938.05	B2-1P21 B3-2R20	
8929	50	133.4485	74935.23	B3-5R5 B14-8P13 C5-12P2	
8930	38	133.4501	74934.39	B0-3P10 B1-3P12	+ CII 133.45323
8931	16	133.4529	74932.80	B12-8R10	CII 133.45323
8932					
8933	0	133.4668	74925.02	B3-0P25	
8934	1	133.4691	74923.69	B20-13P5	
8935	0	133.4782	74918.61	B15-9R12	
8936	1	133.4829	74915.96	B6-4R17	
8937	6	133.4943	74909.56	B13-9P5	
8938	7	133.4977	74907.66	C5-12P3	
8939	2	133.5007	74905.95	C0-7R5	
8940	6	133.5098	74900.85	B8-7P4	
8941	320	133.5131	74898.99	B0-4R2	
8942	12	133.5186	74895.94	B3-2P19 B11-6R17	
8943	2	133.5315	74888.69	C5-12P4 B0-2R16	
8944	5	133.5383	74884.88	B11-7R13	
8945	2	133.5421	74882.77	B16-10P8 B6-6R7	
8946	3	133.5506	74878.00	C5-12P5	
8947	86	133.5549	74875.56	B3-5P4	
8948	12	133.5581	74873.77	B1-4P6	
8949	104	133.5604	74872.51	B5-6R2	
8950	21	133.5642	74870.34	B10-8R3 B10-7R11 B2-4R10 B16-9P13 B20-13P6	
8951					CII 133.56627
8952	30	133.5707	74866.70	C5-12P6	+ CII 133.57077
8953					CII 133.57077
8954	1344	133.5866	74857.82	B0-4P1 B12-8P9 B2-3R15 B9-4R21	
8955	392	133.5948	74853.20	B5-6P1	
8956	4	133.6008	74849.87	B5-5R11	
8957	22	133.6097	74844.88	B10-8P2 B17-10P11 B11-3P25	
8958	24	133.6131	74842.98	B3-4P11	
8959	6	133.6183	74840.07	B4-4R14 B8-6R12	
8960	11	133.6311	74832.90	B6-5R13 B20-13P7	
8961	1	133.6425	74826.48	B9-4P20	
8962	0	133.6487	74823.02	B18-11R11	
8963	0	133.6557	74819.11	B18-11P9	
8964	0	133.6726	74809.66	B1-1R21	
8965	2	133.6844	74803.02	B11-7P12 B4-0R27	
8966	5	133.6916	74799.01	B12-7R15	
8967	6	133.6941	74797.61	B17-11P5 B10-7P10 B19-12P7	
8968	2	133.7055	74791.25	B15-9P11	
8969	14	133.7370	74773.64	B0-2P15 B1-1P20	
8970	624	133.7469	74768.11	B0-4R3	
8971	14	133.7519	74765.30	B5-5P10 B8-6P11 B9-6R14	
8972	0	133.7594	74761.09	B17-10R13	
8973	11	133.7629	74759.16	B2-4P9	
8974	3	133.7676	74756.53	B15-10P5	
8975	2	133.7690	74755.72	C1-8R4 B14-9R10	
8976	2	133.7737	74753.10	B2-3P14	
8977	4	133.7771	74751.19	B6-5P12	
8978	6	133.7803	74749.41	B4-3P17	
8979	152	133.7849	74746.84	B5-6R3 B4-4P13	
8980	0	133.7878	74745.24	B13-6P19	
8981	0	133.8172	74728.80	C1-8R5	
8982	14	133.8322	74720.41	B12-9R0	
8983	16	133.8355	74718.57	B18-12R0 B2-2R19	
8984	192	133.8379	74717.28	B5-6P2 B18-12R1	
8985	3	133.8491	74711.00	B8-3R23	
8986	720	133.8565	74706.86	B0-4P2	
8987		133.8580	74706.04	B12-9R1	sh
8988	9	133.8710	74698.78	B7-6P9	
8989	0	133.8818	74692.74	B18-12R2	
8990	7	133.8911	74687.55	B9-6P13 B18-11R12	
8991	42	133.8959	74684.87	B10-8P3 B10-8R4	
8992	3	133.9044	74680.13	B16-9R15	
8993	2	133.9065	74679.00	B8-7R6	
8994	1	133.9200	74671.46	B20-14R4	
8995	10	133.9217	74670.52	B7-5R15	
8996	0	133.9382	74661.30	B13-9P6	

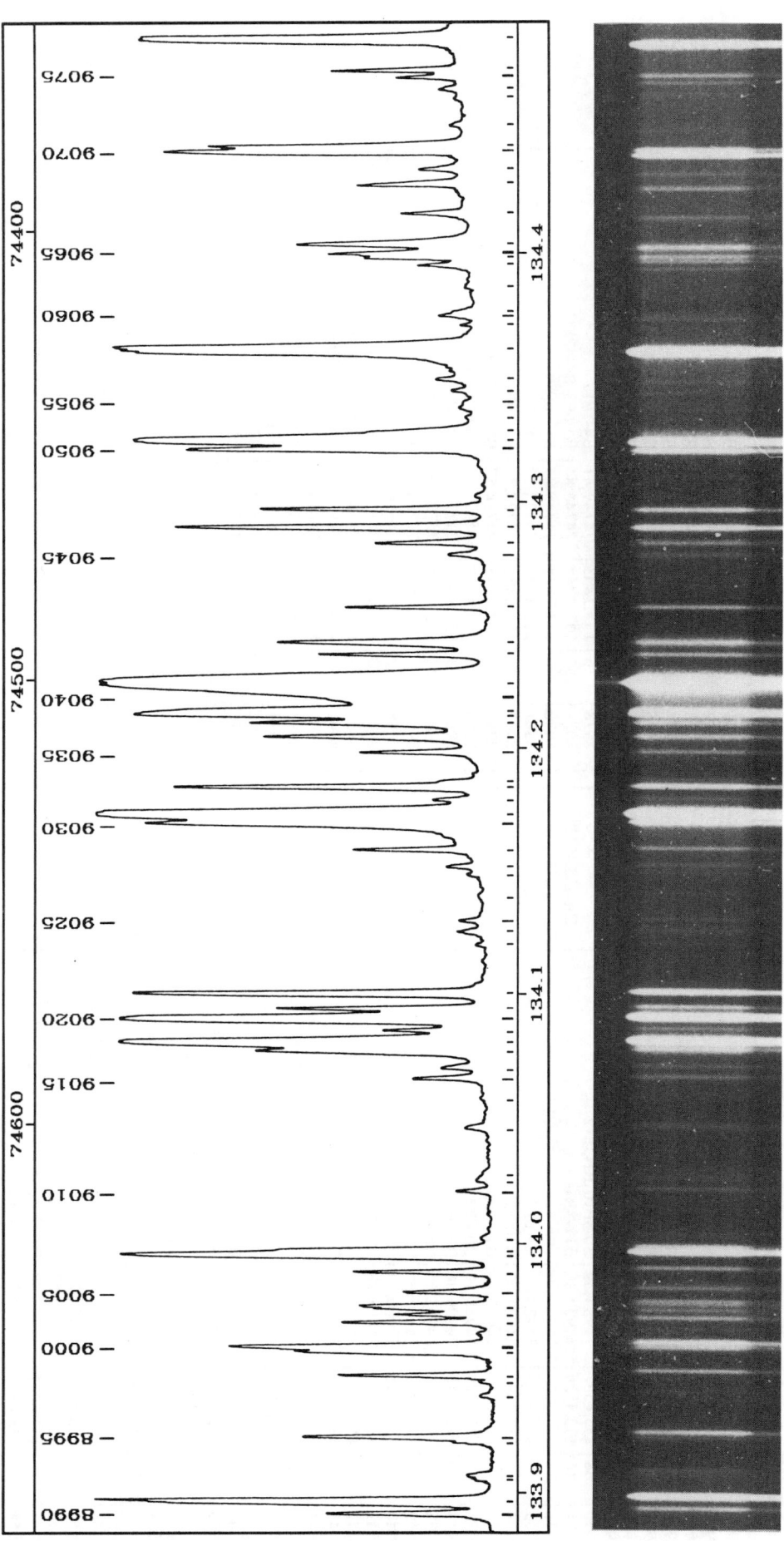

PLATE 98

TABLE 98

Nb	I	λ (nm)	σ (cm⁻¹)	Assignment	Comment
8997	0	133.9442	74657.94	B18-12P1	
8998	6	133.9466	74656.64	B12-9R2	
8999	8	133.9561	74651.30	B9-7P8 B2-2P18	
9000	10	133.9580	74650.28	B3-5R6 B18-12R3 C4-11Q7	
9001	0	133.9636	74647.15	C3-10R3	
9002	8	133.9680	74644.71	B8-4P19 B14-8R15	
9003	2	133.9710	74643.02	B14-9P9	
9004	4	133.9742	74641.23	C4-11Q1 B13-8R13 C2-9R5	b
9005	4	133.9800	74638.01	B8-7P5 C2-9R4 C4-11Q2 C4-11Q6	
9006	4	133.9879	74633.60	B11-8P7 C4-11Q3	
9007	38	133.9944	74629.96	B12-9P1 B20-14R3 B17-11P6 C4-11Q4 C4-11Q5	
9008	6	133.9964	74628.89	B18-11P10 B15-7P19 C2-9R3	
9009	0	134.0017	74625.93	B15-8R17	
9010	1	134.0198	74615.87	B16-10P9	
9011	0	134.0238	74613.62	C3-10R2	
9012	0	134.0253	74612.76	B7-3R22 B9-5P17	
9013	0	134.0450	74601.82	B20-14R2 B4-5P8	
9014	0	134.0571	74595.06	B18-12P2	
9015	4	134.0650	74590.69	B18-12R4 B7-5P14 B7-3P21	
9016	2	134.0693	74588.30	B5-4R16	
9017	8	134.0763	74584.39	B1-4R8 B10-6R16 B12-8R11	
9018	130	134.0795	74582.64	B0-4R4	
9019	2	134.0844	74579.91	B20-14R1	
9020	118	134.0892	74577.20	B3-5P5 C2-9R2	
9021	6	134.0933	74574.93	C3-10R1	
9022	24	134.0995	74571.49	B5-6R4	
9023	0	134.1173	74561.58	B20-14R0	
9024	1	134.1247	74557.48	B0-1P19	
9025	2	134.1291	74555.05	B15-9R13	
9026	0	134.1396	74549.22	B17-10R14	
9027	0	134.1479	74544.58	B15-10P6	
9028	2	134.1512	74542.77	C4-11P2	
9029	4	134.1581	74538.90	B14-10R0	
9030	32	134.1690	74532.88	B12-9R3	
9031	528	134.1723	74531.04	B5-6P3	
9032	2	134.1783	74527.73	B6-6R8 B13-8P12	
9033	17	134.1836	74524.79	B12-9P2 B20-14P1	
9034	0	134.1860	74523.45	C0-7P3	
9035	5	134.1976	74516.98	B16-11R0 B8-2P25 B18-12R5	
9036	8	134.2040	74513.45	B10-8R5 B3-3R17	
9037	11	134.2097	74510.28	C3-10R0 B20-14P2 B12-6P18	
9038	56	134.2132	74508.32	B14-10R1	
9039		134.2142	74507.74	B16-11R1 B20-14P4	sh
9040		134.2226	74503.11	B5-4P15 B20-14P3	sh
9041	1984	134.2257	74501.39	B0-4P3	
9042	6	134.2374	74494.89	C4-11P3 C1-8R11	
9043	8	134.2421	74492.26	B0-3R12	
9044	6	134.2567	74484.19	B10-8P4	
9045	2	134.2781	74472.29	B6-3R21	
9046	4	134.2829	74469.64	B12-8P10 B16-11R2	
9047	20	134.2892	74466.16	B1-4P7	
9048	11	134.2963	74462.20	B14-10R2 B17-11P7	
9049	0	134.3023	74458.91	B6-6P7	
9050	24	134.3203	74448.89	B14-10P1 C4-11P4 B19-13R1	
9051	108	134.3238	74446.97	B7-7R0 B16-11P1 B19-13R2	
9052	4	134.3272	74445.11	B1-2R18	
9053	0	134.3331	74441.82		
9054	0	134.3365	74439.95	B19-13R0	
9055	0	134.3386	74438.80	B19-13R3	
9056	0	134.3438	74435.89	B15-9P12	
9057	2	134.3486	74433.23	B6-3P20	
9058	400	134.3607	74426.54	B7-7R1	
9059	0	134.3705	74421.13		
9060	2	134.3741	74419.10	B10-7R12 B11-7R14	
9061	0	134.3764	74417.81	B7-4P18	
9062	0	134.3861	74412.47	B16-9R16	
9063	3	134.3943	74407.92	B11-5P20 C4-11P5	
9064	5	134.3977	74406.04	B12-9R4 B14-9R11	
9065	6	134.3995	74405.05	B16-11R3	
9066	7	134.4034	74402.91	B1-3R14	
9067	4	134.4152	74396.34	B8-5R17 B14-6P21	
9068	5	134.4272	74389.69	B8-7R7 B19-13P1	
9069	2	134.4330	74386.48	B13-9P7	
9070	28	134.4407	74382.25	B12-9P3	
9071	14	134.4427	74381.14	B14-10R3	
9072	0	134.4504	74376.90	B18-11P11	
9073	0	134.4619	74370.53	C4-11P6	
9074	2	134.4652	74368.70	B5-2R23 B17-10R15	
9075	2	134.4700	74366.01	B16-11P2	
9076	6	134.4731	74364.34	B14-10P2	
9077	106	134.4858	74357.28	B7-7R2 B3-4R13	

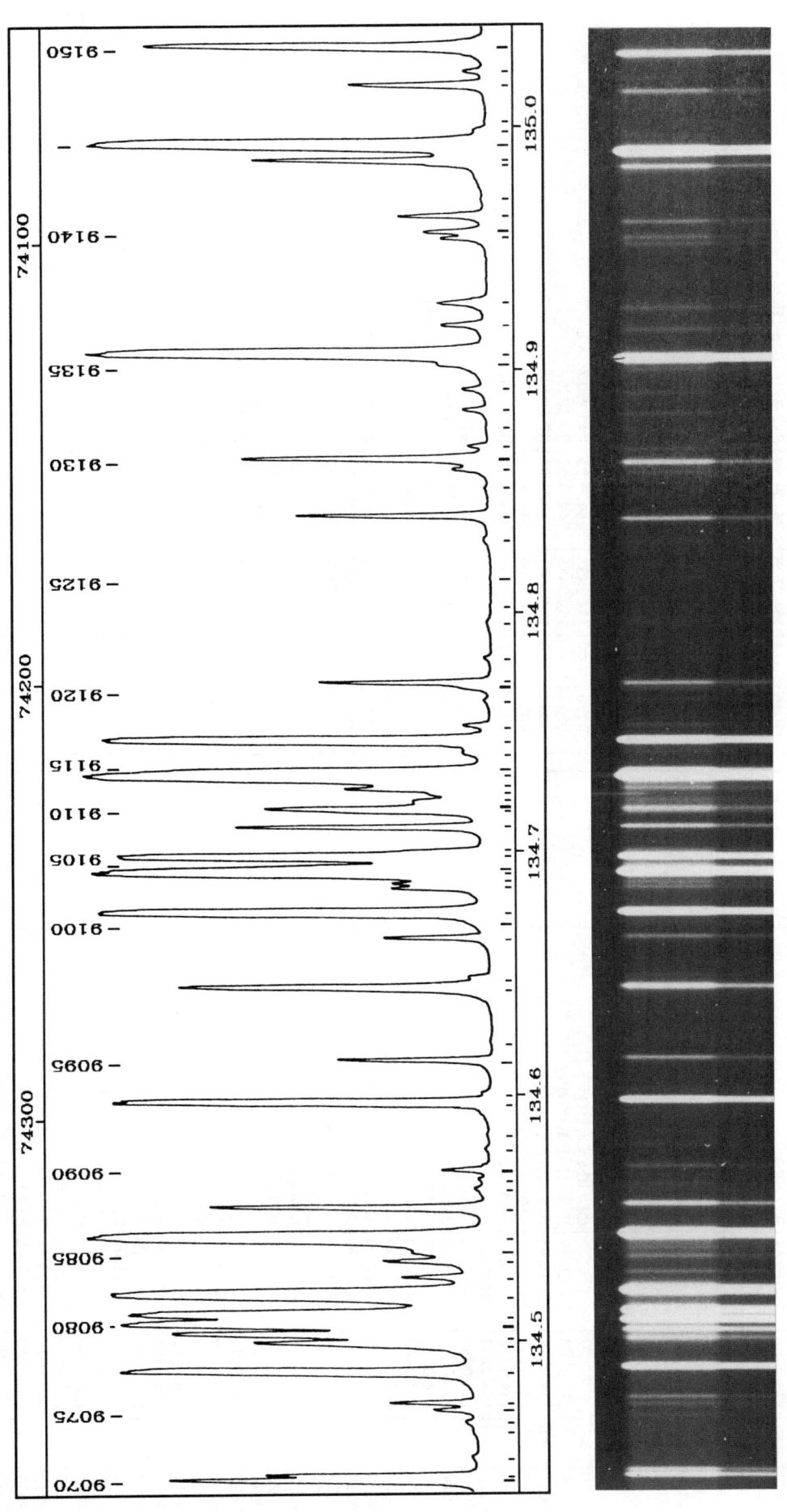

PLATE 99

TABLE 99

Nb	I	λ (nm)	σ (cm⁻¹)	Assignment	Comment
9078	18	134.4977	74350.72	B1-2P17 B2-4R11	
9079	45	134.5010	74348.90	B5-6R5	
9080	240	134.5051	74346.60	B2-5R0	
9081	209	134.5090	74344.49	B0-4R5 B16-10P10	
9082	384	134.5173	74339.87	B7-7P1 B0-3P11 B10-7P11	
9083	3	134.5247	74335.77	C3-10P2 B8-7P6	
9084	6	134.5313	74332.12	B8-6R13 B11-6R18	
9085	0	134.5345	74330.35	C4-11P7	
9086	912	134.5404	74327.10	B2-5R1 B5-5R12	
9087	24	134.5528	74320.25	B3-5R7	
9088	0	134.5605	74316.03	B16-11R4 B12-7R16	
9089	0	134.5645	74313.81	B15-10P7 B7-6R11	
9090	4	134.5681	74311.82	B11-7P13 B17-10P13	
9091	0	134.5769	74306.95	B14-9P10	
9092	0	134.5831	74303.51	B18-12P5	
9093	98	134.5956	74296.61	B5-6P4	
9094	0	134.5994	74294.53	B11-8P8	
9095	11	134.6133	74286.88	C3-10P3 B10-8R6	
9096	0	134.6226	74281.71	B10-4R23	
9097	28	134.6433	74270.32	B1-3P13	
9098	1	134.6476	74267.97	B14-10R4	
9099	8	134.6639	74258.95	B6-5R14 B16-11P3	
9100	0	134.6698	74255.71	B4-2R22	
9101	192	134.6746	74253.07	B2-5R2 B15-9R14	
9102	6	134.6848	74247.41	B10-8P5 B18-11P12	
9103	0	134.6867	74246.40	B9-7P9	
9104	432	134.6911	74243.96	B0-4P4	
9105		134.6932	74242.79	B3-4P12 B8-6P12	sh
9106	134	134.6974	74240.47	B7-7R3	
9107	2	134.6994	74239.38	B12-9R5 B10-4P22	
9108	19	134.7103	74233.35	B3-5P6	
9109		134.7173	74229.54	B5-3P19	sh
9110	16	134.7180	74229.13	B5-5P11 B7-6P10	
9111	4	134.7211	74227.42	B2-4P10	
9112					Cl I 134.72397
9113	10	134.7265	74224.45	B9-6R15 B13-8R14 B4-2P21	
9114	912	134.7317	74221.56	B2-5P1 B6-4R18	
9115		134.7337	74220.50	B14-10P3	
9116	0	134.7413	74216.29	C3-10P4 B17-11P8	
9117	184	134.7465	74213.41	B7-7P2 B5-1P25	
9118	3	134.7527	74210.01	B12-9P4	
9119	0	134.7627	74204.27	B16-11R5 B6-1P26	
9120	2	134.7685	74201.30	B10-5P19	
9121	15	134.7703	74200.30	B4-4R15	
9122	2	134.7805	74194.72	B12-7P15	
9123	0	134.7962	74186.08	B18-12P6 B4-1P24	sh
9124	0	134.8024	74182.64	B16-9P15	
9125	0	134.8137	74176.46	B5-1R26	
9126	1	134.8294	74167.83	B4-1R25	
9127	13	134.8390	74162.51	B6-5P13	
9128	1	134.8508	74156.05	B16-9R17	
9129	4	134.8584	74151.87	B2-3R16 B6-1R27	
9130	16	134.8625	74149.58	B1-4R9 B9-4R22	
9131	2	134.8679	74146.61	C3-10P5	
9132	0	134.8754	74142.49	B6-4P17	
9133	3	134.8827	74138.48	B6-6R9	
9134	2	134.8914	74133.69	B4-5P9	
9135	4	134.9017	74128.03	B9-6P14	
9136	328	134.9061	74125.63	B2-5R3 B14-10R5	
9137	5	134.9179	74119.13	B3-1P23	
9138	4	134.9271	74114.12	B0-2R17	
9139	3	134.9540	74099.35	B13-8P13	
9140	5	134.9566	74097.91	B3-2R21	
9141	8	134.9630	74094.38	B4-4P14 B14-9R12 B9-4P21	
9142	0	134.9691	74091.01	B13-9P8	
9143	0	134.9844	74082.65	B15-9P13	
9144	11	134.9860	74081.78	B5-6R6	
9145	52	134.9924	74078.22	B2-5P2 B7-7R4 B16-10P11 C3-10P6 B15-10P8	
9146	0	134.9981	74075.12	B14-8P15	
9147	0	135.0022	74072.86	B11-3P26	
9148	9	135.0171	74064.70	B8-7R8 B14-10P4 B12-8P11 B18-12P7	
9149	2	135.0228	74061.55	B7-5R16	
9150	42	135.0330	74055.98	B0-4R6 B6-6P8 C5-13Q3	

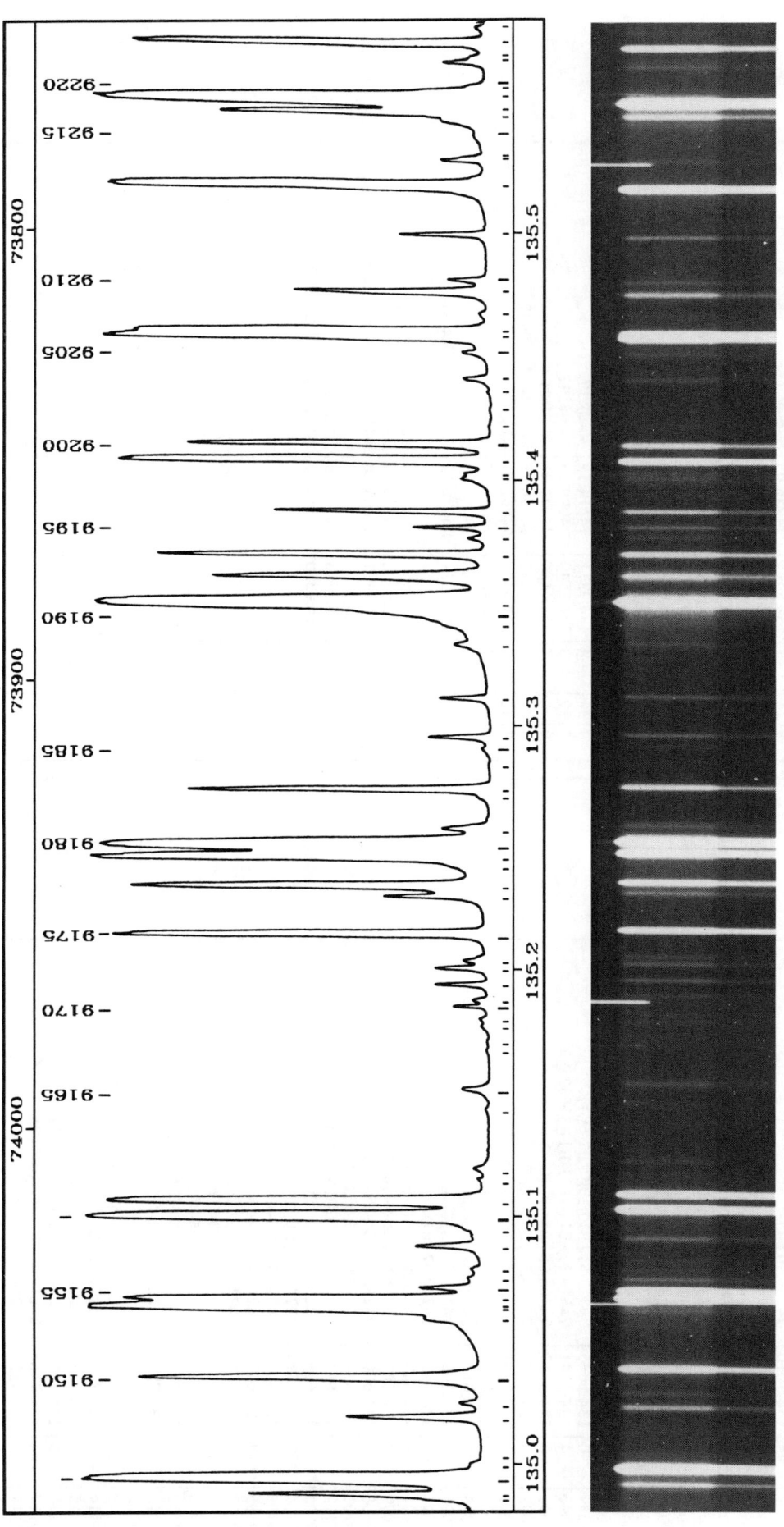

PLATE 100

TABLE 100

Nb	I	λ (nm)	σ (cm⁻¹)	Assignment	Comment
9151	3	135.0571	74042.78	B3-2P20	
9152					CuII 135.05938
9153	512	135.0619	74040.14	B7-7P3 B12-9R6	
9154	102	135.0648	74038.54	B9-8R0	
9155	6	135.0693	74036.07	B2-3P15	
9156	1	135.0736	74033.74	B2-1R23	
9157	6	135.0769	74031.91	B17-11P9	
9158	0	135.0862	74026.82	B10-8R7 B10-6R17	
9159	0	135.0936	74022.75		
9160	320	135.0984	74020.15	B9-8R1	
9161	132	135.1045	74016.77	B5-6P5 B1-4P8 B8-7P7	
9162	0	135.1136	74011.77	C3-10P7	
9163	3	135.1178	74009.50	C1-8P6 B2-1P22	
9164	2	135.1421	73996.21	C5-13P5	
9165	4	135.1500	73991.89	B0-2P16	
9166					ClI 135.16568
9167	0	135.1685	73981.72	B8-4R21	
9168	0	135.1753	73977.99	C5-13Q1	
9169	0	135.1787	73976.16	B4-3P18	
9170					CuII 135.18366
9171	3	135.1861	73972.09	B10-7R13	
9172	4	135.1924	73968.68	B12-9P5	
9173	4	135.1993	73964.89	B19-14R4 B7-5P15	
9174	3	135.2023	73963.25	B14-9P11	
9175	72	135.2134	73957.18	B9-8R2 B14-10R6 B9-5P18	
9176	8	135.2290	73948.66	B3-5R8 B15-9R15 C5-13P4	
9177	54	135.2333	73946.30	B2-5R4 B18-12P8	
9178	272	135.2415	73941.79	B9-8P1	
9179	792	135.2455	73939.59	B0-4P5	
9180	6	135.2504	73936.94	B11-8P9 B11-7R15	
9181	0	135.2571	73933.26	B12-6P19	
9182	27	135.2706	73925.91	B17-12R0 B17-12R1 C5-13P3	
9183	2	135.2733	73924.44	C5-13P2	
9184	0	135.2838	73918.70	B8-4P20 B2-0P25	
9185	8	135.2900	73915.29	B5-4R17	
9186	4	135.2946	73912.77	B17-12R2	
9187	4	135.3104	73904.17	B17-12P1	
9188	0	135.3325	73892.11	B2-2R20	
9189	4	135.3418	73887.01	B4-0P27	sh
9190	0	135.3458	73884.80	B2-5P3	
9191	1360	135.3495	73882.82	B0-3R13	
9192	20	135.3603	73876.91	B7-7R5	
9193	30	135.3693	73872.02	B10-7P12	
9194	2	135.3763	73868.19	B17-12R3	
9195	6	135.3805	73865.91	B17-12R3	
9196	13	135.3870	73862.34	B17-12P1	
9197	3	135.4005	73854.98	B1-1P21	
9198	2	135.4025	73853.87	B14-8R17	
9199	99	135.4079	73850.94	B9-8R3	
9200	29	135.4145	73847.33	B3-5P7	
9201	0	135.4216	73843.49	B17-11P10 B18-12P9	
9202	1	135.4283	73839.80	B19-14R1	
9203	4	135.4356	73835.82	B7-6R12	
9204	4	135.4414	73832.69	B7-3R23 B12-7R17	
9205	174	135.4522	73826.77	B13-8R15	
9206	104	135.4586	73823.29	B9-8P2	
9207	2	135.4600	73822.56	B7-7P4	
9208	16	135.4676	73818.41	B9-7P10 B19-14R0	
9209	4	135.4765	73813.55	B17-12R4 B11-7P14 B2-2P19 B5-4P16	
9210	6	135.4814	73810.85	B12-9R7 B8-6R14	
9211	200	135.5001	73800.67	B17-12P2	
9212		135.5202	73789.72	B4-6R0	CuII 135.53053
9213	3	135.5315	73783.58	B5-5R13	
9214	1	135.5407	73778.56	B11-6R19	
9215	4	135.5471	73775.07	B14-9R13 B12-8R13	
9216	20	135.5502	73773.40	B5-6R7 B3-3R18	
9217	744	135.5557	73770.44	B4-6R1 B8-5R18 B15-10P9 B19-14P4	
9218					OI 135.55977
9219	0	135.5607	73767.68	B14-10R7	
9220	5	135.5707	73762.25	B19-14P3 B17-12R5	
9221	3	135.5729	73761.05	B11-5P21	
9222	52	135.5784	73758.06	B11-9R0	
9224	2	135.5848	73754.61		

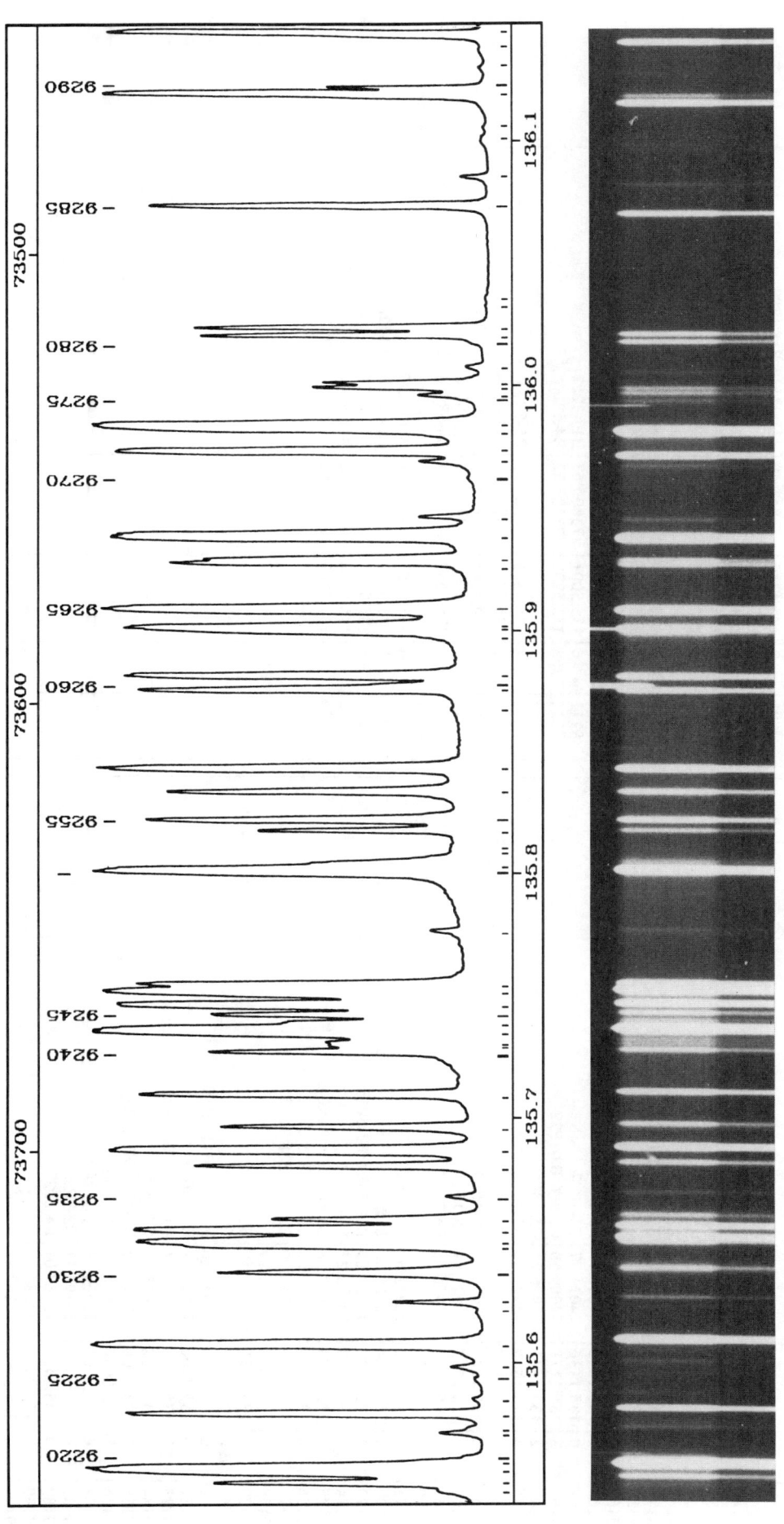

PLATE 101

TABLE 101

Nb	I	λ (nm)	σ (cm-1)	Assignment	Comment
9225	2	135.5930	73750.14	B4-5R11	
9226	3	135.5977	73747.56	B3-4R14	
9227	192	135.6067	73742.65	B11-9R1	
9228	0	135.6201	73735.36	B10-8R8	
9229	7	135.6240	73733.27	B7-6P11	
9230	20	135.6361	73726.70	B1-3R15 B6-0P29	
9231		135.6473	73720.61	B17-12P3	sh
9232	104	135.6488	73719.80	B0-4R7 B6-6R10	
9233	88	135.6534	73717.26	B2-5R5	
9234	15	135.6577	73714.96	B0-3P12	
9235	2	135.6670	73709.92	B8-7R9	
9236	26	135.6792	73703.26	B9-8R4 B8-6P13	
9237	172	135.6857	73699.76	B4-6R2	
9238	20	135.6950	73694.66	B5-6P6 B7-4P19 B6-3R22	
9239	46	135.7084	73687.40	B11-9R2	
9240	20	135.7258	73677.94	B18-13R1 B9-6R16	
9241	8	135.7282	73676.68	B1-4R10	
9242	8	135.7297	73675.82	B18-13R2	
9243	776	135.7345	73673.25	B4-6P1	
9244	14	135.7376	73671.57	B13-8P14 B14-10P6 B6-5R15	
9245	21	135.7412	73669.62	B15-11R0 B18-13R0 B5-5P12	
9246	164	135.7452	73667.45	B11-9P1	
9247	368	135.7509	73664.34	B9-8P3 B2-4P11 B15-9R16 B17-11P11	
9248	58	135.7516	73663.95	B15-11R1	
9249	2	135.7756	73650.94	B12-8P12 B18-13R4	
9250	272	135.8005	73637.43	B2-5P4 B6-3P21	
9251	88	135.8033	73635.90	B4-5P10 B1-2R19 B18-13R5	
9252	0	135.8090	73632.84	B17-12R6	
9253	0	135.8111	73631.67	B18-13R7	
9254	13	135.8172	73628.37	B15-11R2 B14-9P12	
9255	36	135.8216	73626.00	B7-7R6 B13-10R0 B17-12P4 B18-13R6	
9256	29	135.8332	73619.71	B18-13P1 B3-4P13 B6-6P9	
9257	113	135.8428	73614.48	B13-10R1 C1-8P9	
9258	0	135.8662	73601.82		
9259	44	135.8754	73596.86	B15-11P1	
9260					CuII 135.87730
9261	64	135.8812	73593.69	B11-9R3	
9262	0	135.8997	73583.66	B1-3P14	
9263					CuII 135.90091
9264	142	135.9010	73583.00	B0-4P6	
9265	240	135.9087	73578.80	B4-6R3 B18-13P2	
9266	28	135.9278	73568.47	B13-10R2	
9267	22	135.9290	73567.83	B15-11R3	
9268	170	135.9386	73562.65	B11-9P2 B7-7P5 B9-6P15	
9269	3	135.9462	73558.54	B12-9R8 B6-5P14	
9270	0	135.9616	73550.18	B10-5P20	
9271	4	135.9686	73546.42	B4-4R16	
9272	90	135.9728	73544.10	B13-10P1	
9273		135.9823	73539.00	B3-5R9	sh
9274	400	135.9834	73538.38	B4-6P2	CuII 135.99362
9275					
9276	4	135.9955	73531.85	B1-2P18	
9277	11	135.9987	73530.14	B1-4P9	
9278	11	136.0006	73529.10	B18-13P3 B6-4R19	
9279	2	136.0071	73525.58	B16-10P13 B10-4P23	
9280	0	136.0176	73519.92	B17-12P5	
9281	21	136.0202	73518.51	B15-11P2	
9282	22	136.0234	73516.79	B9-8R5 B5-3R21	
9283	0	136.0317	73512.28	B15-10P10	
9284	0	136.0353	73510.32	B10-7R14	
9285	40	136.0734	73489.73	B13-10R3	
9286	2	136.0854	73483.29	B15-11R4	
9287	0	136.1010	73474.87	B18-13P4 B10-6R18 C4-12R3 C5-14P4	
9288	0	136.1050	73472.69	C1-8P10	
9289	68	136.1191	73465.07	B9-8P4	
9290	9	136.1219	73463.56	B11-9R4 B14-9R14	
9291	0	136.1309	73458.72	B11-7R16	
9292	0	136.1392	73454.24	B9-4R23	
9293	45	136.1443	73451.50	B13-10P2	

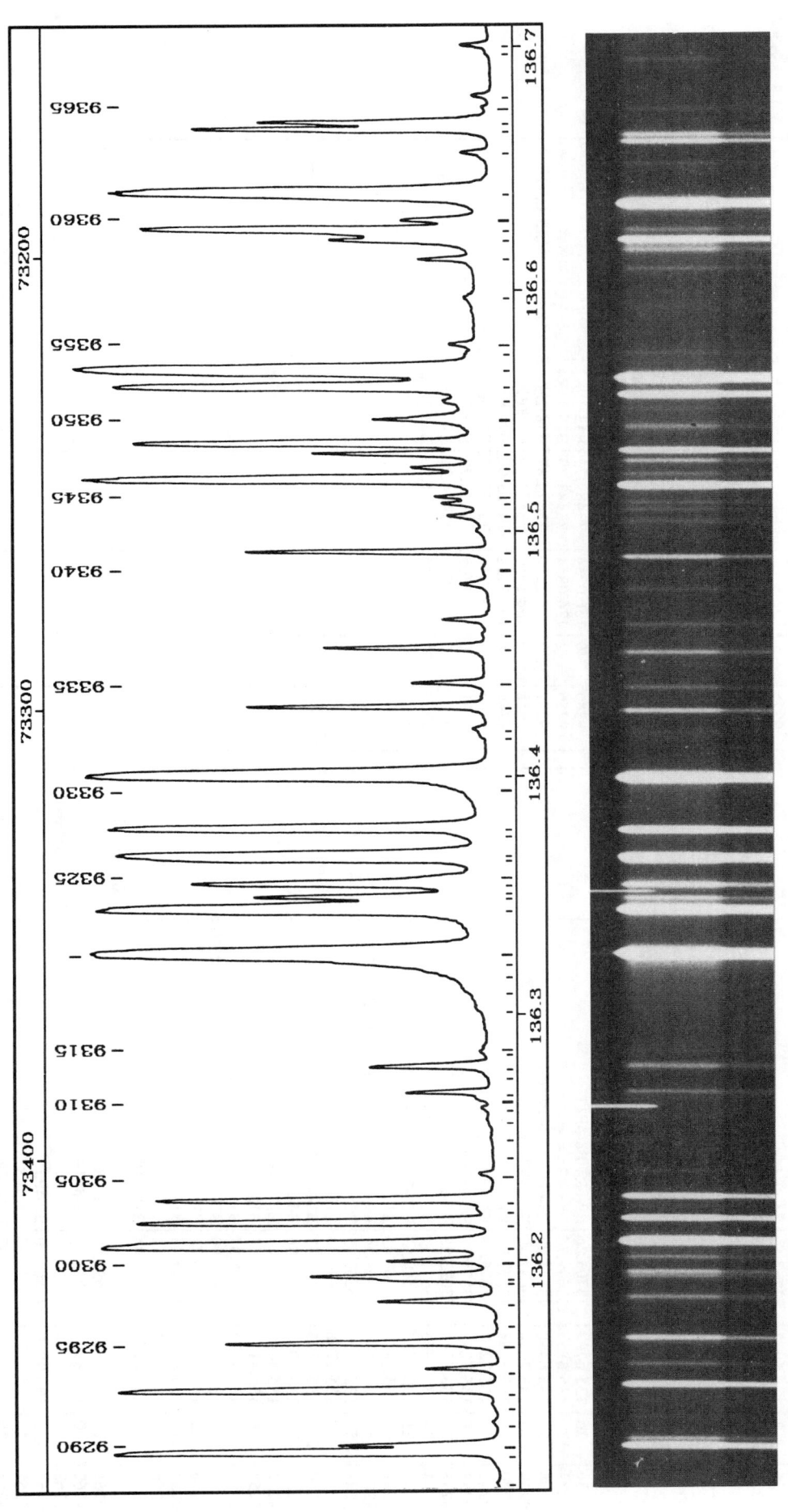

PLATE 102

TABLE 102

Nb	I	λ (nm)	σ (cm⁻¹)	Assignment	Comment
9294	7	136.1527	73446.96	B7-5R17	
9295	19	136.1635	73441.10	B2-5R6 B5-3P20	
9296		136.1721	73436.47	B13-8R16	
9297	8	136.1805	73431.95	B2-3R17 B6-4P18	
9298	6	136.1895	73427.10	B5-6R8	
9299	11	136.1908	73426.39	B4-4P15	
9300	6	136.1972	73422.95	B3-5P8	
9301	216	136.2038	73419.39	B11-9P3 B10-8R9 B13-9P10 B15-9R17	
9302	46	136.2135	73414.19	B15-11P3	
9303	2	136.2183	73411.56	B4-2R23	
9304	38	136.2226	73409.27	B4-6R4	
9305	2	136.2330	73403.67	C4-12R2	
9306	0	136.2402	73399.79		
9307	0	136.2482	73395.48		
9308	0	136.2546	73392.00	B17-11P13	
9309					CuII 136.25997
9310	0	136.2638	73387.07	C5-14P3	
9311	6	136.2660	73385.89	B10-7P13	
9312	0	136.2739	73381.63	B12-8R14	
9313	7	136.2769	73379.99	B13-10R4 B15-11R5	
9314	0	136.2826	73376.91		
9315	0	136.2845	73375.93	B9-4P22	
9316	0	136.2999	73367.63	B12-7R18	
9317	0	136.3083	73363.13	B4-2P22	
9318	0	136.3156	73359.19		
9319	0	136.3206	73356.50		
9320	1120	136.3238	73354.79	B4-6P3	
9321	408	136.3426	73344.67	B2-5P5 B9-7P11	
9322	14	136.3473	73342.12	B7-7R7 B14-10R9	
9323					CuII 136.35031
9324	26	136.3529	73339.09	B0-4R8 B7-6R13 C5-14P2	
9325	0	136.3552	73337.90	C4-12R1	sh
9326	0	136.3634	73333.47	B5-6P7	sh
9327	155	136.3647	73332.77	B6-7R0 B7-5P16 B11-3P27	
9328	0	136.3725	73328.57	B8-7R10	
9329	118	136.3764	73326.48	B13-10P3 B0-2R18	
9330	0	136.3935	73317.30		
9331	536	136.3985	73314.57	B6-7R1 B11-7P15	
9332	0	136.4152	73305.64	B9-5P19	
9333	1	136.4173	73304.51	B2-3P16	
9334	13	136.4265	73299.52	B11-9R5	
9335	5	136.4363	73294.29	B9-8R6	
9336	9	136.4510	73286.39	B15-11P4 B12-9R9 B5-1P26	
9337	0	136.4567	73283.32	B8-4R22 B6-1P27	
9338	5	136.4625	73280.22	B8-6R15	
9339	4	136.4776	73272.12	B6-6R11 B15-11R6	
9340	0	136.4829	73269.24	B11-5R23 B5-1R27	
9341	14	136.4911	73264.85	B7-7P6	
9342	1	136.5004	73259.86	B3-2R22	
9343	2	136.5061	73256.78	C3-11R7 B4-1P25 B4-1R26	
9344	3	136.5112	73254.05	B12-8P13 B6-1R28	
9345	3	136.5141	73252.49	B13-8P15	
9346	120	136.5213	73248.65	B6-7R2	
9347	3	136.5260	73246.11	B15-10P11 B14-9P13	
9348	11	136.5317	73243.08	B13-10R5	
9349	42	136.5360	73240.74	B11-9P4	
9350	7	136.5460	73235.41	B0-3R14	
9351	2	136.5534	73231.44	B5-4R18	
9352	86	136.5593	73228.25	B9-8P5	
9353	403	136.5663	73224.52	B6-7P1	
9354	0	136.5728	73221.04	B5-5R14	
9355	2	136.5767	73218.95	B7-6P12	
9356	0	136.5957	73208.75	B3-1R25	
9357	4	136.6120	73200.01	B4-3P19	
9358	8	136.6202	73195.63	B0-2P17 B8-4P21	
9359	64	136.6242	73193.47	B4-6R5	
9360	6	136.6283	73191.28	B3-2P21 B3-1P24 B12-7P17	
9361	232	136.6394	73185.32	B0-4P7	
9362	3	136.6562	73176.34	B18-14R4 B8-3R25	
9363	24	136.6656	73171.30	B13-10P4	
9364	16	136.6683	73169.86	B1-4R11	
9365	1	136.6748	73166.40	B14-9R15	
9366	2	136.6792	73164.05	C1-8P13	
9367	0	136.6967	73154.65	C4-12P2	
9368	3	136.6998	73153.02	B8-6P14	

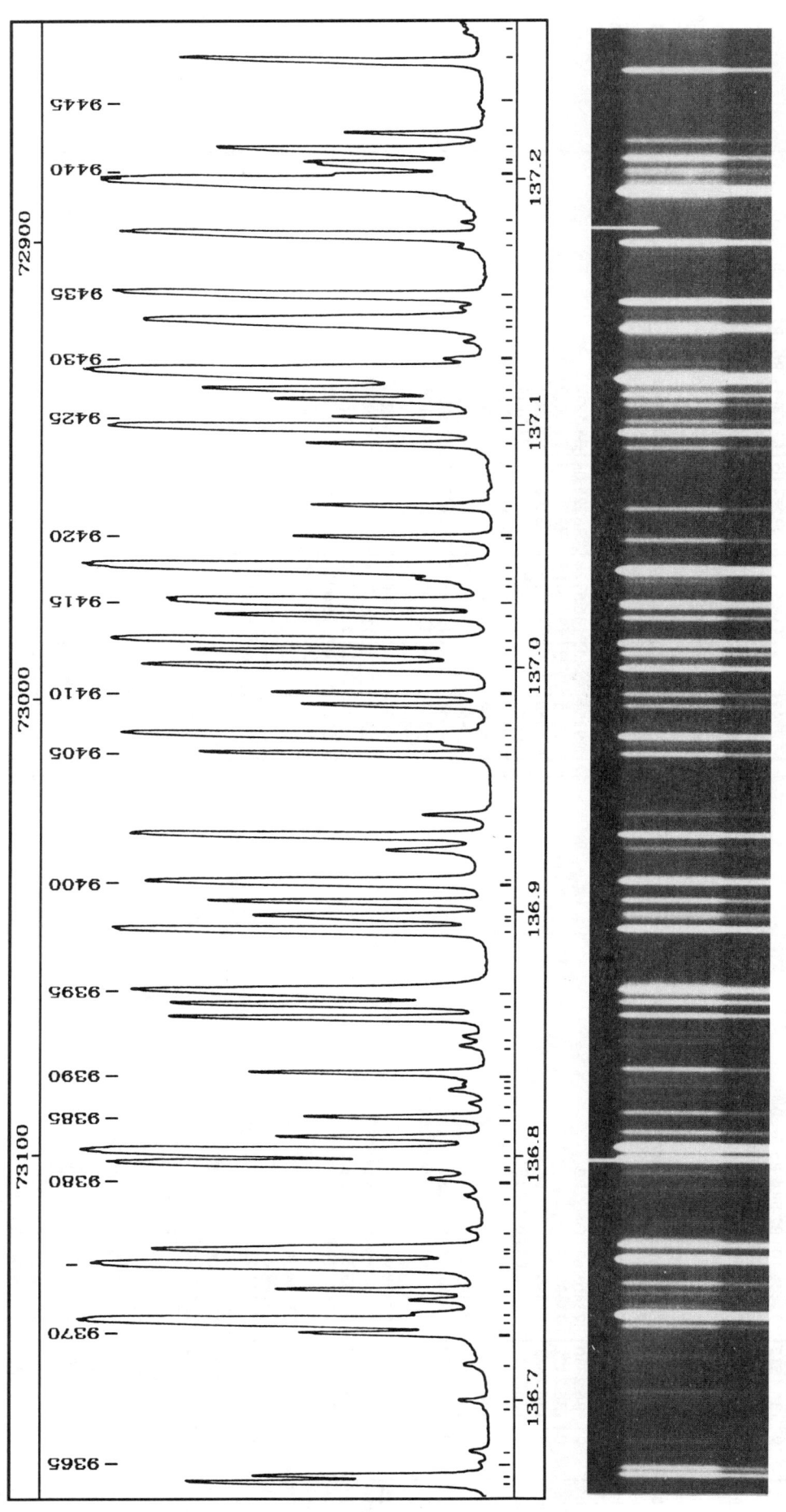

PLATE 103

TABLE 103

Nb	I	λ (nm)	σ (cm-1)	Assignment	Comment
9369	3	136.7143	73145.22	B8-5R19	
9370	8	136.7268	73138.54	B15-11P5 B11-8P11	
9371	200	136.7314	73136.11	B6-7R3	
9372	2	136.7349	73134.23	B11-5P22 C4-12P3	
9373	3	136.7404	73131.30	B9-6R17 C3-11R6	
9374	10	136.7444	73129.16	B18-14R3	
9375	200	136.7541	73123.93	B4-6P4 B2-1R24	
9376	46	136.7602	73120.70	B2-5R7	
9377		136.7613	73120.11	B3-4R15	
9378	1	136.7693	73115.80	B5-4P17	sh
9379	2	136.7831	73108.44	B4-5P11	
9380	4	136.7898	73104.86	B11-9R6	
9381					CuII 136.79509
9382	67	136.7958	73101.64	B16-12R0 B16-12R1	
9383	288	136.8005	73099.15	B6-7P2	
9384	12	136.8067	73095.85	B3-5R10 B18-14R2	
9385	11	136.8149	73091.44	B5-5P13	
9386	1	136.8211	73088.16	B13-9P11	
9387	3	136.8262	73085.40	B15-11R7 B2-1P23	
9388	2	136.8288	73084.04	B11-6P19	
9389	0	136.8312	73082.77	B13-10R6	
9390	11	136.8331	73081.73	B16-12R2 B10-8R10	
9391	2	136.8444	73075.71	B2-4P12	
9392	2	136.8480	73073.78	B6-5R16	
9393	24	136.8554	73069.85	B18-14R1	
9394	30	136.8611	73066.80	B1-5R0	
9395	36	136.8662	73064.05	B0-3P13 B2-2R21 B13-8R17	
9396	96	136.8915	73050.55	B1-5R1	
9397	8	136.8967	73047.79	B18-14R0	
9398	12	136.8976	73047.28	B5-6R9	
9399	17	136.9034	73044.23	B16-12R3 B10-7R15	sh
9400	40	136.9111	73040.08	B16-12P1	
9401		136.9121	73039.56	B9-8R7	
9402	4	136.9251	73032.63	B1-3R16	
9403	42	136.9311	73029.42	B11-9P5	
9404	4	136.9398	73024.80	B7-7R8 B3-3R19	
9405	19	136.9647	73011.52	B18-14P1	
9406	2	136.9690	73009.20	B1-4P10	
9407	52	136.9721	73007.56	B2-5P6	
9408	0	136.9771	73004.93	B7-3P23	
9409	8	136.9848	73000.78	B18-14P4 B12-9R10	
9410	11	136.9897	72998.19	B18-14P2 B1-1R23	
9411	40	137.0006	72992.41	B18-14P3 B16-12R4 B9-6P16 B12-8R15	
9412	18	137.0066	72989.20	B13-10P5 B11-7R17	
9413	106	137.0109	72986.92	B8-8R0 B15-10P12	
9414	20	137.0215	72981.26	B1-5R2	
9415	34	137.0270	72978.30	B6-7R4 B3-4P14 B16-12P2 B7-4P20	
9416	0	137.0337	72974.75	B15-11P6	
9417	1	137.0371	72972.95	B2-2P20	
9418	384	137.0410	72970.84	B8-8R1	CuII 137.05600
9419	10	137.0537	72964.10	B3-5P9	
9420					
9421	9	137.0668	72957.12	B9-8P6 C3-11R3	
9422	0	137.0835	72948.25	B15-11R8	
9423	12	137.0924	72943.53	B6-5P15	
9424	118	137.0987	72940.16	B1-5P1	
9425	4	137.1032	72937.77	B5-6P8 B1-1P22	
9426	8	137.1104	72933.95	B4-6R6	
9427	17	137.1144	72931.80	B7-7P7 B10-6R19	
9428	848	137.1212	72928.20	B6-7P3 B6-3R23	
9429		137.1245	72926.41	B16-12R5	
9430	6	137.1274	72924.88	B8-7R11	sh
9431	2	137.1347	72920.99	B12-7R19	
9432	55	137.1423	72916.99	B0-4R9 B14-9P14	
9433		137.1447	72915.67	B13-10R7	sh
9434	2	137.1486	72913.62	B10-5P21	
9435	77	137.1532	72911.18	B8-8R2	
9436	1	137.1736	72900.35	C3-11R2	
9437	58	137.1778	72898.10	B16-12P3 B9-7P12	CuII 137.18399
9438					
9439	400	137.1991	72886.80	B8-8P1 B17-13R1 B17-13R2	
9440	6	137.2026	72884.92	B11-9R7 B10-7P14	
9441	10	137.2067	72882.75	B4-4R17	
9442	8	137.2076	72882.27	B17-13R3 B14-9R16	
9443	18	137.2129	72879.45	B1-3P15 B17-13R5	
9444	8	137.2201	72875.63	B17-13R0 B17-13R4	
9445	0	137.2321	72869.23	B15-10R15	
9446	26	137.2494	72860.09	B1-5R3	
9447	0	137.2612	72853.80	B16-12R6	

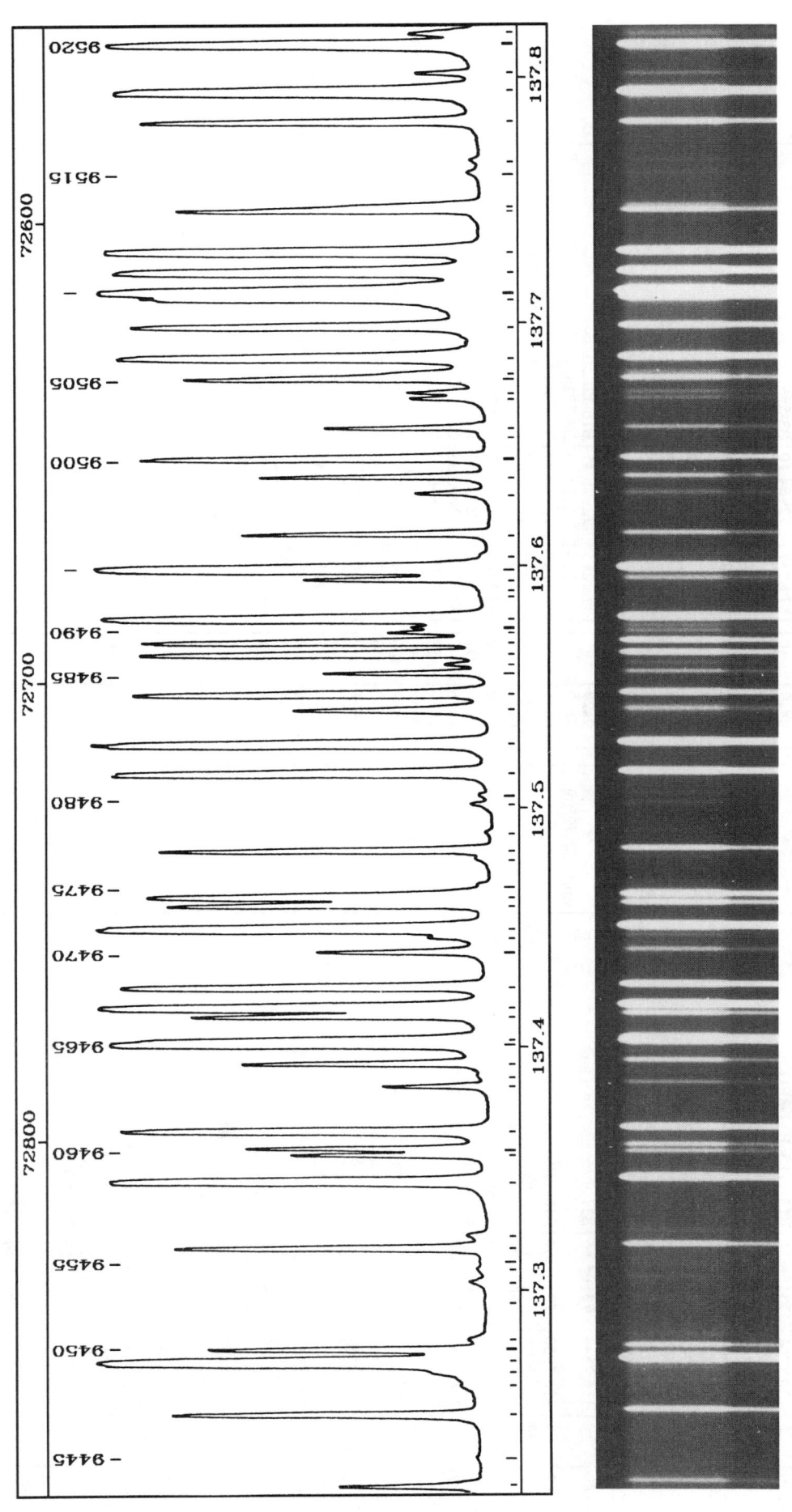

PLATE 104

TABLE 104

Nb	I	λ (nm)	σ (cm⁻¹)	Assignment	Comment
9448	0	137.2666	72850.92	B6-3P22	
9449	272	137.2706	72848.81	B4-6P5	
9450	20	137.2758	72846.06	C3-11R1	
9451	0	137.2801	72843.75	B17-13R7	
9452	0	137.2960	72835.32	B6-4R20	
9453	0	137.3036	72831.29	B7-5R18	
9454	0	137.3090	72828.45	B8-2P27 B12-6P21	
9455	0	137.3121	72826.80	B13-8P16	
9456	23	137.3174	72823.98	B17-13P1 B17-13R6 B11-7P16	
9457	2	137.3229	72821.05	B1-2R20	
9458	98	137.3451	72809.31	B8-8R3	
9459	11	137.3557	72803.68	B16-12P4 B6-6R12	
9460	12	137.3584	72802.24	B14-11R0 B15-11P7	
9461	69	137.3656	72798.45	B1-5P2 B12-8P14	
9462	5	137.3839	72788.71	B11-9P6	
9463	0	137.3885	72786.29	C2-10R7	
9464	14	137.3930	72783.91	B17-13P2 B13-10P6	
9465	160	137.4011	72779.65	B14-11R1	
9466	0	137.4023	72779.00	B6-7R5 B16-12R7	
9467	20	137.4124	72773.62	C3-11R0	sh
9468	208	137.4160	72771.71	B8-8P2	
9469	70	137.4245	72767.24	B10-9R0	
9470	8	137.4391	72759.50	B2-5R8	
9471	2	137.4452	72756.26	B9-8R8	
9472	224	137.4490	72754.26	B10-9R1	
9473	32	137.4588	72749.05	B14-11R2 B13-9P12 C4-13R5	
9474	38	137.4623	72747.22	B0-4P8 B4-4P16 B0-1P21	
9475	0	137.4670	72744.75	B8-6R16	
9476	0	137.4781	72738.87	B15-10P13	
9477	26	137.4814	72737.10	B17-13P3	
9478	0	137.4895	72732.85	B11-8P12	
9479	2	137.5007	72726.90	B10-8R11 B13-8R18	
9480	0	137.5046	72724.83	B12-9R11	
9481	89	137.5136	72720.09	B14-11P1 B6-4P19	
9482	158	137.5256	72713.74	B6-7P4	
9483	12	137.5395	72706.40	B1-2P19 B4-5R13	
9484	48	137.5461	72702.91	B10-9R2 B2-3R18	
9485	10	137.5547	72698.36	B16-12P5	
9486	3	137.5582	72696.52	B7-5P17	
9487	39	137.5626	72694.19	B12-10R0	
9488	33	137.5677	72691.51	B14-11R3	
9489	4	137.5716	72689.45	B7-6P13	
9490	2	137.5737	72688.29	B1-5R4	
9491	120	137.5774	72686.34	B12-10R1 B17-13P4	
9492	0	137.5867	72681.47	C2-10R4	
9493	0	137.5892	72680.11		
9494	8	137.5936	72677.82	B7-7R9	
9495	224	137.5981	72675.41	B10-9P1 B9-4P23	
9496	0	137.6038	72672.42	B13-10R8 B6-6P11	
9497	14	137.6124	72667.87	B8-8R4	
9498	4	137.6290	72659.09	B11-9R8 B5-3P21	
9499	9	137.6361	72655.37	B9-8P7	
9500	32	137.6437	72651.34	B12-10R2 B14-11P2	
9501	0	137.6528	72646.53	B14-9R17	
9502	6	137.6562	72644.75	C2-10R3 B5-5R15	
9503	4	137.6683	72638.38	B17-13P5	
9504	4	137.6707	72637.09	B5-6R10	
9505	16	137.6767	72633.94	B4-6R7	b
9506	4	137.6783	72633.08	B1-4R12	
9507	78	137.6852	72629.43	B2-5P7	
9508	48	137.6983	72622.56	C3-11P2 B3-5R11	
9509	64	137.7102	72616.27	B10-9R3	
9510	520	137.7125	72615.05	B8-8P3 B12-8R16	
9511	146	137.7208	72610.68	B12-10P1 B14-11R4	
9512	192	137.7292	72606.23	B1-5P3	
9513	23	137.7462	72597.28	C3-11P3 B14-9P15	
9514	8	137.7478	72596.45	B8-6P15	
9515	0	137.7611	72589.42	B9-6R18 B17-13P7	
9516	0	137.7666	72586.54	B16-12P6	
9517	40	137.7827	72578.09	B3-6R0 B4-2R24 B10-7R16	
9518	158	137.7950	72571.56	B10-9P2 B0-3R15 C2-10R2	
9519	4	137.8024	72567.70	B7-7P8	
9520	116	137.8143	72561.41	B3-6R1 B2-3P17	
9521	3	137.8186	72559.15	B13-10P7 C3-11P4 B4-5P12	

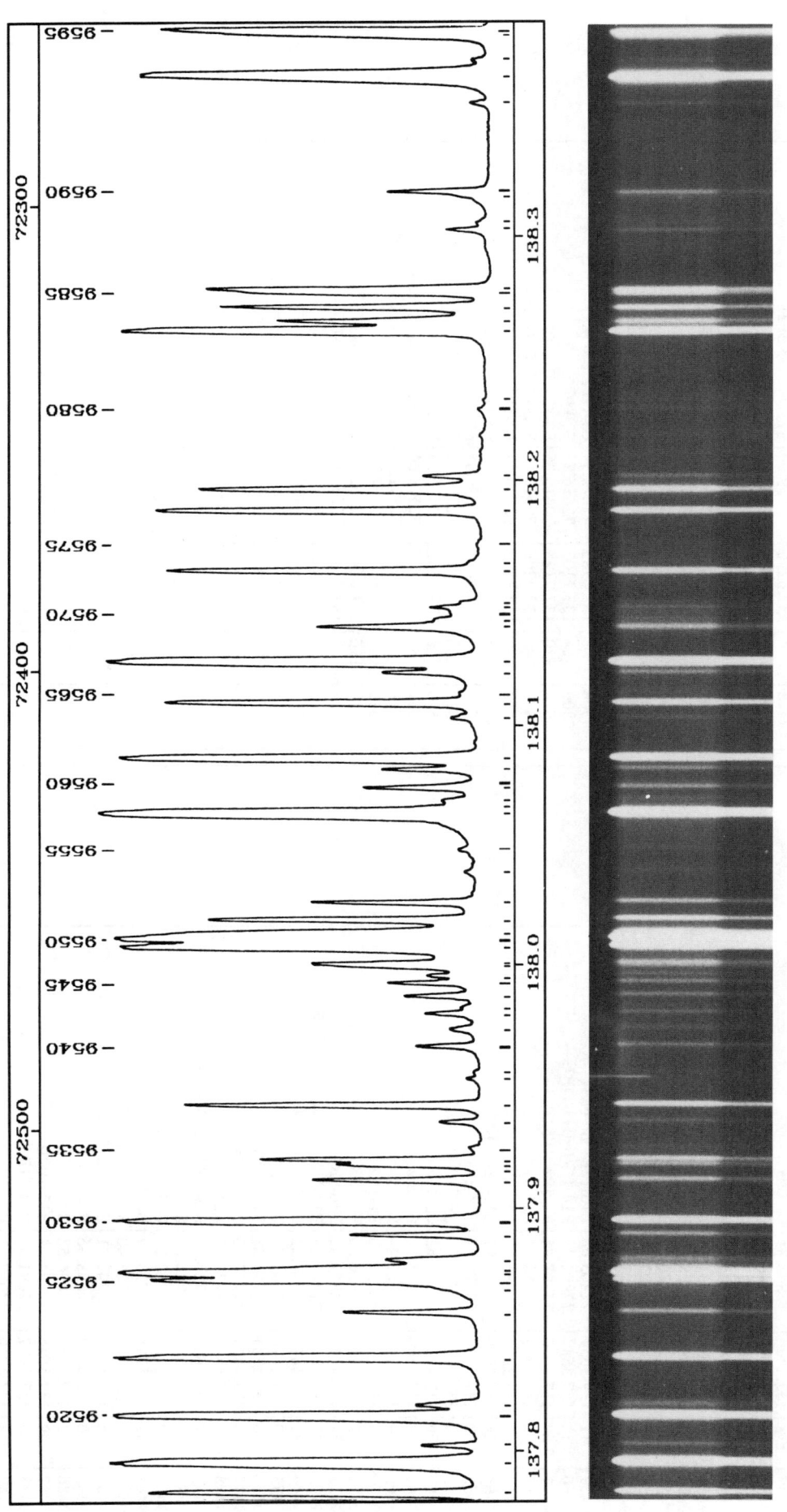

PLATE 105

TABLE 105

Nb	I	λ (nm)	σ (cm⁻¹)	Assignment	Comment
9522	70	137.8380	72548.94	B12-10R3 B5-4R19	
9523	6	137.8564	72539.27	B6-7R6	
9524	0	137.8652	72534.60	B11-5P23 B11-7R18 C1-9R5	
9525	46	137.8696	72532.33	B4-6P6	
9526	192	137.8723	72530.87	B14-11P3	
9527		137.8745	72529.75	C2-10R1 B8-5R20	sh
9528	5	137.8776	72528.08	C1-9R4 C3-11P5 B0-2R19	
9529	6	137.8880	72522.62	B11-9P7	
9530	76	137.8937	72519.65	B12-10P2 B15-10P14	
9531	10	137.9106	72510.76	B5-6P9 B4-2P23	
9532	0	137.9138	72509.08	C3-11P6	
9533	7	137.9170	72507.40	B14-11R5	
9534	10	137.9189	72506.39	B8-8R5	
9535	0	137.9234	72504.04	B8-7R12 B12-7R20	
9536	3	137.9345	72498.19	B5-5P14 C3-11P7	
9537	22	137.9414	72494.57	B3-6R2	
9538					Cl I 137.95278
9539	0	137.9554	72487.21	B8-4P22	
9540	3	137.9657	72481.76	C0-8R9	
9541	2	137.9725	72478.21	B3-4R16	
9542	2	137.9792	72474.68	B3-5P10	
9543	0	137.9819	72473.28	B16-12P7	
9544	6	137.9866	72470.81	B6-5R17	
9545	4	137.9920	72467.95	B1-5R5	
9546	1	137.9951	72466.36	B13-10R9	
9547	7	137.9995	72464.04	C1-9R3 B2-4P13	
9548	7	138.0026	72462.42	C0-8R4	
9549	150	138.0068	72460.22	B3-6P1	
9550	232	138.0102	72458.41	B6-7P5 B10-9R4 B1-4P11	
9551		138.0118	72457.55	B0-4R10	sh
9552	17	138.0180	72454.31	C0-8R5	
9553	10	138.0254	72450.46	B12-10R4	
9554	0	138.0381	72443.76	C0-8R3	
9555	0	138.0471	72439.03	B13-8P17	
9556	344	138.0624	72431.00	B10-9P3	
9557	2	138.0652	72429.53	C1-9R2	
9558		138.0680	72428.10	B3-2R23	sh
9559	7	138.0732	72425.33	B9-7P13 B9-6P17	
9560		138.0745	72424.65	B4-3P20	sh
9561	4	138.0808	72421.36	B9-8R9	
9562	103	138.0853	72419.02	B8-8P4	
9563	2	138.1019	72410.31	B13-9P13	
9564	30	138.1079	72407.15	B14-11P4	
9565	4	138.1117	72405.15	B10-6R20 B13-8R19	
9566	174	138.1205	72400.55	B17-14R4 B10-7P15	
9567	7	138.1248	72398.30	B12-10P3	
9568	0	138.1391	72390.78	B0-3P14	
9569	0	138.1421	72389.25	B0-2P18	
9570	0	138.1448	72387.81	B14-11R6	
9571	2	138.1472	72386.55	B12-9R12 B12-8P15 B6-1P28	
9572	0	138.1492	72385.49	B5-1R28 C1-9R1	
9573	28	138.1622	72378.71	B3-6R3	
9574	0	138.1701	72374.57		
9575	0	138.1740	72372.51		
9576	41	138.1871	72365.64	B1-5P4	
9577	24	138.1960	72360.99	B2-5R9 B4-1R27	
9578	3	138.2014	72358.15	B11-9R9	
9579	0	138.2189	72349.00	B10-8R12	
9580	0	138.2293	72343.56	B3-2P22	
9581	0	138.2327	72341.80	B4-1P26	
9582	90	138.2613	72326.83	B3-6P2 B9-8P8 B11-7P17	
9583	12	138.2652	72324.80	B1-3R17	
9584	18	138.2710	72321.74	B3-4P15 B6-5P16 B13-10P8 B11-8P13	
9585	12	138.2770	72318.61	B12-10R5	
9586	20	138.2780	72318.08	B10-9R5 B6-6R13	
9587	3	138.3027	72305.17	B7-7R10	
9588	0	138.3059	72303.51	B14-9P16	
9589	0	138.3166	72297.89	B10-5P22 B15-10P15	
9590	6	138.3184	72296.97	B4-6R8	
9591	2	138.3549	72277.87	B17-14R1	
9592	90	138.3658	72272.18	B0-4P9 B3-1P25	
9593	2	138.3723	72268.79	B7-4P21	
9594	18	138.3830	72263.19	B6-7R7 B8-8R6	
9595	36	138.3843	72262.55	B14-11P5	

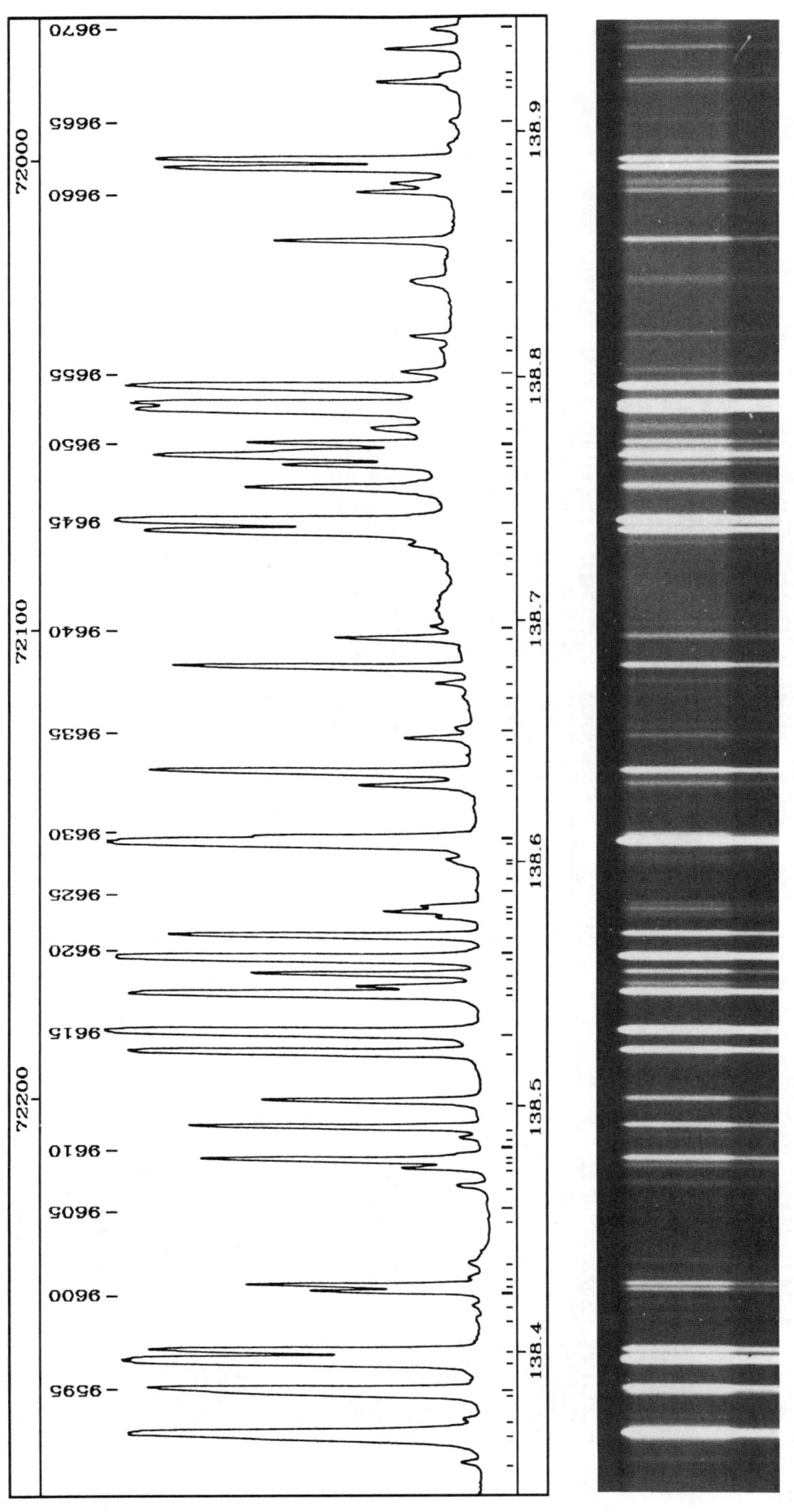

PLATE 106

TABLE 106

Nb	I	λ (nm)	σ (cm-1)	Assignment	Comment
9596	90	138.3957	72256.57	B15-12R1 B10-9P4	
9597	38	138.3999	72254.39	B15-12R0 B14-11R7 B12-10P4 B17-14R0	
9598	0	138.4126	72247.76	B13-10R10	
9599	1	138.4189	72244.47	B12-8R17	
9600	8	138.4245	72241.54	C2-10P3	
9601	13	138.4268	72240.35	B15-12R2	
9602	1	138.4304	72238.48	B2-2R22	
9603	2	138.4361	72235.49	B11-9P8	
9604	0	138.4543	72226.02		
9605	1	138.4600	72223.04	B2-1R25	
9606	4	138.4675	72219.10	B7-5R19	
9607	6	138.4747	72215.38	B3-6R4 B17-14P1	
9608	20	138.4778	72213.76	B2-5P8 B4-4R18	
9609	0	138.4801	72212.56	B17-14P4	
9610	0	138.4844	72210.28		
9611	1	138.4871	72208.90	B8-6R17	
9612	18	138.4912	72206.76	B15-12R3	
9613	13	138.5022	72201.01	B5-6R11 B1-5R6 B17-14P3	
9614	54	138.5218	72190.78	B15-12P1	
9615	160	138.5300	72186.52	B5-7R0 B8-8P5	
9616	58	138.5461	72178.17	B4-6P7	
9617	6	138.5492	72176.50	B7-7P9 B6-3R24	
9618	13	138.5544	72173.83	C2-10P4	
9619	126	138.5608	72170.50	B5-7R1	
9620	0	138.5637	72168.96	B2-1P24	
9621	28	138.5702	72165.59	B6-7P6 B6-6P12	
9622	2	138.5779	72161.59	B12-10R6	
9623	6	138.5804	72160.27	B1-3P16	
9624	2	138.5826	72159.15	B15-12R4	
9625	0	138.5893	72155.65	B4-5R14	
9626	0	138.5944	72152.97	B13-8R20	
9627	2	138.6000	72150.08	C4-13P3 B6-4R21	
9628	2	138.6016	72149.25	B7-6P14	
9629	240	138.6086	72145.61	B3-6P3	
9630	8	138.6109	72144.38	C2-10P5	
9631	8	138.6323	72133.29	B2-2P21	
9632	36	138.6377	72130.45	B15-12P2 B10-9R6	
9633	0	138.6456	72126.33	B12-7R21	
9634	4	138.6519	72123.05	B3-5R12	
9635	0	138.6540	72121.98	B10-7R17	
9636	0	138.6685	72114.45	B14-11R8	
9637	2	138.6745	72111.32	B9-8R10	
9638	22	138.6809	72107.96	B5-7R2	
9639	7	138.6929	72101.73	B14-11P6 B15-12R5	
9640	2	138.6979	72099.15	B1-1R24 B11-7R19	
9641	0	138.7191	72088.13	B13-10P9 B12-9R13	
9642	0	138.7254	72084.86	C2-10P6	
9643	2	138.7316	72081.64	B11-9R10 B13-9P14	
9644	64	138.7366	72079.06	B1-5P5	
9645	156	138.7405	72076.98	B5-7P1 B6-3P23	
9646	14	138.7546	72069.66	B1-4R13 B8-7R13	
9647	10	138.7638	72064.93	B16-13R2	
9648	38	138.7673	72063.09	B16-13R1	
9649	6	138.7690	72062.20	B4-4P17 C1-9P3	
9650	9	138.7726	72060.35	B16-13R3 B5-5R16	
9651	3	138.7789	72057.05	B5-6P10 B9-6R19	
9652	84	138.7863	72053.24	B15-12P3 B16-13R0	
9653	64	138.7885	72052.10	B16-13R4 B10-9P5	
9654	70	138.7957	72048.36	B12-10P5 B14-9P17	
9655	2	138.8024	72044.86	B16-13R5 B16-13R6	
9656	0	138.8119	72039.91	B9-5P21	
9657	2	138.8170	72037.27	B8-6P16 C2-10P7	
9658	2	138.8396	72025.57	B13-10R11 B1-1P23 B8-2P28	b
9659	9	138.8562	72016.96	B8-8R7	
9660	4	138.8761	72006.65	B3-6R5	
9661	2	138.8795	72004.85	B1-2R21	
9662	28	138.8857	72001.63	B16-13P1	
9663	26	138.8891	71999.88	B5-7R3	
9664	0	138.8954	71996.61	B5-3R23	
9665	0	138.9047	71991.82	B4-5P13	
9666	0	138.9181	71984.86	B12-8P16	
9667	4	138.9208	71983.44	B12-10R7 C1-9P4	
9668	0	138.9240	71981.83	B10-8R13	
9669	3	138.9344	71976.40	B9-8P9	
9670	1	138.9424	71972.28	B14-11R9	

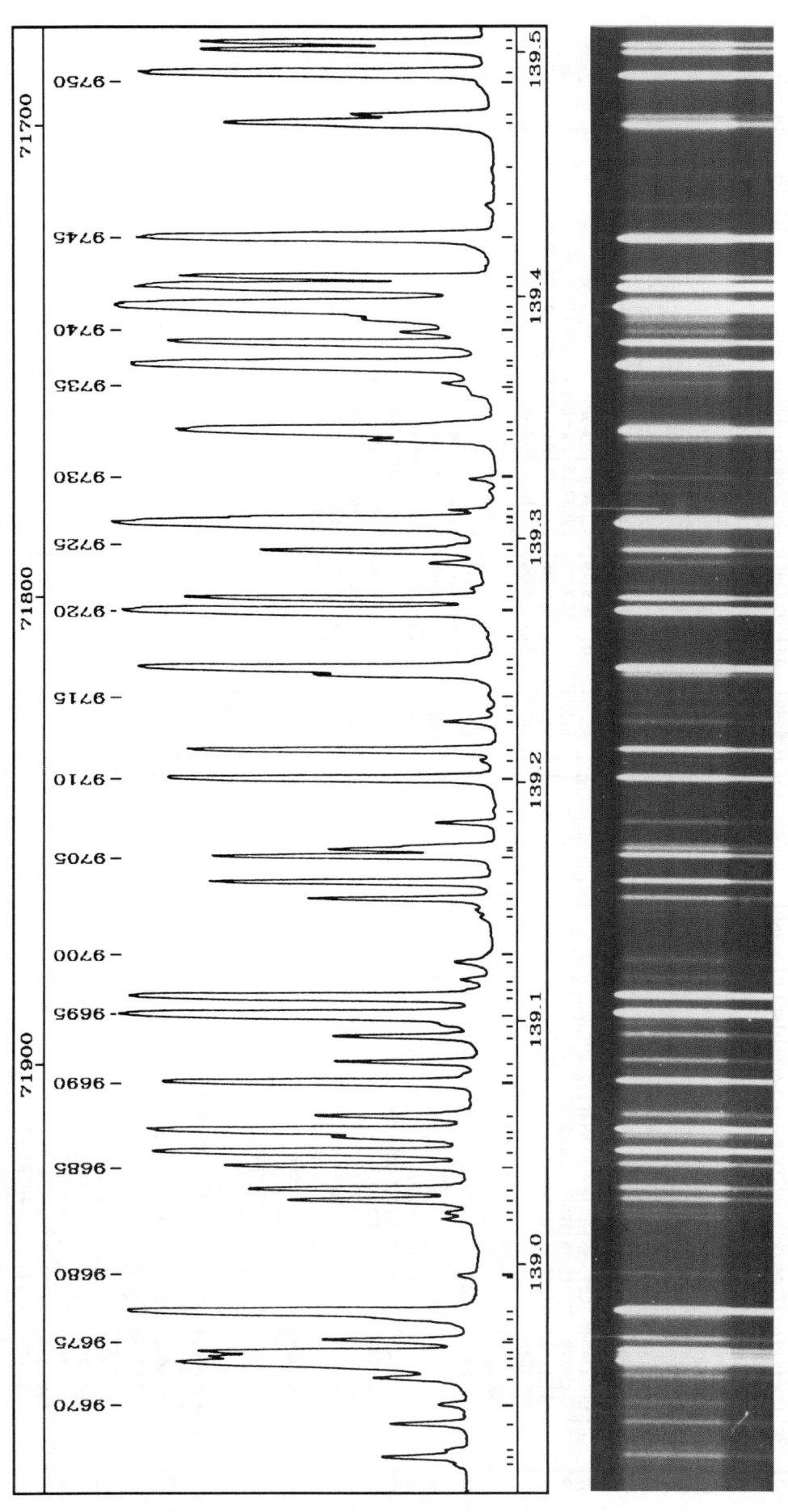

PLATE 107

TABLE 107

Nb	I	λ (nm)	σ (cm⁻¹)	Assignment	Comment
9671	6	138.9533	71966.61	B2-3R19	
9672	40	138.9593	71963.51	B0-4R11	
9673	17	138.9613	71962.50	B15-12P4	
9674	18	138.9636	71961.29	B16-13P2	
9675	8	138.9685	71958.77	B3-5P11	
9676					Cl I 138.96928
9677	90	138.9779	71953.91	B6-7R8	sh
9678	8	138.9799	71952.83	B5-7P2	
9679	2	138.9951	71945.00	B9-7P14	Cl I 138.99569
9680	0				
9681	1	139.0186	71932.84	B11-9P9	
9682	1	139.0211	71931.52	B15-12R7	
9683	10	139.0261	71928.96	B2-5R10 B14-11P7	
9684	15	139.0306	71926.64	B4-6R9 B8-5R21	
9685	18	139.0403	71921.60	B8-8P6	
9686	52	139.0459	71918.71	B3-6P4	
9687	8	139.0521	71915.50	B10-9R7	
9688	52	139.0547	71914.17	B16-13P3 B10-7P16	
9689	8	139.0607	71911.04	B7-7R11 B11-8P14	
9690	35	139.0746	71903.84	B7-8R0	
9691	0	139.0778	71902.21	B14-10P13	
9692	6	139.0830	71899.50	B10-6R21 C1-9P5	
9693	0	139.0934	71894.14	B5-5P15	
9694	0	139.0986	71891.47	B1-5R7	
9695	146	139.1023	71889.53	B7-8R1 B13-11R0 B0-3R16	
9696	84	139.1099	71885.61	B13-11R1	
9697	0	139.1130	71884.01	B5-3P22	
9698	2	139.1173	71881.81	B1-4P12	
9699	0	139.1246	71878.03	B1-2P20	
9700	0	139.1272	71876.67	B15-12R8	
9701	0	139.1434	71868.32	B5-4R20	
9702	0	139.1465	71866.72	B9-6P18 B6-5R18	
9703	10	139.1508	71864.49	B16-13P4	
9704	18	139.1577	71860.91	B15-12P5	
9705	22	139.1685	71855.36	B13-11R2	
9706	10	139.1715	71853.81	B12-10P6 B11-7P18	
9707	4	139.1726	71853.23	C1-9P6	
9708	4	139.1828	71847.96	B5-7R4	
9709	0	139.1879	71845.31	B14-11R10	
9710	44	139.2009	71838.59	B6-7P7	
9711	0	139.2083	71834.78	B2-4P14	
9712	11	139.2127	71832.53	B7-8R2	
9713	5	139.2242	71826.59	B3-4R17	
9714	0	139.2289	71824.17	B15-12R9	
9715	0	139.2350	71821.01	C0-8P5	
9716	10	139.2436	71816.57	B16-13P5 B6-6R14	
9717	88	139.2463	71815.19	B13-11P1	
9718	0	139.2492	71813.69	B2-3P18	
9719	0	139.2600	71808.14	B13-10R12	

Nb	I	λ (nm)	σ (cm⁻¹)	Assignment	Comment
9720	157	139.2701	71802.93	B7-8P1	
9721	29	139.2754	71800.17	B13-11R3	
9722	2	139.2798	71797.94	B8-4P23	CuII 139.31275
9723	3	139.2904	71792.45	B11-9R11	
9724	13	139.2955	71789.83	B4-6P8	
9725	1	139.2981	71788.48	B12-10R8	
9726	256	139.3066	71784.12	B5-7P3	
9727	14	139.3086	71783.09	B10-9P6	
9728					
9729	0	139.3209	71776.75	B16-13P6	
9730	2	139.3256	71774.34	B13-9P15	
9731	6	139.3416	71766.07	B9-8R11	
9732	528	139.3451	71764.30	B2-5P9 B0-4P10	
9733	0	139.3481	71762.72	B7-7P10	sh
9734	0	139.3607	71756.26	C1-9P7 B4-2R25	
9735	0	139.3625	71755.34	B3-6R6	
9736	1	139.3649	71754.08	B15-12P6	
9737	256	139.3719	71750.48	B0-5R0 B14-11P8	sh
9738		139.3732	71749.83	B1-5P6	
9739	40	139.3814	71745.60	B9-9R0	
9740	3	139.3860	71743.21	B5-6R12	
9741	6	139.3912	71740.53	B8-8R8	
9742	1024	139.3961	71738.02	B0-5R1 B13-11P2	
9743	148	139.4042	71733.83	B9-9R1 B7-8R3	
9744	29	139.4082	71731.77	B11-10R0	
9745	114	139.4246	71723.36	B11-10R1 B13-11R4	
9746	0	139.4383	71716.33	C0-8P6 B5-4P19	
9747	0	139.4528	71708.88	B10-5P23	
9748	22	139.4718	71699.08	B0-3P15	
9749	9	139.4749	71697.49	B6-5P17	
9750	0	139.4879	71690.81	C4-14R1	
9751	93	139.4921	71688.64	B7-8P2	
9752	26	139.5016	71683.76	B9-9R2	
9753	26	139.5046	71682.21	B11-10R2 B14-11R11	

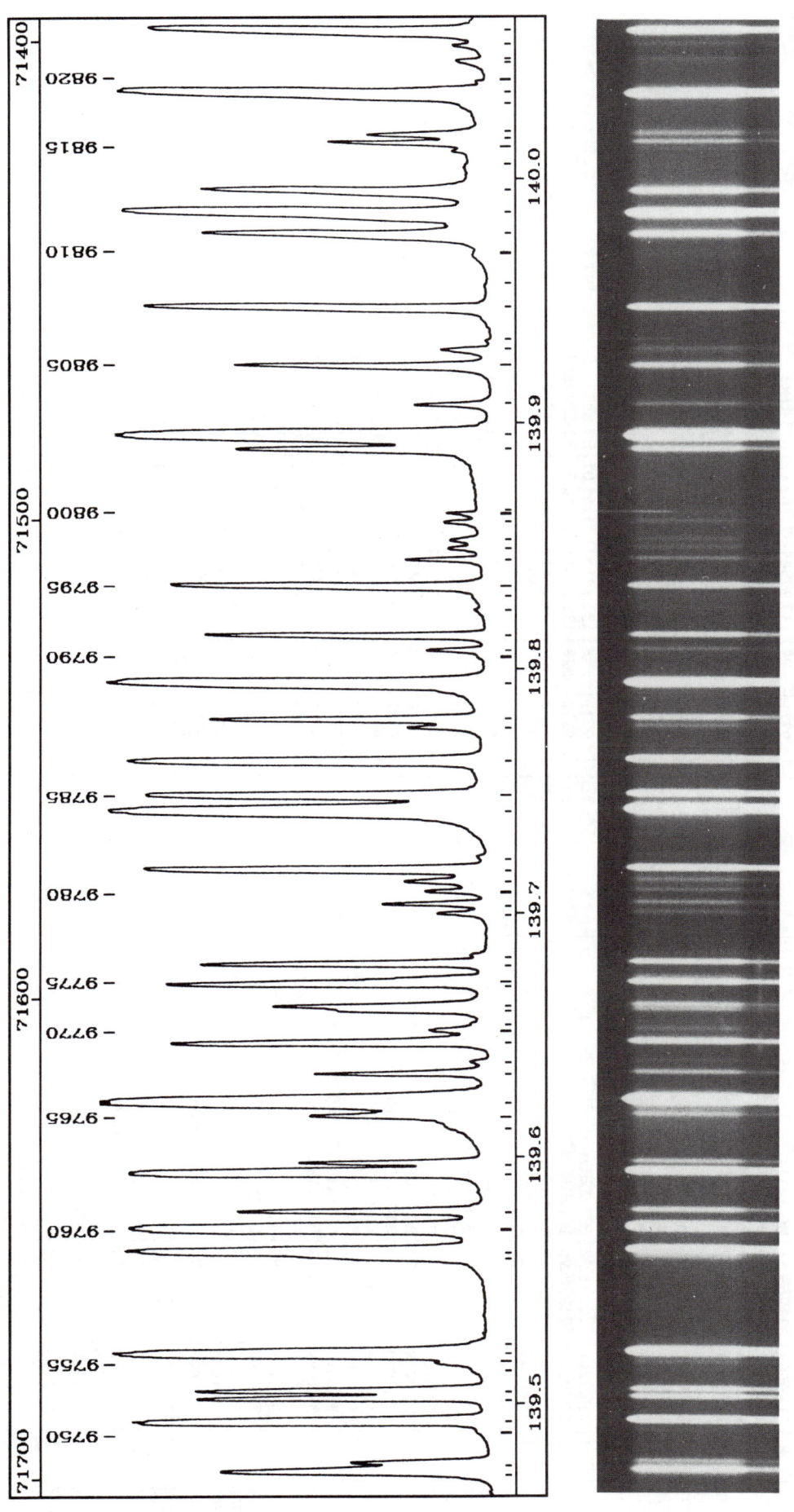

PLATE 108

TABLE 108

Nb	I	λ (nm)	σ (cm-1)	Assignment	Comment
9754	0	139.5127	71678.05	B8-6R18	
9755	2	139.5158	71676.46	B10-9R8 B10-7R18	
9756	264	139.5199	71674.38	B0-5R2	
9757	0	139.5242	71672.15	B4-2P24	
9758	12	139.5588	71654.38	B5-7R5 B3-4P16	
9759	140	139.5610	71653.24	B9-9P1 B4-3P21	
9760	173	139.5704	71648.41	B11-10P1 B3-6P5 B15-12P7 B6-6P13	
9761	18	139.5773	71644.91	B8-8P7	
9762	148	139.5929	71636.86	B13-11P3 B11-9P10	
9763	12	139.5968	71634.88	B12-10P7	
9764	0	139.6111	71627.56	B8-7R14	
9765	10	139.6162	71624.93	B13-11R5	
9766	928	139.6221	71621.88	B0-5P1	
9767	10	139.6335	71616.05	B6-7R9 B7-5R20	
9768	0	139.6382	71613.63	C4-14P3	
9769	34	139.6460	71609.64	B11-10R3 B9-8P10 B13-10R13	
9770	2	139.6509	71607.11	B1-3R18 B10-8R14	
9771	9	139.6532	71605.93	B3-2R24	
9772	9	139.6597	71602.62	B7-6P15 B12-8P17	
9773	13	139.6611	71601.88	B3-5R13	
9774	34	139.6706	71597.02	B9-9R3	
9775		139.6723	71596.17	B7-8R4	sh
9776	23	139.6788	71592.84	C0-8P7	
9777	0	139.6819	71591.24	B4-5R15	
9778	3	139.6995	71582.22	B12-10R9	
9779	6	139.7033	71580.27	B5-6P11	
9780	4	139.7084	71577.64	B16-14R4	
9781	5	139.7125	71575.58	B0-2P19	
9782	58	139.7173	71573.11	B5-7P4 B7-4P22	
9783	0	139.7230	71570.18	B14-11P9	
9784	512	139.7420	71560.43	B0-5R3 B15-12P8	
9785	75	139.7483	71557.20	B11-10P2	
9786	100	139.7626	71549.92	B9-9P2	
9787	6	139.7765	71542.79	B4-4R19 B9-6R20	
9788	24	139.7798	71541.11	B10-9P7 B1-5R8	
9789	272	139.7942	71533.75	B7-8P3	
9790	0	139.8052	71528.12	B3-3R21	
9791	3	139.8078	71526.76	B4-6R10	
9792	23	139.8137	71523.74	B16-14R3	
9793	0	139.8245	71518.22	B14-10P15	
9794	0	139.8291	71515.86	C0-8P8	
9795	32	139.8340	71513.38	B13-11P4 B13-11R6	
9796	4	139.8450	71507.76	B11-10R4	
9797	2	139.8496	71505.37	B11-8P15 B6-1P29	
9798	1	139.8530	71503.66	B3-2P23	
9799	2	139.8603	71499.94	B7-7R12	
9800	0	139.8639	71498.08	B11-9R12	
9801					
9802	18	139.8900	71484.75	B16-14R2 B1-4R14	CuII 139.86419
9803	480	139.8954	71481.99	B0-5P2 B6-7P8 B8-6P17	CuII 139.93527 Ke
9804	3	139.9084	71475.36	B9-9R4 B13-9P16	
9805	19	139.9242	71467.26	B2-5R11 B5-5R17	
9806	3	139.9309	71463.83	B9-7P15 B3-6R7	d
9807					
9808	52	139.9483	71454.95	B16-14R1	
9809	0	139.9579	71450.04	B11-5P25	
9810	0	139.9700	71443.86	B4-1P27	
9811	101	139.9781	71439.73	B0-4R12 B8-8R9 B6-3R25 C3-12R1	
9812	200	139.9867	71435.38	B11-10P3 B10-7P17 B9-5P22	
9813	24	139.9958	71430.71	B16-14R0 B1-3P17	
9814	0	140.0081	71424.45	B15-12P9	
9815	0	140.0129	71421.98	B5-7R6	
9816	12	140.0159	71420.47	B7-8R5 B3-5P12	
9817	9	140.0192	71418.78	B10-9R9 B2-2R23	
9818	0	140.0306	71412.98	B13-10R14	
9819	240	140.0355	71410.46	B9-9P3 B4-5P14	
9820	0	140.0387	71408.83	B9-8R12	
9821	0	140.0482	71404.01	B11-7P19	
9822	2	140.0495	71403.35	B13-11R7	
9823	2	140.0557	71400.14	B12-10P8	
9824	67	140.0612	71397.35	B0-5R4	

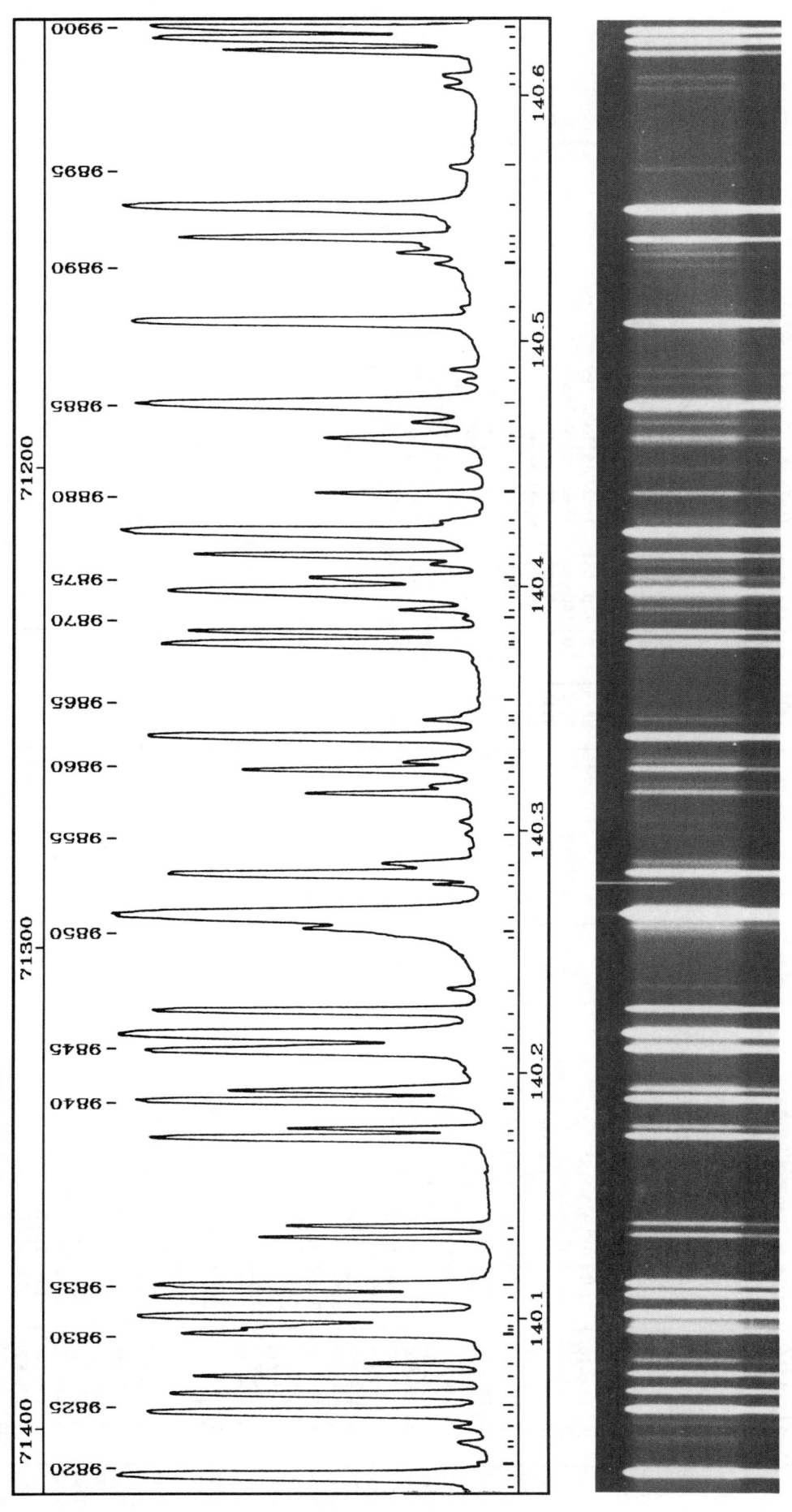

PLATE 109

TABLE 109

Nb	I	λ (nm)	σ (cm-1)	Assignment	Comment
9825	0	140.0650	71395.42	B10-8P13	sh
9826	44	140.0688	71393.50	B16-14P1	
9827	25	140.0758	71389.90	B16-14P4	
9828	6	140.0812	71387.17	B14-12R0	
9829	0	140.0879	71383.75	C3-13R7	
9830	34	140.0933	71381.00	B16-14P2	
9831	17	140.0950	71380.11	B1-5P7	
9832	9	140.0963	71379.47	B11-10R5	
9833	117	140.1001	71377.56	B16-14P3	
9834	80	140.1082	71373.43	B14-12R1 B4-4P18	
9835	53	140.1126	71371.17	B4-6P9 B13-11P5 B12-10R10 B3-1P26	
9836	16	140.1326	71360.99	B14-12R2	
9837	13	140.1373	71358.57	C3-12R0	
9838	62	140.1731	71340.37	B7-8P4	
9839	10	140.1768	71338.46	B3-6P6	
9840	99	140.1879	71332.82	B2-6R0	
9841	16	140.1922	71330.65	B14-12R3	
9842	0	140.1937	71329.89	B7-7P11	sh
9843	0	140.1996	71326.89	B11-7R21	
9844	80	140.2086	71322.28	B5-7P5	
9845		140.2102	71321.49	B9-9R5	sh
9846	376	140.2151	71319.00	B2-6R1 B6-3P24 B9-6P19	
9847	53	140.2250	71313.97	B14-12P1	
9848	2	140.2345	71309.12	B6-6R15	
9849	2	140.2573	71297.55	B2-2P22	
9850	10	140.2594	71296.49	B8-8P8	
9851	1576	140.2648	71293.72	B0-5P3	CuII 140.27770
9852					
9853	49	140.2817	71285.12	B2-5P10 B11-10P4 B14-12R4	
9854	5	140.2861	71282.89	B1-4P13 B5-5P16	
9855	0	140.2988	71276.46	B11-9P11	
9856	0	140.3039	71273.84	C2-11R7	
9857	11	140.3154	71268.02	B5-6R13	
9858	3	140.3182	71266.60	B6-5R19	
9859	16	140.3249	71263.17	B14-12P2 B2-1P25	
9860	2	140.3281	71261.54	B10-9P8	
9861	2	140.3297	71260.77	B7-6R17	
9862	90	140.3381	71256.48	B2-6R2	
9863	3	140.3452	71252.86	B6-7R10 B12-8P18	
9864	0	140.3478	71251.55	B10-7R19	
9865	0	140.3534	71248.73	B8-2P29	
9866	0	140.3691	71240.74	B15-13R6	
9867	55	140.3764	71237.06	B9-9P4	
9868		140.3771	71236.70	C3-12P3 B10-8R15	sh
9869	28	140.3813	71234.55	C3-12P2	
9870		140.3875	71231.43	B13-9P17	
9871	2	140.3909	71229.69	B14-12R5 B2-3R20	
9872	4	140.3956	71227.29	B13-11R8 B11-10R6	
9873	54	140.3982	71225.99	B0-4P11	
9874	12	140.4034	71223.37	B15-13R2	
9875		140.4039	71223.07	B15-13R3	
9876	2	140.4096	71220.20	B15-13R4 B15-13R5	
9877	28	140.4131	71218.43	B15-13R1	
9878	320	140.4227	71213.54	B2-6P1 B13-11P6	
9879	0	140.4272	71211.30	B7-8R6 B1-1R25 B15-13R7	
9880	9	140.4388	71205.39	B15-13R0 B11-9R13	
9881	0	140.4488	71200.33	B9-8P11	
9882	4	140.4601	71194.61	B5-4R21	
9883	12	140.4614	71193.96	B0-3R17	
9884	4	140.4679	71190.62	B2-4P15 B1-2R22	
9885	138	140.4750	71187.04	B0-5R5	
9886	2	140.4846	71182.20	B8-7R15	
9887	2	140.4893	71179.82	B12-10R11	
9888	136	140.5087	71169.97	B14-12P3 B3-4R18	
9889	0	140.5147	71166.94	B14-12R6	
9890	2	140.5327	71157.82	B8-6R19	
9891	3	140.5373	71155.50	B1-5R9 B12-10P9	
9892	1	140.5402	71154.01	B5-7R7	
9893	27	140.5434	71152.41	B15-13P1	
9894	144	140.5558	71146.14	B2-6R3 B10-9R10	
9895	2	140.5723	71137.78	B9-9R6 B3-6R8	
9896	2	140.6053	71121.05	B1-1P24 B11-8P16 B5-3P23	
9897	2	140.6100	71118.71	B8-8R10 B6-6P14	
9898	15	140.6199	71113.70	B15-13P2	
9899	83	140.6246	71111.33	B7-8P5	
9900	58	140.6289	71109.16	B11-10P5	

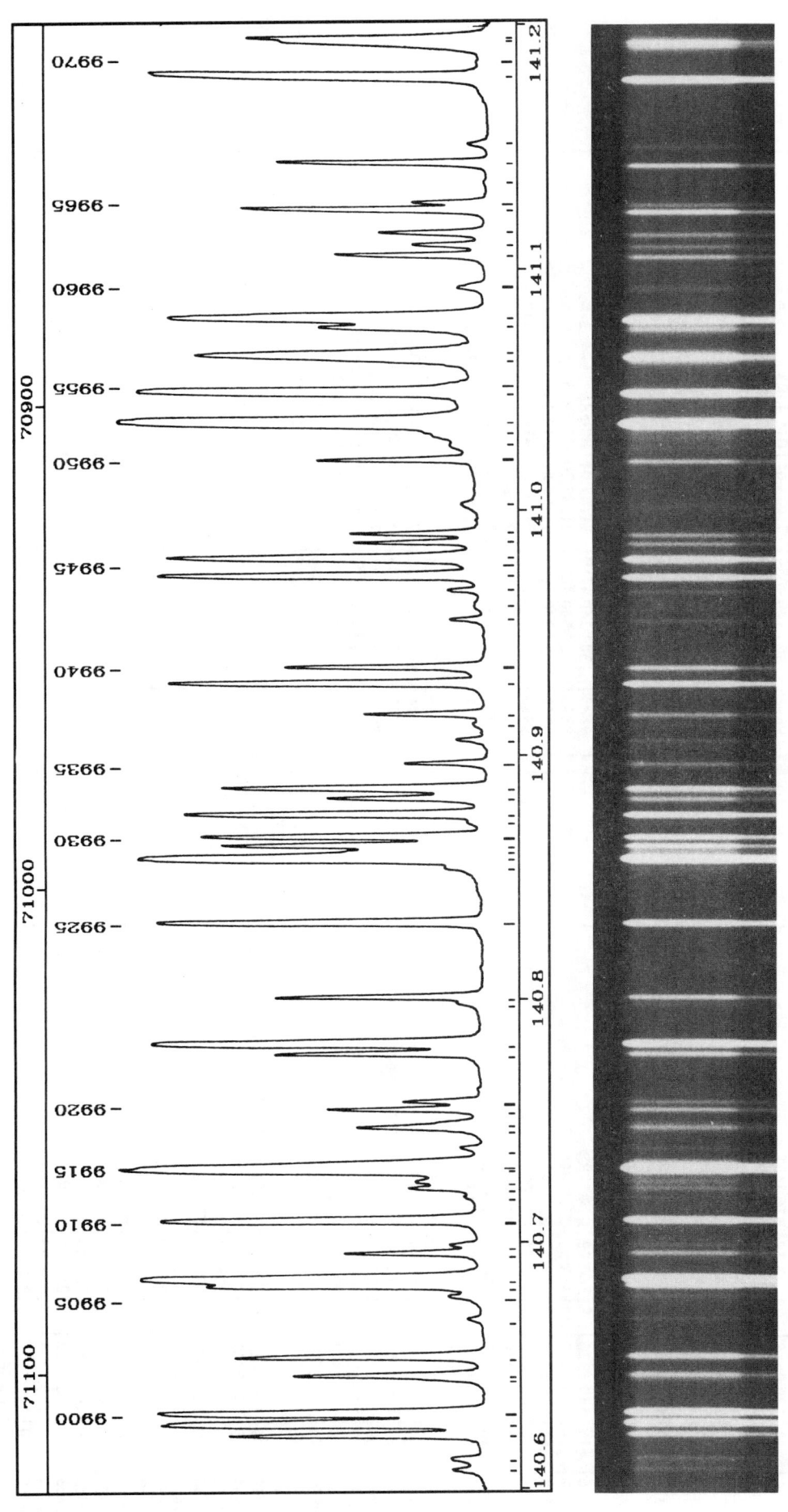

PLATE 110

TABLE 110

Nb	I	λ (nm)	σ (cm⁻¹)	Assignment	Comment
9901	10	140.6429	71102.04	B14-12R7	sh
9902	10	140.6441	71101.43	B4-6R11 C3-13R6	
9903	14	140.6518	71097.55	B6-7P9 B13-11R9	
9904	1	140.6672	71089.77	C1-10R11	
9905	2	140.6764	71085.14	B5-6P12	
9906	27	140.6808	71082.91	B14-12P4	
9907	160	140.6833	71081.62	B2-6P2	
9908	8	140.6938	71076.34	B7-7R13	
9909	2	140.6971	71074.66	B6-5P18	
9910	51	140.7074	71069.48	B15-13P3	
9911					CuII 140.71689
9912	3	140.7205	71062.87	B3-5R14	
9913	4	140.7233	71061.44	B2-3P19	
9914	320	140.7287	71058.71	B0-5P4	
9915		140.7310	71057.54	B11-10R7	sh
9916	1	140.7369	71054.55	B7-6P16	
9917	6	140.7455	71050.22	B1-2P21	
9918	1	140.7471	71049.42	B9-6R21	
9919	7	140.7527	71046.58	B13-11P7	
9920	5	140.7559	71044.97	B9-8R13	
9921	12	140.7759	71034.90	B5-7P6	
9922	67	140.7801	71032.78	B9-9P5	
9923	2	140.7967	71024.41	B7-5R21	
9924	10	140.7993	71023.10	B15-13P4	
9925	38	140.8299	71007.65	B4-7R0	
9926	0	140.8527	70996.17	B14-12R9	
9927	164	140.8566	70994.20	B4-7R1	
9928	7	140.8594	70992.76	B0-3P16	
9929	15	140.8619	70991.50	B3-6P7	
9930	23	140.8658	70989.55	B2-6R4	
9931	0	140.8718	70986.55	B9-7P16	
9932	29	140.8752	70984.82	B14-12P5	
9933	9	140.8819	70981.45	B3-4P17	
9934	16	140.8860	70979.38	B15-13P5 B2-5R12	
9935	4	140.8962	70974.23	B1-5P8 B10-7P18	
9936	2	140.9059	70969.33	B7-8K7	
9937	0	140.9126	70966.00	B13-11R10	
9938	5	140.9167	70963.89	B10-9P9	
9939	34	140.9295	70957.46	B12-11R0 B12-11R1	
9940	9	140.9361	70954.14	B8-8P9	
9941	2	140.9553	70944.48	B15-13P6	
9942	1	140.9609	70941.66	B0-1P23	
9943	2	140.9676	70938.31	B12-11R2	
9944	40	140.9735	70935.31	B4-7R2 B12-10R12	
9945	0	140.9773	70933.39	B8-6P18 B12-8P19	
9946	25	140.9808	70931.65	B0-5R6	
9947	4	140.9872	70928.41	B9-9R7 B15-13P7	
9948	5	140.9908	70926.59	B4-6P10 B11-9R14	
9949	0	141.0028	70920.59	B0-2R21	

Nb	I	λ (nm)	σ (cm⁻¹)	Assignment	Comment
9950	10	141.0214	70911.23	B11-10P6	
9951	1	141.0275	70908.15	B12-9P15	
9952	1	141.0317	70906.02	B12-10P10	d
9953	420	141.0371	70903.35	B2-6P3	
9954	131	141.0496	70897.05	B4-7P1	
9955	0	141.0524	70895.62	B7-4P23	
9956		141.0627	70890.44	B4-3P22	sh
9957	19	141.0647	70889.48	B0-4R13 B11-10R8	
9958	8	141.0765	70883.55	B1-3R19 B7-7P12	
9959	47	141.0799	70881.82	B12-11P1 B1-4R15 B14-12P6	
9960	1	141.0929	70875.31	B13-11P8 B4-4R20 B5-5R18	b
9961	6	141.1060	70868.70	B6-7R11	
9962	3	141.1103	70866.53	B10-9R11	
9963	5	141.1151	70864.14	B3-5P13	
9964	15	141.1246	70859.39	B12-11R3	
9965	4	141.1273	70857.97	C2-11R2 B9-5P23	
9966	0	141.1357	70853.78	B5-7R8	
9967	11	141.1438	70849.72	B7-8P6	
9968	1	141.1514	70845.89	B13-11R11 B4-2P25	
9969	59	141.1792	70831.96	B4-7R3	
9970	0	141.1863	70828.42	B9-8P12	
9971		141.1933	70824.91		NI 141.19318 Ke
9972		141.1945	70824.28		NI 141.19483

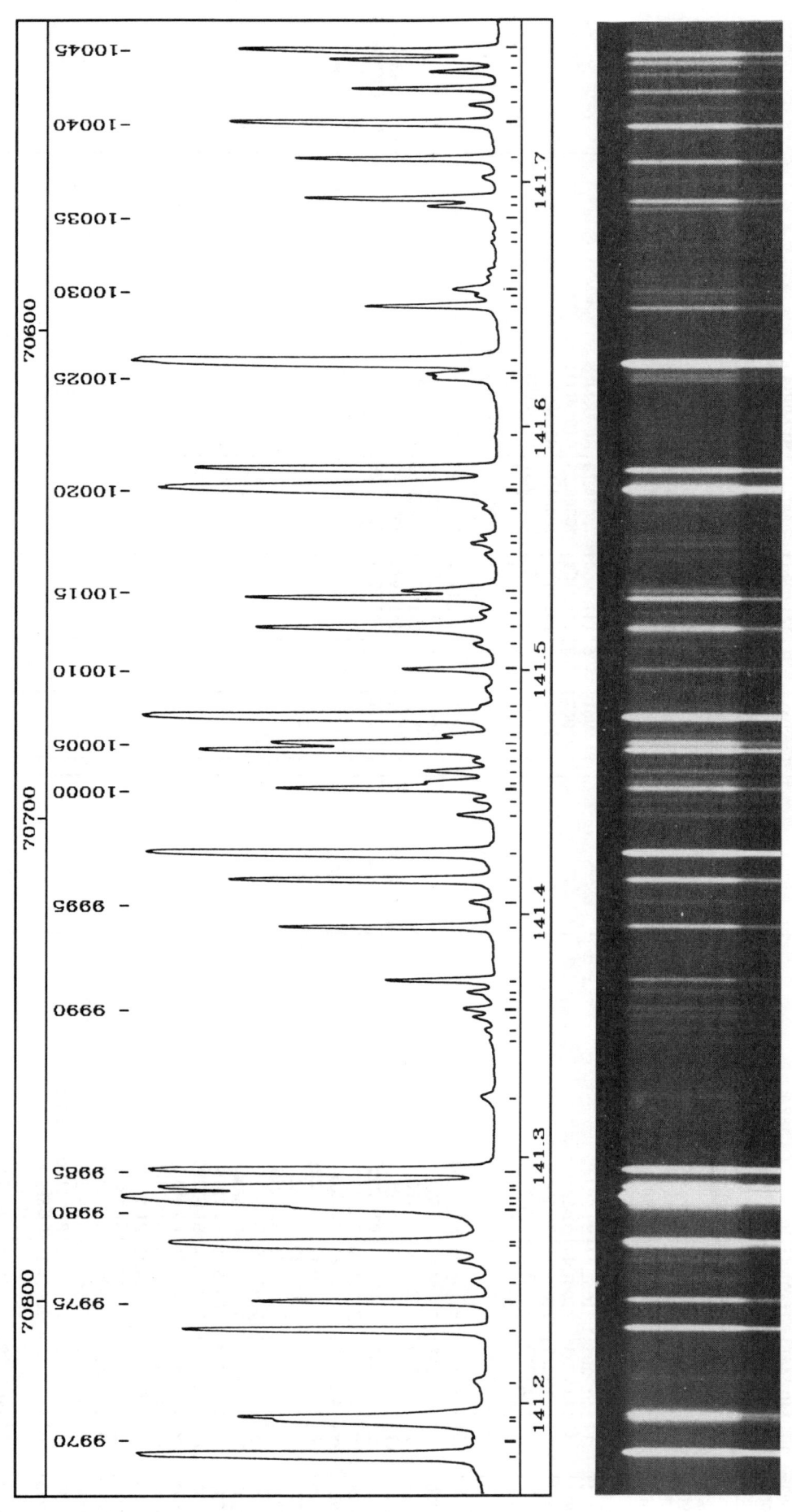

PLATE 111

TABLE 111

Nb	I	λ (nm)	σ (cm-1)	Assignment	Comment
9973	1	141.2092	70816.94	B4-5P15	
9974	27	141.2300	70806.50	B12-11P2	
9975	13	141.2415	70800.70	C2-11R1 B9-9P6	
9976	1	141.2499	70796.50	B3-2R25 B6-6R16	d
9977	1	141.2580	70792.43	B12-11R4 C1-10R8	
9978	12	141.2642	70789.34	B6-8R0 B9-6P20	
9979	33	141.2653	70788.78	B2-6R5	
9980	2	141.2797	70781.56	B8-8R11	
9981		141.2824	70780.20	B2-5P11 B5-6R14	
9982	428	141.2838	70779.50	B0-5P5	sh
9983	58	141.2876	70777.61	B6-8R1	
9984		141.2896	70776.63	B14-12P7 B3-6R9	sh
9985	62	141.2946	70774.10	B4-7P2	
9986	1	141.3246	70758.99	B0-2P20	
9987	0	141.3478	70747.50	B13-11R12	
9988	0	141.3520	70745.37	B15-14R4	
9989	1	141.3572	70742.76	B12-10R13	
9990	2	141.3609	70740.93	B10-10R0	
9991	0	141.3656	70738.58	B8-7R16 C1-10R7	
9992	7	141.3727	70737.45	B11-8P17	
9993	12	141.3949	70735.00	B10-10R1	
9994	1	141.4052	70723.89	B6-8R2	
9995	14	141.4142	70718.78	B13-11P9	
9996	65	141.4253	70714.28	B5-7P7	
9997	3	141.4412	70708.73	B12-11P3	
9998	1	141.4475	70700.76	B7-8R8 B12-11R5	
9999	10	141.4521	70697.59	B9-9R8 B10-10R2	
10000	3	141.4544	70695.29	B11-10P7 B13-10P15	
10001	1	141.4592	70694.15	B1-3P18 B8-9R0	
10002	1	141.4634	70691.78	B6-7P10	
10003	2	141.4678	70689.67	B13-11R13	
10004	30	141.4707	70687.45	B6-8P1	
10005	11	141.4738	70685.99	B4-7R4 B15-14R3 B4-4P19	
10006	3	141.4815	70684.45	B8-9R1	
10007	80	141.4858	70680.63	B2-6P4 B9-8R14	
10008	0	141.4931	70678.45	B14-12P8	
10009	0	141.5013	70674.83	B3-2P24 B6-5R20	
10010	5	141.5115	70670.73	B5-5P17	
10011	2	141.5182	70665.63	B1-4P14	
10012	10	141.5246	70662.31	B0-4P12	
10013	1	141.5305	70659.07	B12-10P11	
10014	17	141.5335	70656.13	B10-10P1	
10015	4	141.5483	70654.66	B4-6R12 B10-9P10 B11-10R9 B8-6R20	b
10016	0	141.5530	70647.25	B11-9R15	
10017	2	141.5559	70644.92	B7-7R14	
10018	1	141.5674	70643.49	B15-14R2	
10019	0	141.5755	70637.73	B8-9R2	
10020	55	141.5779	70633.67	B0-5R7	
10021		141.5779	70632.49	B10-10R3 B12-9P16	sh

Nb	I	λ (nm)	σ (cm-1)	Assignment	Comment
10022	21	141.5835	70629.69	B6-8R3	
10023	0	141.5964	70623.27	B5-1P29	
10024	4	141.6206	70611.21	B3-6P8	
10025	4	141.6223	70610.34	B15-14R1	
10026	155	141.6273	70607.85	B4-7P3 B2-2R24	
10027	0	141.6411	70600.97	B12-11P4	
10028	6	141.6504	70596.36	B14-12P9	
10029	0	141.6544	70594.34	B8-8P10 B12-11R6 B10-8P15	
10030	2	141.6571	70592.98	C2-11P9	
10031	0	141.6619	70590.59	B9-6R22	
10032	0	141.6650	70589.07	B15-14R0	
10033	0	141.6764	70583.40	B6-6P15	
10034	0	141.6798	70581.67	B6-3P25	
10035	0	141.6865	70578.33	B5-6P13	
10036	3	141.6913	70575.98	B6-8P2	
10037	11	141.6944	70574.43	B10-9R12 B12-10R14	
10038	1	141.7031	70570.11	B10-10P2 B3-1R28	
10039	13	141.7105	70566.38	B7-8P7	
10040	14	141.7252	70559.08	B8-9R3	
10041	2	141.7324	70555.49	C2-11P3	
10042	8	141.7393	70552.07	B15-14P4	
10043	4	141.7465	70548.50	B2-6R6	
10044	7	141.7511	70546.18	B9-9P7 B15-14P1	
10045	13	141.7547	70544.40		

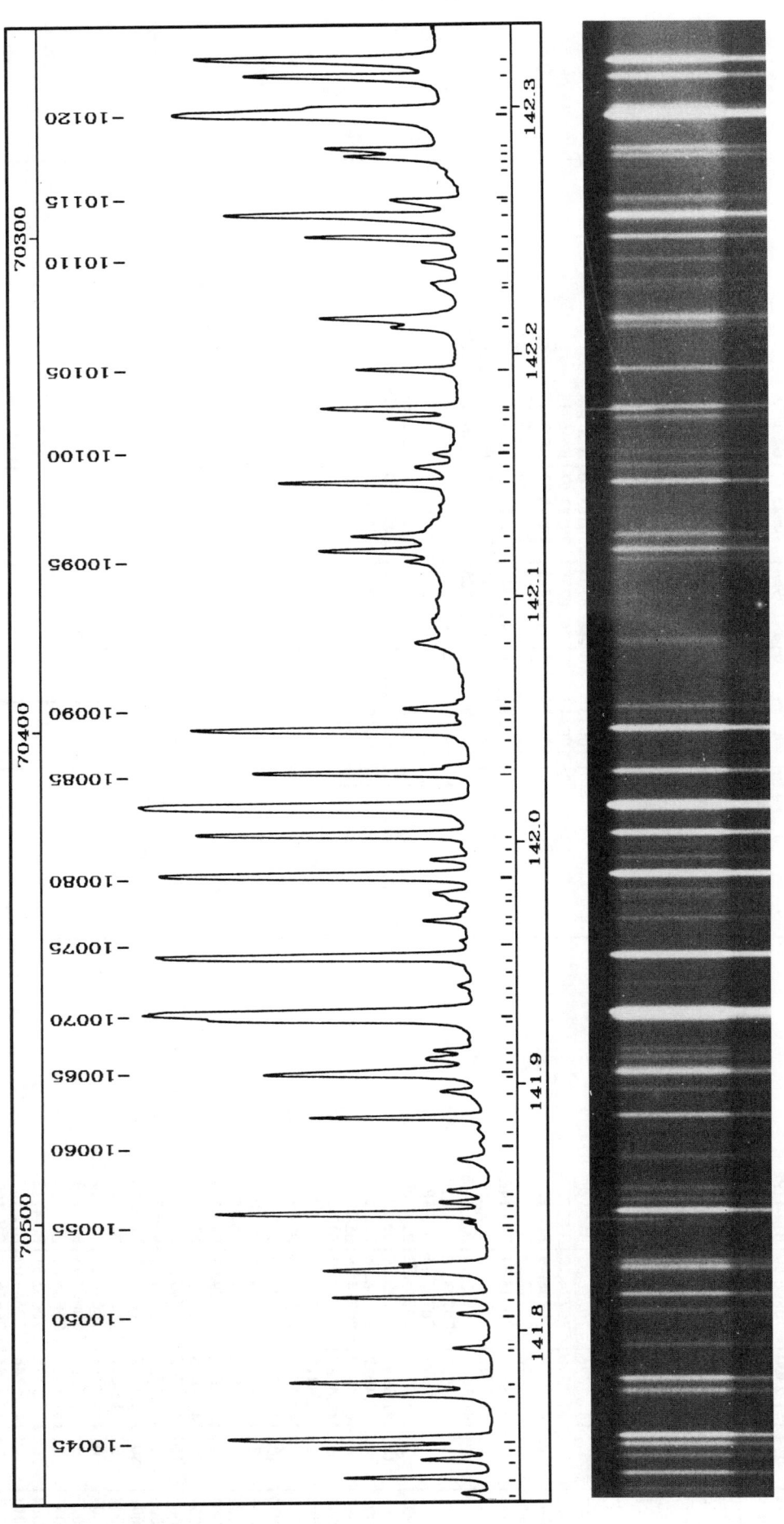

PLATE 112

TABLE 112

Nb	I	λ (nm)	σ (cm⁻¹)	Assignment	Comment
10046	6	141.7728	70535.40	B1-5P9 B2-4P16 B10-7P19	
10047	12	141.7777	70532.94	B15-14P2 B15-14P3 B5-4R22	
10048	2	141.7914	70526.13	C2-11P5	
10049	0	141.7939	70524.92	B5-7R9	
10050	2	141.8052	70519.28	B9-7P17	
10051	8	141.8119	70515.95	C2-11P4 C2-11P6	
10052	10	141.8229	70510.48	B3-5R15 B3-4R19	
10053	5	141.8253	70509.29	B13-11P10 B7-6P17	
10054	0	141.8405	70501.75	B10-10R4 B10-7R21	
10055					CuII 141.84265
10056	14	141.8460	70499.00	B4-7R5	
10057	2	141.8505	70496.75	B6-8R4	
10058	2	141.8553	70494.38	B2-3R21 B8-4P25	
10059	1	141.8679	70488.09	B0-3R18	
10060	0	141.8748	70484.70	B3-1P27	
10061	0	141.8805	70481.86	C0-9R9	
10062	9	141.8851	70479.57	B13-12R1 C1-10R3	
10063	2	141.8957	70474.30	B13-12R0	
10064	9	141.9030	70470.69	B2-5R13 B12-11R7	
10065	0	141.9043	70470.01	B2-2P23	sh
10066	2	141.9089	70467.74	B13-12R2 B6-7R12	
10067	1	141.9124	70466.02	B11-10P8	
10068	0	141.9157	70464.35	B2-1R27	
10069	0	141.9254	70459.55	B4-6P11	sh
10070	73	141.9273	70458.60	B0-5P6 B6-5P19	
10071	0	141.9354	70454.61		
10072	0	141.9390	70452.78	B11-10R10	
10073	0	141.9490	70447.86		
10074	44	141.9508	70446.96	B10-10P3	
10075	0	141.9612	70441.76		b
10076	0	141.9659	70439.43	B8-9R4 B13-12R3	
10077	0	141.9690	70437.91	B4-5R17	
10078	0	141.9749	70434.99	B9-8P13 B12-10R15	
10079	2	141.9772	70433.86	B8-8R12	
10080	40	141.9842	70430.35	B12-11P5	
10081	2	141.9930	70426.02	B7-7P13	
10082	0	141.9964	70424.33	C1-10R2	
10083	24	142.0016	70421.76	B6-8P3 B9-9R9 B12-10P12	
10084	100	142.0129	70416.12	B2-6P5	
10085	15	142.0270	70409.15	B13-12P1	
10086	0	142.0296	70407.84	B7-8R9	
10087	0	142.0410	70402.19	B11-9R16	
10088	23	142.0446	70400.41	B4-7P4	
10089	0	142.0496	70397.93	B13-12R4 B10-10R5	
10090	3	142.0537	70395.91	B8-6P19 B12-9P17	
10091	0	142.0566	70394.47	B11-8P18	
10092	1	142.0803	70382.76	B1-2R23	
10093	1	142.0887	70378.57	B9-7R19	
10094	0	142.0995	70373.24	B2-1P26	
10095	2	142.1139	70366.11	C1-10R1	
10096	5	142.1180	70364.07	B5-7P8 B13-11P11	
10097	3	142.1240	70361.08	B8-9P3	
10098	10	142.1458	70350.29	B13-12P2	
10099	2	142.1524	70347.06	B13-12R5 B14-13R3 B14-13R4 B14-13R5	
10100	1	142.1581	70344.24	B14-13R2	
10101	0	142.1616	70342.47	B12-11R8	
10102	3	142.1722	70337.22	B10-9P11 B14-13R0 B5-4P21	
10103					CuII 142.17589
10104	7	142.1764	70335.17	B1-1R26 B14-13R1	
10105	4	142.1923	70327.28	B6-8R5	
10106	2	142.2099	70318.59	B9-8R15	
10107	3	142.2132	70316.97	B0-4R14 B9-5P24	
10108	0	142.2260	70310.63	B2-3P20 B8-9R5	
10109	2	142.2280	70309.62	C3-13R0	
10110	2	142.2369	70305.22	B3-4P18	
10111	0	142.2417	70302.86	B8-7R17	
10112	9	142.2465	70300.52	B10-10P4	
10113	10	142.2553	70296.17	B0-5R8	
10114		142.2613	70293.18	B3-5P14	sh
10115	3	142.2621	70292.78	B13-12R6 B10-9R13	
10116	0	142.2732	70287.33	B5-5R19	
10117	0	142.2766	70285.62	B9-6P21	
10118	3	142.2797	70284.10	B12-11P6 B6-6R17	
10119	3	142.2827	70282.60	B5-6R15	
10120	48	142.2963	70275.88	B14-13P1 B13-12P3 B0-3P17	
10121	3	142.2990	70274.55	B4-7R6	
10122	7	142.3122	70268.06	B9-9P8 B6-7P11	
10123	16	142.3189	70264.74	B2-6R7 B1-4R16	

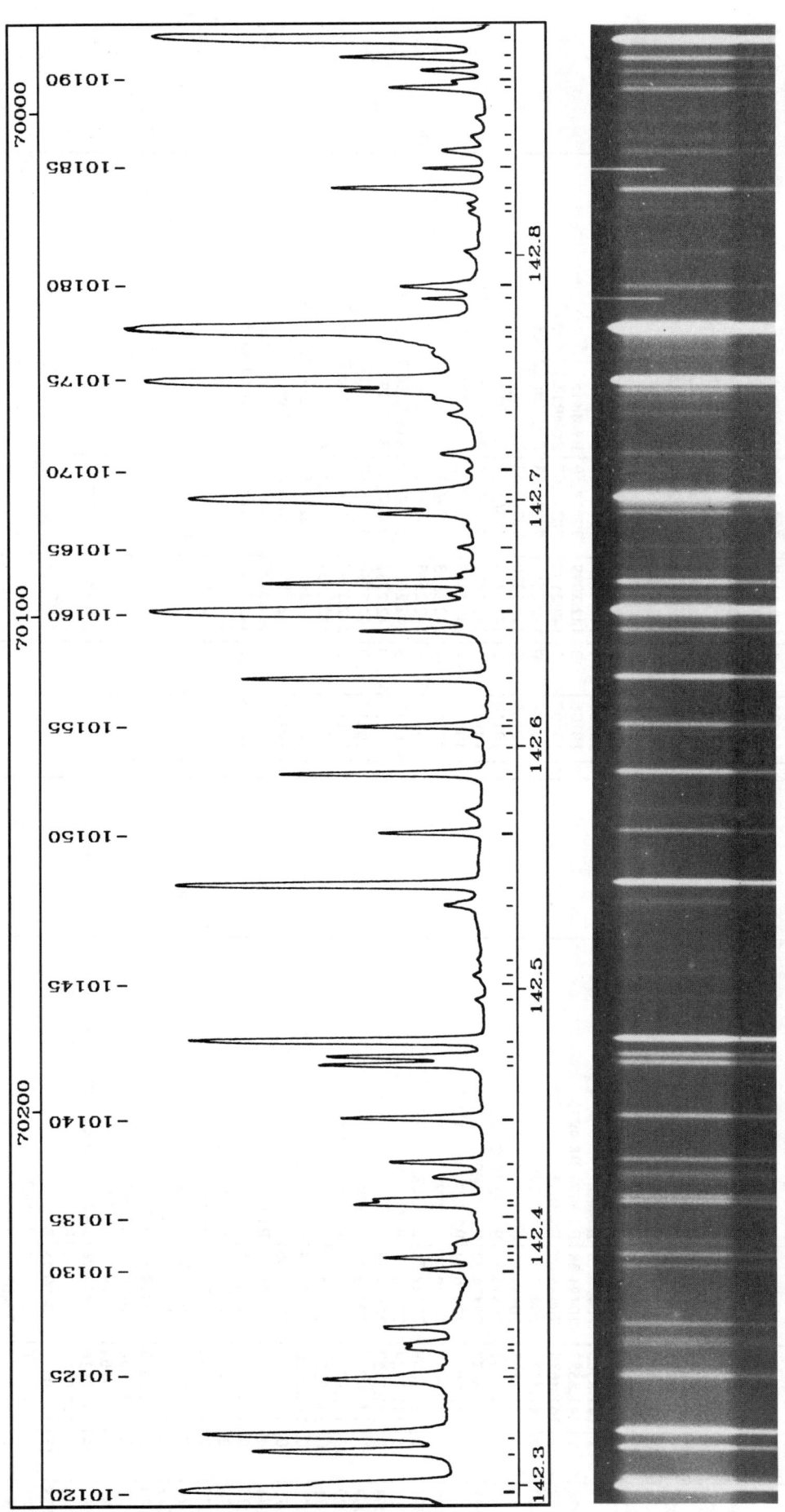

PLATE 113

TABLE 113

Nb	I	λ (nm)	σ (cm⁻¹)	Assignment	Comment
10124	3	142.3414	70253.62	B2-5P12 B10-10R6	
10125	0	142.3434	70252.63	C3-13P3	
10126	0	142.3485	70250.11	B13-12R7	
10127	2	142.3541	70247.38	B11-10R11	
10128	2	142.3559	70246.47	B14-13P2	
10129	2	142.3624	70243.25	B7-8P8	
10130	2	142.3859	70231.68	B6-8P4 B7-6R19	
10131	3	142.3906	70229.35	B11-10P9	
10132	0	142.3936	70227.88	B1-1P25	
10133	0	142.3960	70226.70	B1-2P22	
10134	0	142.4034	70223.04	B12-9P18	
10135	0	142.4078	70220.89	B12-10P13	
10136	3	142.4121	70218.74	B8-8P11	
10137	4	142.4137	70217.94	C3-13P2 B4-5P16	
10138	1	142.4229	70213.43	B4-4R21 B12-11R9	
10139	3	142.4289	70210.45	B7-7R15 B10-8P16	
10140	4	142.4468	70201.67	B3-6P9	
10141	7	142.4686	70190.90	B8-9P4 B4-6R13	
10142	8	142.4719	70189.28	B13-12P4	
10143	33	142.4784	70186.09	B14-13P3	
10144	0	142.4957	70177.56	B9-9R10	
10145	0	142.5034	70173.77		
10146	0	142.5055	70172.74	B8-6R21 B5-7R10	
10147	0	142.5126	70169.25	B4-2R27 B9-6R23	
10148	1	142.5349	70158.25	B1-3R20	b
10149	25	142.5430	70154.29	B4-7P5 B13-12R8	
10150	6	142.5643	70143.79	B14-13P4	
10151	0	142.5731	70139.47	B4-3P23	
10152	10	142.5886	70131.84	B10-10P5	
10153	0	142.5993	70126.57	B10-7P20	
10154	0	142.6041	70124.23	B6-8R6	
10155	3	142.6081	70122.26	B12-11P7 C2-12R7	
10156	0	142.6107	70120.98	B13-12R9	
10157	0	142.6168	70117.95	B0-2R22	
10158	13	142.6276	70112.64	B2-6P6	
10159	8	142.6470	70103.11	B14-13P5	
10160	73	142.6552	70099.11	B0-5P7	
10161	1	142.6621	70095.68	B7-8R10 B6-5R21	
10162	11	142.6668	70093.41	B13-12P5	
10163	0	142.6703	70091.69	B11-8P19	
10164	0	142.6753	70089.24	B12-11R10 B10-10R7	
10165	0	142.6810	70086.40	C0-9R1	
10166	0	142.6893	70082.36		
10167	4	142.6952	70079.46	B8-8R13	
10168	4	142.6988	70077.69	C1-10P3	
10169	17	142.7013	70076.46	B0-4P13	
10170	0	142.7122	70071.09	B14-13P6	
10171	1	142.7193	70067.63	B1-5P10 B9-7P18	
10172	1	142.7357	70059.54	B5-5P18	
10173	2	142.7420	70056.45	B5-6P14 B14-13P7 B3-3R23	
10174	4	142.7453	70054.86	B6-7R13	
10175	185	142.7491	70053.00	B1-6R0	sh
10176	0	142.7602	70047.52	B11-10R12 B0-1P24	
10177	2	142.7700	70044.04	B6-6P16	
10178	620	142.7706	70042.42	B1-6R1	CuII 142.78290
10179					
10180	3	142.7877	70034.03	B1-4P15	
10181	0	142.8020	70027.03	C1-10P4	b
10182	0	142.8180	70019.18	B10-9P12	
10183	0	142.8212	70017.62	B10-9R14	
10184	6	142.8274	70014.56	B4-7R7	CuII 142.83580
10185					
10186	1	142.8428	70007.01	B6-8P5 B11-10P10	
10187	0	142.8485	70004.23	B4-4P20 B3-2R26	
10188	0	142.8563	70000.43	B12-11R11	
10189	7	142.8683	69994.53	B11-11R0 B11-11R1	
10190	1	142.8711	69993.14	B13-12P6	
10191	3	142.8754	69991.04	B8-9P5	
10192	5	142.8810	69988.33	B5-7P9	
10193	124	142.8885	69984.65	B1-6R2	

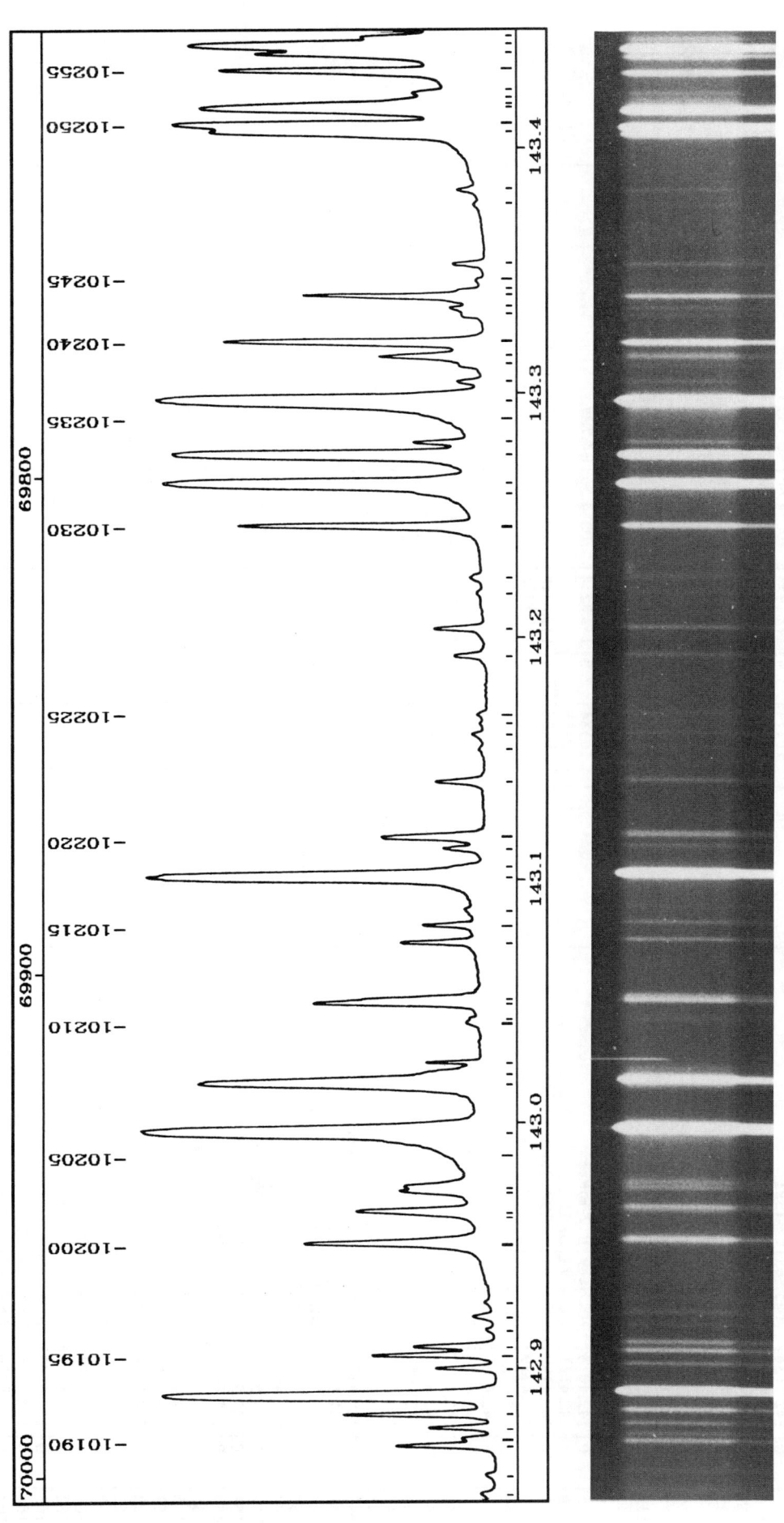

PLATE 114

TABLE 114

Nb	I	λ (nm)	σ (cm⁻¹)	Assignment	Comment
10194	2	142.8997	69979.17	C1-10P5	
10195	4	142.9051	69976.53	B9-9P9	
10196	2	142.9084	69974.89	C1-10P6 B4-6P12	
10197	0	142.9153	69971.50	B7-6P18	
10198	1	142.9206	69968.90	B11-11R2 B9-8R16	
10199	0	142.9267	69965.94	B7-7P14	
10200	5	142.9506	69954.26	B1-3P19 B12-11P8	
10201		142.9628	69948.26	B3-5R16	sh
10202	3	142.9638	69947.79	B2-6R8	
10203	1	142.9718	69943.85	B2-5R14	
10204	1	142.9737	69942.93	B0-2P21	
10205		142.9928	69933.60		sh
10206	640	142.9960	69932.03	B1-6P1	
10207	27	143.0160	69922.27	B0-5R9	
10208	0	143.0202	69920.20	B11-11R3 B11-11P1	
10209	0	143.0407	69910.17	C1-10P10 C1-11R11	CuII 143.02428
10210	0	143.0423	69909.39		
10211	0	143.0488	69906.20		
10212	7	143.0500	69905.61	B7-8P9 B8-9R7 B9-9R11 B10-10R8	
10213	3	143.0738	69894.00	B10-10P6	
10214	2	143.0811	69890.43	B13-12P7	
10215	1	143.0874	69887.35	B6-8R7	
10216	0	143.1014	69880.54	B5-4R23	
10217	175	143.1050	69878.78	B1-6R3	
10218	0	143.1132	69874.76	B8-6P20	
10219	1	143.1181	69872.37	B12-11R12 B2-4P17	
10220	3	143.1405	69861.45	B4-7P6 B14-14R4	
10221	2	143.1538	69854.96	B11-10R13 B3-2P25	
10222	0	143.1598	69852.02	B3-4R20	
10223	0	143.1648	69849.59	B6-5P20	
10224	0	143.1679	69848.06	B11-11R4	
10225	0	143.1920	69836.29	B10-8P17	
10226	1	143.2034	69830.75	B8-8P12	
10227	2	143.2176	69823.84	B6-7P12	
10228	0	143.2243	69820.56	B12-11R13	
10229	0	143.2456	69810.18	B9-5P25 C3-14P4	
10230	17	143.2594	69803.44	B14-14R3 B2-2R25	
10231		143.2629	69801.75	B13-12P8	sh
10232	360	143.2749	69795.90	B1-6P2	
10233	210	143.2801	69793.36	B3-7R0	
10234	4	143.2934	69786.90	C3-14R1	
10235		143.2970	69785.14	B12-11P9	sh
10236	720	143.3048	69781.31	B3-7R1	
10237	1	143.3116	69778.00	B5-6R16	
10238	0	143.3154	69776.16	B7-7R16	
10239	2	143.3211	69773.42	B0-3R19 B6-6R18	
10240	17	143.3322	69768.01	B2-6P7	
10241	0	143.3353	69766.48	B7-8R11	
10242	0	143.3369	69765.30	B3-6P10 B8-9P6 B2-3R22	
10243	11	143.3402	69764.10	B14-14R2	
10244	0	143.3424	69763.02	C0-9P3	
10245	0	143.3467	69760.92	B11-11R5 B10-7P21	sh
10246	1	143.3534	69757.68	B10-9R15	
10247	0	143.3778	69745.82	B11-11P3	
10248	0	143.3841	69742.73	B13-12P9	
10249	30	143.4074	69731.39	B1-6R4 B11-10P11	
10250	137	143.4101	69730.08	B3-7R2	
10251	40	143.4170	69726.74	B14-14R1	
10252	0	143.4189	69725.83	B0-4R15	
10253	0	143.4208	69724.87	B8-8R14	
10254	0	143.4239	69723.41	B4-7R8 B8-6R22	
10255	20	143.4324	69719.24	B9-10R0	
10256	15	143.4392	69715.94	B14-14R0 C3-14P3 B10-10R9	
10257	85	143.4424	69714.39	B9-10R1 B4-6R14 B10-10P7 B4-1P29	
10258	2	143.4463	69712.52	B3-5P15	

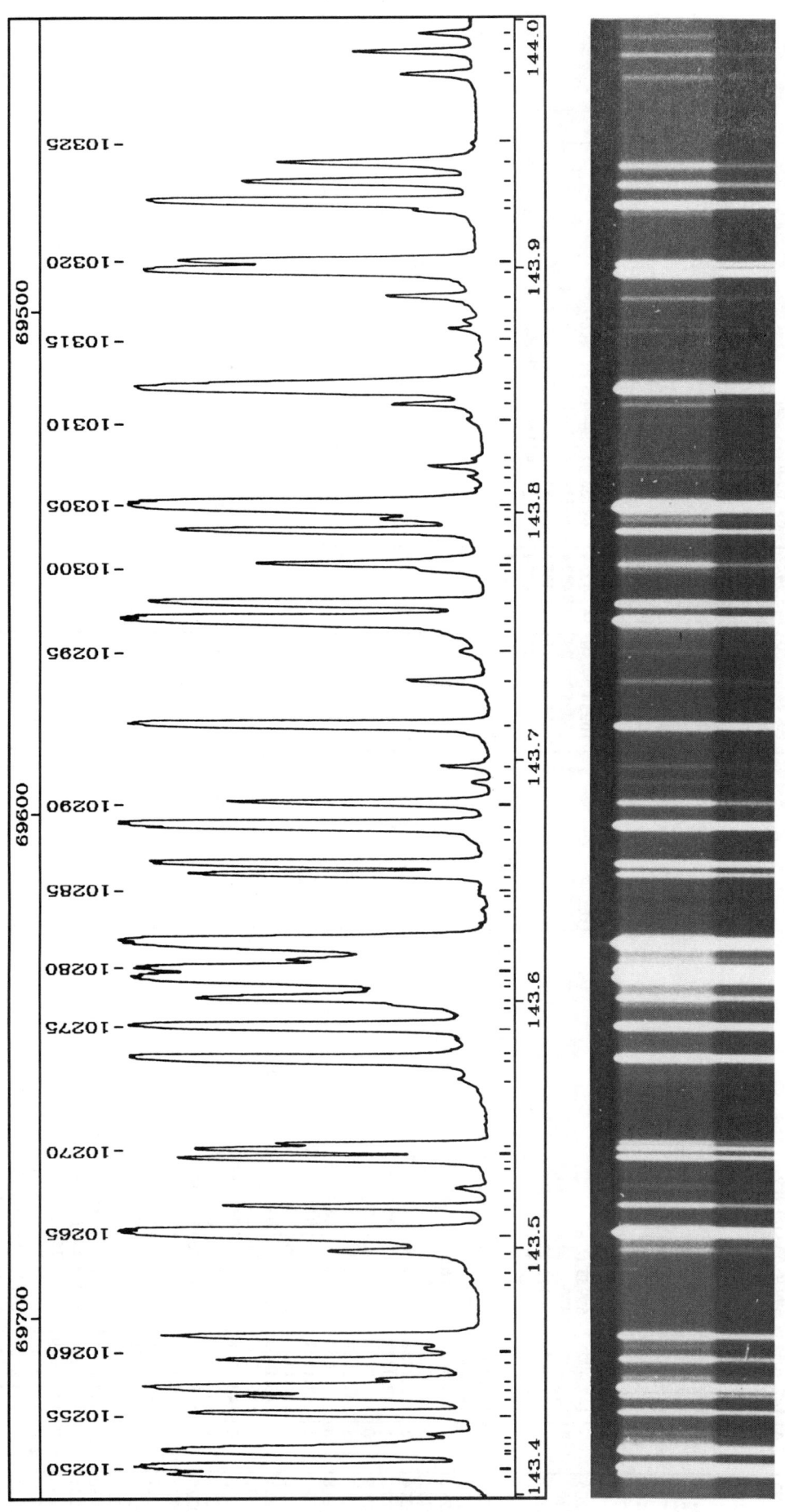

PLATE 115

TABLE 115

Nb	I	λ (nm)	σ (cm⁻¹)	Assignment	Comment
10259	10	143.4539	69708.82	B2-5P13 B5-5R20	
10260	1	143.4588	69706.42	B10-9P13	
10261	16	143.4631	69704.34	B0-5P8 B11-10R14	
10262	0	143.4854	69693.50	C0-9P4	
10263					CuII 143.49037
10264	10	143.4978	69687.50	C3-14R0	
10265	760	143.5049	69684.05	B3-7P1	
10266	19	143.5160	69678.62	B9-10R2	
10267	1	143.5238	69674.86	B9-9P10	
10268					CuII 143.53155
10269	37	143.5354	69669.21	B14-14P4	
10270	34	143.5391	69667.41	B14-14P1	
10271	10	143.5414	69666.31	B14-14P2	
10272	0	143.5684	69653.19	B2-2P24	
10273	177	143.5760	69649.50	B14-14P3	
10274	0	143.5796	69647.77	B5-3P25	
10275	150	143.5889	69643.25	B5-8R0	
10276	0	143.5980	69638.86	B9-7P19	
10277	30	143.6006	69637.59	C3-14P2	
10278	2	143.6018	69637.03	B1-4R17 B9-8R17	
10279	575	143.6090	69633.54	B5-8R1 B9-10P1 B6-7R14	
10280	215	143.6125	69631.82	B3-7R3	
10281	11	143.6163	69630.00	B6-8R8 B3-4P19 B9-9R12	
10282	1095	143.6230	69626.75	B1-6P3 B11-11P4 B12-11P10 C0-9P5	
10283	0	143.6370	69619.95		
10284	0	143.6434	69616.85	B4-5P17 B2-1R28	
10285	0	143.6456	69615.80	B7-4P25	
10286	25	143.6513	69613.00	B9-10R3	
10287	77	143.6556	69610.94	B7-9R0	
10288	0	143.6660	69605.90		
10289	275	143.6718	69603.10	B7-9R1	
10290	15	143.6811	69598.57	B2-6R9	
10291	1	143.6898	69594.35	C2-12R3	
10292	2	143.6963	69591.21	B5-7P10	
10293	111	143.7128	69583.21	B5-8R2 B1-2R24	
10294	3	143.7311	69574.35	B1-5P11 B8-7P17	
10295	1	143.7433	69568.47	B11-10R15	
10296		143.7549	69562.84	B2-3P21	sh
10297	380	143.7558	69562.40	B3-7P2	
10298	63	143.7627	69559.05	B7-9R2 B4-7P7 C0-9P6	
10299	0	143.7764	69552.44	B7-8P10	
10300	12	143.7789	69551.24	B0-3P18 B12-12R1	
10301		143.7807	69550.37	B12-12R2	sh
10302	35	143.7925	69544.65	B9-10P2	
10303	7	143.7975	69542.25	B12-12R0	
10304	500	143.8025	69539.80	B5-8P1	sh
10305	0	143.8053	69538.47	B1-6R5	
10306	0	143.8095	69536.45	B6-5R22	
10307	1	143.8145	69534.01	B8-9P7	
10308	4	143.8187	69531.98	B5-6P15	
10309	0	143.8221	69530.36	B13-13R7	
10310	0	143.8381	69522.59	B10-9R16	
10311	4	143.8443	69519.60	B9-10R4	
10312	230	143.8509	69516.41	B7-9P1	sh
10313	0	143.8524	69515.71	B0-5R10 B10-8P18	
10314	0	143.8644	69509.91	B6-6P17	
10315	0	143.8702	69507.08		
10316	2	143.8750	69504.78	B7-7P15	
10317	1	143.8784	69503.12	B11-10P12	
10318	4	143.8880	69498.49	B12-12R3 B2-1P27	
10319	160	143.8985	69493.45	B5-8R3	
10320	36	143.9018	69491.82	B3-7R4 B10-10P8 B11-11P5	
10321	1	143.9228	69481.69	C0-9P7 B13-13R6	
10322	83	143.9261	69480.10	B7-9R3 B12-11P11	
10323	15	143.9343	69476.15	B4-6P13 B12-12P1 B1-1R27	
10324	9	143.9422	69472.33	B0-4P14	s
10325	0	143.9502	69468.46	B12-12R4	
10326	5	143.9784	69454.86	B13-13R3 B5-5P19	
10327	7	143.9877	69450.40	B13-13R2 B8-8P13	
10328	2	143.9953	69446.71	B7-6P19	

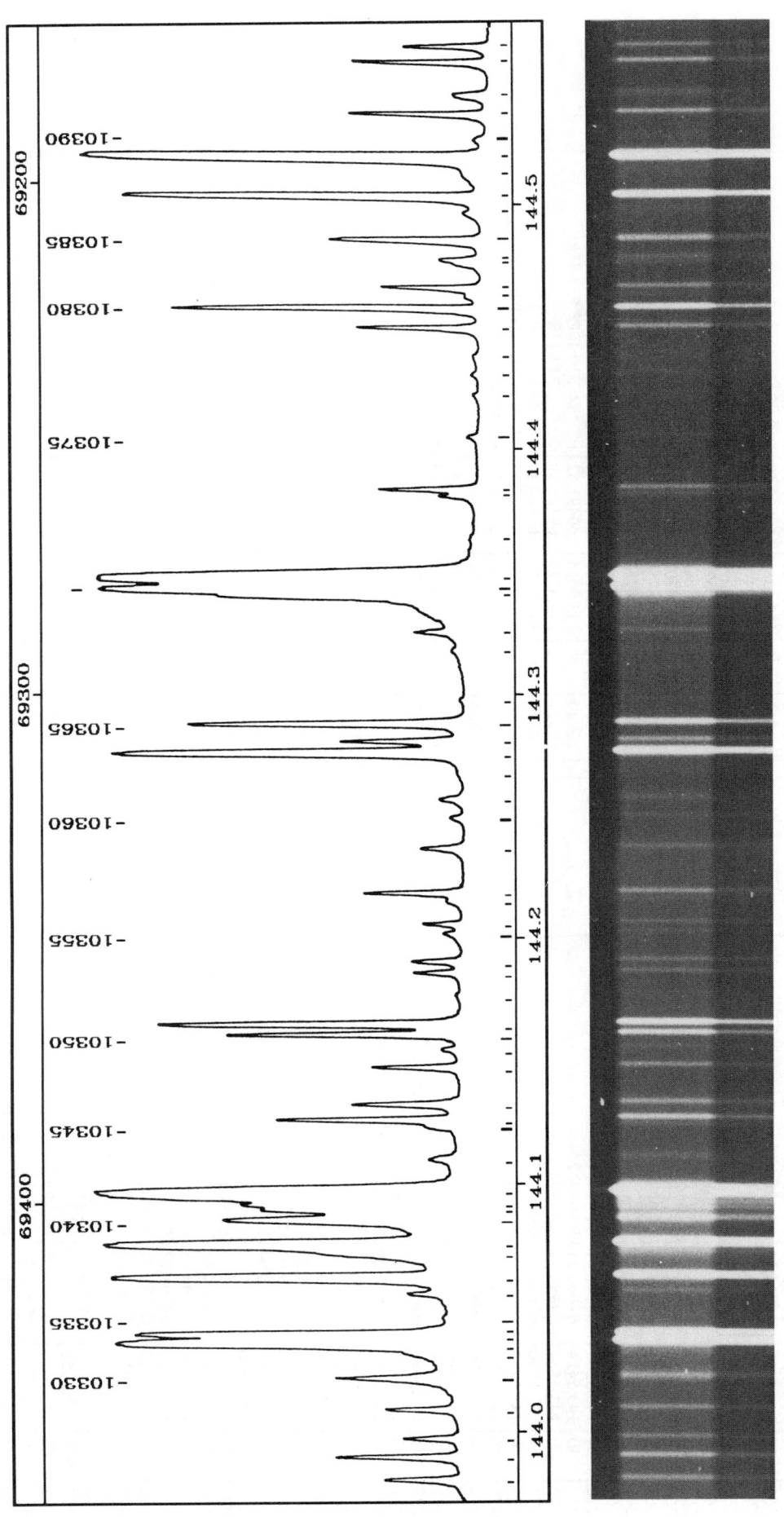

PLATE 116

TABLE 116

Nb	I	λ (nm)	σ (cm⁻¹)	Assignment	Comment
10329	4	144.0069	69441.10	B13-13R1	
10330	5	144.0199	69434.87	B1-3R21	d
10331		144.0309	69429.52	B7-8R12	sh
10332	250	144.0341	69427.98	B5-8P2	
10333	56	144.0374	69426.40	B9-10P3	
10334	0	144.0400	69425.14	B13-13R0	
10335	0	144.0443	69423.11	B12-12R5	
10336	1	144.0543	69418.25	B12-12P2	
10337	122	144.0607	69415.19	B7-9P2	
10338	5	144.0709	69410.27	B1-2P23	
10339	230	144.0740	69408.79	B1-6P4	
10340	14	144.0841	69403.90	B4-7R9 B2-5R15	
10341	12	144.0884	69401.84	B2-6P8	
10342	14	144.0906	69400.80	B9-10R5 C0-9P11	
10343	1020	144.0949	69398.70	B3-7P3	
10344	1	144.1086	69392.10	B1-4P16 B10-9P14	
10345	0	144.1216	69385.84	B8-6P21	
10346	10	144.1244	69384.49	B6-7P13	
10347	7	144.1307	69381.46	B3-5R17	
10348	7	144.1458	69374.19	B8-8R15	
10349	0	144.1533	69370.57	B13-13P1 B12-12R6	
10350	12	144.1589	69367.92	B7-9R4	
10351	20	144.1631	69365.88	B5-8R4	
10352	2	144.1755	69359.92	B12-11P12	
10353	2	144.1843	69355.66	B9-9R13	
10354	3	144.1890	69353.41	B7-7R17	
10355	0	144.2007	69347.79	B12-12P3 B1-1P26	d
10356	1	144.2045	69345.96	B6-8R9	
10357	0	144.2143	69341.26	B11-11P6	
10358	4	144.2172	69339.86	B9-9P11	
10359	2	144.2355	69331.07	B4-4P21	
10360	0	144.2481	69325.00	B10-9R17	
10361	1	144.2558	69321.31	B0-2R23 B10-10R11	
10362	0	144.2660	69316.38	B8-6R23	
10363	47	144.2747	69312.20	B3-7R5	
10364	6	144.2794	69309.94	B3-6P11	
10365	13	144.2867	69306.48	B1-6R6	
10366	0	144.2965	69301.77	B12-11P13	
10367	0	144.3165	69292.13	B13-13P3	
10368	2	144.3243	69288.42	B11-10P13	
10369	15	144.3396	69281.08	B9-10P4 B5-6R17	
10370	300	144.3424	69279.70	B7-9P3 B6-6R19	
10371	730	144.3467	69277.64	B0-5P9 B5-8P3	
10372	1	144.3621	69270.27	B7-6R21	
10373	1	144.3803	69261.55	B6-5P21	
10374	3	144.3828	69260.31	B10-10P9 B9-10R6	
10375	0	144.4042	69250.07	B13-13P4	
10376	0	144.4220	69241.54	B9-7P20	
10377	0	144.4301	69237.63	B8-9P8 B12-12R9	
10378	0	144.4376	69234.05	B3-2R27	d
10379	8	144.4486	69228.76	B4-6R15 B12-12R10	
10380	17	144.4566	69224.93	B7-9R5	
10381	1	144.4619	69222.41	B10-8P19	
10382	5	144.4652	69220.81	B2-6R10	
10383	0	144.4754	69215.94	B4-7P8	
10384	3	144.4770	69215.16	B1-3P20	
10385	5	144.4849	69211.40	B13-13P5 B2-4P18 B6-7R15	
10386	1	144.4967	69205.73	B3-4R21	
10387	33	144.5027	69202.84	B5-8R5	
10388	0	144.5118	69198.48		
10389	202	144.5187	69195.19	B3-7P4	
10390	0	144.5266	69191.40	C2-12P3 C2-12P4	
10391	6	144.5365	69186.66	B7-8P11	
10392	2	144.5443	69182.95	B13-13P6 B11-11P7	
10393	5	144.5573	69176.74	B5-7P11	
10394	4	144.5636	69173.72	B13-13P7	

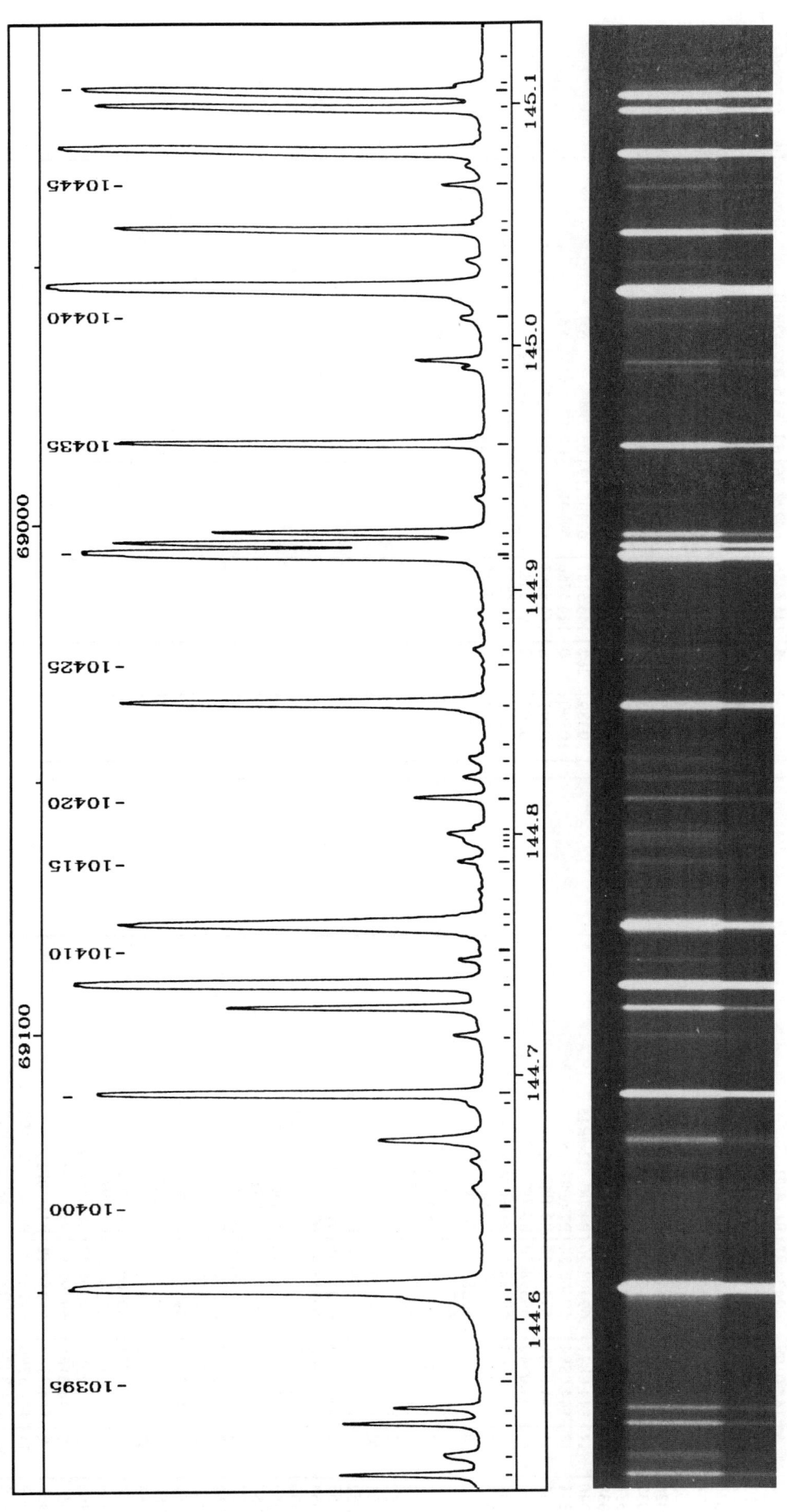

PLATE 117

TABLE 117

Nb	I	λ (nm)	σ (cm⁻¹)	Assignment	Comment
10395	0	144.5809	69165.42	B0-1P25	
10396	0	144.5862	69162.89		
10397	1	144.6085	69152.25	B12-12P5	
10398	340	144.6121	69150.50	B1-6P5 B2-5P14	
10399	0	144.6327	69140.66	B5-5R21	
10400	0	144.6467	69133.95	B8-7P18	
10401	0	144.6536	69130.65	B0-2P22	
10402	6	144.6646	69125.40	B3-5P16	
10403	0	144.6730	69121.39	B0-4R16	
10404	0	144.6882	69114.14	B10-9P15	
10405	60	144.6918	69112.40	B7-9P4 B9-10P5	
10406	1	144.7158	69100.97	B9-10R7	
10407	11	144.7270	69095.61	B3-7R6	
10408	122	144.7366	69091.04	B5-8P4 B9-9R14	
10409	2	144.7469	69086.12	B7-8R13	
10410	0	144.7514	69083.96		
10411	32	144.7613	69079.23	B0-5R11	
10412	0	144.7634	69078.26		
10413	0	144.7764	69072.02		
10414	0	144.7852	69067.82	B11-11R11	
10415	1	144.7879	69066.55	B8-8P14 B3-2P26	
10416	0	144.7948	69063.27	B12-12P6	
10417	0	144.7967	69062.36	B0-3R20	
10418	1	144.7994	69061.08	B4-7R10	
10419	0	144.8026	69059.51	B1-5P12	
10420	3	144.8143	69053.95	B7-9R6	
10421	1	144.8232	69049.71	B9-9P12 B7-7P16	
10422	0	144.8307	69046.15	B2-3R23	
10423	0	144.8371	69043.07	B6-8R10	
10424	27	144.8527	69035.64	B1-6R7 B8-8R16	
10425	0	144.8684	69028.19	B7-4P26	
10426	0	144.8754	69024.84	B10-10P10	
10427	0	144.8859	69019.84	B11-11P8	
10428	0	144.8899	69017.94	B4-5P18	d
10429	0	144.9138	69006.55	B5-8R6	
10430	115	144.9188	69006.04	B10-11R1 B5-6P16	
10431	30	144.9188	69004.16	B10-11R0 B1-4R18	sh
10432	10	144.9233	69002.00	B2-6P9 B6-5R23	
10433	0	144.9380	68995.04	C1-11R5	
10434	30	144.9464	68991.03	B10-8P20	
10435	30	144.9602	68984.46	B10-11R2 B13-14R4 B6-6P18	
10436	0	144.9737	68978.02		
10437	2	144.9913	68969.66	B12-12P7	
10438	4	144.9944	68968.16	B4-6P14	
10439	0	144.9994	68965.79		
10440	1	145.0119	68959.85	B3-4P20	
10441	270	145.0238	68954.19	B3-7P5 B8-9P9	
10442	1	145.0354	68948.68	B11-10P15 B10-10R13	
10443	25	145.0479	68942.74	B10-11R3 B7-7R18	
10444	0	145.0522	68940.72	B7-6P20	
10445	3	145.0667	68933.82	B6-7P14	
10446	1	145.0748	68929.96	B4-4R23	
10447	101	145.0806	68927.20	B10-11P1 B8-6P22 B9-10R8	
10448	0	145.0897	68922.88	B9-10P6	
10449	35	145.0983	68918.79	B13-14R3	
10450	56	145.1046	68915.79	B7-9P5	
10451	1	145.1079	68914.24	C1-11R4	
10452	0	145.1196	68908.67		

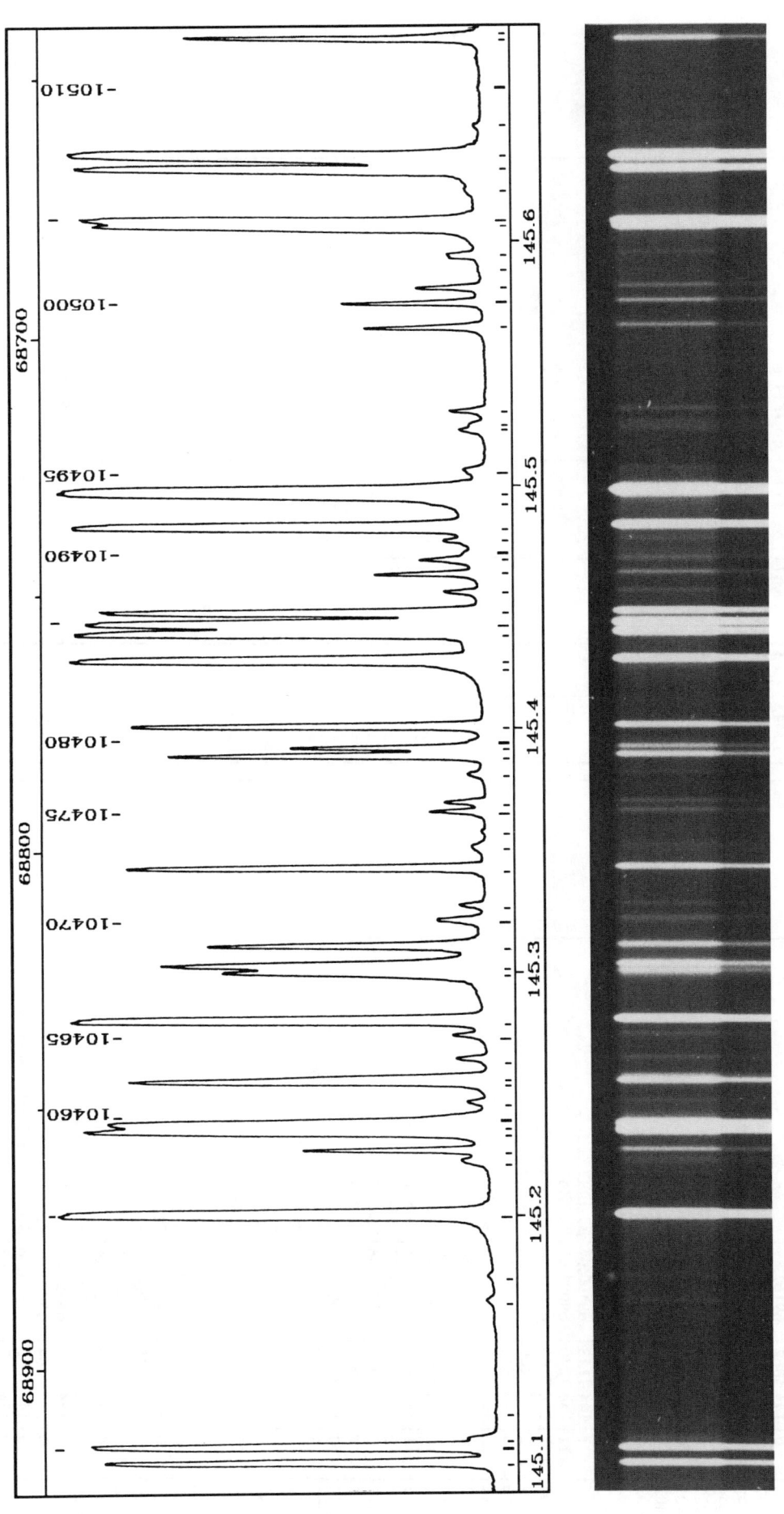

PLATE 118

TABLE 118

Nb	I	λ (nm)	σ (cm⁻¹)	Assignment	Comment
10453	0	145.1647	68887.28	B9-7P21	
10454	0	145.1750	68882.39	B12-12P8	
10455	180	145.1997	68870.65	B5-8P5 B13-14R2	
10456	2	145.2219	68860.16	B11-11P9 B5-5P20	
10457	6	145.2259	68858.25	B7-9R7	
10458	58	145.2339	68854.43	B1-6P6 B2-5R16 B0-4P15	
10459	58	145.2362	68853.34	B10-11P2	
10460		145.2393	68851.88	B2-2P25	
10461	2	145.2456	68848.88	B4-7P9	sh
10462	22	145.2542	68844.84	B3-7R7	
10463		145.2557	68844.12	B10-11R4	sh
10464	4	145.2635	68840.44	B9-9R15	
10465	2	145.2732	68835.81	B3-6P12	
10466	112	145.2794	68832.89	B13-14R1 B7-6R22	
10467	10	145.2996	68823.30	B0-3P19	
10468	17	145.3022	68822.07	B0-5P10	
10469	11	145.3101	68818.34	B2-6R11	
10470	2	145.3209	68813.25	B7-8P12 B3-5R18	
10471	1	145.3271	68810.28	B12-12P9	
10472	30	145.3422	68803.15	B13-14R0	
10473	0	145.3511	68798.92	B1-2R25	
10474	0	145.3584	68795.49	B10-10R14	
10475	2	145.3657	68792.04	B10-10P11	
10476	2	145.3696	68790.19	C1-11R3 B6-7R16	
10477	1	145.3811	68784.72	B5-6R18	
10478	16	145.3886	68781.18	B5-8R7	
10479	7	145.3919	68779.64	B10-11R5 B3-1P29	
10480	0	145.3941	68778.59	C0-10R9	
10481	30	145.4007	68775.46	B13-14P4	
10482		145.4251	68763.90	B12-12P10	sh
10483	100	145.4280	68762.54	B13-14P1	
10484	108	145.4391	68757.30	B10-11P3	
10485	160	145.4428	68755.54	B13-14P3	
10486	58	145.4475	68753.32	B13-14P2	
10487	2	145.4559	68749.36	B5-7P12	
10488	5	145.4634	68745.80	B9-9P13	
10489	3	145.4693	68743.05	B1-4P17	
10490	0	145.4719	68741.78	B7-8R14	
10491	2	145.4772	68739.30	B4-6R16	
10492	150	145.4829	68736.60	B0-6R0	
10493	580	145.4960	68730.43		sh
10494		145.4972	68729.86	B0-6R1 B1-6R8	
10495	0	145.5059	68725.74	B6-8R11 B11-11P10	
10496	1	145.5227	68717.80	B1-3R22 B9-10R9	
10497	0	145.5248	68716.82	B9-10P7 B8-7P19	
10498	3	145.5303	68714.19	B8-8R17	
10499	7	145.5649	68697.89	B4-7R11	
10500	6	145.5749	68693.18	B7-9P6 B6-5P22	
10501	5	145.5810	68690.28	B8-8P15 B4-3P25	

Nb	I	λ (nm)	σ (cm⁻¹)	Assignment	Comment
10502	0	145.5867	68687.60	B10-11R6	
10503	2	145.5948	68683.78	B3-7P6 B10-10R15	
10504	55	145.6059	68678.55	B0-6R2	
10505	145	145.6076	68677.74	B4-4P22	
10506	0	145.6213	68671.29	B8-10R0	
10507	90	145.6291	68667.58	B8-10R1	
10508	300	145.6344	68665.08	B8-9P10	
10509	0	145.6470	68659.14		sh
10510	0	145.6793	68643.92	B10-11P4	
10511	15	145.6823	68642.53	B1-1R28 B7-9R8	
10512		145.6839	68641.77		

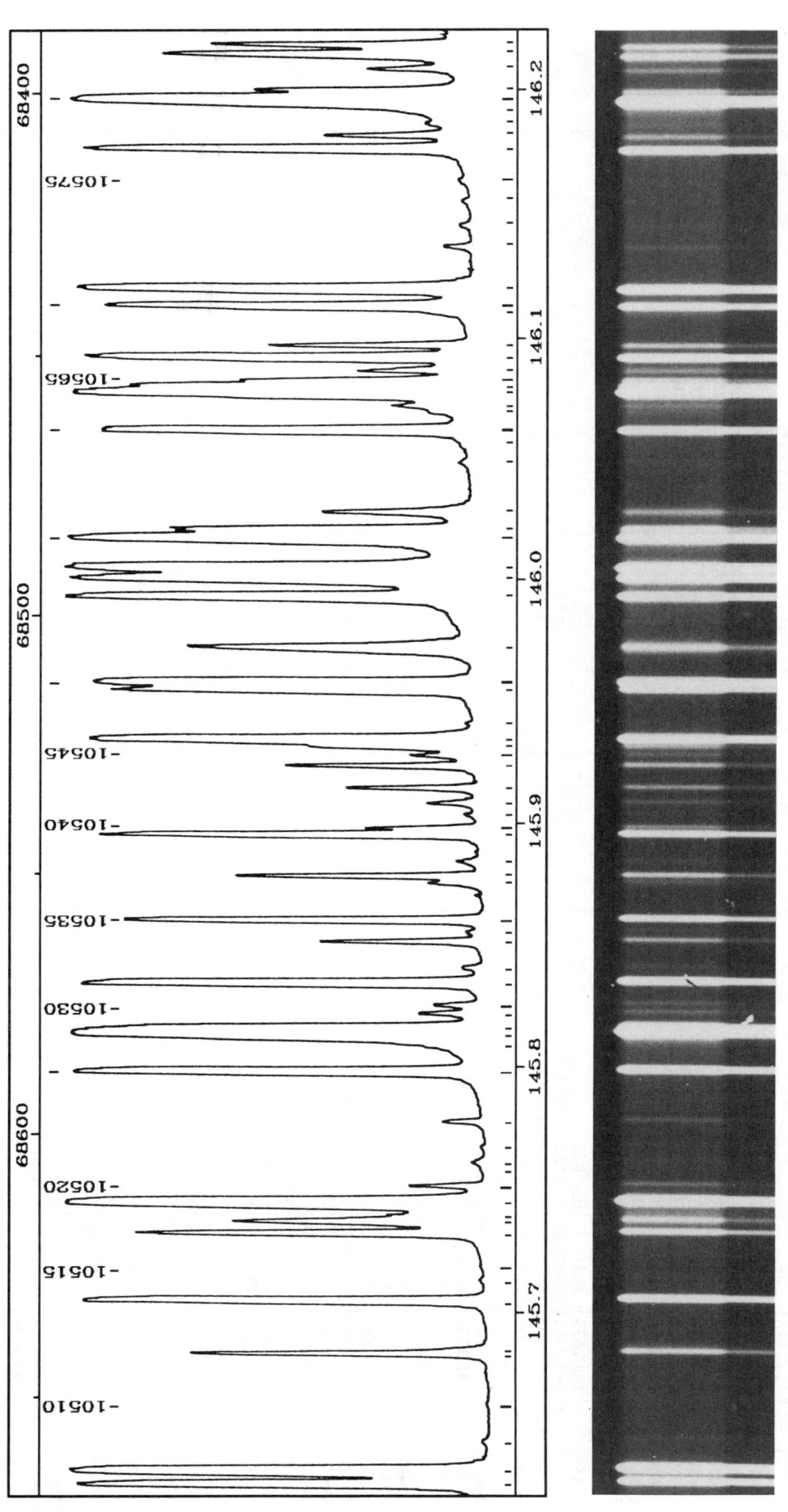

PLATE 119

TABLE 119

Nb	I	λ (nm)	σ (cm⁻¹)	Assignment	Comment
10513	68	145.7036	68632.50	B8-10R2	
10514	0	145.7125	68628.31	C1-11P10	
10515	0	145.7186	68625.42		
10516	20	145.7309	68619.65	B5-8P6	
10517	10	145.7357	68617.36	B0-5R12	
10518	0	145.7382	68616.18	B9-9R16	
10519	500	145.7434	68613.74	B0-6P1	
10520	3	145.7504	68610.46	C2-13R3	
10521	0	145.7564	68607.64		
10522	0	145.7602	68605.81	B1-2P24	
10523	0	145.7672	68602.56	B7-7P17	
10524	1	145.7767	68598.05	B12-13R7	
10525	103	145.7972	68588.40	B11-12R1	
10526		145.8119	68581.50	B8-10P1 B2-5P15	sh
10527	580	145.8132	68580.90	B0-6R3 B11-12R2	
10528		145.8157	68579.71	B11-12R0	sh
10529	3	145.8208	68577.30	B2-6P10	
10530	2	145.8244	68575.64	B10-11R7	
10531	88	145.8335	68571.34	B8-10R3	
10532	2	145.8397	68568.44	B10-10P12 B3-4R22	
10533	6	145.8505	68563.34	B3-7R8	
10534	0	145.8545	68561.48	B12-13R6	
10535	29	145.8593	68559.22	B11-12R3	
10536	2	145.8749	68551.90	B11-11P11 B7-7R19	
10537	12	145.8775	68550.65	B2-7R0	
10538	3	145.8835	68547.85	B2-4P19 B9-10R10	
10539	43	145.8944	68542.71	B2-7R1	
10540	4	145.8972	68541.43	B12-13R5 C1-11P9	
10541	0	145.9031	68538.63	C0-10R7	
10542	3	145.9080	68536.32	B3-5P17	
10543	7	145.9142	68533.44	B12-13R2 B0-2R24	
10544	3	145.9232	68529.19	B12-13R4 B5-8R8	
10545	3	145.9277	68527.07	B1-5P13	
10546	6	145.9316	68525.27	B11-12R4	
10547	95	145.9341	68524.06	B1-6P7	
10548					CuII 145.94117
10549	33	145.9546	68514.47	B12-13R3	
10550	130	145.9575	68513.12	B11-12P1 B12-13R1 B10-11P5 B8-6P23	
10551	13	145.9719	68506.32	B0-4R17	
10552	127	145.9927	68496.59	B6-9R0	
10553	174	146.0003	68493.02	B8-10P2 B12-13R0 B3-2R28	
10554	450	146.0044	68491.08	B6-9R1 B2-7R2	
10555	280	146.0167	68485.32	B0-6P2	
10556	17	146.0199	68483.80	B8-10R4 B11-12R5 B6-7P15 B5-6P17	
10557	8	146.0275	68480.25	B1-3P21	
10558	0	146.0483	68470.52	B6-6P19	
10559	0	146.0557	68467.01	C1-11P8	
10560	98	146.0613	68464.41	B4-8R0	
10561	0	146.0687	68460.95	B4-7P10 B10-11R8	
10562	1	146.0714	68459.68	B7-6P21	
10563	340	146.0770	68457.07	B4-8R1	
10564	45	146.0792	68456.00	B11-12P2	
10565	11	146.0816	68454.89	B4-6P15	
10566	4	146.0858	68452.90	B9-9P14 C2-13R2	
10567	100	146.0913	68450.36	B6-9R2 C0-10R6	
10568	7	146.0962	68448.06	B7-9P7	sh
10569		146.1096	68441.76	B11-11P12	
10570	50	146.1123	68440.50	B0-6R4	
10571	127	146.1193	68437.22	B2-7P1 B12-13P1 B11-12R6 B7-8P13 B7-6R23	
10572	1	146.1365	68429.17	B9-9R17	
10573	0	146.1455	68424.97	B4-5P19	
10574	0	146.1551	68420.47	B8-8R18	
10575	0	146.1636	68416.48	C1-11P7	
10576	69	146.1765	68410.46	B4-8R2	
10577	5	146.1819	68407.90	B7-9R9	
10578	0	146.1869	68405.58	C1-11P6	
10579	0	146.1955	68401.58	B12-13P2 B7-8R15	sh
10580	440	146.1969	68400.90	B6-9P1	
10581	12	146.2005	68399.24	B2-7R3 B6-8R12	
10582	3	146.2093	68395.10	B2-6R12	
10583	21	146.2152	68392.36	B1-6R9 B11-12R7	
10584	9	146.2190	68390.57	B8-10R5	

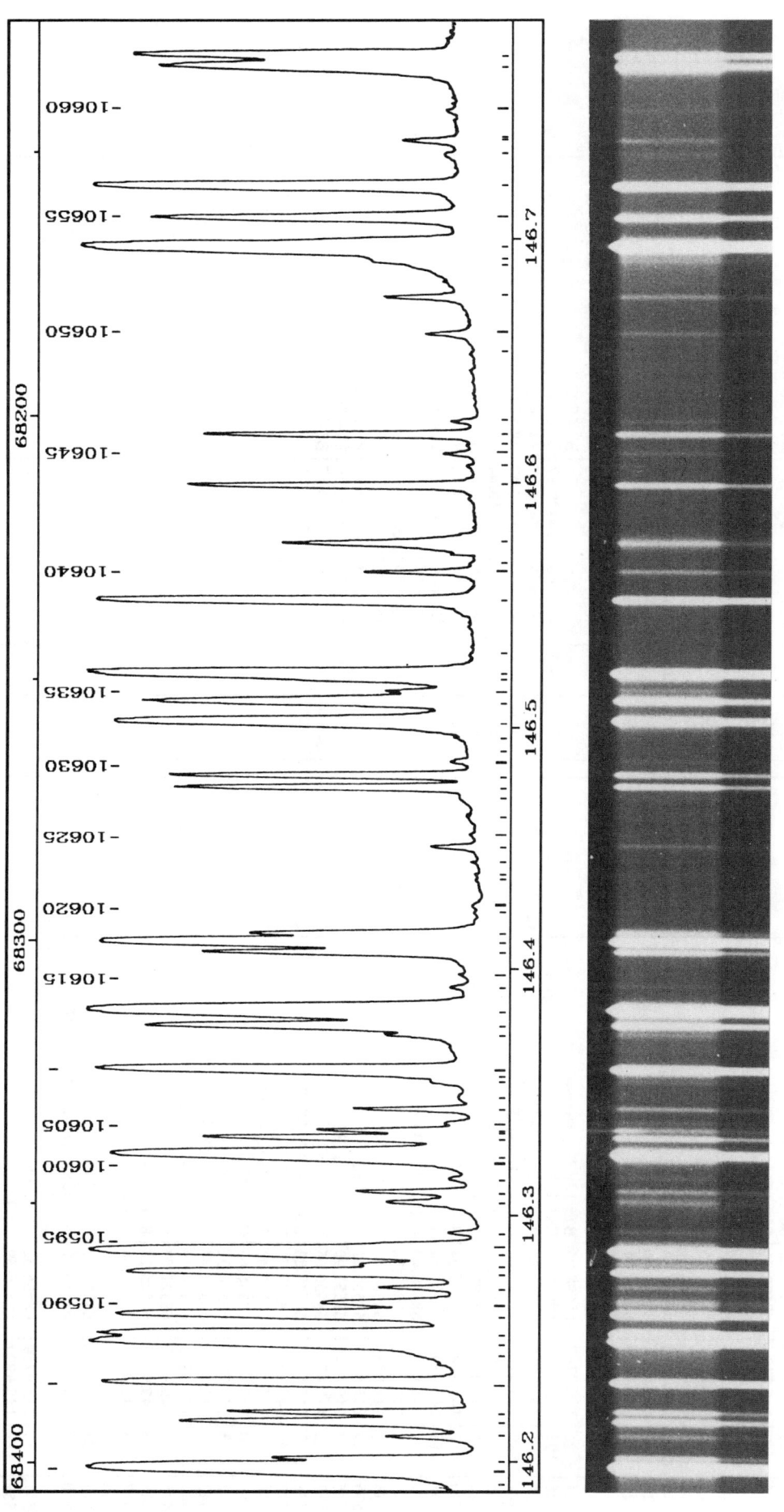

PLATE 120

TABLE 120

Nb	I	λ (nm)	σ (cm⁻¹)	Assignment	Comment
10585	100	146.2312	68384.87	B11-12P3	
10586	0	146.2434	68379.14		
10587	420	146.2481	68376.95	B8-10P3 B6-7R17	
10588	162	146.2508	68375.68	B6-9R3	
10589	63	146.2588	68371.95	B3-7P7 B11-12R8	
10590	7	146.2636	68369.71	B1-4R19 B11-11P13	
10591	3	146.2699	68366.75	C2-13P5	
10592	59	146.2760	68363.90	B12-13P3	
10593	6	146.2791	68362.45	B10-10P13	
10594	400	146.2845	68359.92	B4-8P1 B9-10R11	
10595		146.2873	68358.63	B8-9P11	sh
10596	2	146.2925	68356.19	C2-13R1	
10597	2	146.3046	68350.53	B0-3R21	
10598	3	146.3090	68348.51	B3-6P13	
10599	1	146.3143	68346.03	B10-11R9	
10600	65	146.3220	68342.40	B0-5P11 B2-3R24	sh
10601	65	146.3240	68341.51	B5-8P7	
10602	11	146.3306	68338.40	C0-10R5	CI 146.33367
10603					+CI 146.33367
10604	6	146.3336	68336.99	B10-11P6	
10605	0	146.3367	68335.55	B12-13P4	
10606	4	146.3424	68332.90	B8-7P20	
10607	0	146.3471	68330.71	B8-8P16	
10608	0	146.3537	68327.64	B0-2P23	sh
10609	92	146.3566	68326.25	B4-8R3	
10610	1	146.3584	68325.42	B4-7R12 B4-4R24	
10611	42	146.3729	68318.67	B2-7P2	
10612	820	146.3764	68317.03	B0-6P3 B5-7P13	
10613	0	146.3826	68314.11	B11-12R9	
10614	0	146.3928	68309.36		
10615	18	146.3986	68306.67	B11-12P4	
10616	260	146.4068	68302.82	B6-9P2 B5-6R19	
10617	10	146.4110	68300.86	B2-5R17 B3-4P21	
10618	0	146.4143	68299.34	B3-2P27	
10619	0	146.4267	68293.57		
10620	0	146.4306	68291.73		
10621	0	146.4400	68287.37		
10622	2	146.4420	68286.44	C2-14R4	
10623	0	146.4458	68284.67		
10624	0	146.4507	68282.36	B5-5P21	
10625	0	146.4553	68280.23	C0-10R4	
10626	18	146.4637	68276.30	B5-3P27	
10627	18	146.4713	68272.77	B2-2R27	
10628	1	146.4753	68270.89	B12-13P5	
10629	0	146.4802	68268.63	B6-9R4	
10630	0	146.4859	68265.96	B2-7R4	
10631	82	146.4904	68263.85		
10632		146.4959	68261.32	B4-3R27	
10633		146.5022	68258.38	B0-6R5	

Nb	I	λ (nm)	σ (cm⁻¹)	Assignment	Comment
10634	28	146.5104	68254.53	B3-7R9 B5-8R9 B12-13P6	
10635	3	146.5146	68252.57	C2-13P4	
10636	226	146.5199	68250.10	B12-13P7 B3-5R19 B4-6R17	sh
10637		146.5218	68249.21	B4-8P2	s
10638	75	146.5296	68245.62	B8-10P4	
10639	3	146.5518	68235.24	B8-10R6	
10640	0	146.5636	68229.77		
10641	7	146.5673	68228.05	B0-4P16	
10642	15	146.5754	68224.26	B11-12P5	
10643		146.5990	68213.31		
10644					CuII 146.60702
10645	1	146.6119	68207.28	C2-13P3	
10646	0	146.6157	68205.53	C0-10R3	
10647	15	146.6197	68203.65	B4-8R4	
10648	1	146.6250	68201.19	B10-11P7	
10649	0	146.6530	68188.15	B7-9P8 B5-7R15	
10650	1	146.6610	68184.45	B9-9P15	
10651	3	146.6762	68177.37	B10-10P14	
10652	3	146.6907	68170.63		
10653	4	146.6927	68169.71	B6-9P3	
10654	720	146.6969	68167.77	B1-6P8 B7-9R10	
10655	21	146.7086	68162.31	B2-7P3	
10656	118	146.7215	68156.32	B6-5P23	
10657	0	146.7345	68150.32		
10658					
10659	0	146.7402	68147.64	B10-11R11	CI 146.74024
10660		146.7524	68141.99	B0-5R13	CI 146.74024
10661	27	146.7703	68133.69	B6-9R5 B2-6P11	
10662	31	146.7746	68131.68		

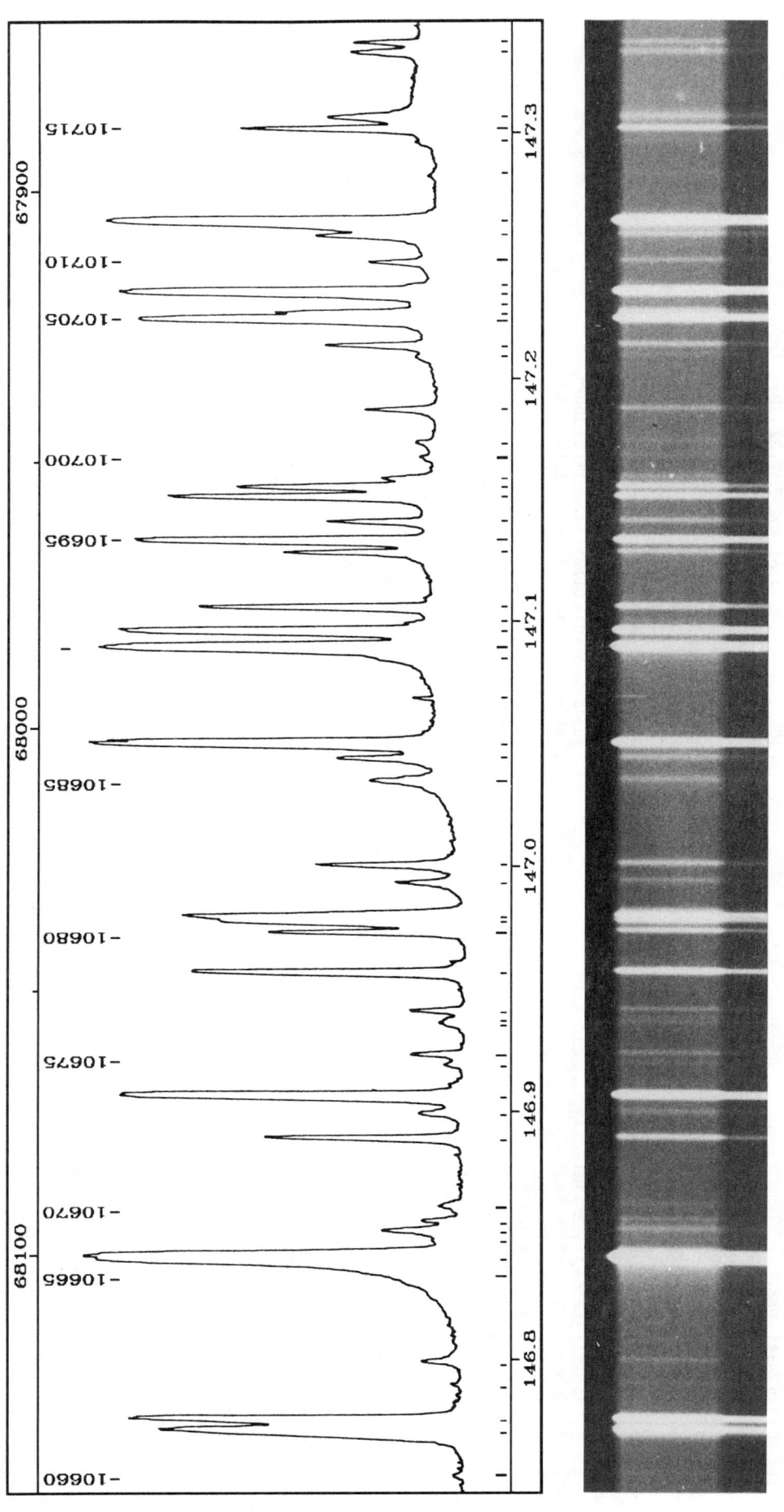

PLATE 121

TABLE 121

Nb	I	λ (nm)	σ (cm-1)	Assignment	Comment
10663	0	146.7891	68124.94	B6-8P11	
10664	1	146.7985	68120.61	B11-12P6	
10665	700	146.8360	68103.21	B7-6R24	sh d
10666	0	146.8396	68101.51	B0-6P4 B4-8P3	
10667	0	146.8461	68098.51	B2-3P23	
10668	2	146.8513	68096.10	B0-3P20	
10669	1	146.8553	68094.24	B2-7R5	
10670	0	146.8618	68091.21	B1-4P18	
10671	8	146.8889	68078.67	B8-10R7	
10672	1	146.8993	68073.85	C2-14R3 B7-8R16	
10673	82	146.9060	68070.74	B8-10P5 B5-5R23	
10674	0	146.9191	68064.68	B6-8R13	
10675	2	146.9232	68062.76	B7-8P14	
10676	0	146.9356	68057.02	B8-9P12	
10677	0	146.9376	68056.08	B4-7P11	
10678	1	146.9410	68054.53	B12-14R4	
10679	18	146.9566	68047.31	B4-8R5	
10680	8	146.9733	68039.59	B5-8P8 B6-7P16	
10681	15	146.9779	68037.44	B3-7P8	
10682	20	146.9796	68036.67	B0-6R6 B10-10P15	
10683	2	146.9939	68030.02	B4-4P23 B11-12P7	
10684	3	147.0009	68026.80	B1-6R10	
10685	2	147.0357	68010.69	B1-3R23	
10686	3	147.0449	68006.45	B2-5P16	
10687	137	147.0508	68003.70	B6-9P4	
10688					CuII 147.06974
10689	0	147.0861	67987.41	B8-7P21	
10690	260	147.0900	67985.57	B9-11R1 B8-8P17	
10691	83	147.0966	67982.53	B9-11R0	
10692	1	147.1003	67980.83	B1-5P14	
10693	17	147.1065	67977.96	B12-14R3 B6-7R18 B6-6P20	
10694	5	147.1291	67967.52	B6-9R6	
10695	55	147.1338	67965.36	B9-11R2	
10696	3	147.1416	67961.74	B5-8R10	
10697	28	147.1519	67957.01	B2-7P4	
10698	10	147.1559	67955.16	B2-6R13	
10699	2	147.1596	67953.42	B12-14R2	
10700	0	147.1683	67949.42	B3-5P18	
10701	0	147.1745	67946.55	B3-4R23	
10702	2	147.1878	67940.41	B4-6P16	
10703	0	147.2101	67930.14	B9-9P16	
10704	4	147.2141	67928.29	B4-7R13	
10705	73	147.2247	67923.39	B9-11R3	
10706	4	147.2272	67922.23	B3-7R10	
10707	0	147.2311	67920.44		
10708	126	147.2357	67918.32	B4-8P4	
10709		147.2388	67916.90		s h
10710	1	147.2483	67912.52	B8-10R8	
10711	5	147.2592	67907.47	B7-9R11 B7-9P9	
10712	250	147.2646	67904.99	B9-11P1 B12-14R1	
10713	0	147.2838	67896.11	B8-9R15	
10714	0	147.2966	67890.24	B2-4P20	
10715	10	147.3025	67887.53	B8-10P6	
10716	4	147.3073	67885.29	B0-4R18 B2-7R6	
10717	3	147.3336	67873.17	B5-7P14 C2-14R2	
10718	3	147.3378	67871.26	B12-14R0	

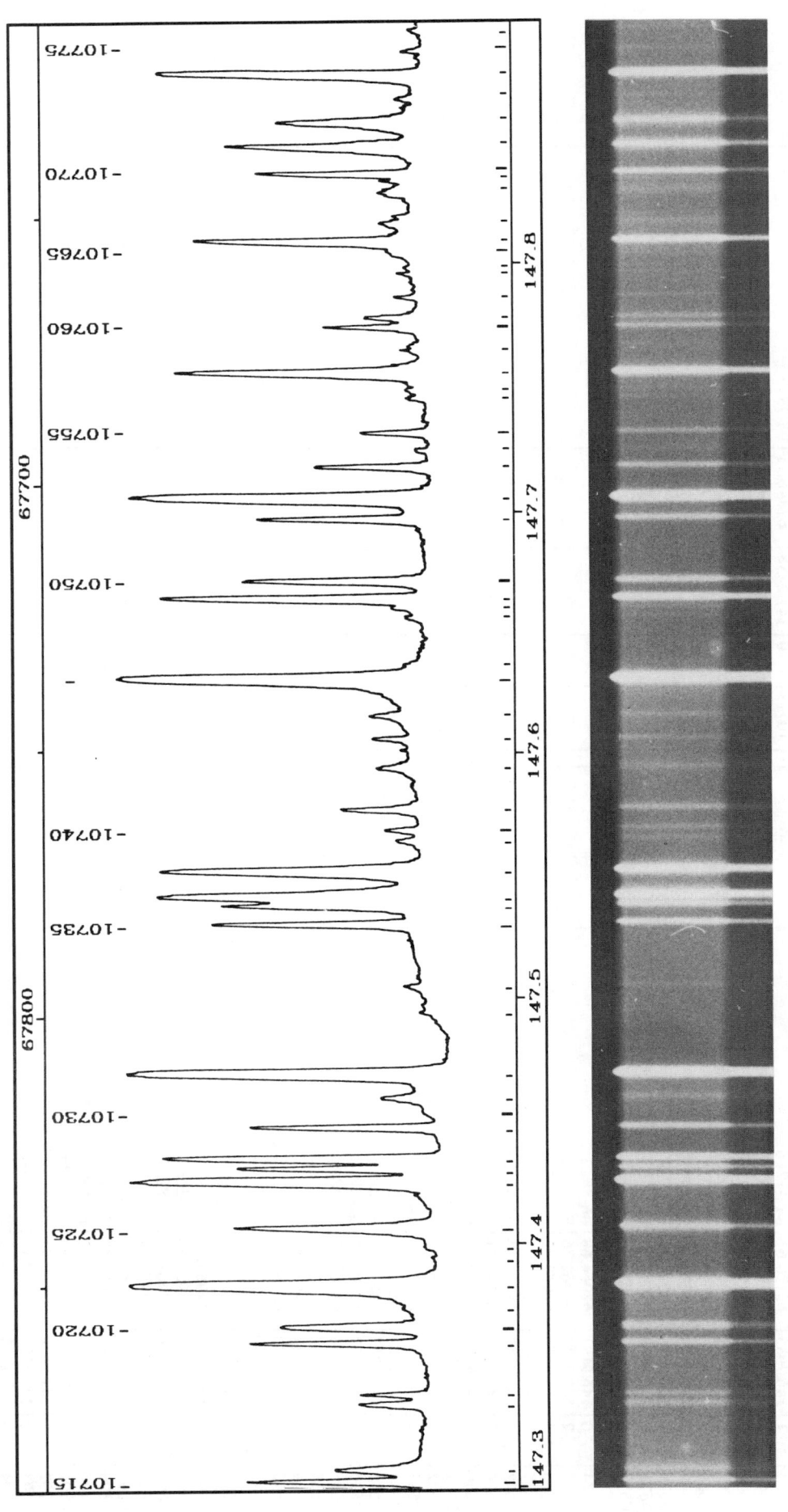

PLATE 122

TABLE 122

Nb	I	λ (nm)	σ (cm-1)	Assignment	Comment
10719	10	147.3586	67861.67	B9-11R4	
10720	7	147.3652	67858.63	B4-8R6 B12-14P4	
10721	0	147.3756	67853.84	C0-10P11	
10722	290	147.3822	67850.82	B0-6P5 B3-6P14	
10723					CuII 147.39785
10724	0	147.4000	67842.61	B4-5P20	
10725	14	147.4056	67840.04	B0-5P12	
10726	165	147.4242	67831.45	B9-11P2 B5-6R20	
10727	16	147.4294	67829.08	B12-14P1	
10728	45	147.4333	67827.30	B12-14P3	
10729	15	147.4461	67821.41	B12-14P2	
10730	1	147.4553	67817.14	B2-1P29	
10731	1	147.4581	67815.86	B1-2P25	
10732	204	147.4679	67811.35	B6-9P5	
10733					CuII 147.49348
10734	0	147.5038	67794.87	B9-10P11	
10735	13	147.5299	67782.88	B9-11R5	
10736	10	147.5375	67779.36	B6-9R7	
10737	43	147.5412	67777.69	B0-6R7 C2-14P4	
10738	38	147.5516	67772.91	B1-6P9	
10739	0	147.5647	67766.89	B4-6R18	
10740	1	147.5691	67764.86	B7-8R17	
10741	2	147.5773	67761.10	B8-9P13 B0-2R25 B6-8P12	
10742	1	147.5940	67753.41	B1-3P22	
10743	0	147.6056	67748.11	C2-14R1	
10744	0	147.6153	67743.64	B2-5R18	
10745	360	147.6302	67736.83	B9-11P3 B8-10R9 B1-4R20	
10746	0	147.6355	67734.37	B4-4R25	
10747	0	147.6559	67725.02	B5-5P22	
10748	0	147.6599	67723.18	B9-9P17	
10749	40	147.6634	67721.58	B2-7P5	
10750	10	147.6707	67718.25	B5-8P9 C1-12R5	
10751	9	147.6962	67706.57	B8-10P7	
10752	175	147.7046	67702.69	B4-8P5	
10753	6	147.7174	67696.82	B7-8P15	
10754	0	147.7241	67693.77	B3-5R20	
10755	2	147.7314	67690.43	B9-11R6	
10756	0	147.7457	67683.86	B8-9R16	
10757	0	147.7492	67682.24	C0-10P6	
10758	27	147.7558	67679.23	B3-7P9	
10759	0	147.7655	67674.81	B8-8P18	
10760	2	147.7746	67670.63	C2-14P3	
10761	1	147.7784	67668.90	B2-6P12	
10762	0	147.7872	67664.85		d
10763		147.7968			
10764	0	147.7996	67659.18	B2-3R25	
10765	0	147.8059	67656.29	B5-8R11 C0-10P7	
10766	21	147.8099	67654.47	B3-4P22 B7-9R12 B1-1P28	
10767	0	147.8172	67651.13		

Nb	I	λ (nm)	σ (cm-1)	Assignment	Comment
10768	0	147.8306	67644.99	B2-7R7 B0-3R22	s
10769	0	147.8350	67643.00		
10770	7	147.8378	67641.71	B4-8R7	
10771	14	147.8487	67636.70	B1-6R11	
10772	7	147.8585	67632.22	B0-5R14	
10773	0	147.8686	67627.63	B11-13R6	
10774	66	147.8778	67623.41	B9-11P4	
10775	0	147.8876	67618.90	B7-9P10	
10776	0	147.8963	67614.96		

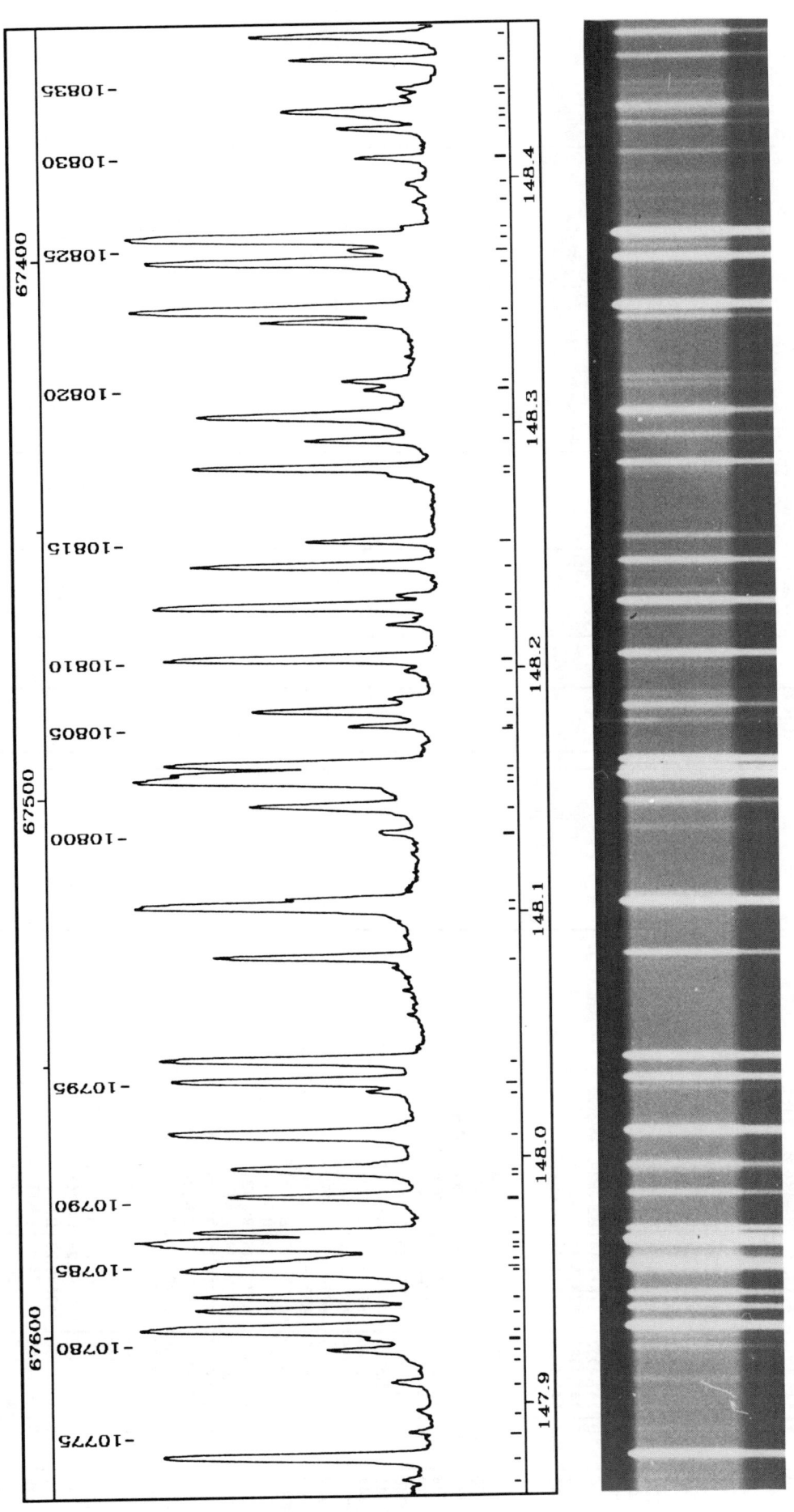

PLATE 123

TABLE 123

Nb	I	λ (nm)	σ (cm-1)	Assignment	Comment
10777	1	147.9079	67609.65	B7-6P23	
10778	0	147.9169	67605.53	C2-14P2	d
10779	5	147.9208	67603.74	B6-7P17	
10780	0	147.9252	67601.75	B11-13R5	
10781	132	147.9291	67599.96	B10-12R1 B9-10P12 B6-7R19	
10782	25	147.9367	67596.46	B10-12R2	
10783	20	147.9423	67593.90	B6-9P6	
10784	40	147.9531	67588.97	B10-12R0 C1-12R4	
10785	17	147.9549	67588.17	B0-4P17 B9-11R7	
10786		147.9574	67587.01	B11-13R4	sh
10787		147.9584	67584.40	B7-10R0	sh
10788	210	147.9647	67583.70	B7-10R1	
10789	28	147.9685	67581.94	B10-12R3	
10790	12	147.9830	67575.32	B11-13R3	
10791		147.9930	67570.77	B6-9R8	sh
10792	12	147.9944	67570.11	B3-7R11	
10793	46	148.0085	67563.69	B0-6P6 B11-13R2	
10794	1	148.0260	67555.71	B8-10R10	
10795	40	148.0301	67553.83	B7-10R2	
10796	56	148.0386	67549.96	B11-13R1 B3-2P28	
10797	20	148.0806	67530.78	B11-13R0 B4-7R14	
10798	140	148.1019	67521.08	B10-12P1	
10799	6	148.1046	67519.85	B10-12R4	
10800	1	148.1329	67506.97	B8-9R17 B6-6P21	
10801	7	148.1438	67502.00	B10-12R5 B2-6R14	
10802	220	148.1545	67497.12	B7-10P1	
10803	57	148.1568	67496.08	B7-10R3	
10804	70	148.1606	67494.34	B9-11P5	
10805					CI 148.17631
10806	10	148.1764	67487.15	B0-6R8	CI 148.17631
10807	1	148.1825	67484.35	B9-11R8 B7-8R18	
10808	1	148.1873	67482.18	B8-9P14	
10809	1	148.1988	67476.95	B11-13P1	
10810	57	148.2036	67474.76	B8-10P8 B5-6P19	
10811	1	148.2179	67468.23	B10-12P2	
10812	77	148.2250	67465.00	B10-12R6	
10813	1	148.2294	67463.00	B4-8P6	
10814	23	148.2417	67457.40	B2-7P6	
10815	5	148.2518	67452.79	B1-4P19	
10816	0	148.2791	67440.41	B11-13P2	
10817	29	148.2819	67439.11	B5-7P15	
10818	7	148.2933	67433.95	B2-5P17 C1-12R3 B4-6P17	
10819	19	148.3030	67429.52	B1-5P15	
10820	1	148.3138	67424.59	B10-12R7	
10821	10	148.3172	67423.06	B7-10R4 B4-4P24	
10822	145	148.3413	67412.10	B7-10P2 C0-11R9	
10823	64	148.3461	67409.91	B11-13P3 B9-10P13	
10824	2	148.3658	67400.99	B4-8R8	
10825		148.3706	67398.82		

Nb	I	λ (nm)	σ (cm-1)	Assignment	Comment
10826	195	148.3758	67396.43	B10-12P3 B7-9R13	
10827	0	148.3803	67394.37	B6-8P13	
10828	0	148.3910	67389.54	B10-12R8	
10829	0	148.3984	67386.19	B5-6R21	
10830	2	148.4089	67381.41	B5-8P10	
10831	3	148.4206	67376.08	B8-10R11	
10832		148.4259	67373.70	B2-7R8 B5-7R17	sh
10833	4	148.4278	67372.84	B0-3P21	
10834	0	148.4340	67370.01	B10-12R9	
10835	0	148.4371	67368.58	B3-5P19	
10836	9	148.4481	67363.62	B11-13P4	
10837	10	148.4575	67359.35	B1-6P10	

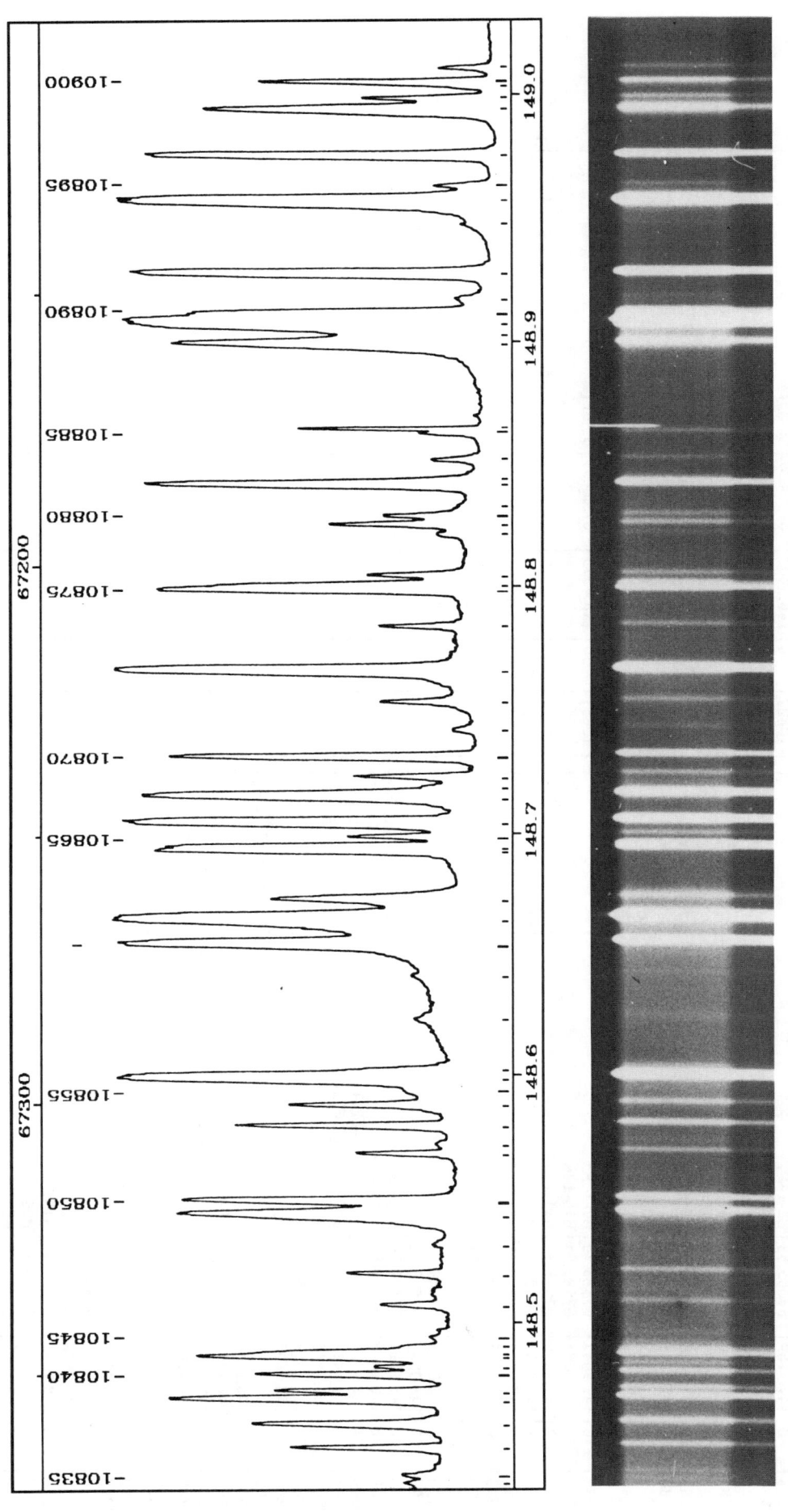

PLATE 124

TABLE 124

Nb	I	λ (nm)	σ (cm⁻¹)	Assignment	Comment
10838	30	148.4677	67354.74	B6-9P7	
10839	9	148.4710	67353.24	B9-11P6	
10840	11	148.4777	67350.18	B5-9R0 B3-6P15	
10841	2	148.4810	67348.67	B9-11R9	
10842	25	148.4849	67346.92	B5-9R1 B3-4R24	
10843	8	148.4863	67346.27	B6-9R9	
10844	1	148.4886	67345.23	B7-8P16	
10845	0	148.4936	67342.99	B4-3P27	
10846	3	148.5061	67337.32	B5-8R12	
10847	3	148.5186	67331.63	B11-13P5	
10848	3	148.5312	67325.91	B7-9P11	
10849	27	148.5426	67320.77	B0-5P13	
10850	36	148.5479	67318.35	B1-3R24 B10-12P4	
10851	4	148.5670	67309.71	B5-9R2 B11-13P7	
10852	0	148.5713	67307.76	B2-2P27	
10853	10	148.5782	67304.62	B7-10R5	
10854	6	148.5867	67300.77	B3-7P10	
10855	0	148.5926	67298.11		
10856	400	148.5979	67295.69	B7-10P3	
10857	0	148.6012	67294.20	B4-6R19	sh
10858	0	148.6218	67284.90	B1-2R27	d
10859	0	148.6398	67276.74	B4-5P21	
10860	270	148.6529	67270.79	B1-7R0	
10861	1020	148.6631	67266.18	B1-7R1	
10862	10	148.6712	67262.51	B0-4R19	
10863	48	148.6915	67253.33	B5-9P1	
10864	48	148.6925	67252.91	B3-8R0	
10865	2	148.6973	67250.72	B8-10P9	
10866	170	148.7032	67248.05	B3-8R1	
10867	70	148.7141	67243.13	B0-6P7	
10868	0	148.7172	67241.70	B2-4P21	
10869	2	148.7220	67239.57	B5-9R3 B7-8R19	
10870	30	148.7298	67236.01	B10-12P5	
10871	0	148.7411	67230.91	B9-10P14 C1-12R1	
10872	2	148.7521	67225.94	B1-6R12	
10873	226	148.7650	67220.13	B1-7R2	
10874	4	148.7830	67212.00	B8-9P15	
10875	40	148.7981	67205.16	B3-8R2	
10876		148.7995	67204.54	B9-11P7 B8-10R12	sh
10877	4	148.8038	67202.58	B3-7R12	
10878	1	148.8207	67194.95	B5-5P23	
10879	4	148.8246	67193.19	B2-6P13	
10880	2	148.8280	67191.66	B2-5R19	
10881	0	148.8321	67189.82		sh
10882	36	148.8408	67185.88	B4-8P7 B6-7P18	sh
10883		148.8427	67185.01	C1-12P6	
10884	2	148.8509	67181.30	B9-11R11	
10885	3	148.8618	67176.40	B7-10R6	
10886					CuII 148.88311
10887	30	148.8987	67159.77	B0-6R9	
10888	1000	148.9054	67156.71	B7-10P4	sh
10889	42	148.9077	67155.71	B1-7P1	
10890	0	148.9105	67154.43	B2-7P7 B5-9P2	
10891	123	148.9170	67151.48	B3-5R21 B7-9R14	
10892	0	148.9272	67146.88	B3-8P1	
10893	350	148.9472	67137.88	B5-9R4	
10894	2	148.9566	67133.66	B1-7R3 B4-8R9	sh
10895	60	148.9627	67130.92	B4-7R15	
10896	17	148.9751	67125.29	B3-8R3	
10897	3	148.9939	67116.85	B0-5R15	
10898		148.9983	67114.85	B10-12P6	
10899	10	149.0033	67112.59	B1-4R21	
10900	3	149.0049	67111.88	B11-14R4	
10901		149.0106	67109.31	B6-9R10	

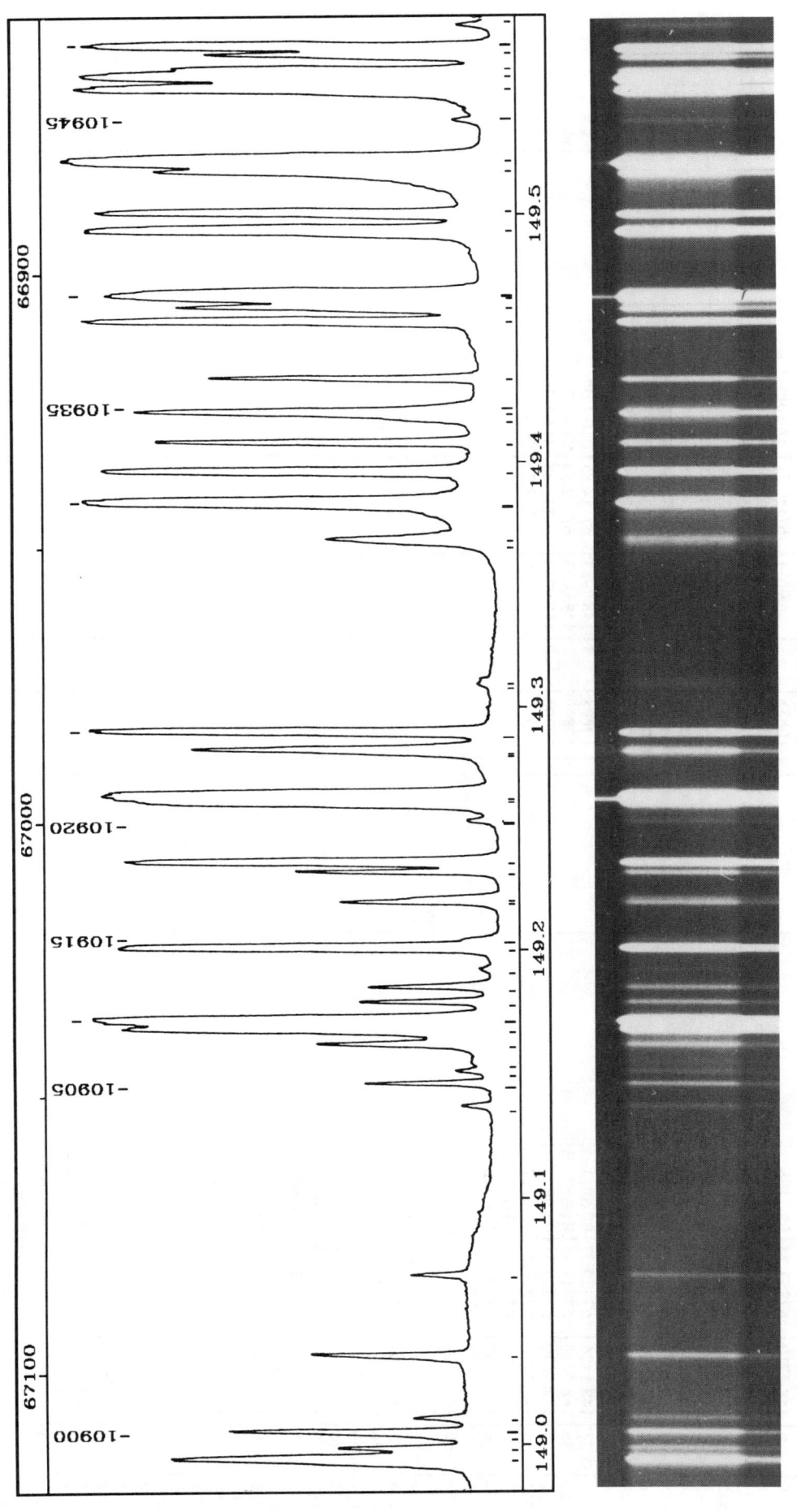

PLATE 125

TABLE 125

Nb	I	λ (nm)	σ (cm⁻¹)	Assignment	Comment
10902	5	149.0358	67097.99	B6-9P8 B9-10P15	b
10903	2	149.0682	67083.39	C1-12P5	
10904	1	149.1364	67052.69	B9-11P8	
10905	4	149.1456	67048.57	B10-12P7	
10906	1	149.1506	67046.31	B8-10R13	
10907	0	149.1542	67044.73	B1-2P26 C1-12P4	
10908	7	149.1615	67041.44	B2-6R15	
10909	100	149.1677	67038.63	B1-3P23 B11-14R3	
10910	540	149.1708	67037.24	B3-8P2 B1-7P2	
10911	5	149.1788	67033.65	B5-8P11	
10912	5	149.1850	67030.86	B7-10R7 B8-10P10 B6-8P14	
10913	0	149.1923	67027.59	C1-12P3	
10914	100	149.2010	67023.66	B5-9P3	
10915	0	149.2040	67022.34	B3-4P23	
10916	8	149.2200	67015.15	B5-8R13	
10917	5	149.2213	67014.58	B7-8P17	
10918	10	149.2322	67009.67	B3-8R4	
10919	65	149.2362	67007.87	B1-7R4 B5-9R5	NI 149.26254 +NI 149.26254
10920	3	149.2537	67000.00	B5-7P16	NI 149.28195 +NI 149.28195
10921					
10922	198	149.2625	66996.06	B7-10P5 B2-3R26	
10923					
10924	20	149.2819	66987.37	B5-7R18	
10925	68	149.2894	66984.00	B11-14R2	
10926	0	149.3093	66975.05	B8-9P16	
10927	0	149.3111	66974.26	B10-12P8	
10928		149.3671	66949.13	B0-3R23	sh
10929	4	149.3685	66948.53	B0-4P18	
10930	300	149.3835	66941.78	B11-14R1	
10931	73	149.3962	66936.10	B8-11R1	
10932	22	149.4080	66930.82	B8-11R0	
10933	0	149.4176	66926.54	B4-6P18	
10934	24	149.4201	66925.41	B1-6P11	
10935		149.4227	66924.24	B7-9R15	sh
10936	16	149.4339	66919.23	B8-11R2	
10937	83	149.4571	66908.83	B11-14R0	
10938	17	149.4627	66906.34	B3-7P11	NI 149.46751 +NI 149.46751
10939					
10940	95	149.4675	66904.19	B9-11P9	
10941	167	149.4942	66892.23	B0-6P8 B3-8P3 B4-8P8	
10942	76	149.5012	66889.09	B11-14P4	
10943	32	149.5178	66881.66	B8-11R3	
10944	1440	149.5221	66879.76	B1-7P3	
10945	0	149.5394	66872.03	B7-10R8	
10946	280	149.5516	66866.55	B11-14P1	
10947	460	149.5565	66864.37	B11-14P3 B6-9R11	
10948	26	149.5593	66863.13	B5-9P4 B1-5P16	
10949	13	149.5651	66860.50	B3-8R5	
10950	170	149.5686	66858.95	B11-14P2	
10951	2	149.5778	66854.84	B2-5P18	

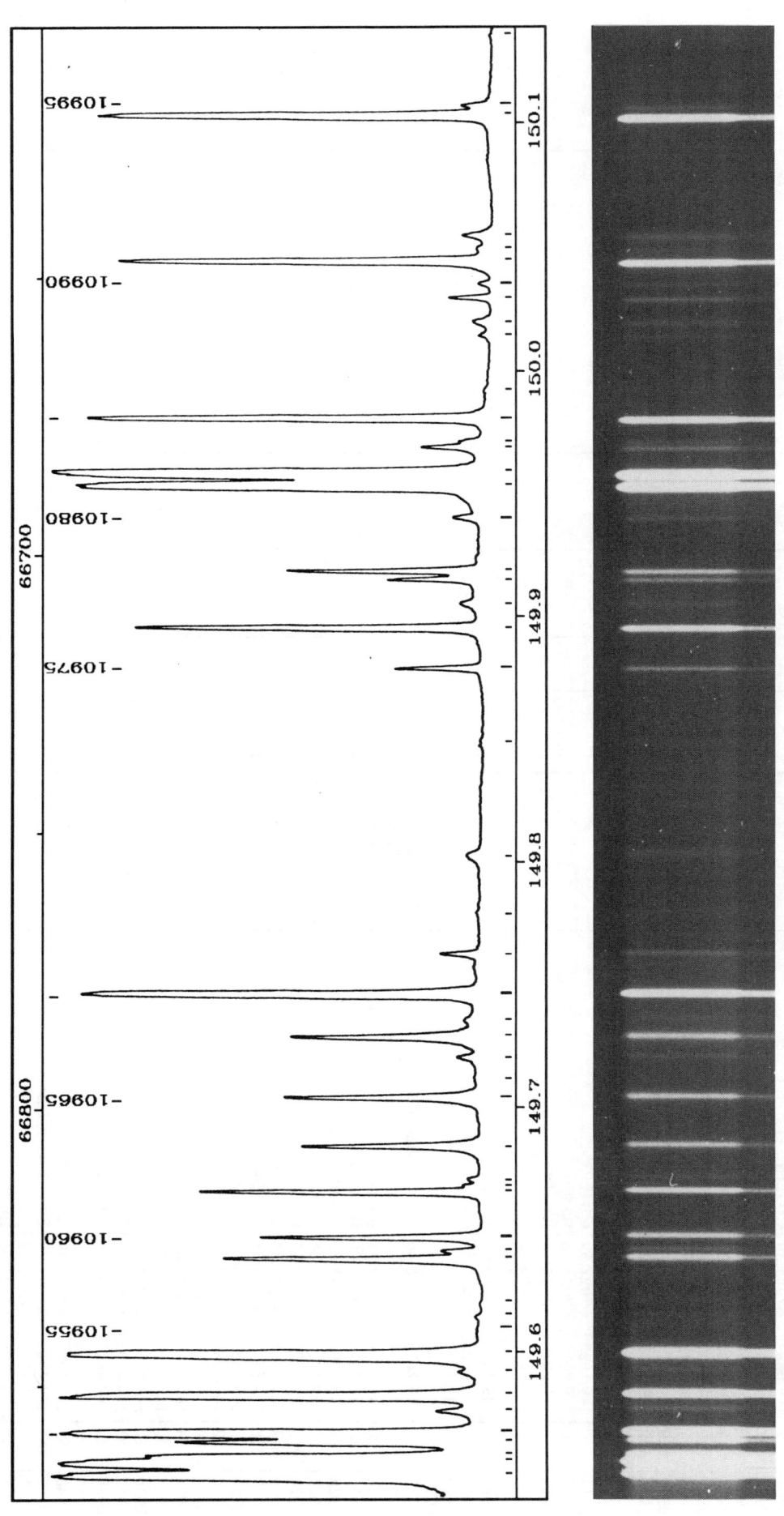

PLATE 126

TABLE 126

Nb	I	λ (nm)	σ (cm-1)	Assignment	Comment
10952	102	149.5839	66852.13	B8-11P1 B4-8R10	
10953	0	149.5938	66847.67	B3-6P16	
10954	90	149.6002	66844.84	B1-7R5	
10955	0	149.6119	66839.62		
10956	0	149.6160	66837.78	B4-6R20	
10957	0	149.6208	66835.62	C1-13R5 B3-2P29	
10958	10	149.6387	66827.63	B6-9P9 B2-7P8	
10959	3	149.6419	66826.21	B8-11R4	
10960	10	149.6474	66823.74	B3-7R13 B4-4P25	
10961	15	149.6658	66815.55	B7-10P6	CuII 149.66867
10962					
10963	0	149.6711	66813.16	B8-10P11	
10964	8	149.6843	66807.27	B0-6R10	
10965	9	149.7040	66798.48	B1-6R13 B3-5P20	
10966	0	149.7123	66794.77	B1-4P20	
10967	3	149.7206	66791.09	B6-7P19	
10968	7	149.7284	66787.58	B0-5P14	
10969	0	149.7359	66784.24		
10970	65	149.7460	66779.76	B8-11P2 B8-9P17	
10971	1	149.7624	66772.44	B8-11R5	
10972	0	149.7796	66764.75	B9-11P10	
10973	0	149.8029	66754.37	B0-2P25	
10974	0	149.8500	66733.40	B4-7R16 B4-5P22	
10975	3	149.8793	66720.34	C0-11R5	d
10976	25	149.8957	66713.07	B3-8P4 B7-8P18	
10977	0	149.9056	66708.63	B2-6P14	
10978	5	149.9153	66704.33	B7-10R9	
10979	7	149.9187	66702.82	B10-13R7	
10980	1	149.9405	66693.10	B5-8R14	
10981	210	149.9535	66687.36	B8-11P3	
10982	300	149.9583	66685.19	B1-7P4	
10983	0	149.9691	66680.40	B3-8R6	
10984	4	149.9715	66679.35	B5-8P12	
10985	38	149.9810	66675.13	B5-9P5 B6-8P15	
10986	0	149.9923	66670.07	B5-9R7	
10987	0	150.0150	66659.98	B8-11R6	
10988	1	150.0205	66657.55	B0-3P22	
10989	3	150.0301	66653.29	B10-13R6	
10990	0	150.0360	66650.69	C1-13R4	
10991	28	150.0450	66646.69	B1-7R6	
10992	0	150.0502	66644.38	B1-3R25	
10993	1	150.0555	66642.02	B0-4R20	
10994	33	150.1039	66620.51	B7-10P7 B10-13R5 B5-7R19 C0-11R4	
10995	2	150.1081	66618.66	B6-9R12	
10996	0	150.1373	66605.70	B2-4P22	

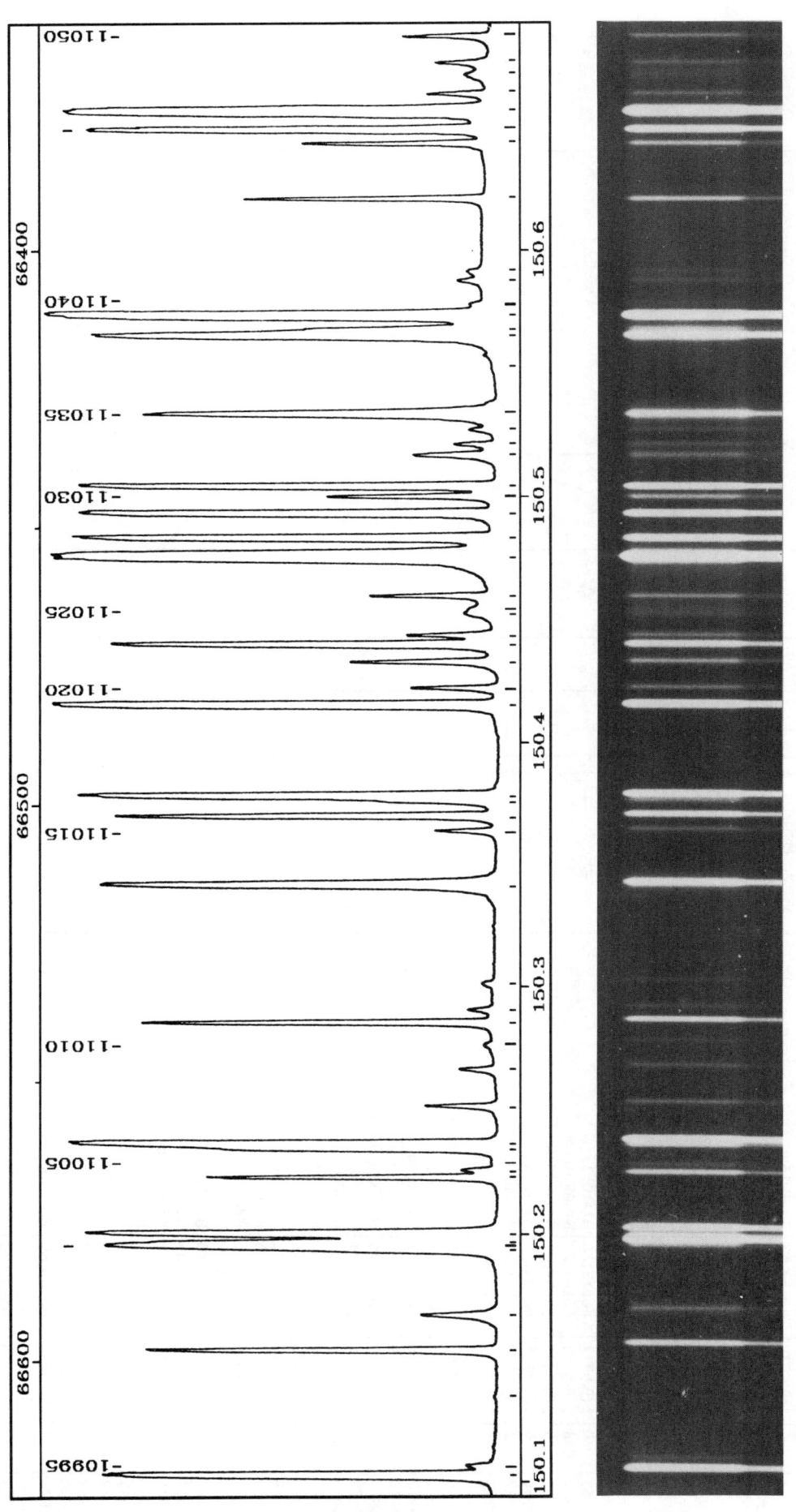

PLATE 127

TABLE 127

Nb	I	λ (nm)	σ (cm⁻¹)	Assignment	Comment
10997	25	150.1548	66597.96	B10-13R3	
10998	4	150.1687	66591.76	B0-5R16	
10999	0	150.1947	66580.26	B10-13R4	sh
11000	74	150.1968	66579.31	B10-13R2 B9-12R1	
11001		150.1979	66578.83	B4-8P9	sh
11002	62	150.2014	66577.26	B8-11P4 B9-12R2 B2-6R16 B5-7P17 B2-2P28	
11003	20	150.2237	66567.40	B9-12R0 B1-2R28	
11004	2	150.2266	66566.09	B8-11R7	
11005	0	150.2302	66564.52		
11006	19	150.2352	66562.28	B9-12R3	
11007	95	150.2370	66561.50	B10-13R1	
11008	7	150.2521	66554.81	B4-8R11	
11009	2	150.2670	66548.22	B6-9P10	
11010	0	150.2773	66543.66	C1-13P7	
11011	26	150.2858	66539.89	B10-13R0	
11012	2	150.2912	66537.48	B9-12R4	
11013	1	150.3022	66532.61	B7-10R10	
11014	39	150.3422	66514.91	B0-6P9 C0-11R3	
11015	5	150.3642	66505.18	B9-12R5	
11016	37	150.3702	66502.52	B3-8P5	
11017	7	150.3764	66499.78	B3-7P12 B1-4R22	
11018	68	150.3785	66498.87	B9-12P1	
11019	119	150.4152	66482.66	B10-13P1	
11020	6	150.4221	66479.61	B2-7P9	
11021	9	150.4328	66474.83	B1-6P12	
11022	36	150.4397	66471.81	B6-10R1 B3-8R7	
11023	6	150.4433	66470.22	B5-9R8 B9-12R6 B6-10R0	
11024	1	150.4519	66466.45	B8-11R8	
11025	7	150.4547	66465.19		
11026	1	150.4594	66463.13	B5-9P6	
11027	440	150.4756	66455.97	B1-7P5 B7-8P19	
11028	65	150.4831	66452.65	B8-11P5	
11029	73	150.4931	66448.23	B10-13P2	
11030	10	150.4999	66445.23	B6-10R2 C1-13R3	
11031	55	150.5041	66443.37	B9-12P2	
11032	7	150.5168	66437.76	B3-7R14 B4-6P19	
11033	4	150.5213	66435.76	B9-12R7	
11034	0	150.5275	66433.05	C0-12R9	
11035	26	150.5336	66430.34	B0-6R11	
11036	0	150.5530	66421.79	C0-11R2	
11037	56	150.5658	66416.15	B1-7R7 B3-4P24	
11038	6	150.5681	66415.13	B7-10P8	
11039	225	150.5740	66412.54	B10-13P3	
11040	0	150.5787	66410.46	B9-12R8	
11041	1	150.5882	66406.27	B9-12R10	
11042	1	150.5919	66404.65	B4-6R21	
11043	11	150.6215	66391.59	B6-10R3	
11044	11	150.6441	66381.63	B6-10P1	

Nb	I	λ (nm)	σ (cm⁻¹)	Assignment	Comment
11045	49	150.6497	66379.15	B10-13P4	
11046	165	150.6571	66375.89	B9-12P3 C1-13P6 B6-9R13 B5-8R15	d
11047	4	150.6640	66372.85	B9-12R9	
11048	1	150.6723	66369.21		
11049	3	150.6767	66367.26	B8-11R9	
11050	6	150.6876	66362.45	B7-10R11	

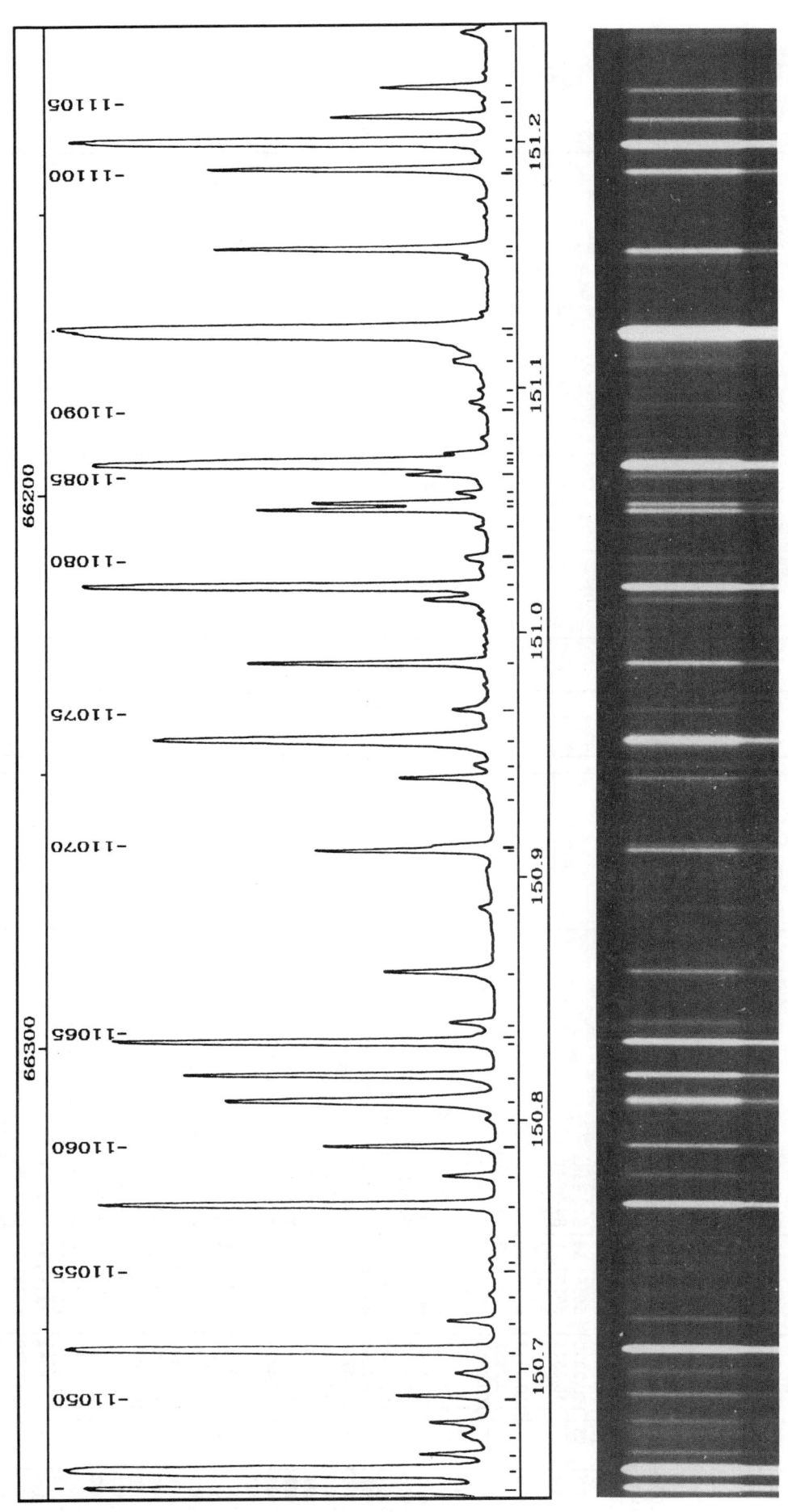

PLATE 128

TABLE 128

Nb	I	λ (nm)	σ (cm⁻¹)	Assignment	Comment
11051	2	150.6969	66358.35	B1-6R14	
11052	69	150.7060	66354.35	B10-13P5	
11053	4	150.7182	66349.00	B3-6P17	
11054	1	150.7291	66344.20	B4-7R17	
11055	0	150.7391	66339.77	B1-3P24	
11056	0	150.7431	66338.02	C0-11R1	
11057	0	150.7509	66334.61	B6-8P16	
11058	30	150.7652	66328.31	B10-13P7	
11059	4	150.7769	66323.17	B5-8P13	
11060	10	150.7891	66317.77	B8-11P6	
11061	0	150.8004	66312.84	B6-10R4	
11062	15	150.8075	66309.72	B0-4P19	
11063	19	150.8181	66305.05	B10-13P6 C1-13R2	
11064	37	150.8314	66299.21	B9-12P4	
11065	0	150.8339	66298.11	B1-5P17	
11066	2	150.8391	66295.81	B1-2P27 B6-10P2	
11067	8	150.8604	66286.45	B2-5P19 B6-8R19	
11068	1	150.8869	66274.82	B8-11R10	
11069	9	150.9104	66264.47	B6-9P11 C0-11R0	
11070	4	150.9125	66263.58	B3-8P6	
11071	0	150.9319	66255.05	B5-9R9	
11072	5	150.9407	66251.17	B4-8P10	
11073	1	150.9460	66248.84	B4-8R12	
11074	23	150.9563	66244.34	B0-5P15 B3-5P21	
11075	2	150.9686	66238.95	B3-8R8	
11076	13	150.9879	66230.46	B5-9P7	
11077	4	151.0141	66218.99	B2-6P15 B7-9P15	
11078	56	151.0193	66216.69	B9-12P5	
11079	0	151.0260	66213.78	B4-5P23	
11080	1	151.0310	66211.56	B6-10R5	
11081	0	151.0430	66206.30		
11082	11	151.0506	66202.97	B7-10P9	
11083	10	151.0533	66201.80	C1-13P5	
11084	2	151.0572	66200.07	B7-10R12	
11085	5	151.0651	66196.64	B8-11R11	
11086	82	151.0694	66194.75	B1-7P6	
11087	1	151.0702	66194.41	B8-11P7	sh
11088		151.0737	66192.88	C1-13R1	
11089	0	151.0800	66190.12		
11090	0	151.0914	66185.11		
11091	0	151.0946	66183.71	B6-10P3	
11092	0	151.1001	66181.29	B8-11R14	sh
11093	0	151.1117	66176.20	C1-14R4	
11094		151.1225	66171.48	B4-9R0 B5-7P18	
11095	570	151.1244	66170.63	B4-9R1	
11096	16	151.1535	66157.90	B1-4P21	
11097		151.1574	66156.22	B1-7R8	
11098	0	151.1703	66150.58	B4-3P29	
11099	0	151.1772	66147.53		s
11100	15	151.1887	66142.51	B8-11R12 B6-9R14	sh
11101	0	151.1901	66141.92	C0-11P7	
11102	121	151.1963	66139.18		
11103	10	151.2013	66136.99	B4-9R2	
11104	0	151.2115	66132.54	B9-12P6	
11105	8	151.2166	66130.29	B3-5R23	
11106	2	151.2235	66127.29	B8-11R13	
11107		151.2458	66117.55	B2-5R21	

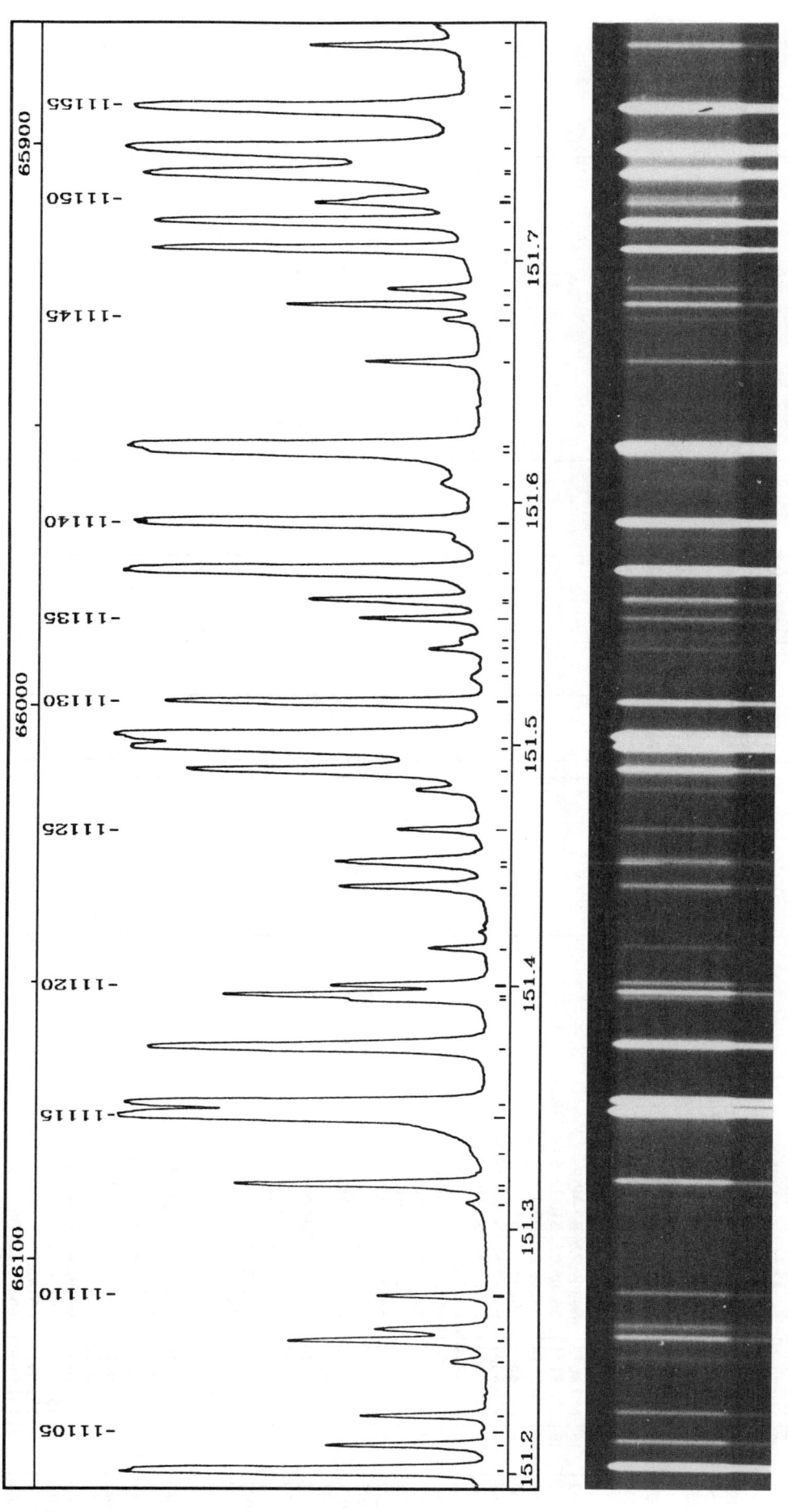

PLATE 129

TABLE 129

Nb	I	λ (nm)	σ (cm⁻¹)	Assignment	Comment
11108	11	151.2542	66113.88	B0-6P10	
11109	7	151.2587	66111.89	B2-7P10 B2-6R17 C0-11P6	
11110	7	151.2723	66105.94	B10-14R4 C1-13P4	
11111	1	151.3102	66089.39	B6-10R6 C0-11P3	
11112	0	151.3162	66086.76	C0-11P5	
11113	17	151.3181	66085.96	B3-7P13	
11114	0	151.3311	66080.26	C0-11P4	
11115	505	151.3467	66073.45	B4-9P1	
11116	178	151.3513	66071.46	B4-9R3 B5-8R16	
11117	50	151.3745	66061.32	B0-5R17 B10-14R3	
11118	8	151.3941	66052.79	B7-10R13	
11119	16	151.3962	66051.85	B9-12P7	
11120	10	151.3999	66050.24	B3-7R15	
11121	3	151.4147	66043.80	C1-13P3	
11122	6	151.4401	66032.70	B0-6R12	
11123	7	151.4502	66028.32		CuII 151.44924
11124	4	151.4640	66022.29	B0-4R21 B5-9R10	
11125	3	151.4804	66015.13	B8-11P8	
11126	22	151.4885	66007.43	B6-8P17	
11127	238	151.4981	66005.64	B1-6P13	
11128	845	151.5022	65999.45	B2-8R0	
11129	37	151.5164	65994.72	B2-8R1	
11130	0	151.5273	65992.35	B10-14R2 B3-8P7	
11131	0	151.5327	65989.77	B1-3R26 B0-2P26	
11132	2	151.5386	65987.07	B7-9P16	
11133	1	151.5448	65984.36	B7-10P10 B2-4P23	
11134	6	151.5511	65981.17	B5-7R21	d
11135		151.5584	65980.95	B3-8R9	sh
11136	8	151.5589	65975.74	B6-9P12 B9-12P8	
11137	300	151.5709	65970.13	B5-9P8	
11138	205	151.5838	65967.09	B4-9P2 B4-9R4	
11139	0	151.5907	65959.97	B5-8P14	
11140	420	151.6071	65954.25	B2-8R2 B4-6P20	d
11141		151.6203	65953.42		sh
11142	7	151.6222	65938.06	B0-7R0 B0-3P23	
11143	1	151.6575	65930.36	B0-7R1 B10-14R1 B6-10R7	
11144	10	151.6752	65927.77	B4-8R13	
11145	7	151.6812	65924.93	B7-10R14	
11146	57	151.6877	65917.93	B9-12P9	
11147	80	151.7038	65913.12	B6-9R15	
11148	8	151.7149	65909.76	B10-14R0	
11149	3	151.7226	65908.99	B0-7R2 B4-8P11	
11150	152	151.7244	65904.56	B1-6R15	
11151		151.7346	65903.92	C1-14R3	
11152		151.7360	65900.20	B1-7P7 B10-14P4	sh
11153		151.7446	65892.71	B9-12P10	
11154	980	151.7619	65891.39	B2-8P1	
11155	300	151.7649		B2-8R3	
11156	0			B6-10P5	
11157	10	151.7878	65881.44	B8-11P9	

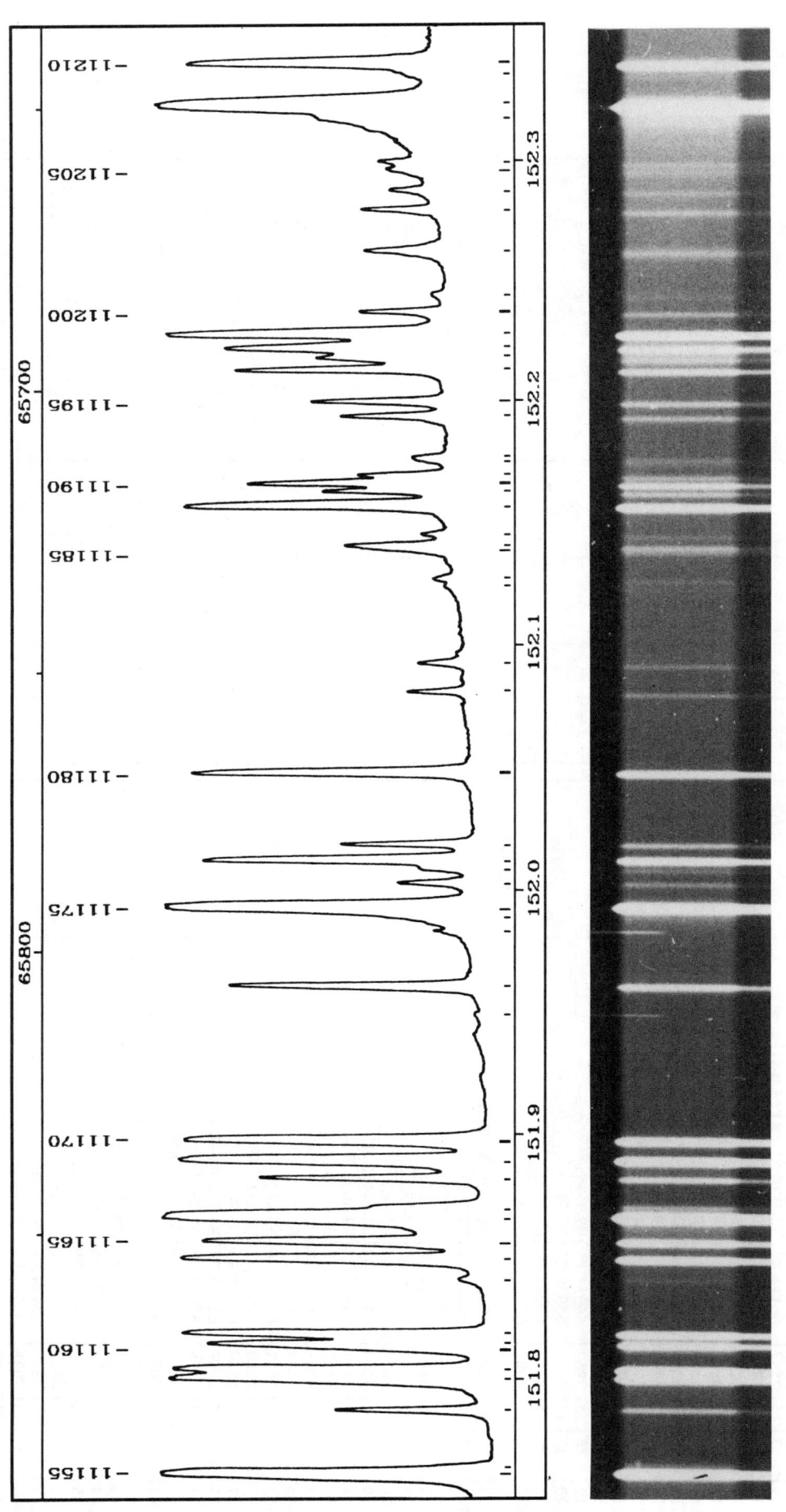

PLATE 130

TABLE 130

Nb	I	λ (nm)	σ (cm⁻¹)	Assignment	Comment
11207	1460	152.3180	65652.12	B2-6R18	sh
11208		152.3232	65649.90	B2-8P3	sh d
11209	71	152.3365	65644.16	B4-6R23	
11210		152.3403	65642.53	B2-8R5	

Nb	I	λ (nm)	σ (cm⁻¹)	Assignment	Comment
11158	185	151.8005	65875.92	B10-14P3	
11159	163	151.8038	65874.51	B10-14P1 B2-2P29	
11160		151.8121	65870.89		sh
11161	38	151.8142	65869.99	B1-7R9	
11162	75	151.8185	65868.12	B10-14P2	
11163	2	151.8410	65858.37	B3-6P18	
11164	90	151.8491	65854.84	B7-11R1	
11165	39	151.8562	65851.78	B4-9R5	
11166	700	151.8662	65847.42	B4-9P3 B7-11R0	
11167	8	151.8702	65845.69	B7-10R15	
11168	21	151.8819	65840.61	B7-11R2 B3-4P25	
11169	212	151.8894	65837.40	B0-7P1	
11170	117	151.8974	65833.90	B0-7R3	
11171					CuII 151.94918
11172	28	151.9607	65806.47	B7-11R3 B7-9P17	
11173					CuII 151.98371
11174	0	151.9898	65793.91	B5-9R11	
11175	500	151.9936	65792.26	B2-8P2	
11176	5	152.0034	65788.01	B5-7P19 B2-4R25	
11177	2	152.0091	65785.52	B5-8R17	
11178	50	152.0129	65783.87	B2-8R4	
11179	7	152.0194	65781.08	B7-10P11	
11180	60	152.0491	65768.23	B7-11P1	
11181	5	152.0815	65754.21	B7-11R4	
11182	2	152.0934	65749.08	B8-11P10	
11183	0	152.1252	65735.35	B1-5P18	
11184	0	152.1283	65734.01	B4-5P24 B6-9R16	
11185		152.1407	65728.61	B2-6P16 B2-5P20	sh
11186	5	152.1420	65728.09	B2-7P11	
11187	1	152.1468	65726.01	B6-8P18 C1-14R2	
11188	115	152.1588	65720.80	B0-7P2	
11189	8	152.1644	65718.39	B5-9P9	sh
11190	22	152.1679	65716.90	B0-7R4 B6-10P6	
11191	3	152.1709	65715.58	B9-13R7	
11192		152.1766	65713.11	B3-8P8	
11193	0	152.1781	65712.49	B3-8R10	
11194	6	152.1954	65705.00	B6-9P13	
11195	10	152.2014	65702.43	B4-9R6	
11196	27	152.2146	65696.72	B7-11P2	
11197	7	152.2195	65694.61	B0-5P16	
11198	25	152.2237	65692.81	B0-6P11	
11199	128	152.2295	65690.29	B4-9P4	
11200	4	152.2382	65686.54	B7-11R5	
11201	0	152.2452	65683.50	B7-9P18	
11202	2	152.2633	65675.72	B0-4P20	
11203	4	152.2800	65668.49	B3-7P14	d
11204	1	152.2881	65665.03	B3-7R16	
11205	0	152.2967	65661.30	B1-3P25	
11206	0	152.2997	65659.99	B9-13R6	

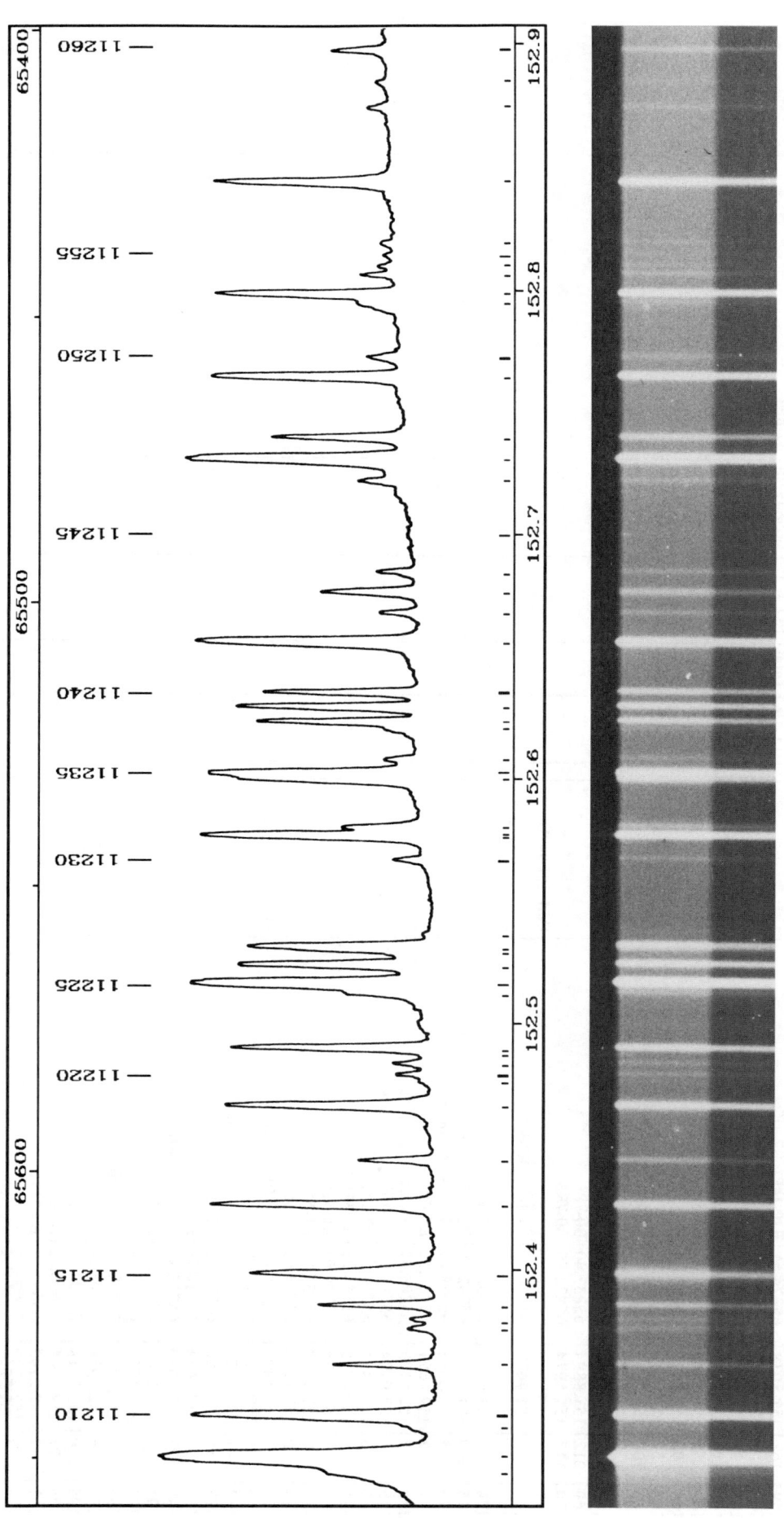

PLATE 131

TABLE 131

Nb	I	λ (nm)	σ (cm-1)	Assignment	Comment
11260	4	152.8970	65403.49	B7-10P13	

Nb	I	λ (nm)	σ (cm-1)	Assignment	Comment
11211	6	152.3607	65633.71	B8-11P11	
11212	1	152.3751	65627.52	B4-8R14	
11213	1	152.3787	65625.95	B5-8P15	
11214	7	152.3845	65623.49	B9-13R5	
11215	20	152.3969	65618.12	B0-6R13	d
11216	49	152.4249	65606.08	B0-3R25 B7-11P3 B7-11R6 B2-5R22	
11217	5	152.4431	65598.25	B9-13R4	
11218	30	152.4655	65588.63	B1-7P8	
11219	0	152.4753	65584.41	C1-14R1	
11220	1	152.4783	65583.12	B7-10P12	
11221	2	152.4827	65581.23	B6-9R17	
11222					CuII 152.48601
11223	37	152.4889	65578.56	B9-13R3	
11224	7	152.5117	65568.76	B4-8P12	
11225	350	152.5155	65567.11	B0-7P3	
11226	40	152.5231	65563.84	B0-7R5	
11227	27	152.5297	65560.99	B1-7R10	sh
11228	0	152.5307	65560.59	B9-13R2	
11229	0	152.5362	65558.20	B5-9R12	
11230	2	152.5665	65545.21	B8-11P12	
11231					CuII 152.57645
11232	110	152.5764	65540.94	B9-13R1	
11233	8	152.5799	65539.43	B1-6P14	
11234	30	152.6000	65530.81	B4-9R7 B8-12R2	
11235	92	152.6019	65529.99	B8-12R1 B0-5R18	
11236	2	152.6080	65527.38	B6-10P7	
11237	1	152.6202	65522.12	B4-6P21 B8-11P14	
11238	24	152.6233	65520.79	B8-12R3	
11239	35	152.6293	65518.22	B9-13R0 B7-11R7	
11240	22	152.6352	65515.70	B8-12R0	
11241	160	152.6552	65507.09	B4-9P5	
11242	3	152.6673	65501.91	B8-12R4	
11243	11	152.6757	65498.30	B7-11P4 B8-11P13	
11244	2	152.6841	65494.72	B8-12R5	
11245					CuIII 152.6998 Ke
11246	2	152.7214	65478.70	B6-8P19	
11247	280	152.7303	65474.91	B2-8P4	
11248	16	152.7391	65471.11	B2-8R6	
11249	87	152.7641	65460.41	B9-13P1	
11250	1	152.7721	65456.98	B1-6R16	
11251	0	152.7941	65447.55	B5-9P10	
11252	69	152.7977	65446.00	B8-12P1	
11253	0	152.8056	65442.61	C0-12R5	
11254	1	152.8091	65441.12	B6-9P14	
11255	0	152.8138	65439.10	B8-12R6	
11256	0	152.8187	65437.02	B5-7P20	
11257	58	152.8431	65426.56	B9-13P2 B0-4R22 B3-8R11 B7-11R8	
11258	1	152.8736	65413.53	B8-12R7 B8-12R10	
11259	0	152.8842	65408.98	B3-8P9	

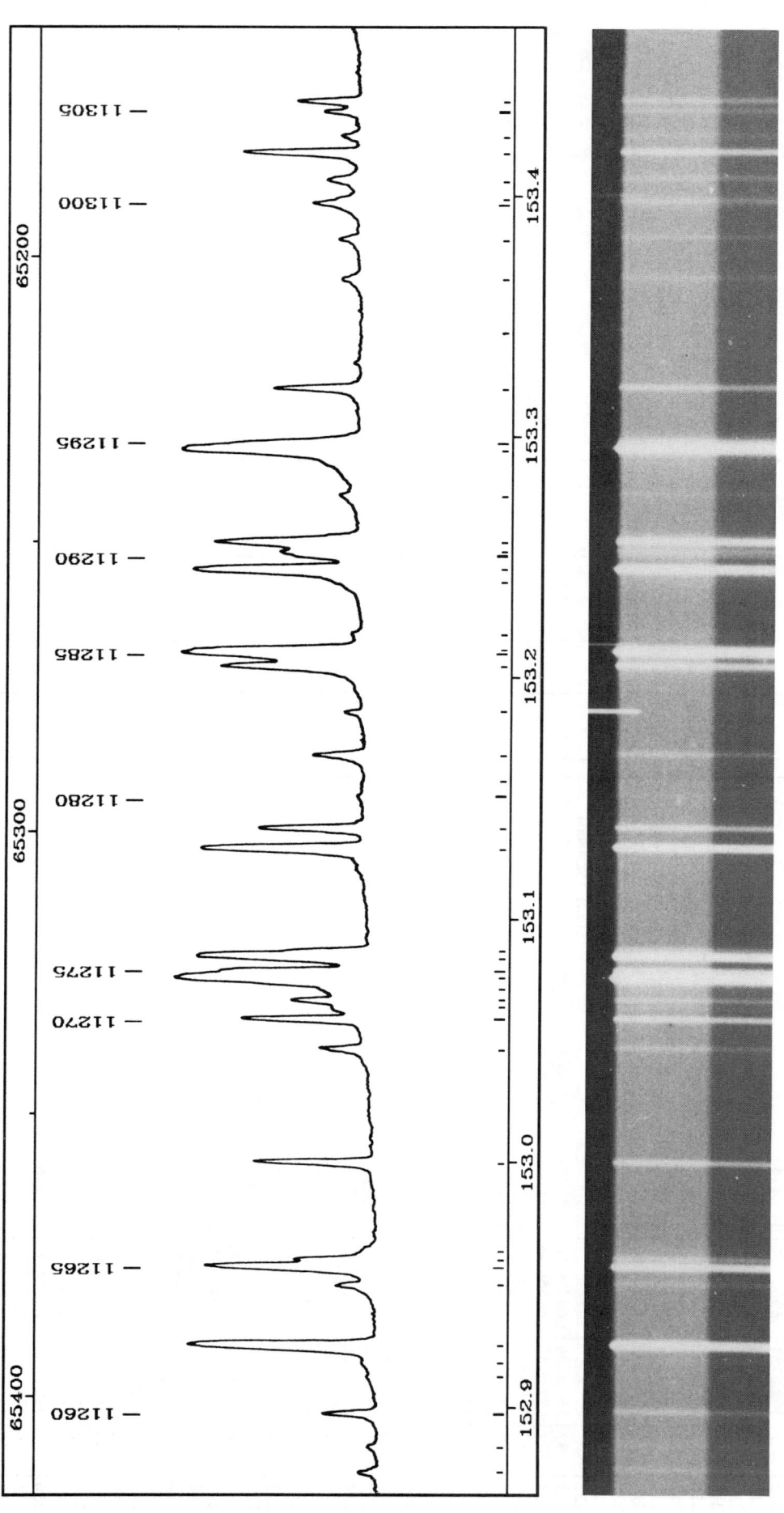

PLATE 132

TABLE 132

Nb	I	λ (nm)	σ (cm⁻¹)	Assignment	Comment
11261	0	152.9126	65396.84	B3-6R21	
11262	0	152.9184	65394.34	B8-12R8 B2-4P24	
11263	334	152.9241	65391.90	B9-13P3 B8-12P2 B8-12R9	
11264	4	152.9493	65381.15	B3-6P19	
11265	70	152.9562	65378.18	B0-7P4 B7-11P5	
11266	10	152.9592	65376.89	B0-7R6	
11267	0	152.9625	65375.52	B1-3R27	
11268	18	152.9988	65359.97	B9-13P4	
11269	3	153.0455	65340.03	B4-9R8	
11270	28	153.0575	65334.92	B9-13P5 B7-11R9	
11271	0	153.0621	65332.94	B2-7P12	
11272	7	153.0655	65331.52	B9-13P7	
11273	0	153.0699	65329.63	B6-10P8	
11274	640	153.0745	65327.65	B5-10R1	
11275	65	153.0771	65326.55	B8-12P3 B5-9R13 B1-4R24	
11276	176	153.0836	65323.80	B5-10R0	
11277	10	153.0858	65322.84	B9-13P6 B4-8R15	
11278	155	153.1289	65304.45	B5-10R2	
11279	25	153.1374	65300.82	B4-9P6	
11280	0	153.1510	65295.04	B4-5P25	d
11281	0	153.1567	65292.61	B6-8P20	d
11282	10	153.1678	65287.88	B3-7R17	
11283					CuII 153.18559
11284	45	153.2040	65272.43	B2-8R7	
11285	490	153.2099	65269.92	B2-8P5	
11286					CuII 153.21306
11287	0	153.2168	65266.98	B0-3P24	
11288	0	153.2395	65257.34	B0-2P27	
11289	220	153.2440	65255.39	B5-10R3 B0-6P12	
11290	8	153.2500	65252.86	B8-12P4	
11291	14	153.2516	65252.19	B3-7P15 B7-10P14	
11292	71	153.2556	65250.47	B1-7P9	
11293	0	153.2750	65242.23	B2-6P17	
11294	660	153.2940	65234.11	B5-10P1	
11295	0	153.2959	65233.33	B1-7R11	sh
11296	20	153.3191	65223.46	B4-8P13	
11297					SiII 153.34318
11298	2	153.3644	65204.20	B2-6R19	
11299	3	153.3811	65197.10	B6-9P15	
11300	6	153.3961	65190.69	B0-6R14	CuII 153.39865
11301					
11302	5	153.4057	65186.64	B2-5P21	
11303	33	153.4163	65182.11	B5-10R4	
11304	4	153.4232	65179.21	B1-5P19	
11305	4	153.4334	65174.87	B8-12P5	
11306	7	153.4374	65173.16	B5-9P11	

Nb	I	λ (nm)	σ (cm⁻¹)	Assignment	Comment

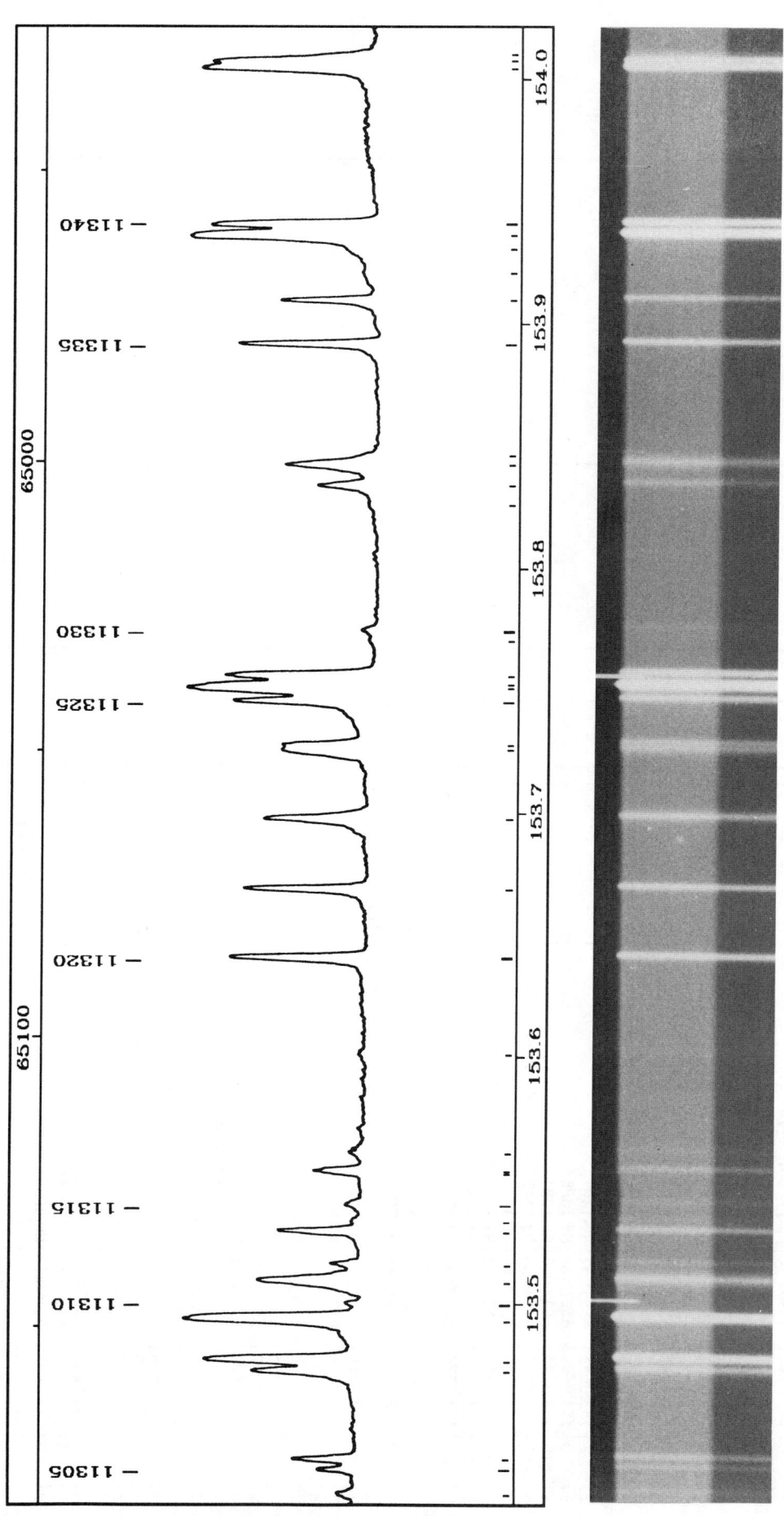

PLATE 133

TABLE 133

Nb	I	λ (nm)	σ (cm⁻¹)	Assignment	Comment
11307	20	153.4726	65158.19	B0-7R7	
11308	128	153.4771	65156.31	B0-7P5	
11309	320	153.4935	65149.33	B5-10P2	CuII 153.50023
11310					d
11311	24	153.5089	65142.80	B0-5P17	
11312	5	153.5156	65139.95	B7-10P15	
11313	13	153.5291	65134.24	B4-9R9	
11314	1	153.5331	65132.55	B3-8R12	
11315	2	153.5402	65129.51	B4-7P18 B5-7P21	
11316					CuII 153.55238
11317	5	153.5536	65123.83	B9-14R4 B6-10P9	
11318	0	153.5612	65120.60	B2-5R23	
11319	0	153.6010	65103.74	B5-9R14	
11320	47	153.6405	65087.00	B5-10R5	
11321	30	153.6692	65074.87	B4-9P7	
11322	21	153.6986	65062.41	B1-6P15	
11323	12	153.7266	65050.55	B0-4P21	
11324	12	153.7286	65049.69	B2-8R8	
11325	45	153.7468	65041.99	B9-14R3	
11326	890	153.7524	65039.65	B5-10P3	
11327					CuII 153.75590
11328	76	153.7572	65037.61	B2-8P6	
11329	0	153.7705	65031.97	B6-10R13	
11330	0	153.7759	65029.69	B4-8R16	
11331	0	153.8274	65007.93	B1-3P26 B3-6R22	
11332	8	153.8351	65004.67	B1-6R17	d
11333	11	153.8436	65001.10	B0-5R19	CuII 153.84795
11334					
11335	33	153.8927	64980.33	B9-14R2	
11336	12	153.9107	64972.76	B5-10R6	
11337	0	153.9198	64968.89	B7-11P8 B0-3R26	
11338	0	153.9339	64962.97		
11339	395	153.9371	64961.58	B3-9R1	
11340	116	153.9415	64959.76	B3-9R0	
11341	157	154.0054	64932.79	B9-14R1	
11342	95	154.0077	64931.82	B3-9R2	
11343	0	154.0101	64930.81	B2-7P13	

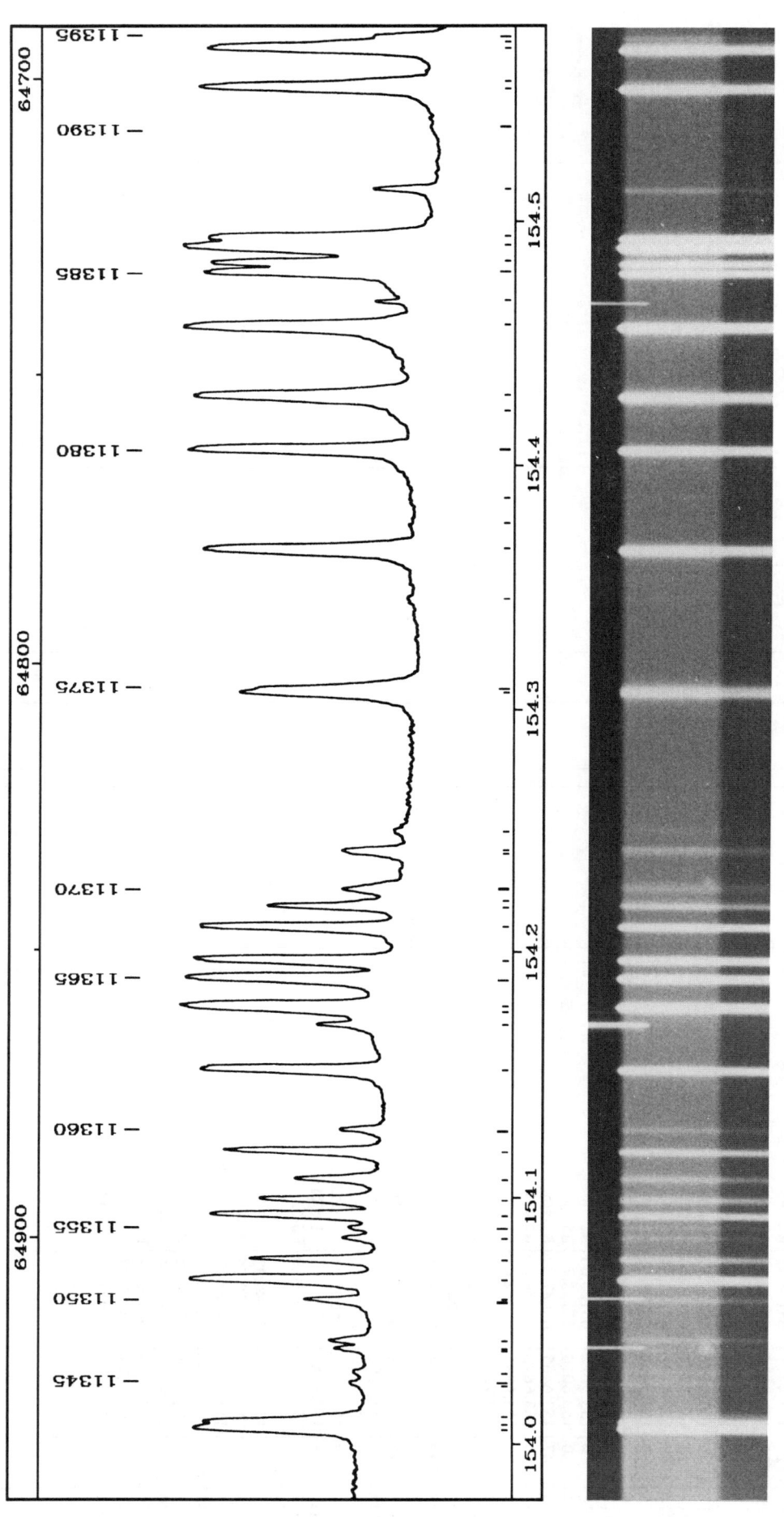

PLATE 134

TABLE 134

Nb	I	λ (nm)	σ (cm⁻¹)	Assignment	Comment
11344					
11345	1	154.0242	64924.85	B3-7R18	CuII 154.02394
11346	0	154.0286	64923.00	B3-6P20	
11347	0	154.0383	64918.91	B6-10P10	
11348					
11349	2	154.0416	64917.52	B4-9R10	CuII 154.03887
11350	6	154.0580	64910.60	B0-7R8	
11351					CuII 154.05883
11352	157	154.0659	64907.29	B5-10P4	
11353	24	154.0742	64903.80	B0-7P6	
11354	4	154.0830	64900.09	B5-9P12	
11355	3	154.0872	64898.30	B5-9R15	
11356	48	154.0920	64896.30	B9-14R0	
11357	18	154.0983	64893.66	B1-7P10	
11358	12	154.1068	64890.06	B1-7R12	
11359	35	154.1178	64885.42	B9-14P4 B1-2P29	
11360	5	154.1268	64881.64	B4-8P14	
11361	125	154.1513	64871.32	B3-9R3	CuII 154.17032
11362					CuII 154.17560 Ke
11363					
11364	460	154.1772	64860.42	B3-9P1	
11365	240	154.1891	64855.42	B9-14P3	
11366	160	154.1966	64852.26	B9-14P1	
11367	115	154.2104	64846.45	B9-14P2	
11368	18	154.2191	64842.81	B5-10R7	
11369		154.2212	64841.91	B3-7P16	sh
11370	3	154.2264	64839.75	B0-4R23	d
11371		154.2410	64833.60	B7-11P9 B3-8R13	sh
11372	6	154.2422	64833.09	B4-9P8	
11373		154.2502	64829.74	B2-4P25	
11374	36	154.3069	64805.91	B2-8R9	
11375	26	154.3082	64805.38	B0-6P13	
11376	0	154.3458	64789.57		
11377	117	154.3657	64781.24	B3-9R4 B2-8P7	
11378	0	154.3763	64776.78	B1-4R25	
11379	0	154.3878	64771.96	B2-6R20	
11380	256	154.4066	64764.06	B3-9P2 B2-6P18	
11381	0	154.4208	64758.12	B4-7P19	
11382	200	154.4285	64754.88	B5-10P5 B0-6R15 B4-8R17	
11383	500	154.4572	64742.84	B6-11R1	
11384					CuII 154.46771
11385	140	154.4795	64733.49	B6-11R0	
11386	115	154.4833	64731.92	B6-11R2	
11387	720	154.4902	64729.03	B1-8R1	
11388	198	154.4936	64727.61	B1-8R0	
11389	5	154.5135	64719.28	B6-10P11 B5-9R16	
11390	0	154.5388	64708.66	B7-11P10	
11391	180	154.5548	64701.98	B6-11R3	
11392	180	154.5578	64700.72	B5-10R8	sh

Nb	I	λ (nm)	σ (cm⁻¹)	Assignment	Comment
11393	174	154.5709	64695.25	B1-8R2	sh
11394		154.5727	64694.47	B4-9R11	
11395	5	154.5754	64693.34	B8-13R7	

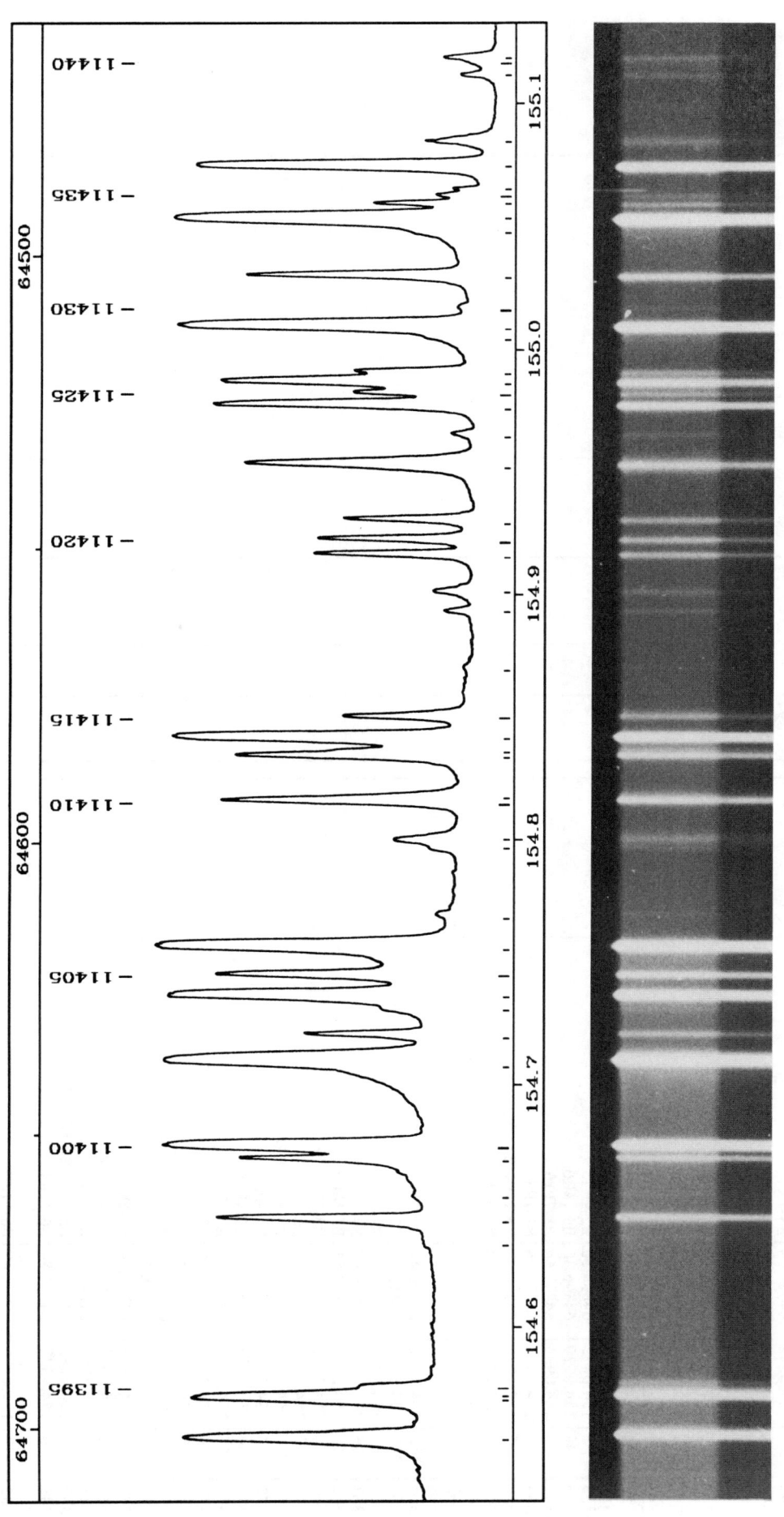

PLATE 135

TABLE 135

Nb	I	λ (nm)	σ (cm-1)	Assignment	Comment
11396	0	154.6354	64668.25	B2-5R24	
11397	35	154.6431	64665.04	B3-9R5 B2-5P22	
11398	0	154.6528	64660.96	B3-6R23	
11399	25	154.6673	64654.92	B6-11R4	
11400	500	154.6725	64652.74	B6-11P1	
11401	800	154.7072	64638.23	B3-9P3 B0-7R9	
11402	11	154.7177	64633.86	B5-9P13 B1-5P20	
11403	1	154.7291	64629.10	B8-13R6	
11404	300	154.7338	64627.11	B1-8R3	
11405	35	154.7422	64623.60	B0-7P7	
11406	700	154.7540	64618.70	B1-8P1	
11407	1	154.7677	64612.96	B8-13R5	
11408	1	154.7953	64601.46	B7-11P11	
11409	2	154.7990	64599.90	B0-3P25	d
11410	30	154.8150	64593.22	B6-11R5 B0-5P18	
11411		154.8163	64592.69		sh
11412	29	154.8334	64585.54	B5-10P6	
11413	8	154.8352	64584.78	B1-6P16	
11414	240	154.8408	64582.46	B6-11P2 B3-7R19	
11415	7	154.8494	64578.87	B4-9P9 B5-9R17	
11416	0	154.8694	64570.53	B2-7R16	
11417	2	154.8924	64560.94	C0-13R5	
11418	3	154.9004	64557.62	B1-6R18	
11419	11	154.9155	64551.32	B5-10R9	
11420	10	154.9217	64548.74	B4-8P15	
11421	9	154.9298	64545.38	B2-8R10	
11422	15	154.9525	64535.91	B1-7R13 B3-8R14	
11423	1	154.9672	64529.79	B6-10P12	
11424	32	154.9768	64525.78	B1-8R4 B2-7P14	
11425	2	154.9816	64523.80	B3-9R6	
11426	23	154.9862	64521.87	B1-7P11	
11427	3	154.9899	64520.32	B6-11R6	
11428	0	155.0039	64514.51	B8-13R2	
11429	350	155.0090	64512.40	B1-8P2	
11430	0	155.0165	64509.25	B4-8R18	
11431	12	155.0295	64503.85	B2-8P8	
11432	0	155.0468	64496.67	B5-9R18	d
11433	710	155.0529	64494.11	B6-11P3	
11434	8	155.0589	64491.63	B8-13R1	
11435	5	155.0624	64490.16	B3-6P21	
11436					CuII 155.06533
11437	140	155.0745	64485.12	B3-9P4	
11438	2	155.0847	64480.91	B0-5R20	
11439	1	155.1118	64469.63	B4-9R12	
11440	0	155.1167	64467.59		
11441	2	155.1188	64466.70	B8-13R0	

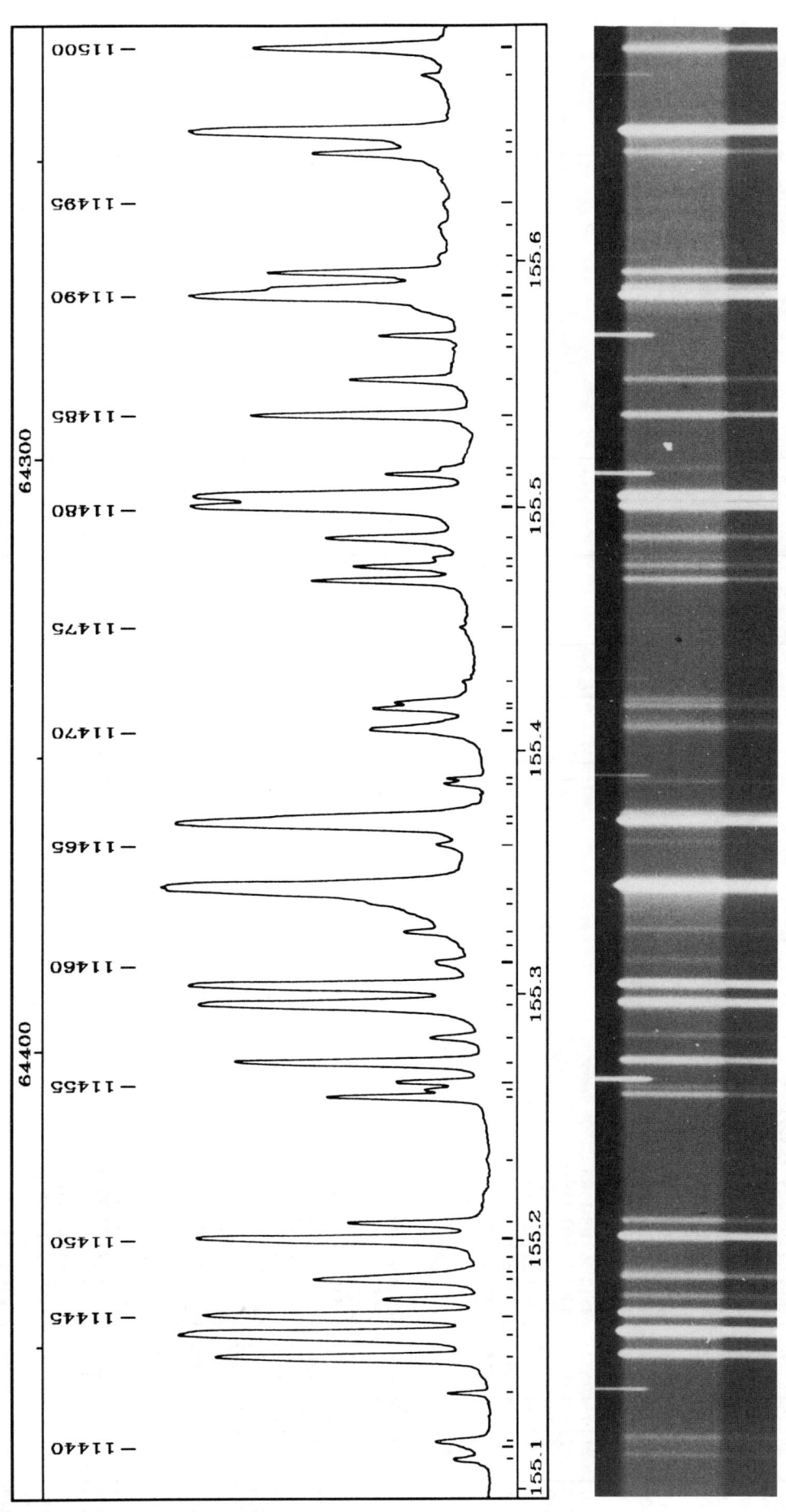

PLATE 136

TABLE 136

Nb	I	λ (nm)	σ (cm⁻¹)	Assignment	Comment
11442	55	155.1529	64452.56	B7-12R2	CuII 155.13890
11443	260	155.1620	64448.76	B7-12R1	
11444	68	155.1696	64445.61	B7-12R3	
11445	7	155.1764	64442.79	B3-7P17	
11446	9	155.1845	64439.41	B6-11R7	
11447	1	155.1858	64438.87	B0-4P22	
11448	0	155.1932	64435.83		sh
11449	72	155.2006	64432.73	B7-12R0	
11450	8	155.2071	64430.05	B7-12R4	
11451	11	155.2328	64419.38	B4-7P20	
11452	4	155.2581	64408.87	B7-12R5	
11453		155.2607	64407.79	B8-13P1	
11454					
11455					CuII 155.26464
11456	30	155.2721	64403.07	B5-10P7	
11457	3	155.2822	64398.88	B5-10R10	
11458	85	155.2958	64393.23	B1-8R5	
11459	125	155.3036	64389.99	B6-11P4 B7-12R10	
11460	0	155.3129	64386.15	B1-3P27 B7-12R6	
11461	3	155.3194	64383.46	B5-9P14	
11462		155.3261	64380.67	B8-13P2	sh
11463	1040	155.3374	64376.01	B1-8P3	
11464	240	155.3438	64373.35	B7-12R7	
11465	16	155.3624	64365.64	B7-12P1 B2-6R21	
11466	1	155.3709	64362.13	B3-9R7 B0-3R27 B6-10P13	
11467		155.3731	64361.21	B6-11R8	
11468		155.3870	64355.43		
11469					CuII 155.38962
11470	7	155.4096	64346.10	B0-6P14	
11471	0	155.4124	64344.95	B8-13P3	
11472	1	155.4181	64342.58	B0-7R10	
11473	3	155.4204	64341.63		
11474		155.4293			NiII 155.4293 Ke
11475	0	155.4507	64329.08	B8-13P7	
11476	12	155.4698	64321.16	B0-7P8	
11477	7	155.4755	64318.83	B4-9P10	
11478	1	155.4790	64317.36	B0-6R16 B8-13P4	
11479	10	155.4871	64314.02	B7-12P2	
11480	125	155.5000	64308.68	B3-9P5	
11481	240	155.5039	64307.05		
11482					CuII 155.51344
11483	1	155.5153	64302.37	B4-8R19 B2-4P26	
11484	0	155.5326	64295.19		
11485	29	155.5372	64293.30	B8-13P5	
11486	8	155.5517	64287.29	B8-13P6	
11487		155.5646	64281.98		s
11488					CuII 155.57030
11489	0	155.5812	64275.11	B0-4R24	d
11490	173	155.5864	64272.98	B6-11P5 B6-11R9	
11491	25	155.5891	64271.84	B2-8R11	
11492	17	155.5957	64269.11	B3-7R20 C0-13P7	
11493	0	155.6033	64265.97	B1-4R26	d
11494	0	155.6146	64261.32	B2-5R25	d
11495	12	155.6248	64257.13	B5-10R11 B4-9R13	
11496	0	155.6448	64248.86		
11497		155.6487	64247.25		
11498	280	155.6535	64245.28	B7-12P3	
11499		155.6766			NiII 155.6766 Ke
11500	22	155.6869	64231.47	B1-8R6 B4-8P16	

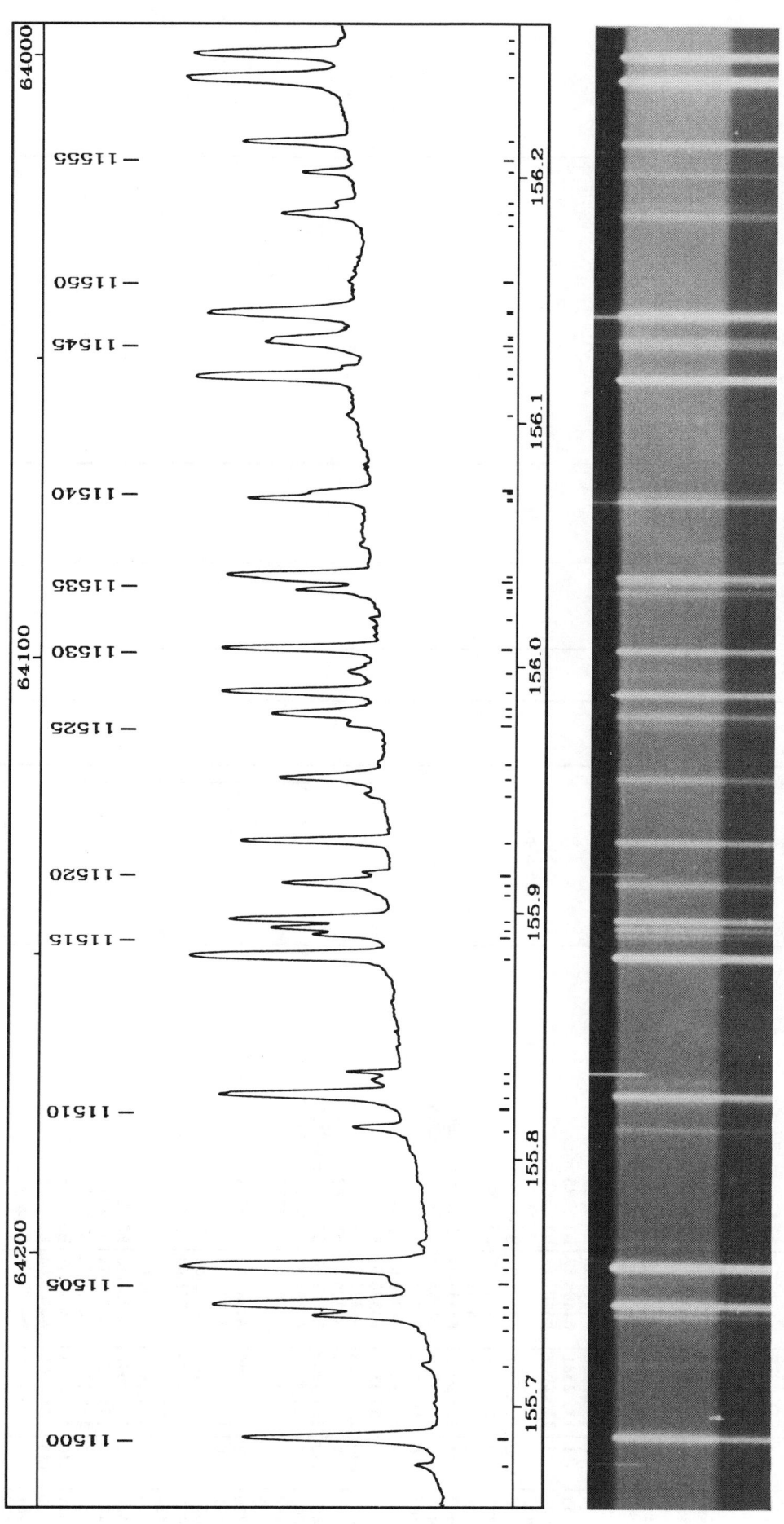

PLATE 137

TABLE 137

Nb	I	λ (nm)	σ (cm-1)	Assignment	Comment
11501	0	155.7166	64219.23	B6-10P14	
11502	0	155.7295	64213.90		
11503	7	155.7359	64211.28	B5-10P8	
11504	50	155.7401	64209.54	B2-8P9	
11505	0	155.7499	64205.49	B2-7R17	
11506	212	155.7555	64203.18	B1-8P4	
11507					CuII 155.75867
11508	0	155.7649	64199.30	B6-11R10	
11509	6	155.8116	64180.09	B3-9R8	
11510		155.8232	64175.30	B1-7R14	sh
11511	45	155.8250	64174.56	B7-12P4	
11512	3	155.8306	64172.26	B2-5P23	
11513					CuII 155.83447
11514	126	155.8813	64151.39	B4-10R1	
11515	7	155.8901	64147.76	B5-9P15	
11516	15	155.8929	64146.61	B6-11P6	
11517	38	155.8965	64145.12	B4-10R0	
11518	0	155.9074	64140.63	B6-11R11	
11519	12	155.9112	64139.06	B1-7P12	
11520		155.9160			NiII 155.9159 Ke
11521	30	155.9288	64131.83	B4-10R2	
11522	0	155.9481	64123.91	B4-7P21	
11523	15	155.9548	64121.16	B1-6R19 B2-7P15	
11524		155.9609	64118.65	B6-10P15	
11525	3	155.9763	64112.30	B6-11R13	
11526	14	155.9804	64110.60	B1-6P17	
11527		155.9823	64109.83	B3-9P6 B5-10R12 B6-11R12	
11528	41	155.9892	64107.01		sh
11529	0	155.9971	64103.76	B1-5P21	
11530	35	156.0068	64099.76	B7-12P5	
11531	0	156.0189	64094.79		
11532	0	156.0288	64090.72	B3-6P22	
11533					CI 156.03092
11534	17	156.0309	64089.88		CI 156.03092
11535	17	156.0351	64088.15	B8-14R4	
11536	42	156.0367	64087.50	B4-10R3	
11537					CI 156.06822
11538		156.0682	64074.54		CI 156.06822
11539					CI 156.07090
11540		156.0706	64073.55	B3-7P18	CI 156.07090
11541	0	156.1024	64060.52	B4-10P1	
11542	182	156.1179	64054.17	B4-9P11	
11543	0	156.1214	64052.71		
11544		156.1301	64049.16		sh
11545	15	156.1321	64048.31	B0-5P19	
11546					CI 156.13402
11547		156.1339	64047.60		CI 156.13402
11548					CI 156.14384
11549	88	156.1439	64043.50	B1-8R7	+CI 156.14384

Nb	I	λ (nm)	σ (cm-1)	Assignment	Comment
11550	0	156.1570	64038.11	B4-9R14	
11551	0	156.1820	64027.88	B0-7R11	
11552	10	156.1845	64026.86	B7-12P6	
11553	2	156.1884	64025.23	B4-10R4	
11554	6	156.2012	64019.99		
11555	24	156.2052	64018.36		
11556	320	156.2136	64014.90	B6-11P7 B5-10P9	
11557	143	156.2392	64004.42	B1-8P5	
11558	0	156.2490	64000.40	B8-14R3	
11559		156.2540	63998.35	B3-7R21	

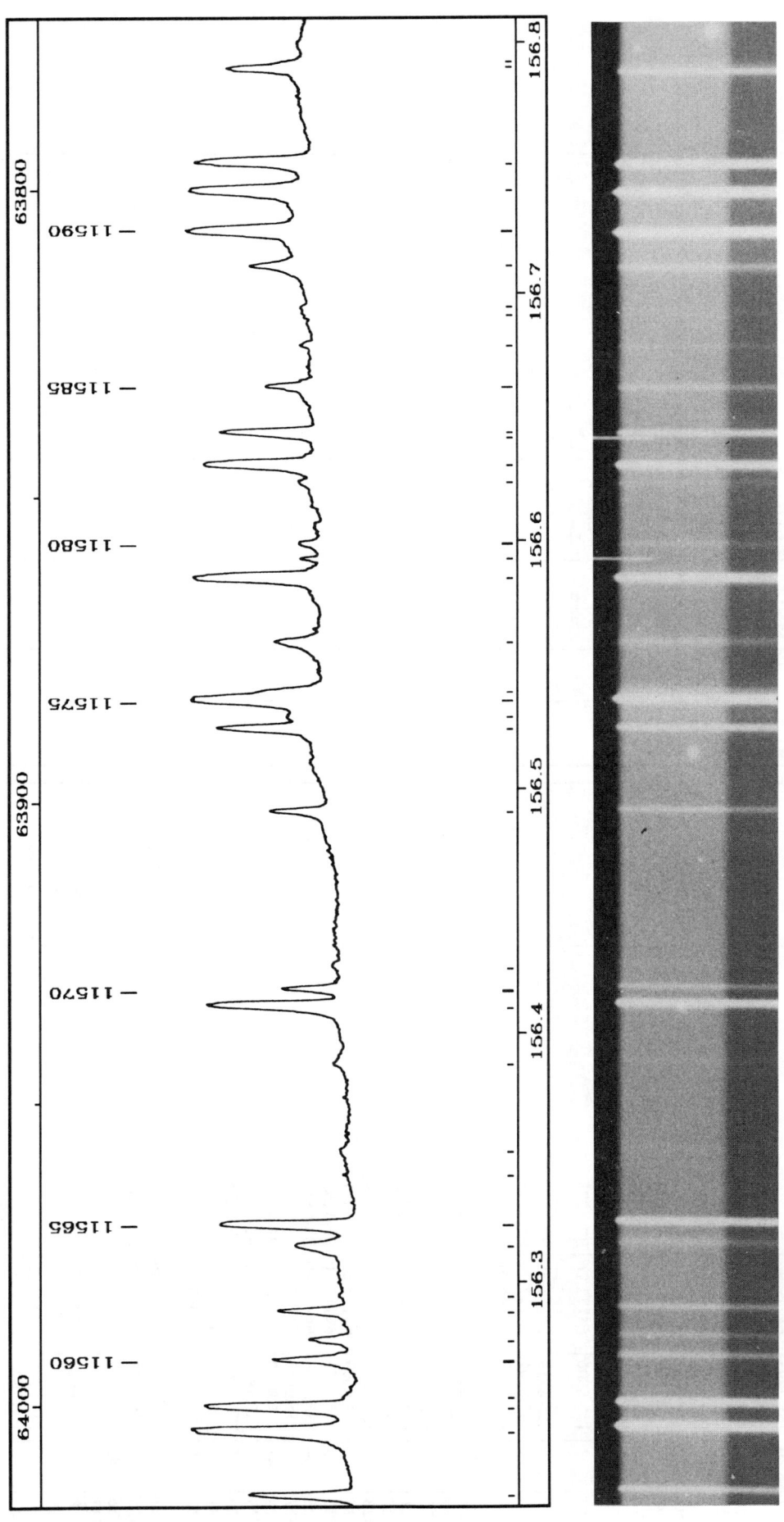

PLATE 138

TABLE 138

Nb	I	λ (nm)	σ (cm⁻¹)	Assignment	σ (cm⁻¹)	Assignment	Comment

Nb	I	λ (nm)	σ (cm⁻¹)	Assignment	Comment
11560	15	156.2681	63992.58	B0-7P9	
11561	8	156.2762	63989.28	B2-8R12	
11562	15	156.2876	63984.59	B3-9R9	
11563	0	156.2935	63982.20	B5-10R13	
11564	6	156.3136	63973.98	B0-5R21	
11565	115	156.3218	63970.59	B4-10P2	d
11566	0	156.3389	63963.59		
11567	0	156.3514	63958.48	B0-3P26	
11568	0	156.3872	63943.86	B5-9P16	
11569	144	156.4104	63934.38	B8-14R2 B4-8P17	
11570	12	156.4172	63931.58	B4-10R5	
11571	0	156.4266	63927.75	C0-13P5	
11572	17	156.4899	63901.90	B2-8P10	
11573	68	156.5234	63888.21	B3-9P7	
11574	0	156.5288	63886.00		
11575	740	156.5350	63883.49	B8-14R1	
11576	15	156.5380	63882.25	B0-6P15 B6-11P8	
11577	10	156.5583	63873.96	B0-6R17	
11578	380	156.5842	63863.42	B4-10P3	
11579					CuII 156.59243
11580	1	156.5980	63857.78	B7-12P9	
11581	0	156.6229	63847.64	B7-12P10	
11582	220	156.6300	63844.75	B8-14R0 B0-4P23	
11583					CuII 156.64148
11584	69	156.6430	63839.45	B8-14P4	
11585	15	156.6617	63831.82	B1-8R8	
11586	4	156.6784	63824.99	B4-10R6	
11587	0	156.6911	63819.83	B5-10R15	
11588	0	156.6940	63818.65	B5-10P10	
11589	20	156.7106	63811.90	B1-7R15	
11590	710	156.7247	63806.15	B8-14P3	
11591	740	156.7409	63799.56	B8-14P1 B1-4R27	
11592	350	156.7523	63794.90	B8-14P2	
11593	60	156.7899	63779.63	B1-8P6	
11594		156.7919	63778.79	B3-9R10	sh

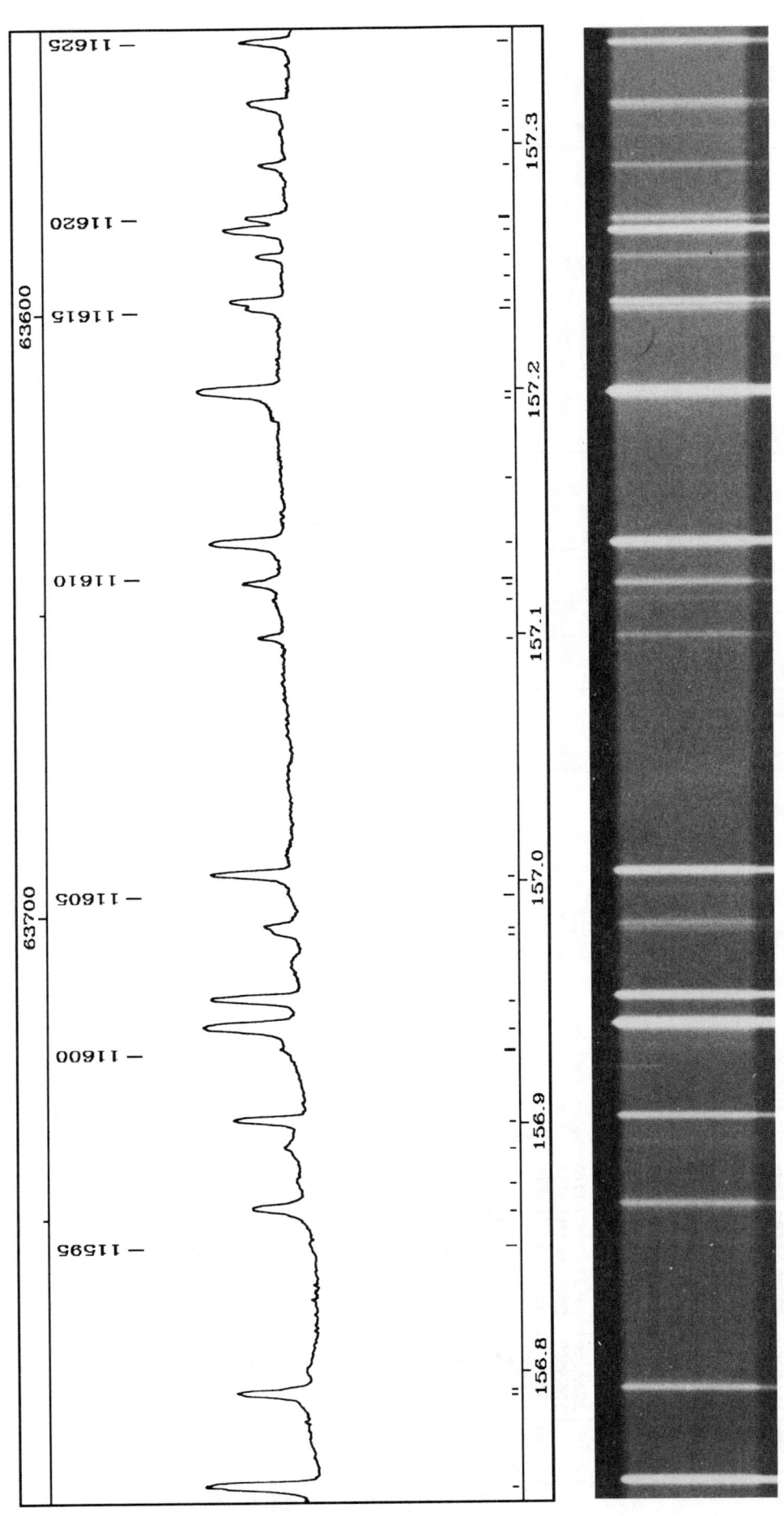

PLATE 139

TABLE 139

Nb	I	λ (nm)	σ (cm⁻¹)	Assignment	Comment
11595	0	156.8507	63754.91	B6-11P9	
11596	37	156.8649	63749.12	B1-7P13	
11597	0	156.8760	63744.61	C0-13P3	
11598	0	156.8901	63738.90	B0-4R25	d
11599	87	156.9006	63734.62	B4-10P4	
11600	5	156.9296	63722.86	B2-7P16	
11601	1195	156.9383	63719.29	B2-9R1	b
11602	290	156.9500	63714.55	B2-9R0 B2-5P24	
11603	13	156.9775	63703.38	B4-10R7	
11604	22	156.9799	63702.43	B2-8R13 B1-6R20 B3-7P19	
11605	5	156.9935	63696.92	B0-7R12	
11606	265	157.0013	63693.73	B2-9R2	
11607	21	157.0984	63654.39	B3-9P8	
11608	5	157.1134	63648.28	B0-7P10	
11609	32	157.1202	63645.54	B7-13R7	
11610		157.1218	63644.88	B1-6P18	
11611	400	157.1367	63638.85	B2-9R3	sh
11612	0	157.1633	63628.10	B5-10P11	
11613		157.1961	63614.81		sh
11614	1050	157.1985	63613.85	B2-9P1	
11615	35	157.2328	63599.95	B1-8R9	
11616	84	157.2354	63598.89	B5-11R1	
11617	0	157.2473	63594.08	B1-5P22	
11618	20	157.2540	63591.37	B5-11R2	
11619	163	157.2649	63586.98	B4-10P5 B5-11R0	
11620	40	157.2697	63585.05	B2-8P11	
11621	18	157.2914	63576.25	B7-13R6	
11622	4	157.3052	63570.70	B4-10R8	
11623	18	157.3155	63566.54	B3-9R11	
11624	36	157.3169	63565.96	B5-11R3	
11625	78	157.3416	63555.99	B2-9R4	

Nb	I	λ (nm)	σ (cm⁻¹)	Assignment	Comment

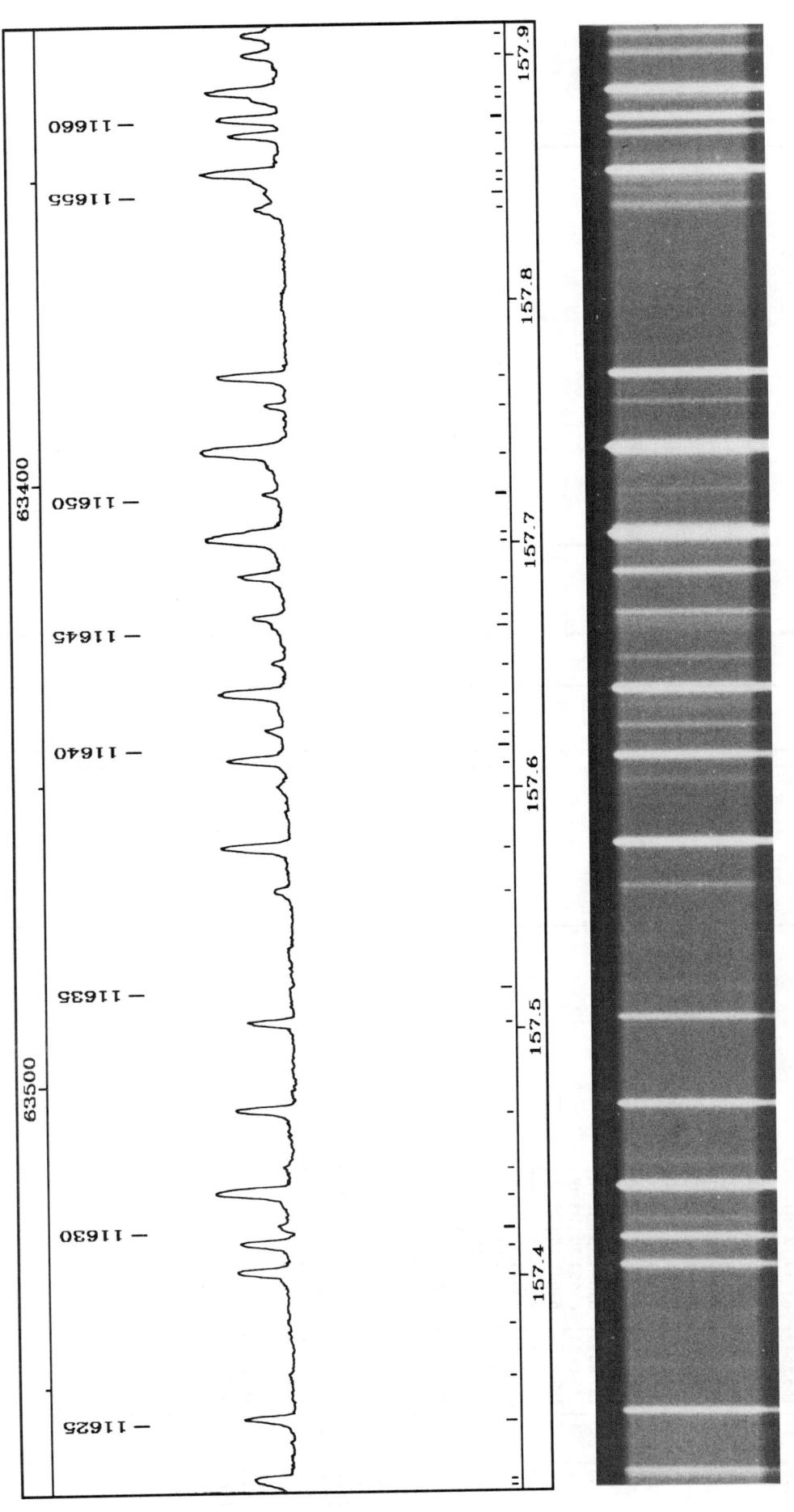

PLATE 140

TABLE 140

Nb	I	λ (nm)	σ (cm-1)	Assignment	Comment
11626	0	157.3601	63548.52	B4-9R17	
11627	0	157.3813	63539.95	B6-11P11	
11628	80	157.4013	63531.87	B1-8P7	
11629	80	157.4126	63527.30	B7-13R5	
11630	3	157.4198	63524.42	B5-11R4 B2-7R19	
11631	510	157.4334	63518.91	B2-9P2	
11632	1	157.4437	63514.75	B0-5P20	
11633	125	157.4670	63505.35	B5-11P1	
11634	44	157.5029	63490.91	B7-13R4	
11635	0	157.5147	63486.12	B0-5R22 B6-11P12	d
11636	12	157.5572	63469.02	B5-11R5 B6-11P12	
11637	300	157.5752	63461.76	B7-13R3	
11638	4	157.6013	63451.27	B1-7R16	
11639	83	157.6116	63447.11	B2-9R5	
11640	0	157.6183	63444.41	B3-8P15	
11641	12	157.6240	63442.11	B6-11P13	
11642	0	157.6335	63438.31	B0-6R18	
11643	240	157.6393	63435.96	B5-11P2 B7-13R2	
11644	11	157.6521	63430.81	B4-10R9	
11645	1	157.6669	63424.84		d
11646	28	157.6705	63423.42	B4-10P6	
11647	55	157.6875	63416.56	B0-8R1 B2-8R14 B0-6P16	
11648	840	157.7028	63410.43	B7-13R1 B0-8R0	sh
11649	0	157.7052	63409.45	B3-9P9	b
11650	3	157.7209	63403.12	B5-11R6	
11651	1520	157.7387	63395.99	B2-9P3	
11652	14	157.7575	63388.41	B0-8R2	
11653	210	157.7694	63383.66	B7-13R0	
11654	15	157.8384	63355.93	B1-7P14 B0-7R13	
11655	3	157.8448	63353.36	B3-9R12	
11656	6	157.8499	63351.34	B1-8R10	
11657	270	157.8529	63350.13	B5-11P3	
11658	0	157.8574	63348.33	B0-3P27	
11659	55	157.8683	63343.92	B6-12R2	
11660	97	157.8751	63341.21	B6-12R3	
11661	13	157.8831	63338.01	B6-12R10	
11662	240	157.8861	63336.79	B6-12R1 B2-7P17	
11663	22	157.9015	63330.63	B6-12R4 B5-11R7	
11664	27	157.9093	63327.48	B0-8R3	

Nb	I	λ (nm)	σ (cm-1)	Assignment	Comment

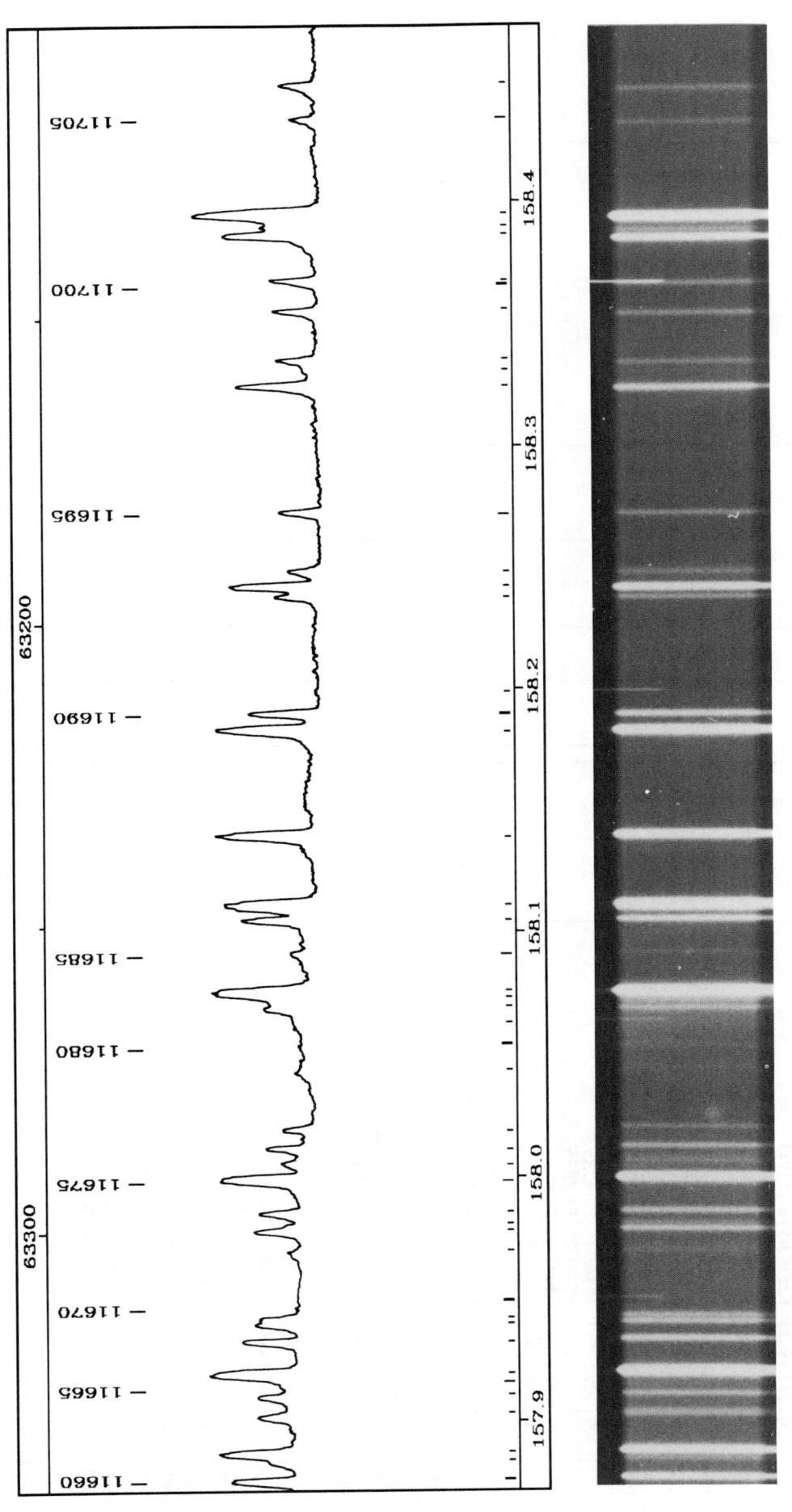

PLATE 141

TABLE 141

Nb	I	λ (nm)	σ (cm^{-1})	Assignment	Comment

Nb	I	λ (nm)	σ (cm^{-1})	Assignment	Comment
11665	0	157.9119	63326.47	B7-13P1	
11666	880	157.9185	63323.82	B6-12R0	
11667	63	157.9318	63318.48	B6-12R5	
11668	32	157.9389	63315.64	B6-12R9	
11669	22	157.9411	63314.74	B2-9R6	
11670					CuII 157.94918
11671	5	157.9682	63303.88	B1-6R21	d
11672	44	157.9768	63300.42	B0-8P1	
11673	8	157.9790	63299.57	B6-12R6	
11674	23	157.9843	63297.43	B7-13P2	
11675	510	157.9979	63292.00	B0-7P11 B4-10R10 B5-10P13	
11676	8	158.0045	63289.34	B6-12R7	b
11677	20	158.0104	63286.96	B6-12R8	
11678	9	158.0184	63283.76	B0-4P24	
11679	0	158.0429	63273.96		d
11680	0	158.0519	63270.37	B3-8R19	d
11681					CuII 158.06257
11682	23	158.0672	63264.23	B1-8P8	
11683	1680	158.0713	63262.59	B2-8P12	
11684	1	158.0741	63261.46	B7-13P3	sh
11685	78	158.0904	63254.96	B5-11R8	
11686	480	158.1043	63249.39	B5-11P4	
11687	400	158.1103	63247.00	B2-9P4 B6-12P1 B4-10P7	
11688	460	158.1394	63235.35	B7-13P4 B7-13P7 B0-8R4	
11689	70	158.1825	63218.11	B7-13P5	
11690	23	158.1893	63215.41	B7-13P6	
11691					CuII 158.19953
11692	196	158.2378	63196.02	B0-8P2	
11693	10	158.2418	63194.43	B6-12P2	
11694	11	158.2481	63191.93	B1-6P19	
11695	42	158.2725	63182.16	B5-11R9	
11696	6	158.3242	63161.53	B2-9R7	
11697	12	158.3304	63159.06	B5-10P14	
11698	15	158.3347	63157.36	B3-9P10	
11699	128	158.3550	63149.27	B4-10R11	
11700	19	158.3674	63144.30	B3-9R13	
11701					CuII 158.36823
11702	650	158.3862	63136.82	B5-11P5	
11703	7	158.3892	63135.61	B2-8R15	
11704	8	158.3949	63133.33	B6-12P3	
11705	7	158.4337	63117.87	B5-11R10	
11706	8	158.4475	63112.38	B0-8R5 B1-5P23	

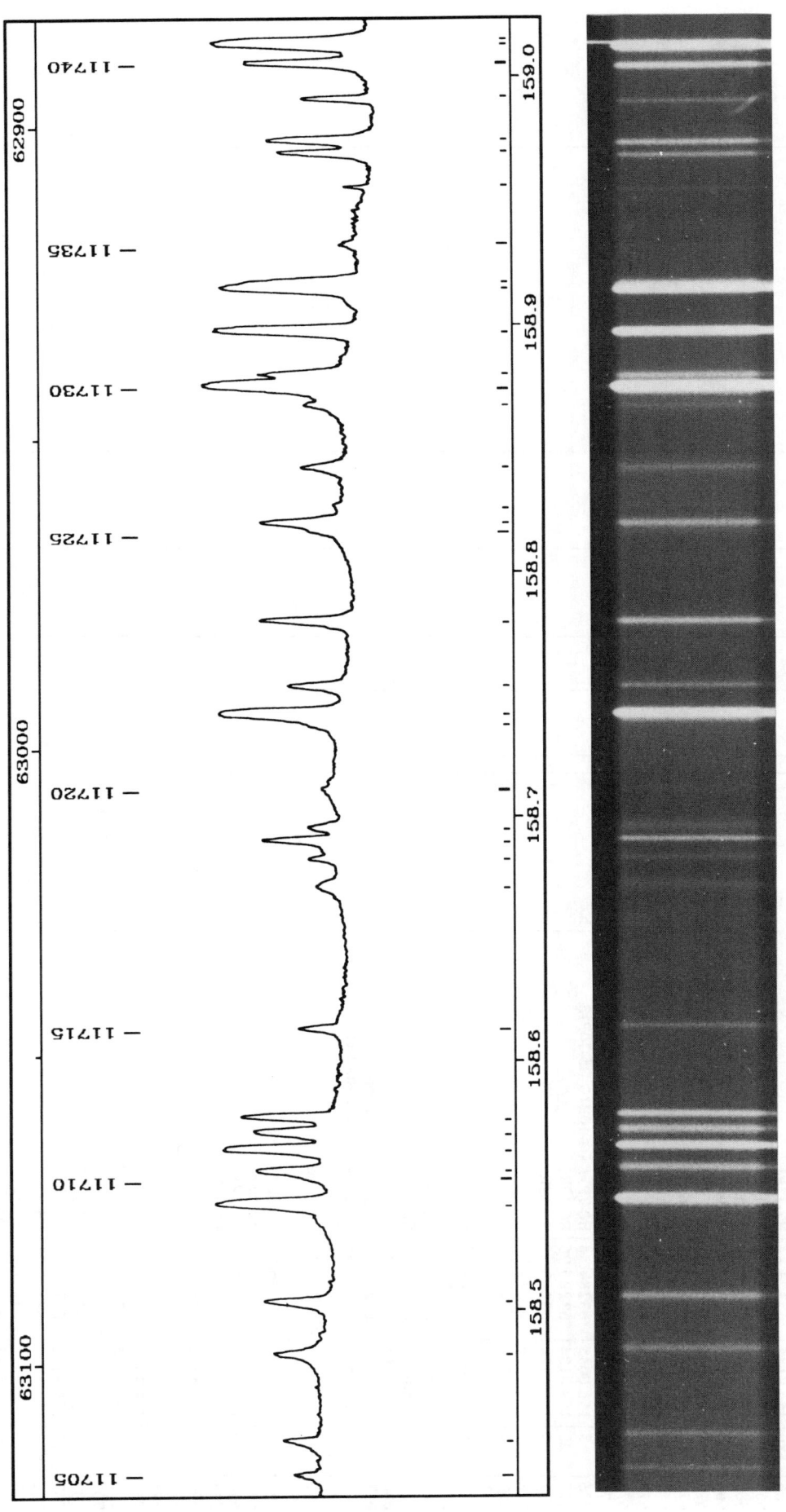

PLATE 142

TABLE 142

Nb	I	λ (nm)	σ (cm-1)	Assignment	Comment
11707	12	158.4822	63098.57	B1-7R17	
11708	17	158.5033	63090.16	B1-8R11	
11709	390	158.5427	63074.50	B2-9P5	
11710	0	158.5528	63070.48	B4-9P15	
11711	29	158.5556	63069.37	B5-11R11 B5-10P15	
11712	150	158.5645	63065.82	B6-12P4	
11713	32	158.5711	63063.20	B4-10P8 B5-11R13	
11714	58	158.5772	63060.75	B0-8P3	
11715	10	158.6132	63046.44	B5-11R12	
11716	2	158.6711	63023.46	B0-5R23	d
11717	5	158.6825	63018.93	B4-10R12	
11718	22	158.6903	63015.84	B5-11P6	
11719	5	158.6955	63013.75	B0-6R19	
11720	0	158.7113	63007.49	B0-7R14	sh
11721		158.7399	62996.14	B0-5P21	
11722	320	158.7420	62995.32	B6-12P5	
11723	11	158.7532	62990.84	B2-9R8	
11724	28	158.7795	62980.44	B1-8P9	
11725	0	158.8164	62965.79	B2-7P18	
11726	20	158.8196	62964.53	B1-7P15 B2-7R21	
11727	0	158.8266	62961.75	B0-8R6	
11728	6	158.8424	62955.49	B0-6P17	
11729	8	158.8678	62945.41	B3-9R14	
11730	1160	158.8756	62942.33	B3-10R1	
11731	30	158.8801	62940.53	B2-8P13	
11732	320	158.8981	62933.43	B3-10R0	
11733	250	158.9154	62926.55	B3-10R2	
11734		158.9171	62925.90	B6-12P6 B7-14R3	sh
11735	0	158.9331	62919.56	B0-7P12	s
11736	0	158.9562	62910.41		
11737	24	158.9701	62904.92	B4-10R13	
11738	29	158.9752	62902.89	B3-9P11	
11739	14	158.9921	62896.20	B0-8P4	
11740	44	159.0066	62890.48	B5-11P7	
11741	376	159.0146	62887.31	B3-10R3	CuII 159.01649
11742					

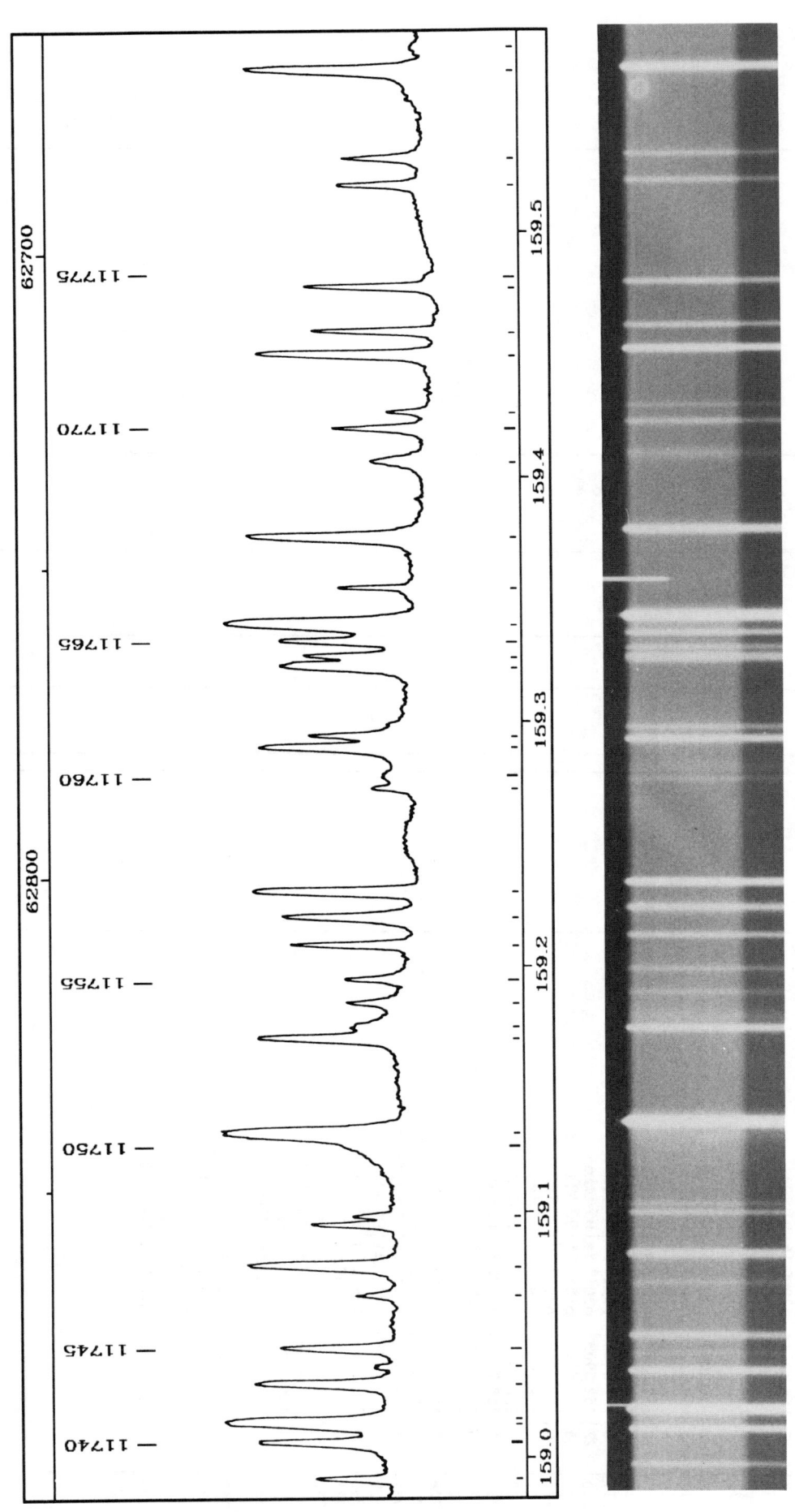

PLATE 143

TABLE 143

Nb	I	λ (nm)	σ (cm-1)	Assignment	Comment
11743	58	159.0302	62881.12	B2-9P6	
11744	1	159.0373	62878.33	B4-9P16	
11745	23	159.0446	62875.45	B4-10P9	
11746	2	159.0661	62866.95	B2-8R16	
11747	84	159.0779	62862.29	B6-12P7	
11748	14	159.0947	62855.64	B7-14R2	
11749	4	159.0982	62854.25	B3-8P17	
11750		159.1262	62843.20		sh
11751	1140	159.1313	62841.20	B3-10P1	
11752	61	159.1698	62826.00	B3-10R4	d
11753	5	159.1739	62824.37		
11754	6	159.1847	62820.11	B1-8R12	
11755	8	159.1944	62816.28	B4-10R14	
11756	18	159.2084	62810.77	B6-12P8	
11757	27	159.2199	62806.22	B2-9R9	
11758	93	159.2302	62802.17	B7-14R1	
11759	1	159.2728	62785.35	B0-8R7	
11760	0	159.2777	62783.43	B0-4R27	d
11761	43	159.2893	62778.86	B6-12P9	
11762	13	159.2940	62776.99	B6-12P10	
11763	28	159.3230	62765.58	B5-11P8 B4-10R15	
11764	18	159.3269	62764.03	B3-9R15	
11765	28	159.3331	62761.61	B7-14R0 B7-14P4	
11766	560	159.3401	62758.86	B3-10P2 B1-6P20 B1-7R18	CuII 159.35556
11767					
11768	88	159.3758	62744.78	B3-10R5	
11769	1	159.4069	62732.56	B0-4P25	d
11770	10	159.4200	62727.40	B4-9P17	
11771	3	159.4267	62724.75	B7-14P3	
11772	63	159.4501	62715.55	B7-14P1	
11773	16	159.4598	62711.75	B7-14P2	
11774	21	159.4777	62704.68	B0-8P5	
11775	0	159.4816	62703.15		
11776	8	159.5194	62688.30	B4-10P10	
11777	7	159.5278	62684.99	B1-8P10	
11778	97	159.5656	62670.15	B2-9P7	
11779	0	159.5751	62666.43	B1-5P24	

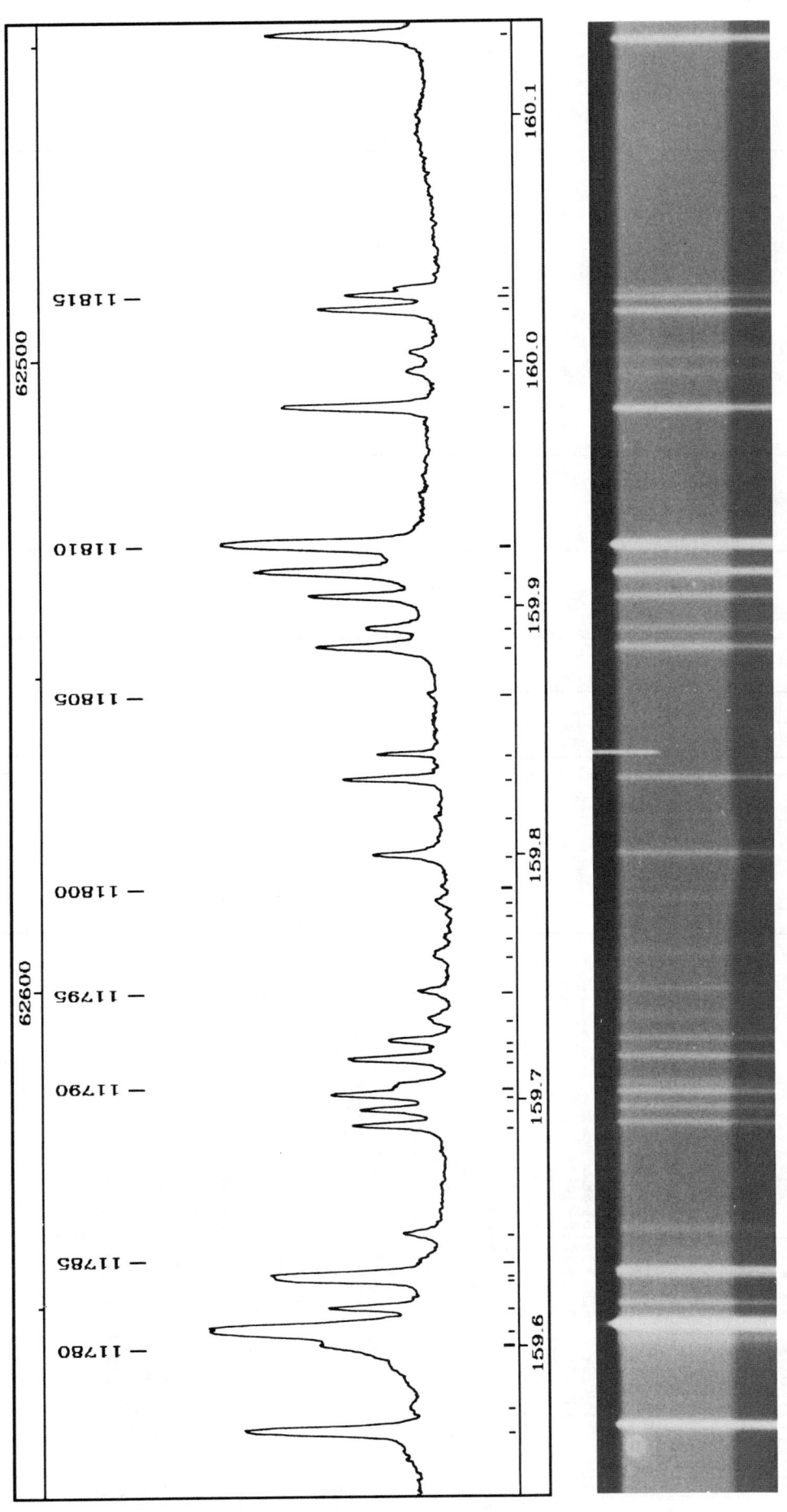

PLATE 144

TABLE 144

Nb	I	λ (nm)	σ (cm⁻¹)	Assignment	Comment
11780	14	159.6003	62656.52	B0-7R15	
11781	1650	159.6059	62654.32	B3-10P3	
11782	10	159.6148	62650.84	B3-9P12	
11783	30	159.6267	62646.16	B3-10R6	
11784	33	159.6278	62645.72	B5-11P9	
11785	0	159.6345	62643.09		
11786	1	159.6450	62638.99	B4-9P18	d
11787	8	159.6880	62622.10	B2-8P14	
11788	5	159.6944	62619.61	B2-7P19	
11789	6	159.7007	62617.15	B2-8R17	
11790	1	159.7040	62615.84	B1-6R23	
11791	8	159.7148	62611.59	B2-9R10	
11792	0	159.7183	62610.22		
11793	5	159.7221	62608.75	B3-9R16	
11794	0	159.7314	62605.11	B0-6R20	
11795	1	159.7424	62600.78	B3-8P18	
11796	0	159.7576	62594.82	B0-5R24	
11797	0	159.7645	62592.14		d
11798	0	159.7739	62588.43		
11799	0	159.7794	62586.28	B0-8R8	
11800	0	159.7857	62583.82		
11801	5	159.7983	62578.90	B1-7P16	
11802	0	159.8133	62573.01		
11803	6	159.8296	62566.63	B6-13R7	
11804					CuII 159.84023
11805	0	159.8636	62553.33	B1-8R13	
11806	10	159.8830	62545.74	B0-7P13	
11807	3	159.8906	62542.76	B5-11P10	
11808	10	159.9038	62537.62	B3-10R7	
11809	40	159.9137	62533.74	B3-10P4	
11810	300	159.9248	62529.39	B4-10P11	
11811	25	159.9808	62507.50	B0-6P18	
11812	0	159.9955	62501.74	B0-5P22	
11813	0	160.0032	62498.76	B3-9R17	
11814	14	160.0205	62491.98	B6-13R6	
11815	6	160.0263	62489.73	B0-8P6	
11816	3	160.0290	62488.68	B5-11P11	
11817	28	160.1315	62448.66		

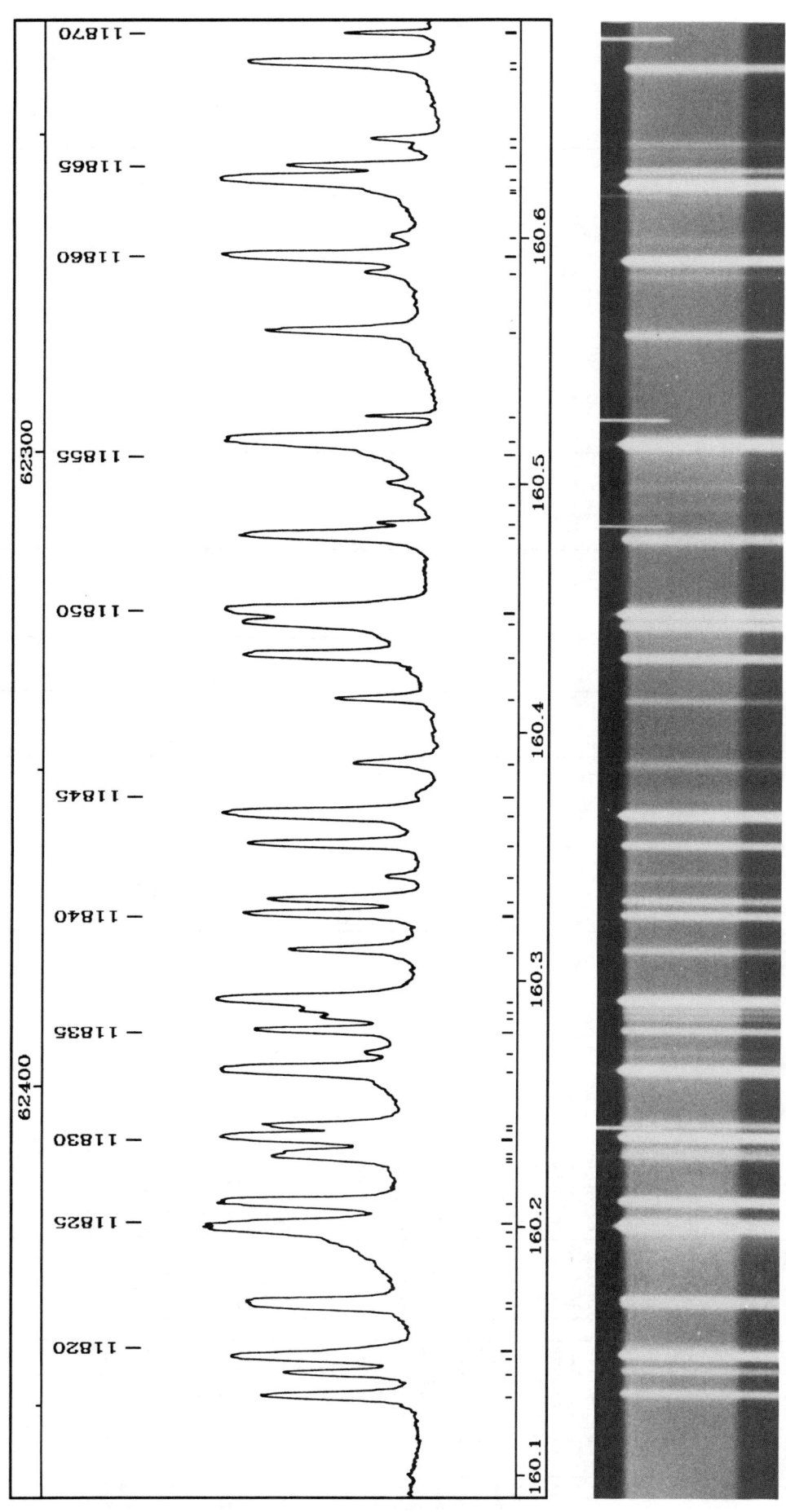

PLATE 145

TABLE 145

Nb	I	λ (nm)	σ (nm)	Assignment	Comment
11818	15	160.1407	62445.09	B2-9P8	
11819	169	160.1471	62442.60	B1-9R1	
11820		160.1496	62441.60	B1-7R19	sh
11821	51	160.1679	62434.50	B1-9R0 B3-9R18	
11822	51	160.1692	62433.99	B6-13R5	sh
11823		160.1945	62424.12		
11824	1140	160.1987	62422.46	B1-9R2 B4-11R1	sh
11825		160.2021	62421.17		
11826	245	160.2087	62418.58	B4-11R2	
11827					CuII 160.22729
11828	26	160.2274	62411.30	B2-9R11	
11829	12	160.2287	62410.80	B3-10R8	
11830	320	160.2348	62408.40	B4-11R0	
11831					CuII 160.23880
11832	23	160.2396	62406.56	B3-9P13	
11833	370	160.2619	62397.87	B4-11R3	
11834	2	160.2689	62395.13	B2-8R18	
11835	42	160.2780	62391.60	B6-13R4	
11836	12	160.2836	62389.43	B3-8P19	
11837	16	160.2864	62388.32	B5-11P12	
11838	410	160.2904	62386.75	B3-10P5	
11839	16	160.3108	62378.83	B1-8P11	
11840	66	160.3254	62373.16	B1-9R3	
11841	26	160.3311	62370.95	B5-11P13	
11842	1	160.3404	62367.31	B0-8R9	
11843	60	160.3541	62362.00	B4-11R4	
11844	385	160.3662	62357.29	B6-13R3	
11845	0	160.3734	62354.47	B1-6R24	d
11846	8	160.3869	62349.22	B1-6P21	
11847	9	160.4130	62339.10	B4-10P12	
11848	155	160.4306	62332.25	B1-9P1	
11849	250	160.4439	62327.07	B6-13R2	
11850	1215	160.4486	62325.27	B4-11P1	
11851	104	160.4792	62313.39	B4-11R5 B2-8P15	
11852					CuII 160.48475
11853	1	160.4925	62308.20	B0-7R16	
11854	3	160.5005	62305.10	B2-7P20	
11855		160.5130	62300.24		sh
11856	1170	160.5178	62298.37	B6-13R1 B1-9R4	
11857					CuII 160.52813
11858	31	160.5625	62281.05	B3-10R9	
11859	3	160.5862	62271.84	B1-8R14	
11860	310	160.5928	62269.31	B6-13R0	
11861	1	160.6007	62266.23	B1-5P25	d
11862	0	160.6184	62259.35		
11863		160.6196			
11864	625	160.6236	62257.35	B4-11P2 B5-12R10	
11865	21	160.6292	62255.19	B4-11R6	
11866	0	160.6367	62252.29	B0-3P29	d
11867	4	160.6403	62250.88	B0-8P7	sh
11868	77	160.6690	62239.77	B3-8P20	CuII 160.68341
11869		160.6710	62238.98	B1-9P2	
11870					

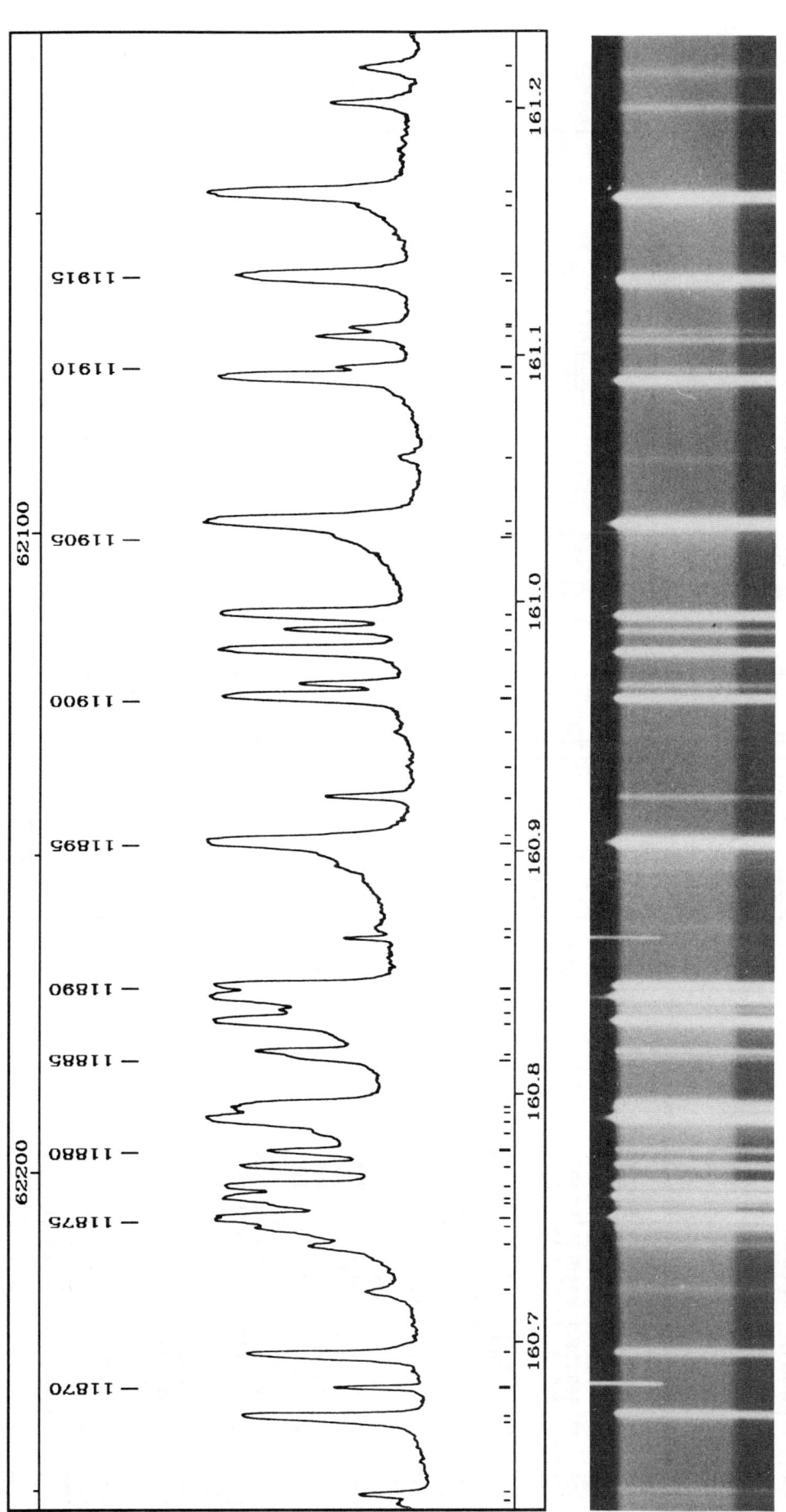

PLATE 146

TABLE 146

Nb	I	λ (nm)	σ (cm⁻¹)	Assignment			Comment
11871	64	160.6957	62229.40	B3-10P6	B0-4P26		
11872	1	160.7213	62219.51	B0-6R21			
11873	11	160.7395	62212.45	B2-8R19			
11874	50	160.7467	62209.68	B2-9R12	B2-9P9		
11875	1085	160.7504	62208.26	B6-13P1			
11876	31	160.7563	62205.96	B5-12R9			
11877	420	160.7588	62204.98	B5-12R3	B1-7P17		
11878	280	160.7633	62203.24	B5-12R2			
11879	65	160.7716	62200.04	B5-12R4			
11880	22	160.7776	62197.71	B1-9R5			
11881	8	160.7855	62194.66				
11882	1340	160.7906	62192.70	B5-12R1			
11883	117	160.7948	62191.07	B5-12R5	B4-11R7		
11884	73	160.7970	62190.20	B4-10P13			
11885	13	160.8155	62183.06	B5-12R8			
11886	31	160.8175	62182.30	B5-12R6			
11887	563	160.8295	62177.63	B6-13P2	B5-12R7		
11888	23	160.8339	62175.95	B3-9P14			
11889	1710	160.8396	62173.75	B4-11P3			
11890	350	160.8439	62172.09	B5-12R0			
11891							CuII 160.86393
11892	0	160.8679	62162.79	B0-7P14			
11893	0	160.8935	62152.93	B1-7R20			
11894	5	160.8979	62151.21				
11895	1320	160.9026	62149.40	B6-13P3	B3-10R10		
11896		160.9058	62148.17				sh
11897	8	160.9216	62142.05	B6-13P7			
11898		160.9342					
11899	0	160.9476	62132.02	B0-8R10			
11900	218	160.9616	62126.63	B6-13P4			
11901	11	160.9668	62124.62	B4-11R8			
11902	200	160.9804	62119.35	B1-9P3			
11903	15	160.9888	62116.11	B6-13P6			
11904	240	160.9952	62113.66	B6-13P5			
11905		161.0278	62101.06				sh
11906							CuII 161.02964
11907	1300	161.0325	62099.25	B5-12P1			
11908	0	161.0592	62088.98	B2-8R20			d
11909	320	161.0912	62076.65	B4-11P4			
11910	5	161.0952	62075.11	B1-9R6			
11911	9	161.1074	62070.39	B4-10P14			
11912	3	161.1109	62069.05	B1-8P12			
11913							CuII 161.11181
11914	115	161.1308	62061.37	B4-11R9	B0-6P19		sh
11915		161.1321	62060.88	B3-10P7			
11916	0	161.1603	62050.03				
11917	700	161.1653	62048.09	B5-12P2			
11918	11	161.2020	62033.98	B2-7P21			
11919	3	161.2162	62028.51	B0-5P23			d

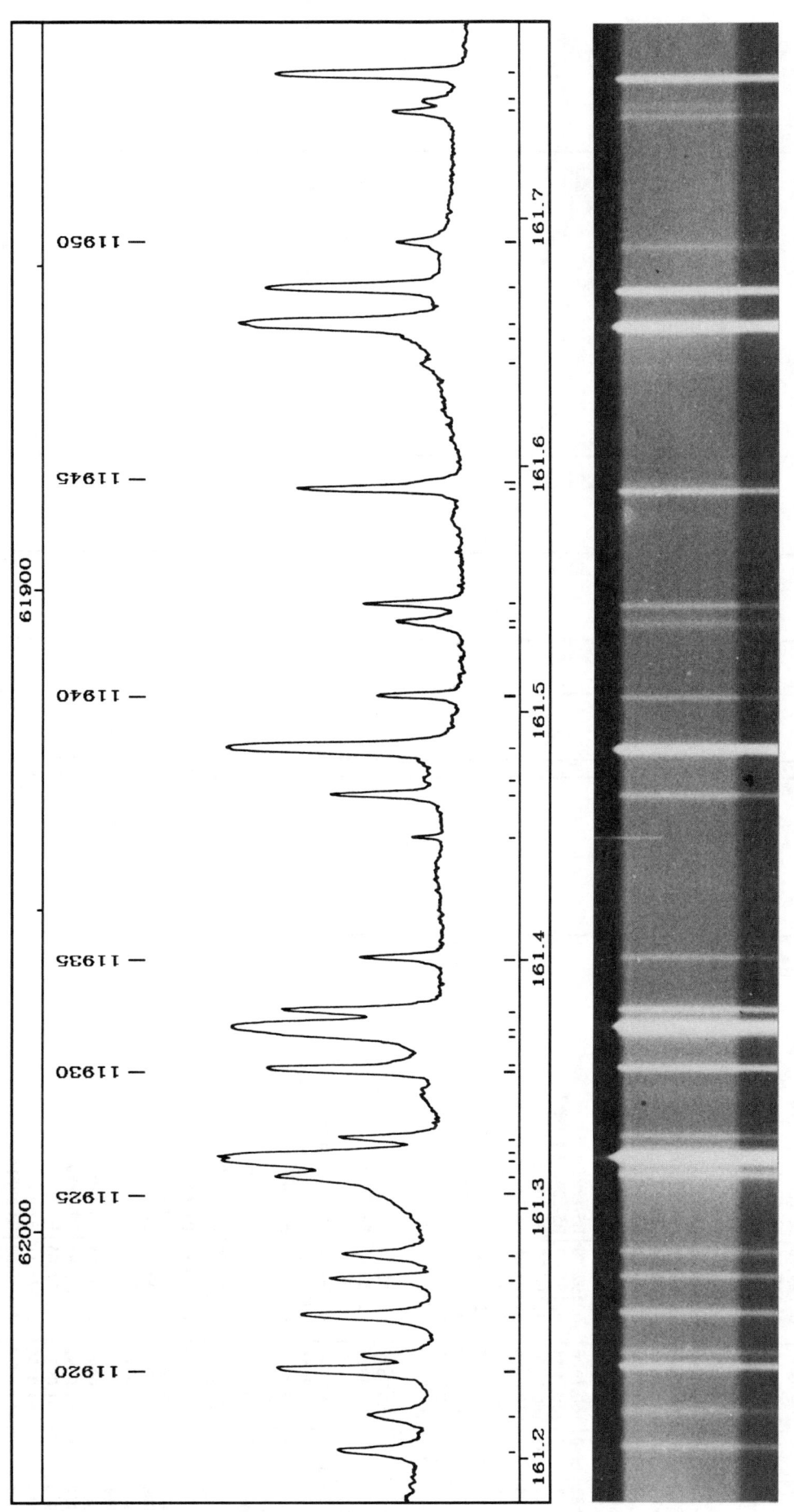

PLATE 147

TABLE 147

Nb	I	λ (nm)	σ (cm-1)	Assignment	Comment
11920	29	161.2345	62021.48	B3-10R11	
11921	5	161.2399	62019.38	B2-8P16	
11922	17	161.2561	62013.16	B2-9R13	
11923	10	161.2709	62007.48	B4-11R10	
11924	8	161.2803	62003.84	B1-8R15	
11925	5	161.3054	61994.20	B0-8P8	
11926	38	161.3115	61991.87	B4-10P15	
11927	2060	161.3183	61989.26	B5-12P3	
11928		161.3220	61987.84		sh
11929	9	161.3270	61985.89	B4-11R13	
11930	40	161.3550	61975.16	B1-9P4	
11931	0	161.3573	61974.26	B1-6P22	
11932	0	161.3694	61969.63	B4-11R11	sh
11933	480	161.3716	61968.76	B4-11P5 B2-9P10 B0-7R17	
11934	26	161.3786	61966.07	B3-9P15	
11935	6	161.4000	61957.88	B4-11R12	
11936		161.4495			NiII 161.4495 Ke
11937	9	161.4664	61932.40	B1-9R7	
11938	0	161.4729	61929.88	B1-5P26	d
11939	396	161.4852	61925.19	B5-12P4	
11940	6	161.5065	61917.00	B6-14R4	
11941		161.5341	61906.43		sh
11942	3	161.5365	61905.51	B1-7R21	
11943	8	161.5439	61902.69	B3-10R12	d
11944	20	161.5904	61884.86	B3-10P8	
11945	0	161.5929	61883.89	B0-8R11	
11946	0	161.6410	61865.50	B0-6R22	
11947	0	161.6527	61861.00		
11948	440	161.6576	61859.13	B5-12P5	
11949	66	161.6721	61853.59	B4-11P6	
11950	1	161.6903	61846.64	B1-7P18	
11951	2	161.7429	61826.52	B2-9R14	
11952	1	161.7572	61821.05		
11953	42	161.7582	61820.66	B6-14R3	

Nb	I	λ (nm)	σ (cm-1)	Assignment	Comment

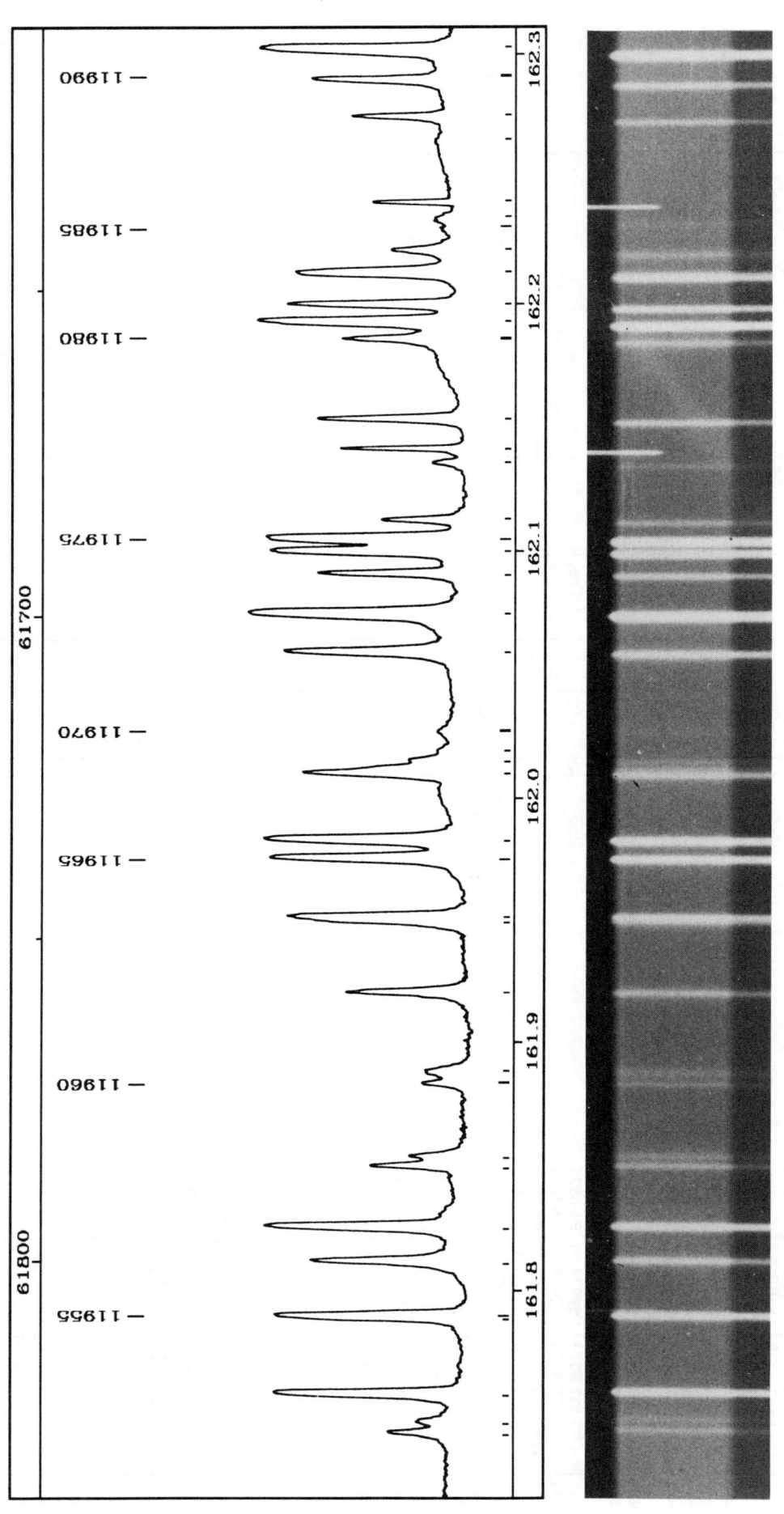

PLATE 148

TABLE 148

Nb	I	λ (nm)	σ (cm⁻¹)	Assignment	Comment

Nb	I	λ (nm)	σ (cm⁻¹)	Assignment	Comment
11954	60	161.7891	61808.85	B1-9P5	CuII 161.79154
11955					
11956	16	161.8113	61800.38	B3-10R13	
11957	61	161.8252	61795.06	B5-12P6	
11958	7	161.8499	61785.65	B3-9P16	
11959	3	161.8537	61784.20	B0-7P15	
11960	2	161.8832	61772.92	B1-9R8	
11961	1	161.8880	61771.11	B0-4P27	d
11962	9	161.9202	61758.81	B1-8P13	
11963	33	161.9498	61747.53	B2-8P17 B1-8R16	sh
11964	77	161.9511	61747.05	B6-14R2	
11965	103	161.9751	61737.90	B5-12P7	
11966	20	161.9828	61734.95	B4-11P7	
11967	4	162.0102	61724.51	B2-9P11	
11968	0	162.0122	61723.73	B3-10R14	
11969	0	162.0154	61722.52	B0-8P9	d
11970	38	162.0279	61717.75	B1-7R22	
11971	320	162.0594	61705.76	B3-10P9	
11972	15	162.0748	61699.91	B2-10R1	
11973	115	162.0914	61693.59	B5-12P8	
11974	142	162.1005	61690.14	B6-14R1	
11975	5	162.1053	61688.29	B2-10R2 B2-10R0	
11976	1	162.1132	61685.30	B3-10R15	
11977	15	162.1365	61676.41	B5-12P10	
11978	10	162.1543	61669.67	B5-12P9	CuII 162.14256
11979	116	162.1868	61657.29	B2-9R15	
11980	37	162.1936	61654.71	B2-10R3	
11981	34	162.2006	61652.04	B6-14P4	
11982	2	162.2134	61647.18	B6-14R0 B3-9P17	b
11983	0	162.2227	61643.64	B1-6P23	
11984	0	162.2308	61640.59	B0-7R18	
11985	0	162.2355	61638.78	B0-6P20	CuII 162.24278
11986					
11987					
11988	0	162.2672	61626.74	B0-8R12	
11989	9	162.2769	61623.06	B1-9P6	
11990	18	162.2918	61617.40	B4-11P8	
11991	210	162.3040	61612.76	B6-14P3	

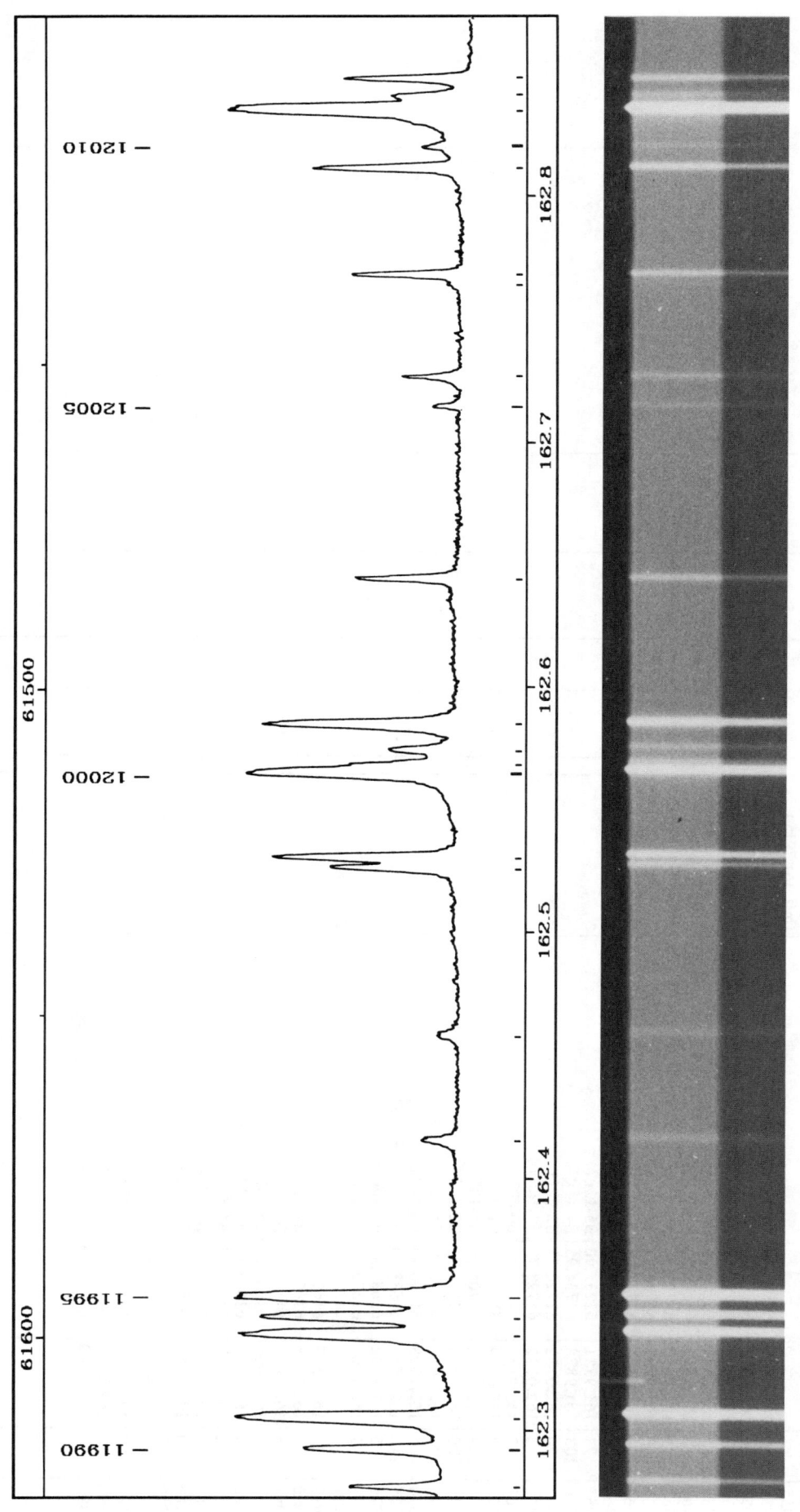

PLATE 149

TABLE 149

Nb	I	λ (nm)	σ (cm-1)	Assignment	Comment
11992	137	162.3374	61600.10	B6-14P1 B2-10R4 B1-9R9	CuII 162.31732 Ke
11993	70	162.3443	61597.48	B6-14P2	
11994	276	162.3524	61594.39	B2-10P1 B0-5P24	
11995	1	162.4154	61570.53	B3-9P18	d
11996	0	162.4582	61554.28	B0-6R23	d
11997	12	162.5269	61528.29	B3-10P10	
11998	23	162.5309	61526.76	B2-10R5	
11999	140	162.5658	61513.57	B2-10P2 B2-9R16	
12000	8	162.5687	61512.47	B1-7P19	
12001	3	162.5752	61510.01	B1-8R17	
12002	45	162.5853	61506.17	B4-11P9 B2-8P18	
12003	6	162.6443	61483.86	B2-9P12	
12004	1	162.7143	61457.40	B5-13R7	
12005	2	162.7262	61452.93	B1-8P14	
12006	0	162.7647	61438.37	B0-8P10	
12007	7	162.7681	61437.09	B2-10R6	
12008	16	162.8112	61420.82	B1-9P7	
12009	1	162.8197	61417.63	B1-9R10	
12010	360	162.8349	61411.88	B2-10P3 B0-7P16	
12011	4	162.8402	61409.89	B2-9R17	
12012	8	162.8472	61407.26	B4-11P10	
12013					

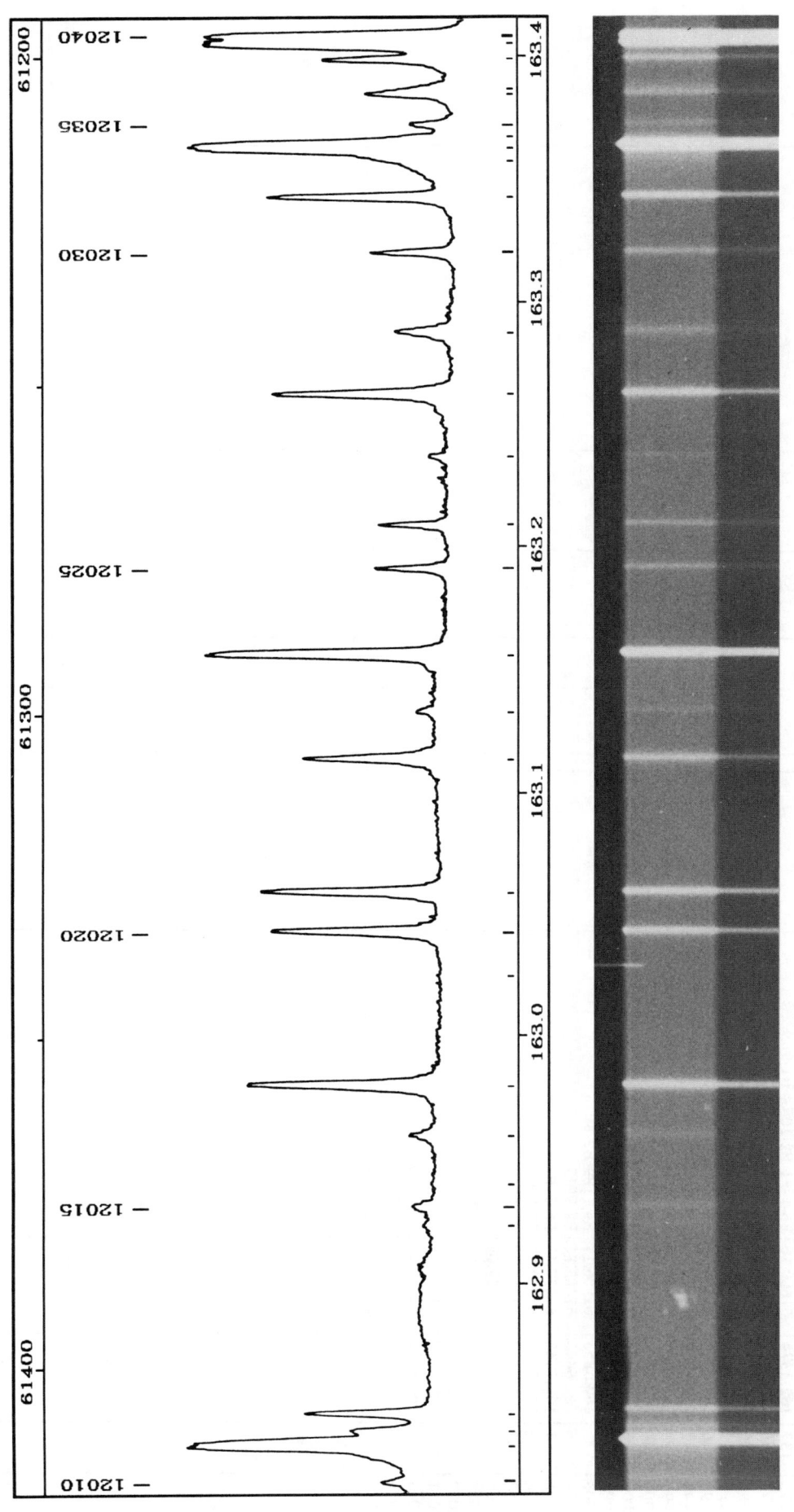

PLATE 150

TABLE 150

Nb	I	λ (nm)	σ (cm⁻¹)	Assignment	Comment
12014	0	162.9220	61379.07		
12015	1	162.9295	61376.24	B1-6P24	
12016	0	162.9390	61372.64	B5-13R6	
12017	1	162.9584	61365.36	B0-8R13	
12018	25	162.9785	61357.79	B3-10P11	d
12019					CuII 163.02681
12020	15	163.0413	61334.17	B2-10R7 B0-7R19	
12021	17	163.0575	61328.06	B4-11P11	
12022	7	163.1127	61307.31	B2-8P19	
12023	1	163.1320	61300.05	B1-8R18	
12024	71	163.1554	61291.25	B2-10P4	
12025	2	163.1905	61278.06	B4-11P12	
12026	2	163.2084	61271.34	B4-11P13	
12027	0	163.2367	61260.72	B5-13R4	
12028	16	163.2619	61251.28	B2-9P13	
12029	0	163.2870	61241.85	B0-6P21	
12030	3	163.3194	61229.70	B1-9R11	
12031	15	163.3422	61221.16	B2-10R8 B5-13R3	
12032		163.3585	61215.06		sh
12033	480	163.3629	61213.41	B3-11R1 B3-11R2	
12034	0	163.3667	61212.00		
12035	3	163.3717	61210.10	B1-7P20	
12036	3	163.3843	61205.40	B1-9P8	
12037		163.3861	61204.70	B0-5P25	sh d
12038	7	163.3983	61200.14	B3-10P12	
12039	152	163.4048	61197.70	B3-11R3	
12040	112	163.4075	61196.70	B3-11R0	

Nb	I	λ (nm)	σ (cm⁻¹)	Assignment	Comment

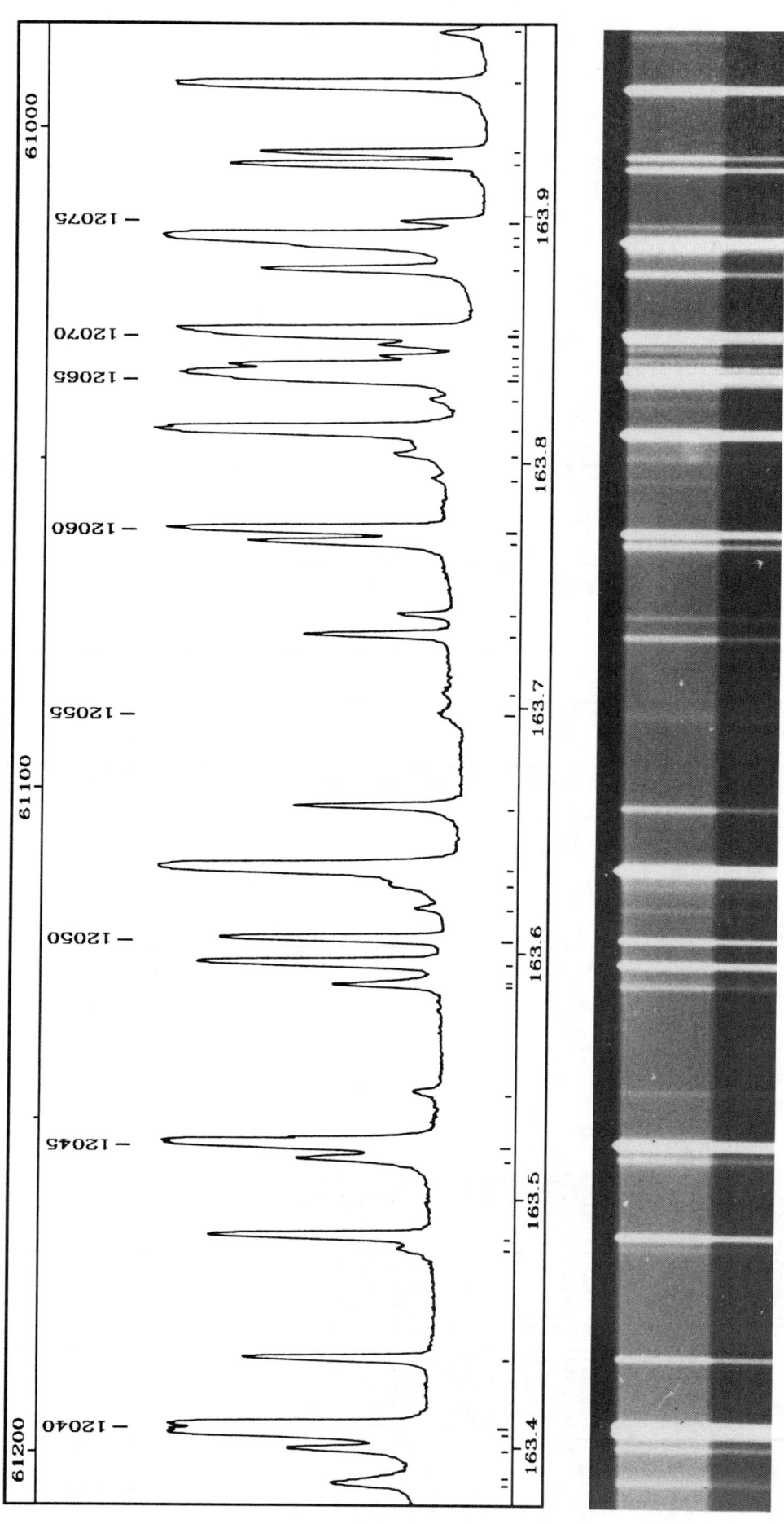

PLATE 151

TABLE 151

Nb	I	λ (nm)	σ (cm⁻¹)	Assignment		Comment
12041	12	163.4348	61186.49	B5-13R2		
12042	1	163.4793	61169.81	B2-8P20		
12043	24	163.4840	61168.07	B3-11R4		
12044	7	163.5150	61156.47	B1-8P15		
12045	163	163.5210	61154.22	B5-13R1	B2-10P5	
12046	1	163.5423	61146.27	B0-8P11		
12047	5	163.5858	61130.00	B0-9R1		
12048		163.5873	61129.45	B1-8R19		sh
12049	36	163.5946	61126.71	B3-11R5		
12050	18	163.6046	61122.96	B5-13R0		
12051	1	163.6170	61118.35	B0-9R0		
12052		163.6268	61114.68	B0-9R2		
12053	380	163.6333	61112.26	B3-11P1		
12054	8	163.6589	61102.71	B2-10R9		
12055	0	163.6969	61088.52			d
12056	0	163.7051	61085.45	B4-12R9		
12057	7	163.7289	61076.59	B3-11R6		
12058	3	163.7371	61073.53	B0-9R3		
12059	11	163.7664	61062.59	B3-10P13		
12060	48	163.7713	61060.77	B5-13P1		
12061	0	163.7922	61052.96	B4-12R8		
12062	1	163.8023	61049.19	B0-7P17		
12063	178	163.8118	61045.65	B3-11P2		
12064	0	163.8247	61040.87	B1-9R12		
12065	13	163.8330	61037.76	B4-12R4	B4-12R7	
12066	94	163.8351	61036.97	B4-12R3		
12067	14	163.8384	61035.75	B4-12R5		
12068	1	163.8419	61034.45	B4-12R6		
12069	1	163.8469	61032.60	B2-9P14		
12070	75	163.8508	61031.13	B5-13P2		
12071	0	163.8526	61030.48	B4-12R2		
12072	12	163.8777	61021.12	B3-11R7		
12073	6	163.8877	61017.41	B5-13P7		
12074	340	163.8905	61016.35	B4-12R1		
12075	2	163.8969	61013.96	B0-9P1		
12076	18	163.9202	61005.30	B5-13P3		
12077	13	163.9249	61003.55	B2-10P6		
12078	97	163.9525	60993.29	B4-12R0		
12079	2	163.9738	60985.36	B5-13P6	B5-13P4	

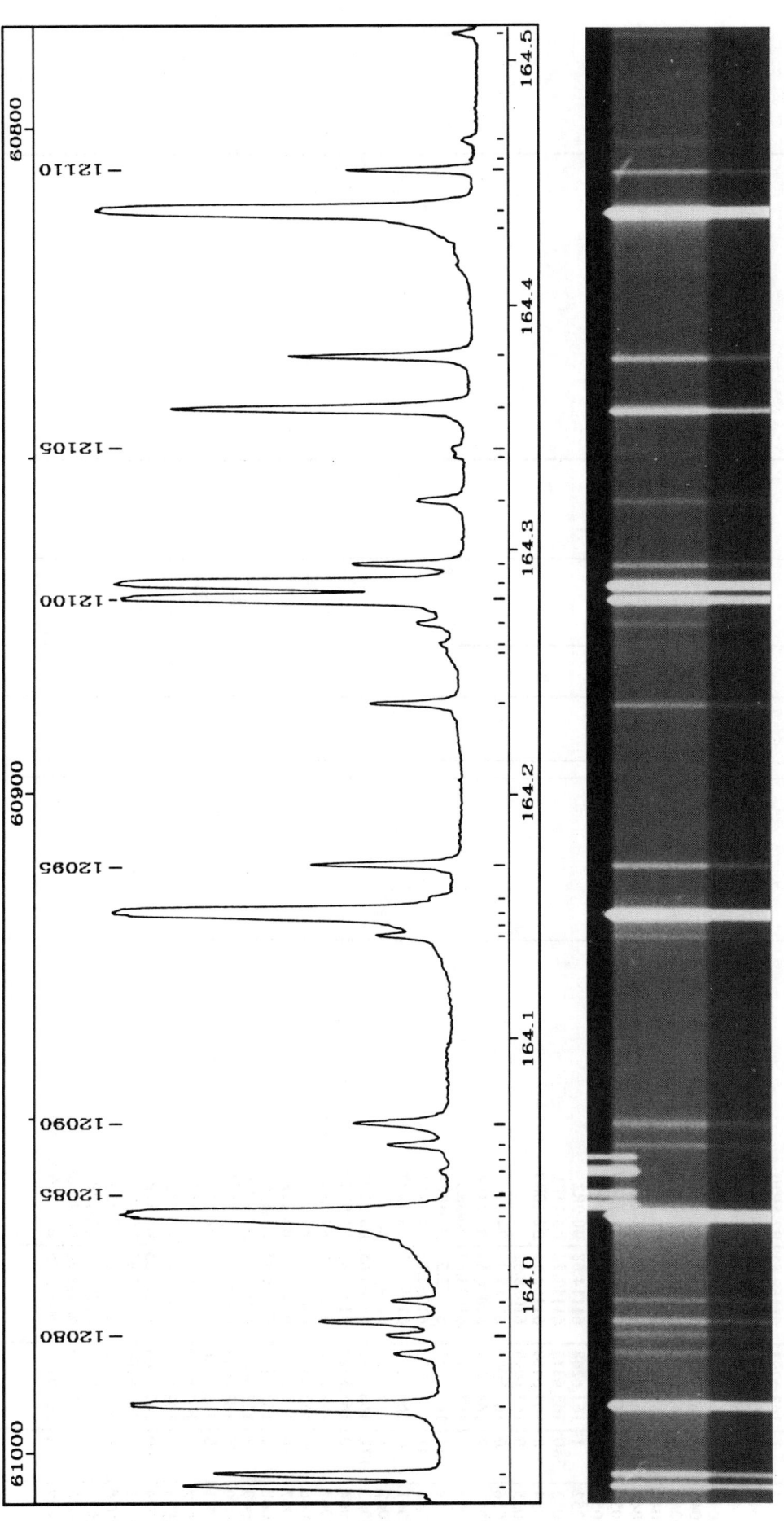

PLATE 152

TABLE 152

Nb	I	λ (nm)	σ (cm⁻¹)	Assignment	Comment
12080	1	163.9812	60982.60	B2-10R10	
12081	5	163.9865	60980.63	B1-9P9	
12082	1	163.9951	60977.43	B5-13P5	
12083	480	164.0295	60964.65	B3-11P3 B3-11R8	
12084		164.0337			HeII 164.0345 Ke
12085		164.0380			
12086		164.0390			
12087		164.0477			HeII 164.0474 Ke P
12088		164.0532			
12089	1	164.0579	60954.10	B3-10P14	
12090	3	164.0665	60950.89	B1-7P21	
12091	1	164.1429	60922.53	B0-9P2	
12092		164.1472	60920.92		sh
12093	320	164.1520	60919.16	B4-12P1	
12094	0	164.1578	60916.99	B0-9R5	
12095	5	164.1719	60911.78	B3-11R9	
12096	2	164.2377	60887.38	B3-10P15	
12097	0	164.2575	60880.03	B0-5P26	
12098	0	164.2621	60878.31	B0-6P22	
12099	1	164.2707	60875.14	B1-8P16	
12100	87	164.2807	60871.42	B3-11P4	d
12101	155	164.2865	60869.27	B4-12P2 B3-11R10	
12102	3	164.2947	60866.24	B2-10R11	
12103	2	164.3209	60856.54	B1-9R13	
12104	0	164.3398	60849.55	B0-8R15 B0-8P12	
12105	0	164.3427	60848.44		
12106	18	164.3580	60842.79	B2-10P7 B3-11R11 B3-11R12	
12107	6	164.3793	60834.90	B2-9P15	
12108	0	164.4293	60816.42		
12109	340	164.4388	60812.88	B4-12P3	
12110	5	164.4554	60806.75	B0-9P3	
12111	0	164.4598	60805.13	B0-9R6	
12112	0	164.4676	60802.24		
12113	0	164.5108	60786.29	B5-14R4	

Nb	I	λ (nm)	σ (cm⁻¹)	Assignment	Comment

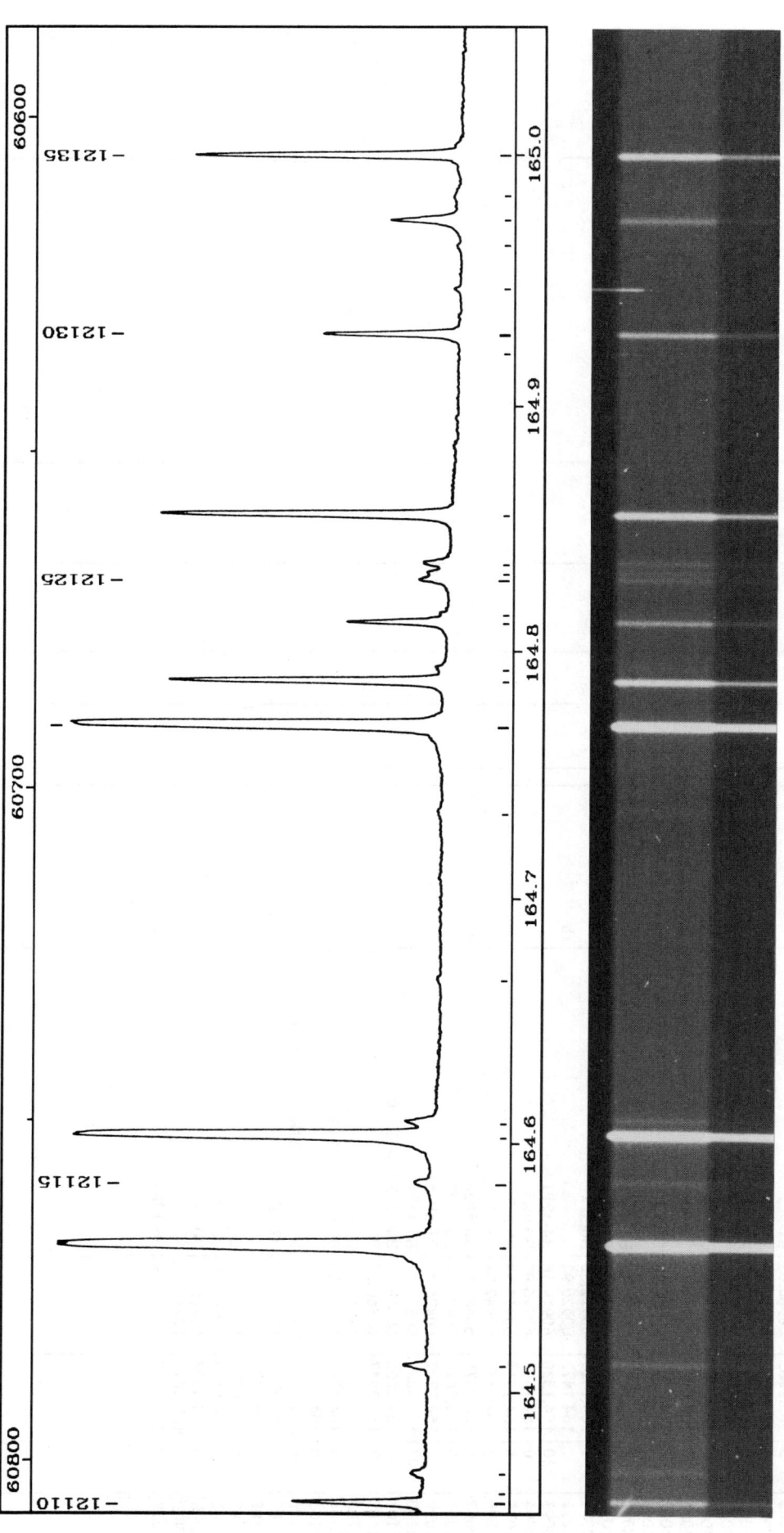

PLATE 153

TABLE 153

Nb	I	λ (nm)	σ (cm-1)	Assignment	Comment

Nb	I	λ (nm)	σ (cm-1)	Assignment	Comment
12114	110	164.5585	60768.65	B3-11P5	
12115	1	164.5836	60759.41	B2-10R12	
12116	55	164.6025	60752.42	B4-12P4	
12117	1	164.6084	60750.25	B1-9P10	
12118	0	164.6668	60728.69		
12119	0	164.7347	60703.68	B0-7P18	
12120	57	164.7688	60691.09	B4-12P5	
12121	12	164.7868	60684.48	B5-14R3	
12122	0	164.7923	60682.43	B1-9R14	
12123	3	164.8108	60675.63	B2-10P8	
12124	0	164.8145	60674.26	B0-9R7	
12125	1	164.8283	60669.20	B2-10R13	
12126	0	164.8308	60668.26	B0-9P4	
12127	1	164.8349	60666.78	B2-9P16	
12128	14	164.8543	60659.61	B3-11P6	
12129					GeII 164.91942
12130	6	164.9272	60632.80	B4-12P6	
12131					CuII 164.94575
12132	0	164.9622	60619.95		
12133	2	164.9734	60615.82	B1-8P17	
12134	0	164.9824	60612.53		
12135	10	164.9994	60606.28	B5-14R2	

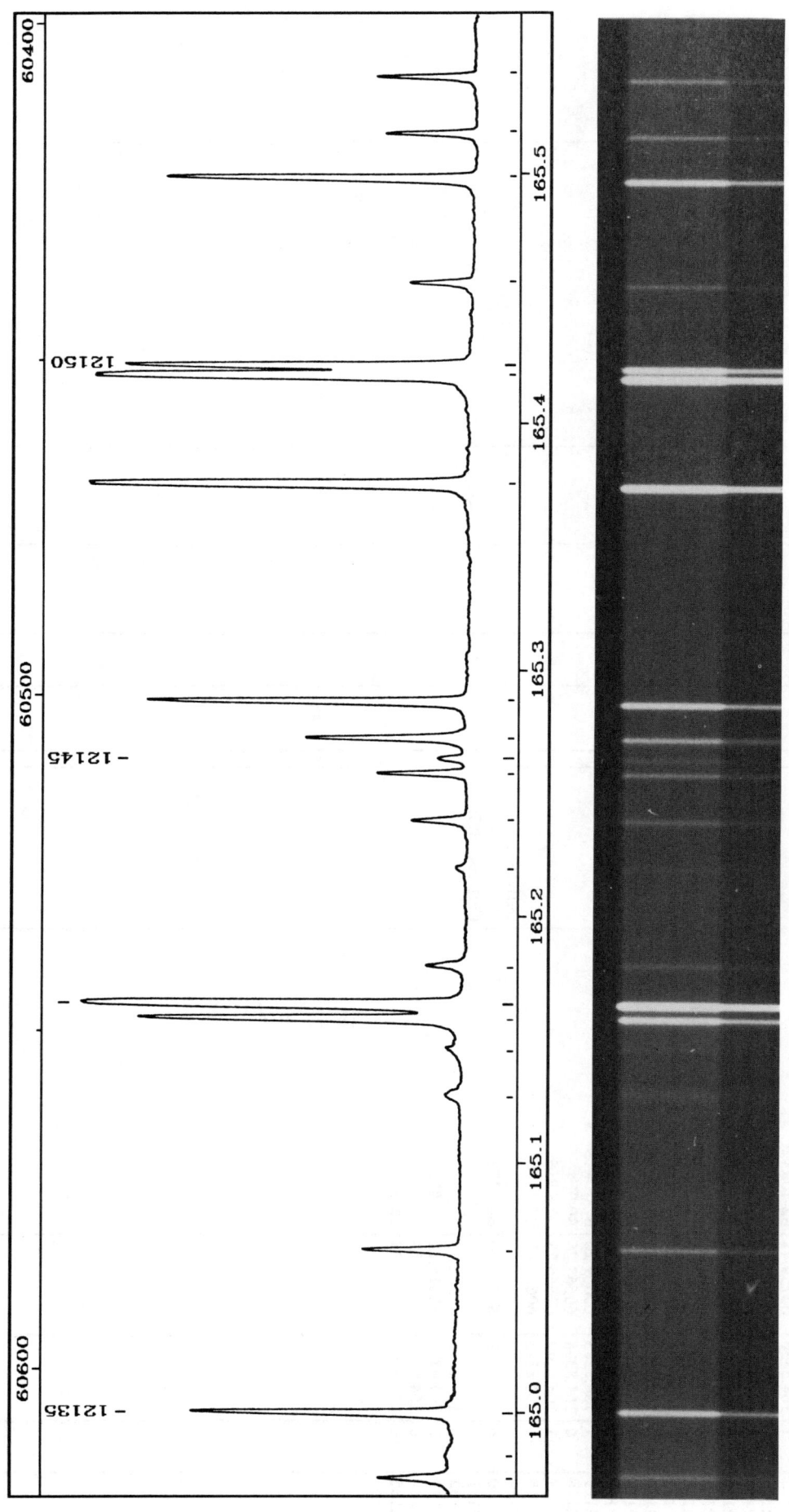

PLATE 154

TABLE 154

Nb	I	λ (nm)	σ (cm⁻¹)	Assignment		Comment
12136	3	165.0645	60582.37	B4-12P7		
12137	0	165.1273	60559.33	B0-6P23		d
12138	0	165.1461	60552.45	B0-8P13		
12139	18	165.1577	60548.18	B3-11P7		
12140	55	165.1635	60546.08	B5-14R1	B4-12P8 B4-12P10	
12141	2	165.1797	60540.14	B2-9P17		
12142	0	165.2196	60525.50	B1-9R15		
12143	2	165.2391	60518.35	B1-9P11		
12144	2	165.2583	60511.35	B5-14P4		
12145	1	165.2643	60509.13	B0-9P5		
12146	6	165.2725	60506.15	B2-10P9		
12147	15	165.2867	60500.93	B5-14R0		
12148	45	165.3749	60468.67	B5-14P3		
12149	50	165.4191	60452.51	B5-14P1		
12150	23	165.4230	60451.09	B5-14P2		
12151	1	165.4567	60438.78	B3-11P8		
12152	15	165.4992	60423.26	B1-10R1		
12153	3	165.5174	60416.60	B1-10R2		
12154	4	165.5403	60408.26	B1-10R0		

Nb	I	λ (nm)	σ (cm⁻¹)	Assignment	Comment

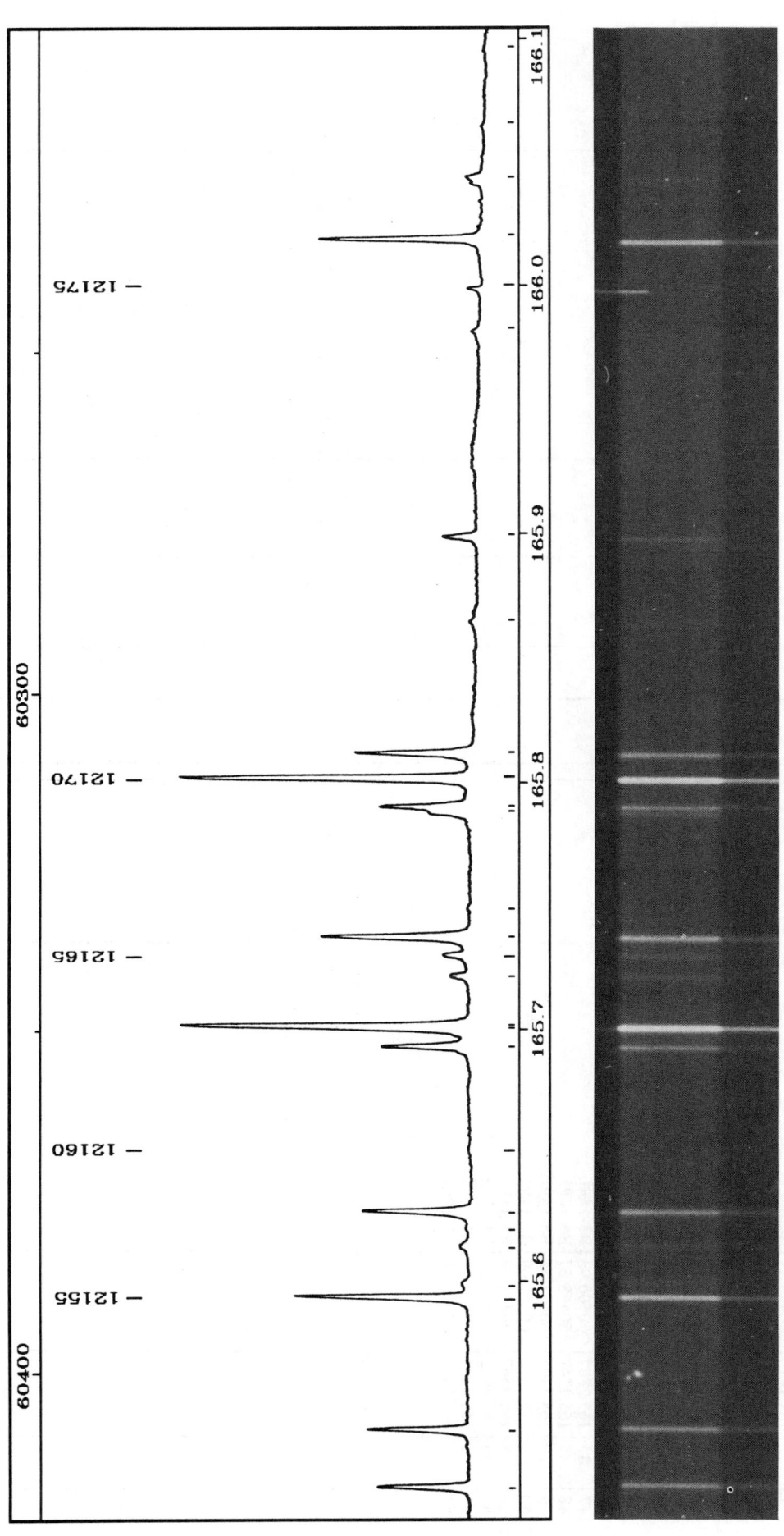

PLATE 155

TABLE 155

Nb	I	λ (nm)	σ (cm⁻¹)	Assignment	Comment
12155	5	165.5929	60389.09	B1-10R3	
12156	0	165.5987	60386.94	B1-8P18	
12157	0	165.6136	60381.51	B0-7P19	
12158	0	165.6246	60377.52		
12159		165.6267	60376.74		CI 165.62672
12160	0	165.6523	60367.40	B0-9R9	
12161		165.6928	60352.64		CI 165.69283
12162					CI 165.70082
12163		165.7010	60349.68		CI 165.70082
12164	1	165.7218	60342.09	B1-10R4	
12165	1	165.7302	60339.05	B2-10P10	
12166	6	165.7376	60336.33	B3-11P9	+CI 165.73792
12167	0	165.7494	60332.06	B0-9P6	
12168	1	165.7884	60317.87	B4-13R7	
12169	14	165.7909	60316.95		CI 165.79068
12170		165.8021	60312.87	B1-10P1	
12171		165.8123	60309.15		CI 165.81212
12172	0	165.8657	60289.73	B1-9P12	
12173	1	165.8997	60277.38	B1-10R5	
12174	0	165.9829	60247.17	B3-11P10	CuII 166.00009
12175					
12176	6	166.0199	60233.73	B1-10P2	
12177	0	166.0449	60224.69	B4-13R6	
12178	0	166.0650	60217.38		
12179	0	166.0978	60205.51		

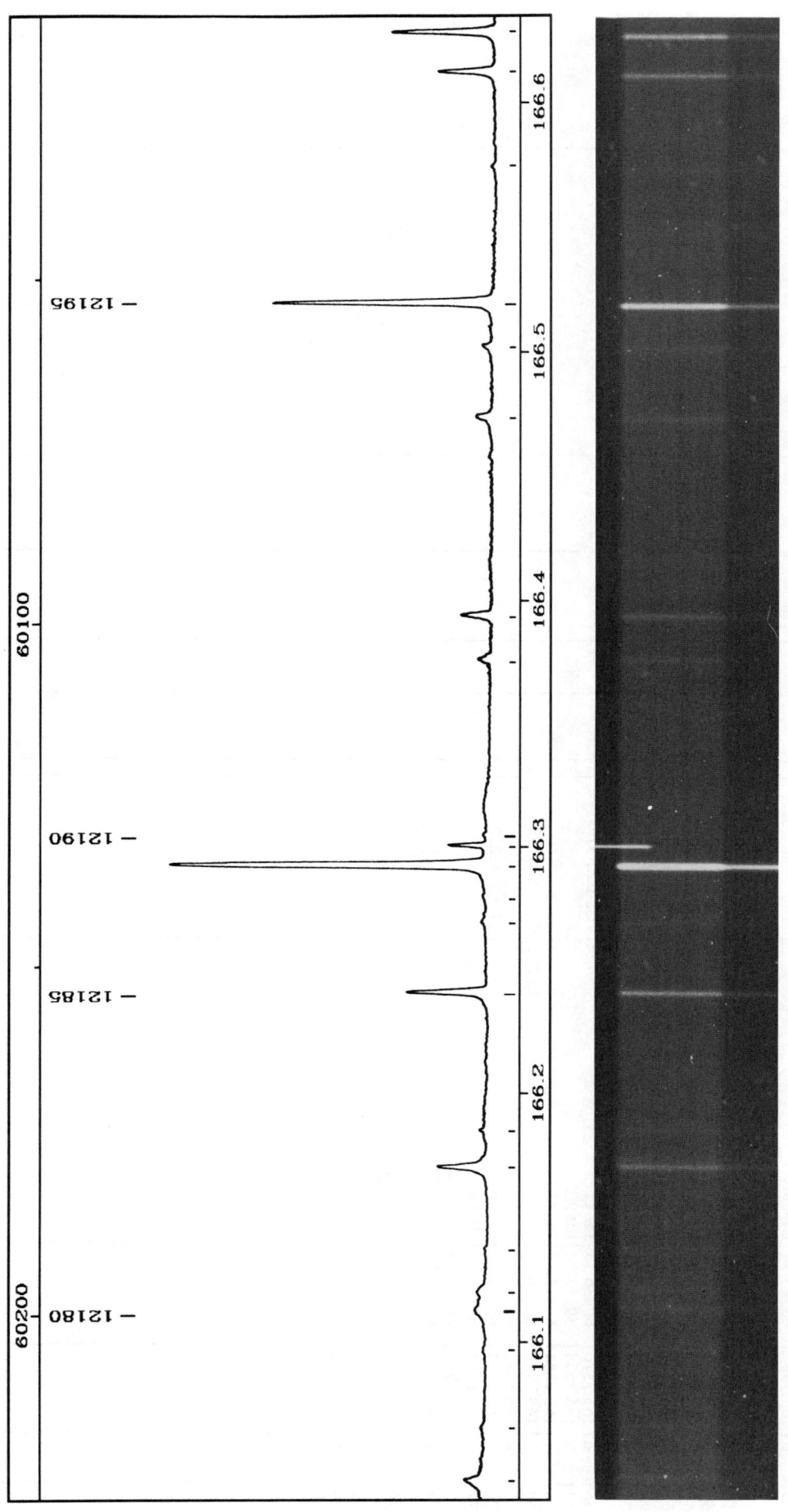

PLATE 156

TABLE 156

Nb	I	λ (nm)	σ (cm-1)	Assignment	Comment
12180	0	166.1129	60200.03	B1-8P19	
12181	0	166.1213	60196.97	B1-10R6	
12182	0	166.1375	60191.11		
12183	1	166.1698	60179.40	B2-10P11 B3-11P11	
12184	0	166.1802	60175.65		s
12185	2	166.2398	60154.06	B4-13R5	
12186	0	166.2678	60143.95	B3-11P13	
12187	0	166.2780	60140.25	B0-9P7	
12188	16	166.2917	60135.28	B1-10P3	CuII 166.30020
12189					s
12190	0	166.3042	60130.76	B1-10R7	
12191	0	166.3755	60105.01	B4-13R4	
12192	1	166.3932	60098.62	B1-9P13	
12193	0	166.4735	60069.63		
12194	0	166.5021	60059.29	B4-13R3	
12195	8	166.5194	60053.06	B2-10P12	
12196	0	166.5750	60033.02	B1-10P4	
12197	3	166.6127	60019.42	B4-13R2	
12198	5	166.6286	60013.72		

Nb	I	λ (nm)	σ (cm-1)	Assignment	Comment

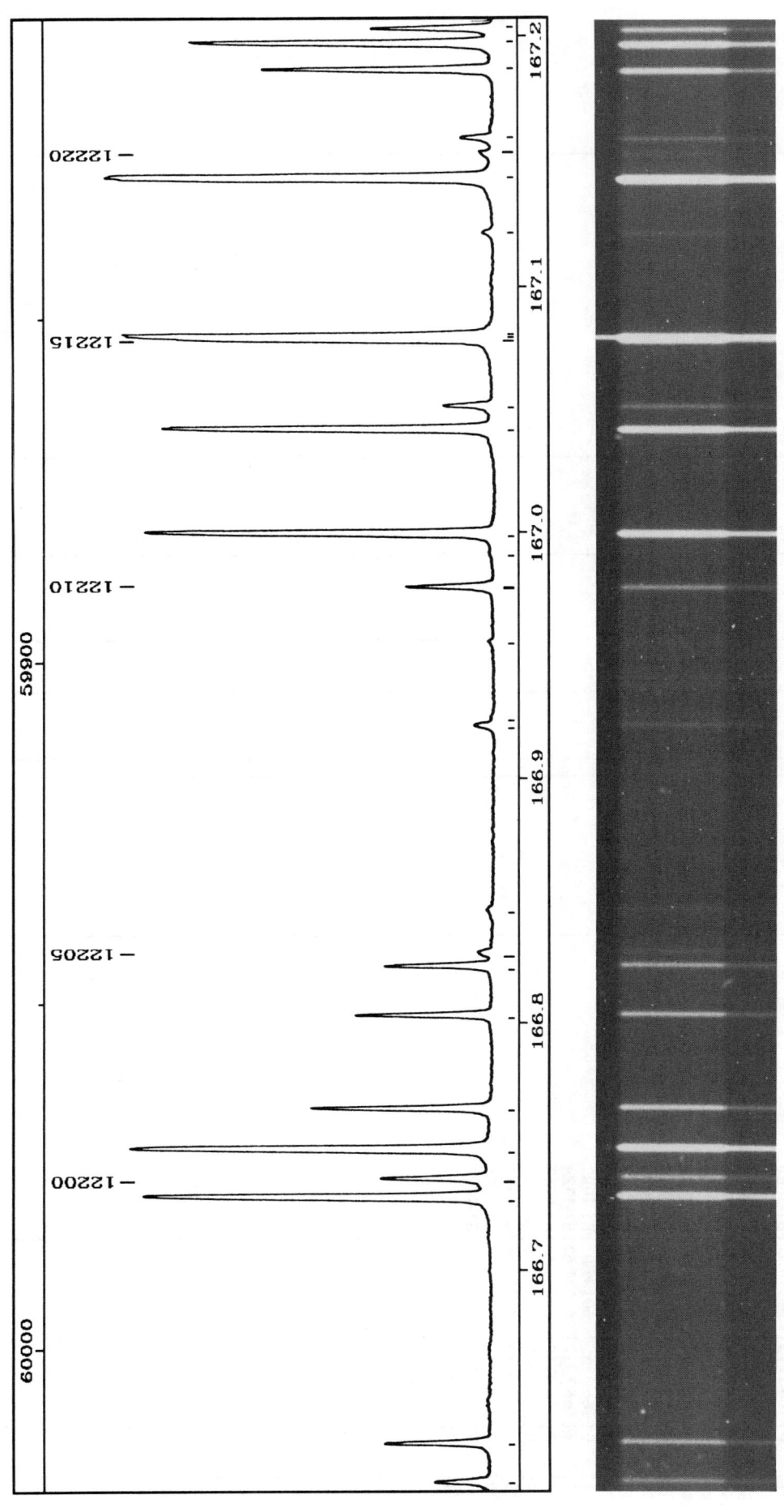

PLATE 157

TABLE 157

Nb	I	λ (nm)	σ (cm⁻¹)	Assignment	Comment
12199	16	166.7279	59977.95	B4-13R1	
12200	3	166.7354	59975.25	B2-11R2	
12201	20	166.7473	59970.99	B2-11R1	
12202	6	166.7636	59965.12	B2-11R3	
12203	5	166.8017	59951.43	B2-11R0	
12204	5	166.8220	59944.15	B4-13R0	
12205	5	166.8274	59942.19	B2-11R4	
12206	0	166.8453	59935.75	B0-9P8 B3-12R9	
12207	1	166.9207	59908.69	B2-11R5	
12208	0	166.9245	59907.34	B2-10P13	
12209	0	166.9541	59896.71	B1-10R9	
12210	4	166.9767	59888.58	B1-10P5	
12211	0	166.9921	59883.06		
12212	16	166.9986	59880.73	B4-13P1	
12213	16	167.0410	59865.53	B2-11P1	
12214	1	167.0505	59862.14	B4-13P7	
12215		167.0778	59852.34	B4-13P2	sh
12216		167.0788	59852.01		Al II 167.07867
12217	1	167.1212	59836.80	B3-12R3	Al II 167.07867
12218	1	167.1432	59828.95	B4-13P3	
12219	37	167.1534	59825.28	B3-12R2	
12220	1	167.1595	59823.10	B4-13P6	
12221	1	167.1870	59813.26	B4-13P4	
12222	9	167.1977	59809.44	B4-13P5	
12223	12	167.2037	59807.29	B3-12R1	
12224	4				

Nb	I	λ (nm)	σ (cm⁻¹)	Assignment	Comment

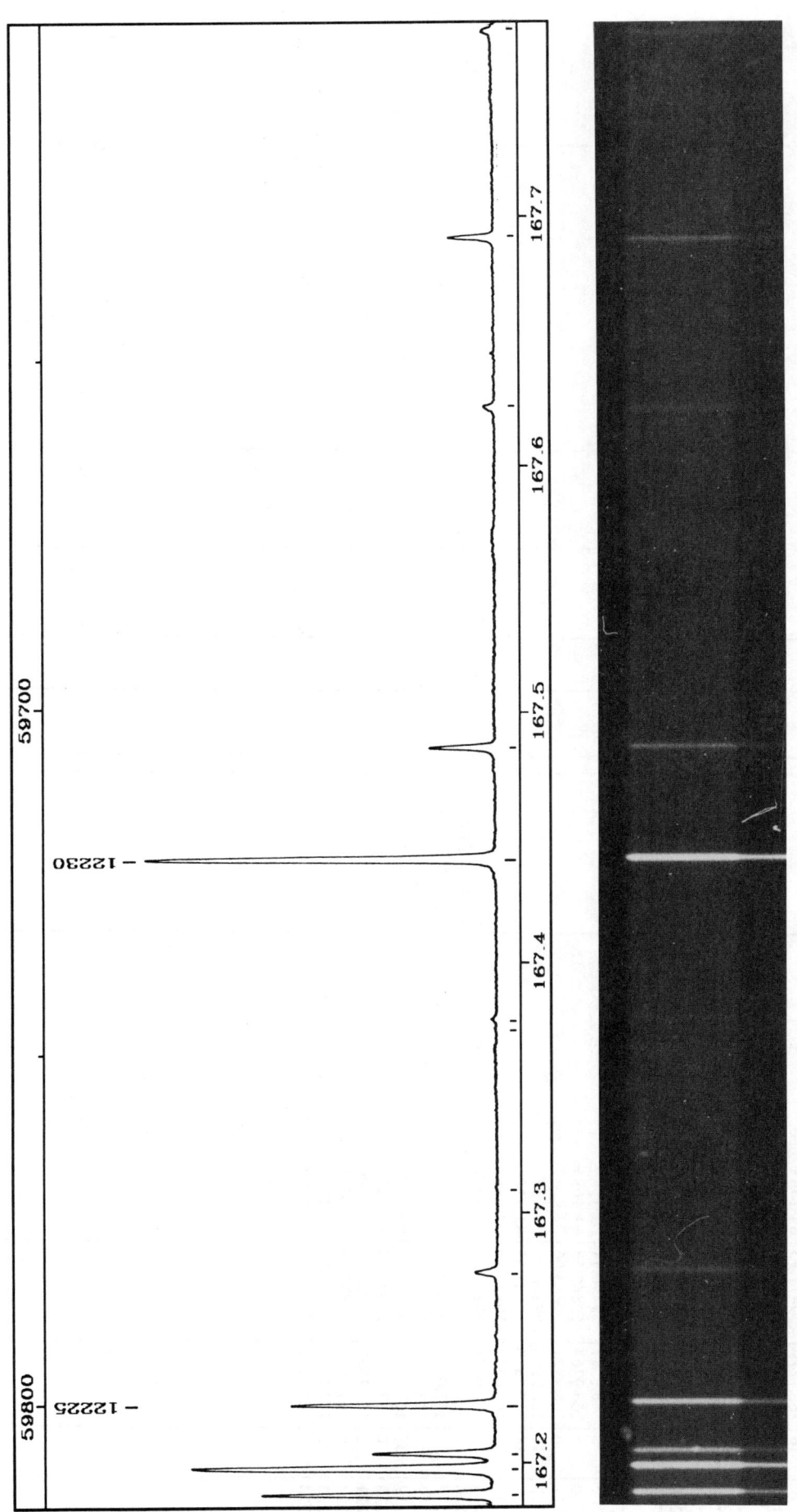

PLATE 158

TABLE 158

Nb	I	λ (nm)	σ (cm⁻¹)	Assignment		Comment
12225	8	167.2230	59800.39	B2-11P2		
12226	1	167.2758	59781.52	B3-12R0		
12227	0	167.3123	59768.46			
12228	0	167.3727	59746.88	B1-10P6		
12229	0	167.3769	59745.41	B2-11P3 B0-9P9		
12230	18	167.4417	59722.28	B3-12P1		
12231	3	167.4870	59706.12	B3-12P2		
12232	0	167.6235	59657.50	B2-11P4		
12233	2	167.6916	59633.27	B3-12P3		
12234	0	167.7747	59603.74			

Nb	I	λ (nm)	σ (cm⁻¹)	Assignment	Comment

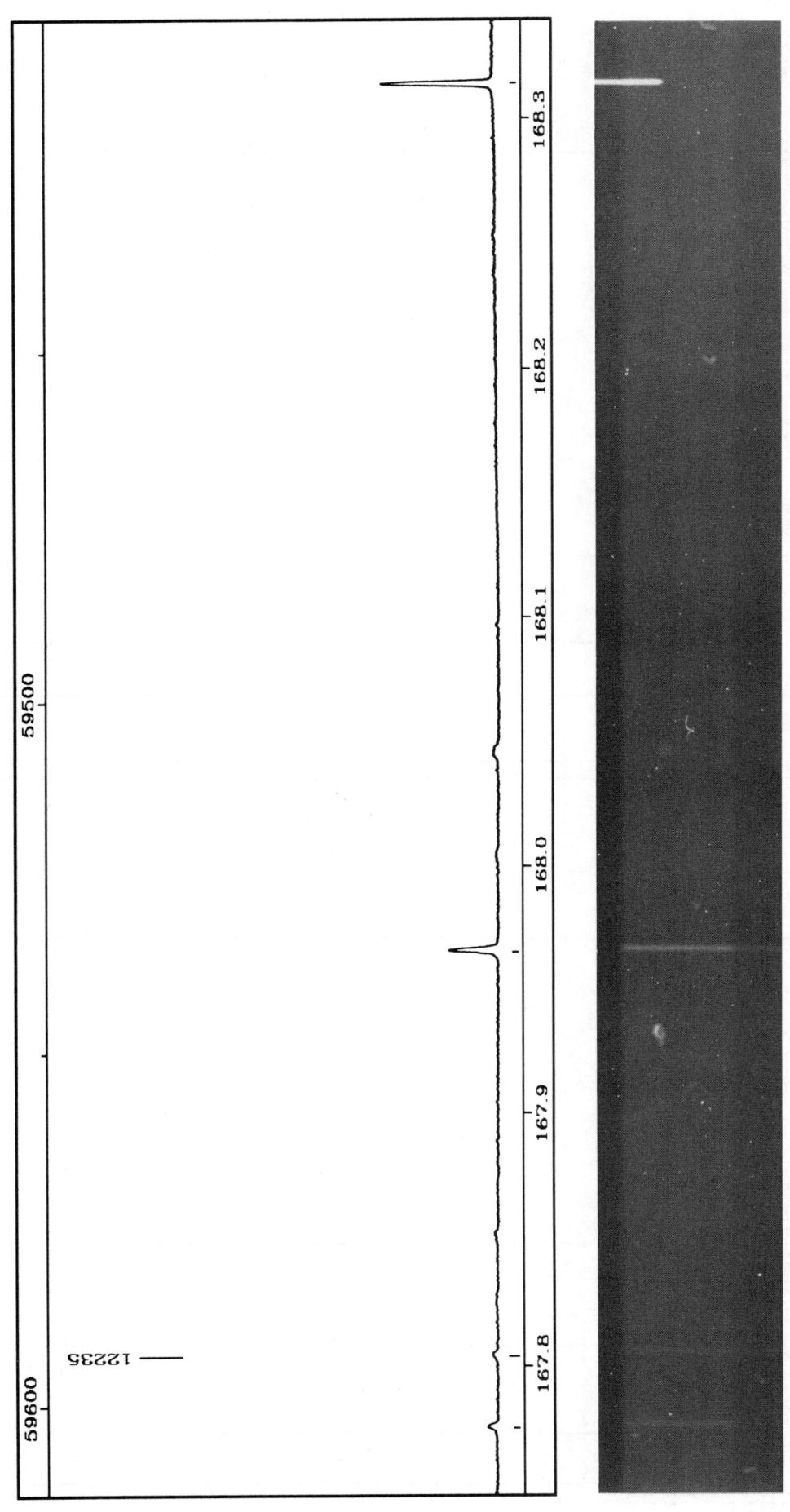

PLATE 159

TABLE 159

Nb	I	λ (nm)	σ (cm-1)	Assignment	Comment
12235	0	167.8037	59593.44	B1-10P7	CuII 168.31585 Ke
12236	3	167.9655	59536.04	B2-11P5	
12237					

Nb	I	λ (nm)	σ (cm-1)	Assignment	Comment

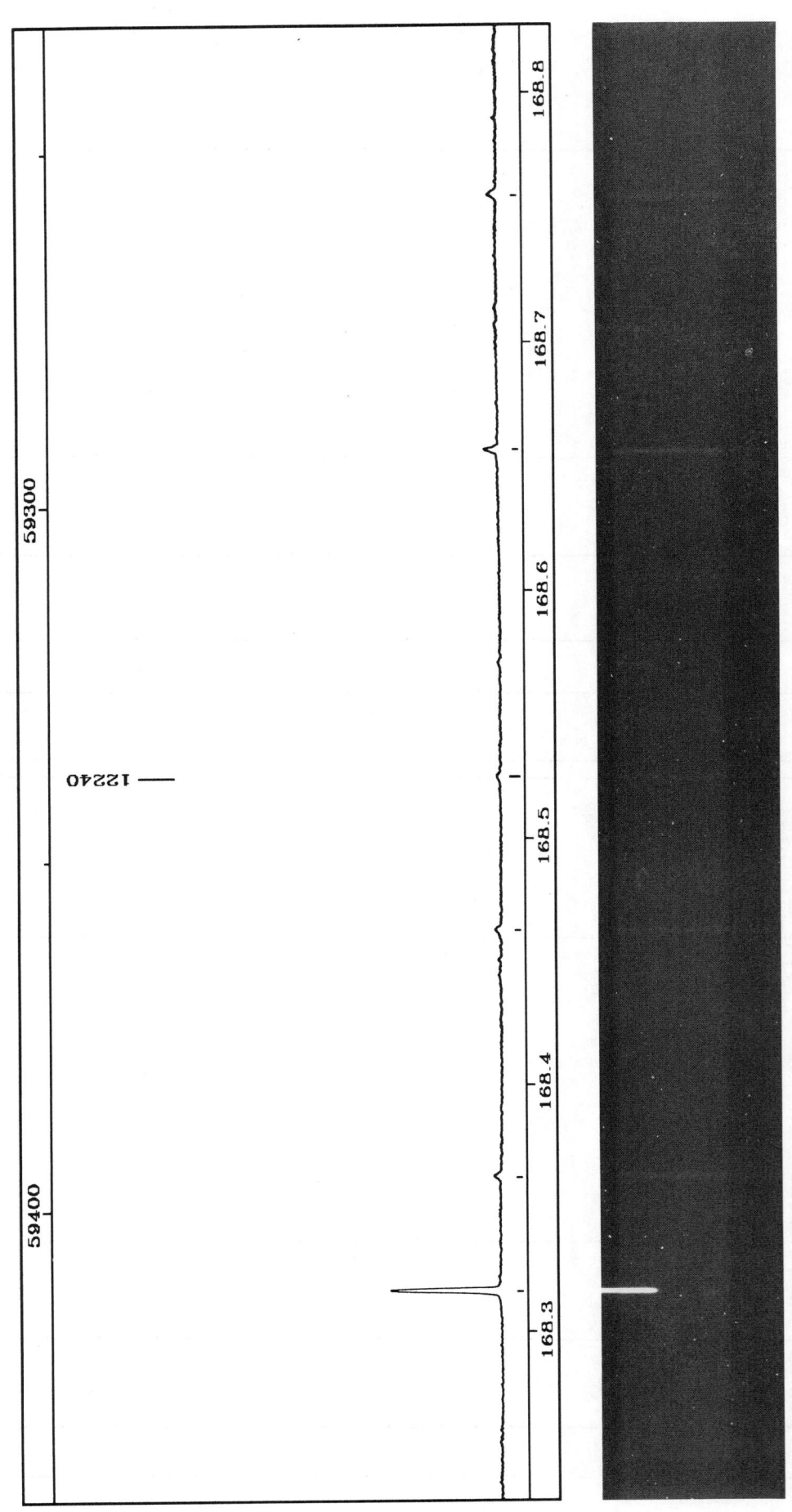

PLATE 160

TABLE 160

Nb	I	λ (nm)	σ (cm-1)	Assigment	Comment
12238	0	168.3627	59395.58	B3-12P7	
12239	0	168.4619	59360.60	B3-12P9	
12240	0	168.5251	59338.35	B4-14P4	
12241	1	168.6560	59292.29	B4-14P3	
12242	0	168.7589	59256.15		

Nb	I	λ (nm)	σ (cm-1)	Assigment	Comment

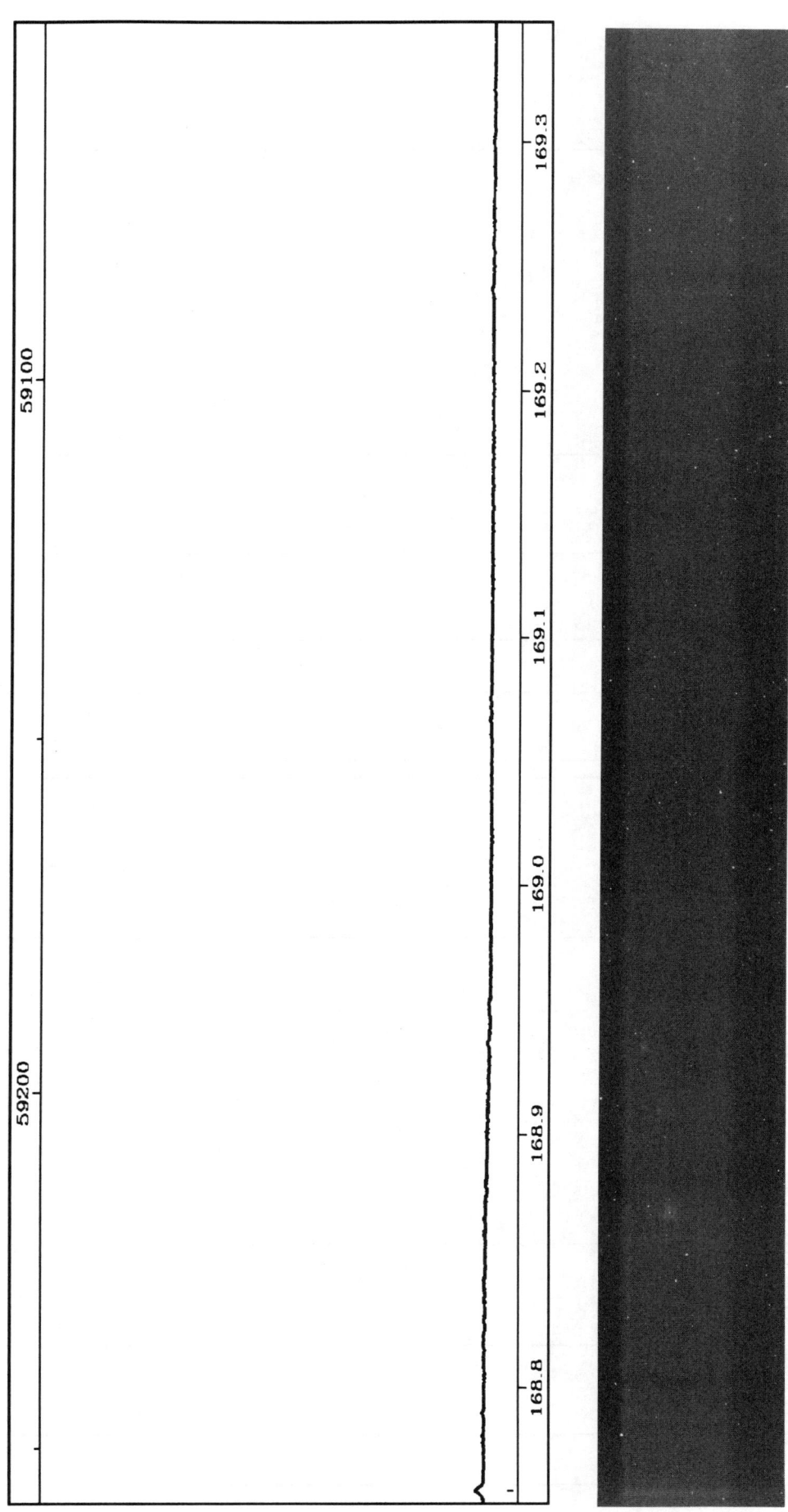

PLATE 161

TABLE 161

Nb	I	λ (nm)	σ (cm⁻¹)	Assignment	Comment

Nb	I	λ (nm)	σ (cm⁻¹)	Assignment	Comment

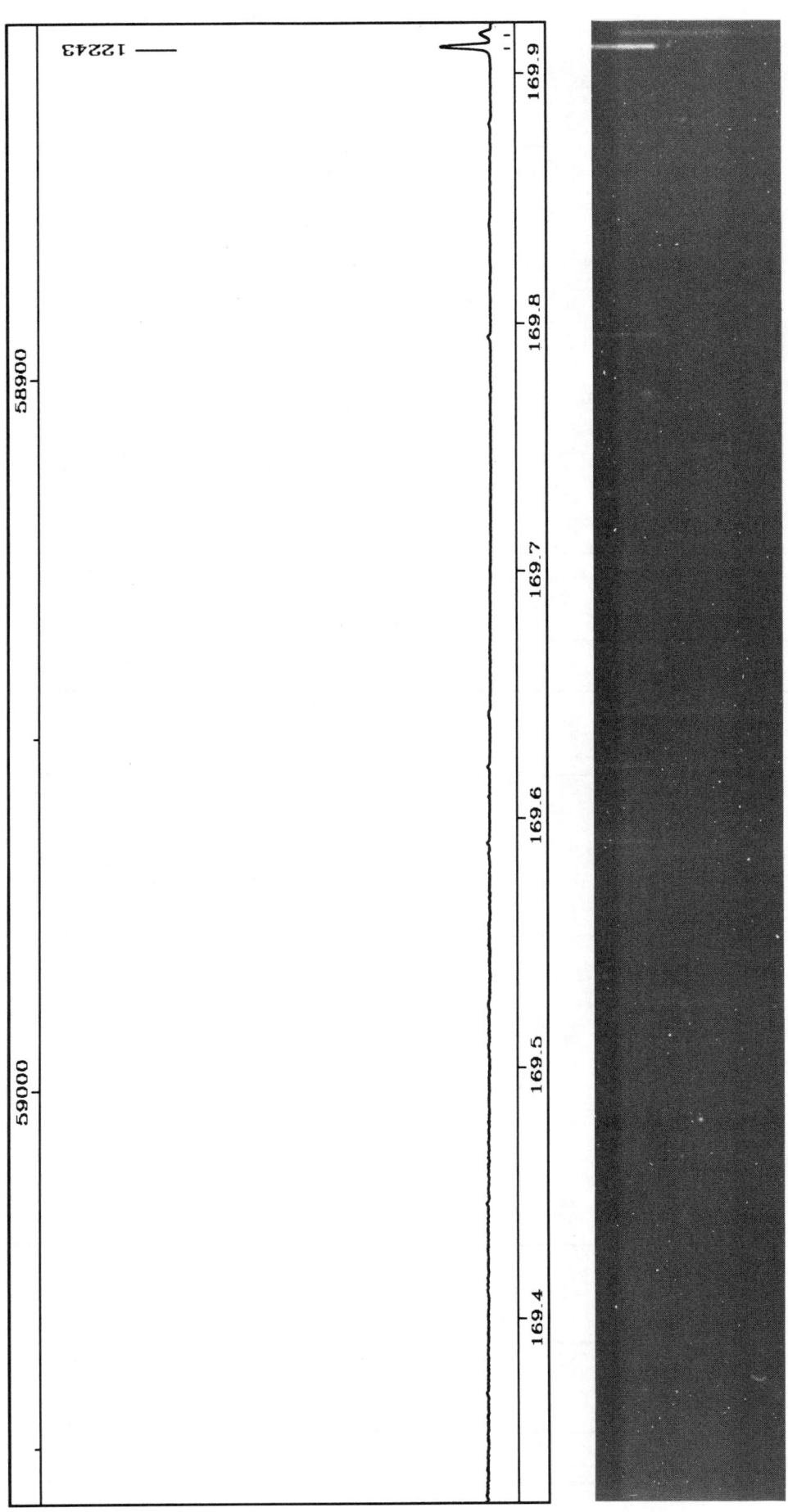

PLATE 162

TABLE 162

Nb	I	λ (nm)	σ (cm⁻¹)	Assignment	Comment

Nb	I	λ (nm)	σ (cm⁻¹)	Assignment	Comment
12243		169.9099			CuII 169.90953 Ke
12244	0	169.9151	58852.92	B3-13R3	+CuII 169.91023 Ke

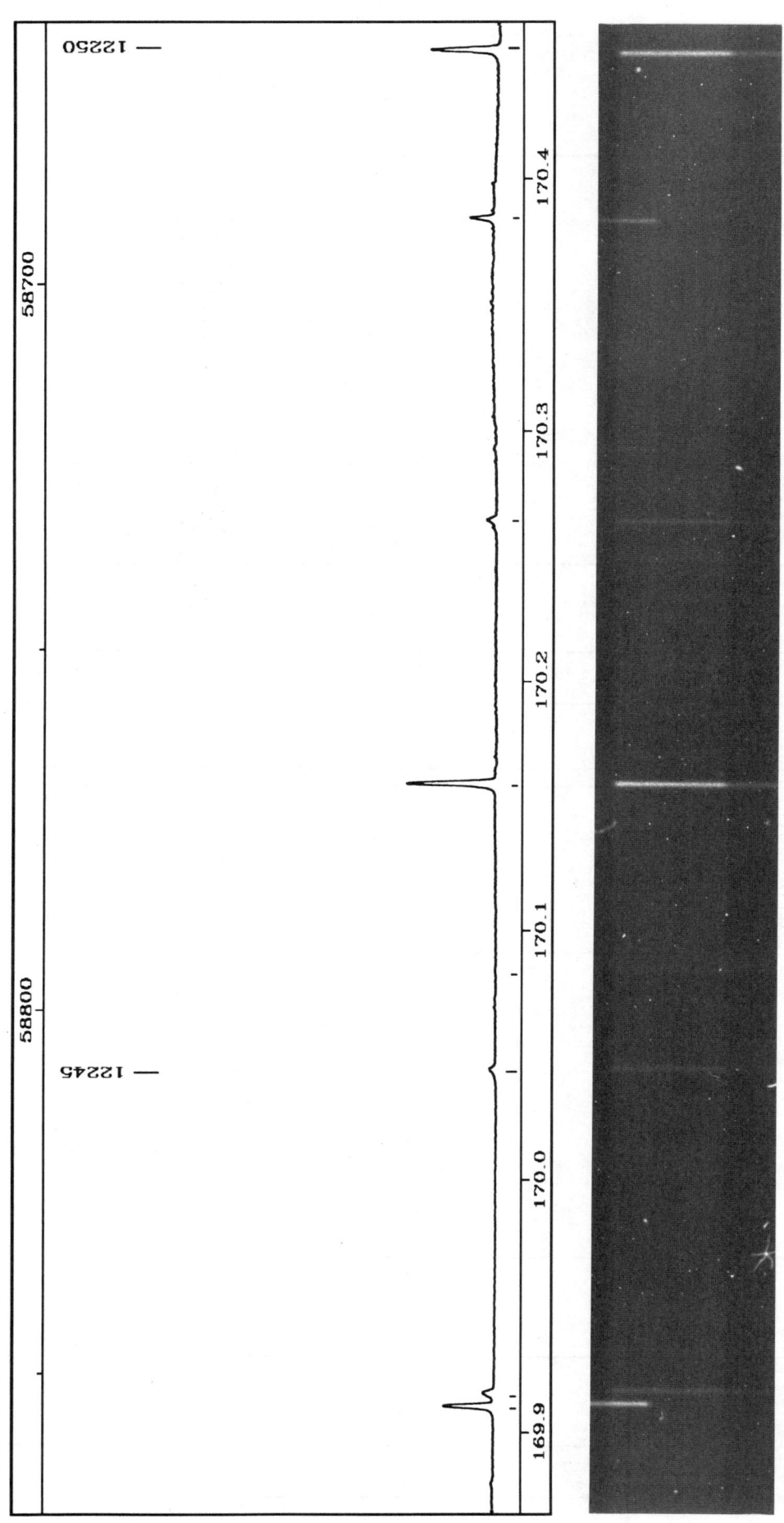

PLATE 163

TABLE 163

Nb	I	λ (nm)	σ (cm⁻¹)	Assignment	Comment
12245	0	170.0437	58808.42	B3-13R2	
12246	0	170.0840	58794.47		
12247	1	170.1582	58768.83	B3-13R1	
12248	0	170.2639	58732.35	B3-13R0	CuI 170.3843 Ke
12249		170.3846			
12250	1	170.4516	58667.67	B3-13P1	

Nb	I	λ (nm)	σ (cm⁻¹)	Assignment	Comment

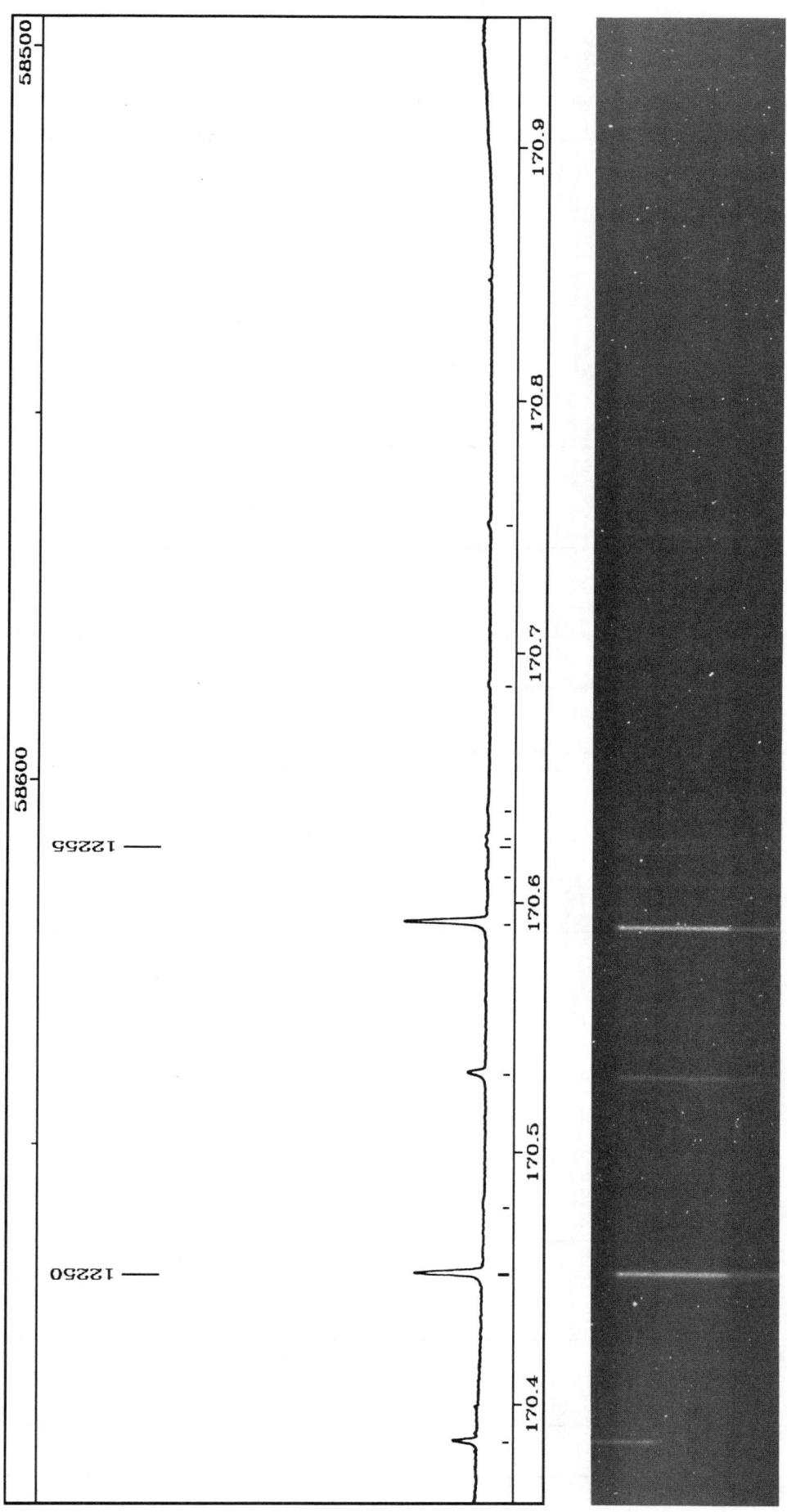

PLATE 164

TABLE 164

Nb	I	λ (nm)	σ (cm⁻¹)	Assignment	Comment
12251	0	170.4784	58658.47		
12252	0	170.5299	58640.74	B3-13P2	
12253	1	170.5907	58619.83	B3-13P3	
12254	0	170.6107	58612.98		
12255	0	170.6218	58609.17	B3-13P5	
12256	0	170.6255	58607.89	B3-13P4	
12257	0	170.6367	58604.05	B2-12R3	
12258	0	170.6863	58587.00	B2-12R2	
12259	0	170.7494	58565.36	B2-12R1	

Nb	I	λ (nm)	σ (cm⁻¹)	Assignment	Comment

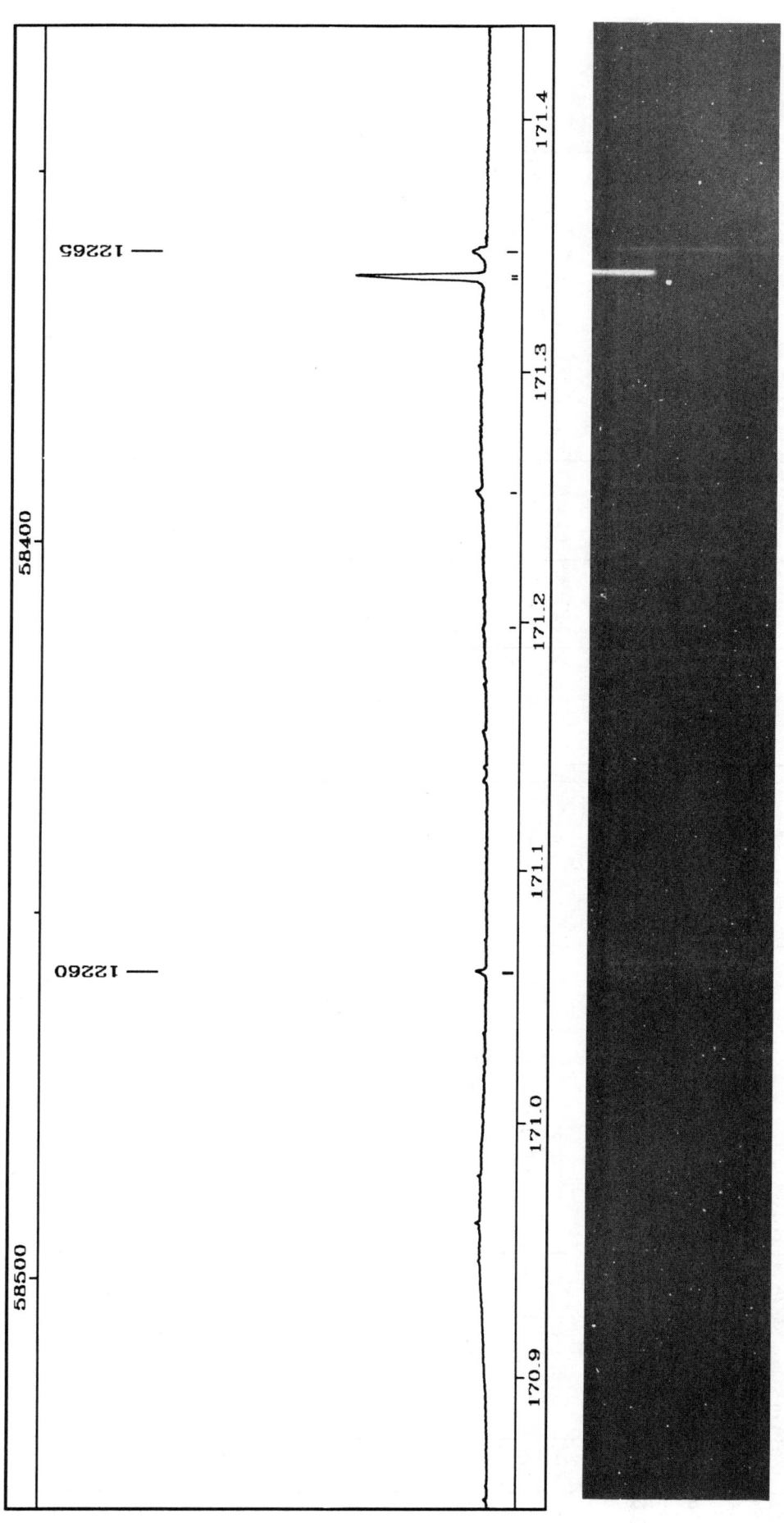

PLATE 165

TABLE 165

Nb	I	λ (nm)	σ (cm⁻¹)	Assignment	Comment

Nb	I	λ (nm)	σ (cm⁻¹)	Assignment	Comment
12260	0	171.0591	58459.32	B2-12P1	
12261	0	171.1967	58412.33	B2-12P2	
12262	0	171.2523	58393.37		
12263		171.3365			CuI 171.3365 Ke
12264		171.3374			
12265	0	171.3464	58361.30	B2-12P3	

Journal of Physical and Chemical Reference Data
Cumulative Listing of Reprints

Reprints from Volume 1

1. Gaseous Diffusion Coefficients, *T.R. Marrero and E.A. Mason,* Vol. 1, No. 1, pp. 1–118 (1972) $14.00

2. Selected Values of Critical Supersaturation for Nucleation of Liquids from the Vapor, *G.M. Pound,* Vol. 1, No. 1, pp. 119–134 (1972) $6.00

3. Selected Values of Evaporation and Condensation Coefficients for Simple Substances, *G.M. Pound,* Vol. 1, No. 1, pp. 135–146 (1972) $6.00

4. Atlas of the Observed Absorption Spectrum of Carbon Monoxide between 1060 and 1900 Å, *S.G. Tilford and J.D. Simmons,* Vol. 1, No. 1, pp. 147–188 (1972) $9.00

5. Tables of Molecular Vibrational Frequencies, Part 5, *T. Shimanouchi,* Vol. 1, No. 1, pp. 189–216 (1972) (superseded by No.103) $8.00

6. Selected Values of Heats of Combustion and Heats of Formation of Organic Compounds Containing the Elements C, H, N, O, P, and S, *Eugene S. Domalski,* Vol. 1, No. 2, pp. 221–278 (1972) $10.00

7. Thermal Conductivity of the Elements, *C.Y. Ho, R.W. Powell, and P.E. Liley,* Vol. 1, No. 2, pp. 279–422 (1972) $15.00

8. The Spectrum of Molecular Oxygen, *Paul H. Krupenie,* Vol. 1, No. 2, pp. 423–534 (1972) $13.00

9. A Critical Review of the Gas-Phase Reaction Kinetics of the Hydroxyl Radical, *Wm. E. Wilson, Jr.,* Vol. 1, No. 2, pp. 535–574 (1972) (superseded by Monograph 1) $9.00

10. Molten Salts: Volume 3, Nitrates, Nitrites, and Mixtures, Electrical Conductance, Density, Viscosity, and Surface Tension Data, *G.J. Janz, Ursula Krebs, H.F. Siegenthaler, and R.P.T. Tomkins,* Vol. 1, No. 3, pp. 581–746 (1972) $17.00

11. High Temperature Properties and Decomposition of Inorganic Salts—Part 3. Nitrates and Nitrites, *Kurt H. Stern,* Vol. 1, No. 3, pp. 747–772 (1972) $8.00

12. High-Pressure Calibration: A Critical Review, *D.L. Decker, W.A. Bassett, L. Merrill, H.T. Hall, and J.D. Barnett,* Vol. 1, No. 3, pp. 773–836 (1972) $10.00

13. The Surface Tension of Pure Liquid Compounds, *Joseph J. Jasper,* Vol. 1, No. 4, pp. 841–1009 (1972) $17.00

14. Microwave Spectra of Molecules of Astrophysical Interest, I. Formaldehyde, Formamide, and Thioformaldehyde, *Donald R. Johnson, Frank J. Lovas, and William H. Kirchhoff,* Vol. 1, No. 4, pp. 1011–1046 (1972) $9.00

15. Osmotic Coefficients and Mean Activity Coefficients of Uni-univalent Electrolytes in Water at 25° C, *Walter J. Hamer and Yung-Chi Wu,* Vol. 1, No. 4, pp. 1047-1099 (1972) $10.00

16. The Viscosity and Thermal Conductivity Coefficients of Gaseous and Liquid Fluorine, *H.J.M. Hanley and R. Prydz,* Vol. 1, No. 4, pp. 1101-1113 (1972) $6.00

Reprints from Volume 2

17. Microwave Spectra of Molecules of Astrophysical Interest, II. Methylenimine, *William H. Kirchhoff, Donald R. Johnson, and Frank J. Lovas,* Vol. 2, No. 1, pp. 1-10 (1973) $6.00

18. Analysis of Specific Heat Data in the Critical Region of Magnetic Solids, *F.J. Cook,* Vol. 2, No. 1, pp. 11-24 (1973) $6.00

19. Evaluated Chemical Kinetic Rate Constants for Various Gas Phase Reactions, *Keith Schofield,* Vol. 2, No. 1, pp. 25–84 (1973) $10.00

20. Atomic Transition Probabilities for Forbidden Lines of the Iron Group Elements. (A Critical Data Compilation for Selected Lines), *M.W. Smith and W.L. Wiese,* Vol. 2, No. 1, pp. 85–120 (1973) (superseded by Vol. 17, Suppl. 4) $9.00

21. Tables of Molecular Vibrational Frequencies, Part 6, *T. Shimanouchi,* Vol. 2, No. 1, pp. 121–162 (1973) (superseded by No. 103) $9.00

22. Compilation of Energy Band Gaps in Elemental and Binary Compound Semiconductors and Insulators, *W.H. Strehlow and E.L. Cook,* Vol. 2, No. 1, pp. 163–200 (1973) $9.00

23. Microwave Spectra of Molecules of Astrophysical Interest, III. Methanol, *R.M. Lees, F.J. Lovas, W.H. Kirchhoff, and D.R. Johnson,* Vol. 2, No. 2, pp. 205–214 (1973) $6.00

Journal of Physical and Chemical Reference Data
Reprint and Supplements/Monographs Orders

To: American Chemical Society
Distribution Office
1155 Sixteenth Street, N.W.
Washington, DC 20036-9976

Name: _____

Title: _____

Organization: _____

Address: _____

City: _____ State: _____

Country: _____ Zip: _____

I am a member of _____
(ACS, AIP, or Affiliated Society)

ORDERS FOR REPRINTS AND SUPPLEMENTS MUST BE PREPAID.
*Foreign orders for Reprints, add $2.50 for each reprint for postage and handling. Foreign orders for Reprint Packages, add $5.00 for each Reprint Package for postage and handling. Make checks payable to the American Chemical society.

*Bulk Rates: Substract 20% from the listed price for orders of 50 or more of any one item.

PRICES ARE SUBJECT TO CHANGE WITHOUT NOTICE.

Please ship the following Reprints and Supplements/Monographs:

Reprint No./Package _____ , _____ copies $ _____

Reprint No./Package _____ , _____ copies $ _____

Reprint No./Package _____ , _____ copies $ _____

Reprint No./Package _____ , _____ copies $ _____

Reprint No./Package _____ , _____ copies $ _____

Reprint No./Package _____ , _____ copies $ _____

Vol. ____ , Suppl. ____ ☐ Hardcover _____ copies $ _____

Vol. ____ , Suppl. ____ ☐ Hardcover _____ copies $ _____

Vol. ____ , Suppl. ____ ☐ Hardcover _____ copies $ _____

Vol. ____ , Suppl. ____ ☐ Hardcover _____ copies $ _____

Vol. ____ , Suppl. ____ ☐ Hardcover _____ copies $ _____

Vol. ____ , Suppl. ____ ☐ Hardcover _____ copies $ _____

Mono. No. 1, 1989 ☐ Hardcover _____ copies $ _____

Mono. No. 2, 1994 ☐ Hardcover _____ copies $ _____

Mono. No. 3, 1994 ☐ Hardcover _____ copies $ _____

Total Enclosed $ _____

24. Microwave Spectra of Molecules of Astrophysical Interest, IV. Hydrogen Sulfide, *Paul Helminger, Frank C. De Lucia, and William H. Kirchhoff,* Vol. 2, No. 2, pp. 215–224 (1973) ... $6.00

25. Tables of Molecular Vibrational Frequencies, Part 7, *T. Shimanouchi,* Vol. 2, No. 2, pp. 225–256 (1973) (superseded by No. 103) ... $8.00

26. Energy Levels of Neutral Helium (^{4}He I), *W.C. Martin,* Vol. 2, No. 2, pp. 257–266 (1973) ... $6.00

27. Survey of Photochemical and Rate Data for Twenty-eight Reactions of Interest in Atmospheric Chemistry, *R.F. Hampson, Editor, W. Braun, R.L. Brown, D. Garvin, J.T. Herron, R.E. Huie, M.J. Kurylo, A.H. Laufer, J.D. McKinley, H. Okabe, M.D. Scheer, W. Tsang, and D.H. Stedman,* Vol. 2, No. 2, pp. 267–312 (1973) (superseded by Nos. 159, 203, 264, 366) ... $9.00

28. Compilation of the Static Dielectric Constant of Inorganic Solids, *K.F. Young and H.P.R. Frederikse,* Vol. 2, No. 2, pp. 313–410 (1973) (superseded by Nos. 159, 203, 264, 366) ... $13.00

29. Soft X-Ray Emission Spectra of Metallic Solids: Critical Review of Selected Systems, *A.J. McAlister, R.C. Dobbyn, J.R. Cuthill, and M.L. Williams,* Vol. 2, No. 2, pp. 411–426 (1973) ... $6.00

30. Ideal Gas Thermodynamic Properties of Ethane and Propane, *J. Chao, R.C. Wilhoit, and B.J. Zwolinski,* Vol. 2, No. 2, pp. 427–438 (1973) ... $6.00

31. An Analysis of Coexistence Curve Data for Several Binary Liquid Mixtures Near Their Critical Points, *A. Stein and G.F. Allen,* Vol. 2, No. 3, pp. 443–466 (1973) ... $8.00

32. Rate Constants for the Reactions of Atomic Oxygen (O ^{3}P) with Organic Compounds in the Gas Phase, *John T. Herron and Robert E. Huie,* Vol. 2, No. 3, pp. 467–518 (1973) ... $10.00

33. First Spectra of Neon, Argon, and Xenon 136 in the 1.2-4.0 μm Region, *Curtis J. Humphreys,* Vol. 2, No. 3, pp. 519–530 (1973) ... $6.00

34. Elastic Properties of Metals and Alloys, I. Iron, Nickel, and Iron-Nickel Alloys, *H.M. Ledbetter and R.P. Reed,* Vol. 2, No. 3, pp. 531–618 (1973) ... $12.00

35. The Viscosity and Thermal Conductivity Coefficients of Dilute Argon, Krypton, and Xenon, *H.J.M. Hanley,* Vol. 2, No. 3, pp. 619–642 (1973) ... $8.00

36. Diffusion in Copper and Copper Alloys, Part I. Volume and Surface Self-Diffusion in Copper, *Daniel B. Butrymowicz, John R. Manning, and Michael E. Read,* Vol. 2, No. 3, pp. 643–656 (1973) ... $6.00

37. The 1973 Least-Squares Adjustment of the Fundamental Constants, *E. Richard Cohen and B.N. Taylor,* Vol. 2, No. 4, pp. 663–734 (1973) ... $11.00

38. The Viscosity and Thermal Conductivity Coefficients of Dilute Nitrogen and Oxygen, *H.J.M. Hanley and James F. Ely,* Vol. 2, No. 4, pp. 735–756 (1973) ... $8.00

39. Thermodynamic Properties of Nitrogen Including Liquid and Vapor Phases from 63 K to 2000 K with Pressures to 10,000 Bar, *Richard T. Jacobsen and Richard B. Stewart,* Vol. 2, No. 4, pp. 757–922 (1973) ... $17.00

40. Thermodynamic Properties of Helium 4 from 2 to 1500 K at Pressures to 10^8 Pa, *Robert T. McCarty,* Vol. 2, No. 4, pp. 923–1042 (1973) ... $14.00

Reprints from Volume 3

41. Molten Salts: Volume 4, Part 1, Fluorides and Mixtures, Electrical Conductance, Density, Viscosity, and Surface Tension Data, *G.J. Janz, G.L. Gardner, Ursula Krebs, and R.P.T. Tomkins,* Vol. 3, No. 1, pp. 1–115 (1974) ... $14.00

42. Ideal Gas Thermodynamic Properties of Eight Chloro- and Fluoromethanes, *A.S. Rodgers, J. Chao, R. C. Wilhoit, and B.J. Zwolinski,* Vol. 3, No. 1, pp. 117–140 (1974) ... $8.00

43. Ideal Gas Thermodynamic Properties of Six Chloroethanes, *J. Chao, A.S. Rodgers, R.C. Wilhoit, and B.J. Zwolinski,* Vol. 3, No. 1, pp. 141–162 (1974) ... $8.00

44. Critical Analysis of Heat-Capacity Data and Evaluation of Thermodynamic Properties of Ruthenium, Rhodium, Palladium, Iridium, and Platinum from 0 to 300 K. A Survey of the Literature Data on Osmium, *George T. Furukawa, Martin L. Reilly, and John S. Gallagher,* Vol. 3, No. 1, pp. 163–209 (1974) ... $9.00

45. Microwave Spectra of Molecules of Astrophysical Interest, V. Water Vapor, *Frank C. De Lucia, Paul Helminger, and William H. Kirchhoff,* Vol. 3, No. 1, pp. 211–219 (1974) ... $6.00

46. Microwave Spectra of Molecules of Astrophysical Interest, VI. Carbonyl Sulfide and Hydrogen Cyanide, *Arthur G. Maki,* Vol. 3, No. 1, pp. 221–244 (1974) ... $8.00

47. Microwave Spectra of Molecules of Astrophysical Interest, VII. Carbon Monoxide, Carbon Monosulfide, and Silicon Monoxide, *Frank J. Lovas and Paul H. Krupenie,* Vol. 3, No. 1, pp. 245–257 (1974) ... $6.00

48. Microwave Spectra of Molecules of Astrophysical Interest, VIII. Sulfur Monoxide, *Eberhard Tiemann,* Vol. 3, No. 1, pp. 259–268 (1974) ... $6.00

49. Tables of Molecular Vibrational Frequencies, Part 8, *T. Shimanouchi,* Vol. 3, No. 1, pp. 269–308 (1974) (superseded by No. 103) ... $9.00

50. JANAF Thermochemical Tables, 1974 Supplement, *M.W. Chase, J.L. Curnutt, A.T. Hu, H. Prophet, A.N. Syverud, and L.C. Walker,* Vol. 3, No. 2, pp. 311–480 (1974) (superseded by Vol. 14, Suppl. 1) ... $17.00

51. High Temperature Properties and Decomposition of Inorganic Salts, Part 4. Oxy-Salts of the Halogens, *Kurt H. Stern,* Vol. 3, No. 2, pp. 481–526 (1974) ... $9.00

52. Diffusion in Copper and Copper Alloys, Part II. Copper-Silver and Copper-Gold Systems, *Daniel B. Butrymowicz, John R. Manning, and Michael E. Read,* Vol. 3, No. 2, pp. 527–602 (1974) ... $11.00

53. Microwave Spectral Tables I. Diatomic Molecules, *Frank J. Lovas and Eberhard Tiemann,* Vol. 3, No. 3, pp. 609–770 (1974) ... $17.00

54. Ground Levels and Ionization Potentials for Lanthanide and Actinide Atoms and Ions, *W.C. Martin, Lucy Hagan, Joseph Reader, and Jack Sugar,* Vol. 3, No. 3, pp. 771–780 (1974) ... $6.00

55. Behavior of the Elements at High Pressures, *John Francis Cannon,* Vol. 3, No. 3, pp. 781–824 (1974) ... $9.00

56. Reference Wavelengths from Atomic Spectra in the Range 15 Å to 25000 Å, *Victor Kaufman and Bengt Edlén,* Vol. 3, No. 4, pp. 825–895 (1974) ... $11.00

57. Elastic Properties of Metals and Alloys. II. Copper, *H.M. Ledbetter and E.R. Naimon,* Vol. 3, No. 4, pp.897–935 (1974) ... $9.00

58. A Critical Review of H-Atom Transfer in the Liquid Phase: Chlorine Atom, Alkyl, Trichloromethyl, Alkoxy, and Alkylperoxy Radicals, *D.G. Hendry, T. Mill, L. Piszkiewicz, J.A. Howard, and H.K. Eigenmann,* Vol. 3, No. 4, pp. 937–978 (1974) ... $9.00

59. The Viscosity and Thermal Conductivity Coefficients for Dense Gaseous and Liquid Argon, Krypton, Xenon, Nitrogen, and Oxygen, *H.J.M. Hanley, R.D. McCarty, and W.M. Haynes,* Vol. 3, No. 4, pp. 979–1017 (1974) ... $9.00

(Continuation of Cumulative Listing of Reprints)

Reprints from Volume 4

60. JANAF Thermochemical Tables, 1975 Supplement, *M.W. Chase, J.L. Curnutt, H. Prophet, R.A. McDonald, and A.N. Syverud,* Vol. 4, No. 1, pp. 1–175 (1975) (superseded by Vol. 14, Suppl. 1) — **$17.00**

61. Diffusion in Copper and Copper Alloys, Part III. Diffusion in Systems Involving Elements of the Groups IA, IIA, IIIB, IVB, VB, VIB, and VIIB, *Daniel B. Butrymowicz, John R. Manning, and Michael E. Read,* Vol. 4, No. 1, pp. 177–249 (1975) — **$12.00**

62. Ideal Gas Thermodynamic Properties of Ethylene and Propylene, *Jing Chao and Bruno J. Zwolinski,* Vol. 4, No. 1, pp. 251–261 (1975) — **$6.00**

63. Atomic Transition Probabilities for Scandium and Titanium (A Critical Data Compilation of Allowed Lines), *W.L. Wiese and J.R. Fuhr,* Vol. 4, No. 2, pp. 263–352 (1975) (superseded by Vol. 17, Suppl. 3) — **$12.00**

64. Energy Levels of Iron, Fe I through Fe XXVI, *Joseph Reader and Jack Sugar,* Vol. 4, No. 2, pp. 353–440 (1975) (superseded by Vol. 14, Suppl. 2) — **$12.00**

65. Ideal Gas Thermodynamic Properties of Six Fluoroethanes, *S.S. Chen, A.S. Rodgers, J. Chao, R.C. Wilhoit, and B.J. Zwolinski,* Vol. 4, No. 2, pp. 441–456 (1975) — **$6.00**

66. Ideal Gas Thermodynamic Properties of the Eight Bromo- and Iodomethanes, *S.A. Kudchadker and A.P. Kudchadker,* Vol. 4, No. 2, pp. 457–470 (1975) — **$6.00**

67. Atomic Form Factors, Incoherent Scattering Functions, and Photon Scattering Cross Sections, *J.H. Hubbell, Wm.J. Veigele, E.A. Briggs, R.T. Brown, D.T. Cromer, and R.J. Howerton,* Vol. 4, No. 3, pp. 471–538 (1975) — **$11.00**

68. Binding Energies in Atomic Negative Ions, *H. Hotop and W.C. Lineberger,* Vol. 4, No. 3, pp. 539–576 (1975) — **$9.00**

69. A Survey of Electron Swarm Data, *J. Dutton,* Vol. 4, No. 3, pp. 577–856 (1975) — **$24.00**

70. Ideal Gas Thermodynamic Properties and Isomerization of *n*-Butane and Isobutane, *S.S. Chen, R.C. Wilhoit, and B.J. Zwolinski,* Vol. 4, No. 4, pp. 859–869 (1975) — **$6.00**

71. Molten Salts: Volume 4, Part 2, Chlorides and Mixtures, Electrical Conductance, Density, Viscosity, and Surface Tension Data, *G.J. Janz, R.P.T. Tomkins, C.B. Allen, J.R. Downey, Jr., G.L. Gardner, U. Krebs, and S.K. Singer,* Vol. 4, No. 4, pp. 871–1178 (1975) — **$26.00**

72. Property Index to Volumes 1–4 (1972–1975), Vol. 4, No. 4, pp. 1179–1192 (1975) (superseded by Reprint --- 194, Vol. 1–10) — **$6.00**

Reprints from Volume 5

73. Scaled Equation of State Parameters for Gases in the Critical Region, *J.M.H. Levelt Sengers, W.L. Greer, and J.V. Sengers,* Vol. 5, No. 1, pp. 1–51 (1976) — **$10.00**

74. Microwave Spectra of Molecules of Astrophysical Interest, IX. Acetaldehyde, *A. Bauder, F.J. Lovas, and D.R. Johnson,* Vol. 5, No. 1, pp. 53–77 (1976) — **$8.00**

75. Microwave Spectra of Molecules of Astrophysical Interest, X. Isocyanic Acid, *G. Winnewisser, W.H. Hocking, and M.C.L. Gerry,* Vol. 5, No. 1, pp. 79–101 (1976) — **$8.00**

76. Diffusion in Copper and Copper Alloys, Part IV. Diffusion in Systems Involving Elements of Group VIII, *Daniel B. Butrymowicz, John R. Manning, and Michael E. Read,* Vol. 5, No. 1, pp. 103–200 (1976) — **$13.00**

77. A Critical Review of the Stark Widths and Shifts of Spectral Lines from Non-Hydrogenic Atoms, *N. Konjevic and J.R. Roberts,* Vol. 5, No. 2, pp. 209–257 (1976) — **$10.00**

78. Experimental Stark Widths and Shifts for Non-Hydrogenic Spectral Lines of Ionized Atoms (A Critical Review and Tabulation of Selected Data), *N. Konjevic and W.L. Wiese,* Vol. 5, No. 2, pp. 259–308 (1976) — **$10.00**

79. Atlas of the Absorption Spectrum of Nitric Oxide (NO) between 1420 and 1250 Å, *E. Miescher and F. Alberti,* Vol. 5, No. 2, pp. 309–317 (1976) — **$6.00**

80. Ideal Gas Thermodynamic Properties of Propanone and 2-Butanone, *Jing Chao and Bruno J. Zwolinski,* Vol. 5, No. 2, pp. 319–328 (1976) — **$6.00**

81. Refractive Index of Alkali Halides and Its Wavelength and Temperature Derivatives, *H.H. Li,* Vol. 5, No. 2, pp. 329–528 (1976) — **$19.00**

82. Tables of Critically Evaluated Oscillator Strengths for the Lithium Isoelectronic Sequence, *G.A. Martin and W.L. Wiese,* Vol. 5, No. 3, pp. 537–570 (1976) — **$9.00**

83. Ideal Gas Thermodynamic Properties of Six Chlorofluoromethanes, *S.S. Chen, R. C. Wilhoit, and B.J. Zwolinski,* Vol. 5, No. 3, pp. 571–580 (1976) — **$6.00**

84. Survey of Superconductive Materials and Critical Evaluation of Selected Properties, *B.W. Roberts,* Vol. 5, No. 3, pp. 581–821 (1976) — **$25.00**

85. Nuclear Spins and Moments, *Gladys H. Fuller,* Vol. 5, No. 4, pp. 835–1092 (1976) — **$23.00**

86. Nuclear Moments and Moment Ratios as Determined by Mössbauer Spectroscopy, *J.G. Stevens and B.D. Dunlap,* Vol. 5, No. 4, pp. 1093–1121 (1976) — **$8.00**

87. Rate Coefficients for Ion-Molecule Reactions, I. Ions Containing C and H, *L. Wayne Sieck and Sharon G. Lias,* Vol. 5, No. 4, pp. 1123–1146 (1976) — **$8.00**

88. Microwave Spectra of Molecules of Astrophysical Interest, XI. Silicon Sulfide, *Eberhard Tiemann,* Vol. 5, No. 4, pp. 1147–1156 (1976) — **$6.00**

89. Property Index and Author Index to Volumes 1–5 (1972–1976), Vol. 5, No. 4, pp. 1161–1183 (superseded by Reprint --- 194, Vol. 1–10) — **$8.00**

Reprints from Volume 6

90. Diffusion in Copper and Copper Alloys, Part V. Diffusion in Systems Involving Elements of Group VA, *Daniel B. Butrymowicz, John R. Manning, and Michael E. Read,* Vol. 6, No. 1, pp. 1–50 (1977) — **$10.00**

91. The Calculated Thermodynamic Properties of Superfluid Helium-4, *James S. Brooks and Russell J. Donnelly,* Vol. 6, No. 1, pp. 51–104 (1977) — **$10.00**

92. Thermodynamic Properties of Normal and Deuterated Methanols, *S.S. Chen, R.C. Wilhoit, and B.J. Zwolinski,* Vol. 6, No. 1, pp. 105–112 (1977) — **$6.00**

93. The Spectrum of Molecular Nitrogen, *Alf Lofthus and Paul H. Krupenie,* Vol. 6, No. 1, pp. 113–307 (1977) — **$19.00**

94. Energy Levels of Chromium, Cr I through Cr XXIV, *Jack Sugar and Charles Corliss,* Vol. 6, No. 2, pp. 317–383 (1977) (superseded by Vol. 14, Suppl. 2) — **$11.00**

95. The Activity and Osmotic Coefficients of Aqueous Calcium Chloride at 298.15 K, *Bert R. Staples and Ralph L. Nuttall,* Vol. 6, No. 2, pp. 385–407 (1977) — **$8.00**

96. Molten Salts: Volume 4, Part 3, Bromides and Mixtures; Iodides and Mixtures–Electrical Conductance, Density, Viscosity, and Surface Tension Data, *G.J. Janz, R.P.T. Tomkins, C.B. Allen, J.R. Downey, Jr., and S.K. Singer,* Vol. 6, No. 2, pp. 409–596 (1977) — **$18.00**

97. The Viscosity and Thermal Conductivity Coefficients for Dense Gaseous and Liquid Methane, *H.J.M. Hanley, W.M. Haynes, and R.D. McCarty,* Vol. 6, No. 2, pp. 597–609 (1977) — **$6.00**

98. Phase Diagrams and Thermodynamic Properties of Ternary Copper-Silver Systems, *Y. Austin Chang, Daniel Goldberg, and Joachim P. Neumann,* Vol. 6, No. 3, pp. 621–673 (1977) — **$10.00**

(Continuation of Cumulative Listing of Reprints)

99. Crystal Data Space-Group Tables, *Alan D. Mighell, Helen M. Ondik, and Bettijoyce Breen Molino,* Vol. 6, No. 3, pp. 675–829 (1977) $16.00

100. Energy Levels of One-Electron Atoms, *Glen W. Erickson,* Vol. 6, No. 3, pp. 831–869 (1977) $9.00

101. Rate Constants for Reactions of ClO_x of Atmospheric Interest, *R.T. Watson,* Vol. 6, No. 3, pp. 871–917 (1977) $9.00

102. NMR Spectral Data: A Compilation of Aromatic Proton Chemical Shifts in Mono- and Di-Substituted Benzenes, *B.L. Shapiro and L.E. Mohrmann,* Vol. 6, No. 3, pp. 919–991 (1977) $11.00

103. Tables of Molecular Vibrational Frequencies. Consolidated Volume II. *T. Shimanouchi,* Vol. 6, No. 3, pp. 993–1102 (1977) (supersedes Nos. 5, 21, 25, 49) $13.00

104. Effects of Isotopic Composition, Temperature, Pressure, and Dissolved Gases on the Density of Liquid Water, *George S. Kell,* Vol. 6, No. 4, pp. 1109–1131 (1977) $8.00

105. Viscosity of Water Substance–New International Formulation and Its Background, *A. Nagashima,* Vol. 6, No. 4, pp. 1133–1166 (1977) $9.00

106. A Correlation of the Existing Viscosity and Thermal Conductivity Data of Gaseous and Liquid Ethane, *H.J.M. Hanley, K.E. Gubbins, and S. Murad,* Vol. 6, No. 4, pp. 1167–1180 (1977) $6.00

107. Elastic Properties of Zinc: A Compilation and a Review, *H.M. Ledbetter,* Vol. 6, No. 4, pp. 1181–1203 (1977) $8.00

108. Behavior of the AB-Type Compounds at High Pressures and High Temperatures, *Leo Merrill,* Vol. 6, No. 4, pp. 1205–1252 (1977) $9.00

109. Energy Levels of Manganese, Mn I through Mn XXV, *Charles Corliss and Jack Sugar,* Vol. 6, No. 4, pp. 1253–1329 (1977) (superseded by Vol. 14, Suppl. 2) $11.00

Reprints from Volume 7

110. Tables of Atomic Spectral Lines for the 10 000 Å to 40 000 Å Region, *Michael Outred,* Vol. 7, No. 1, pp. 1–262 (1978) $23.00

111. Evaluated Activity and Osmotic Coefficients for Aqueous Solutions: The Alkaline Earth Metal Halides, *R.N. Goldberg and R.L. Nuttall,* Vol. 7, No. 1, pp. 263–310 (1978) $9.00

112. Microwave Spectra of Molecules of Astrophysical Interest XII. Hydroxyl Radical, *Robert A. Beaudet and Robert L. Poynter,* Vol. 7, No. 1, pp. 311–362 (1978) $10.00

113. Ideal Gas Thermodynamic Properties of Methanoic and Ethanoic Acids, *Jing Chao and Bruno J. Zwolinski, Vol. 7, No. 1, pp. 363–377 (1978)* $6.00

114. Critical Review of Hydrolysis of Organic Compounds in Water Under Environmental Conditions, *W. Mabey and T. Mill,* Vol. 7, No. 2, pp. 383–415 (1978) $9.00

115. Ideal Gas Thermodynamic Properties of Phenol and Creosols, *S.A. Kudchadker, A.P. Kudchadker, R.C. Wilhoit, and B.J. Zwolinski,* Vol. 7, No. 2, pp. 417–423 (1978) $6.00

116. Densities of Liquid $CH_{4-a}X_a$ (X=Br,I) and $CH_{4-(a+b+c+d)}F_aCl_bBr_cI_d$ Halomethanes, *A.P. Kudchadker, S.A.Kudchadker, P.R. Patnaik, and P.P. Mishra,* Vol. 7, No. 2, pp. 425–439 (1978) $6.00

117. Microwave Spectra of Molecules of Astrophysical Interest XIII. Cyanoacetylene, *W.J. Lafferty and F.J. Lovas,* Vol. 7, No. 2, pp. 441–493 (1978) $10.00

118. Atomic Transition Probabilities for Vanadium, Chromium, and Manganese (A Critical Data Compilation of Allowed Lines), *S.M. Younger, J.R. Fuhr, G.A. Martin, and W.L. Wiese,* Vol. 7, No. 2, pp. 495–629 (1978) (superseded by Vol. 17, Suppl. 3) $15.00

119. Thermodynamic Properties of Ammonia, *Lester Haar and John S. Gallagher,* Vol. 7, No. 3, pp. 635–792 (1978) $16.00

120. JANAF Thermochemical Tables, 1978 Supplement, *M.W. Chase, Jr., J.L. Curnutt, R.A. McDonald, and A.N. Syverud,* Vol. 7, No. 3, pp. 793–940 (1978) (superseded by Vol. 14, Suppl. 1) $16.00

121. Viscosity of Liquid Water in the Range −8 °C to 150 °C, *Joseph Kestin, Mordechai Sokolov, and William A. Wakeham,* Vol. 7, No. 3, pp. 941–948 (1978) $6.00

122. The Molar Volume (Density) of Solid Oxygen in Equilibrium with Vapor, *H.M. Roder,* Vol. 7, No. 3, pp. 949–957 (1978) $6.00

123. Thermal Conductivity of Ten Selected Binary Alloy Systems, *C.Y. Ho, M.W. Ackerman, K.Y. Wu, S.G. Oh, and T.N. Havill,* Vol. 7, No. 3, pp. 959–1177 (1978) $20.00

124. Semi-Empirical Extrapolation and Estimation of Rate Constants for Abstraction of H from Methane by H, O, HO, and O_2, *Robert Shaw,* Vol. 7, No. 3, pp. 1179–1190 (1978) $6.00

125. Energy Levels of Vanadium, V I through V XXIII, *Jack Sugar and Charles Corliss,* Vol. 7, No. 3, pp. 1191–1262 (1978) (superseded by Vol. 14, Suppl. 2) $11.00

126. Recommended Atomic Electron Binding Energies, $1s$ to $6p_{3/2}$, for the Heavy Elements, $Z = 84$ to 103, *F.T. Porter and M.S. Freedman,* Vol. 7, No. 4, pp. 1267–1284 (1978) $8.00

127. Ideal Gas Thermodynamic Properties of $CH_{4-(a+b+c+d)}F_aCl_bBr_cI_d$ Halomethanes, *Shanti A. Kudchadker and Arvind P. Kudchadker,* Vol. 7, No. 4, pp. 1285–1307 (1978) $8.00

128. Critical Review of Vibrational Data and Force Field Constants for Polyethylene, *John Barnes and Bruno Fanconi,* Vol. 7, No. 4, pp. 1309–1321 (1978) $6.00

129. Tables of Molecular Vibrational Frequencies, Part 9, *Takehiko Shimanouchi, Hiroatsu Matsuura, Yoshiki Ogawa, and Issei Harada,* Vol. 7, No. 4, pp. 1323–1443 (1978) $14.00

130. Microwave Spectral Tables. II. Triatomic Molecules, *Frank J. Lovas,* Vol. 7, No. 4, pp. 1445–1750 (1978) $26.00

Reprints from Volume 8

131. Energy Levels of Titanium, Ti I through Ti XXII, *Charles Corliss and Jack Sugar,* Vol. 8, No. 1, pp. 1–62 (1979) (superseded by Vol. 14, Suppl. 2) $10.00

132. The Spectrum and Energy Levels of the Neutral Atom of Boron (B I), *G.A. Odintzova and A.R. Striganov,* Vol. 8, No. 1, pp. 63–67 (1979) $6.00

133. Relativistic Atomic Form Factors and Photon Coherent Scattering Cross Sections, *J.H. Hubbell and I. Øverbø,* Vol. 8, No. 1, pp. 69–105 (1979) $9.00

134. Microwave Spectra of Molecules of Astrophysical Interest. XIV. Vinyl Cyanide (Acrylonitrile), *M.C.L. Gerry, K. Yamada, and G. Winnewisser,* Vol. 8, No. 1, pp. 107–123 (1979) $8.00

135. Molten Salts: Volume 4, Part 4, Mixed Halide Melts. Electrical Conductance, Density, Viscosity, and Surface Tension Data, *G.J. Janz, R.P.T. Tomkins, and C.B. Allen,* Vol. 8, No. 1, pp. 125–302 (1979) $18.00

136. Atomic Radiative and Radiationless Yields for K and L Shells, *M.O. Krause,* Vol. 8, No. 2, pp. 307–327 (1979) $8.00

137. Natural Widths of Atomic K and L Levels, $K\alpha$ X-ray Lines and Several KLL Auger Lines, *M.O. Krause and J.H. Oliver,* Vol. 8, No. 2, pp. 329–338 (1979) $6.00

138. Electrical Resistivity of Alkali Elements, *T.C. Chi,* Vol. 8, No. 2, pp. 339–438 (1979) $13.00

139. Electrical Resistivity of Alkaline Earth Elements, *T.C. Chi,* Vol. 8, No. 2, pp. 439–497 (1979) $10.00

(Continuation of Cumulative Listing of Reprints)

140. Vapor Pressures and Boiling Points of Selected Halomethanes, *A.P. Kudchadker, S.A. Kudchadker, R.P. Shukla, and P.R. Patnaik,* Vol. 8, No. 2, pp. 499–517 (1979) $8.00

141. Ideal Gas Thermodynamic Properties of Selected Bromoethanes and Iodoethane, *S.A. Kudchadker and A.P. Kudchadker,* Vol. 8, No. 2, pp. 519–526 (1979) $6.00

142. Thermodynamic Properties of Normal and Deuterated Naphthalenes, *S.S. Chen, S.A. Kudchadker, and R.C. Wilhoit,* Vol. 8, No. 2, pp. 527–535 (1979) $6.00

143. Microwave Spectra of Molecules of Astrophysical Interest. XV. Propyne, *A. Bauer, D. Boucher, J. Burie, J. Demaison, and A. Dubrulle,* Vol. 8, No. 2, pp. 537–558 (1979) $8.00

144. A Correlation of the Viscosity and Thermal Conductivity Data of Gaseous and Liquid Propane, *P.M. Holland, H.J.M. Haniey, K.E. Gubbins, and J.M. Haile,* Vol. 8, No. 2, pp. 559–575 (1979) $8.00

145. Microwave Spectra of Molecules of Astrophysical Interest. XVI. Methyl Formate, *A. Bauder,* Vol. 8, No. 3, pp. 583–618 (1979) $9.00

146. Molecular Structures of Gas-Phase Polyatomic Molecules Determined by Spectroscopic Methods, *Marlin D. Harmony, Victor W. Laurie, Robert L. Kuczkowski, R.H. Schwendeman, D.A. Ramsay, Frank J. Lovas, Walter J. Lafferty, and Arthur G. Maki,* Vol. 8, No. 3, pp. 619–721 (1979) $13.00

147. Critically Evaluated Rate Constants for Gaseous Reactions of Several Electronically Excited Species, *Keith Schofield,* Vol. 8, No. 3, pp. 723–798 (1979) $11.00

148. A Review, Evaluation, and Correlation of the Phase Equilibria, Heat of Mixing, and Change in Volume on Mixing for Liquid Mixtures of Methane + Ethane, *M.J. Hiza, R.C. Miller, and A.J. Kidnay,* Vol. 8, No. 3, pp. 799–816 (1979) $8.00

149. Energy Levels of Aluminum, Al ɪ through Al xɪɪɪ, *W.C. Martin and Romuald Zalubas,* Vol. 8, No. 3, pp. 817–864 (1979) $9.00

150. Energy Levels of Calcium, Ca ɪ through Ca xx, *Jack Sugar and Charles Corliss,* Vol. 8, No. 3, pp. 865–916 (1979) (superseded by Vol. 14, Suppl. 2) $10.00

151. Evaluated Activity and Osmotic Coefficients for Aqueous Solutions: Iron Chloride and the Bi-univalent Compounds of Nickel and Cobalt, *R.N. Goldberg, R.L. Nutall, and B.R. Staples,* Vol. 8, No. 4, pp. 923–1003 (1979) $12.00

152. Evaluated Activity and Osmotic Coefficients for Aqueous Solutions: Bi-univalent Compounds of Lead, Copper, Manganese, and Uranium, *R.N. Goldberg,* Vol. 8, No. 4, pp. 1005–1050 (1979) $9.00

153. Microwave Spectra of Molecules of Astrophysical Interest. XVII. Dimethyl Ether, *F.J. Lovas, H. Lutz, and H. Dreizler,* Vol. 8, No. 4, pp. 1051–1107 (1979) $10.00

154. Energy Levels of Potassium, K ɪ through K xɪx, *Charles Corliss and Jack Sugar,* Vol. 8, No. 4, pp. 1109–1145 (1979) (superseded by Vol. 14, Suppl. 2) $9.00

155. Electrical Resistivity of Copper, Gold, Palladium, and Silver, *R.A. Matula,* Vol. 8, No. 4, pp. 1147–1298 (1979) $16.00

Reprints from Volume 9

156. Energy Levels of Magnesium, Mg ɪ through Mg xɪɪ, *W.C. Martin and Romuald Zalubas,* Vol 9, No. 1, pp. 1–58(1980) $12.00

157. Microwave Spectra of Molecules of Astrophysical Interest. XVIII. Formic Acid, *Edmond Willemot, Didier Dangoisse, Nicole Monnanteuil, and Jean Bellet,* Vol. 9, No. 1, pp. 59–160 (1980) $15.00

158. Refractive Index of Alkaline Earth Halides and Its Wavelength and Temperature Derivatives, *H.H. Li,* Vol. 9, No. 1, pp. 161–289 (1980). $17.00

159. Evaluated Kinetic and Photochemical Data for Atmospheric Chemistry, *D.L. Baulch, R.A. Cox, R.F. Hampson, Jr., J.A. Kerr, J. Troe, and R.L. Watson,* Vol. 9, No. 2, pp. 295–471 (1980) $20.00

160. Energy Levels of Scandium, Sc ɪ through Sc xxɪɪ, *Jack Sugar and Charles Corliss,* Vol. 9, No. 2, pp. 473–511 (1980) (superseded by Vol. 14, Suppl. 2) $11.00

161. A Compilation of Kinetic Parameters for the Thermal Degradation of n-Alkane Molecules, *D.L. Allara and Robert Shaw,* Vol. 9, No. 3, pp. 523–559 (1980) $11.00

162. Refractive Index of Silicon and Germanium and Its Wavelength and Temperature Derivatives, *H.H. Li,* Vol. 9, No. 3, pp. 561–658 (1980) $15.00

163. Microwave Spectra of Molecules of Astrophysical Interest XIX. *Methyl Cyanide, D. Boucher, J. Burie, A. Bauer, A. Dubrulle, and J. Demaison,* Vol. 9, No. 3, pp. 659–719 (1980). $12.00

164. A Review, Evaluation, and Correlation of the Phase Equilibria, Heat of Mixing, and Change in Volume on Mixing for Liquid Mixtures of Methane + Propane, *R.C. Miller, A.J. Kidnay, and M.J. Hiza,* Vol. 9, No. 3, pp. 721–734 (1980) $8.00

165. Saturation States of Heavy Water, *P.G. Hill and R.D. Chris MacMillan,* Vol. 9, No. 3, pp. 735–749 (1980) $8.00

166. The Solubility of Some Sparingly Soluble Lead Salts: An Evaluation of the Solubility in Water and Aqueous Electrolyte Solution, *H. Lawrence Clever and Francis J. Johnston,* Vol. 9, No. 3, pp. 751–784 (1980) $11.00

167. Molten Salts Data as Reference Standards for Density, Surface Tension, Viscosity, and Electrical Conductance: KNO_3 and NaCl, *George J. Janz,* Vol. 9, No. 4, pp. 791–829 (1980) $11.00

168. Molten Salts: Volume 5, Part 1, Additional Single and Multi-Component Salt Systems. Electrical Conductance, Density, Viscosity, and Surface Tension Data, *G.J. Janz and R.P. Tomkins,* Vol. 9, No. 4, pp. 831–1021 (1980) $21.00

169. Pair, Triplet, and Total Atomic Cross Sections (and Mass Attenuation Coefficients) for 1 MeV–100 GeV Photons in Elements Z=1 to 100, *J.H. Hubbell, H.A. Gimm, and I. Øverbø,* Vol. 9, No. 4, pp. 1023–1147 (1980) $16.00

170. Tables of Molecular Vibrational Frequencies, Part 10, *Takehiko Shimanouchi, Hiroatsu Matsuura, Yoshiki Ogawa, and Issei Harada,* Vol. 9, No. 4, pp. 1149–1254 (1980) $15.00

171. An Improved Representative Equation for the Dynamic Viscosity of Water Substance, *J.T.R. Watson, R.S. Basu, and J.V. Sengers,* Vol. 9, No. 4, pp. 1255–1290 (1980) $11.00

172. Static Dielectric Constant of Water and Steam, *M. Uematsu and E. U. Franck,* Vol. 9, No. 4, pp. 1291–1306 (1980) $8.00

173. Compilation and Evaluation of Solubility Data in the Mercury (I) Chloride-Water System, *Y. Marcus,* Vol. 9, No. 4, pp. 1307–1329 (1980) $10.00

Reprints from Volume 10

174. Evaluated Activity and Osmotic Coefficients for Aqueous Solutions: Bi-Univalent Compounds of Zinc, Cadmium, and Ethylene Bis(Trimethylammonium) Chloride and Iodide, *R. N. Goldberg,* Vol. 10, No. 1, pp. 1–55 (1981) $12.00

(Continuation of Cumulative Listing of Reprints)

175. Tables of the Dynamic and Kinematic Viscosity of Aqueous KCl Solutions in the Temperature Range 25–150 °C and the Pressure Range 0.1–35 MPa, *Joseph Kestin, H. Ezzat Khalifa, and Robert J. Correia*, Vol. 10, No. 1, pp. 57–70 (1981) $8.00

176. Tables of the Dynamic and Kinematic Viscosity of Aqueous NaCl Solutions in the Temperature Range 20–150 °C and the Pressure Range 0.1–35 MPa, *Joseph Kestin, H. Ezzat Khalifa, and Robert J. Correia, Vol. 10, No. 1, pp. 71–87 (1981)* $10.00

177. Heat Capacity and Other Thermodynamic Properties of Linear Macromolecules. I. Selenium, *Umesh Gaur, Hua-Cheng Shu, Aspy Mehta, and Bernhard Wunderlich*, Vol. 10, No. 1, pp. 89–117 (1981) $10.00

178. Heat Capacity and Other Thermodynamic Properties of Linear Macromolecules. II. Polyethylene, *Umesh Gaur and Bernhard Wunderlich*, Vol. 10, No. 1, pp. 119–152 (1981) $11.00

179. Energy Levels of Sodium, Na I through Na XI, *W. C. Martin and Romuald Zalubas*, Vol. 10, No. 1, pp. 153–195 (1981) $11.00

180. Energy Levels of Nickel, Ni I through Ni XXVIII, *Charles Corliss and Jack Sugar*, Vol. 10, No. 1, pp. 197–289 (1981) (superseded by Vol. 14, Suppl. 2) $14.00

181. Ion Product of Water Substance, 0–1000 °C, 1–10,000 bars New International Formulation and Its Background, *William L. Marshall and E. U. Franck*, Vol. 10, No. 2, pp. 295–304 (1981) $8.00

182. Atomic Transition Probabilities for Iron, Cobalt, and Nickel (A Critical Data Compilation of Allowed Lines), *J. R. Fuhr, G. A. Martin, W. L. Wiese, and S. M. Younger*, Vol. 10, No. 2, pp. 305–565 (1981) (superseded by Vol. 17, Suppl. 4) $25.00

183. Thermodynamic Tabulations for Selected Phases in the System CaO-Al$_2$O$_3$SiO$_2$-H$_2$O at 101.325 kPa (1 atm) between 273.15 and 1800 K, *John L. Haas, Jr., Gilpin R. Robinson, Jr., and Bruce S. Hemingway*, Vol. 10, No. 3, pp. 575–669 (1981) $14.00

184. Evaluated Activity and Osmotic Coefficients for Aqueous Solutions: Thirty-Six Uni-Bivalent Electrolytes, R. N. Goldberg, Vol. 10, No. 3, pp. 671–764 (1981) $14.00

185. Activity and Osmotic Coefficients of Aqueous Alkali Metal Nitrites, *Bert R. Staples*, Vol. 10, No. 3, pp. 765–778 (1981) $8.00

186. Activity and Osmotic Coefficients of Aqueous Sulfuric Acid at 298.15 K, *Bert R. Staples*, Vol. 10, No. 3, pp. 779–798 (1981) $10.00

187. Rate Constants for the Decay and Reactions of the Lowest Electronically Excited Singlet State of Molecular Oxygen in Solution, *Francis Wilkinson and James G. Brummer*, Vol. 10, No. 4, pp. 809–999 (1981) $20.00

188. Heat Capacity and Other Thermodynamic Properties of Linear Macromolecules. III. Polyoxides, *Umesh Gaur and Bernhard Wunderlich*, Vol. 10, No. 4, pp. 1001–1049 (1981) $11.00

189. Heat Capacity and Other Thermodynamic Properties of Linear Macromolecules. IV. Polypropylene, *Umesh Gaur and Bernhard Wunderlich*, Vol. 10, No. 4, pp. 1051–1064 (1981) $8.00

190. Tables of N$_2$O Absorption Lines for the Calibration of Tunable Infrared Lasers from 522 cm^{-1} to 657 cm^{-1} and from 1115 cm^{-1} to 1340 cm^{-1}, *W. B. Olson, A. G. Maki, and W. J. Lafferty*, Vol. 10, No. 4, pp. 1065–1084 (1981) $10.00

191. Microwave Spectra of Molecules of Astrophysical Interest. XX. Methane, *I. Ozier, M. C. L. Gerry, and A. G. Robiette*, Vol. 10, No. 4, pp. 1085–1095 (1981) $8.00

192. Energy Levels of Cobalt, Co I through Co XXVII, *Jack Sugar and Charles Corliss*, Vol. 10, No. 4, pp. 1097–1174 (1981) (superseded by Vol. 14, Suppl. 2) $13.00

193. A Critical Review of Henry's Law Constants for Chemicals of Environmental Interest, *Donald Mackay and Wan Ying Shiu*, Vol. 10, No. 4, pp. 1175–1199 (1981) $10.00

194. Property, Materials, and Author Indexes to the Journal of Physical and Chemical Reference Data, Vol. 1–10, pp. 1205–1225 (1972–1981) $10.00

Reprints from Volume 11

195. A Fundamental Equation of State for Heavy Water, *P. G. Hill, R. D. Chris MacMillan, and V. Lee*, Vol. 11, No. 1, pp. 1–14 (1982) $10.00

196. Volumetric Properties of Aqueous Sodium Chloride Solutions, *P. S. Z. Rogers and Kenneth S. Pitzer*, Vol. 11, No. 1, pp. 15–81 (1982) $18.00

197. Ideal Gas Thermodynamic Properties of CH$_3$, CD$_3$, CD$_4$, C$_2$D$_2$, C$_2$D$_4$, C$_2$D$_6$, C$_2$H$_6$, CH$_3$N$_2$CH$_3$, and CD$_3$N$_2$CD$_3$, *Krishna M. Pamidimukkala, David Rogers, and Gordon B. Skinner*, Vol. 11, No. 1, pp. 83–99 (1982) $12.00

198. Peak Absorption Coefficients of Microwave Absorption Lines of Carbonyl Sulphide, *Z. Kisiel and D. J. Millen*, Vol. 11, No. 1, pp. 99–116 (1982) $12.00

199. Vibrational Contributions to Molecular Dipole Polarizabilities, *David M. Bishop and Lap M. Cheung*, Vol. 11, No. 1, pp. 119–133 (1982) $10.00

200. Energy Levels of Iron, Fe I through Fe XXVI, *Charles Corliss and Jack Sugar*, Vol. 11, No. 1, pp. 135–241 (1982) (superseded by Vol. 14, Suppl. 2) $22.00

201. Microwave Spectra of Molecules of Astrophysical Interest. XXI. Ethanol(C$_2$H$_5$OH) and Propionitrile (C$_2$H$_5$CN), *Frank J. Lovas*, Vol. 11, No. 2, pp. 251–312 (1982) $16.00

202. Heat Capacity and Other Thermodynamic Properties of Linear Macromolecules, V. Polystyrene, *Umesh Gaur and Bernhard Wunderlich*, Vol. 11, No. 2, pp. 313–325 (1982) $10.00

203. Evaluated Kinetic and Photochemical Data for Atmospheric Chemistry: Supplement 1, CODATA Task Group on Chemical Kinetics, *D. L. Baulch, R. A. Cox, P. J. Crutzen, R. F. Hampson, Jr., J. A. Kerr (Chairman), J. Troe, and R. T. Watson*, Vol. 11, No. 2, pp. 327–496 (1982) $30.00

204. Molten Salts Data: Diffusion Coefficients in Single and Multi-Component Salt Systems, *G. J. Janz and N. P. Bansal*, Vol. 11, No. 3, pp. 505–693 (1982) $32.00

205. JANAF Thermochemical Tables, 1982 Supplement, *M. W. Chase, Jr., J. L. Curnutt, J. R. Downey, Jr., R. A. McDonald, A. N. Syverud, and E. A. Valenzuela*, Vol. 11, No. 3, pp. 695–940 (1982) (superseded by Vol. 14, Suppl. 1) $40.00

206. Critical Evaluation of Vapor-Liquid Equilibrium, Heat of Mixing, and Volume Change of Mixing Data. General Procedures, *Buford D. Smith, Ol Muthu, Ashok Dewan, and Matthew Gierlach*, Vol. 11, No. 3, pp. 941–951 (1982) $10.00

207. Rate Coefficients for Vibrational Energy Transfer Involving the Hydrogen Halides, *Stephen R. Leone*, Vol. 11, No. 3, pp. 953–996 (1982) $14.00

208. Behavior of the AB$_2$-Type Compounds at High Pressures and High Temperatures, *Leo Merrill*, Vol. 11, No. 4, pp. 1005–1064 (1982) $16.00

209. Heat Capacity and Other Thermodynamic Properties of Linear Macromolecules. VI. Acrylic Polymers, *Umesh Gaur, Suk-fai Lau, Brent B. Wunderlich, and Bernhard Wunderlich*, Vol. 11, No. 4, pp. 1065–1089 (1982) $12.00

210. Molecular Form Factors and Photon Coherent Scattering Cross Sections of Water, *L. R. M. Morin*, Vol. 11, No. 4, pp. 1091–1098 (1982) $10.00

211. Evaluation of Binary *PTxy* Vapor–Liquid Equilibrium Data for C_6 Hydrocarbons. Benzene + Cyclohexane, *Buford D. Smith, Ol Muthu, Ashok Dewan, and Matthew Gierlach*, Vol. 11, No. 4, pp. 1099–1126 (1982) $12.00

212. Evaluation of Binary Excess Enthalpy Data for C_6 Hydrocarbons. Benzene + Cyclohexane, *Buford D. Smith, Ol Muthu, Ashok Dewan, and Matthew Gierlach*, Vol. 11, No. 4, pp. 1127–1149 (1982) $12.00

213. Evaluation of Binary Excess Volume Data for C_6 Hydrocarbons. Benzene + Cyclohexane, *Buford D. Smith, Ol Muthu, Ashok Dewan, and Matthew Gierlach*, Vol. 11, No. 4, pp. 1151–1169 (1982) $12.00

Reprints from Volume 12

214. Thermodynamic Properties of Steam in the Critical Region, *J. M. H. Levelt Sengers, B. Kamgar-Parsi, F. W. Balfour, and J. V. Sengers*, Vol. 12, No. 1, pp. 1–28 (1983) $12.00

215. Heat Capacity and Other Thermodynamic Properties of Linear Macromolecules. VII. Other Carbon Backbone Polymers, *Umesh Gaur, Brent B. Wunderlich, and Bernhard Wunderlich*, Vol. 12, No. 1, pp. 29–63 (1983) $14.00

216. Heat Capacity and Other Thermodynamic Properties of Linear Macromolecules. VIII. Polyesters and Polyamides, *Umesh Gaur, Suk-fai Lau, Brent B. Wunderlich, and Bernhard Wunderlich*, Vol. 12, No. 1, pp. 65–89 (1983) $12.00

217. Heat Capacity and Other Thermodynamic Properties of Linear Macromolecules. IX. Final Group of Aromatic and Inorganic Polymers, *Umesh Gaur, Suk-fai Lau, and Bernhard Wunderlich*, Vol. 12, No. 1, pp. 91–108 (1983) $12.00

218. An Annotated Compilation and Appraisal of Electron Swarm Data in Electronegative Gases, *J. W. Gallagher, E. C. Beaty, J. Dutton, and L. C. Pitchford*, Vol. 12, No. 1, pp. 109–152 (1983) $14.00

219. The Solubility of Oxygen and Ozone in Liquids, *Rubin Battino, Timothy R. Rettich, and Toshihiro Tominaga*, Vol. 12, No. 2, pp. 163–178 (1983) $10.00

220. Recommended Values for the Thermal Expansivity of Silicon from 0 to 1000 K, *C. A. Swenson*, Vol. 12, No. 2, pp. 179–182 (1983) $10.00

221. Electrical Resistivity of Ten Selected Binary Alloy Systems, *C. Y. Ho, M. W. Ackerman, K. Y. Wu, T. N. Havill, R. H. Bogaard, R. A. Matula, S. G. Oh, and H. M. James*, Vol. 12, No. 2, pp. 183–322 (1983) $26.00

222. Energy Levels of Silicon, Si I through Si XIV , *W. C. Martin and Romuald Zalubas*, Vol. 12, No. 2, pp. 323–380 (1983) $16.00

223. Evaluation of Binary *PTxy* Vapor–Liquid Equilibrium Data for C_6 Hydrocarbons. Benzene + Hexane, *Buford D. Smith, Ol Muthu, and Ashok Dewan*, Vol. 12, No. 2, pp. 381–387 (1983) $10.00

224. Evaluation of Binary Excess Enthalpy Data for C_6 Hydrocarbons. Benzene + Hexane, *Buford D. Smith, Ol Muthu, and Ashok Dewan*, Vol. 12, No. 2, pp. 389–393 (1983) $10.00

225. Evaluation of Binary Excess Volume Data for C_6 Hydrocarbons. Benzene + Hexane, *Buford D. Smith, Ol Muthu, and Ashok Dewan*, Vol. 12, No. 2, pp. 395–401 (1983) $10.00

226. Atlas of the High-Temperature Water Vapor Spectrum in the 3000 to 4000 cm^{-1} Region, *A. S. Pine, M. J. Coulombe, C. Camy-Peyret, and J-M. Flaud*, Vol. 12, No. 3, pp. 413–465 (1983) $16.00

227. Small-Angle Rayleigh Scattering of Photons at High Energies: Tabulations of Relativistic HFS Modified Atomic Form Factors, *D. Schaupp, M. Schumacher, F. Smend, P. Rullhusen, and J. H. Hubbell*, Vol. 12, No. 3, pp. 467–512 (1983) $14.00

228. Thermodynamic Properties of D_2O in the Critical Region, *B. Kamgar-Parsi, J. M. H. Levelt Sengers, and J. V. Sengers*, Vol. 12, No. 3, pp. 513–529 (1983) $12.00

229. Chemical Kinetic Data Sheets for High-Temperature Chemical Reactions, *N. Cohen and K. R. Westberg*, Vol. 12, No. 3, pp. 531–590 (1983) $16.00

230. Molten Salts: Volume 5, Part 2. Additional Single and Multi-Component Salt Systems. Electrical Conductance, Density, Viscosity and Surface Tension Data, *G. J. Janz and R. P. T. Tomkins*, Vol. 12, No. 3, pp. 591–815 (1983) $38.00

231. International Tables of the Surface Tension of Water, *N. B. Vargaftik, B. N. Volkov, and L. D. Voljak*, Vol. 12, No. 3, pp. 817–820 (1983) $10.00

232. Evaluated Theoretical Cross Section Data for Charge Exchange of Multiply Charged Ions with Atoms. I. Hydrogen Atom-Fully Stripped Ion Systems, *R. K. Janev, B. H. Bransden, and J. W. Gallagher*, Vol. 12, No. 4, pp. 829–872 (1983) $14.00

233. Evaluated Theoretical Cross Section Data for Charge Exchange of Multiply Charged Ions with Atoms. II. Hydrogen Atom-Partially Stripped Ion Systems, *J. W. Gallagher, B. H. Bransden, and R. K. Janev*, Vol. 12, No. 4, pp. 873–890 (1983) $12.00

234. Recommended Data on the Electron Impact Ionization of Light Atoms and Ions, *K. L. Bell, H. B. Gilbody, J. G. Hughes, A. E. Kingston, and F. J. Smith*, Vol. 12, No. 4, pp. 891–916 (1983) $12.00

235. A Correlation of the Viscosity and Thermal Conductivity Data of Gaseous and Liquid Ethylene, *P. M. Holland, B. E. Eaton, and H. J. M. Hanley*, Vol. 12, No. 4, pp. 917–932 (1983) $10.00

236. Transport Properties of Liquid and Gaseous D_2O over a Wide Range of Temperature and Pressure, *N. Matsunaga and A. Nagashima*, Vol. 12, No. 4, pp. 933–966 (1983) $14.00

237. Thermochemical Data for Gaseous Monoxides, *J. B. Pedley and E. M. Marshall*, Vol. 12, No. 4, pp. 967–1031 (1983) $18.00

238. Vapor Pressure of Coal Chemicals, *J. Chao, C. T. Lin, and T. H. Chung*, Vol. 12, No. 4, pp. 1033–1063 (1983) $12.00

Reprints from Volume 13

239. Thermodynamic Properties of Aqueous Sodium Chloride Solutions, *Kenneth S. Pitzer, J. Christopher Peiper, and R. H. Busey*, Vol. 13, No. 1, pp. 1–102 (1984) $22.00

240. Refractive Index of ZnS, ZnSe, and ZnTe and Its Wavelength and Temperature Derivatives, *H. H. Li*, Vol. 13, No. 1, pp. 103–150 (1984) $14.00

241. High Temperature Vaporization Behavior of Oxides. I. Alkali Metal Binary Oxides, *R. H. Lamoreaux and D. L. Hildenbrand*, Vol. 13, No. 1, pp. 151–173 (1984) $12.00

242. Thermophysical Properties of Fluid H_2O, *J. Kestin, J. V. Sengers, B. Kamgar-Parsi, and J. M. H. Levelt Sengers*, Vol. 13, No. 1, pp. 175–183 (1984) $10.00

243. Representative Equations for the Viscosity of Water Substance, *J. V. Sengers and B. Kamgar-Parsi*, Vol. 13, No. 1, pp. 185–205 (1984) $12.00

244. Atlas of the Schumann–Runge Absorption Bands of O_2 in the Wavelength Region 175–205 nm, *K. Yoshino, D. E. Freeman, and W. H. Parkinson*, Vol. 13, No. 1, pp. 207–227 (1984) $12.00

245. Equilibrium and Transport Properties of the Noble Gases and Their Mixtures at Low Density, *J. Kestin, K. Knierim, E. A. Mason, B. Najafi, S. T. Ro, and M. Waldman*, Vol. 13, No. 1, pp. 229–303 (1984) $18.00

246. Evaluation of Kinetic and Mechanistic Data For Modeling of Photochemical Smog, *Roger Atkinson and Alan C. Lloyd*, Vol. 13, No. 2, pp. 315–444 (1984) $26.00

247. Rate Data for Inelastic Collision Processes in the Diatomic Halogen Molecules, *J. I. Steinfeld*, Vol. 13, No. 2, pp. 445–553 (1984) $22.00

248. Water Solubilities of Polynuclear Aromatic and Heteroaromatic Compounds, *Robert S. Pearlman, Samuel H. Yalkowsky, and Sujit Banerjee*, Vol. 13, No. 2, pp. 555–562 (1984) $10.00

249. The Solubility of Nitrogen and Air in Liquids, *Rubin Battino, Timothy R. Rettich, and Toshihiro Tominaga*, Vol. 13, No. 2, pp. 563–600 (1984) $14.00

250. Thermophysical Properties of Fluid D_2O, *J. Kestin, J. V. Sengers, B. Kamgar-Parsi, and J. M. H. Levelt Sengers*, Vol. 13, No. 2, pp. 601–609 (1984) $10.00

251. Experimental Stark Widths and Shifts for Spectral Lines of Neutral Atoms (A Critical Review of Selected Data for the Period 1976 to 1982), *N. Konjević, M. S. Dimitrijević, and W. L. Wiese*, Vol. 13, No. 3, pp. 619–647 (1984) $12.00

252. Experimental Stark Widths and Shifts for Spectral Lines of Positive Ions (A Critical Review and Tabulation of Selected Data for the Period 1976 to 1982), *N. Konjević, M. S. Dimitrijević, and W. L. Wiese*, Vol. 13, No. 3, pp. 649–686 (1984) $14.00

253. A Review of Deuterium Triple-Point Temperatures, *L. A. Schwalbe and E. R. Grilly*, Vol. 13, No. 3, pp. 687–693 (1984) $10.00

254. Evaluated Gas Phase Basicities and Proton Affinities of Molecules; Heats of Formation of Protonated Molecules, *Sharon G. Lias, Joel F. Liebman, and Rhoda D. Levin*, Vol. 13, No. 3, pp. 695–808 (1984) $24.00

255. Isotopic Abundances and Atomic Weights of the Elements, *Paul De Bièvre, Marc Gallet, Norman E. Holden, and I. Lynus Barnes*, Vol. 13, No. 3, pp. 809–891 (1984) $20.00

256. Representative Equations for the Thermal Conductivity of Water Substance, *J. V. Sengers, J. T. R. Watson, R. S. Basu, B. Kamgar-Parsi, and R. C. Hendricks*, Vol. 13, No. 3, pp. 893–933 (1984) $14.00

257. Ground-State Vibrational Energy Levels of Polyatomic Transient Molecules, *Marilyn E. Jacox*, Vol. 13, No. 4, pp. 945–1068 (1984) $24.00

258. Electrical Resistivity of Selected Elements, *P. D. Desai, T. K. Chu, H. M. James, and C. Y. Ho*, Vol. 13, No. 4, pp. 1069–1096 (1984) $12.00

259. Electrical Resistivity of Vanadium and Zirconium, *P. D. Desai, H. M. James, and C. Y. Ho*, Vol. 13, No. 4, pp. 1097–1130 (1984) $14.00

260. Electrical Resistivity of Aluminum and Manganese, *P. D. Desai, H. M. James, and C. Y. Ho*, Vol. 13, No. 4, pp. 1131–1172 (1984) $14.00

261. Standard Chemical Thermodynamic Properties of Alkane Isomer Groups, *Robert A. Alberty and Catherine A. Gehrig*, Vol. 13, No. 4, pp. 1173–1197 (1984) $12.00

262. Evaluated Theoretical Cross-Section Data for Charge Exchange of Multiply Charged Ions with Atoms. III. Nonhydrogenic Target Atoms, *R. K. Janev and J. W. Gallagher*, Vol. 13, No. 4, pp. 1199–1249 (1984) $16.00

263. Heat Capacity of Reference Materials: Cu and W, *G. K. White and S. J. Collocott*, Vol. 13, No. 4, pp. 1251–1257 (1984) $10.00

264. Evaluated Kinetic and Photochemical Data for Atmospheric Chemistry: Supplement II. CODATA Task Group on Gas Phase Chemical Kinetics, *D. L. Baulch, R. A. Cox, R. F. Hampson, Jr., J. A. Kerr (Chairman), J. Troe, and R. T. Watson*, Vol. 13, No. 4, pp. 1259–1380 (1984) $24.00

Reprints from Volume 14

265. Thermodynamic Properties of Key Organic Oxygen Compounds in the Carbon Range C_1 to C_4. Part 1. Properties of Condensed Phases, *Randolph C. Wilhoit, Jing Chao, and Kenneth R. Hall*, Vol. 14, No. 1, pp. 1–175 (1985) $30.00

266. Standard Chemical Thermodynamic Properties of Alkylbenzene Isomer Groups, *Robert A. Alberty*, Vol. 14, No. 1, pp. 177–192 (1985) $10.00

267. Assessment of Critical Parameter Values for H_2O and D_2O, *J. M. H. Levelt Sengers, J. Straub, K. Watanabe, and P. G. Hill*, Vol. 14, No. 1, pp. 193–207 (1985) $10.00

268. The Viscosity of Nitrogen, Oxygen, and Their Binary Mixtures in the Limit of Zero Density, *Wendy A. Cole and William A. Wakeham*, Vol. 14, No. 1, pp. 209–226 (1985) $12.00

269. The Thermal Conductivity of Fluid Air, *K. Stephan and A. Laesecke*, Vol. 14, No. 1, pp. 227–234 (1985) $10.00

270. The Electronic Spectrum and Energy Levels of the Deuterium Molecule, *Robert S. Freund, James A. Schiavone, and H. M. Crosswhite*, Vol. 14, No. 1, pp. 235–383 (1985) $28.00

271. Microwave Spectra of Molecules of Astrophysical Interest. XXII. Sulfur Dioxide (SO_2), *F. J. Lovas*, Vol. 14, No. 2, pp. 395–488 (1985) $20.00

272. Evaluation of the Thermodynamic Functions for Aqueous Sodium Chloride from Equilibrium and Calorimetric Measurements below 154 °C, *E. Colin W. Clarke and David N. Glew*, Vol. 14, No. 2, pp. 489–610 (1985) $24.00

273. The Mark–Houwink–Sakurada Equation for the Viscosity of Linear Polyethylene, *Herman L. Wagner*, Vol. 14, No. 2, pp. 611–617 (1985) $10.00

274. The Solubility of Mercury and Some Sparingly Soluble Mercury Salts in Water and Aqueous Electrolyte Solutions, *H. Lawrence Clever, Susan A. Johnson, and M. Elizabeth Derrick*, Vol. 14, No. 3, pp. 631–680 (1985) $16.00

275. A Review and Evaluation of the Phase Equilibria, Liquid-Phase Heats of Mixing and Excess Volumes, and Gas-Phase *PVT* Measurements for Nitrogen +Methane, *A. J. Kidnay, R. C. Miller, E. D. Sloan, and M. J. Hiza*, Vol. 14, No. 3, pp. 681–694 (1985) $10.00

276. The Homogeneous Nucleation Limits of Liquids, *C. T. Avedisian*, Vol. 14, No. 3, pp. 695–729 (1985) $14.00

277. Binding Energies in Atomic Negative Ions: II, *H. Hotop and W. C. Lineberger*, Vol. 14, No. 3, pp. 731–750 (1985) $12.00

278. Energy Levels of Phosphorus, P I through P XV, *W. C. Martin, Romuald Zalubas, and Arlene Musgrove*, Vol. 14, No. 3, pp. 751–802 (1985) $16.00

279. Standard Chemical Thermodynamic Properties of Alkene Isomer Groups, *Robert A. Alberty and Catherine A. Gehrig*, Vol. 14, No. 3, pp. 803–820 (1985) $12.00

280. Standard Chemical Thermodynamic Properties of Alkylnaphthalene Isomer Groups, *Robert A. Alberty and Theodore M. Bloomstein*, Vol. 14, No. 3, pp. 821–837 (1985) $12.00

281. Carbon Monoxide Thermophysical Properties from 68 to 1000 K at Pressures to 100 MPa, *Robert D. Goodwin*, Vol. 14, No. 4, pp. 849–932 (1985) $20.00

(Continuation of Cumulative Listing of Reprints)

282. Refractive Index of Water and Its Dependence on Wavelength, Temperature, and Density, *I. Thormählen, J. Straub, and U. Grigull,* Vol. 14, No. 4, pp. 933–945 (1985) — $10.00

283. Viscosity and Thermal Conductivity of Dry Air in the Gaseous Phase, *K. Kadoya, N. Matsunaga, and A. Nagashima,* Vol. 14, No. 4, pp. 947–970 (1985) — $12.00

284. Charge Transfer of Hydrogen Ions and Atoms in Metal Vapors, *T. J. Morgan, R. E. Olson, A. S. Schlachter, and J. W. Gallagher,* Vol. 14, No. 4, pp. 971–1040 (1985) — $18.00

285. Reactivity of HO_2/O_2^- Radicals in Aqueous Solution, *Benon H. J. Bielski, Diane E. Cabelli, Ravindra L. Arudi, and Alberta B. Ross,* Vol. 14, No. 4, pp. 1041–1100 (1985) — $16.00

286. The Mark–Houwink–Sakurada Equation for the Viscosity of Atactic Polystyrene, *Herman L. Wagner,* Vol. 14, No. 4, pp. 1101–1106 (1985) — $10.00

287. Standard Chemical Thermodynamic Properties of Alkylcyclopentane Isomer Groups, Alkylcyclohexane Isomer Groups, and Combined Isomer Groups, *Robert A. Alberty and Young S. Ha,* Vol. 14, No. 4, pp. 1107–1132 (1985) — $12.00

Reprints from Volume 15

288. Triplet–Triplet Absorption Spectra of Organic Molecules in Condensed Phases, *Ian Carmichael and Gordon L. Hug,* Vol. 15, No. 1, pp. 1–250 (1986) — $40.00

289. Recommended Rest Frequencies for Observed Interstellar Molecular Microwave Transitions—1985 Revision, *F. J. Lovas,* Vol. 15, No. 1, pp. 251–303 (1986) — $16.00

290. New International Formulations for the Thermodynamic Properties of Light and Heavy Water, *J. Kestin and J. V. Sengers,* Vol. 15, No. 1, pp. 305–320 (1986) — $10.00

291. Forbidden Lines in ns^2np^k Ground Configurations and $nsnp$ Excited Configurations of Beryllium through Molybdenum Atoms and Ions, *Victor Kaufman and Jack Sugar,* Vol. 15, No. 1, pp. 321–426 (1986) — $22.00

292. Thermodynamic Properties of Twenty-One Monocyclic Hydrocarbons, *O. V. Dorofeeva, L. V. Gurvich, and V. S. Jorish,* Vol. 15, No. 2, pp. 437–464 (1986) — $12.00

293. Evaluated Kinetic Data for High-Temperature Reactions. Volume 5. Part 1. Homogeneous Gas Phase Reactions of the Hydroxyl Radical with Alkanes, *D. L. Baulch, M. Bowers, D. G. Malcolm, and R. T. Tuckerman,* Vol. 15, No. 2, pp. 465–592 (1986) — $24.00

294. Thermodynamic Properties of Ethylene from the Freezing Line to 450 K at Pressures to 260 MPa, *Majid Jahangiri, Richard T Jacobsen, Richard B. Stewart, and Robert D. McCarty,* Vol. 15, No. 2, pp. 593–734 (1986) — $26.00

295. Thermodynamic Properties of Nitrogen from the Freezing Line to 2000 K at Pressures to 1000 MPa, *Richard T Jacobsen, Richard B. Stewart, and Majid Jahangiri,* Vol. 15, No. 2, pp. 735–909 (1986) — $30.00

296. A Critical Review of Aqueous Solubilities, Vapor Pressures, Henry's Law Constants, and Octanol–Water Partition Coefficients of the Polychlorinated Biphenyls, *Wan Ying Shiu and Donald Mackay,* Vol. 15, No. 2, pp. 911–929 (1986) — $12.00

297. Computer Methods Applied to the Assessment of Thermochemical Data. Part I. The Establishment of a Computerized Thermochemical Data Base Illustrated by Data for $TiCl_4(g)$, $TiCl_4(l)$, $TiCl_3(cr)$, and $TiCl_2(cr)$, *S. P. Kirby, E. M. Marshall, and J. B. Pedley,* Vol. 15, No. 3, pp. 943–965 (1986) — $12.00

298. Thermodynamic Properties of Iron and Silicon, *P. D. Desai,* Vol. 15, No. 3, pp. 967–983 (1986) — $12.00

299. Cross Sections for Collisions of Electrons and Photons with Nitrogen Molecules, *Y. Itikawa, M. Hayashi, A. Ichimura, K. Onda, K. Sakimoto, K. Takayanagi, M. Nakamura, H. Nishimura, and T. Takayanagi,* Vol. 15, No. 3, pp. 985–1010 (1986) — $12.00

300. Thermochemical Data on Gas-Phase Ion-Molecule Association and Clustering Reactions, *R. G. Keesee and A. W. Castleman, Jr.,* Vol. 15, No. 3, pp. 1011–1071 (1986) — $16.00

301. Standard Reference Data for the Thermal Conductivity of Liquids, *C. A. Nieto de Castro, S. F. Y. Li, A. Nagashima, R. D. Trengove, and W. A. Wakeham,* Vol. 15, No. 3, pp. 1073–1086 (1986) — $10.00

302. Chemical Kinetic Data Base for Combustion Chemistry. Part I. Methane and Related Compounds, *W. Tsang and R. F. Hampson,* Vol. 15, No. 3, pp. 1087–1279 (1986) — $34.00

303. Improved International Formulations for the Viscosity and Thermal Conductivity of Water Substance, *J. V. Sengers and J. T. R. Watson,* Vol. 15, No. 4, pp. 1291–1314 (1986) — $12.00

304. The Viscosity and Thermal Conductivity of Normal Hydrogen in the Limit of Zero Density, *M. J. Assael, S. Mixafendi, and W. A. Wakeham,* Vol. 15, No. 4, pp. 1315–1322 (1986) — $10.00

305. The Viscosity and Thermal Conductivity Coefficients of Gaseous and Liquid Argon, *B. A. Younglove and H. J. M. Hanley,* Vol. 15, No. 4, pp. 1323–1337 (1986) — $10.00

306. Standard Chemical Thermodynamic Properties of Alkyne Isomer Groups, *Robert A. Alberty and Ellen Burmenko,* Vol. 15, No. 4, pp. 1339–1349 (1986) — $10.00

307. Recent Progress in Deuterium Triple-Point Measurements, *L. A. Schwalbe,* Vol. 15, No. 4, pp. 1351–1356 (1986) — $10.00

308. Rate Constants for Reactions of Radiation-Produced Transients in Aqueous Solutions of Actinides, *S. Gordon, J. C. Sullivan, and Alberta B. Ross,* Vol. 15, No. 4, pp. 1357–1367 (1986) — $10.00

309. Thermodynamic Properties of Key Organic Oxygen Compounds in the Carbon Range C_1 to C_4. Part 2. Ideal Gas Properties, *Jing Chao, Kenneth R. Hall, Kenneth N. Marsh, and Randolph C. Wilhoit,* Vol. 15, No. 4, pp. 1369–1436 (1986) — $18.00

Reprints from Volume 16

310. Thermochemical Data on Gas Phase Compounds of Sulfur, Fluorine, Oxygen, and Hydrogen Related to Pyrolysis and Oxidation of Sulfur Hexafluoride, *John T. Herron,* Vol. 16, No. 1, pp. 1–6 (1987) — $10.00

311. The Thermochemical Measurements on Rubidium Compounds: A Comparison of Measured Values with Those Predicted from the NBS Tables of Chemical and Thermodynamic Properties, *V. B. Parker, W. H. Evans, and R. L. Nuttall,* Vol. 16, No. 1, pp. 7–59 (1987) — $16.00

312. Standard Thermodynamic Functions of Gaseous Polyatomic Ions at 100–1000 K, *Aharon Loewenschuss and Yitzhak Marcus,* Vol. 16, No. 1, pp. 61–89 (1987) — $12.00

313. Thermodynamic Properties of Manganese and Molybdenum, *P. D. Desai,* Vol. 16, No. 1, pp. 91–108 (1987) — $12.00

314. Thermodynamic Properties of Selected Binary Aluminum Alloy Systems, *P. D. Desai,* Vol. 16, No. 1, pp. 109–124 (1987) — $10.00

315. ^{13}C Chemical Shielding in Solids, *T. M. Duncan,* Vol. 16, No. 1, pp. 125–151 (1987) — $12.00

316. The Mark–Houwink–Sakurada Relation for Poly(Methyl Methacrylate), *Herman L. Wagner,* Vol. 16, No. 2, pp. 165–173 (1987) — $10.00

(Continuation of Cumulative Listing of Reprints)

317. The Viscosity of Carbon Dioxide, Methane, and Sulfur Hexafluoride in the Limit of Zero Density, *R. D. Trengove and W. A. Wakeham,* Vol. 16, No. 2, pp. 175–187 (1987) $10.00

318. The Viscosity of Normal Deuterium in the Limit of Zero Density, *M. J. Assael, S. Mixafendi, and W. A. Wakeham,* Vol. 16, No. 2, pp. 189–192 (1987) $10.00

319. Standard Chemical Thermodynamic Properties of Alkanethiol Isomer Groups, *Robert A. Alberty, Ellen Burmenko, Tae H. Kang, and Michael B. Chung,* Vol. 16, No. 2, pp. 193–208 (1987) $10.00

320. Evaluation of Binary Excess Volume Data for the Methanol+Hydrocarbon Systems, *R. Srivastava and B. D. Smith,* Vol. 16, No. 2, pp. 209–218 (1987) $10.00

321. Evaluation of Binary Excess Enthalpy Data for the Methanol+Hydrocarbon Systems, *R. Srivastava and B. D. Smith,* Vol. 16, No. 2, pp. 219–237 (1987) $12.00

322. Extinction Coefficients of Triplet–Triplet Absorption Spectra of Organic Molecules in Condensed Phases: A Least-Squares Analysis, *Ian Carmichael, W. P. Helman, and G. L. Hug,* Vol. 16, No. 2, pp. 239–260 (1987) $12.00

323. Evaluated Chemical Kinetic Data for the Reactions of Atomic Oxygen $O(^3P)$ with Unsaturated Hydrocarbons, *R. J. Cvetanović,* Vol. 16, No. 2, pp. 261–326 (1987) $18.00

324. Spectral Data for Molybdenum Ions, Mo VI–Mo XLII, *Toshizo Shirai, Yohta Nakai, Kunio Ozawa, Keishi Ishii, Jack Sugar, and Kazuo Mori,* Vol. 16, No. 2, pp. 327–377 (1987) $16.00

325. Standard Chemical Thermodynamic Properties of Alkanol Isomer Groups, *Robert A. Alberty, Michael B. Chung, and Theresa M. Flood,* Vol. 16, No. 3, pp. 391–417 (1987) $12.00

326. High-Temperature Vaporization Behavior of Oxides II. Oxides of Be, Mg, Ca, Sr, Ba, B, Al, Ga, In, Tl, Si, Ge, Sn, Pb, Zn, Cd, and Hg, *R. H. Lamoreaux, D. L. Hildenbrand, and L. Brewer,* Vol. 16, No. 3, pp. 419–443 (1987) $12.00

327. Equilibrium and Transport Properties of Eleven Polyatomic Gases at Low Density, *A. Boushehri, J. Bzowski, J. Kestin, and E. A. Mason,* Vol. 16, No. 3, pp. 445–466 (1987) $12.00

328. The Thermochemistry of Inorganic Solids IV. Enthalpies of Formation of Compounds of the Formula MX_aY_b, *Mohamed W. M. Hisham and Sidney W. Benson,* Vol. 16, No. 3, pp. 467–470 (1987) $10.00

329. Chemical Kinetic Data Base for Combustion Chemistry. Part 2. Methanol, *Wing Tsang,* Vol. 16, No. 3, pp. 471–508 (1987) $14.00

330. Phase Diagrams and Thermodynamic Properties of the 70 Binary Alkali Halide Systems Having Common Ions, *James Sangster and Arthur D. Pelton,* Vol. 16, No. 3, pp. 509–561 (1987) $16.00

331. Thermophysical Properties of Fluids. II. Methane, Ethane, Propane, Isobutane, and Normal Butane, *B. A. Younglove and J. F. Ely,* Vol. 16, No. 4, pp. 577–798 (1987) $36.00

332. Methanol Thermodynamic Properties from 176 to 673 K at Pressures to 700 Bar, *Robert D. Goodwin,* Vol. 16, No. 4, pp. 799–892 (1987) $20.00

333. International Equations for the Saturation Properties of Ordinary Water Substance, *A. Saul and W. Wagner,* Vol. 16, No. 4, pp. 893–901 (1987) $10.00

334. Rate Data for Inelastic Collision Processes in the Diatomic Halogen Molecules. 1986 Supplement, *J. I. Steinfeld,* Vol. 16, No. 4, pp. 903–910 (1987) $10.00

335. Critical Survey of Data on the Spectroscopy and Kinetics of Ozone in the Mesosphere and Thermosphere, *Jeffrey I. Steinfeld, Steven M. Adler-Golden, and Jean W. Gallagher,* Vol. 16, No. 4, pp. 911–951 (1987) $14.00

336. Critical Compilation of Surface Structures Determined by Low-Energy Electron Diffraction Crystallography, *Philip R. Watson,* Vol. 16, No. 4, pp. 953–992 (1987) $14.00

337. Viscosity and Thermal Conductivity of Nitrogen for a Wide Range of Fluid States, *K. Stephan, R. Krauss, and A. Laesecke,* Vol. 16, No. 4, pp. 993–1023 (1987) $12.00

Reprints from Volume 17

338. Pressure and Density Series Equations of State for Steam as Derived from the Haar–Gallagher–Kell Formulation, *R. A. Dobbins, K. Mohammed, and D. A. Sullivan,* Vol. 17, No. 1, pp. 1–8 (1988) $10.00

339. Absolute Cross Sections for Molecular Photoabsorption, Partial Photoionization, and Ionic Photofragmentation Processes, *J. W. Gallagher, C. E. Brion, J. A. R. Samson, and P. W. Langhoff,* Vol. 17, No. 1, pp. 9–153 (1988) $28.00

340. Energy Levels of Molybdenum, Mo I through Mo XLII, *Jack Sugar and Arlene Musgrove,* Vol. 17, No. 1, pp. 155–239 (1988) $20.00

341. Standard Chemical Thermodynamic Properties of Polycyclic Aromatic Hydrocarbons and Their Isomer Groups I. Benzene Series, *Robert A. Alberty and Andrea K. Reif,* Vol. 17, No. 1, pp. 241–253 (1988) $10.00

342. Electronic Energy Levels of Small Polyatomic Transient Molecules, *Marilyn E. Jacox,* Vol. 17, No. 2, pp. 269–511 (1988) $40.00

343. Critical Review of Rate Constants for Reactions of Hydrated Electrons, Hydrogen Atoms and Hydroxyl Radicals ($\cdot OH/\cdot O^-$) in Aqueous Solution, *George V. Buxton, Clive L. Greenstock, W. Phillip Helman, and Alberta B. Ross,* Vol. 17, No. 2, pp. 513–886 (1988) $56.00

344. Chemical Kinetic Data Base for Combustion Chemistry. Part 3. Propane, *Wing Tsang,* Vol. 17, No. 2, pp. 887–951 (1988) $18.00

345. Evaluated Chemical Kinetic Data for the Reactions of Atomic Oxygen $O(^3P)$ with Saturated Organic Compounds in the Gas Phase, *John T. Herron,* Vol. 17, No. 3, pp. 967–1026 (1988) $16.00

346. Rate Constants for Reactions of Inorganic Radicals in Aqueous Solution, *P. Neta, Robert E. Huie, and Alberta B. Ross,* Vol. 17, No. 3, pp. 1027–1284 (1988) $42.00

347. Recommended Data on the Electron Impact Ionization of Atoms and Ions: Fluorine to Nickel, *M. A. Lennon, K. L. Bell, H. B. Gilbody, J. G. Hughes, A. E. Kingston, M. J. Murray, and F. J. Smith,* Vol. 17, No. 3, pp. 1285–1363 (1988) $18.00

348. Evaluated Chemical Kinetic Data for the Reactions of Atomic Oxygen $O(^3P)$ with Sulfur Containing Compounds, *D. L. Singleton and R. J. Cvetanović,* Vol. 17, No. 4, pp. 1377–1437 (1988) $16.00

349. New International Skeleton Tables for the Thermodynamic Properties of Ordinary Water Substance, *H. Sato, M. Uematsu, K. Watanabe, A. Saul, and W. Wagner,* Vol. 17, No. 4, pp. 1439–1540 (1988) $22.00

350. Benzene Thermophysical Properties from 279 to 900 K at Pressures to 1000 Bar, *Robert D. Goodwin,* Vol. 17, No. 4, pp. 1541–1636 (1988) $20.00

351. Estimation of the Thermodynamic Properties of Hydrocarbons at 298.15, K, *Eugene S. Domalski, and Elizabeth D. Hearing,* Vol. 17, No. 4, pp. 1637–1678 (1988) $14.00

352. Wavelengths and Energy Level Classifications of Scandium Spectra for All stages of Ionization, *V. Kaufman and J. Sugar,* Vol. 17, No. 4, pp. 1679–1789 (1988) $22.00

353. Atomic Weights of the Elements 1987, *J. R. De Laeter,* Vol. 17, No. 4, pp. 1791–1793 (1988) $10.00

(Continuation of Cumulative Listing of Reprints)

354. The 1986 CODATA Recommended Values of the Fundamental Physical Constants, *E. Richard Cohen and Barry N. Taylor,* Vol. 17, No. 4, pp. 1795–1803 (1988) $10.00

Reprints from Volume 18

355. Standard Electrode Potentials and Temperature Coefficients in Water at 298.15 K, *Steven G. Bratsch,* Vol. 18, No. 1, pp. 1–21 (1989) $12.00

356. Cross Sections for Collisions of Electrons and Photons with Oxygen Molecules, *Y. Itikawa, A. Ichimura, K. Onda, K. Sakimoto, K. Takayanagi, Y. Hatano, M. Hayashi, H. Nishimura, and S. Tsurubuchi,* Vol. 18, No. 1, pp. 23–42 (1989) $12.00

357. Thermal Conductivity of Refrigerants in a Wide Range of Temperature and Pressure, *R. Krauss and K. Stephan,* Vol. 18, No. 1, pp. 43–76 (1989) $14.00

358. Standard Chemical Thermodynamic Properties of Polycyclic Aromatic Hydrocarbons and Their Isomer Groups. II. Pyrene Series, Naphthopyrene Series, and Coronene Series, *Robert A. Alberty, Michael B. Chung, and Andrea K. Reif,* Vol. 18, No. 1, pp. 77–109 (1989) $14.00

359. Cross Sections for K-Shell X-ray Production by Hydrogen and Helium Ions in Elements from Beryllium to Uranium, *G. Lapicki,* Vol 18. No. 1, pp. 111–218 (1989) $22.00

360. Rate Constants for the Quenching of Excited States of Metal Complexes in Fluid Solution, *Morton Z. Hoffman, Fabrizio Bolletta, Luca Moggi, and Gordon L. Hug,* Vol. 18. No. 1, pp. 219–543 (1989). $50.00

361. The Thermal Conductivity of Nitrogen and Carbon Monoxide in the Limit of Zero Density, *J. Millat and W. A. Wakeham,* Vol. 18, No. 2, pp. 565–581 (1989) $12.00

362. Thermophysical Properties of Methane, *Daniel G. Friend, James F. Ely, and Hepburn Ingham,* Vol. 18, No. 2, pp. 583–638 (1989) $16.00

363. Thermodynamic Properties of Argon from the Triple Point to 1200 K with Pressures to 1000 MPa, *Richard B. Stewart and Richard T. Jacobsen,* Vol. 18, No. 2, pp. 639–798 (1989) $28.00

364. Thermodynamic Properties of Dioxygen Difluoride (O_2F_2) and Dioxygen Fluoride (O_2F), *John L. Lyman,* Vol. 18, No. 2, pp. 799–807 (1989) $10.00

365. Thermodynamic and Transport Properties of Carbohydrates and their Monophosphates: The Pentoses and Hexoses, *Robert N. Goldberg and Yadu B. Tewari,* Vol. 18, No. 2, pp. 809–880 (1989) $18.00

366. Evaluated Kinetic and Photochemical Data for Atmospheric Chemistry: Supplement III. *IUPAC Subcommittee on Gas Kinetic Data Evaluation for Atmospheric Chemistry, R. Atkinson, D. L. Baulch, R. A. Cox, R. F. Hampson, Jr., J. A. Kerr (Chairman), and J. Troe,* Vol. 18, No. 2, pp. 881–1097 (1989) $36.00

367. Octanol–Water Partition Coefficients of Simple Organic Compounds, *James Sangster,* Vol. 18, No. 3, pp. 1111–1229 (1989) $24.00

368. Evaluation of Data on Solubility of Simple Apolar Gases in Light and Heavy Water at High Temperature, *Roberto Fernández Prini and Rosa Crovetto,* Vol. 18, No. 3, pp. 1231–1243 (1989) $10.00

369. Microwave Spectral Tables. III. Hydrocarbons, CH to $C_{10}H_{10}$, *F. J. Lovas an R. D. Suenram,* Vol. 18, No. 3, pp. 1245–1524 (1989) $44.00

370. A Fundamental Equation for Water Covering the Range from the Melting Line to 1273 K at Pressures up to 25 000 MPa, *A. Saul and W. Wagner,* Vol. 18, No. 4, pp. 1537–1564 (1989) $12.00

371. Toluene Thermophysical Properties from 178 to 800 K at Pressures to 1000 Bar, *Robert D. Goodwin,* Vol. 18, No. 4, pp. 1565–1636 (1989) $18.00

372. Reduction Potentials of One-Electron Couples Involving Free Radicals in Aqueous Solution, *Peter Wardman,* Vol. 18, No. 4, pp. 1637–1755 (1989) $24.00

373. Photoemission Cross Sections for Atomic Transitions in the Extreme Ultraviolet due to Electron Collisions with Atoms and Molecules, *P. J. M. van der Burgt, W. B. Westerveld, and J. S. Risley,* Vol. 18, No. 4, pp. 1757–1805 (1989) $16.00

Reprints from Volume 19

374. Chemical Kinetic Data Base for Combustion Chemistry. Part 4. Isobutane, *Wing Tsang,* Vol. 19, No. 1, pp. 1–68 (1990) $18.00

375. Thermodynamic Functions and Properties of MgO at High Compression and High Temperature, *Orson L. Anderson and Keshan Zou,* Vol. 19, No. 1, pp. 69–83 (1990) $10.00

376. Critical Compilation of Surface Structures Determined by Ion Scattering Methods, *Philip R. Watson,* Vol. 19, No. 1, pp. 85–111 (1990) $12.00

377. Benzene: A Further Liquid Thermal Conductivity Standard, *M. J. Assael, M. L. V. Ramires, C. A. Nieto de Castro, and W. A. Wakeham,* Vol. 19, No. 1, pp. 113–117 (1990) $10.00

378. Energy Levels of Atomic Aluminum with Hyperfine Structure, *Edward S. Chang,* Vol. 19, No. 1, pp. 119–125 (1990) $24.00

379. Spectral Data and Grotrian Diagrams for Highly Ionized Iron, Fe VIII–XXVI, *Toshizo Shirai, Yoshio Funatake, Kazuo Mori, Jack Sugar, Wolfgang L. Wiese, and Yohta Nakai,* Vol. 19, No. 1, pp. 127–275 (1990) $28.00

380. Updated Excitation and Ionization Cross Section for Electron Impact on Atomic Oxygen, *Russ R. Laher and Forrest R. Gilmore,* Vol. 19, No. 1, pp. 277–305 (1990) $12.00

381. Standard Chemical Thermodynamic Properties of Isomer Groups of Monochloroalkanes, *Robert A. Alberty and Michael B. Chung,* Vol. 19, No. 2, pp. 321–348 (1990) $12.00

382. Standard Chemical Thermodynamic Properties of Polycyclic Aromatic Hydrocarbons and Their Isomer Groups. III. Napthocoronene Series, Ovalene Series, and First Members of Some Higher Series, *Robert A. Alberty, Michael B. Chung, and Andrea K. Reif,* Vol. 19, No. 2, pp. 349–370 (1990) $12.00

383. The Dielectric Constant of Water and Debye-Huckel Limiting Law Slopes, *Donald G. Archer and Peiming Wang,* Vol. 19, No. 2, pp. 371–411 (1990) $14.00

384. Rate Constants for Reactions of Peroxyl Radicals in Fluid Solutions, *P. Neta, Robert E. Huie, and Alberta B. Ross,* Vol. 19, No. 2, pp. 413–513 (1990) $22.00

385. Energy Levels of Copper, Cu I through Cu XXIX, *Jack Sugar and Arlene Musgrove,* Vol. 19, No. 3, pp. 527–616 (1990) $20.00

386. Cross Sections and Related Data for Electron Collisions with Hydrogen Molecules and Molecular Ions, *H. Tawara, Y. Itikava, H. Nishimura, and M. Yoshino,* Vol. 19, No. 3, pp. 617–636 (1990) $12.00

387. Cross Sections for Collisions of Electrons and Photons with Atomic Oxygen, *Y. Itikawa and A. Ichimura,* Vol. 19, No. 3, pp. 637–651 (1990). $10.00

388. Cross Sections and Swarm Coefficients for H^+, H_2^+, H_3^+, H, H_2, and H^- in H_2 for Energies from 0.1 eV to 10 keV, *A. V. Phelps,* Vol. 19, No. 3, pp. 653–675 (1990) $12.00

389. Refractive Index of Water and Steam as Function of Wavelength, Temperature and Density, *P. Schiebener, J. Straub, J. M. H. Levelt Sengers, and J. S. Gallagher,* Vol. 19, No. 3, pp. 677–717 (1990) $14.00

(Continuation of Cumulative Listing of Reprints)

390. Heat Capacities of Organic Compounds in the Liquid State I. C_1 to C_{18} 1-Alkanols, *Milan Zábranský, Vlastimil Růžička, Jr., and Vladimír Majer*, Vol. 19, No. 3, pp. 719–762 (1990) $14.00

391. The Transport Properties of Carbon Dioxide, *V. Vesovic, W. A. Wakeham, G. A. Olchowy, J. V. Sengers, J. T. R. Watson, and J. Millat*, Vol. 19, No. 3, pp. 763–808 (1990) $14.00

392. Energy Levels of Sulfur, S I Through S XVI, *W. C. Martin, Romuald Zalubas, and Arlene Musgrove*, Vol. 19, No. 4, pp. 821–880 (1990) $16.00

393. Heat Capacities and Entropies of Organic Compounds in the Condensed Phase, Volume II, *Eugene S. Domalski and Elizabeth D. Hearing*, Vol. 19, No. 4, pp. 881–1047 (1990) $30.00

394. The Thermodynamics of the Krebs Cycle and Related Compounds, *Stanley L. Miller and David Smith-Magowan*, Vol. 19, No. 4, pp. 1049–1073 (1990) $12.00

395. Transport Properties of Fluid Oxygen, *A. Laesecke, R. Krauss, K. Stephen, and W. Wagner*, Vol. 19, No. 5, pp. 1089–1122 (1990) $14.00

396. Thermal Conductivity of Nine Polyatomic Gases at Low Density, *F. J. Uribe, E. A. Mason, and J. Kestin*, Vol. 19, No. 5, pp. 1123–1136 (1990) $10.00

397. The Thermal Conductivity of Methane and Tetrafluoromethane in the Limit of Zero Density, *M. J. Assael, J. Millat, V. Vesovic, and W. A. Wakeham*, Vol. 19, No. 5, pp. 1137–1147 (1990) $10.00

398. Coupled Phase Diagram-Thermodynamic Analysis of the 24 Binary Systems, A_2CO_3–AX and A_2SO_4–AX where A = Li, Na, K and X = Cl, F, NO_3, OH, *Yves Dessureault, James Sangster, and Arthur D. Pelton*, Vol. 19, No. 5, pp. 1149–1178 (1990) $12.00

399. Equilibrium and Transport Properties of Gas Mixtures at Low Density: Eleven Polyatomic Gases and Five Noble Gases, *J. Bzowski, J. Kestin, E. A. Mason, and F. J. Uribe*, Vol. 19, No. 5, pp. 1179–1232 (1990) $16.00

400. A Unified Fundamental Equation for the Thermodynamic Properties of H_2O, *Philip G. Hill*, Vol. 19, No. 5, pp. 1233–1274 (1990) $14.00

401. The Viscosity and Thermal Conductivity of Pure Monatomic Gases From Their Normal Boiling Point up to 5000 K in the Limit of Zero Density and at 0.101325 MPa, *E. Bich, J. Millat, and E. Vogel*, Vol. 19, No. 6, pp. 1289–1305 (1990) $12.00

402. Experimental Stark Widths and Shifts for Spectral Lines of Neutral and Ionized Atoms (A Critical Review of Selected Data for the Period 1983 through 1988), *N. Konjević and W. L. Wiese*, Vol. 19, No. 6, pp. 1307–1385 (1990) $18.00

403. Vibrational and Electronic Energy Levels of Polyatomic Transient Molecules. Supplement 1, *Marilyn E. Jacox*, Vol. 19, No. 6, pp. 1387–1546 (1990) $28.00

404. Thermodynamic and Thermophysical Properties of Organic Nitrogen Compounds. Part I. Methanamine, Ethanamine, 1- and 2-Propanamine, Benzenamine, 2-, 3-, and 4-Methylbenzenamine, *J. Chao, N. A. M. Gadella, B. E. Gammon, K. N. Marsh, A. S. Rodgers, G. R. Somayajulu, and R. C. Wilhoit*, Vol. 19, No. 6, pp. 1547–1615 (1990) $18.00

Reprints from Volume 20

405. Spectral Data and Grotrian Diagrams for Highly Ionized Copper, Cu X-Cu XXIX, *T. Shirai, T. Nakagaki, Y. Nakai, J. Sugar, K. Ishii, and K. Mori*, Vol. 20, No. 1, pp. 1–81 (1991) $20.00

406. Wavelengths and Energy Level Classifications of Magnesium Spectra for All Stages of Ionization (Mg I thru Mg XII), *V. Kaufman and W. C. Martin*, Vol. 20, No. 1, pp. 83–152 (1991) $18.00

407. Spectroscopy and Structure of the Alkali Hydride Diatomic Molecules and Their Ions, *W. C. Stwalley, W. T. Zemke, and S. C. Yang*, Vol. 20, No. 1, pp. 153–187 (1991) $14.00

408. Critical Evaluation of Liquid Crystal Transition Temperature I: 4,4'-Alkyl/Alkoxyphenylbenzoates, *T. T. Blair, M. E. Neubert, M. Tsai, and C. Tsai* Vol. 20, No. 1, pp. 189–204 (1991) $10.00

409. Chemical Kinetic Data Base for Combustion Chemistry. Part V. Propene, *Wing Tsang*, Vol. 20, No. 2, pp. 221–273 (1991) $16.00

410. Thermophysical Properties of Ethane, *Daniel G. Friend, Hepburn Ingham, and James F. Ely*, Vol. 20, No. 2, pp. 275–347 (1991) $18.00

411. Heat Capacity and Other Thermodynamic Properties of Linear Macromolecules. X. Update of the ATHAS 1980 Data Bank, *Manika Varma-Nair and Bernhard Wunderlich*, Vol. 20, No. 2, pp. 349–404 (1991) $16.00

412. Heat Capacities of Organic Compounds in Liquid State. II. C_1 to C_{18} n-Alkanes, *Vlastimil Růžička, Jr., Milan Zábranský, and Vladimír Majer*, Vol. 20, No. 2, pp. 405–444 (1991) $14.00

413. Kinetics and Mechanisms of the Gas-Phase Reactions of the NO_3 Radical with Organic Compounds, *Roger Atkinson*, Vol. 20, No. 3, pp. 459–507 (1991) $16.00

414. Thermodynamic Properties of the NaBr + H_2O System, *Donald G. Archer*, Vol. 20, No. 3, pp. 509–555 (1991) $14.00

415. Cross Sections and Swarm Coefficients for Nitrogen Ions and Neutrals in N_2 and Argon Ions and Neutrals in Ar for Energies from 0.1 eV to 10 keV, *A. V. Phelps*, Vol. 20, No. 3, pp. 557–573 (1991) $12.00

416. Evaluation of Solubility Data of the System CO_2–H_2O from 273 K to the Critical Point of Water, *Rosa Crovetto*, Vol. 20, No. 3, pp. 575–589 (1991) $10.00

417. Chemical Kinetic Data Base for Propellant Combustion. I. Reactions Involving NO, NO_2, HNO, HNO_2, HCN, and N_2O, *Wing Tsang and John T. Herron*, Vol. 20, No. 4, pp. 609–663 (1991) $16.00

418. Ab-Initio Calculations and Ideal Gas Thermodynamic Functions of Cyclopentadiene and Cyclopentadiene Derivatives, *Miriam Karni, Izhack Oref, and Alexander Burcat*, Vol. 20, No. 4, pp. 665–683 (1991) $12.00

419. Improved Fits for the Vibrational and Rotational Constants of Many States of Nitrogen and Oxygen, *Russ R. Laher and Forrest R. Gilmore*, Vol. 20, No. 4, pp. 685–712 (1991) $12.00

420. Solubilities of Solids and Liquids of Low Volatility in Supercritical Carbon Dioxide, *K. D. Bartle, A. A. Clifford, S. A. Jafar, and G. F. Shilstone*, Vol. 20, No. 4, pp. 713–756 (1991) $14.00

421. Wavelengths and Energy Level Classifications for the Spectra of Aluminum (Al I through Al XIII), *Victor Kaufman and W. C. Martin*, Vol. 20, No. 5, pp. 775–858 (1991) $20.00

422. Energy Levels of Krypton, Kr I through Kr XXXVI, *Jack Sugar and Arlene Musgrove*, Vol. 20, No. 5, pp. 859–915 (1991) $16.00

423. Thermodynamic Properties of Oxygen from the Triple Point to 300 K with Pressures to 80 MPa, *Richard B. Stewart, Richard T. Jacobsen, and W. Wagner*, Vol. 20, No. 5, pp. 917–1021 (1991) $22.00

424. Sixteen Thousand Evaluated Experimental Thermodynamic Property Data for Water and Steam, *H. Sato, K. Watanabe, J. M. H. Levelt Sengers, J. S. Gallagher, P. G. Hill, J. Straub, and W. Wagner*, Vol. 20, No. 5, pp. 1023–1044 (1991) $12.00

(Continuation of Cumulative Listing of Reprints)

425. A New Equation of State and Tables of Thermodynamic Properties for Methane Covering the Range from the Melting Line to 625 K at Pressures up to 1000 MPa, *U. Setzmann and W. Wagner*, Vol. 20, No. 6, pp. 1061–1155 (1991) $20.00

426. Thermodynamic Properties of the Aqueous Sulfuric Acid System to 350 K, *Frank J. Zeleznik*, Vol. 20, No. 6, pp. 1157–1200 (1991) $14.00

427. The Solubility of Carbon Dioxide in Water at Low Pressure, *John J. Carroll, John D. Slupsky, and Alan E. Mather*, Vol. 20, No. 6, pp. 1201–1209 (1991) $10.00

428. Chemical Kinetic Data Sheets for High-Temperature Reactions. Part II, *N. Cohen and K. R. Westberg*, Vol. 20, No. 6, pp. 1211–1311 (1991) $22.00

429. Atomic Weights of the Elements 1989, *J. R. De Laeter and K. G. Heumann*, Vol. 20, No. 6, pp. 1313–1325 (1991) $10.00

430. Isotopic Compositions of the Elements 1989, *J. R. De Laeter, K. G. Heumann, and K. J. R. Rosman*, Vol. 20, No. 6, pp. 1327–1337 (1991) $10.00

431. Property, Materials, and Author Indexes to the Journal of Physical and Chemical Reference Data, Vols. 11–20, pp. 1343–1394 (1982–1991) $16.00

Reprints from Volume 21

432. Thermodynamic Properties of the NaCl+H₂O System. I. Thermodynamic Properties of NaCl(cr), *Donald G. Archer*, Vol. 21, No. 1, pp. 1–21 (1992) $12.00

433. Spectral Data and Grotrian Diagrams for Highly Ionized Cobalt, Co VIII through Co XXVII, *Toshizo Shirai, Alberto Mengoni, Yohta Nakai, Jack Sugar, Wolfgang L. Wiese, Kazuo Mori, and H. Sakai*, Vol. 21, No. 1, pp. 23–121 (1992) $22.00

434. Critical Compilation of Surface Structures Determined by Surface Extended X-Ray Absorption Fine Structure (SEXAFS) and Surface Extended Electron Energy Loss Spectroscopy (SEELFS), *Philip R. Watson*, Vol. 21, No. 1, pp. 123–156 (1992) $14.00

435. Laser-Induced Kerr Constants for Pure Liquids, *N. J. Harrison and B. R. Jennings*, Vol. 21, No. 1, pp. 157–163 (1992) $10.00

436. Recommended Rest Frequencies for Observed Interstellar Molecular Microwave Transitions—1991 Revision, *Frank J. Lovas*, Vol. 21, No. 2, pp. 181–272 (1992) $20.00

437. Spectral Data and Grotrian Diagrams for Highly Ionized Vanadium, V VI through V XXIII, *Toshizo Shirai, Toshiaki Nakagaki, Jack Sugar, and Wolfgang L. Wiese*, Vol. 21, No. 2, pp. 273–390 (1992) $24.00

438. Evaluated Kinetic Data for Combustion Modelling, *D. L. Baulch, C. J. Cobos, R. A. Cox, C. Esser, P. Frank, Th. Just, J. A. Kerr, M. J. Pilling, J. Troe, R. W. Walker, and J. Warnatz*, Vol. 21, No. 3, pp. 411–734 (entire issue) (1992) $70.00

439. Chemical Kinetic Data Base for Propellant Combustion. II. Reactions Involving CN, NCO, and HNCO, *Wing Tsang*, Vol. 21, No. 4, pp. 753–791 (1992) $14.00

440. Thermodynamic Properties of the NaCl+H₂O System. II. Thermodynamic Properties of NaCl(aq), NaCl·2H₂O(cr), and Phase Equilibria, *Donald G. Archer*, Vol. 21, No. 4, pp. 793–829 (1992) $14.00

441. Vibrational Bands of HₓNᵧOᵤ Molecules, *F. Mélen and M. Herman*, Vol. 21, No. 4, pp. 831–881 (1992) $16.00

442. Collisions of H⁺, H₂⁺, H₃⁺, ArH⁺, H⁻, H, and H₂ with Ar and of Ar⁺ and ArH⁺ with H₂ for Energies from 0.1 eV to 10 keV, *A. V. Phelps*, Vol. 21, No. 4, pp. 883–897 (1992). $10.00

443. A Critical Compilation of Atomic Transition Probabilities for Singly Ionized Argon, *V. Vujnović and W. L. Wiese*, Vol. 21, No. 5, pp. 919–939 (1992) $12.00

444. The Solubility of Some Sparingly Soluble Salts of Zinc and Cadmium in Water and in Aqueous Electrolyte Solutions, *H. Lawrence Clever, M. Elizabeth Derrick, and Susan A. Johnson*, Vol. 21, No. 5, pp. 941–1004 (1992) $16.00

445. Franck–Condon Factors, r-Centroids, Electronic Transition Moments, and Einstein Coefficients for Many Nitrogen and Oxygen Band Systems, *Forrest R. Gilmore, Russ R. Laher, and Patrick J. Espy*, Vol. 21, No. 5, pp. 1005–1107 (1992) $22.00

446. Evaluated Kinetic and Photochemical Data for Atmospheric Chemistry. Supplement IV, *IUPAC Subcommittee on Gas Kinetic Data Evaluation for Atmospheric Chemistry, R. Atkinson, D. L. Baulch, R. A. Cox, R. F. Hampson, Jr., J. A. Kerr (Chairman), and J. Troe*, Vol. 21, No. 6, pp. 1125–1568 (entire issue) (1992) $70.00

Reprints from Volume 22

447. Thermodynamic Properties of the Group IIA Elements, *C. B. Alcock, M. W. Chase, and V. Itkin*, Vol. 22, No. 1, pp. 1–85 (1993) $20.00

448. Spectroscopy and Structure of the Lithium Hydride Diatomic Molecules and Ions, *William C. Stwalley and Warren T. Zemke*, Vol. 22, No. 1, pp. 87–112 (1993) $12.00

449. Quantum Yields for the Photosensitized Formation of the Lowest Electronically Excited Singlet State of Molecular Oxygen in Solution, *Francis Wilkinson, W. Phillip Helman, and Alberta B. Ross*, Vol. 22, No. 1, pp. 113–262 (1993) $28.00

450. Wavelengths and Energy Level Classifications for the Spectra of Sulfur (S I through S XVI), *Victor Kaufman and W. C. Martin*, Vol. 22, No. 2, pp. 279–375 (1993) $22.00

451. Thermodynamic Properties of Alkenes (Mono-Olefins Larger than C₄), *W. V. Steele and R. D. Chirico*, Vol. 22, No. 2, pp. 377–430 (1993) $16.00

452. The Thermodynamic Behavior of the CO₂–H₂O System from 400 to 1000 K, up to 100 MPa and 30% Mole Fraction of CO₂, *J. S. Gallagher, R. Crovetto, and J. M. H. Levelt Sengers*, Vol. 22, No. 2, pp. 431–513 (1993) $20.00

453. Thermodynamics of Enzyme-Catalyzed Reactions: Part 1. Oxidoreductases, *Robert N. Goldberg, Yadu B. Tewari, Donna Bell, Karl Fazio, and Ellen Anderson*, Vol. 22, No. 2, pp. 515–582 (1993) $18.00

454. Estimation of the Heat Capacities of Organic Liquids as a Function of Temperature Using Group Additivity. I. Hydrocarbon Compounds, *Vlastimil Růžička, Jr., and Eugene S. Domalski*, Vol. 22, No. 3, pp. 597–618 (1993) $12.00

455. Estimation of the Heat Capacities of Organic Liquids as a Function of Temperature Using Group Additivity. II. Compounds of Carbon, Hydrogen, Halogens, Nitrogen, Oxygen, and Sulfur, *Vlastimil Růžička, Jr., and Eugene S. Domalski*, Vol. 22, No. 3, pp. 619–657 (1993) $14.00

456. Thermodynamic and Thermophysical Properties of Organic Nitrogen Compounds. Part II. 1- and 2-Butanamine, 2-Methyl-1-Propanamine, 2-Methyl-2-Propanamine, Pyrrole, 1-, 2-, and 3-Methylpyrrole, Pyridine, 2-, 3-, and 4-Methylpyridine, Pyrrolidine, Piperidine, Indole, Quinoline, Isoquinoline, Acridine, Carbazole, Phenanthridine, 1- and 2-Naphthalenamine, and 9-Methylcarbazole, *A. Das, M. Frenkel, N. A. M. Gadalla, S. Kudchadker, K. N. Marsh, A. S. Rodgers, and R. C. Wilhoit*, Vol. 22, No. 3, pp. 659–782 (1993) $24.00

(Continuation of Cumulative Listing of Reprints)

457. International Equations for the Saturation Properties of Ordinary Water Substance. Revised According to the International Temperature Scale of 1990. Addendum to J. Phys. Chem. Ref. Data 16, 893 (1987), *W. Wagner and A. Pruss*, Vol. 22, No. 3, pp. 783–787 (1993) — $10.00

458. Estimation of the Thermodynamic Properties of C–H–N–O–S–Halogen Compounds at 298.15 K, *Eugene S. Domalski and Elizabeth D. Hearing*, Vol. 22, No. 4, pp. 805–1159 (entire issue) (1993) — $80.00

459. A Compilation of Energy Levels and Wavelengths for the Spectrum of Singly-Ionized Oxygen (O II), *W. C. Martin, Victor Kaufman, and Arlene Musgrove*, Vol. 22, No. 5, pp. 1179–1212 (1993) — $14.00

460. Energy Levels of Germanium, Ge I through Ge XXXII, *Jack Sugar and Arlene Musgrove*, Vol. 22, No. 5, pp. 1213–1278 (1993) — $18.00

461. Spectral Data and Grotrian Diagrams for Highly Ionized Chromium, Cr V through Cr XXIV, *Toshizo Shirai, Yohta Nakai, Toshiaki Nakagaki, Jack Sugar, and Wolfgang L. Wiese*, Vol. 22, No. 5, pp. 1279–1423 (1993) — $28.00

462. Thermodynamic Properties of Synthetic Sapphire (α-Al_2O_3), Standard Reference Material 720 and the Effect of Temperature-Scale Differences on Thermodynamic Properties, *Donald G. Archer*, Vol. 22, No. 6, pp. 1441–1453 (1993) — $10.00

463. Thermodynamic Properties of Gaseous Silicon Monotelluride and the Bond Dissociation Enthalpy D_m°(SiTe) at $T \rightarrow 0$, *P. A. G. O'Hare*, Vol. 22, No. 6, pp. 1455–1458 (1993) — $10.00

464. The Disilicides of Tungsten, Molybdenum, Tantalum, Titanium, Cobalt, and Nickel, and Platinum Monosilicide: A Survey of Their Thermodynamic Properties, *M. S. Chandrasekharaiah, J. L. Margrave, and P. A. G. O'Hare*, Vol. 22, No. 6, pp. 1459–1468 (1993) — $10.00

465. Evaluated Bimolecular Ion-Molecule Gas Phase Kinetics of Positive Ions for Use in Modeling Planetary Atmospheres, Cometary Comae, and Interstellar Clouds, *Vincent G. Anicich*, Vol. 22, No. 6, pp. 1469–1569 (1993) — $22.00

466. Atomic Weights of the Elements 1991, *IUPAC Commission on Atomic Weights and Isotopic Abundances*, Vol. 22, No. 6, pp. 1571–1584 (1993) — $10.00

JPCRD Supplements/Monographs

When the topic demands it, and the quantity and quality of the data justifies it, the **Journal of Physical and Chemical Reference Data** issues a separate hard-cover-bound compilation. Monographs (before 1989 known as Supplements) are a collection of tables of highly significant physical or chemical property data in one complete volume. Listed below are the Monographs/Supplements to **JPCRD** which have been published. Each one is a valuable resource for the physicist and chemist.

PHYSICAL AND THERMODYNAMIC PROPERTIES OF ALIPHATIC ALCOHOLS, by R. C. Wilhoit and B. J. Zwolinski. (Supplement No. 1 to Volume 2) 1973, 420 pages.*

U.S. & Canada	$66/$60
Abroad:	$80/$72

THERMAL CONDUCTIVITY OF THE ELEMENTS: A COMPREHENSIVE REVIEW, by C. Y. Ho, R. W. Powell, and P. E. Liley. (Supplement No. 1 to Volume 3) 1974, 796 pages.*

U.S. & Canada	$120/$110
Abroad:	$144/$132

ENERGETICS OF GASEOUS IONS, by H. M. Rosenstock, K. Draxl, B. W. Steiner, and J. T. Herron. (Supplement No. 1 to Volume 6) 1977, 783 pages.*

U.S. & Canada	$140/$130
Abroad:	$168/$156

EVALUATED KINETIC DATA FOR HIGH TEMPERATURE REACTIONS: VOLUME 4, HOMOGENEOUS GAS PHASE REACTIONS OF HALOGEN- AND CYANIDE-CONTAINING SPECIES by D. L. Baulch, J. Duxbury, S. J. Grant, and D. C. Montague. (Supplement No. 1 to Volume 10) 1981, 721 pages. Hardcover.

U.S. & Canada	$160.00
Abroad:	$192.00

THERMOPHYSICAL PROPERTIES OF FLUIDS. 1. ARGON, ETHYLENE, PARAHYDROGEN, NITROGEN, NITROGEN TRIFLUORIDE, AND OXYGEN by B. A. Younglove (Supplement No. 1 to Volume 11) 1982, 368 pages. Hardcover.

U.S. & Canada	$80.00
Abroad:	$96.00

THE NBS TABLES OF CHEMICAL THERMODYNAMIC PROPERTIES. SELECTED VALUES FOR INORGANIC AND C_1 AND C_2 ORGANIC SUBSTANCES IN SI UNITS, by D. D. Wagman, W. H. Evans, V. B. Parker, R. H. Schumm, I. Halow, S. M. Bailey, K. L. Churney, and R. L. Nuttall. (Supplement No. 2 to Volume 11) 1982, 394 pages. Hardcover.

U.S. & Canada	$80.00
Abroad:	$96.00

HEAT CAPACITIES AND ENTROPIES OF ORGANIC COMPOUNDS IN THE CONDENSED PHASE by E. S. Domalski, W. H. Evans, and E. D. Hearing (Supplement No. 1 to Volume 13) 1984, 288 pages. Hardcover.

U.S. & Canada	$80.00
Abroad:	$96.00

JANAF THERMOCHEMICAL TABLES, Third Edition, by M. W. Chase, Jr., C. A. Davies, J. R. Downey, Jr., D. J. Frurip, R. A. McDonald, and A. N. Syverud. (Supplement No. 1 to Volume 14) 1985, 1896 pages, 2 volumes. Hardcover.

U.S. & Canada	$260.00
Abroad:	$312.00

ATOMIC ENERGY LEVELS OF THE IRON-PERIOD ELEMENTS: POTASSIUM THROUGH NICKEL, by J. Sugar and C. Corliss. (Supplement No. 2 to Volume 14) 1985, 664 pages. Hardcover.

U.S. & Canada	$100.00
Abroad:	$116.00

ATOMIC AND IONIC SPECTRUM LINES BELOW 2000 ANGSTROMS: HYDROGEN THROUGH KRYPTON, by Raymond L. Kelly. (Supplement No. 1 to Volume 16) 1987, 1689 pages, 3 volumes. Hardcover.

U.S. & Canada	$150.00
Abroad:	$180.00

GAS-PHASE ION AND NEUTRAL THERMOCHEMISTRY, by S. G. Lias, John E. Bartmess, J. F. Liebman, J. L. Holmes, R. D. Levin, and W. G. Mallard (Supplement No. 1 to Volume 17) 1988, 872 pages. Hardcover.

U.S. & Canada	$140.00
Abroad:	$168.00

THERMODYNAMIC AND TRANSPORT PROPERTIES FOR MOLTEN SALTS: CORRELATED EQUATIONS FOR CRITICALLY EVALUATED DENSITY, SURFACE TENSION ELECTRICAL CONDUCTANCE, AND VISCOSITY DATA, by George J. Janz (Supplement No. 2 to Volume 17) 1988, 320 pages. Hardcover.

U.S. & Canada	$50.00
Abroad:	$60.00

ATOMIC TRANSITION PROBABILITIES SCANDIUM THROUGH MANGANESE, by G. A. Martin, J. R. Fuhr, and W. L. Wiese. (Supplement No. 3 to Volume 17) 1988, 523 pages. Hardcover.

U.S. & Canada	$130.00
Abroad:	$156.00

ATOMIC TRANSITION PROBABILITIES IRON THROUGH NICKEL, by J. R. Fuhr, G. A. Martin, and W. L. Wiese. (Supplement No. 4 to Volume 17) 1988, 504 pages. Hardcover.

U.S. & Canada	$130.00
Abroad:	$156.00

KINETICS AND MECHANISMS OF THE GAS-PHASE REACTIONS OF THE HYDROXYL RADICAL WITH ORGANIC COMPOUNDS, by Roger Atkinson. (Monograph No. 1) 1989, 246 pp. Hardcover.

U.S. & Canada	$110.00
Abroad:	$132.00

GAS-PHASE TROPOSPHERIC CHEMISTRY OF ORGANIC COMPOUNDS, by Roger Atkinson. (Monograph No. 2) 1994, 216 pp. Hardcover.

U.S. & Canada	$120.00
Abroad:	$144.00

VIBRATIONAL AND ELECTRONIC ENERGY LEVELS OF POLYATOMIC TRANSIENT MOLECULES, by Marilyn E. Jacox. (Monograph No. 3) 1994, 461 pp. Hardcover.

U.S. & Canada	$140.00
Abroad:	$168.00

*Prices for hardcover/softcover

JPCRD Special Issues

EVALUATED KINETIC DATA FOR COMBUSTION MODELLING. D. L. Baulch, C. J. Cobos, R. A. Cox, C. Esser, P. Frank, Th. Just, J. A. Kerr, M. J. Pilling, J. Troe, R. W. Walker, and J. Warnatz, Vol. 21 No. 3, pp. 411–734 (entire issue) (1992) $70.00

EVALUATED KINETIC AND PHOTOCHEMICAL DATA FOR ATMOSPHERIC CHEMISTRY, SUPPLEMENT IV, R. Atkinson, D. L. Baulch, R. A. Cox, R. F. Hampson, Jr., J. A. Kerr, and J. Troe, Vol. 21, No. 6, pp. 1125–1568 (entire issue) (1992) $70.00

ESTIMATION OF THE THERMODYNAMIC PROPERTIES OF C–H–N–O–S–HALOGEN COMPOUNDS AT 298.15 K. Eugene S. Domalski and Elizabeth D. Hearing, Vol. 22, No. 4, pp. 805–1159 (entire issue) (1993) $80.00

SPECIAL REPRINT PACKAGES

JPCRD special reprint packages contain selected articles on specific subjects offered at a more economical package rate than aggregate individual purchase. Your specific requirements are available in a complete library of literature, at a fractional cost of purchasing previous journal issues.

The reprint packages are listed by subject area. The corresponding reprint numbers and titles are given in the Cumulative Reprints List. Build your information bank in this thorough and economical manner!

PACKAGE 1: ATOMIC PHYSICS

Package 1A (13 Parts) ENERGY LEVELS AND SPECTRA. Consisting of Reprints Nos. 378, 324, 379, 385, 392, 402, 405, 406, 421, 422, 433, 437, 443.

| | If purchased individually: | $254.00 |
| | **PACKAGE:** | **$203.00** |

Package 1B (8 Parts) COLLISION CROSS SECTIONS. Consisting of Reprint Nos. 380, 356, 299, 386, 387, 388, 415. 442.

| | If purchased individually: | $92.00 |
| | **PACKAGE:** | **$73.00** |

PACKAGE 2: MOLECULAR PROPERTIES

Package 2A (6 Parts) MICROWAVE SPECTRA. Consisting of Reprint Nos. 289, 198, 53, 130, 369, 436.

| | If purchased individually: | $135.00 |
| | **PACKAGE:** | **$108.00** |

Package 2B (6 Parts) OPTICAL SPECTRA. Consisting of Reprint Nos. 244, 270, 288, 93, 8, 441.

| | If purchased individually: | $128.00 |
| | **PACKAGE:** | **$101.00** |

Package 2C (10 Parts) MOLECULAR CONSTANTS. Consisting of Reprint Nos. 257, 342, 103, 129, 170, 403, 407, 419, 445, 448.

| | If purchased individually: | $194.00 |
| | **PACKAGE:** | **$155.00** |

PACKAGE 3: CONDENSED MATTER PHYSICS

Package 3A (8 Parts) ELECTRICAL PROPERTIES. Consisting of Reprint Nos. 258, 259, 260, 221, 138, 139, 28, 84.

| | If purchased individually: | $127.00 |
| | **PACKAGE:** | **$102.00** |

Package 3B (6 Parts) OPTICAL PROPERTIES. Consisting of Reprint Nos. 240, 162, 81, 158, 282, 389.

| | If purchased individually: | $89.00 |
| | **PACKAGE:** | **$71.00** |

PACKAGE 4: THERMOPHYSICS

Package 4A (8 Parts) WATER AND STEAM. Consisting of Reprint Nos. 333, 370, 383, 290, 400, 338, 349, 424.

| | If purchased individually: | $104.00 |
| | **PACKAGE:** | **$83.00** |

Package 4B (7 Parts) CRYOGENIC FLUIDS. Consisting of Reprint Nos. 295, 337, 281, 304, 305, 363, 395.

| | If purchased individually: | $124.00 |
| | **PACKAGE:** | **$99.00** |

Package 4C (10 Parts) HYDROCARBONS AND DERIVATIVES. Consisting of Reprint Nos. 331, 362, 294, 350, 371, 332, 377, 397, 410, 425.

| | If purchased individually: | $194.00 |
| | **PACKAGE:** | **$155.00** |

PACKAGE 5: THERMOCHEMISTRY

Package 5A (16 Parts) ORGANIC COMPOUNDS. Consisting of Reprint Nos. 265, 309, 292, 381, 382, 391, 393, 394, 404, 254, 197, 351, 397, 404, 412, 418.

| | If purchased individually: | $272.00 |
| | **PACKAGE:** | **$217.00** |

Package 5B (6 Parts) INORGANIC COMPOUNDS. Consisting of Reprint Nos. 310, 375, 398, 237, 297, 183.

| | If purchased individually: | $76.00 |
| | **PACKAGE:** | **$60.00** |

PACKAGE 6: PHYSICAL PROPERTIES

Package 6A (12 Parts) SOLUBILITY. Consisting of Reprint Nos. 296, 248, 193, 219, 246, 274, 166, 368, 416, 420, 427, 444.

| | If purchased individually: | $143.00 |
| | **PACKAGE:** | **$114.00** |

Package 6B (3 Parts) VAPOR PRESSURE. Consisting of Reprint Nos. 140, 241, 326.

| | If purchased individually: | $32.00 |
| | **PACKAGE:** | **$25.00** |

PACKAGE 7: CHEMICAL KINETICS

Package 7A (7 Parts) ATMOSPHERIC CHEMISTRY. Consisting of Reprint Nos. 413, 159, 203, 264, 366, 246, 335.

| | If purchased individually: | $168.00 |
| | **PACKAGE:** | **$134.00** |

Package 7B (10 Parts) COMBUSTION CHEMISTRY. Consisting of Reprint Nos. 302, 329, 344, 374, 345, 348, 409, 417, 428, 438.

| | If purchased individually: | $240.00 |
| | **PACKAGE:** | **$192.00** |

PACKAGE 8: SURFACE CHARACTERIZATION

Package 8A (3 Parts) Consisting of Reprint Nos. 336, 376, 434.

| | If purchased individually: | $40.00 |
| | **PACKAGE:** | **$32.00** |

American Institute of Physics

500 Sunnyside Boulevard
Woodbury, NY 11797-2999

Roland W. Schmitt	Chair, Governing Board
Marc H. Brodsky	Executive Director and CEO
Roderick M. Grant	Secretary
Arthur T. Bent	Treasurer and CFO
John S. Rigden	Director of Physics Programs
Darlene Carlin	Director of Publishing

The American Institute of Physics was founded in 1931; it is chartered as a membership corporation with leading societies in the fields of physics and astronomy as members. The Institute combines into one operating agency those functions which can best be done by the societies jointly. Its purpose is the advancement and diffusion of the knowledge of physics and its application to human welfare. About 100 000 physicists belong to its ten member societies: The American Physical Society, Optical Society of America, Acoustical Society of America, Society of Rheology, American Association of Physics Teachers, American Crystallographic Associa-

tion, American Astronomical Society, American Association of Physicists in Medicine, American Vacuum Society, American Geophysical Union. In addition, there are 20 Affiliated Societies, 84 Corporate Associates, and 6500 members of the Society of Physics Students.

For itself and its member societies, AIP publishes 20 archival journals, 17 translated Soviet journals, a general interest journal, conference proceedings, handbooks, and bulletin-programs, as well as an integrated package of current awareness journals, magnetic tapes, and microfilm. AIP also serves the public by making available to the press and other channels of public information reliable communications on physics and astronomy and their progress; carries on extensive career planning and placement activities; encourages and assists in the documentation and study of the history and philosophy of recent physics; cooperates with local, national, and international organizations devoted to physics and the related sciences; and fosters the relations of the science of physics to other sciences and to the arts and industries.

American Chemical Society

1155 Sixteenth Street, NW
Washington, DC 20036-9976

Paul H. L. Walter	Chairman of the Board
Ned D. Heindel	President
John K. Crum	Executive Director
Brian A. Bernstein	Treasurer
D.H. Michael Bowen	Deputy Executive Director and Secretary
Robert J. Massie	Director, Chemical Abstracts Service
Robert H. Marks	Director, Publications Division*
Halley A. Merrell	Director, Membership Division
Sylvia A. Ware	Director, Education Division

The American Chemical Society was founded in 1876 as a scientific and educational society, and in 1937 was granted a national charter by an act of Congress. Its objectives include the advancement of chemistry, the promotion of research in science and industry, improvement of qualifications and usefulness of chemists, increased diffusion of chemical knowledge, and the promotion of scientific interest and inquiry.

ACS publishes 26 magazines, journals, and other scientific periodicals in addition to its participation in publication of the *Journal of Physical and Chemical Reference Data*. It also publishes books and other nonperiodical reports and operates the Chemical Abstracts Service, the key to the world's chemical literature. It pursues its objectives through its services to chemists and chemical engineers and to education, industry, and government.

*Inquiries regarding ACS participation in JPCRD should be directed to this division.

National Institute of Standards and Technology

U.S. Department of Commerce
Gaithersburg, MD 20899-0001

U.S. Department of Commerce	Ronald H. Brown, Secretary
Technology Administration	Mary L. Good, Under Secretary for Technology
National Institute of Standards and Technology	Arati Prabhakar, Director

The National Institute of Standards and Technology (NIST), formerly the National Bureau of Standards, was established by an act of Congress 3 March 1901. NIST works to strengthen U.S. industry's international competitiveness, advance science and engineering, and improve public health, safety, and the environment. With an annual budget of more than $350 million and about 3100 staff members, NIST conducts science and engineering research in commercially important fields such as advanced materials, information systems, bio-

technology, optoelectronics, computer-integrated manufacturing, and sensor technology.

NIST's laboratory research is designed to support development of critical emerging technologies and the new measurement methods and standards necessary to make them commercially viable. The ability of U.S.-based industries to exploit these new technologies determines in large part the health of the U.S. economy.

Several new NIST programs are designed to help spur innovation at U.S. businesses through seed money for development of generic technologies; grants to states for support of technology transfer programs; and financial and technical assistance to help small and mid-sized companies adopt more efficient manufacturing methods.

Throughout its history, NIST has helped industry to develop high-quality products and processes by supplying standard reference materials and data; by continuously improving technologies for precision measurements of mass, time, length, voltage, radiation, and other units; and through active participation in voluntary standards-setting bodies.